D1687933

Handbook of Petroleum Processing

Steven A. Treese • Peter R. Pujadó
David S. J. Jones
Editors

Handbook of Petroleum Processing

Second Edition

Volume 2

With 571 Figures and 395 Tables

Springer Reference

Editors
Steven A. Treese
Puget Sound Investments LLC
Katy, TX, USA

Peter R. Pujadó
UOP LLC (A Honeywell Company)
Kildeer, IL, USA

David S. J. Jones
Calgary, AB, Canada

ISBN 978-3-319-14528-0   ISBN 978-3-319-14529-7 (eBook)
ISBN 978-3-319-14530-3 (print and electronic bundle)
DOI 10.1007/978-3-319-14529-7

Library of Congress Control Number: 2015937620

Springer Cham Heidelberg New York Dordrecht London
1st edition: © Springer Science + Business Media B.V. 2006
2nd edition: © Springer International Publishing Switzerland 2015
This work is subject to copyright. All rights are reserved by the Publisher, whether the whole or part of the material is concerned, specifically the rights of translation, reprinting, reuse of illustrations, recitation, broadcasting, reproduction on microfilms or in any other physical way, and transmission or information storage and retrieval, electronic adaptation, computer software, or by similar or dissimilar methodology now known or hereafter developed.
The use of general descriptive names, registered names, trademarks, service marks, etc. in this publication does not imply, even in the absence of a specific statement, that such names are exempt from the relevant protective laws and regulations and therefore free for general use.
The publisher, the authors and the editors are safe to assume that the advice and information in this book are believed to be true and accurate at the date of publication. Neither the publisher nor the authors or the editors give a warranty, express or implied, with respect to the material contained herein or for any errors or omissions that may have been made.

Printed on acid-free paper

Springer International Publishing AG Switzerland is part of Springer Science+BusinessMedia (www.springer.com)

Preface to the Second Edition

This Second Edition of the *Handbook of Petroleum Processing* follows in the footsteps and traditions of the First Edition. We have greatly updated and expanded the information and scope in this edition. We have included many new processes and unconventional feedstocks that are encountered in today's facilities. The book has been reorganized to better facilitate its use as a reference.

This Handbook provides a basic knowledge of petroleum refining and processing as a foundation. This general knowledge may be sufficient for many users, but we have continued to provide detailed, practical approaches to designing units and solving the most common challenges facing today's processing and design professionals. Where possible, we have included additional information for those who must operate the process units. Not all the information is at the same level of detail, but we have provided more detail in this edition about operation-related issues.

Most of the calculation techniques described here are manual. We authors and editors grew up using many of these methods; however, we recognize that computer process simulation and applications play the major roles in designing and operating plants today. Many of the methods provided here are behind the simulation software or can be converted easily to computer applications. These methods can quickly provide rough checks on computer results, or they can be used if you do not have a computer or simulation package readily available.

In different chapters, you will find the use of different units of measure. The units reflect those familiar to the chapter authors. Some chapters provide both English and metric units and some only English or only metric. The petroleum industry is accustomed to mixing units and converting between units. In fact, many product specifications use mixed systems. Several common conversion factors are provided in Appendix D.

This edition of the Handbook was structured with both electronic and print publication in mind. It is primarily conceived as a comprehensive book, and the organization of the chapters reflects that premise. For electronic publication, however, we have made the individual chapters as self contained as possible. This means that the information in a chapter generally includes the needed reference materials. The self-contained approach only takes us so far. Lest we have to repeat large portions of other chapters or resources, individual chapters will still contain references to other chapters or appendices in this Handbook and to other resources.

The book is organized into six parts:

Part I: General Refinery Processes and Operations

This part describes crude oil and the processes used to convert it to finished products along with the design and operation of these processes.

Included here are
- A general understanding of crude and crude properties
- An understanding of products and their important properties
- How a refinery is configured to match the desired crudes and markets
- Detailed discussions of the processes, design, and operation of crude and vacuum distillation units, light ends processing, catalytic reforming, fluid catalytic cracking, hydroprocessing, alkylation, olefin condensation, isomerization, gas treating, residual oil upgrading, and hydrogen production
- How product qualities are specified, measured, and controlled
- Techniques for planning refinery operations and economic analyses
- How a petroleum processing project is executed

Part II: Variations

Today's petroleum processing facility typically runs or handles more than just traditional crude oil and may make more than just fuel products.

This part of the Handbook explores facilities that make
- Lube oils
- Petrochemicals
- Other chemicals

It also provides insights into nonconventional feedstocks which find their way into the processing scheme or are processed somewhat like petroleum, e.g., coal and gas liquids, shale oil, shale crude, bitumens, and renewable stocks.

One chapter discusses biorefineries, which are increasingly providing fuels and fuel blend stocks.

Part III: Support Systems

All process facilities need certain common support systems. These are discussed in some detail in Part III, along with many design and operating techniques.

Included here are discussions of
- Instrumentation and control systems
- Utilities (steam, condensate, fuel, water, air, power, nitrogen, others)
- Off-site facilities (storage, blending, loading, waste hydrocarbons, wastewater)
- Environmental controls and practices (air emissions, aqueous effluents, solid wastes, noise)

Part IV: Safety Systems

Today, it is an absolute requirement that process facilities operate safely. Facilities must not endanger employees or the community. Facilities that cannot do this are usually not allowed to operate at all. While a large part of safety derives from the way a facility is operated, many features can be designed into the plant to enable safer operations.

This part of the Handbook focuses on the systems and practices that allow for excellent safety. Included here, you will find discussions of
- Process safety management (PSM) basics
- Safety systems for pressure and temperature
- Oxygen deficiency protection practices
- Confined space entry practices
- Facility siting considerations
- Hazardous materials
- Fire prevention and protection

Part V: Reference
The Reference part of the book provides detailed discussion of several general equipment types and how to design them:
- Vessels
- Fractionation towers
- Pumps
- Compressors
- Heat exchangers
- Fired heaters
- Piping and pressure drop

This part also has a comprehensive (and in some cases fairly detailed) dictionary of abbreviations, acronyms, expressions, and terms you will hear around refineries and petroleum processing facilities. We have added many terms and eliminated most of the repetition in the first edition. Many of these terms we use in the industry without even realizing they have specific connotations in our business. Having them all in one place helps. Where appropriate, the definitions provide references back to specific chapters in the book for additional information.

Part VI: Appendices
The appendices include reference materials that either did not fit elsewhere or were needed in multiple chapters. The appendices are divided into four sections:

Appendix "Examples of Working Flow Sheets in Petroleum Refining" provides examples of various flow sheets and documents used in petroleum processing.

Appendix "General Data for Petroleum Processing" contains general data on petroleum properties and some equipment properties. Included here, you will find relationships for viscosity, specific gravity, °API gravity, boiling points, freezing points, and tray geometry.

Appendix "Selection of Crude Oil Assays for Petroleum Processing" has several examples of crude and other oil properties in the form of simple assays. This appendix has been expanded to include many of the newer stocks, such as shale crudes, bitumens, and synthetic crudes. Some of these assays are dated, so for design or planning work, it is best to always get an updated assay from your supplier.

Appendix "Conversion Factors Used in Petroleum Processing" contains conversion factors. No Handbook worth its salt can get away without some common conversion factors. While we have not included all the factors you will need here, we have included those we found most useful or hardest to find in the literature. The tables include general factors plus pressure and viscosity interconversions.

We have tried to make this work as comprehensive as possible, but it is not feasible to conceive of or include everything you might want to know. There are numerous outside references provided, which can lead you to more references. Online searches often provide good information but can sometimes be misleading or wrong as well. Always apply your own judgment when looking online, or even when using information in this book for that matter. Things should make sense – processing of petroleum is not a great mystery.

This book has been edited by Steven A. Treese (retired from Phillips 66 Company) and Peter R. Pujadó (retired from UOP LLC, a Honeywell Company). Our fellow editor emeritus was David S.J. Jones (retired from Fluor Corporation), who passed away a few years ago. His contributions are found throughout this book, especially in the detailed techniques for calculation. He is missed.

We appreciate the help of Karin Bartsch with Springer Reference, who provided good guidance in making this an "electronic-friendly" work, as well as a useful reference book.

We would like to also acknowledge our many contributing authors for lending their excellent and invaluable expertise: Mark P. Lapinski, Stephen M. Metro, Mark Moser, Warren Letzch, Maureen Bricker, Vasant Thakkar, John Petri, Peter Kokayeff, Steven Zink, Pamela Roxas, Douglas A. Nafis, Kurt A. Detrick, Robert L. Mehlberg, Dennis J. Ward, Dana K. Sullivan, Bipin Vora, Greg Funk, Andrea Bozzano, Stanley J. Frey, and Geoffrey W. Fichtl. Hopefully, we have not missed anyone.

As a final disclaimer, we have made every effort to provide accurate information in this work, but we offer no warrantees in any specific application. The user assumes all responsibility when applying the information contained herein.

We hope you find this Handbook useful. It has been an interesting adventure for us (and a good review) in compiling it. Use it in safety and good health!

Peter R. Pujadó and Steven A. Treese
Editors and Authors

Contents

Volume 1

Part I General Refinery Processes and Operations 1

Introduction to Crude Oil and Petroleum Processing 3
David S. J. Jones

Petroleum Products and a Refinery Configuration 53
David S. J. Jones and Steven A. Treese

Atmospheric and Vacuum Crude Distillation Units in Petroleum
Refineries 125
David S. J. Jones

Distillation of the "Light Ends" from Crude Oil in Petroleum
Processing 199
David S. J. Jones

Catalytic Reforming in Petroleum Processing 229
Mark P. Lapinski, Stephen Metro, Peter R. Pujadó, and Mark Moser

Fluid Catalytic Cracking (FCC) in Petroleum Refining 261
Warren Letzsch

Hydrocracking in Petroleum Processing 317
Maureen Bricker, Vasant Thakkar, and John Petri

Hydrotreating in Petroleum Processing 361
Peter Kokayeff, Steven Zink, and Pamela Roxas

Alkylation in Petroleum Processing 435
Douglas A. Nafis, Kurt A. Detrick, and Robert L. Mehlberg

Olefin Condensation 457
Robert L. Mehlberg, Peter R. Pujadó, and Dennis J. Ward

Isomerization in Petroleum Processing 479
Dana Sullivan, Stephen Metro, and Peter R. Pujadó

Refinery Gas Treating Processes 499
David S. J. Jones and Steven A. Treese

Upgrading the Bottom of the Barrel 531
David S. J. Jones

Hydrogen Production and Management for Petroleum Processing ... 565
Steven A. Treese

Quality Control of Products in Petroleum Refining 649
David S. J. Jones

Petroleum Refinery Planning and Economics 685
David S. J. Jones and Peter R. Pujadó

Petroleum Processing Projects 787
David S. J. Jones and Steven A. Treese

Part II Variations 841

Non-energy Refineries in Petroleum Processing 843
David S. J. Jones and Steven A. Treese

Chemicals from Natural Gas and Coal 883
Bipin Vora, Gregory Funk, and Andrea Bozzano

Unconventional Crudes and Feedstocks in Petroleum Processing 905
Steven A. Treese

Biorefineries 965
Stanley J. Frey and Geoffrey W. Fichtl

Volume 2

Part III Support Systems 1007

Process Controls in Petroleum Processing 1009
David S. J. Jones and Steven A. Treese

Utilities in Petroleum Processing 1093
David S. J. Jones and Steven A. Treese

Off-Site Facilities for Petroleum Processing 1167
David S. J. Jones and Steven A. Treese

Environmental Control and Engineering in Petroleum Processing 1215
David S. J. Jones and Steven A. Treese

Part IV Safety Systems 1305

Safety Systems for Petroleum Processing 1307
David S. J. Jones and Steven A. Treese

Hazardous Materials in Petroleum Processing 1351
David S. J. Jones and Steven A. Treese

Fire Prevention and Firefighting in Petroleum Processing 1415
David S. J. Jones and Steven A. Treese

Part V Reference 1437

Process Equipment for Petroleum Processing 1439
David S. J. Jones and Steven A. Treese

**Dictionary of Abbreviations, Acronyms, Expressions, and
Terms Used in Petroleum Processing and Refining** 1685
David S. J. Jones, Peter R. Pujadó, and Steven A. Treese

Part VI Appendices 1827

Examples of Working Flow Sheets in Petroleum Refining 1829
Steven A. Treese

General Data for Petroleum Processing 1835
Steven A. Treese

Selection of Crude Oil Assays for Petroleum Refining 1845
Steven A. Treese

Conversion Factors Used in Petroleum Processing 1881
Steven A. Treese

Index 1885

About the Editors

Steven A. Treese retired from Phillips 66 in 2013 as the Hydroprocessing Team Lead after 40 years but continues to take on the occasional consulting assignment in process engineering and refining as a consultant with Puget Sound Investments, L.L.C. He started his professional career with Union Oil Company of California in 1973 as a Research Engineer after obtaining a B.S. in Chemical Engineering from Washington State University. He followed company heritages through Unocal, Tosco, Phillips, ConocoPhillips, and Phillips 66. Steve's range of experience includes catalyst development, hydroprocessing, hydrogen production, utilities, sulfur recovery, geothermal, shale oil, nitrogen fertilizers, process design, procurement, and licensing. He is a licensed Professional Engineer. Steve has several publications, a few patents, and was on the 1994 NPRA Question and Answer Panel. He is a member of the American Institute of Chemical Engineers. Steve's hobbies include woodworking, boating, fermentation, and photography. He is a mentor for FIRST Robotics Team 624, CRyptonite, in Katy, Texas.

Peter Pujadó retired from UOP LLC (a Honeywell subsidiary) in 2005 as Senior Manager/R&D Fellow responsible for the development and commercialization of technologies for the production of light olefins (ethylene and propylene) by the catalytic conversion of methanol. He started his career as a lecturer at the University of Manchester Institute of Science and Technology (UMIST) in Manchester, England; he then worked as a process engineer for SA Cros in Barcelona, Spain, in areas as diverse as chlorine, caustic, chlorinated hydrocarbons, ammonia, urea, nitric acid, and NPK fertilizers; he joined UOP LLC as an R&D process coordinator responsible for the production of cumene, phenol/acetone, aromatics isomerization, aromatics disproportionation and transalkylation, terephthalic acid, acrylonitrile, acetic acid, etc. After retirement from UOP LLC, he again worked as a lecturer at Northwestern University, Evanston, Illinois, and has done some consulting in the petrochemicals area. Peter had graduated with an M.S. in Chemical and Petroleum Refining Engineering from the Colorado School of Mines, a Ph.D. in Chemical Engineering from the University of Minnesota, and an MBA from the University of Chicago. He is a licensed Professional Engineer and a Fellow Member of the American Institute of Chemical Engineers. He is the author of over 95 papers and publications and of 44 patents. Peter's hobbies include travel, mountain hiking, and reading.

David S. J. Jones was from a small coal mining village (Ynystawe) in South Wales, UK. He left school at the age of 16 and joined the army, where he ended up in India as a Regimental Sergeant Major. After that, he returned to Wales and worked for BP in a quality control lab.

He had ambition, a remarkable ability, and dogged determination to study Chemical Engineering in order to improve on his lot in this world. He studied Chemical Engineering at night school and obtained a Bachelor of Science degree. He had a striking career and made it to the top of the tree, spending many years with Fluor as well as consulting.

When he retired, he occupied himself with the writing of several publications on chemical engineering, such as the *Elements of Petroleum Processing* (Wiley and Sons 1995) and the *Handbook of Petroleum Processing* (Springer 2006), mainly aimed at students and young graduate engineers with an emphasis on problem-solving.

Stan passed away in 2005.

Contributors

Andrea Bozzano UOP LLC, A Honeywell Company, Des Plaines, IL, USA

Maureen Bricker UOP LLC, A Honeywell Company, Des Plaines, IL, USA

Kurt A. Detrick UOP, A Honeywell Company, Des Plaines, IL, USA

Geoffrey W. Fichtl UOP, A Honeywell Company, Des Plaines, IL, USA

Stanley J. Frey UOP, A Honeywell Company, Des Plaines, IL, USA

Gregory Funk UOP LLC, A Honeywell Company, Des Plaines, IL, USA

David S. J. Jones Deceased. Formerly at: Calgary, AB, Canada

Peter Kokayeff UOP, A Honeywell Company, Des Plaines, IL, USA

Mark P. Lapinski UOP LLC, A Honeywell Company, Des Plaines, IL, USA

Warren Letzsch Technip-Stone & Webster, Houston, TX, USA

Consulting PC, Ellicott City, MD, USA

Robert L. Mehlberg UOP, A Honeywell Company, Des Plaines, IL, USA

Stephen Metro UOP LLC, A Honeywell Company, Des Plaines, IL, USA

Mark Moser UOP LLC, A Honeywell Company, Des Plaines, IL, USA

Douglas A. Nafis UOP, A Honeywell Company, Des Plaines, IL, USA

John Petri UOP LLC, A Honeywell Company, Des Plaines, IL, USA

Peter R. Pujadó UOP LLC, A Honeywell Company, Kildeer, IL, USA

Pamela Roxas UOP, A Honeywell Company, Des Plaines, IL, USA

Dana Sullivan UOP LLC, A Honeywell Company, Des Plaines, IL, USA

Vasant Thakkar UOP LLC, A Honeywell Company, Des Plaines, IL, USA

Steven A. Treese Puget Sound Investments LLC, Katy, TX, USA

Bipin Vora R&D, UOP LLC, A Honeywell Company, Des Plaines, IL, USA

Dennis J. Ward Deceased. Formerly at: Ft. Myers, FL, USA

Steven Zink UOP, A Honeywell Company, Des Plaines, IL, USA

Part III

Support Systems

Process Controls in Petroleum Processing

David S. J. Jones and Steven A. Treese

Contents

Introduction	1010
Definitions	1011
Control System Types and Architectures	1014
Evolution of Control System Types	1014
Control System Architecture	1017
Special Control Systems	1018
Flow Measurement and Control	1019
Flow Elements and Their Characteristics	1019
Typical Arrangements	1028
Final Control Elements	1031
Flow Measurement and Control Conclusion	1039
Temperature Measurement and Control	1039
Temperature Measurement Elements	1039
Typical Temperature Measurement Arrangements	1046
Control Elements	1049
Concluding Comments	1049
Pressure Measurement and Control	1050
Pressure Measurement Elements and Characteristics	1051
Typical Installations	1051
Pressure Control Elements	1054
Concluding Comments on Pressure	1055
Level Measurement and Control	1063
Level Measurement Elements and Characteristics	1063
Level Instrument Installations	1067

David S. J. Jones: deceased.
Steven A. Treese has retired from Phillips 66.

D.S.J. Jones
Calgary, AB, Canada

S.A. Treese (✉)
Puget Sound Investments, LLC, Katy, TX, USA
e-mail: streese256@aol.com

Level Control	1069
Specific Level Considerations	1069
Concluding Comments on Level Measurement and Control	1075
Composition Measurement and Control	1075
Composition Measurement Elements	1076
Online Analyzer Installations	1077
Inferred Properties	1084
Controllers	1084
Types of Control Actions	1085
Control Loop Tuning	1086
Conclusion	1087
Appendix 1: Control Valve Sizing	1087
Process Flow Coefficient (C_v) and Valve Sizing	1087
Pressure Drop	1088
Flashing Liquids	1088
Two-Phase Flow	1088
Valve Rangeability	1088
Valve Flow Coefficient ($C_{v'}$)	1089
Control Valve Sizing	1089
Valve Action on Air Failure	1090
References	1090
Flow Measurement and Control	1090
Temperature Measurement and Control	1091
Pressure Measurement and Control	1091
Level Measurement and Control	1092
Composition Measurement and Control	1092

Abstract

This chapter focuses on the control systems which support the refining processes. This is not a definitive work on control systems, but paints the applications of control systems in a modern refinery with a broad brush. The material in this chapter starts with a discussion of control system architecture and continues with detailed descriptions of the major types of parameters controlled: flow, temperature, pressure, level, and composition. Additional material is provided on specific, common control situations. There is a brief discussion of control theory and there are procedures provided for sizing control valves. Numerous references are available for further information.

Keywords

Refinery • Process • Control • Distributed control system • Advanced control • SIS • DCS

Introduction

This chapter focuses on the control systems which support the refining processes. This is not a definitive work on control systems, but paints the applications of control systems in a modern refinery with a broad brush. More detailed treatment of process controls can be found in instrumentation and controls references, some of which are noted at the end of each section of this chapter.

The discussion includes:

- Definitions
- Control system types and architectures
- Specific discussion of each type of parameter monitoring and controls:
  - Flow
  - Temperature
  - Pressure
  - Level
  - Composition
- Discussion of special control situations commonly encountered:
  - Surge
  - Reflux
  - Steam generation
  - Column pressure control
  - Safety instrumented or integrated systems

Definitions

We will start with a few basic definitions that will be encountered frequently in control systems discussions.

Advanced process control (APC). A control system overlaying the DCS or basic process control system that performs more complex control functions for a process unit. It analyzes and coordinates several loops to obtain an optimum operation with the limitations of the refinery unit. The solution and changes are passed back to the DCS or other control system for implementation.

Basic process control (BPC). This is the workhorse control system that takes input from the process variables, compares the input to the set points, and outputs the appropriate control moves to the control elements in the field.

Cascade control. A control system with two or more controllers. The "master controller(s)" provides a set point to the "slave controller."

Control element. A device that performs a control function, such as a control valve or variable-speed controller.

Control loop. The basic device arrangement from process sensing element in the field to the process controller to the final field control element (like a control valve).

Control system architecture. Refers to essentially a map of how the process control system is assembled. Architecture would include what devices are tied into the process control system or network.

Control valve. A final control element that introduces a variable pressure drop into a flowing line. The pressure drop causes the flow rate in the line to change in a characterized and predictable way.

Control valve response. The minimum time that should be allowed between HLL and LLL to permit the control valve to respond effectively to changes in level. Typical values are:

CV size, inches	Response, sec
1	6
2	15
3	25
4	35
6	40
8	45
10	50

The above times allow for air signal lags and for operating at a proportional band of about 50 %.

Data historian. An electronic data storage system that collects data periodically from the process control systems and laboratory for archiving and analysis.

Dead time (dead band). The amount of time before a process starts changing after a disturbance or control action in the system.

Derivative. A control mode where the controller output is determined by the rate of change of the deviation from the set point. Acts like a dampening term. Normally used as a fine-tuning parameter in a proportional control mode.

Distributed control system (DCS). This is a computer-based control system used for basic process control. The basic control loops are in the background. The DCS displays the process and control loop information for the operator. It also records the information and communicates with any advanced control systems. The operator and APC change the settings in the DCS, which then sends the moves on to the process loops. DCS is the most common type of control system today in refineries.

dP cell. A diaphragm device for measuring small differences in pressure with the higher pressure on one side of the diaphragm and the lower pressure on the other side. The difference may be directly indicated or may be transmitted.

Enterprise control. An overall control system that factors in economics, market, and other factors to optimize and coordinate several refinery units simultaneously.

Error. This is usually defined in process control as the difference between the set point and the process variable or measurement.

Feedforward control. A control scheme where changes in an upstream variable initiate changes in downstream controls in anticipation of upcoming deviations. For example, if the feed rate to a unit changes, the appropriate controllers can also be changed downstream so they are set correctly for the new feed rate.

Input. The measured parameter or parameters provided to a controller. These may be direct readings, 3–15 psi pneumatic signals, 4–20 ma electrical signals, or digital.

Instrumentation. Typically, this term applies to the physical elements of the control system. It is usually considered separately from the actual controls.

Integral. A control mode where the controller output is determined by the deviation from the set point and how long that deviation has existed. Acts like an amplification term. Normally used as a fine-tuning parameter in a proportional control mode.

Level control range. This is the distance between the high liquid level (HLL) and the low liquid level (LLL) in the vessel. When using a level controller, the signal to the control valve at HLL will be to fully open the valve. At LLL the signal will be to fully close the valve.

Output. The signal coming from a controller to the field control device. These may be 3–15 psi pneumatic signals, 4–20 ma electrical signals, digital, or some other form.

Process control. Management of process operating conditions to meet product objectives safely and efficiently within equipment limitations.

Programmable logic controller (PLC). An electronic control system that uses programmed logic to carry out repetitive, sequential process tasks. Outputs from the PLC are used to control field elements. It may interface with other control systems but is intended to perform its function stand-alone.

Proportional. A control mode where the controller output is determined by the deviation from set point times a constant. See proportional band.

Proportional band. This determines the response time of the controller. Normally a proportional band is adjustable between 5 % and 150 %. The wider the proportional band, the less sensitive is the control. If a slower response time is required, a wider proportional band is used.

Reset. The rate at which a control system makes changes to a process. A high reset rate will make more frequent changes to the system. In a PID or PI controller, this may result in windup of controller output which causes it to overshoot the set point. A low reset rate may never reach its target.

Safety integrated system (SIS). AKA safety integrity system. This is a system that takes secure process information parallel to other control systems and applies specific limits to the measured variables. If one of the variables goes outside the predefined limits, the system first alarms at low deviation and then takes a predefined set of emergency actions at high or sustained deviation. The system is similar to a PLC but is normally triple redundant to ensure high integrity.

Sensing element. The field device that measures a physical or chemical property to be used in controlling the process. The most common elements measure things like flow, temperature, pressure, and level.

Set point. The desired value of a process variable that is provided to a controller. The controller tries to make the process measurement match the set point.

Supervisory control. Essentially basic process control.

Surge volumes. This is the volume of liquid between the normal liquid level (NLL) and the bottom (usually the tangent line) of a vessel.

Control System Types and Architectures

Evolution of Control System Types

Control systems can be viewed from several perspectives. Controls have evolved from simple control valves that were manually manipulated to sophisticated, enterprise-wide systems found in some companies today. Most refineries have some semblance of each evolution of control systems in operation today.

A control system can be viewed as a hierarchy of control layers, one on top of the other. Each layer has increasing amounts of information available to it and can perform increasingly complex control functions. This hierarchy is illustrated in Fig. 1, with the most basic level system at the bottom and progressing upward in sophistication.

The most basic process control is accomplished by simple hand valves manually controlled by the operator. Until the early twentieth century, this was the primary control method for chemical processes. We still see manual control used in refineries in devices such as "hand valves" in large steam turbines. These manual valves control the number of steam orifices through which steam impacts the turbine blades.

In the next control evolution, the manual valves were automated to control a specific variable directly and continuously. This is a simple loop control and it is common in refineries. An example would be a pressure regulator. A relief valve is

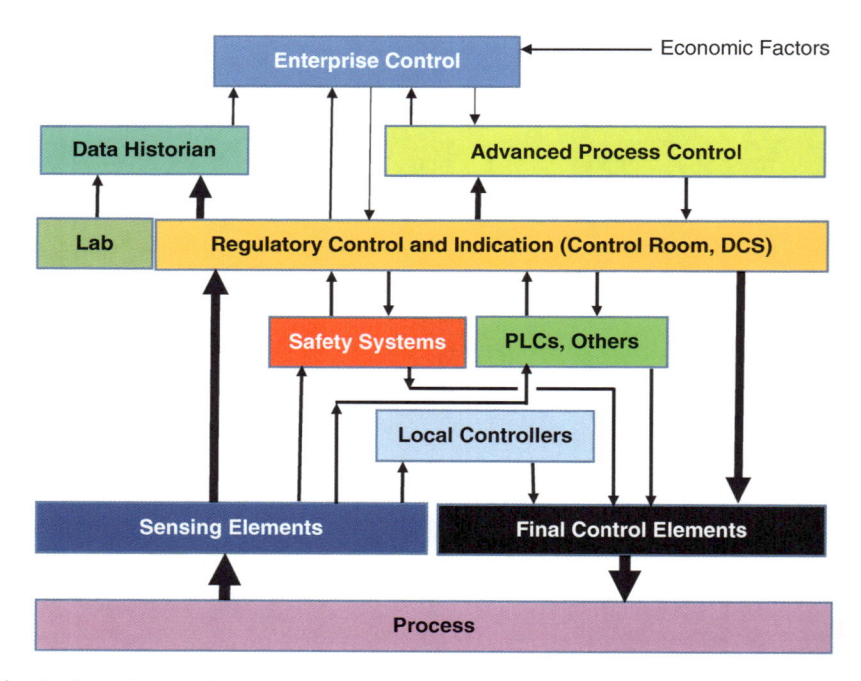

Fig. 1 Control system hierarchy

also a special case of this type of control with an on/off action. In a simple loop, illustrated by Fig. 2, a sensing element directly controls a process variable at a set value. The controls may be mechanical or pneumatic.

Simple loops lead to single loop controls, Fig. 3, where a sensing element sends a signal to a controller (local or in a control room). The controller has a set point target. The controller compares the set point to the measured variable and outputs a signal to a field control device to adjust the device and bring the variable closer to the set point. This type of controller normally has additional logic beyond the simple loop to enable it to keep the process variable under tighter control over a wider range of variation than a simple controller and to enable an operator to change set points easily. Single control loops are among the most common controls you will encounter in a refinery. In single loops here, we will also include indicator

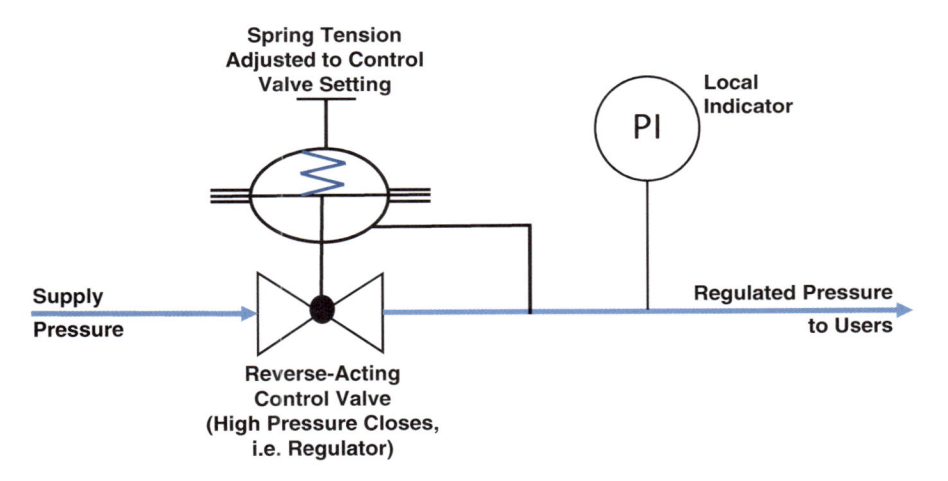

Fig. 2 Simple direct control – pressure regulation

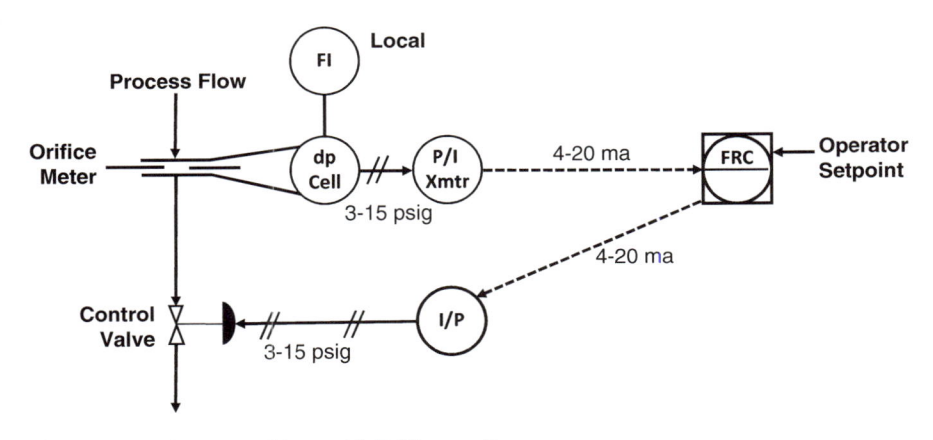

Fig. 3 Single-flow control loop with DCS controller

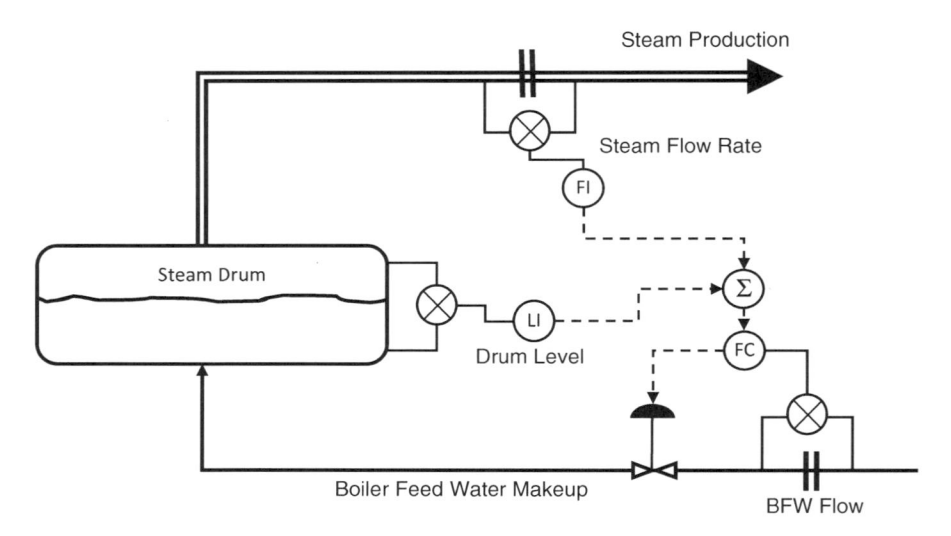

Fig. 4 Basic three element boiler control

loops, which may have no control function but provide data to the operator. More sophisticated controls are built upon a foundation of loop controls.

Multiple control loops can be coordinated using several schemes, such as using one control loop to look at more than one variable and then provide set points to other control loops based on its evaluation. See Fig. 4. Examples of this include cascading one loop to another and 3-element boiler control.

In the example in Fig. 4, the steam flow rate and the steam drum level changes are both added together to determine the total water demand for the steam generation system. The control system then sends a set point to the boiler feed water flow controller indicating how much water the system wants. The controller compares the set point with the current BFW flow rate and adjusts the BFW control valve to provide the required amount of water.

Multiple single loop controls are normally pulled together into a control room. The control room may have individual loop controllers (that may be pneumatic or electronic) or may have a distributed control system (DCS), which is computer based. You will find numerous mixtures and variations between these extremes. The layer of individual control or sensing loops, however, remains as a foundation. More sophisticated controls rely on this foundation layer. The compilation of data from all the loops in one system enables operating data to be retained and archived for analysis, opening up a whole range of operation improvements. The archived data is stored in a "data historian" for access by plant personnel.

Above the loops, DCS and other control systems enable better coordination of the controls and more sophisticated control schemes. Having a DCS can be viewed as the investment required for smarter controls. A common example would be advanced process control (APC), found in most refineries today. An APC application, for instance, can be used to control several loops around a fractionator to target specific product properties off the fractionator and optimize the energy consumed making these products all at the same time. APC can be used in a hydrotreater to control the multiple individual bed temperatures to optimize catalyst and energy use.

The "highest" level of control currently seen we will call "enterprise" control, which factors market considerations into control decisions. A facility using this approach economically optimizes the entire facility within facility and market constraints. The system knows what each product is worth and what each feed or utility costs. Then it finds a solution that makes the most money. This is like a refinery LP but is run more frequently and can more immediately impact the operation.

From a practical standpoint, APC and higher control applications are "slow loops"; that is, they are normally "tweaking" controls or are providing information on a much lower frequency than the normal loop controllers. They may make a move once an hour or once a day. The operator is usually given the option to accept or reject the control solution from the more sophisticated systems. This interface is needed because the advanced applications may not have current information about a refinery's capability and status at any given time.

Control System Architecture

Control system architecture refers to essentially a map of how the process control system is assembled. Architecture would include what devices are tied into the process control system or network. Figure 1 indicates the general architecture layers possible.

In a simple refinery control architecture, the loops are connected to unit control rooms where they are displayed, monitored, and controlled. In this configuration, the system ends with the "Regulatory Control and Indication" layer. Data from the process is recorded, at some frequency, manually on a data sheet for review and analysis by the refinery staff. Up until the advent of computer or DCS systems, this was the refinery norm.

A modern refinery architecture would still look somewhat like the simple architecture, with a control loop level, but the loops are tied to a DCS with advanced control overlaying the control loops and optimizing the loops. Data from the control system and the laboratory information system are collected in a data historian. The refinery staff, and remote users, can access the data for analysis. There may still be some manual readings taken to cover in instruments that are not part of the system and to help ensure operators monitor specific equipment on some frequency. This is the architecture you would find most commonly today.

The enterprise layer of the system is not yet common, but growth in this area is expected.

Special Control Systems

There are two categories of devices that are common in facilities today and have special purposes within control systems. These are programmable logic controllers (PLCs) and safety integrated systems (SIS). While these may interface with the normal process control systems, they have dedicated functions and operate much like very complex control loops.

Programmable Logic Controllers

Programmable logic controllers are used where a clearly defined complex control must be implemented continuously and repeatedly. A simple example of this type of control would be an instrument air-dryer which senses the moisture in the instrument air and switches to its regeneration cycle based on the moisture content. The controller then sequences the dryer regeneration cycle. A more sophisticated example would be the antisurge controller for a large centrifugal compressor, which analyzes the compressor performance and distance away from surge continuously and limits or shuts down the compressor if it approaches surge.

Another application for PLCs is found in pressure swing adsorption units where a PLC controls and monitors the valve sequences and pressures required to operate the PSA. More sophisticated PSA systems will also contain diagnostics and fallback operations if something goes amiss in the PSA. The system prevents damage to the PSA.

Safety Integrated Systems

Also known as safety integrity systems, these are normally triple-redundant logic solvers (essentially parallel PLCs) that monitor specific variables and take specific override actions if the data indicate that a process unit is headed toward a major process safety incident. The requirement for an SIS originates in Process Hazards Analysis of a process unit. A Layers of Protection Analysis (LOPA) is used to define the requirements and adequacy for an SIS design. The types of events the system is targeted to avoid would include fire, explosion, loss of containment, or any other serious incident.

An example where an SIS is employed is monitoring a hydrocracker for potential loss of positive reactor temperature control (runaway), which can lead to rapid overheating of equipment and subsequent failure accompanied by fire and/or explosion. An SIS in this service can detect the incipient runaway and, after pre-alarms, would automatically shutdown the unit and vent the hydrogen pressure to the flare at a high rate. These actions are designed to eliminate runaway risk. While the unit is shutdown by the actions, this is safer than a loss of containment incident.

SIS systems normally have their own, separate, dedicated, multiple sensing elements in the process unit. They may communicate information to the DCS but operate outside and separately from the DCS. There are specific requirements for design, analysis, installation, redundancy, modification review and approvals, testing frequency, and maintenance of the SISs. How rigorous the requirements are is determined by the severity of the possible consequence.

Flow Measurement and Control

Flow is one of the most common and important measurements in a refinery. From an overall standpoint, crude oil, feedstocks, and refined products are normally bought and sold by volume, hence volumetric flow directly affects a facility's economics. On a day-to-day basis, flows in and around each unit must be measured and managed. Material balancing is important for monitoring and controlling units and flow measurements are required to develop a balance.

This section is intended to be survey of flow measurement and control. More detailed information on this topic is available from many sources, including those listed as references.

We will review the basics of:

- Flow sensing elements and their characteristics
- Typical arrangements
- Control elements

Flow Elements and Their Characteristics

Rate gauging by changes in tank or drum levels is probably the most basic and direct method for determining flow rates. This method relies on changes in level within a known, calibrated vessel. While this is technically a determination of flow rate, it is not continuous and is not our subject in this section. Our focus here is on elements that continuously measure flow rate for use in controlling a process.

Table 1, 2 and 3 provides a listing of many common types of flow metering elements currently commercially available. The table summarizes the measurement principle, application characteristics, advantages, disadvantages, typical locations used, and other factors relevant to refinery applications.

Note that there is no truly successful method for metering two-phase flows, so this should generally be avoided.

Within the chemical industry, the prevalence of meters by type is represented in the following, after G. Livelli of ABB, Inc. Refining would be a subset of the chemical industry in this regard.

Meter type/principle	Percentage of meters
Differential pressure	29
Magnetic	20
Direct mass	18
Positive displacement	9
Turbine	8
Ultrasonic	7
Anemometer	5
Swirl, vortex, other	4

Table 1 Common types of flow elements and their characteristics

Principle	Differential pressure			Magnetic
Meter type	Orifice plate	Venturi tube	Pitot tube	Magnetic
Applications	Clean liq, gas, steam	Clean liq, gas, some dirt, slurry	Clean gas or liquid	Clean or dirty liquids, sulfur
Sizes, inches	>1	>2	>3	>0.1
Accuracy, ±%	0.5–3	0.5–1.5	3	0.5
Turndown	3:1	3:1	3:1	10:1
Min Reynolds # or vise limits	10,000	>100,000	>100,000	None
Max psig	H	L–H	L	M
Δp	H	L	L	L
Meter cost	L	H	L	H
Installation cost	M–H	M	M	H
Maintenance cost	M–H	L	L	M
U/S pipe, dias	10–30	5–10	20–30	5
Output form	Sq.Rt.	Sq.Rt.	Sq.Rt.	Linear
Some advantages	Low cost, excellent history, very common in refinery	Low dP, dirty fluids, short pipe run	Similar to pitot venturi	Good range, zero Δp, insensitive to fluid props
Some disadvantages	High Δp, limited range, wear, density effects	Expensive, limited range, high R_d required	Similar to pitot venturi + very low Δp, taps plug	Expensive, minimum conductivity required

Of the types of flow meters available, the orifice meter is, by far, the most common in refineries. It is simple and reliable for most applications. We will spend most of this section reviewing the equations used for sizing orifice meters and how they are applied.

We will also briefly discuss applications of the other meter types in a refinery. We will specifically discuss custody transfer metering.

Orifice Meters

An orifice meter is simply a designed restriction plate (a plate with a hole through it) in a pipe that produces a pressure drop with a known relationship to the flow rate. The pressure is measured upstream and immediately downstream of the plate where the flow is at the highest velocity (at the vena contracta). Using Bernoulli's principle, the flow rate is a function of the pressure drop.

Table 2 Common types of flow elements and their characteristics

Principle	Oscillatory		Positive displacement		
Meter type	Coriolis	Vortex shedding	Positive displc.	Oval gear	Nutating disc
Applications	Clean or dirty liquids	Clean liq, gas, steam, some dirt	Clean liq, gas, higher viscosity	Clean liq, gas, high viscosity	Clean liq
Sizes, inches	<6	>0.5	>12	1/8–3	5/8–2
Accuracy, ±%	0.25	1	0.5–1	0.25	2
Turndown	>25:1	10–20:1	20:1	High	Med
Min Reynolds # or visc limits	None	>10,000	>10 cSt	<50,000 cP	<5,000 cP
Max psig			L	M	L
Δp	M	M	H	H	H
Meter cost	H	M	H	H	M
Installation cost	H	L–M	H	H	M
Maintenance cost	L–M	M	H	H	H
U/S pipe, dias	None	15–25	None	None	None
Output form	Linear	Linear	Linear	Linear	Linear
Some advantages	Direct mass reading, open channel, high accuracy, wide range	Accuracy unaffected by props, high range, high turndown, moderate dP	Custody transfer, high accuracy, wide range, viscous liquids	Similar to PD	Similar to PD
Some disadvantages	Difficult installation, not for gases, expensive	High Reynolds no., visc. <20 cP, not for dirty, corrosive or erosive services	Mechanical, expensive, not good for rates, output affected by viscosity	Similar to PD, small sizes only	Similar to PD, small sizes only

Orifice meters normally used in refineries are the sharp-edged, concentric-type devices. Figure 5 shows the key parts of a typical orifice meter installation using orifice taps. Specific application may use eccentric or segmental-type orifice plates, but these are not common. If another type of plate is used, appropriate changes to the calculations are needed. The taps may also be in other locations around the orifice plate, such as in the pipe upstream and downstream, but these arrangements are not common.

The basic, generalized equation for an orifice meter is:

$$Q = (\pi/48) C \ Y \ d_o^2 \left[(2\rho_f g_c \Delta p)/(1 - \beta^4) \right]^{1/2} \qquad (1)$$

Table 3 Common types of flow elements and their characteristics

Principle	Turbine	Ultrasonic		Variable area	Thermal
Meter type	Turbine	Doppler	Transit time	Rotameter	Thermal dispersion
Applications	Clean liq, gas, steam	Dirty/corrosive liq or slurry	Clean/corrosive liq	Clean liq or gas, some contaminants	Clean/dirty liq or gas, slurry
Sizes, inches	>0.25	>0.5	>0.5	<3	<3
Accuracy, ±%	0.5	2–5	2–5	0.5–5	1–5
Turndown	10–35:1	10:1	10:1	10:1	100:1
Min Reynolds # or visc. limits	2–15 cSt	None	None	<3 cP	None
Max psig	M	M	M	L	M
Δp	M	L	L	M	L
Meter cost	M	M	M	L–M	L–M
Installation cost	M	L	L	L	L
Maintenance cost	M–H	L	L	L	L–M
U/S pipe, dias	None	5–20	5–20	None	10–20
Output form	Linear	Linear	Linear	Linear	Log
Some advantages	Good range, low flow, easy to install	Nonintrusive, low cost, bidirectional	Same as Doppler	Low cost, direct reading	High range, almost any fluid
Some disadvantages	Accuracy affected by wear and fluid props, clean fluids, high cost, gas meters tend high	Needs particles to work, liquids only, low accuracy	Clean liquids only, low accuracy	Expensive larger sizes, fluid props affect accuracy	Sensor fouling affects performance, low accuracy

where:

Q = flow rate for compressible fluids, lb_m/s
C = orifice discharge coefficient (correction factor)
Y = expansion factor (correction factor for compressible fluids)
d_o = orifice diameter in inches
ρ_f = fluid density at upstream flow conditions, lb_m/cft
g_c = gravity constant, 32.174 ft*lb_m/(lb_f s^2)
Δp = pressure drop across the orifice, psi
β = ratio of the diameter of the orifice to the pipeline ID, dimensionless

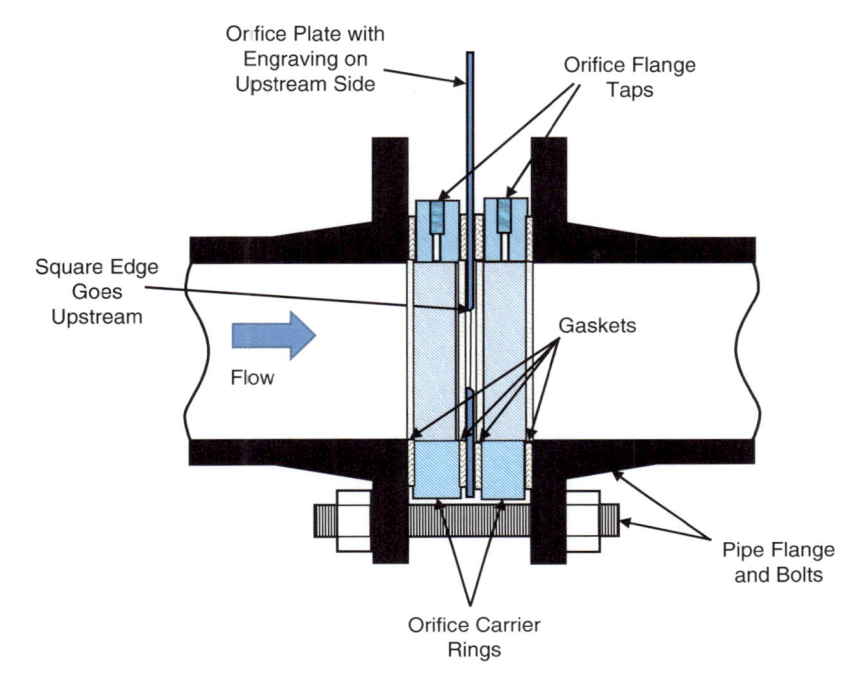

Process Controls in Petroleum Processing 1023

Fig. 5 Orifice meter detail

For incompressible (or nearly incompressible fluids, like liquids), Y = 1. For compressible fluids:

$$Y = 1 - (Y_0 d_p)/(p_1 k) \qquad (2)$$

where:

Y_0 = function depending on tap location. For the normal flange taps
$Y_0 = 0.41 + 0.35\beta^4$
p_1 = absolute pressure of upstream fluid, *psia*
k = ratio of fluid specific heats = C_p/C_v

Now,

$$C = C_\infty + b/R_d^n \qquad (3)$$

where:

C_∞ = orifice coefficient at infinite Reynolds number

For normal orifice flange taps:

$$\text{Small pipes} (\sim 2''): C_\infty = 0.5959 + 0.0312\beta^{2.1} - 0.184\beta^8$$
$$+ 0.039\beta^4/(1-\beta^4) - 0.0337\beta^3/d_p$$

$$\text{Large pipes}: C_\infty = 0.5959 + 0.0312\beta^{2.1} - 0.184\beta^8$$
$$+ 0.09\beta^4/(1-\beta^4) - 0.0337\beta^3/d_p$$

$b = 91.71\ \beta^{2.5}$
R_d = Reynolds number of orifice $= 12 * d_o * v_o * \rho_f/\mu$
n = 0.75
v_o = fluid velocity through orifice in ft/s
μ = absolute viscosity in lb/(ft*s)

Solving these equations is primarily accomplished using computer programs today, so the task is not as onerous as it might appear.

From a practical operating standpoint, the key observation from the orifice equations is that the mass flow rate through an existing orifice is proportional to the square root of the pressure drop and the fluid density. Flow meters for orifice taps often make the square root conversion for you. The density correction may also be incorporated into the control system but is normally done separately. If density does not change much, the correction is usually ignored except in test runs.

To correct a liquid flow rate from flowing conditions to a standard, base set of conditions:

$$Q_{base} = Q_{meas}(\rho_{base}/\rho_{meas})^{0.5} \qquad (4)$$

where:

Q_{base}, Q_{meas} = flow rate, base conditions and measured, respectively
ρ_{base}, ρ_{act} = liquid density at base and actual conditions, respectively

This correction is normally minor, but it may be done for detailed test runs.

To correct a compressible fluid flow rate to a base condition density from actual density, then:

$$Q_{corr} = Q_{meas}(P_{base}/P_{act})^{0.5}(M_{base}/M_{act})^{0.5}(T_{act}/T_{base})^{0.5}(Z_{act}/Z_{base})^{0.5} \qquad (5)$$

where:

Q_{corr}, Q_{meas} = flow rate, base conditions and measured, respectively
P_{base}, P_{act} = absolute pressure, base and actual, respectively
M_{base}, M_{act} = molecular weight of fluid, base and actual, respectively

T_{base}, T_{act} = absolute temperature of fluid, base and actual, respectively
z_{corr}, z_{act} = compressibility factor of fluid, base and actual, respectively

Note that by expressing the corrections as ratios, you don't need to deal with conversion of units. You just need to be sure the numerator and denominator of each correction factor are in the same units. Also, note that the pressure and temperature units for the compressible gas corrections are in *absolute* units.

Orifice meters have several cousins:

- Venturi flow meters – These apply the same measurement principles. They have a more smooth flow pattern and recover more head; hence, they have lower pressure drop. Venturis are often used in low-pressure applications where loss of head is expensive.
- Nozzle meters – These are like orifices but use a smooth tube in place of an orifice.
- Wedge or V meters – This type of meter is often used where slurries are processed. The meter consists of a roughly triangular wedge that projects into the flow, with pressure drop measure between the upstream fluid and the narrowest part of the wedge restriction.

Orifice meters, and other meters that depend on pressure drop, enjoy a long history with many applications in refineries. They are handicapped by their limited 3:1 turndown and the pressure drop they impose. They require straight runs of pipe upstream and downstream for good accuracy, which can make the installation physically large, but they can be used for very large flows.

Magnetic Flow Meters

Mag flow meters or magmeters make use of Faraday's law. Faraday discovered that a conductive material moving through a magnetic field will generate a voltage that is directly proportional to the fluid velocity. In a magmeter, a magnetic field is imposed perpendicular to a flowing stream. The voltage generated through the flowing stream is then measured perpendicular to the magnetic field and the stream flow.

The meter calibration is specific to the fluid handled. Density corrections for differences in fluid density are required for the best accuracy. Accuracy can be ± 0.2 % of full scale.

Obviously, this method of measurement requires a conductive fluid to operate. Mag flow meters are "nonintrusive" to the process, i.e., they do not interfere with flow, and they can be built with corrosion-resistant liners to handle corrosive materials. You will find them in refineries when metering molten sulfur, concentrated acids, and other onerous services. They are generally limited to smaller-sized lines by the ability to produce an adequate magnetic field. Installations are fairly small, requiring only short runs of straight pipe.

Oscillatory Meters

There are two common types of oscillatory flow meters in refineries: mass flow and vortex-shedding meters. These meters are expensive but have value in some applications. Mass flow meters require no special pipe runs, but vortex meters require substantial meter runs length. Other types of oscillatory meters include Coanda and momentum exchange meters.

Mass Flow Meters

Mass flow meters are finding increasing use in refinery liquid services. They are generally not suitable for gases. The most common mass flow meter makes use of the Coriolis effect. In this meter, the fluid flows through one or two tubes (usually looped) that are vibrated at their natural resonant frequency by electromagnetic drivers. Fluid flow creates a secondary vibration in the tubes that is a linear function of the fluid flow rate and density. The meter sensors measure the secondary vibration and convert it to flow rate. Accuracy can be ± 0.2 % and these meters have an extremely wide operating rate (up to 100:1 turndown). The flow is measured directly as mass.

A volumetric reading from a mass flow meter can be corrected to a base condition using the ratio of base density to measured density. The mass flow reading needs no corrections.

Vortex-Shedding Meters

The vortex-shedding meter depends on the phenomenon that swirls are created downstream of anything inserted into the middle of a flowing stream and the swirl direction oscillates from one side to the other on a frequency that is proportional to the flow rate. This oscillation can be sensed and converted to flow rate. Flow rates from a vortex-shedding meter can be corrected for density to determine actual flow. These meters have ± 1 % accuracy and good turndown. They offer little flow resistance.

A high Reynolds number (i.e., turbulent flow) is required for vortex shedding to work best. The fluid needs to be clean and noncorrosive to avoid erosion or corrosion of the "bluff body" that creates the swirls.

Positive Displacement (PD) Meters

PD meters directly measure the volumetric flow of a fluid. Meters in this category include nutating discs, rotary pistons, rotary vanes, reciprocating pistons, lobes, and oval gears. Positive displacement meters can be used for both liquids and vapors. They can handle viscous fluids but may lose accuracy at low viscosities. The standard gas meters on homes are PD meters.

These meters are essentially positive displacement pumps forced to run backward. The rate at which they run is sensed and converted to volumetric flow. Correction of the flow to a base or standard condition requires simple adjustment for density.

These meters tend to be mechanically complex (i.e., reliability issues) and expensive. They are used often for custody transfer because they have high accuracy and wide rangeability. Sizes are limited to about 1,000 gpm or less.

The fluid should be clean to prevent undue wear and tear. They do not like two-phase flows. No pipe meter runs are required for PD meters.

Turbine Meters
Another type of meter commonly applied in refineries, especially for custody transfer, is the turbine meter. Accuracy is good and the meters operate over a wide range of turndown. Cost is moderate. The meters require clean, non-plugging or eroding fluids. Gas turbine meters tend to read high. Corrections for density are required for most accurate work. The meters usually have their own straightening vanes incorporated into them.

Essentially these meters operate by spinning a small turbine or multiblade propeller using the flowing fluid. The turbine speed is linear with the flow rate.

Ultrasonic Meters
Two types of ultrasonic meters are common today: Doppler and transit time (or time of flight). The meters are nonintrusive, relatively cheap, and have no pressure drop. They do require a meter run of straight pipe. Accuracy is 2–5 % at best, so they are used mostly for difficult services or where you need a temporary reading on a line without any other meters. They are often "strap-on"-type meters. They are not for gas streams, but liquids or slurries only.

Doppler Meters
A Doppler meter determines the flow rate by sensing the Doppler frequency shift of an ultrasonic signal imposed on a stream and reflected off particles moving in the stream. This means that the Doppler meter only works on dirty liquids or slurries or liquids with entrained bubbles. The resulting flow signal is linear.

Transit Time or Time of Flight Meters
Converse to a Doppler meter, a transit time meter needs very clean fluid to work properly. In this type of meter, the time for an imposed ultrasonic signal to reach a sensor that is a fixed distance away across the flowing stream is determined. The change in this time is converted to a flow rate. The result is linear with flow rate.

Variable Area Meters
A variable area meter imposes a fixed force set against a flowing stream in an expanding flow tube. The flow pushes against the force and moves it in the tube, increasing the flow area until the flowing force equals the resistance. The most common application of this is the rotameter.

These meters are generally limited in size, with variable accuracy. They can be very small but are limited to about $3''$ maximum size. The fluid used must be fairly clean, but some contamination is tolerable. The tubes can be armored to allow higher pressures, but these devices are usually limited to <700 psig.

They are low cost and read flow directly. Fluid properties affect the accuracy. Corrections for fluid density are required for highest accuracy. No meter runs are needed, although a rotameter must normally be installed in a vertical pipe run.

Thermal Meters

There are two common varieties of thermal meters: anemometers and thermal dispersion. These meters are some of the most tolerant of fluid conditions, working with liquids, gases, or slurries. They have a wide turndown range. The primary issue with them would be fouling of the elements. Accuracy is only in the ± 1–5 % range, however, and they do need metering runs, which can be expensive in large scale.

Both types of meters determine the heat loss into a flowing stream by various methods. The resulting output is proportional to the log of the mass flow rate. Corrections for density and stream properties would be required.

You might find this type of meter in a flare header, for instance.

Typical Arrangements

Flow Element Installation

The installation of the flow sensing elements in the process depends on the type of element being used.

For the most common elements, a straight section of pipe upstream and downstream of the meter element is required. This is often referred to as the meter run. For large lines, the meter run may require a great deal of space and cost a lot. For instance, a typical orifice meter run in a 24-in. pipe would require 40 f. of undisturbed 24″ pipe upstream and about 8 f. downstream, plus the orifice installation for a total of about 50 f. of straight pipe. Some types of meters do not require meter runs, so have an advantage.

The metering run does not have to be horizontal in most cases. A vertical run will work just as well. In some cases, such as two phase or slurry, a vertical run is preferred.

If there is insufficient room for a good meter run in the piping, straightening vanes can be used to reduce the length of pipe required, but there will be some compromise in accuracy. Poor incoming flow to an orifice meter can result in the meter being off 15–20 %.

Figure 6 illustrates a normal orifice meter installation, including the meter run. We will spend some time reviewing this installation, since it is most common.

Following the inlet meter run, the orifice is installed in specific orifice flanges that have pressure taps incorporated into the flanges. The taps sense pressure upstream of the orifice and immediately downstream of the orifice. These are normally called "flange taps."

The orifice itself is installed between the orifice flanges, using gaskets upstream and downstream. The orifice will have a tab that protrudes from the flange to indicate the orifice number and size. *The lettering stamped into the tab must point toward the incoming flow for the orifice to correctly measure flow rate*. Errors of 3–10 % are introduced if the orifice is pointed the wrong way.

Downstream of the orifice flanges, there should be about five diameters of straight, undisturbed flow. This is not as important as the upstream meter run, but will introduce inaccuracy if the flow runs immediately into an elbow, for instance.

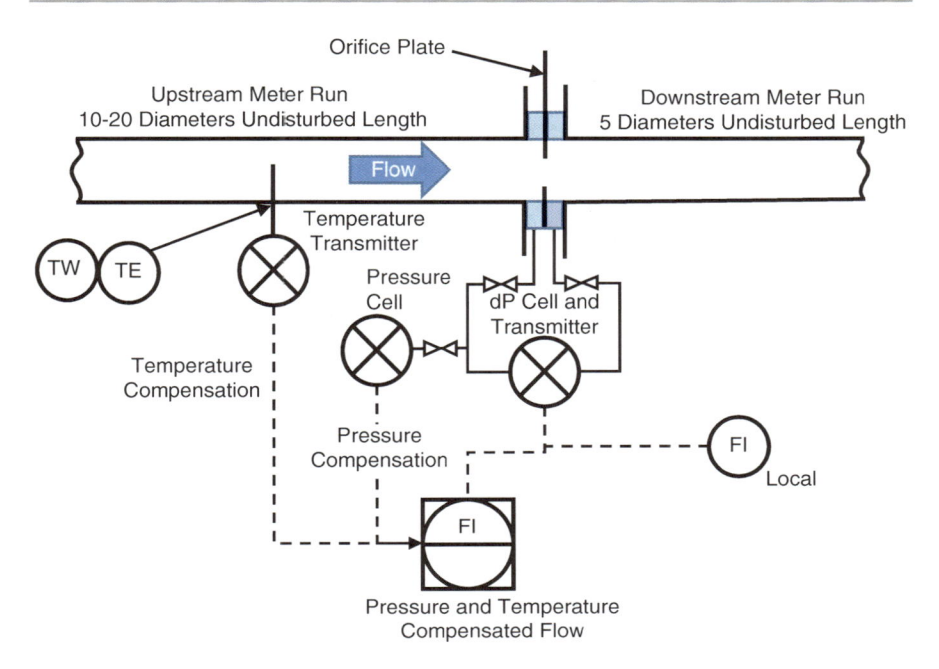

Fig. 6 Orifice meter installation

The taps on the orifice flanges are close-coupled to root valves that comply with the same specification as the piping. These root valves can be used to isolate the flange from the Δp cell that senses the pressure drop for maintenance or if there is a loss of containment from the small piping and tubing connected to the flange. The small piping and tubing are especially susceptible to damage from activity in the area and from vibration.

Piping or tubing from the root valves leads to a differential pressure cell through an inlet instrument manifold. The manifold has its own isolation valves and a bypass valve, as well as calibration connections. It allows the dP cell to be zeroed, spanned, and otherwise maintained. Standard instrument manifolds are available or one can be built from piping, tubing, and instrument valves.

The dP cell is a sensitive instrument that detects the pressure differential between the upstream and downstream orifice taps by the position or strain on a diaphragm with upstream pressure on one side and downstream pressure on the other. The cell may be direct indicating of the pressure drop (or may have a square root scale) or the cell may transmit the dP to the control system. The signal may be pneumatic or electronic and analog or digital.

For flow elements other than orifices, the element installation is somewhat simpler:

- Magnetic and vortex-shedding meters are installed between flanges after appropriate meter runs.

- Coriolis, positive displacement (all kinds), and turbine meters are installed between flanges without any significant meter runs.
- Ultrasonic meters are strapped onto the pipe with an appropriate meter run. They need to be mechanically coupled to the pipe wall to ensure good contact.
- Rotameters are installed into a vertical pipe run.
- Thermal dispersion and pitot tube meters are inserted into the pipe or ducting after an appropriate meter run through a flange using a root valve and packing gland arrangement. They can usually be removed from the line in service and isolated for maintenance.

Most of the non-orifice elements themselves incorporate a sensing system that translates the flow sensor reading into, normally, an electronic signal for transmission to the controller. The signal may be analog or digital and pneumatic or electronic, depending on the control system architecture. For pitot meters, a dP cell is used, similar to an orifice.

While most of the flow sensing elements are connected to the main stream, there are cases where it is impractical to meter the entire stream due to size. In these cases, sidestream metering is an option. For this approach, a smaller branch is created off the main flow line and a known, small pressure drop is incorporated into the main line to cause flow in the branch. The branch flow is metered. The branch flow will be a function of the main line flow rate. In this way, a smaller meter can be used to measure a large stream.

Each refinery and engineering company has its own standards for instrument installation, but most are similar to the above description.

Flow Signal to the Controller

In most cases, the signal from the flow element is transmitted to the controller where the actual conversion to flow rate and any monitoring and control actions are determined. The transmitter may also have its own local flow indication for field operator information.

In some cases, a local controller or indicator is used, but the flow signal is normally sent to the control room for display, recording, and decisions on control action.

Controller Processing

The controller takes the flow signal, compares it to the desired flow rate (set point), and determines the required response, if any. The resulting output from the controller is sent back to the final control element in the field. Controllers are discussed more fully in the section entitled "Controllers".

Output Signal to the Control Element

The output signal from the flow controller is returned to the appropriate flow control element in the field. The output signal may be electronic or pneumatic.

The most common flow control element in a refinery is the control valve. These will be discussed in detail later.

Another control element that is seeing more application in refineries is the variable-speed controller. More on this element in a while.

The control signal is usually received at the final control element by some sort of signal converter that translates the signal into something the control element can use.

Final Control Elements

There are three final control elements we will discuss here:

- Control valves
- Variable-speed drives
- Variable volume

Control Valves

The most common final control element for flow is the control valve. There are many types of control valves available. The different characteristics of these valves offer the opportunity for optimization. Selection of the valve type appropriate to a service is critical.

By way of definition, a control valve is "a power-operated device forming the final element in a process control system. It consists of a body subassembly containing an internal means for changing the flow rate of the process fluid. The body is linked to one or more actuators which respond to a signal transmitted from a controlling element (controller)."

The major types of control valves and actuators used in refining today are described in Table 4. A detailed discussion of the most common types of control valves and key considerations follows.

Figure 7 shows a conventional control valve which in this case is taken as a double-seated plug-type valve. Like most control valves, it is operated pneumatically by an air stream exerting a pressure on a diaphragm which in turn allows the movement of a spring-loaded valve stem. One or two plugs (the diagram shows two) are attached to the bottom of the valve stem which, when closed, fit into valve seats, thus providing tight shutoff. The progressive opening and closing of the plugs on the valve seats due to the movement of the stem determines the amount of the controlled fluid flowing across the valve. The pressure of the air to the diaphragm controlling the stem movement is varied by a control parameter, such as a temperature measurement, or flow measurement, or the like. In many of the more modern refineries, many control valves are operated electronically.

Figure 8 shows two other types of control valves in common service in the industry. These are the venturi type and the butterfly type. Both these types when pneumatically operated (which they usually are) work in the same way as described above for a plug-type valve. The major difference in these two types is in the valve system itself. In the case of the venturi, the fluid being controlled is subjected to a 90° angle change in direction within the valve body. The inlet and outlet

Table 4 Control valve and actuator options

Control valve categories			
Type of action	Valve type	General description	Typical flow pattern past plug/seat
Linear or rising stem	Single port	Globe-type valve with single plug and seat arrangement. Opens as stem rises	Normally upward. Plug pushes against flow
	Double port	Globe-type valve with two plugs and seats on common stem, one above the other. As stem rises, both seats are opened. Forces balanced by flow action	Both upward and downward past plugs and seats to balance forces
	Balanced plug, cage guided	Plug is contained inside a cage with multiple holes in the cage. As plug rises, more holes are uncovered to increase flow. Balance obtained by arrangement of internals	Flow horizontal through cage walls and then downward. From outside of cage inward
	High capacity, cage guided	Similar to balanced plug, cage-guided principle. Rising stem increases flow	Flow upward toward plug and then horizontally outward through cage
	Reverse acting, cage guided	Similar to balanced plug, cage-guided principle. Rising stem *decreases* flow as holes in cage are covered	Flow horizontal from outside cage through the cage wall and then downward and out through passages in the plug
	Three way	Single-plug proportions flow to two different destinations. May be cage guided. Position of stem changes proportion of flow to different destinations	Normally up flow to one destination and downflow to the other destination
	Angle	Similar to a single-port valve. Rising stem increases flow. For very high-pressure services, this type of valve may have several plugs and seats with part of the dP across each set	Downflow past plug and through each seat
Rotary shaft	Butterfly	Similar to damper. A near-line-sized valve disc is rotated with a center shaft located along the disc diameter. Rotation is 90° from full closed to full open	Flow is through the seat and past the disc. Forces on the disc are roughly balanced
	V-notch ball	This is a ball valve with a notch removed from the ball. Greater rotation of the ball exposes more of the wedge-shaped notch, allowing increased flow. Rotation is 90° from full closed to full open	Flow passes through the notch in the ball. The seat for the ball surrounds it
	Eccentric plug	Similar to a butterfly valve, but the axis of rotation is offset away from a smaller disc, so that the disc rotates in or out of a seat ring	Flow may go either direction past the eccentric plug or the seat

(*continued*)

Table 4 (continued)

Actuator categories		
Principle	Actuator type	General description
Pneumatic	Direct acting	Control air pressure pushes down on a diaphragm against a spring to move a linear stem down. The stem is connected to the valve. Loss of air supply causes stem to rise
	Reverse acting	Control air pressure pushes up on a diaphragm against a spring to move a linear stem upward. The stem is connected to the valve. Loss of air supply causes stem to drop
	Rotary acting	Linear motion from a diaphragm or piston actuator is converted to rotary motion similar to "locomotion" acting through a cam eccentrically on the rotary stem
	Piston	Air pressure acting on a piston causes linear movement of the piston rod which moves the valve stem. The piston may have a spring return or may be double acting, using air pressure to move in either direction
Electrically operated	Electrohydraulic	An electrically driven hydraulic pump provides oil pressure to actuate a valve. This type of actuator is often used where rapid action is required, like power generation steam turbine controls. Significant amounts of oil can be stored in high-pressure accumulators to provide rapid action. Can be used on very large valves
	Electric motor	Actuation is achieved by a gearbox connected to an electric motor. This type of valve will fail in place normally if no power is available
Manually operated	Manual	Actuation is normally by the operator turning a handwheel. This type of actuator can be used in any type of valve but is hard to use on very large valves. Most other types of actuators can also have handwheels on them to enable control if the primary control actuator fails. A manually actuated valve has no failure position; essentially it fails in place

dimensions are also different, with the inlet having the larger diameter. The valve itself is plug type but seats in the bend of the valve body. Venturi type or angle valves are used in cases where there exists a high-pressure differential between the fluid at the inlet side of the valve and that required at the outlet side.

Butterfly valves operate at very low-pressure drop across them. They can operate quite effectively at only inches of water gauge pressure drop and where the operation of the conventional plug-type valves would be unstable. The action of this valve is by means of a flap in the process line. The movement of this flap from open to shut is made by a valve stem outside the body itself. This stem movement, as in the case of the other pneumatically operated valves, is provided by air and spring loads onto the stem from a diaphragm chamber. The only major disadvantage in this type of valve is the fact that very tight shutoff is difficult to obtain due to the flap-type action of the valve.

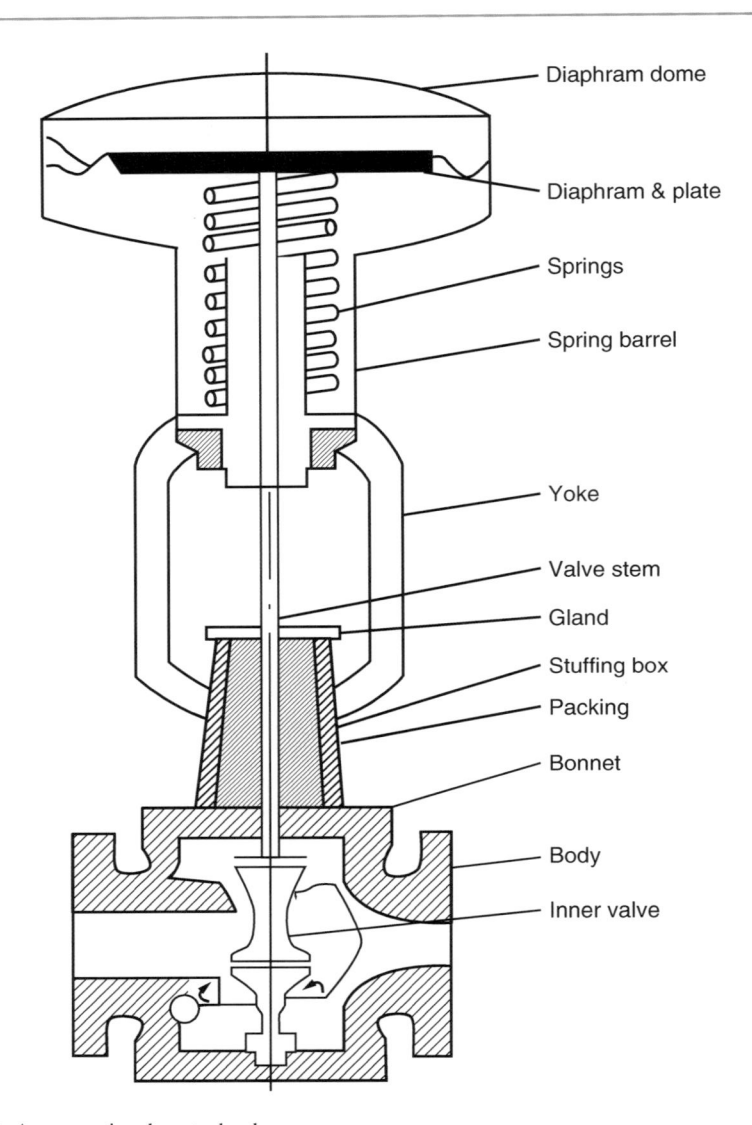

Fig. 7 A conventional control valve

The Valve Plug

There are two types of valve plugs used for the conventional control valve function. These are:

- Single-seated valves
- Double-seated valves

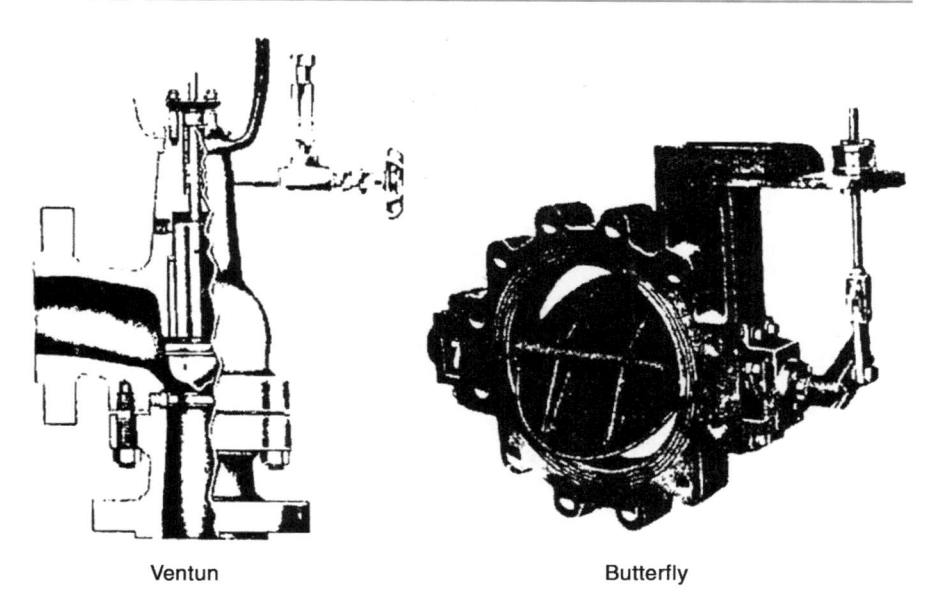

Fig. 8 The venturi and butterfly control valves

Single-seated valves are inherently unbalanced so that the pressure drop across the plug affects the force required to operate the valve. Double-seated valves are inherently balanced valves and are the first choice in many services.

The conventional control valves predominantly in use can have either an equal percentage, or linear characteristics. The difference between these two is given in Fig. 9.

With an equal percentage characteristic, equal incremental changes in valve stem lift result in equal percentage changes in the flow rate. For example, if the lift were to change from 20 % to 30 % of maximum lift, the flow at 30 % would be about 50 % more than the flow at 20 %. Likewise, if the lift increases from 40 % to 50 %, the flow at 50 % would be about 50 % more than at 40 %.

With a linear valve having a constant pressure drop across it, equal incremental changes in stem lift result in equal incremental changes in flow rate. For example, if the lift increases from 40 % to 50 % of maximum, the flow rate changes from 40 % to 50 % of maximum. Thus, equal percentage is the more desirable characteristic for most applications and is the one most widely used.

Pressure Drop Across Control Valves

In sizing or specifying the duty of the control valve, the pressure drop across the control valve must be determined for the design or maximum flow rate. In addition, if it is known that a valve must operate at a flow rate considerably lower than the maximum rate, the pressure drop at this lower flow rate must also be calculated. This will be required to establish the rangeability of the valve.

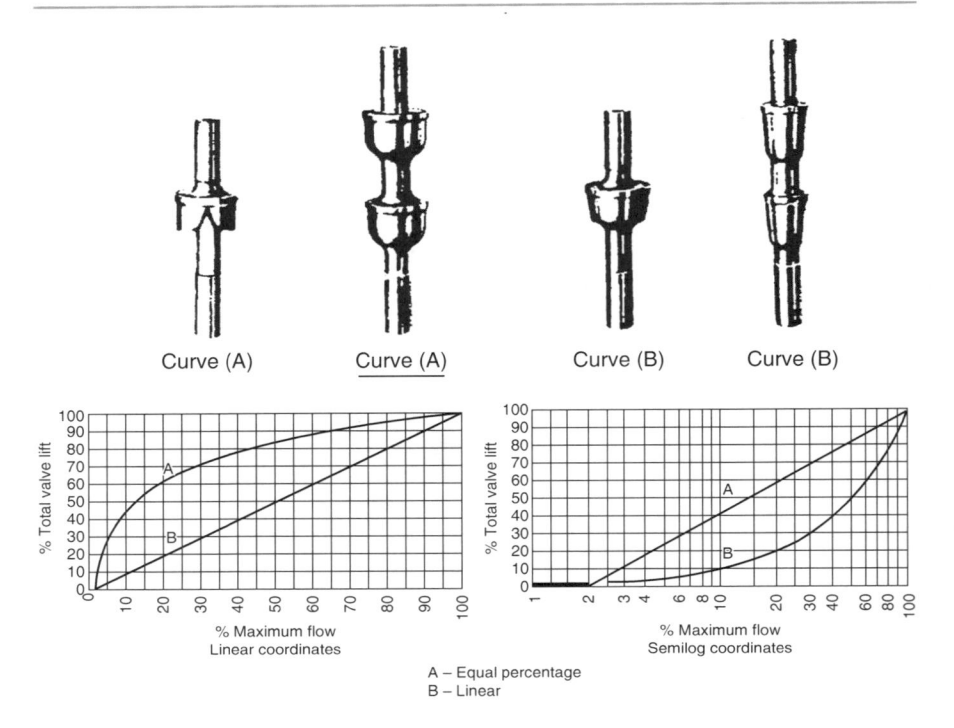

Fig. 9 Control valve characteristics

As a general rule of thumb, the sum of following pressure drops at maximum flow may be used for this purpose:

(a) 20 % of the friction drop in the circuit[1] (excluding the valve).
(b) 10 % of the static pressure of the vessel into which the circuit discharges up to pressures of 200 psig, 20 psig from 200 to 400 psig, and 5 % above 400 psig.

The static pressure is included to allow for possible changes in the pressure level in the system (i.e., by changing the set point on the pressure controller on a vessel). The percentage included for static pressure can be omitted in circuits such as recycle and reflux circuits in which any change in pressure level in the receiver will be reflected through the entire circuit. In some circuits the control valve will have to take a much greater pressure drop than calculated from the percentages listed above. This occurs in circuits where the control valve serves to bleed down fluid from a high-pressure source to a low-pressure source. Examples are pressure control valves releasing gas from a tower or streams going out to tankage from

---

[1] A circuit generally includes all equipment between the discharge of a pump, compressor or vessel, and the next point downstream of which pressure is controlled. In most cases this latter point is a vessel.

Process Controls in Petroleum Processing 1037

vessels operating at high pressure. These are the circumstances where venturi or angle valves are used, as described earlier.

Valve Action on Air Failure
In the analysis of the design and operation of any process or utility system, the question always arises on the action of control valves in the system on instrument air failure.

Should the control valve fail open or closed is the judgment decision based principally on evaluating all aspects of safety and damage in each event. For example: control valves on fired heater tube inlets should always fail open to prevent damage to the tubes from low or no flow through them when they are hot. On the other hand control valves controlling fuel to the heaters should fail closed on air failure to avoid overheating of the heater during the air failure.

The failure position of the valve is established by introducing the motive air to either above the diaphragm for a failed open requirement or below the diaphragm for a failed shut situation. The air failure to the valve above the diaphragm allows the spring to pull up the plugs from the valve seats. Air failure to valves below the diaphragm forces the spring to seat the valves in the closed position. It is also possible to fail a valve in place by blocking in the air to both the top and bottom of the valve, but this option is only used selectively.

Sizing a Control Valve
The sizing procedure for sizing control valves is described in the Appendix of this chapter.

Variable-Speed Control
Of increasing importance in flow control are variable-speed controllers. By controlling the speed of a compressor or pump, the delivered flow rate can be changed directly. The flow delivered will depend on the flow curve and characteristics of the specific device as speed changes.

In a simple example, a reciprocating pump, which delivers a fixed volume for each stroke, can vary flow by changing the number of strokes per minute. This type of pump is used for controlling the injection rate of sulfiding chemicals into a hydrotreating unit during catalyst activation.

In a centrifugal compressor application, varying the speed of a large compressor can be used to reduce power requirements and keep the compressor away from its surge point and the potential for damage to the machine.

If we look around for potential uses of variable-speed drives, there are many possibilities.

There are four main types of variable-speed drives you will find in refineries:

- Turbines
- Engines
- Electronic
- Mechanical

Turbines

One of the oldest and most common variable drives is the steam turbine. Speed of the turbine is controlled by a governor valve that admits a variable steam flow rate to the turbine steam chest. More steam results in higher turbine speed. The upper speed on the turbine is limited by the overspeed trip. Larger turbines incorporate more sophisticated controls for the governor but essentially operate the same way.

Turbines are normally used for their high reliability, but their variable speed and high-power output capabilities recommend them especially for large compressors.

Engines

The internal combustion engine is another traditional, variable-speed driver. Environmental regulations and economics are slowly phasing out engine drivers, but they are still used in remote locations.

Jet engine-derivative drives (really turbojets) are also variable-speed devices that are considered here in the same category as internal combustion engines. These are not common in refineries outside of cogeneration units, which make both power and steam.

Electronic Speed Controls

Electronic speed controllers or variable-speed drives have been steadily increasing in capability over the past few decades. Controllers capable of supplying over 5,000 hp are available now, where as they were limited to 1,000 hp only a few years ago.

Variable-speed drives normally control the frequency of electrical power to an AC motor to generate a variable speed. Higher frequency increases the speed. There are also variable voltage, DC drives.

An important advantage of variable electronic drives are in power (and operating cost) savings. You only use as much power as you need, plus a small loss as heat. The drives have become very reliable and efficient compared to early variable-speed drives. A few decades ago, a 5,000 hp variable-speed drive was the size of a semitrailer and needed an almost equally large cooling system to dissipate the heat. New drives are much smaller and do not generate nearly as much heat.

Mechanical Speed Control

Speed for smaller pumps and drivers can also be controlled mechanically. This is common on many reciprocating and progressing cavity pumps, for instance. A crank is turned to change the speed of the controller. Various mechanisms are used for these controllers, including belts and gears. Essentially the drive is akin to an automotive transmission.

Variable Volume Control

The final common method for flow control in a refinery is variable volume control. In this approach the volume displaced by a positive displacement device is changed mechanically to regulate the flow rate. The mechanism usually changes the stroke length. Variable volume devices encompass both piston and diaphragm types.

Variable volume devices are used extensively in metering pumps. The unloaders on reciprocating compressors are also variable volume devices, which change the effective compressor displacement to reduce the net flow and power required.

Flow Measurement and Control Conclusion

Flow metering and control is the most fundamental operation conducted by a control system. It is critical to modern, continuous operations when controlling pressure, level, and temperature, in addition to direct flow rates. There are many options when choosing the primary flow sensing elements, final control elements, and actuators. These must all be coordinated to produce the necessary control in the refinery processes.

Temperature Measurement and Control

Temperature measurement and control of refinery processes is a second key consideration. There are fewer methods for measuring temperature than flow, but these are no less important. In fact, from a safety standpoint, temperature is one of the most critical factors in avoiding loss of containment.

In the following discussion:

- The primary temperature sensing elements will be reviewed, along with their characteristics and limitations.
- The most common installation arrangements will be described, along with specific considerations.
- Implementation of various temperature controls will be discussed.
- Some special temperature monitoring and control cases will be reviewed, including reactor temperatures, skin and surface temperatures, and furnace tube wall temperatures.

Temperature Measurement Elements

Table 5 provides a survey of the most common temperature measurement elements in refineries and petrochemical facilities.

The most common elements and their measurement principles are:

Thermocouples
These devices make use of the fact that, when heated, a solid-state junction of two dissimilar metals generates a small voltage relative to a fixed reference junction. The voltage is linear over a wide temperature range. The most common thermocouples used in petroleum refining are J and K types.

Table 5 Range of common refinery temperature measurement elements

Property change with temperature (range)	Reading method	Element	Operating range		Accuracy	Annual drift	Additional notes
			English, °F	Metric, °C			
Voltage between dissimilar metals (−330 to +4,000 °F, −200 to +1,700 °C)	High-impedance volt meter with cold reference junction	Thermocouple, J	Continuous: +32 to +1,382 Short term: −350 to +1,472	Continuous: 0 to +750 Short term: −210 to +800	±4 °F at 32 °F, ±17 °F at 2,300 °F	<±9 °F	Iron-Constantan. More sensitive than type K
		Thermocouple, K	Continuous: +32 to +2,012 Short term: −450 to +2,500	Continuous: 0 to +1,100 Short term: −270 to +1,372			Chromel-Alumel. General use inexpensive
		Thermocouple, N	Continuous: +32 to +2,012 Short term: −454 to +2,372	Continuous: 0 to +1,100 Short term: −270 to +1,300			Nicrosil-Nisil. Stable and oxidation resistant
		Thermocouple, R	Continuous: +32 to +2,912 Short term: −58 to +3,210	Continuous: 0 to +1,600 Short term: −50 to +1,768			Pt/Rh 87/13 % − Pt
		Thermocouple, S	Continuous: +32 to +2,912 Short term: −58 to +3,210	Continuous: 0 to +1,600 Short term: −50 to +1,768			Pt/Rh 90/10 % − Pt
		Thermocouple, T	Continuous: −301 to +572 Short term: −450 to +752	Continuous: −185 to +300 Short term: −270 to +400			Copper-Constantan
		Thermocouple, E	Continuous: +32 to +1,472 Short term: −450 to +1,830	Continuous: 0 to +800 Short term: −270 to +1,000			Chromel-Constantan. Cryogenic uses
	Many other options are available						

Resistance (−420 to +1,200 °F, −250 to +650 °C)	Wheatstone bridge	Platinum resistance temperature detector (RTD)	to +1,764 depending on type of element	to +962 depending on type of element	±0.2 °F at 32 °F, ±2 °F at 1,200 °F	<±0.2 °F	Available in various coiled or wound-wire forms: strain free, thin film, wire-wound, and coiled. High accuracy and repeatability below 600 °C (1,112 °F)
		Other detector material options are available					
Expansion – filled systems (−320 to +1,400 °F, −195 to +760°C)	Direct	Bulb thermometer	Typical: +5 to +248; Available up to 930 °F	Typical: −15 to +138; Available to 500 °C	±5 °F	None	Expansion of a liquid in a sealed tube creates a force that drives a mechanical indicator or controller. Highest cost, lowest accuracy
	Direct	Bimetal strip	Linear: −4 to +392 Maximum: ~842	Linear: −20 to +200 Maximum: ~450			Difference in expansion between two dissimilar metal strips bonded together bends the strips one direction, creating a force that drives a mechanical indicator or controller, e.g., home thermostat and overtemperature protection of circuits

(*continued*)

Table 5 (continued)

Property change with temperature (range)	Reading method	Element	Operating range		Accuracy	Annual drift	Additional notes
			English, °F	Metric, °C			
Infrared radiation (−40 to 5,400 °F, −40 to 3,000 °C)	Direct or image enhanced	Infrared thermometer	−76 to +5,432	−60 to +3,000	±0.5–1.0 % of reading		Good range. Works for heater tubes. Requires corrections for reflection and hot gas luminosity. Requires correct emissivity
		Infrared camera	32–1,832	0 to 1,000+			Portable and fixed available. Good for more comprehensive furnace tube monitoring

A J-type thermocouple has a junction between iron and Constantan. Constantan is an alloy of about 55 % copper and 45 % nickel. The normal, stable operating temperature range for this type of thermocouple is about 32–1,382 °F (0–750 °C).

The most common, general purpose thermocouple is K-type. This has a junction between Chromel (90 % Ni, 10 % Cr) and Alumel (95 % Ni, 2 % Mn, 2 % Al, and 1 % Si). Stable operating range is 32–2,012 °F (0–1,100 °C).

There are many other types of thermocouples available in varying temperature ranges.

Thermocouples are the workhorses of temperature measurement in a refinery. They are located throughout the process systems and supply the bulk of the temperature data to the control systems.

Thermocouples have three issues that may affect their accuracy and repeatability:

1. Gradual, cumulative drift on long exposure to elevated temperatures. This is mostly seen in base-metal thermocouples (like J and K). It is due to oxidation, carburization, or metallurgy changes within the metals in contact.
2. Short-term cyclic change when heated into the 500–1,200 °F (250–650 °C) range. Notable in types J, K, T, and E. Again this is due to metallurgical changes.
3. Time-temperature changes due to magnetic transformations. For K couples this occurs in the 77–437 °F (25–225 °C) range and for J couples above 1,350 °F (730 °C).

The issues may dictate the preference for one thermocouple type over another for certain applications. Noble metal thermocouples (like R and S) and some nonmagnetic types of couples help mitigate these issues.

Resistance Temperature Detectors (RTDs)

When very accurate, repeatable temperatures are needed, you look to RTDs. This type of temperature element consists of a length of very fine (usually platinum) wire wound around a core. The wound element is surrounded by a sheath to protect it from direct exposure.

There are also carbon film RTDs that can be used at temperatures down to absolute zero, but these are not common in refineries.

The usual platinum RTDs are limited to applications below about 1,100 °F (600 °C). Above this temperature, the platinum can become contaminated by metal from the RTD sheath and lose accuracy.

There are six factors to consider in comparing RTDs with thermocouples for an application:

1. *Temperature range*: An RTD upper temperature is limited as noted above, but below that temperature, the RTD will be more accurate than a thermocouple. Above 932 °F (500 °C), a thermocouple is the only option for contact temperature measurement.

2. *Response time*: Thermocouples are better for fast temperature response times of fractions of a second. RTDs are slower.
3. *Size*: RTDs, including their sheaths, are usually two to four times larger in diameter than thermocouples (1/8–¼ in. or 3.2–6.4 mm).
4. *Accuracy*: Thermocouple accuracy is limited to about ±3–4 °F (±2 °C). This is normally good enough for most applications. RTDs are more accurate; so if an exact temperature is required, an RTD is indicated.
5. *Stability*: A thermocouple will drift some for the first few hours of use and will age at high temperatures. An RTD is stable for many years within its operating limits.
6. *Cost*: A thermocouple installation is less expensive than an RTD.

Normally, a thermocouple will be used in a refinery, but RTDs will be used in specific applications.

Thermal Expansion Devices

Bulb Thermometers

An industrial bulb thermometer is a metal tube with a fluid trapped inside. This is similar to the classic bulb thermometers we all have at home in our water heaters. The fluid may be volatile.

When heated, the trapped fluid expands or vaporizes, increasing the internal pressure and moving a connected sensing tube, producing a mechanical force which can cause a control action or just be connected to an indicator. This type of element is often seen in low-cost temperature control systems. The bulb of the thermometer is immersed in the material being measured. The long coil of attached tubing is connected to the controller or indicator.

A typical application for this type of thermometer would be in a temperature-controlled cooling water (or tempered water) system to control the amount of water bypassing the cooler. Some dial thermometers for local indication may have bulb sensors.

The biggest argument in favor of these devices is that they are cheap and adequate for many applications. They have limited temperature range, however.

Bimetal Strip Temperature Measurement

A bimetal strip is made from two thin strips of different metals with different coefficients of expansion that are brazed or otherwise attached side by side over their length. When heated, this "bimetal strip" bends as the metal on one side expands more than the other. The phenomenon is repeatable.

The device is almost always local to the temperature measurement point.

The bimetal strip can be connected to an indicator for temperature or to a control means, usually a pair of on/off electrical contacts. The bimetal strip may actually be one of the contacts. This type of instrument can be used in heating system thermostats and in electrical circuit breakers. Dial thermometers for local temperature indication may be bimetal.

Infrared Radiation

Any material emits infrared radiation that is related to the temperature of the material. Infrared thermometers make use of this phenomenon to determine the temperature without having to physically contact the material. The temperatures determined by infrared devices are not as accurate as contact methods, but they can measure the temperatures of materials that are far hotter than anything a contact thermometer can manage.

The high temperature limits for IR devices make them ideal for monitoring furnaces and furnace tube temperatures. Handheld IR "guns" are used to "shoot the tubes" in a furnace to determine the tube wall temperature. IR cameras are used to provide an overall survey of furnace tubes for hot spots, scan for overloaded circuits, look for insulation failures, and many other tasks. The IR images from a camera are translated into a false-color image to accentuate temperature differences.

Because of the cost, IR devices are not normally permanently installed in a refinery, but are instead used on an as needed basis.

The accuracy of measurements from IR devices is limited by several factors, among these are:

- Difficulty in determining the correct emissivity for the material being measured.
- Reflections from the surroundings, like other tubes and refractory walls in a furnace.
- Temperature is most accurate if the view angle is straight on or 0 degrees.
- When measuring furnace temperatures, you are looking through the firebox gases, which are emitting their own radiation. A narrow measurement frequency of IR helps reduce this impact, but it still influences the measurement.

The limitations in IR devices means that the accuracy is normally limited to about ± 4 °F (± 2 °C) in low range with good calibration. In a furnace at 2,000 F and several interferences, the accuracy is closer to ± 25–50 °F (± 15–30 °C).

Additional Temperature Measurement Options

There are a few additional temperature measurement options coming onto the scene in recent years, although these are not widely used in refineries to date.

One of the most interesting options uses the light reflecting properties of fiber optics. These new temperature detectors enable many temperature points to be incorporated into one fiber-optic cable, which can be very long. One possible use of such a system would be to provide an indication of the hottest wall temperature on a reactor shell with much greater resolution than other methods. The upper temperature limit for this technology so far is only about 500 °F (260 °C), but it does offer another possible measurement development to watch.

Typical Temperature Measurement Arrangements

Here we will focus on the application of thermocouples and RTDs in refinery process temperature measurement. There are four types of applications discussed, which cover the most commonly encountered situations.

Thermowells

A thermocouple or RTD is seldom inserted into process fluids directly. They are inserted into the fluid housed within a thermowell. This is essentially a small, capped pipe designed for the process pressure, temperature, and chemical environment. The pipe is essentially open to atmosphere at the end opposite the cap. The temperature measuring element is inserted into the open end with the thermocouple junction or RTD positioned inside the thermowell at the desired location. The thermowells may be threaded or flanged, as dictated by the piping design conditions and specifications.

There are different types of thermowells, as illustrated in Table 6. The main types are as follows:

- *Straight thermowells* are commonly used where vibration is not a potential problem. Straight thermowells may contain more than one thermocouple, as in reactor thermowells. Each thermocouple can be independently positioned in the thermowell. Clips may be used to hold the thermocouple points against the thermowell wall and position them.
- *Tapered thermowells* are like straight thermowells but taper toward a smaller diameter toward the capped end. This helps them resist vortex-shedding-induced vibration damage. These are common in many fast-flowing process lines.
- *Flexible thermowells* are becoming the preferred type of thermowell in fixed bed reactors. These thermowells are small diameter and can be bent to place the

Table 6 Common thermowell types

Type of thermowell	Straight	Tapered	Flexible	Flexible multipoint
Picture	Threaded style	Flanged style	Individual small tube for each TC	4 TC points in same thermowell tube
Applications	General use	Flowing streams where flow-induced vibration could be a problem	Normally reactors	Normally reactors

thermocouple points in specific locations, not limited by the normal straight thermowells. There are multi-tube versions, where each thermocouple has its own tube, and multipoint versions, where all the thermocouples are packed in a small flexible tube with the thermocouple junction locations marked. These thermowells must have a support system installed to prevent damage or movement in service.

Furnace Tubeskin Temperatures

Infrared temperature measurement of furnace tubes and other similar services is fine for general temperature trending, but for more accurate temperatures, direct contact measurement is required. This is a challenge because the furnace firebox environment is very hostile toward materials when they are not cooled by external means, such as flow. It is also necessary to install the skin TIs on the side of the tube facing the flame, which is much hotter. Successful placement of the skin TI at the hottest point on the tube wall is also problematical. This can be calculated, but changes in fouling and firing tend to move the hottest point in the tube wall around. At best a tubeskin TI is just an indication.

While tubeskin TIs have a hard time with reliability in the firebox, they work fine in the convection section, where the environment is less rigorous.

Systems have been developed to attach thermocouples to furnace tubes that overcome many of the obstacles of the furnace environment. They can't help the location limitations. The two most common installations for skin TIs are direct welded pads and refractory-shielded pads:

Welded Pads

In a welded pad installation, the thermocouple is routed through a flexible stainless steel sheath with MgO insulation to a high-alloy (like Hastelloy-X) weld pad and peened into the pad. The pad is then welded to the tube on the side of the tube facing the fire. There is adequate length and flexibility designed into the sheath for differential thermal expansion.

This type of installation has been somewhat successful, but the welds of the pad to the tube can break, the thermocouple can break off the pad, or the sheath and thermocouple may burn up. Variations of this installation have also been used, with similar issues.

The welded pads work well in the furnace convection section.

Refractory-Shielded Pads

This approach overcomes some of the issues of a welded pad. The thermocouple is routed through a stainless steel sheath insulated with MgO to a weld pad where it is attached. The pad is welded to the tube wall on the side pointing toward the tube. So far it looks just like a pad installation.

Now a 310 stainless steel shield is wrapped over the shielded TC and welded to the tube wall. The shield is packed with refractory and extends to cover the sheathed TC over the whole length exposed directly to the radiant heat.

This type of installation, and similar approaches available from manufacturers, has been somewhat more successful in improving skin TI reliability. They are still limited by the fixed location.

Reactor and Vessel Wall Temperatures

It is often necessary to attach skin temperature instruments directly to the wall of a reactor or other hot vessel. These are most often used to ensure minimum pressurization temperature requirements are met and to detect possible hot spots or maldistribution inside the vessel near the wall. The situation is somewhat parallel to the furnace piping, but welding onto the pressure vessel after fabrication and stress relief is a problem.

Note that this is one case where a thermocouple is not used in a thermowell.

One might ask why a magnetic thermocouple holder isn't used. The difficulty with the magnetic holders is that they become demagnetized at reactor operating conditions and then fall off. They are satisfactory for short-term measurements, however.

A number of approaches have been used for installing these skin thermocouples. A couple of them are described below, but each refiner has preferences.

Welded and Threaded Stud

During vessel fabrication, a threaded stainless steel stud can be welded to the vessel at the desired location. The thermocouple is then clamped down to wall using an appropriate washer and nut on the stud. This approach works somewhat but requires preplanning. The studs sometimes break off or the thermocouple comes loose in service. Since the location is usually heavily insulated, it is hard to detect the failure. Backup temperature measurements may be needed if a skin TI does not change along with the other skins.

Drilled Well

Another method used by some refiners is to carefully drill a hole into the vessel shell a specific distance, with a special taper to receive the thermocouple. The thermocouple is inserted and retained in the hole with the TC located at approximately the middle of the vessel wall. Many refiners are unwilling to drill into the vessel walls, but this method does appear to work.

Peened Thermocouple

In this approach, a smaller, shorter hole is drilled into the reactor wall and the thermocouple junction is peened into the vessel wall. These have a tendency to pop off from thermal expansion and contraction, so are have limited reliability.

Temperature Mixing Distances

In many cases, the objective is to measure the temperature of a combined stream that includes both hot and cold fluids. If you measure the temperature just downstream of the mix point, assuming flow is turbulent, you will find the temperature varies significantly.

Process Controls in Petroleum Processing   1049

Just as with flow metering, the location for temperature measurement of a stream when two fluids are mixed must be measured at a point far enough downstream that the fluids are again at a steady state. If the mixing involves reaching a new phase equilibrium, the problem is further complicated.

As a rule of thumb, a temperature measurement for the combination of two streams should be made at least 10–20 pipe diameters downstream of the mix point for turbulent flow. If flow is laminar or stratified, the measurement point may need to be much further downstream. The addition of an in-line static mixer can significantly shorten this distance. Manufacturers can help with the design and computerized fluid dynamic (CFD) modeling of the flow resulting from a static mixer and enable you to place the downstream TI at the best location.

Control Elements

Temperatures in refining processes are controlled by heat exchange, furnace firing, and flow of the reactants for the most part. Temperature control is thus reduced essentially to flow control of feeds, internal fluid flow, or fuels – all flow control issues.

Hence, the control of temperatures mimics the flow control discussion. A typical temperature control scheme for the inlet to a reactor using a fired heater might look like Fig. 10.

In this scheme, there are three TIs at the top of the reactor plus an inlet TI. All the reactor influent passes through the fired heater. The inlet TI can control the reactor inlet temperature by controlling the fuel gas pressure to the heater (providing a set point to the fuel pressure controller). The center TI at the top of the reactor bed acts as a backup controller through a selector switch on the DCS. The other two bed TIs are for calculation and indication.

TIs that will be used only for indication (TI 3 and TI 4) are sent directly to a thermocouple input unit that is part of the distributed control system (DCS). The TIs that may be used for control (TC 1 and TC 2) are transmitted to the DCS using smart, digital transmitters to improve accuracy and provide diagnostics. The cost of a smart or transmitted TI loop is higher than a simpler TI loop, so instruments do not always use a transmitter.

Overall, temperature cascades to fuel gas pressure control and the fuel gas flow is available for monitoring and alarming. As an aside, some people like to cascade temperature control to fuel gas flow control and then to fuel gas pressure control. This does produce smoother control, but response is slow if there is a major temperature change needed.

Concluding Comments

There are several options for temperature measurement, but the most common elements for temperature are thermocouples, with other instruments a remote second. The temperature instruments are usually installed in thermowells inserted

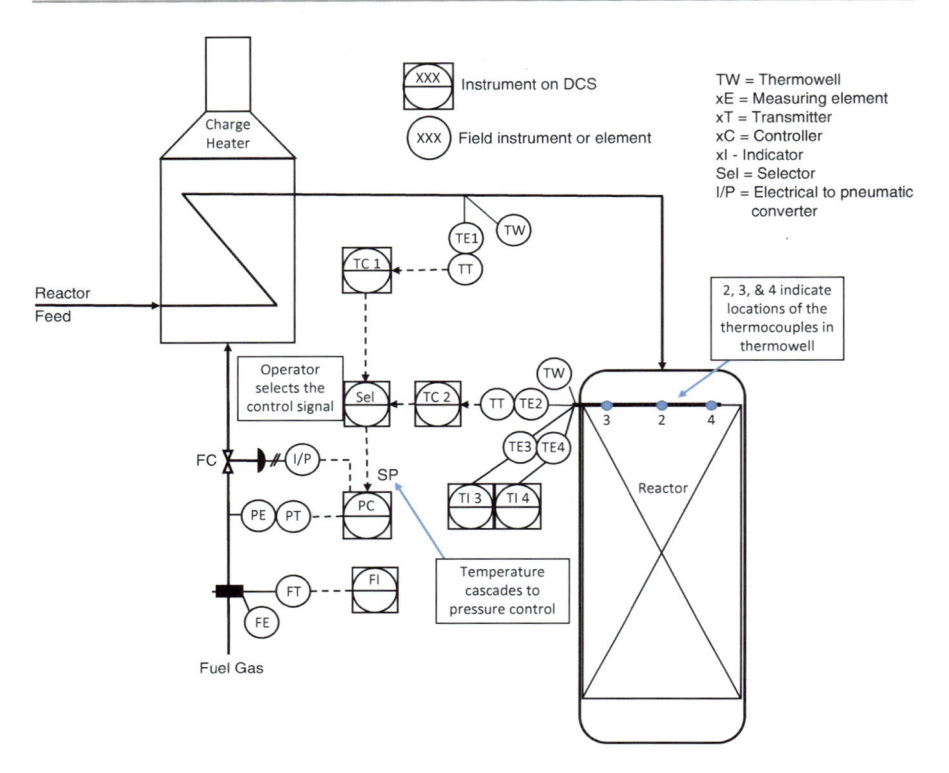

Fig. 10 Reactor inlet temperature control scheme

into the process fluids, although some types of temperature measurement (like infrared) do not require direct contact and some are inserted directly into the material being measured.

When mixing two fluids, be sure to allow adequate mixing distance and time or provide devices to accelerate mixing so the resulting temperature correctly represents the combined fluid conditions.

Remember that temperature control frequently translates into flow control to effect action.

This has only been a superficial treatment of temperature control. Working with your instrument supplier and experts within a facility can clarify the options and allow you to make better decisions on specific applications.

Pressure Measurement and Control

The third parameter that must be regularly measured and controlled in refinery processing is pressure. Aside from the process implications of having the wrong pressure for the physical and chemical processes in the units, pressure control is critical to safety.

Pressure Measurement Elements and Characteristics

Table 7 provides a survey of the general range of commonly applied pressure instruments in petroleum processing.

The most common pressure indicators in processing are the Bourdon tube gauges you will see throughout a facility. These are classically what we think of as a pressure gauge. They usually provide only local indication, although the principles can be adapted for local pressure control or the pressure can be transmitted to a remote control system.

More common for control systems today are electromechanical pressure sensors. These readily provide data to a remote control system as an electrical signal. The electronics for the sensor can be packaged with the sensor housing, so no separate transmitter is needed. The most common remote pressure sensors are normally capacitive.

There are three types of pressures that are measured in general:

- *Gauge pressure* – This is the pressure of the process compared to atmospheric pressure. A positive gauge pressure means pressure is above atmospheric. A negative gauge pressure indicates vacuum.
- *Absolute pressure* – This is the total pressure of the process compared to total vacuum. It is always positive and equals the gauge pressure plus atmospheric pressure. We are normally only concerned about measuring absolute pressures in vacuum systems.
- *Differential pressure* – Sometimes we are looking for small differences in pressure between two streams. In this case we do not want to compare against atmospheric pressure, but want to compare the two pressures directly. This is done in a dP or differential pressure cell. Essentially this is similar to other pressure sensors, except that the two sides of the sensor can be connected to the two pressures of interest in the dP rather than having one side connected to atmosphere. As an alternative, modern, high accuracy electronic pressure sensors can also be used for this service by taking the difference between two gauge pressures, if the difference is large enough.

Typical Installations

Some typical instrument installations, in simplified P&ID format, are in Fig. 11.

Pressure sensors are normally tapped directly into the process at the point being measured.

For simple, local Bourdon tube pressure gauges, there would be a root valve meeting the piping line class at the location point and a bleeder valve to allow servicing of the gauge. If the gauge is close enough to the tap or the pressure is low, the bleed valve may be omitted.

For a control pressure installation, there is normally a piping-class root valve installed for maintenance followed by tubing impulse lines leading to the pressure

Table 7 Pressure sensing instruments

Element type	Sub-categories	Pressure range, psia	Typical accuracy	Advantages	Limitations
Mechanical	Bellows elements	0–30	±0.5 %	Moderate cost. High force output. Absolute or differential pressures. Low-pressure capable	Require spring characterization. Need temperature compensation. Low ranges only. Work hardening of bellows
	Diaphragms and capsules	0–200	±0.5 %	Moderate cost. Close-couple to process. Small. Capable of high overpressures. Linear response	Limited range. Hard to repair. Susceptible to shock and vibration. More complex output transmission
	General purpose Bourdon tube gauge	15–10,000	±2 %	Low cost. Simple. Extensively used. High pressures. Good accuracy versus cost. Adaptable to transducers	Poor below 50 psig. Mechanical movement. Susceptible to shock. Hysteresis effects over range
	High accuracy test gauge	Vac to 3,000	±0.25 %	Similar to general purpose gauge, but higher accuracy	Similar to general purpose gauge and higher cost
Electromechanical	Strain gauge transducers	0–200,000	±0.25 %	Good accuracy. Range to 200,000 psig. Stable. High output signal. Small. No moving parts. Overload capacity	Need power supply. Electrical only. Ambient temps below 600 °F. Limited corrosion-resistant materials. Cost
	Piezoresistive sensors	0–6,000	±0.1–0.5 %	High accuracy. Linear. Repeatable. Smart sensors available. Small size	Higher cost. Electrical systems only. Temperature compensation. For dP, also compensate pressure
	Capacitive sensors	1 in.Wtr to 6,000	±0.2 %	Good accuracy and linearity. Easy to adapt to digital. Moderate cost	Temperature compensation preferred. Limited corrosion resistance

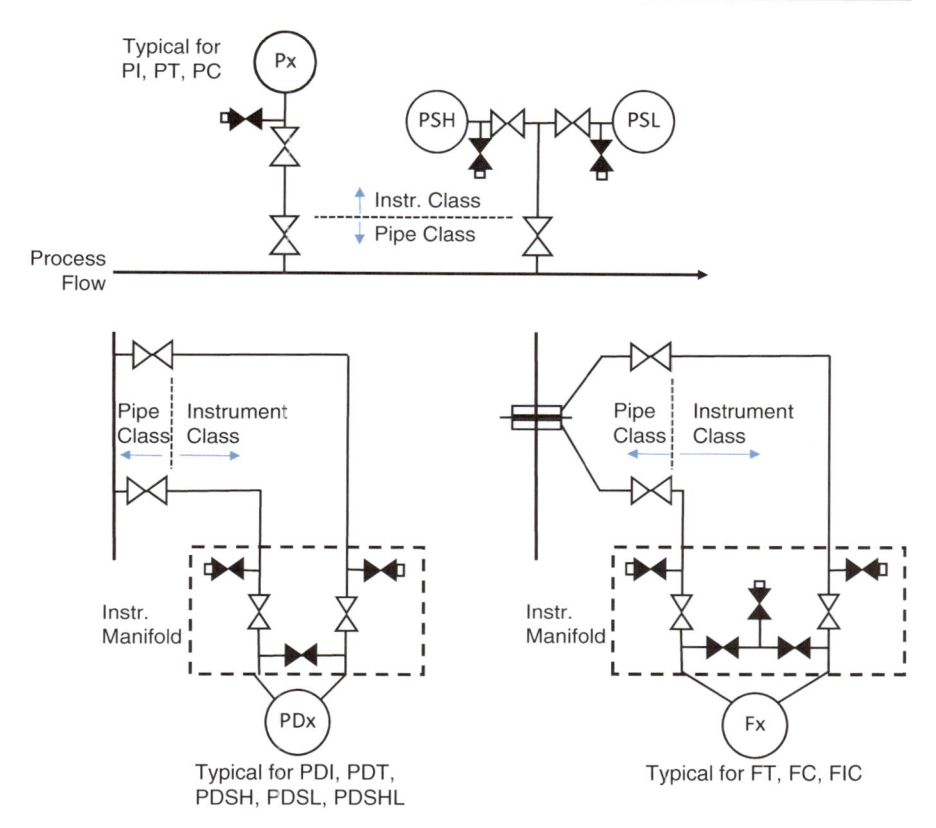

Fig. 11 Some typical pressure instrument installations

instrument manifold. The manifold on the pressure instrument is pretty much the same as those used for flow instruments. It allows for calibration of the sensor and bleeding the sensor impulse lines.

It is preferred that the impulse lines free drain to the sensor for liquids. If the impulse lines are very short, the liquid pressure instrument may be on the top half of the line also. A sensor has to be calibrated with the impulse line(s) full of liquid if this will be the normal condition.

For vapors, the pressure instrument should be located above the tap and the impulse lines should free drain into the process. These instruments can be located below the line but must then be calibrated with the lines full of liquid.

Many fluids can damage or corrode the sensitive parts of the pressure sensor. In these cases, isolation is provided between the process and the instrument. Isolation can take the form of filled taps (with an isolation fluid like glycol) or by installation of a high alloy or gold diaphragm seal at the pressure sensing point with a filled tube connecting the seal to the instrument. The pressure instrument must be calibrated with whatever sealing device is used. Care must be taken to avoid breaking or losing a seal fluid. Also, if a seal diaphragm is used against a hot hydrogen stream, the diaphragm must be gold plated to avoid intrusion of hydrogen and subsequent rupture of the diaphragm seals.

Steam pressure measurement presents a special case for installation. In steam lines, the impulse lines will often have a coil (or pigtail or siphon) in them which sets up a liquid condensate seal between the steam and the pressure sensor. This acts as an isolator between steam and instrument. The same approach can also be used for other condensing vapors, if desired.

Often pressure must be measured on pulsating flows. The shock created by each pulsation will damage pressure sensors if they are not protected. There are several devices available to dampen the pulses. Options include throttle screws, pulsation dampeners, snubbers, needle valves, and other devices. These slow down response to pressure but avoid damage.

Some process fluids can freeze, resulting in plugged or even ruptured impulse lines. This is particularly an issue with water or steam. In these cases, the lines must be "winterized." This can be accomplished by sealing the impulse lines with glycol solution or another, nonfreezing fluid, or by heat tracing the impulse line with steam or electric tracing. The pressure sensor also has to be winterized if it sees process fluid, although care has to be taken to avoid damaging the sensor with heat.

The same precautions listed above apply regardless of the pressure sensor service: indicating, controlling, differential, or flow metering.

Pressure Control Elements

As with flow and temperature, the primary pressure control elements are essentially controlling flows using control valves or one of the other control mechanisms discussed under flow.

The devices controlling pressures receive their signals from pressure controllers that are either field mounted or remote (in a control room or DCS), so the pressure control loops look like other loops.

We will, however, examine a few special cases where pressure control entails specific requirements.

Pressure Regulators

A pressure regulator is a self-contained pressure sensor/controller combination that maintains a fixed pressure. The set pressure can usually be changed manually by adjusting tension in a spring.

Tank Blanket Gas Control

Because of their large size, most storage tanks are not designed for pressures above a few inches of water. In many cases, tanks have to be blanketed with inert gases to avoid undesired reactions. At other times, they may have to vent vaporized gases to avoid overpressure. Special pressure control valves are used to accomplish these tasks. The valves can maintain a tank gas blanket pressure at a set point within a couple of inches of water. These valves use large diaphragms that are controlled by spring, pilot, or weights to achieve the fineness of control necessary.

Heater Draft

As with tanks, heater draft requires very tight control to within a few inches of water pressure. This is accomplished by using a narrow-range pressure transmitter that feeds a draft pressure controller which sends its output to a damper at the base of the heater stack or an induced draft fan to manage the pressure in the firebox. The control system is often redundant and has checks on its operation, such as excess oxygen monitoring and unburned hydrocarbons. These checks can override the controls and shut the heater down.

Most heaters do not have a draft control system, but have local draft meters or gauges instead. Sometimes these are actually water manometers but they can be special Bourdon tube draft gauges. The damper is manually controlled. The draft gauge taps are located at several points in the heater, including:

- At the burner level
- At the arch or convection inlet
- Ahead of the damper
- Above the damper

The normal point at which draft is controlled is the arch or convection inlet. A slightly negative pressure is required at this point to ensure air flow into the heater at the burners.

Distillation Column Pressure

The old adage that "there is more than one way to skin a cat" certainly applies to distillation column pressure control. Many ways of controlling the tower pressure at the overhead have been developed. One of the best summaries and discussions available is probably that by A.W. Sloley of The Distillation Group, Inc., published in Chemical Engineering Progress, January 2001 (see references). Table 8 provides a summary of the 19 methods described in that article.

Some techniques work better than others. Simplicity works much better than elegance. Operators must be able to understand and control a tower using whatever method is chosen or the control will be unsuccessful.

Concluding Comments on Pressure

Several pressure measurement options are available, with the most common being Bourdon tube gauges, pressure transducers, piezoresistive sensors, and capacitive sensors.

Installation of these sensors requires care to protect the sensor from damage due to freezing process fluids, corrosives, pulsations, high temperatures, and fouling.

Table 8 Distillation column pressure control options

Method	Description	Sketch	Advantages	Disadvantages	Applications	Variants	Configuration	Concept	Precautions
Tower net vapor rate always > zero									
1	Direct control of product vapor rate from OH accumulator		Simple and direct	Generally none. May be slow with low vapor rate	Usually best choice for positive vapor rates	Condenser outlet to bottom of drum, but adds dP	Works best with condenser above drum. Gravity flow to drum from condenser. Free drain two-phase exchanger outlet	Controller directly affects column vapor inventory and pressure	
2	Control recycle vapor rate to compressor (spill back)		Simple and rapid control	May need to oversize compressor to add control margin. Not particularly energy efficient	When vapor goes to higher pressure equipment	Vapor may return upstream of condenser, to drum, or to compressor suction	Bypass to flare can be added to operate when compressor is down	Material is recycled to maintain constant suction pressure	
3	Control recycle vapor rate to ejector – ejector discharge recycle; ejector pulls vapor from OH accumulator		Ejector discharge available for recycle. Stable	Difficult to implement with ejector close-coupled to condenser. Large recycle line (low pressure) and control valve	Vacuum systems. Best for ejectors that don't like zero load. Often most economical	Recycle upstream of condenser. Recycle condenser vent upstream of ejector	Vapor may be a product or just noncondensables. In multistage ejectors, recycle loops around first stage (unstable otherwise)	Recycle gas loads up the ejector. More recycle means higher suction pressure	

4	Control makeup ejector load; ejector pulls vapor from OH accumulator, load control using steam injection to process		Easy to retrofit steam injection. Small flows needed	Lower energy efficiency. More steam and cooling water loads	Add on after the fact. Improves control. Useful in batch systems	Inert, noncondensable gas can also be used instead of steam	Inert gas use loads up condenser and any subsequent ejectors	Added load moves ejector along operating curves. Higher load = higher suction pressure	Injected steam should be dry to avoid erosion
5	Control ejector motive steam pressure; ejector pulls vapor from OH accumulator		Minimizes steam demand and condensate	Not normally a good choice. May only have a small control range	Added on after the fact. Poor choice			Lower pressure steam supply reduces ejector flow rate, reducing pressure ratio across ejector	Sometimes does not work. Narrow control range
6	Control recycle vapor rate to ejector – ejector discharge recycle; ejector pulls vapor from tower OH		All ejector discharge available for recycle. Often most stable	Hard to do with ejector close-coupled to condenser. Large recycle line (low pressure) and control valve. OH must be compatible with water	Best scheme for ejectors connected to tower that are unstable at zero load. Often most economical for stability	Can be used without overhead reflux (external reflux supply)	In multistage ejectors, recycle loops around first stage (unstable otherwise)	Recycle loads up ejector, increasing suction pressure	

(*continued*)

Table 8 (continued)

Method	Description	Sketch	Advantages	Disadvantages	Applications	Variants	Configuration	Concept	Precautions
Tower net vapor rate > or = zero									
7	Control vapor product rate in conjunction with a secondary method				Good for systems requiring inert gas venting	Numerous	Can be used with other methods that vary condenser duty, except for flooded drum (12)	Direct overhead pressure control. Changes heat transfer conditions in condenser	
8	Control makeup vapor supply (inert gas)		Simple. Fast response	Pressurizing gas make/vent arrangement more complex. Tuning issues	Pressure towers	Can add the inert upstream of condenser, but slower response	Pressurizing gas must be compatible. Does not require equalizing line. Small overlap where makeup and vent are both active works best	Gas addition or venting directly controls vapor inventory	
Tower net vapor rate zero (total condensing)									
9	Control flow to condenser with condenser at lower pressure		Simple	Drum level affects pressure. Need large control valves and extra exchanger capacity	Stable and effective for many applications		Need equalizing line for stability. Condensed liquid must enter below liquid in drum. If use air cooler, slope to drain. Cool to lowest temp	dP from control valve creates semi-flooded condition in condenser. Effectively variable area	

Process Controls in Petroleum Processing

10	Control flow to condenser with condenser at higher pressure		Simple. Smaller control valve	Gravity drain required from condenser to drum. Extra exchanger area needed. Careful design of elevations	Stable and effective for many applications	Liquid may enter top of drum	Need equalizing line for stability. Condensed liquid must enter below liquid in drum. If use air cooler, slope to drain	Control valve causes liquid backup into condenser. Variable condensing area	
11	Control bypass flow to condenser receiver – hot vapor bypass scheme		Condenser may be mounted below drum	Complex concept hard to understand. Condenser must be able to subcool	Used with large, heavy condensers at grade	Can use drum pressure instead of tower pressure. Condenser does not count as a stage	Must have liquid level in condenser at all times. Liquid must enter drum subsurface. Need adequate bypass size to make it work	Varies surface area in condenser. Drum has superheated gas over subcooled liquid. L and V not in equilibrium in drum	Sometimes does not work. Largely empirical design. Problem with high-purity products, self-refluxing condensers, corrosion of internal pipe
12	Direct control of liquid product rate		Liquid control of drum level not required	Changing exchanger area varies distillate product rate. May cause upsets downstream. Hard to reject noncondensables from upsets	Used with large, heavy condensers at grade	No condensate drum. Control of reflux rate instead of product rate	Must have liquid level in condenser at all times. Drum sometimes used to decant immiscible liquids. Sometimes no drum	Varies condenser surface area	If condenser is below the drum, subcooling is required for stability

(continued)

Table 8 (continued)

Method	Description	Sketch	Advantages	Disadvantages	Applications	Variants	Configuration	Concept	Precautions
13	Dual pressure control of bypass and condensate		Condenser may be below drum	Requires more control valves and subcooling area in condenser	Used with large, heavy condensers at grade. Used if condenser capacity cannot be easily varied (e.g., water boxes)		Must have liquid in condenser at all times	Varies condenser surface area	
14	Control vaporizing coolant level			Hard to blow down the vaporizing coolant if required	Heat recovery where one tower reboils another or for steam generators	Direct flow control of coolant, without level controller	Common in cryogenic plants	Varies level of coolant in the condenser to change heat transfer	
15	Control condenser inlet coolant pressure		Easier to blow down vaporizing coolant if required	More equipment	Heat recovery. Works with multiple shells in parallel		Unusual	Changes LMTD on coolant side of the condenser, changing capacity	

16	Control condenser inlet coolant temperature – tempered water concept		Gets the most out of coolant stream	More equipment. Slow response	Heat recovery. Helps control potential for overcooling of products and freezing	Pump in exchanger outlet instead of inlet	Called a tempered water system when used with cooling water. Common in steam generation heat recovery arrangements	Cooling stream temperature varies, changing LMTD	
17	Control coolant rate		Simple	May overheat coolant. Promotes cooling water fouling	Not common in new plants	Can pinch the outlet valve from exchanger and do this manually. Generally not a good idea	Need an override or limits to avoid fouling	Cooling stream temperature varies, changing LMTD	Causes exchanger fouling on coolant side
18	Control air cooler coolant rate (fans)		Simple	Mechanical issues with air cooler controls. Variable speed not used often due to cost	Most common method with air coolers	Can use blade pitch, louvers, or fan speed to control	Can combine types of control (pitch, speed, louvers) for fine-tuning. Slope to drain to drum	Varies air (coolant) flow rate to change LMTD and exchanger capacity	

(continued)

Table 8 (continued)

Method	Description	Sketch	Advantages	Disadvantages	Applications	Variants	Configuration	Concept	Precautions
19	Control condensation temperature or pressure		Simple, fast response	Big control valve. Condenser and drum pressure vary, causing problems for reflux pump	Pressure towers		Cooling is done to lowest temperature (i.e., overhead pressure is lower)	Direct overhead pressure control	Hard to do in vacuum systems, unless system has vacuum pump downstream. Variable reflux pump suction pressure

Process Controls in Petroleum Processing   1063

Pressure control ultimately boils down to flow control and management of fluids entering and leaving the process system. Pressure control for gases is similar to level control for liquids in that it is really a material balance control. If gas production rate varies, which is the usual case, a pressure controller will relieve the system of gas by holding a given pressure in a drum or tower. In this case the entire space above the liquid level actually constitutes a surge volume.

Level Measurement and Control

Level instrumentation is the next type we will explore.

Level monitoring and control is critical to avoid overflow or overfilling equipment. The principle applications of level control cover many key areas, including:

- Crude, intermediate, and production tank level and inventory change monitoring
- Custody transfer
- Management of process surge drum and tower levels and inventories
- Avoiding liquid carryover into rotating equipment resulting in damage
- Avoiding level carryover from vaporizing liquids
- Managing fluidized catalyst levels and densities
- Avoiding damaging pumps from running dry
- Avoiding spills and loss of containment from tanks, vessels, flares, and other equipment

We will start this examination by looking at the types of level-sensing instruments available and their characteristics. Then some common installation details and critical factors will be described. The discussion will move on to control elements and considerations for level, which will look a lot like flow control. This section will conclude with discussion, in some detail, of specific level control application issues, like minimum recommended surge times for various services.

Level Measurement Elements and Characteristics

Table 9 provides a survey of most of the types of level instruments found in petroleum processing today. The list is divided into three primary types of services:

- Continuous measurement – usually for control or monitoring
- Point measurement – usually for alarming and safety actions
- Liquid inventory gauging – tank levels

The most common level instruments in refineries depend on pressure differences, floats/displacers, or gauge glasses. These cover the majority of the continuous services and point applications in process systems. In tank farms, float and tape or head gauges dominate the applications.

Table 9 Range of level-sensing instrument types

Technology	Application phases	Operating conditions limits	Advantages	Disadvantages
Continuous level				
Pressure, differential pressure	Liquid	6,000 psig, 200 ft	Reliable. Simple. Easy to use. Flexible. Optional diaphragm seals. Unaffected by foam. Tolerable of agitation	Problems with high temp or vacuum. Affected by density. Can plug. Process corrosion. Hydrogen infiltration. Requires two process taps
Capacitance	Liquid, solid	1,000 psig, 500 °F, 100 ft	High temps and pressures. Good for interfaces. Unaffected by density, moderate foam, agitation, vapors	Affected by dielectric constant variation. Not good for high visc. Individual calibration. Fouling of probe. Error in emulsion layers
Ultrasonic	Liquid, solid	3,000 psig, −100 to +300 °F	Noncontact. Easy installation. Unaffected by density, dielectric constant, sludge, contamination, and fouling	Affected by changes in vapor. Temp and press limited. Poor in foam, high agitation, temperature changes
Radar, guided wave	Liquid, some solids		Different options. Unaffected by density, dielectric constant, visc, sludge. Low maintenance	Poor with agitation and foams or start-up. Larger process connections. Complex to configure. Antenna fouling
Displacer	Liquid	10,000 psig, 500 °F, 30 ft	Unaffected by agitation. Small spans possible. Measures interface. Can measure density. Low maintenance. High T and P	Not as good in high visc. Density affects. Sticky materials. Smaller spans (<32 in)
Nuclear	Liquid, solid	Not limited except by wall thickness	Noninvasive. Noncontact. High T and P. Aggressive materials. Does not clog. Unaffected by agitation	Negative perception. Requires licensing. Leak checks required. Expensive
Armored, magnetic	Liquid	to 10,000 psig, 600 °F	Simple. Direct visual reading. High metallurgy possible. Good for hazardous materials. Long ranges available	Float may stick. Fouling services. Mechanical indicators
Gauge glass	Liquid	1,450 psig, 750 °F	Direct reading. Simple	Risk of breakage. Limited height per gauge. Primarily local manual reading. Glass fouling or etching

Process Controls in Petroleum Processing 1065

Point level				
Capacitance	Liquid, solid	1,000 psig, 500 °F, 100 ft	Conductive or nonconductive materials. Interface measurement. Good for slurries. Easy installation. Wide temperature range. High pressures	Time-consuming calibration. Coating on probe
Floats	Liquid	2,000 psig, 750 °F, to 132 in	Unaffected by density changes. High temperatures and pressures. Low cost. Simple	Affected by density. Not for high visc. May stick. Corrosive or turbulent fluid protection. Hard to change set point. Physically large. Damage to float
Ultrasonic	Liquid	3,000 psig, −100 to +300 °F	Reliable. Easy to use. Self-testing. Unaffected by visc, density, dielectric, agitation	Affected by foam. Cannot handle temp extremes well. Not for high vacuum
Microwave (radar)	Liquid, solid		Noncontact. Nonintrusive. Unaffected by sludge, contamination, vapors	High cost
Vibration/tuning fork	Liquid, solid		Slurries. Interface. No moving parts. Unaffected by fouling. Good for low-density materials	No vibrating services. Need significant differences in specific gravity for liquid/vapor interfaces
Conductivity	Liquid	500 psig, 600 °F, 60 in.	Slurries. Interfaces. Low cost. Simple. No moving parts. Easy to install	Conductive fluids only. Fouling. May accelerate corrosion
Nuclear	Liquid, solid	Not limited except by wall thickness	Slurries. Noncontact. Nonintrusive. No moving parts. Unaffected by most factors	Negative perception. Requires licensing. Leak checks required. Expensive. Hard to calibrate

(continued)

Table 9 (continued)

Technology	Application phases	Operating conditions limits	Advantages	Disadvantages
Liquid inventory				
Hydrostatic tank gauging	Liquid	Ambient tanks, normal liquid temperatures for tankage	Mass measurement. Density correction. Install in service. Low maintenance. Very simple	Stratification and agitation affects reading
Float and tape	Liquid		Inexpensive. Many applications. Good accuracy	Intrusive. Mechanical. Reference point movement. Corrections for temp and specific gravity. High maintenance
Servo	Liquid		High accuracy. Can measure density	Intrusive. Reference point movement. Stilling well needed. Mechanical design. More complicated installation. Temperature corrections
Radar, guided wave	Liquid		High accuracy. Noncontact. Nonintrusive. Relatively insensitive to product type. Low maintenance. High reliability. Install in service	Reference point movement. Temperature compensation. Manual density required. Antenna fouling
Magnetostrictive	Liquid		High accuracy	Intrusive. Sensitive to shock, vibration. Moving parts may stick
Weight and cable	Liquid		High accuracy. Interface measurements. Deep tanks. Short contact	Intrusive. Mechanical. Must be manually activated. High maintenance
Hybrid systems	Liquid		High accuracy for level and mass. Direct density. Install in service	Very expensive. Need average tank temperature

Sources: "Selecting a level device based on application needs," Parker (Fisher-Rosemount Inc.) (1998) and "Sighting in on level instruments," Wallace (Fluor E&C) (1976)

Operation of float-type gauges (float chambers, displacers, and magnetic gauges) at high pressures requires special precautions because the floats can be damaged or collapsed by sudden venting or drain flows or sudden changes in pressure. Specific procedures are needed to flush the taps or level bridles around these types of instruments. The procedures are not difficult, but need to be followed. The level instrument supplier can advise on the proper procedures.

Some of the newer types of level gauges, such as guided wave radar (GWR), are becoming more common due to their lower installation costs and relatively low maintenance. Some of these technologies require backup systems for start-up as they are not suitable for the transitional conditions that accompany start-up.

One special application that warrants some discussion is interface gauging among oil/emulsion/water. If the density difference between the oil and water is sufficient and the elevation difference is adequate, then a traditional dP level instrument or displacer with the right specific gravity float works fine. For more difficult emulsions or where gravities are closer, the most reliable probes reported employ microwave (agar or equal) or radio frequency. Sensing by conductivity changes is also feasible as an on/off control. There are other approaches, but these seem to be the most reliable.

Level Instrument Installations

Figure 12 illustrates some of the more common level instrument installations for process systems and tanks.

Process Levels

In process systems, the level instruments may be connected directly to the process vessel or, commonly, they are connected to one or more level bridles that are connected to the vessel. There is normally a root valve between the process and the instrument system that meets the piping specification to allow isolation. A double block and bleed arrangement with blinding capability may be used in high-pressure services.

Outside the root valve, the class changes to instrument line class. If the instruments are hung off a level bridle, each instrument will have its own piping-class root valve and then instrument valves.

Direct connections to the process, without a level bridle, are simple, but if the root valve leaks or can't be moved, the level instrument may end up out of service. This arrangement also exposes the level instrument to risk of fouling, high temperatures, and process corrosion. In some cases, the risks are managed by using sealed diaphragms or glycol-filled impulse lines for dP instruments.

For safety instruments, like high-high level shutdowns, the normal preference is for these to connect directly to the vessel to reduce some of the risks of plugging if a bridle is used.

A level bridle is a vertical section of pipe that is connected to the process with root valves. The bridle normally has vent and drain valves connected to a safe location (like flare) to enable the bridle to be blown free of foulants periodically and to be depressured. The level instruments are then hung off the bridle.

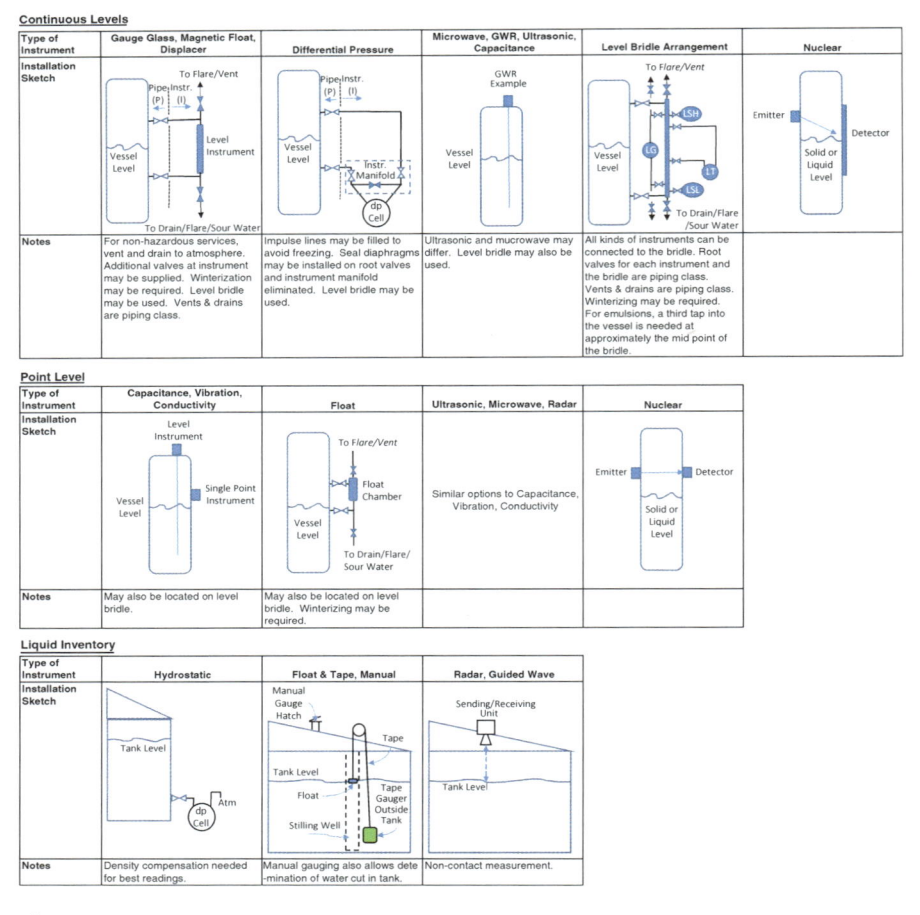

Fig. 12 Some typical level sensor installations

This arrangement has the advantages of (1) acting as a stilling well, dampening out the normal turbulence present in an active process vessel; (2) ability to blow out potential foulants; and (3) lower operating temperature than the process. Conversely, a bridle (1) introduces additional possible leak points, (2) can plug, (3) removes the attached instruments when taken out of service, and (4) readings must be compensated for density differences (especially on steam drums).

Note that a level bridle used for 3-phase separation needs three connections to the process. This is sometimes missed, resulting in poor level indication.

Level instruments, including any bridles, will require steam or electric tracing if the process fluids can freeze. Care must be used, however, to avoid tracing that causes the fluids in the bridle to boil.

Some refiners chose to purge the taps with inert gas or oil to keep them clear instead of periodically blowing down.

Level gauges or sight gauges can be a concern for leakage or failure. The gauge glasses can crack and fail if not correctly assembled and torqued. Steam sight gauges operating above about 500 °F require mica shields to avoid etching of the glass. In higher pressure or hazardous services, traditional gauge glasses are often being replaced by magnetic floats inside stainless or alloy tubes. These magnetic gauges have indicators for level that may consist of a single mag-follower or a section of many flip indicators that change color as the magnetic float passes by them. Magnetic gauges can be fooled by rapid level changes or by damaged floats, so redundant level indicators are a good idea.

Figure 12 does not address some of the level sensor installations, like nuclear. For these installations, the refiner needs to depend on their experience and the instrument vendor for installation details.

Tank Levels

Figure 12 shows a few of the more common tank gauging installations. These are fairly straightforward. If a liquid head-type gauge is used, the level reading will have to be compensated for density differences. Floats and other internal tank level devices, including manual tape gauging, are normally installed in stilling wells. These are just slotted tubes which protect the level devices from the liquid current moving around in the tank as it is filled or drained.

The water/oil interface in a tank is normally determined manually using a gauging tape coated with water indicator paste (water cutting).

Level Control

It is no surprise that the primary elements that control level in process systems and tankage are the same elements used to control flow. Levels that are used for control are connected to their respective level instrument, and the instrument output can be routed directly to the final level control element, a flow controller, or advanced process control. Ultimately, control all comes down to field manipulation of a flow control device as discussed under flow control.

We will not reiterate considerations in flow control here.

Specific Level Considerations

In this section, we will discuss a number of specific level control considerations that affect system design. The primary focus is on surge volumes, which is why we inventory fluids to begin with.

Surge volume is the volume retained in a vessel during operation at a set level. It is used for:

- Protecting equipment from damage caused by flow failures
- Protecting downstream processes from fluctuating flows which could cause poor process performance
- Protecting downstream processing from fluctuations in feed composition or temperature
- Protecting equipment from damage due to coolant failure

Types of Surge Volumes
There are primarily two types of surge volumes. These are:

1. *Upstream protection surge.* This is a surge volume provided to protect the upstream equipment and its associated pump from feed failure.
2. *Downstream flow surge.* This is surge volume provided to protect downstream equipment from feed failures or fluctuations.

Examples of Surge Types

Process Feed
Feed to process units is almost invariably on flow control. Many units also have a feed surge drum, particularly those units that are sensitive to flow fluctuations or where complete flow failure can cause equipment damage. This is an example of "downstream flow surge." The surge volume of the drum will depend on

- Source and reliability of source
- Type of control at source
- Variations and fluctuations in source rate

Column Feed from an Upstream Column
This feed stream will usually be controlled by the level in the source column, hence it will be fluctuating. If surge volume is provided only in the source column, it must be sufficient to cater for "upstream protection" and "downstream protection." The use of a surge vessel would be recommended for this case.

Feed to Fired Heaters or Boilers
The failure of flow through the tubes of fired heaters or steam boilers can cause serious damage through overheating of the tubes. Consequently "downstream protection" is required in this case. Invariably flows to heaters are on flow control.

Reflux Drum Considerations
- When the drum only furnishes reflux or reflux and product to storage, all that is needed in terms of surge volume is sufficient to provide "upstream protection." That is, the surge volume required is only to protect the reflux pump from losing

suction in the case of column feed failure. The pump will be required to circulate reflux and cool the column down during an orderly shutdown period.
- When the reflux drum furnishes reflux and the feed stream to another unit, then the drum must furnish "upstream protection" surge and "downstream protection" surge.
- If there is a vapor product from the drum, additional volume must be provided in the drum to allow vapor/liquid disengaging. This will be such as to retain the same liquid surge capacity as described above.
- Should the vapor phase from the reflux drum be routed to the suction of a compressor, an even larger volume reflux drum will be required. This is to ensure complete disengaging of the vapor/liquid. Internal baffles or screens are also used in the drums vapor outlet section to ensure complete phase separation.

Quantity of Surge Volume

The amount of surge volume will vary with the various types given above and with the specific case in question. Sometimes this amount is set by company specifications or, in the case of engineering contractors, by the client. Generally however the process engineer will be responsible for setting a safe and economic surge volume. In doing this the engineer needs to analyze each case in terms of why the surge volume is being provided, then deciding how much based on this answer. Figure 13 provides some guidelines to the amount of surge that should be applied.

Some useful equations used in setting and handling surge volumes are.
For surge volume:

$$\text{Vol cuft} = (GPM)\ (\text{minutes})/7.48 \tag{6}$$

For vessel size:

$$\text{Diam, } D = \sqrt[3]{(\text{cuft}/2.35)} @ L/D = 3 \tag{7}$$

$$\text{Diam, } D = \sqrt[3]{(\text{cuft}/1.96)} @ L/D = 2.5 \tag{8}$$

$$\text{Diam, } D = \sqrt[3]{(\text{cuft}/1.57)} @ L/D = 2.0 \tag{9}$$

For line size:

$$\text{Diam, } D = \sqrt[3]{(GPM/25)} @ \text{velocity} = 10\,\text{ft/sec.} \tag{10}$$

$$\text{Diam, } D = \sqrt[3]{(GPM/17)} @ \text{velocity} = 7\,\text{ft/sec.} \tag{11}$$

$$\text{Diam, } D = \sqrt[3]{(GPM/12)} @ \text{velocity} = 5\,\text{ft/sec.} \tag{12}$$

For flow rate:

$$\text{ft/sec} = GPM/450 \tag{13}$$

TOWER OVERHEAD CONTROL 1

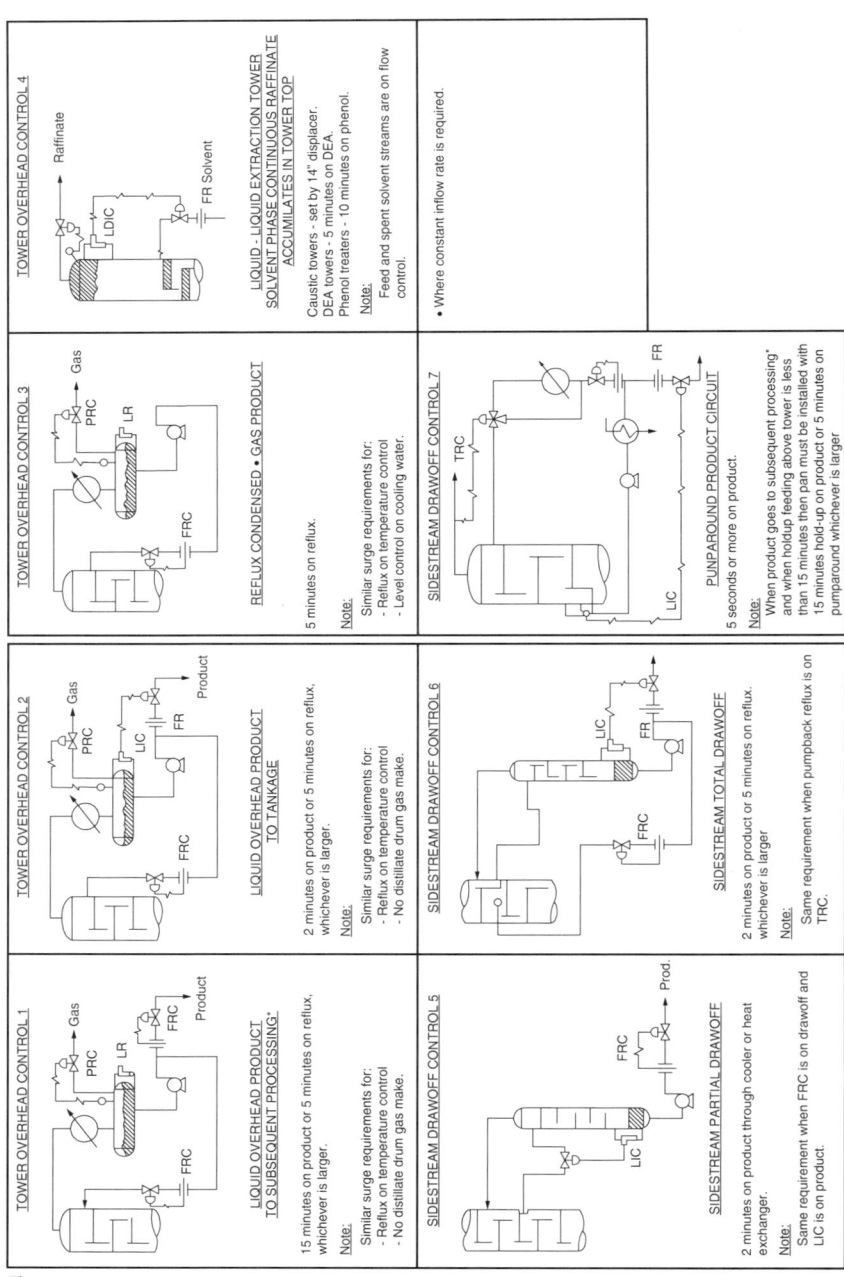

LIQUID OVERHEAD PRODUCT TO SUBSEQUENT PROCESSING*

15 minutes on product or 5 minutes on reflux, whichever is larger.

Similar surge requirements for:
- Reflux on temperature control
- No distillate drum gas make.

TOWER OVERHEAD CONTROL 2

LIQUID OVERHEAD PRODUCT TO TANKAGE

2 minutes on product or 5 minutes on reflux, whichever is larger.

Note:
Similar surge requirements for:
- Reflux on temperature control
- No distillate drum gas make.

TOWER OVERHEAD CONTROL 3

REFLUX CONDENSED - GAS PRODUCT

5 minutes on reflux.

Note:
Similar surge requirements for:
- Reflux on temperature control
- Level control on cooling water.

TOWER OVERHEAD CONTROL 4

LIQUID – LIQUID EXTRACTION TOWER SOLVENT PHASE CONTINUOUS RAFFINATE ACCUMULATES IN TOWER TOP

Caustic towers - set by 14" displacer.
DEA towers - 5 minutes on DEA.
Phenol treaters - 10 minutes on phenol.

Note:
Feed and spent solvent streams are on flow control.

- Where constant inflow rate is required.

SIDESTREAM DRAWOFF CONTROL 5

SIDESTREAM PARTIAL DRAWOFF

2 minutes on product through cooler or heat exchanger.

Note:
Same requirement when FRC is on drawoff and LIC is on product.

SIDESTREAM DRAWOFF CONTROL 6

SIDESTREAM TOTAL DRAWOFF

2 minutes on product or 5 minutes on reflux, whichever is larger.

Note:
Same requirement when pumpback reflux is on TRC.

SIDESTREAM DRAWOFF CONTROL 7

PUMPAROUND PRODUCT CIRCUIT

5 seconds or more on product.

Note:
When product goes to subsequent processing* and when holdup feeding above tower is less than 15 minutes then pan must be installed with 15 minutes hold-up on product or 5 minutes on pumparound whichever is larger

(continued)

a

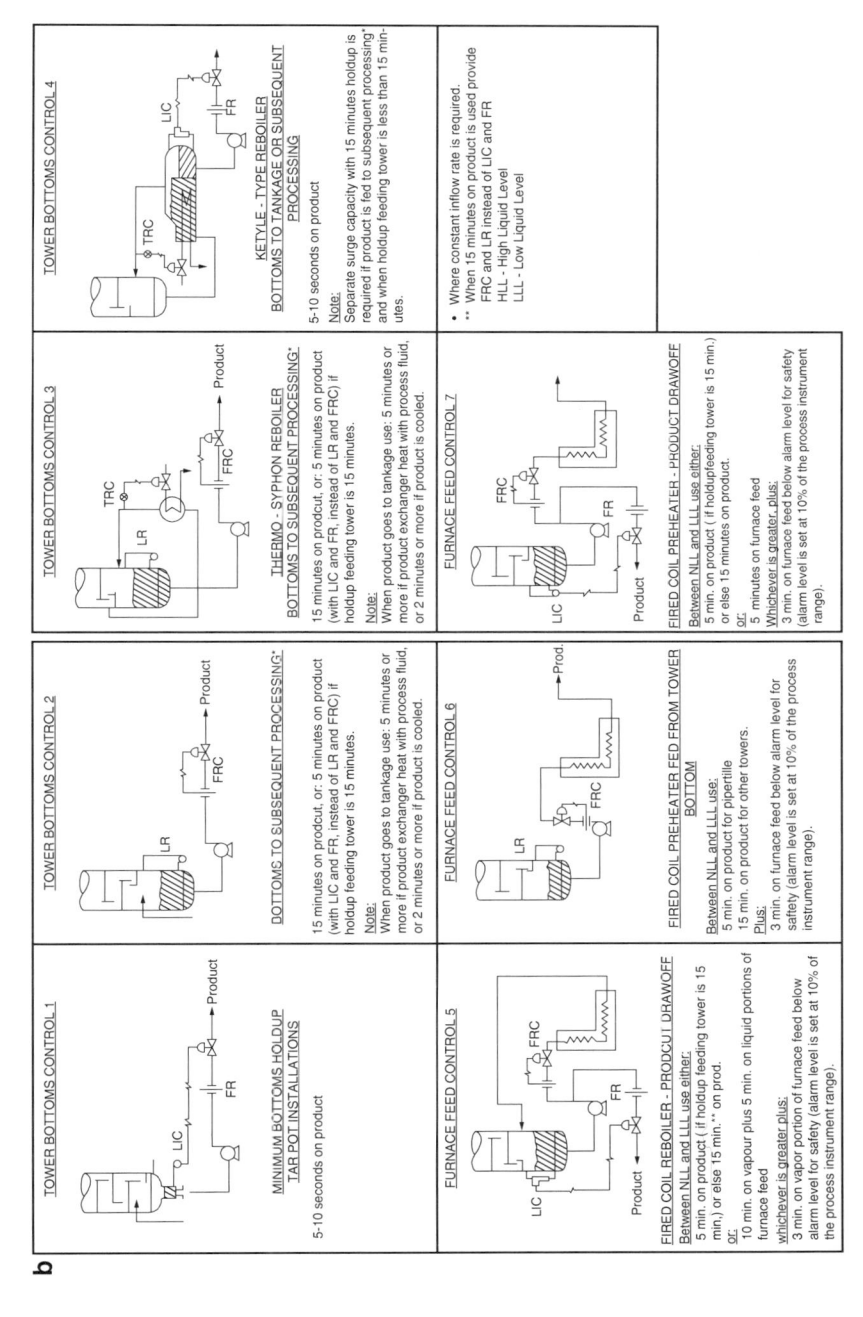

Fig. 13 (a, b) Typical surge requirements

For approximate control valve size:
One size smaller than line size. Thus:

Line	CV
4″	3″
6″	4″
8″	6″
10″	8″

Level Control

Surge volumes are maintained by controlling the amount of liquid entering or leaving the vessel in question. There are several means of accomplishing this and these are described and discussed in the following paragraphs:

- *Control the surge liquid outlet on level control*. This will give a close level control but a fluctuating outlet flow. The level control valve (LCV) will close completely at LLL. Thus, flow through the outlet line will be completely shut off at LLL.
- *Control the surge liquid outlet on flow control and provide a low level alarm*. This will give a fluctuating level but will eliminate flow fluctuations. As there is no LCV to restrict flow at LLL, then operators must physically reset the flow controller to maintain the surge volume. This has the disadvantage that the alarm condition could be missed or even ignored, resulting in possible damage to downstream equipment, and, in extreme cases, result in a fire or explosion hazard.
- *Control the surge liquid outlet on flow control reset by a level control*. This will give a wandering level but a smooth outlet flow. The LCV reset can still however cut off the flow completely on LLL.
- *Control the outlet flow by flow control and the system feed by surge volume level*. This will give close level control and also close outlet flow control. The outlet flow also will not be closed off by a LLL of the surge. In the case of feed to a fractionating tower, level control on the feed stream could cause tower upset conditions. This would be particularly undesirable on fractionators that operate close to critical conditions such as deethanizer.
- *Control the surge liquid on level control to an intermediate surge vessel*. The liquid from the surge vessel may then be flow controlled. This is the ideal method for controlling feed to a fired heater. The only question in this case is one of economics.

Level Control Range

Should the decision now be to use an LC on the surge outlet, the range of the instrument needs to be determined. The range is the vertical distance between the HLL and the LLL. Now if the liquid outlet is feeding another unit which requires a smooth flow, it is possible to achieve this by using a wide proportional band and a large range. The larger the instrument range and the wider the proportional band, the less sensitive is the level controller. Consequently the flow becomes smoother. However, the larger the range, the more expensive is the controller and of course the larger is the tower or vessel in order to accommodate the greater distance between HLL and LLL.

The selection of level control system and the level control range depends therefore on:

- How many outlet streams are there from the surge vessel?
- Which streams cannot tolerate complete shutoff before all the available surge is used?
- The degree to which the outlet stream requires smooth flow.

Surge and Level Control Summary

There are two very general rules to follow in selecting a control system for process inventory control. These are:

- If it is permissible for the product outflow rate to vary, use level control and a relatively small amount of surge capacity.
- If the product goes to a subsequent process where feed rate must be held constant, use flow control and considerably more surge.

Concluding Comments on Level Measurement and Control

There are numerous options for level sensing and control available. Each method has limitations. When selecting a means of measuring level, consider the entire range of operation over which the level instrument is expected to function from start-up, through operation, and to shutdown. How easy is it to maintain? What conditions will fool the level sensor?

Installation of level instruments must consider the process conditions, fluid, and phases to ensure reliable level indication. Maintenance of the level instruments is an important consideration also as they tend to require more attention than some other types of instruments.

Careful attention to the requirements and objectives for surge and level control actions is required. Think about how the level you are controlling fits into the process scheme and what the process needs for good management.

Composition Measurement and Control

Many processes require measurement and control of specific compositions or properties. Most often this is accomplished by periodic sampling of the process fluids followed by laboratory analyses. The lab results are then used to guide changes to the operation.

There are also a number of analyzers that can be installed online to measure specific properties. The online analyzer results can be used immediately for control.

New techniques have arisen in recent years that use calibrated correlations using normal operating parameters to infer properties for streams. We will touch on these, too.

Composition Measurement Elements

A large variety of online analyzers are available for measuring streams common in refineries and petrochemical plants. Table 10 provides a list of many of these options and where they can be applied. The following discussion will highlight a few of the more common online analyzer types.

- *Boiling point analyzer* – This analyzer can be used to determine the 5 % or 95 % points of products for fractionation control purposes, for instance. In this type of analyzer, a fixed, metered flow rate of stock is passed over a heater with vaporizes the sample. The surplus sample is then discharged at a rate appropriate to the desired % recovery. The heat input is varied to balance the target % vaporized. The balanced temperature of the vaporizer determines the boiling point at the desired % recovery.
- *Cloud point, pour point, or freeze point analyzer* – In these types of analyzers, a sample is introduced into the analyzer cell and cooled. The presence of crystals or complete solidification is determined optically by reflection or other means.
- *Distillation analyzer* – This can be an automated batch analyzer that essentially duplicates the ASTM distillation methods. These analyzers are expensive and maintenance intensive.
- *Chemiluminescence analyzer* – This type of analyzer works for compounds like NO, NO_2, NO_x, and O2. The analyzer makes use of the luminescence when the compounds are reacted with ozone. These analyzers are often found in emissions monitoring. They are often automatically calibrated on a set timing using calibration gas.
- *Flash point* – An example of this type of analyzer mixes air and sample together. Then sends them through a heater followed by a flash chamber where the mixture is subjected to a spark. The temperature at the flash chamber inlet is recorded when a pressure sensor sees evidence of a detonation.
- *Gas chromatograph* – This type of analyzer can be applied for many purposes, including tracking of separation quality in distillation columns. An online GC operates much like a lab GC. The preheated sample is periodically introduced into the column and the sample compounds are detected as they leave the column. The results are translated into the analysis. These analyzers tend to be expensive and require a great deal of maintenance.
- *Infrared and near-infrared* – There are several options available in the IR and NIR range. These look at IR transmittance through a sample compared to a reference stream at a specific wavelength (or wavelengths). Many properties can

Process Controls in Petroleum Processing 1077

be inferred from this analysis once calibrated, including octane and cetane. These can be used for continuous blending analysis, for instance. The analyzers are moderately expensive and require good calibration.

- *Ion-specific and pH electrodes* – These electrodes enable online analysis of several aqueous components continuously. The sample flows through where the electrode is immersed in the stream. The result is compared to a reference electrode. The electrodes must be replaced periodically. These systems are not overly expensive, but require regular maintenance and recalibration. The cells themselves are glass, so subject to breakage.
- *Mass spectrometer* – These analyzers are not common in refineries, but have been used to monitor fugitive emissions and gas analyses. The analyzer essentially rips the sample apart into ions and then passes the ions through a strong magnetic field where their deflection is a function of their mass and charge. The different ion concentrations are detected. The matrix of ion concentrations can be translated into composition. These instruments require regular calibration, specific to the service. They tend to be expensive and maintenance intensive.
- *Knock or cetane engine* – The normal laboratory knock or cetane engines can be adapted for online service, such as blending. These tend to be mechanically complex and require a lot of maintenance. They are also expensive. Some refiners have replaced the engines with NIR analyzers, although the final blends still require engine testing for certification.
- *Thermal conductivity* – These are very simple, inexpensive analyzers that use the heat transfer properties of the fluids being measured to infer compositions. They work well for gases like hydrogen in hydrocarbon gases. The analyzers are similar to the corresponding flow meters of the same types.
- *Ultraviolet spectrometer* – UV spectrometers are similar in principle to IR spectrometers using different light wavelengths. They may use absorption or transmittance.
- *X-ray fluorescence or absorption* – In these analyzers, a sample is excited by X-rays which cause it to fluoresce or absorb in specific wavelengths that depend on composition. The fluorescence or absorption is detected and translated to composition. These instruments require calibration and maintenance.

You can see that many of the methods employed in online analyzers are really just versions of laboratory instruments that have been adapted to field use and hardened to prevent damage. Most online instruments require regular calibration and maintenance to keep them running.

Online Analyzer Installations

Analyzers for online applications require appropriate sample taps and conditioning systems to ensure reliability and accuracy. This is critical in capturing the value that an analyzer can help create. If the analyzer is not available and reliable, the operators soon ignore it and all value is lost.

Table 10 A sampling of petroleum processing online analyzer applications

Property	Solid	Liquid	Gas	Alumina capacitance/conductance	Boiling point analyzer	Calorimetry	Chemiluminescence	Cloud point analyzer	Colorimetry	Conductivity analyzer	Capacitance	Density (several options)	Distillation analyzer	Flame photometry	Flash point	Gas chromatography	Hardness	Infrared (various)	Ion – specific electrode or cell
Alkalinity		X																	
Ammonia			X				X											X	
Antimony	X	X																	
Aromatics		X	X																
Benzene		X	X																
Bromine	X	X																	
Cetane		X																X	
Chloride		X																	
Chlorine		X																	X
CO			X															X	
CO$_2$			X															X	
Combustibles			X			X													
Conductivity		X								X									
Dissolved oxygen		X																	X
Dissolved solids		X								X									
Distillation		X											X			X			
Emissions			X															X	
Gas composition			X													X			
H$_2$S		X	X						X										
Hardness		X																	X
HCl			X															X	
Heating value		X	X			X									X				
HF			X															X	X
Hydrocarbons		X	X													X		X	
Hydrogen			X																
Leak detection			X															X	
Metals (general)	X	X																	
Methane																		X	
NH$_4$HS		X																	X
Nitrates, Nitrites		X							X										
Nitrogen purity		X	X																
NO, NOx			X				X											X	

Light transmission	Mass spectrometer	Near – infrared spectrometer	Optical dew point	Oxidation cell	Phosphorous oxide electrolytic cell	Lead sulfide	Engine (knock or cetane)	pH electrolytic cell	Pour point analyzer	Parametric susceptibility	Resistance	Refractometry	Thermal conductivity	Thermal dispersion	Titration	Ultraviolet spectrometer	Viscometer (several kinds)	Vapor pressure analyzer	X – Ray fluorescence/absorption	Zirconium oxide fuelcell	Notes
							X								X						
													X								
																		X			
	X															X					
																X					
																			X		
	X						X														
																			X		
																	X		X		Aqueous
													X								
					X						X		X								
																					Aqueous
																					Aqueous
																X					Air, CEMS
X																					
						X					X					X					
																					Aqueous
X													X								
													X	X							
X																					
																			X		
																					Aqueous
													X								
															X						

(*continued*)

Table 10 (continued)

Property	Phase			Online analysis methods available														
	Solid	Liquid	Gas	Alumina capacitance/conductance	Boiling point analyzer	Calorimetry	Chemiluminescence	Cloud point analyzer	Colorimetry	Conductivity analyzer	Capacitance	Density (several options)	Distillation analyzer	Flame photometry	Flash point	Gas chromatography	Hardness	Infrared (various)	Ion – specific electrode or cell
Octane		X														X			
Oil in water		X																X	
Opacity			X																
Oxygen			X																X
Oxygen purity		X	X																
Petroleum products		X			X			X	X			X	X		X				
PH		X																	
Phosphate		X								X									
Phosphorous	X	X																	
Red-Ox potential		X																	X
Silica		X								X									
SO$_2$			X											X					
Sodium		X								X									X
Sulfur		X																	
Sulfuric acid		X									X								
Suspended solids		X																	
Toluene		X	X																
Toxic gases			X																X
Viscosity		X																	
Waste water		X																	
Water content	X									X	X						X	X	
Water content		X		X							X		X					X	
Water content			X	X						X								X	
Xylene		X	X																

Process Controls in Petroleum Processing

Light transmission	Mass spectrometer	Near-infrared spectrometer	Optical dew point	Oxidation cell	Phosphorous oxide electrolytic cell	Lead sulfide	Engine (knock or cetane)	pH electrolytic cell	Pour point analyzer	Parametric susceptibility	Resistance	Refractometry	Thermal conductivity	Thermal dispersion	Titration	Ultraviolet spectrometer	Viscometer (several kinds)	Vapor pressure analyzer	X – Ray fluorescence/absorption	Zirconium oxide fuelcell	Notes
		X					X														
																X					Aqueous
										X									X		Excess O_2
													X								
									X								X				
								X													
																					Aqueous
																		X			
																					Bleach
																					Aqueous
																	X				
																					Aqueous
																			X		Oil
X												X									
																X					
													X								
																	X				
																X					Aqueous
															X						
			X	X																	
																	X				

Sample Location and Lag Time

Analyzer samples may be drawn directly from the main process line and routed to the sample system through tubing. The tubing may be electrically or steam traced, depending on the fluid properties.

Ideally, the distance from the sample point to the analyzer should be as short as possible if tubing is used. Otherwise, the stream the analyzer is sampling has long since left the process and the sample data is meaningless.

To reduce the time lag between sampling and the analysis, it is common to establish a fast loop that has a much higher flow rate than a small sample tube. The fast loop starts at a high-pressure location and returns to a low-pressure location in the same process stream or another safe location. For instance, if the process has a level control valve on the stream, the fast loop flow could be pulled ahead of the valve and the fast loop return would be downstream of the valve. The fast loop line is routed to a location near the analyzer. A smaller sample line goes from the fast loop line to the analyzer, thus shortening the time from sample to analysis. The return sample can go back to a low-pressure point in the process, to slops or recovered oil, or to flare.

When a fast loop is incorporated into the design, care must be used to ensure the loop does not defeat the purpose of the control system or create unsafe conditions. If the fast loops work around a level control valve and if that valve closes due to a low-low level, the fast loop needs to stop also.

The sample point design needs to consider the flow conditions within the line being sampled:

- For a single-phase liquid or vapor, pull the sample from the top half of the line.
- For two-phase liquid and vapor, pull the sample from a section of line where the flow is turbulent and well mixed. Otherwise, attempts to get a good sample in two-phase lines are problematical.
- For large lines or ducts, it may be necessary to use an isokinetic line to get a representative sample.
- If there are foulants in the line, avoid any samples from the bottom of the line and it may be necessary to install a small extension of the sample tube into the line to avoid catching foulants running along the wall.
- In the case of flue gas excess oxygen analysis with ZrO_2 fuel cells, the probe must be inserted into the flue at high temperature. This is an exception to most sampling.

Obtaining a valid sample is the first critical step in getting a good analytical result. This applies to intermittent samples as well as continuous.

Sample Conditioning

A raw process sample needs to be "conditioned" in many cases to prepare it for analysis. A few of the situations frequently encountered in this regard are:

- *Cooling* – A hot sample will generally have to be cooled. There are high-pressure, high-alloy exchangers available for this purpose. Having to cool a

sample is common for steam and condensate, some hydrogen plant samples, inter-reactor samples, and many others. The sample may be totally condensed as in steam samples or may contain noncondensable gases. If there are noncondensibles, a flash pot downstream of the cooler may be needed. The cooling system needs to prevent damage to the analytical instruments.
- *Liquid knockout/coalescing* – Sometimes you are only interested in the vapor and liquid will damage the analytical instrument. In these cases, as with cooling, a separator is needed ahead of the instrument. You have to be very careful, however, that the liquid removed is not part of what you are trying to analyze. For example, if the stream being analyzed is feed to a hydrogen plant and it has some condensable hydrocarbons that get removed in sample conditioning, the feed will be incorrectly characterized, resulting in too little steam and coking.
- *Filtration* – Most analytical instruments cannot handle incidental incoming solids or fines. These normally are removed by fine ($\ll 1$ μm) filtration. The filters need to be changed out on a routine basis.
- *Sample flow* – Most online analyzers are designed and calibrated for a specific flow rate. Deviations from the design sample flow rate will produce errors. Where critical, the flow is usually metered by a tiny rotameter or other constant flow control device.
- *Pressure reduction* – Many samples are taken at elevated pressures, but most sample analyzers operate at low pressures. Regulators are available to letdown the pressure. These are generally stainless or alloy and have very small C_vs. There should always be filters ahead of the pressure regulator.
- *Manual sample point* – It is normal practice to incorporate a manual sample point into the sample conditioning system. This may be a routine operator sample point or it may only be used as a check sample for the analyzer.

Analyzer Installation

Many online analyzers are designed to be field installed. This would include pH and thermal conductivity analyzers. These instruments are hardened sufficiently that they are simply installed in the field on the process line or a takeoff from the line.

The more sophisticated analyzers, such as NIR, GC, or UV, require a controlled environment. This is especially important for environmental monitoring using continuous emissions monitors (CEMs). These analyzers are normally housed in small buildings with environmental controls. The buildings have positive pressure to prevent possible explosions in the event of a nearby hydrocarbon or hydrogen leak. This also means the analyzer and its associated electronics don't have to be explosion proof. Calibration gases or fluids are housed next to the building. If there is a fast loop from which the sample is drawn, the fast loop runs near the building, with only the smaller sample tubing entering the building. The sample conditioning system and analyzer with its electronics are inside the building. The sample vent or outlet line leaves the enclosure immediately after the instrument.

Sample Disposal

Once a sample is analyzed, you have to do something with the remains. For many samples that undergo nondestructive procedures, the sample can be returned to the process at a lower pressure point. If there is insufficient pressure to get back into the process, flare or slops/recovered oil systems are options. The destination has to be consistent with the hazards of the sample.

Some online analyzers for liquids may discharge the spent samples into a drum where they accumulate for later recovery. Be sure any such drum is well labeled to avoid regulatory problems.

Spent aqueous samples can often be sent to process sewer. If the samples become contaminated during analysis, they may need to be accumulated as a hazardous waste and periodically sent to disposal.

Inferred Properties

The growth of advanced control has brought the need for more continuous process analysis to take advantage of the improved techniques. This has led to the increasing use of inferred analyses for streams.

An inferred analyzer uses normal process data, such as flows, temperatures, and pressures, to estimate the value of a specific stream property that you would normally measure by sampling and lab testing. This enables much of the value of having an online analyzer with the hassle of actually having to buy it, install it, and maintain it.

Inferred properties are really just complex correlations based on analysis of a lot of operating and analytical data collected specifically for this purpose.

Inferred analysis is most often used for distillation-based properties. These could include 5 %, 10 %, 90 %, 95 %, and flash point. Inferred properties have also been used for following the concentration of a heavy key in a tower overhead or a light key in a tower bottoms.

Inferred analyses are periodically checked against the actual analysis by samples sent to the lab. The inferred property estimates are corrected based on the actual sample results. When an inferred property correlation drifts too far (i.e., the corrections are significant), a new correlation will be developed to replace the old.

The use of this approach is expected to expand as control systems become more sophisticated.

Controllers

Regardless of the architecture or type of control system being used, the core of the control system is the controllers themselves. These take the data from the various sensors in the field, the analyzers, the operator, the APC modules, and a variety of other inputs and determine the control moves that need to be made for each stream

Process Controls in Petroleum Processing

to satisfy all the desired operating conditions. The controllers then output the conditions to the field instruments, which make the actual adjustments. This is feedback control.

Types of Control Actions

While this is not a treatise on control instruments, it is appropriate to talk about some of the basic types of simple controllers encountered in a refinery as we draw the controls discussion to a close:

On/Off-Discrete Control

For some control applications, the controlled variable only needs to be kept between limits. For this purpose, we can use on/off control. You may find this type of control on a knockout drum level, where a level sensor signal goes to a controller that monitors the level. When the level reaches a fixed point, the controller turns on a pump and pumps the level down until it reaches a set low level. The controller then shuts off the pump until it is needed again. In some applications, there are only level switches at the high and low points and the action is accomplished with relays only.

PID Control

The most common type of control found in a refinery is PID control or proportional-integral-derivative. There are various forms of this control logic:

- Proportional only
- Proportional-derivative
- Proportional-integral
- Full PID

The terminology comes from the form of the controller equation:

$$u(t) = K_p e(t) + K_i \int_0^t e(\tau)d\tau + K_d(de(t)/dt) \tag{15}$$

where:

$u(t)$ = controller output
K_p = proportional gain, tuning parameter
K_i = integral gain, tuning parameter
K_d = derivative gain, tuning parameter
e = error or difference between set point and measured variable
t = time or instantaneous time
τ = variable of integration between time 0 and time t

In the equation, the first term $[K_p e(t)]$ is called the proportional term because the output from it is proportional to the error from set point.

The second term $[K_i \int_0^t e(\tau) d\tau]$ is called the integral term. The contribution of this term is determined by both the magnitude and the duration of the deviation. If the measured value is far from the set point, the integral term makes a bigger move on output.

The final term $[K_d (de(t)/dt)]$ is the derivative term. This term fine-tunes the output for how quickly and in what direction the actual measurement is changing relative to the set point.

In practice, you will normally find:

- *Proportional control* – where only the first term is used. In proportional control, the output only changes if there is a deviation. If the error is constant, the output is constant, but it may not be at the set point. There is generally an offset between the set point and the process variable to make it work. Proportional band and gain are the tuning parameters. Proportional band is the % change in error that causes a 100 % change in output. Gain or proportional gain is the % change in output divided by the % change in input.
- *PI control* – where proportional and integral are used. This eliminates the error in proportional control but may result in a "reset windup" that drives the process variable too far or overshoots the set point. Then it takes a long time to get back. This is driven by reset rate or the frequency with which the control loop takes action (in resets per minute or minutes per reset). You may get cycling if not properly tuned.
- *PD control* – where proportional and derivative are used. Improves the rate of response of the controller. The derivative acts to boost the proportional controller output based on the rate and direction that the error is changing. This helps in slow processes. Can cause cycling in fast processes. The rate setting for derivative is normally in minutes.
- *PID control* – where all the terms are used. When tuned properly this provides the best basic control for a process, but tuning can be difficult.

Control Loop Tuning

There is automatic control loop tuning software available, but it is helpful to understand how the different types of control terms can be tuned. This is a very simple approach but can be a start:

- *Proportional* – Reduce the proportional band (or increase the gain) until the process begins to swing after a disturbance. Then double the proportional band or cut the gain in half.

- *Integral* – Increase the repeats per minute (or reduce the reset) until the process begins to cycle following a disturbance. Then reduce the reset to about one third of the setting.
- *Derivative* – Increase the rate setting until the process cycles after a disturbance. Then reduce the rate to one third of the value.

Conclusion

This has just been a survey around the various aspects of control systems as applied in refining today. For additional information, consult the listed references of any of the myriad materials available online and in your technical library on the subject.

Appendix 1: Control Valve Sizing

Process Flow Coefficient (C_v) and Valve Sizing

Process flow coefficient C_V is defined as the water flow in GPM through a given restriction for 1 psi pressure drop. These C_Vs can be determined by the following equations:

$$C_v = Q_L \sqrt{\frac{G_L}{\Delta P}} \text{ for liquid} \qquad (16)$$

$$C_v = \frac{Q_s}{82} \sqrt{\frac{T}{\Delta P - P_2}} \text{ for steam} \qquad (17)$$

$$C_v = \frac{Q_G}{1360} \sqrt{\frac{\mu_2 S T}{\Delta P - P_2}} \text{ for gases} \qquad (18)$$

where:

C_v = flow coefficient
Q_L = liquid flow in GPM at conditions
ΔP = pressure drop across valve, psi
G_L = specific gravity of liquid at conditions
Q_S = steam rate in lbs/h
P_2 = pressure downstream of valve psia
Q_G = gas flow in SCFH (60 °F, 14.7 psia)
T = temperature of gas °R(°F + 460)
S = mol weight of gas divided by 29
μ_2 = compressibility factor at downstream conditions

The following are some special considerations that may have to be made in determining process C_V values.

Pressure Drop

For compressible fluids the maximum usable pressure drop in equations (b) and (c) is the critical value. As a rule of thumb and for design purposes, this value is 50 % of the absolute upstream pressure. (The valve can take more than the critical pressure drop, but any pressure drop over the critical takes the form of exit losses.)

Flashing Liquids

In the absence of accurate information, it is recommended that for flashing service the valve body be specified as one nominal size larger than the valve port.

Two-Phase Flow

If two-phase flow exists upstream of the control valve, experience has shown that for fluids below their critical point, a sufficiently accurate process C_V value can be arrived at by adding the process C_V values for the gas and liquid portions of the stream. The calculation is based on the quantities of gas and liquid at upstream conditions. The valve body is specified to be one nominal size larger than the port to allow for expansion.

Valve Rangeability

The rangeability of a control valve is the ratio of the flow coefficient at the maximum flow rate to the flow coefficient at the minimum flow rate ($R = C_V$ Max$/C_V$ Min). Valve rangeability is actually a criterion which is used to judge whether a given valve will be in a controlling position throughout its required range of operation (neither wide open nor fully closed). In practice the selection of the actual valve to be installed is the responsibility of the instrument engineer. As the process engineer is usually the person responsible for the correct operation of the process itself; however, he must be satisfied that the item selected meets the control criteria required. He must therefore satisfy himself that the valve will control over the range of the process flow.

Control valves are usually limited to a rangeability of 10:1. If R is greater than 10:1, then a dual-valve installation should be considered in order to assure good control at the maximum and minimum flow conditions.

In some applications, particularly on compressor or blower suction, butterfly valves have been specified to be line size without considering that as a result the valve may operate almost closed for long periods of time. Under this condition, there have been cases of erosion resulting from this. It is recommended therefore that butterfly valves be sized so that they will not operate below 10 % open for any appreciable period of time and not arbitrarily be made line size.

Valve Flow Coefficient ($C_{v'}$)

In order to ensure that the valve is in a controlling position at the maximum flow rate, the valve $C_{V'}$ is the maximum process C_V value determined above, divided by 0.8. The reasons for using this factor are that:

- It is not desirable to have the valve fully open at maximum flow since it is not then in a controlling position.
- The valves supplied by a single manufacturer often vary as much as 10–20 % in C_V.
- Allowance must be made for pressure drop, flow rate, etc., values which differ from design.

Control Valve Sizing

Control valve sizes are determined by the manufacturers from the process data submitted to them. However, there are available some simple equations to give a good estimate of the required valve sizes to meet a process duty. Three of these are given below: Single-seated control valve sizes may be estimated by:

$$S(\text{inches}) = \left[\frac{\text{Valve}\, C_V}{9}\right]^{1/2} \quad (19)$$

Double-seated control valve sizes may be estimated by:

$$S(\text{inches}) = \left[\frac{\text{Valve}\, C_V}{12}\right]^{1/2} \quad (20)$$

Butterfly valve sizes may be estimated by:

$$S(\text{inches}) = \left[\frac{\text{Valve}\, C_V}{20}\right]^{1/2} \quad (21)$$

The constants (9, 12, or 20) in the denominators of these equations can vary as much as 25 % depending on the valve manufacturer.

A control valve should be no larger than the line size. A control valve size that is calculated to be greater than line size should be carefully checked together with the calculation used for determining line size. Usually, a control valve size should be one size smaller than line size.

Once the valve size is estimated and the valve C_V known, then the percent opening of the valve at minimum flow and maximum flow can be obtained by dividing the respective process C_V values conditions by the selected valve C_V. This information is normally required to check the percent opening of a butterfly at minimum flow. It is not normally necessary to calculate it for any other type of valve.

Valve Action on Air Failure

In the analysis of the design and operation of any process or utility system, the question always arises on the action of control valves in the system on instrument air failure. Should the control valve fail open or closed is the judgment decision of the process engineer after evaluating all aspects of safety and damage in each event. For example, control valves on fired heater tube inlets should always fail open to prevent damage to the tubes through low or no flow through them when they are hot. On the other hand, control valves controlling fuel to the heaters should fail closed on air failure to avoid overheating of the heater during the air failure.

The failure action of the valve is established by introducing the motive air to either above the diaphragm for a failed open requirement or below the diaphragm for a failed shut situation. The air failure to the valve above the diaphragm allows the spring to pull up the plugs from the valve seats. Air failure to valves below the diaphragm forces the spring to seat the valves in the closed position. Failure of a valve in place or locked is also possible, but seldom used.

References

Flow Measurement and Control

Alicat.com, Types of gas mass flow meters (2005). Accessed Jan 2014
W. Chin, Magmeters: how they work and where to use them. Control Mag. 32–34 (Krohne America, 1990)
W.S. Corcoran, J. Honeywell, Practical methods for measuring flows. Chem. Eng. Mag. 86–92 (Fluor Engineers and Constructors, Inc., 1975)
DAC Electric, High voltage variable speed drives, www.dac-electric.com. Accessed 29 Jan 2014
J.W. Dolenc, Choose the right flow meter. Chem. Eng. Progress. 22–32 (Fisher-Rosemount, 1996)
D. Ginesi (Bristol Babcock Inc.), G. Grebe (Cincinnati Test Systems), Flow meters: a performance review. Chem. Eng. Mag. 100–118 (1987)
D. Ginesi, Choices abound in flow measurement. Chem. Eng. Mag. 88–100 (Foxboro Co., 1991)
D. Ginesi, A raft of flowmeters on tap. Chem. Eng. Mag. 146–155 (Foxboro, 1991)
Greyline, Two technologies for flow measurement from outside a pipe, www.greyline.com. Accessed 27 Jan 2014

R.C. Hunt, Oscillatory flowmeters: an effective solution for flow measurement (MycroSENSOR Technologies, 2002), www.fluidicflowmeters.com. Accessed Jan 2014
Instrumart.com, Vortex shedding flow meters, www.instrumart.com. Accessed 27 Jan 2014
R. Kern, How to size flow meters. Chem. Eng. Mag. 161–168 (Hoffmann-La Roche Inc., 1975)
P.K. Khandelwal (Uhde India), V. Gupta (Sycom Consultants Consortium), Make the most of orifice meters. Chem. Eng. Progress 32–37 (1993)
M.D. Kyser, Positive displacement flow measurement. Control Mag. 40–42 (Badger Meter, 1990)
G. Livelli, Selecting flowmeters to minimize energy costs. Chem. Eng. Prog. 34–39 (ABB, 2013)
C.J. O'Brien, Flowmeter terms, types, & successful selection. InTech Mag. 30–33 (Moore Products, 1989)
S. Peramanu, J.C. Wah, Improve material balance by using proper flowmeter corrections (Canadian Natural Resources, 2011), www.hydrocarbonprocessing.com. Accessed Jan 2014
Seekyouranswers.blogspot.com, Turbine flow meter (2013). Accessed Jan 2014
Siemens Moore Process Automation, Inc., Three element feed control system. Accessed 29 Jan 2014
Spirax Sarco, Types of steam flowmeters, Steam engineering tutorials, www.spiraxsarco.com. Accessed 29 Jan 2014
J. Taylor, Selecting the right flow meter for your hydronic system (Wood-Harbinger, 2013), www.woodharbinger.com. Accessed Jan 2014
P.C. Tung, M. Mikasinovic, Sizing orifices for flow of gases and vapors. Chem. Eng. Mag. 83–85 (Ontario Hydro, 1982)
Wikipedia, Mass flow meter. Accessed 27 Jan 2014

Temperature Measurement and Control

R.H. Kennedy, Selecting temperature sensors. Chem. Eng. Mag. 54–71 (The Foxboro Co., 1983)
J.G. Seebold, Tube skin thermocouples. Chem. Eng. Progress 57–59 (Chevron Corp., 1985)
Wikipedia, Resistance thermometer. Accessed 30 Jan 2014
Wikipedia, Thermocouple. Accessed 30 Jan 2014

Pressure Measurement and Control

Ashcroft Inc., Pressure gauge installation, operation, and maintenance, www.ashcroft.com. (2002). Accessed Jan 2014.
R.E. Bicking, Fundamentals of pressure sensor technology. Sens. Mag. (1998)
W.J. Demorest, Jr., Pressure measurement. Chem. Eng. 56–68 (Honeywell Inc., 1985)
K. Hamza, *Pressure Measurement*, http://science-hamza.blogspot.com/ (Cairo, 2011). Accessed Jan 2014.
Instrumentationtoolbox.com, Strain gauge substitute (2011), www.instrumentationtoolbox.com. Accessed Jan 2014
B.G. Liptak, Pressure regulators. Chem. Eng. 69–76 (Bela G. Liptak Assoc., 1987)
R. Repas (Machine Design), C. Dixon (Kavlico Corp.), ABCs of refrigeration pressure sensing (2008), machinedesign.com. Accessed Jan 2014
D.L. Roper, J.L. Ryans, Select the right vacuum gage. Chem. Eng. 125–144 (Tennesee Eastman Co., 1989)
A.W. Sloley of The Distillation Group, Inc. Chem. Eng. Progr., (Jan 2001)
Winters instruments website, http://winters.com/. Accessed Jan 2014

Level Measurement and Control

D. Anderson, Match fit radars. Hydrocarb. Process. 104–109 (Vega Controls Ltd., 2013)
Control, What's the best way to control an interface level when an emulsion tends to form between phases? Control 48–52 (1992)
P. Hagar, Avoid temperature differences when using external chambers. Control 71–73 (Syncrude Canada, 1995)
E.A. Knight, J.R. Pugh, Properly select level-measurement devices for bulk solids. Chem Eng Progress 50–55 (Glasgow Caledonian University, 1996)
B.G. Liptak, Level measurement with problem liquids. Chem Eng. 130–133 (Bela G. Liptak Assoc., 1993)
S. Parker, Selecting a level device based on application needs. Chem. Process., Fluid Flow Annual 75–80 (Fisher-Rosemount, 1998)
B.O. Paul, Seventeen level sensing methods. Chem. Proc. 63–72 (Editor, 1999)
L.M. Wallace, Sighting in on level instruments. Chem. Eng. 95–104 (Fluor E&C, 1976)

Composition Measurement and Control

Controls Wiki, Process control definitions and terminology (2013), https://controls.engin.umich.edu. Accessed Jan 2014
PAControl.com, Process control fundamentals (2006), www.PAControl.com (Excellent and clear). Accessed Jan 2014
F.G. Shinskey, Foxboro, *Process Control Systems: Application, Design, and Tuning* (McGraw Hill Publishing Company, New York, 1988)
Wikipedia, PID controller. Accessed 3 Feb 2014

Utilities in Petroleum Processing

David S. J. Jones and Steven A. Treese

Contents

Introduction	1094
Steam and Condensate Systems	1095
The Refinery Steam System	1095
Components of the Steam System	1097
Boiler Controls	1102
Boiler and Condensate Chemical Treatment	1102
High Reliability Steam Supply	1103
Steam and Condensate System Monitoring	1104
Troubleshooting Steam Systems: A Few Common Issues	1105
Fuel Systems	1107
Fuel Gas System	1108
Fuel Oil System	1111
Burner Management Systems and Safe Firing of Heaters	1114
Fuel Gas Cleanup and Burner Fouling	1115
Troubleshooting	1116
Water Systems	1116
Cooling Water	1118
Boiler Feed Water (BFW) Treating	1128
Plant Water	1138
Potable Water	1139
Fire Water	1140
Compressed Air Systems	1141
Plant Air System	1141
Instrument Air System	1141

David S. J. Jones: deceased.
Steven A. Treese has retired from Phillips 66.

D.S.J. Jones
Calgary, AB, Canada

S.A. Treese (✉)
Puget Sound Investments LLC, Katy, TX, USA
e-mail: streese256@aol.com

© Springer International Publishing Switzerland 2015
S.A. Treese et al. (eds.), *Handbook of Petroleum Processing*,
DOI 10.1007/978-3-319-14529-7_17

Plant Air System 1145
Protection of Air Systems 1145
Electrical Power Supply 1146
Electrical Power Distribution System 1146
Cogeneration and Power Recovery 1147
Power Versus Emergency Steam Loads 1148
Electrical Power Distribution Conclusion 1149
Nitrogen System 1149
General Discussion of Nitrogen Utilities 1149
Specifications 1150
Typical Nitrogen Supply Flow Sheets 1150
Prevention of Nitrogen System Contamination 1152
Other Utility and Utility-Like Systems 1153
Flushing Oil 1153
Ammonia 1154
Oxygen Enrichment Supply 1155
Hydrogen Distribution 1156
Caustic Soda 1156
Sulfuric Acid 1156
Amine Distribution 1157
Sewer Systems 1157
Slops and Recovered Oil 1161
Summary of Practices to Prevent Utility System Contamination 1162
Utilities Conclusion 1163
References 1163
Steam and Condensate Systems 1163
Water Systems 1163
Compressed Air Systems 1164
Electrical Power Supply 1165
Nitrogen System 1165

Abstract

All oil refineries and other petroleum processing facilities need utilities in order to function. The common utility systems include steam, fuel, various waters, air, electrical power, and sewers, among others. This chapter explores the processes, design, reliability, and operation of these critical systems.

Keywords

Utilities • Refinery • Steam • Power • Water • Fuel

Introduction

All oil refining processes require utilities in order to function. The common utility systems are:

- Steam and condensate
- Fuel (gas and oil)

- Water systems (cooling, boiler feed, potable, fire)
- Compressed air (instrument, plant)
- Electrical power supply
- Nitrogen
- Sewers (oily water/plant, chemical, storm)
- Other utilities

The engineering and design of several of these systems is usually the responsibility of a chemical engineer. On operating plants and processes, the process engineer undertakes the responsibility for the correct and efficient operation of utility facilities. The duties associated with the power systems are usually left to the electrical engineer or department, although the process engineer does have an input in the sizing of the system by developing a list of power requirements for all electrical equipment. This includes all motors and electrically operated equipment such as: the desalter, product dehydrators, electrostatic precipitators, and the like.

Because utility systems are so critical to the operation of a processing facility, it is necessary to ensure the security of these systems. Redundancy and other methods are used to ensure reliable supplies. Anti-backflow and anti-cross-contamination provisions are required to ensure the utilities are not contaminated by process fluids or other utilities. Remarks on typical methods to maintain system reliability and potential traps are incorporated into the discussion of each utility.

The following paragraphs describe the typical utility systems found in the oil refining industry. The details of these systems may vary from facility to facility, but their format and general layout will be similar to that described here.

Steam and Condensate Systems

Steam is the most basic utility in a refinery. Many of the uses of steam have been around since ancient times. While it has largely been replaced by electrical power in driving machinery, it still maintains several roles, including process heating, process reactant, driving rotating equipment, and heat tracing/retention.

An important feature of steam for many uses is its extreme reliability. It is not unusual for steam systems to have been in service, without outage, for 40–50 years. No other utility can claim that level of reliability. Of course, the system has to be designed specifically for that sort of reliability and the associated maintenance. We will mention some of the techniques for maintaining steam availability at all times.

The Refinery Steam System

Figure 1 illustrates a configuration and integration of a typical steam system in a modern refinery. The system is centered on steam headers that run throughout the facility to distribute or accept steam at different pressure levels. The highest level (HP Superheated) is usually a high pressure, superheated steam at

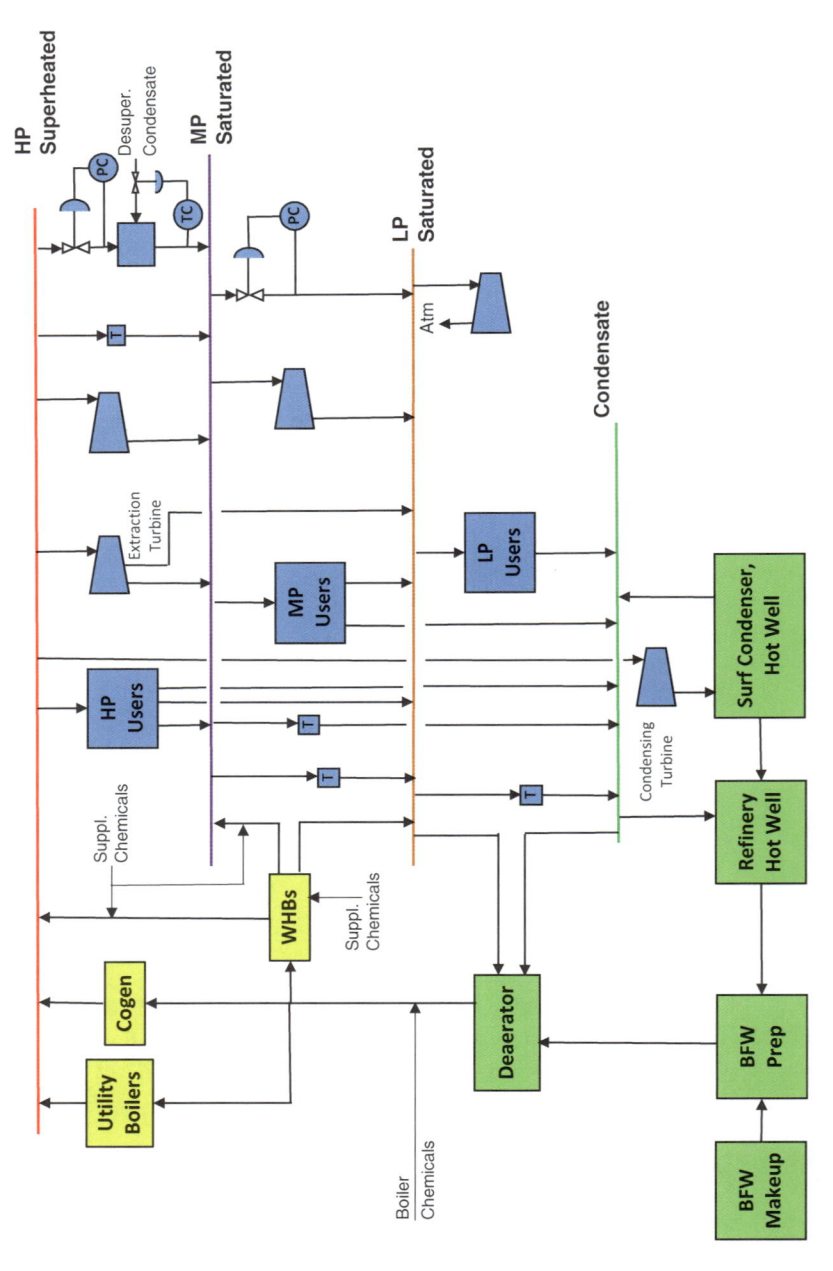

Fig. 1 Typical refinery steam system

about 600 psig (~40 barg) with, perhaps, 180 °F (100 °C) of superheat. This high-pressure steam supplies dry steam to turbines and other users. The high-pressure steam is letdown to lower-pressure levels through turbines and control valves with desuperheating.

The MP-saturated steam header may run about 150 psig (10 barg). It may also feed turbines and process heat users. Often, waste heat boilers supply this header.

The lowest steam pressure header (LP Saturated) usually runs around 50 psig (3–4 barg). It is primarily used for steam tracing, but can also supply heat for low-pressure reboilers, strippers, and the deaerator. There are sometimes emergency generator or compressor steam turbines tied to the LP header that exhaust to atmosphere. They are not normally in service.

All the steam headers cascade eventually to the condensate header(s) which operates at low pressure (5–10 psig or 0.3–0.7 barg). Higher pressure, intermediate condensate headers may also be present.

Some large turbine drivers may also contain surface condensers and operate with deep vacuums at the turbine final stage outlet. The surface condensate has to be pumped to the combined condensate systems for reuse. Because the surface condenser runs at vacuum, it often pulls air into the system, adding load to the boiler feed water deaerator and oxygen scavenger chemicals.

Boiler feed water makeup is very expensive, so most refineries try to maximize the reuse of condensate, adding only supplemental boiler feed water (BFW) makeup from softeners, demineralizers, or reverse osmosis units. We will discuss the production of BFW a little later in this chapter.

Steam generation is critical to the system. The traditional generator is the utility boiler. In addition, refineries today will generate steam from waste heat boilers (AKA heat recovery steam generators) tied to the process units and from cogeneration facilities, which make both steam and power. Another company may own and operate the cogeneration facility in an over-the-fence arrangement.

Of course not all facilities have all these headers, generators, and users, but this is a very common configuration in many refineries and petrochemical facilities.

Components of the Steam System

There are many components in a refinery or petrochemical plant steam system. This section will describe some of the key components.

- Boiler feed water manufacture
- Steam generation
- Waste heat boilers
- Pressure letdown and desuperheating
- Steam traps and condensate headers

Refer to Fig. 2 in the following discussion, which provides additional detail.

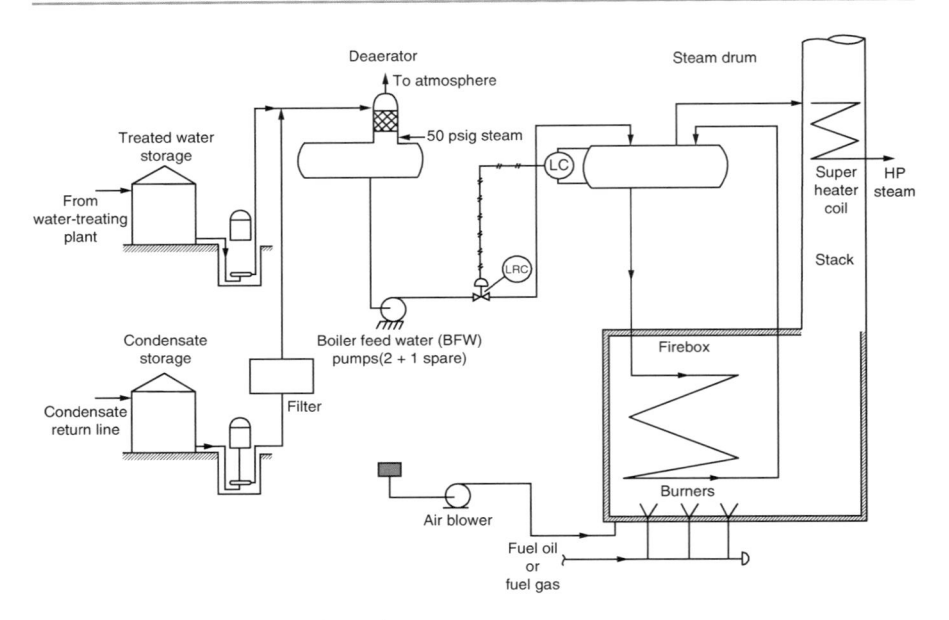

Fig. 2 A typical steam generation unit

Making Boiler Feed Water

In most plants steam condensate, accumulated in the various processes, is collected into a single header and returned to the steam-generating plant. It is stored separately from the treated raw water, because condensates may contain some oil contamination. There is additional discussion of condensate issues later in this section.

A stream of treated water and condensate are taken from the respective storage tanks and pumped to the deaerator drum.

The condensate stream passes through a simple filter en route to the deaerator to remove solid contamination. The combined water and condensate streams enter the top of a packed section of the deaerator called the "heater section." Low-pressure steam is introduced immediately below the packing in the drum to flow upward countercurrent to the liquid streams. Any air or CO_2 entrained in the water is removed by this countercurrent flow of steam to be vented to atmosphere. The combined, deaerated waters are the boiler feed water (BFW).

An oxygen-scavenging chemical is usually injected into the deaerator tank or drum to remove any residual, trace oxygen. The target oxygen is usually less than about 10 ppb. BFW leaving the deaerator tank is usually injected with the boiler chemicals.

Refer to section "Boiler Feed Water (BFW) Treating" for a more detailed discussion on making BFW.

Steam Generation

The deaerated BFW is pumped by the BFW pumps into the steam drum of the steam generator. There will normally be multiple pumps for this service (e.g., $3 \times 60\%$ each).

Two will be operational and one will be on standby. Those pumps normally operating are usually motor driven, while the standby pump is very often driven by an automatically started steam turbine or diesel engine. These pumps are quite large in capacity, operating at high head and discharge pressure. The main steam lines in most plants are high pressure (at least 700 psig at the generator coil outlet), so the pump discharge pressure will be much greater than the HP steam outlet. These pumps are the most important in any refinery or chemical plant. If they fail, no steam can be generated, and the whole complex is in danger of total shutdown or worse. Therefore multiple separate pumps are used to cater for the normal high head and high capacity, and a separate pump driver operating on a completely different power source than electrical power or steam is mandatory to minimize the danger of complete shutdown.

The steam drum is located above the generator's firebox. The liquid in the drum flows through the generator's coils located in the firebox by gravity and thermosiphon. A mixture of steam and water is generated in the coils and flows back to the steam drum. Here the steam and water are separated from the stream. The saturated steam from the drum goes through a non-return (check) valve and flows to a superheater coil.

The steam is heated to the plant's HP steam main temperature in the superheater coil and enters the high-pressure steam header for distribution to the various users. The steam pressure is controlled by a pressure controller on the steam outlet to the header.

Waste Heat Boilers

Most refinery and petrochemical processes occur at greatly elevated temperatures. The process schemes usually preheat feedstocks for the process using heat in the process effluent. This works economically down to a point, but often there is more than enough process heat to supply the preheat needs. In those cases, waste heat boilers (WHBs) are often used. In a hydrogen plant, the waste heat boilers can actually be major sources of steam in a facility. Cogeneration is really just another form of waste heat boiler, where the surplus heat coming from the turbine or engine generator set can be used to raise and superheat steam.

WHBs take two common forms:

- For large heat sources, the WHB will have exchangers or coils in the process or flue gas along with a steam drum. These systems essentially operate like the utility-type boiler described above. They may deliver saturated or superheated steam to the steam headers.
- For many smaller heat sources, the WHB will take the form of a kettle with a steam separator (see Fig. 3). BFW brought into the kettle is boiled off, collected, runs through a steam separator, and then enters the steam header. These systems usually make only saturated steam. The kettle must be periodically blown down to maintain conductivity control.

Some WHBs have their own BFW systems, and it may be necessary to treat the steam from these generators separately for condensate corrosion control.

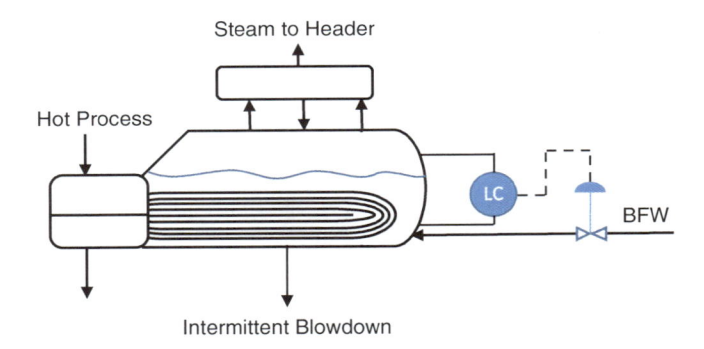

Fig. 3 Kettle-type waste heat boiler arrangement

Pressure Letdown and Desuperheating

Steam to the lower-pressure headers is generated through turbines where possible. Where lower-pressure steam is required and it is not possible to produce it through equipment, then letdown stations are located in suitable places in the system. When steam pressure is reduced to the lower-pressure headers, the associated increase of temperature above that specified for the lower-pressure steam may need to be reduced. Desuperheaters are used for this purpose.

A desuperheater consists of a chamber in the steam line into which cooler condensate is injected. These items are purchased equipment with specially designed injection nozzles for the condensate. The amount of condensate delivered is controlled by the downstream temperature of the steam. Desuperheaters are located at critical locations of the plant where relatively large quantities of high-pressure steam are reduced to low pressures such as the discharge from turbines.

Steam Traps and Condensate

There is continuous heat loss from the steam headers. This causes local condensation of the steam, especially in low flow regions. This condensate will collect at low points in the headers. With flow changes or even normally, the condensate can damage the headers through pipe hammer or the turbines and control valves through erosion. The condensate can also freeze in cold weather, rupturing the lines where it has collected. For these reasons, steam traps are installed on steam lines. The traps are located at low points in the lines and at each drop ahead of a steam turbine. The traps help insure the steam is dry, where required. They also protect the condensate header from overpressure by steam blow-through from the higher-pressure steam headers.

There are several types of condensate traps available from manufacturers. The most commonly found traps in refineries and petrochemical facilities are thermodynamic and bucket types, although others (like continuous drainer float types) are used in specific services.

Sizing of steam traps is an important consideration and is often overlooked. A trap that is too large and does not cycle frequently enough will fail, as will a trap that is too small and cycles all the time. A trap should open and close at least every few minutes. Monitoring and repairing failed traps can save a great deal of energy. Manufacturers have tools and techniques for monitoring trap operation.

Avoid trapping high-pressure steam directly to condensate. It is best to trap high-pressure steam to medium-pressure or low-pressure steam as shown in Fig. 1.

When most types of steam traps fail, they blow steam through into the lower-pressure header. This failure mode is somewhat by design as it continues to maintain dry steam at the expense of higher usage. But, this can overpressure the downstream header, especially if it is the condensate header, and prevent other traps from functioning properly. Monitoring of steam traps can be as simple as listening to them for how often they cycle or as complex as infrared monitoring. Manufacturers have simple tools and techniques for monitoring trap operation. There is a great deal of good training information about traps that is worth your review. Much is available on the Internet.

The traps ultimately cascade down to the condensate header(s). The main condensate return header is usually operated at a positive pressure of between 5 and 10 psig (0.3–0.7 barg). The collected condensate is stored at atmospheric pressure, and very often the small amount of steam flashed from the header pressure to the storage pressure is used in the deaerator instead of the low-pressure steam (the deaerator operates at or near atmospheric pressure).

Note that the recovered condensate will be reused as boiler feed water, hydroprocessing wash water, process makeup water, and so on. It can come from many places, some of which may contaminate the condensate. Ahead of the facilities depicted in Fig. 2, hot condensate should be collected in one or more "hot wells" and then forwarded by pump to the common storage tank(s) or header. The vapor space in the hot well should be monitored for hydrocarbons, which would indicate possible leaks into the condensate system. Small quantities of oils can usually be handled, but larger amounts of oil or oxygen are not allowed.

If a significant leak is identified, that condensate should be dumped to sewer (or a safe location) until the leak can be eliminated; otherwise, the steam system becomes contaminated with hydrocarbon or oxygen, which is not healthy in many of the steam uses. For instance, you cannot have hydrocarbon in your steam when you are trying to gas-free a fired heater before lighting it or when you are trying to gas-free a vessel for maintenance.

Utility Steam Stations

Most plants have utility steam stations that come off the medium- or low-pressure steam headers. These numerous "utility drops" are located strategically throughout the units. The stations have steam hoses for use in cleaning equipment, temporary heating, and many other uses. There is a steam trap located just upstream of the utility station.

Often these utility stations also have plant water and plant air connections. The steam and plant water systems are sometimes connected to allow direct injection heating of the plant water that can be controlled manually or with a self-contained steam valve. There is further discussion of these stations under the appropriate other utilities in this chapter.

Prevention of Steam System Contamination

As with most utilities, it is necessary to ensure that what comes out of the steam lines is steam *and only steam*. Wherever steam is connected to the process, there should be backflow prevention consisting of one or more check valves, one or more block valves, a bleeder, and an isolation point (blind) as a typical minimum.

Connections made for steaming process vessels and other cleanout purposes should be removed immediately after use and especially before pressuring the process system.

These are normally considered best practices, but often get overlooked.

Boiler Controls

Utility boiler systems, and some waste heat boiler services, are designed for reliable, nearly self-contained control of the boiler.

The primary control on the boiler is aimed at maintaining steam pressure at its set point. This is accomplished by controlling the fuel pressure/flow rate to make as much steam as is needed to hold the pressure. Air is normally supplied to the boiler firebox based on demand using a forced draft fan. The air may be preheated by flue gas. If there is a preheater, there will normally also be an induced draft fan that controls the heater draft (firebox pressure) and fine-tunes the draft based on excess oxygen and CO in the flue.

Because the boiler will attempt to make whatever steam is demanded, it could run dry unless the water supply is ensured. Water supply is ensured by use of a three-element control scheme in most systems. This is covered in the handbook chapter entitled "▶ Process Controls in Petroleum Processing" under the section on "Evolution of Control System Types". Using steam rate and steam drum level changes, a 3-element control brings in BFW on flow control so that the boiler has water under all conditions unless the water supply fails completely. This is a material balancing system.

Steam drum or kettle boiler level instrumentation has to be compensated for the difference in densities between the steam drum/kettle and the level bridle/instrument. This adjustment can be significant. Inside the drum/kettle, the water is much hotter than the level instrument, and there will likely be multiple phases present. The liquid is also turbulent, not a quiet level. In the level instrument, the fluid is cooler, more dense, and quieter. If you run the level instrument at the same level you want in the drum, the actual level in the drum will be much higher, leading to carryover. Be sure to properly account for the density differences and fluid dynamics involved.

Boiler and Condensate Chemical Treatment

We alluded to boiler chemicals earlier. There are four treatment objectives for boilers:

- Eliminate contaminants that can deposit downstream
- Prevent corrosion by dissolved oxygen and CO_2

- Prevent deposition and corrosion of the steam generation system
- Prevent corrosion of the condensate system

The elimination of contaminants from the BFW is accomplished by demineralization, softening, or reverse osmosis as noted earlier. These processes will be discussed in more detail later.

Oxygen and CO_2 are removed by steam stripping of the BFW in the deaerator as previously described. In a BFW system, either inorganic or organic oxygen scavengers can be used to eliminate any remaining, trace dissolved oxygen. The most common inorganic scavenger is sodium sulfite (Na_2SO_3). Organic scavengers often take the form of derivatives of erythorbic acid. The scavengers are injected into the deaerated water leaving the heater/stripper as it drops into the drum/storage section of the deaerator.

Corrosion and deposition in the boiler system are managed a couple of different ways. The buildup of dissolved solids and hardness in the boiler water is managed by purging or blowing down a small percentage of the water in the boiler. The lower the purity of the feed water, the more boiler water must be purged. In a large steam generator, there will usually be a small continuous blowdown pulled near the surface of the water level in the steam drum and an intermittent blowdown pulled through a pipe along the bottom of the steam drum and any mud drums (low points).

In addition to blowdown, a chemical mixture is used to manage deposits and corrosion in the boiler systems. These impacts come from accumulation of iron, calcium, and magnesium. Blowdown controls these somewhat, but chemicals help the blowdown to be effective. The common chemicals used are phosphates for precipitating solids, a chelant for complexing the potential foulants, and a polymer dispersant to keep foulants and corrosion products moving into blowdown. High purity caustic may also be injected for pH control. These chemicals are normally injected into the BFW as it heads to the users and may be supplemented at the users.

The condensate system ends up collecting all the CO_2 and other acidic compounds that may be in the steam system. The pH of the condensate system can be quite low. pHs as low as 3.0 have been reported in untreated condensate systems. To control condensate corrosion, amines and oxygen scavengers are normally injected into the steam header at the source leaving the boilers. One or more amines are selected so that they drop out into the steam and condensate headers at different temperatures. An incorrect amine selection will result in corrosion of one of the headers.

Chemical vendors are a valuable resource for understanding boiler treatment chemistry and solutions to corrosion or fouling problems.

More detail on boiler feed water chemistry is in section "Boiler Feed Water (BFW) Treating" of this chapter.

High Reliability Steam Supply

Reliability of steam supply is critical to plant operations. Many facilities depend on backup steam when the primary electrical system fails to supply the most sensitive operations.

Steam reliability starts at the BFW level. The purification equipment that supplies the BFW makeup should be redundant. Multiple trains are generally available to purify the water. A significant amount of purified water is normally held in a storage tank. The tanks should be sized to contain all the water needed by the plant steam system for several hours, typically 12–24 h. This means a big tank. If process water users pull off the tank, the suction lines to these users should be high in the tank so that there is sufficient water reserved below their connections for the boilers.

BFW from the tank is pulled by multiple, redundant transfer pumps (possibly separate connections to the tank) to the deaerator. The transfer pumps will usually be electrically driven with steam or diesel backup. The deaerated water is usually sent to the steam generators via multiple, redundant, high-pressure BFW pumps using at least two different types of drives.

In the most reliable configurations, there are at least two primary steam generators (sometimes several) that have priority over any other generators for BFW. The generators produce into a ring header that allows isolation of parts of the header for boiler turnarounds and maintenance, while the balance of the header remains in service. Hydrogen plants can also be designed to provide emergency steam.

For turbine-driven BFW pump standbys, the emergency steam supply is drawn from the ring header, as are the standby turbines for FD and ID fans. Instrument air compressors are also pulled off the ring header. Essentially, everything needed to run the steam system is internally supplied from the ring header in the event of outside or electrical failure. If necessary, the steam may be vented to atmosphere.

The integration of a cogeneration unit into the steam and electrical systems needs to be carefully considered. The connection of these two systems through the cogen puts both at risk. If the cogen provides too much of the steam in the plant, failure of the cogen will bring the plant down, because the boilers cannot fire up fast enough to cover a loss without tripping. Simultaneously, you have most likely also lost electrical power. The best practice is to ensure you have enough base load on the boilers that the plant can take a cogen trip combined with a steam and electrical load-shedding plan to be implemented immediately in the event of cogen loss. To prevent trips caused by outside electrical disturbances, it is also advisable to have protective relaying that will allow the facility to island or isolate itself if the outside electrical grid trips. Of course, this won't always work, but it helps.

When designing a steam system, always think about achieving absolute reliability of supply. Know when you are accepting a compromise that could affect reliability and be sure you agree with taking that risk.

Steam and Condensate System Monitoring

The steam and condensate systems need to be sampled and monitored regularly to ensure a good, reliable performance. A water treatment chemical supplier can help with the analytical tests and their meaning.

Utilities in Petroleum Processing

Some recommended minimum periodic samples and tests, along with their frequencies, are:

- Boiler feed water makeup streams (daily to weekly): pH, conductivity, hardness, total dissolved solids, oil, and grease (condensates)
- Boiler feed water supply (daily to weekly): pH, conductivity, hardness, dissolved oxygen, total dissolved solids, and boiler treatment chemical concentrations
- Boiler water/blowdown (weekly): pH, conductivity (daily or continuous), hardness, total dissolved solids, boiler treatment chemical concentrations, and iron
- Steam from steam drum (condensed, weekly to monthly): pH and conductivity
- Condensate at various locations in the system (monthly): pH, conductivity, amine inhibitor concentration, and iron

Boilers should be monitored daily by tracking and trending:

- Net steam production per unit of heat fired (e.g., Mlbs/MMBtu)
- Stack excess oxygen
- Stack temperature
- Preheated air temperature
- Stack appearance (steam in the stack indicates a leak)
- Visual verification of steam drum or generator liquid levels
- Visual check of flame patterns and adjustment or cleaning if indicated

Analysis of the results and implementation of actions when results are heading in a bad direction are the keys to maintaining healthy steam and condensate systems.

Troubleshooting Steam Systems: A Few Common Issues

We will touch briefly here on a few common boiler or steam generation issues and provide some hints about where to start looking for problems. While these are couched in terms of boilers for convenience, they apply to all types of steam generators.

Boiler Carryover

Carryover occurs when the steam from the boiler contains entrained water or other material (even volatilized salts or silica) from the steam drum. Allowing carryover for any significant amount of time will foul and damage downstream equipment. Turbines and superheat coils are especially at risk from carryover.

Some places to look when investigating carryover include:

- *Variations in pressure.* Sudden reductions in drum pressure will cause the drum water contents to suddenly expand due to the release of steam bubbles. The level in the drum goes high and allows the carryover. This can be a problem with back pressure control, or one of the users may be pulling the steam header down.

This may also occur during an upset that suddenly increases steam demand, but those situations are usually short-lived. If pressure reductions are large enough, the boiler will shut down on high-high level.
- *Foaming.* Bubbles can build up on the surface of the water in the steam drum, effectively raising the drum level. The bubbles get carried out with the steam. If you have foaming, check your boiler chemistry. Be sure the BFW preparation, especially the demineralizers, softeners, or RO, is actually working. Be sure the drum is being blown down correctly. Increase blowdown frequency and amount to purge out whatever bad actors are present.
- *Drum level instrument problems.* As mentioned under controls, the steam drum level has to be compensated for density difference between the drum and the level instruments. Check the level taps. Ensure they are clear. Verify that the level gauge matches the instruments. Verify that all the instruments agree on the level. Correct level is absolutely critical to avoiding carryover.
- *Internal failure.* A steam drum is not just an empty space. There are baffles, steam purifiers, steam separators, belly plates, etc., that ensure controlled separation of the steam from water. Mechanical failure of any of these can result in carryover. These are hard to diagnose externally, but are easily spotted once the drum is opened. If you've tried everything else, then you need to physically inspect the drum.

Condensate Corrosion

Corrosion in the condensate systems will often show at remote, low-pressure locations. A pattern of corrosion in condensate systems should trigger an investigation. Common problem areas include:

- *Deaerator/degasifier operation.* Ensure the deaerator is stripping correctly. This can be done by testing the BFW at the deaerator for dissolved oxygen (DO). The DO test is very sensitive. Your chemical vendor may be able to help, or you can use an ampoule test tube immersed completely in water. Check the condensate pH and analyze it for CO_2 and other acids which can indicate poor deaerator performance.
- *Amine selection.* Review the condensate treatment program with your treatment chemical vendor. Condensate treatment is not a one-size-fits-all application. Amines of different boiling ranges need to be selected based on your steam and condensate system pressures.

Tube Failures

Boiler tube failures or leaks indicate that deaeration or chemical treatment is inadequate. However, they may also indicate firebox problems. A couple of things to check:

- *Verify deaerator is operating correctly.* Deaerator problems are often indicated by oxygen pitting of the steam drum or deaerator.
- *Verify the boiler chemicals are being applied correctly.* Waterside fouling will result in failures.

- *Verify adequate blowdown.* Again, this is part of waterside management.
- *Flame impingement.* Flames licking tubes may cause local overheating and failures. Look at the flame patterns for issues. Flame instability is another indication of a possible leak. The burner nozzles may need cleaning, or there may be a problem with the air registers.

Decreasing Steam Production Efficiency

Sometimes the steam production may begin decreasing, but the firing stays the same or increases. This can indicate a couple of things:

- *Fouling of the boiler tubes.* Check water chemistry, chemicals, blowdown, and the other related factors.
- *Tube leak.* Trend the steam production vs. makeup BFW or material balance the waterside of the boiler. If you can't account for all the incoming water, it is probably leaking out. Check the stack for steam on a cold morning.

Loud Boiler Hum or Vibration

This phenomenon can occur when the heating value of the fuel gas gets too low, usually because of excessive amounts of hydrogen. The gas is combusting by small explosions. If not corrected, the vibrations will ruin the heater refractory and cause more damage. This problem has to be corrected at the fuel supply source.

Condensate Header Banging

Loud banging of a condensate header is the result of steam collapsing in the header and creating shock waves. This normally means that one or more steam traps have failed or are not primed and steam is blowing through into the header. The failed trap(s) need to be located and repaired. Sometimes all this takes is blocking the trap in temporarily on the condensate side and then unblocking it after sufficient condensate has accumulated. Alternately, the trap can be immersed in cold water or ice to establish a condensate seal. This phenomenon may also occur during sudden rainstorms, but is normal in that case.

Fuel Systems

Oil refineries and petrochemical plants may have three separate fuel systems:

- Fuel gas
- Fuel oil
- Natural gas

Some burners in these plants may be designed to fire any of these fuels. Some normally fire either the fuel gas or the fuel oil stream and can be easily switched over from one to the other. Most modern refineries, limited by environmental regulations, only fire natural gas and fuel gas. In the USA, for instance, few refineries fire fuel oil.

The pilot burners however must be natural gas or fuel gas. Natural gas is preferred as it does not plug or coke the pilot gas tip. If fuel gas is used, the pilot tips require regular cleaning or replacement. The pilot burner fuel supply is kept separate from the main fuel gas supply.

Some systems are designed such that if the pilot flame is lost, the whole burner system is shut down. In other systems, like side wall fired furnaces, there are no pilots, but there are literally hundreds of burners. The furnace design and company safety philosophy dictate the options for a given furnace.

Generally the design of the burner systems in most plants has many safety and shutdown features. After all, in processes that handle flammable material, the heater burners are the one feature in the plant design that can be a major fire or explosion hazard if not properly operated and managed.

Common burner and heater shutdowns include:

- High-high and low-low fuel pressure
- High-high process outlet temperature
- Loss of pilot burner (depending on the design, may also be detected as HH or LL pilot gas pressure)
- Atomizing steam failure (oil burners)
- Low-low process flow to heater
- Low-low heater draft

There are exceptions to all these shutdowns. It is common practice to provide a pre-alarm on any shutdown, so all these parameters also have pre-alarms to allow time for operator action.

Fuel Gas System

This is the simpler of the two systems. A typical fuel gas system is illustrated in Fig. 4.

Waste gas streams from the process plants are gathered and scrubbed free of H_2S with amine solutions (MEA, DEA, DGA, MDEA). Streams that have already been scrubbed may bypass the columns. The rich amine solution is sent to the sulfur plant for regeneration and sulfur recovery. Generally, there will be two or more scrubbers which may handle saturated gases, unsaturated gas, or a mixture. The H_2S is normally reduced to less than 100 ppm by the scrubbers. Recent US regulations have reduced the allowable total sulfur to less than 30 ppm, including H_2S, COS, disulfides, mercaptans, and any other sulfur form. Additional treatment may be used to reach these new targets. Refer to the topic titled "▶ Refinery Gas Treating Processes" in this handbook for additional details on gas treating.

The low sulfur gases are then directed to the plant's fuel gas mix drum. This drum operates at 30–150 psig (2–10 barg) pressure (depending on the system design)

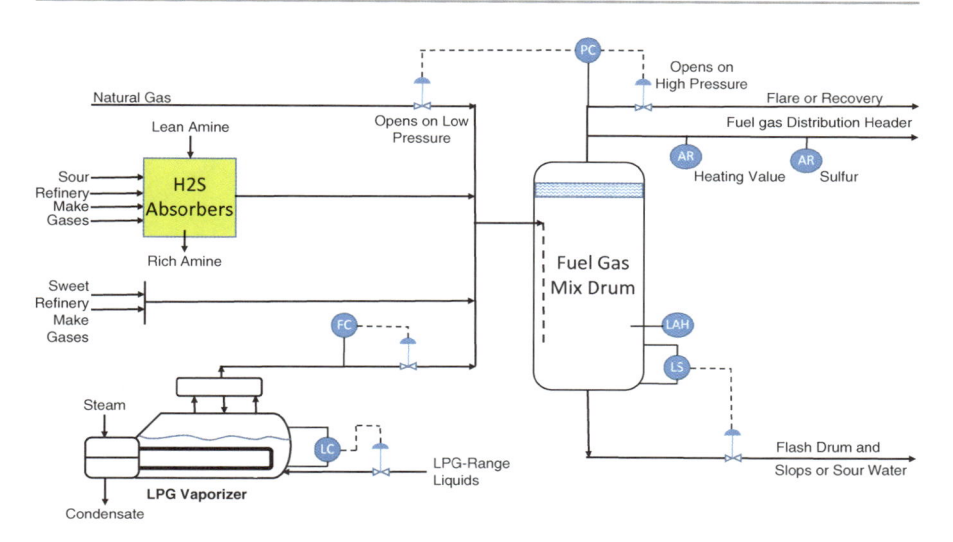

Fig. 4 A typical fuel gas mix system

and somewhat above ambient temperature. Collected hydrocarbons and moisture may be knocked out in this drum and drained to slops, or a small steam coil may be installed in the drum to gasify any "below dew point" material that may have condensed out.

The drum is held at the desired pressure by pressure control valves which allow surplus gas to flow to flare or bring in an external, clean source of gas, usually natural gas or clean LPG, on low pressure. If LPG is used as a secondary fuel source, it is routed to the fuel gas drum via a vaporizer. The vaporizer is usually a kettle-type boiler, heating and vaporizing the LPG at the drum pressure. Medium- or low-pressure steam is used as the heating medium for the vaporizer.

In practice, sending surplus fuel gas to the flare is the least preferable option. Environmental regulations prohibit this type of flaring in many areas. More commonly, the fuel gas pressure control system is operated so that there is always a small, positive makeup gas stream into the system to hold pressure.

Figure 5 depicts the key parts of a fuel gas system within a process unit. There will typically be a local fuel gas knockout drum, which may service several heaters. Any liquid that has condensed in the fuel gas distribution headers (and amine carryover from the scrubbers) is drained to slops/recovered oil from the knockout drum. From the knockout drum, the fuel gas is controlled into the local fuel gas ring header at each heater, and the burners pull their fuel off the heater fuel gas ring header. It is advisable to steam trace and insulate the fuel gas line after it leaves the knockout drum to prevent further condensation and corrosion. Corrosion products will plug the burners. Often, inline Y strainers are installed in each burner line to ensure clean gas to the burner. Some processors also use stainless steel pipe between the knockout drum and each heater.

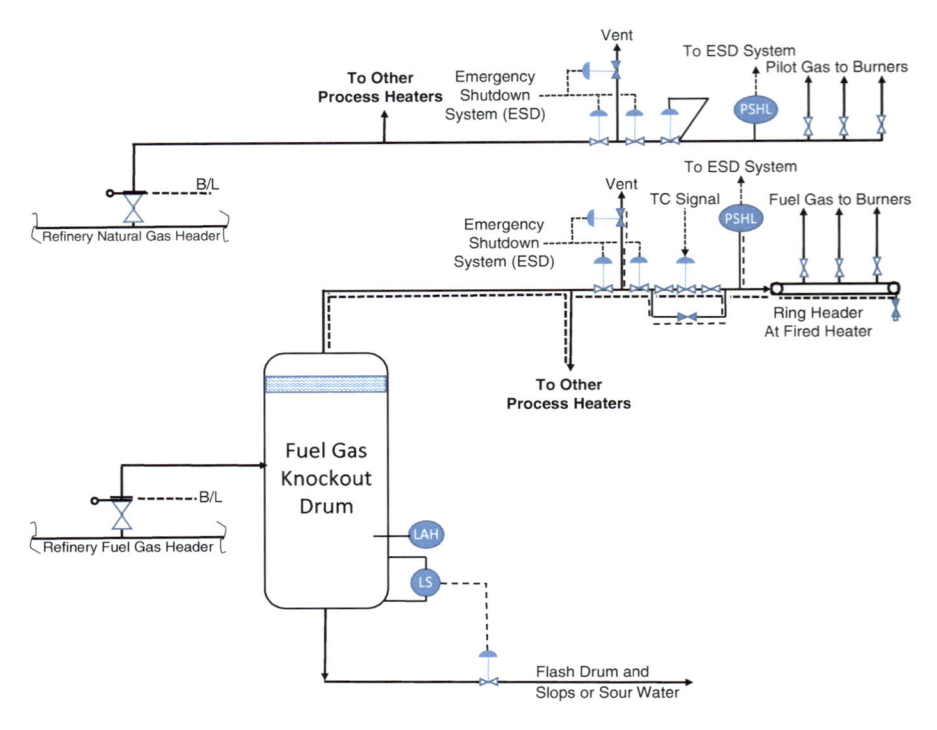

Fig. 5 Local fuel gas system

Parallel to the fuel gas header is the pilot gas header. This header has a simple pressure regulator controlling supply pressure at each heater to normally about 5 psig (0.3 barg). Pilot gas flow is either off or on, with the pilot orifice itself regulating the flow. The pilot gas is often filtered or strained. As noted above, natural gas is preferred for pilots to prevent plugging.

Some heaters contain an automatic switch over from gas firing to fuel oil firing on low gas flows or when manually selected. The fuel gas system is "dead ended." That is, there is no return system to the fuel gas drum, the gas header is pressured up, and gas flows to the burners by means of this differential pressure and intermediate control valves.

Fuel gas flow to the burners is controlled by heater outlet temperature or another temperature downstream of the heater coil process outlet line. Several control configurations may be used for fuel gas to the heater:

- Direct control with TC output
- TC cascade to fuel gas PC which controls the valve
- TC cascade to fuel gas FC which controls the valve
- TC cascade to FC cascade to PC which controls the valve

Each of these approaches has some merits and drawbacks. The control responses get slower, but smoother as you go down the list. We tend to like to have the fuel gas respond quickly, so the first couple of approaches tend to be used the most.

In a dual-fired furnace, the same controller may also regulate the oil firing arrangement when the heater is operated on fuel oil.

Fuel Oil System

Figure 6 is a schematic of a typical fuel oil system. Some plants use petroleum residues as fuel oil. These types of fuels are high in viscosity and very often have a high pour point. For these reasons the fuel oil is stored in insulated and heated cone-roofed tanks. Heating may be accomplished by steam coils located in the base of the tank or by external steam heat exchanger through which the fuel oil is continually circulated.

Positive displacement pumps (usually rotary type) are used to deliver the fuel oil from the tank, through the distribution system to the heater burners. These pumps are always spared, and the spare pump is driven by a steam turbine, while the operating pump is motor driven.

The fuel oil passes through a duplex filter before entering the suction of the pumps. This filter is included to remove any solid contaminants that may be in the oil such as fine coke particles which would foul the fuel oil burner. The discharge pressure of the pumps is controlled by a slipstream routed back to the storage tanks through a pressure control valve. This valve is activated by a pressure control element on the pump discharge header.

The pumps discharge the fuel oil via the pressure controller to a preheater. This preheater may be a simple double-pipe heat exchanger for relatively small units or a regular shell and tube exchanger for the larger systems. Double-pipe type exchangers are favored in this service when economical because they are easier to clean and maintain. The fuel oil leaves the preheater to enter the fuel oil distribution system hot enough to maintain a viscosity low enough for the oil to flow easily and to be easily atomized by steam at the fuel oil "gun" (or burner).

All the piping associated with residual fuel systems is heavily insulated and steam traced. The distribution systems for residual fuel oils are usually the recirculating type. That is, the fuel leaves the preheater to circulate to all the user plants in a loop where the quantity used is taken off the stream and the remainder allowed to return to the system. The return header is routed back to the storage tanks. The circulation system handles between one and three times the quantity of oil that is actually burned.

Fuel oils are introduced into the fire box and ignited through a fuel oil burner sometimes called a fuel oil gun. In order to ensure combustion in a manner suitable for a fire box operation, the fuel oil needs to be dispersed into small droplets or spray at the burner tip. In heavy residual oils this is almost always accomplished by steam. Alternately, compressed air is sometimes used for this purpose.

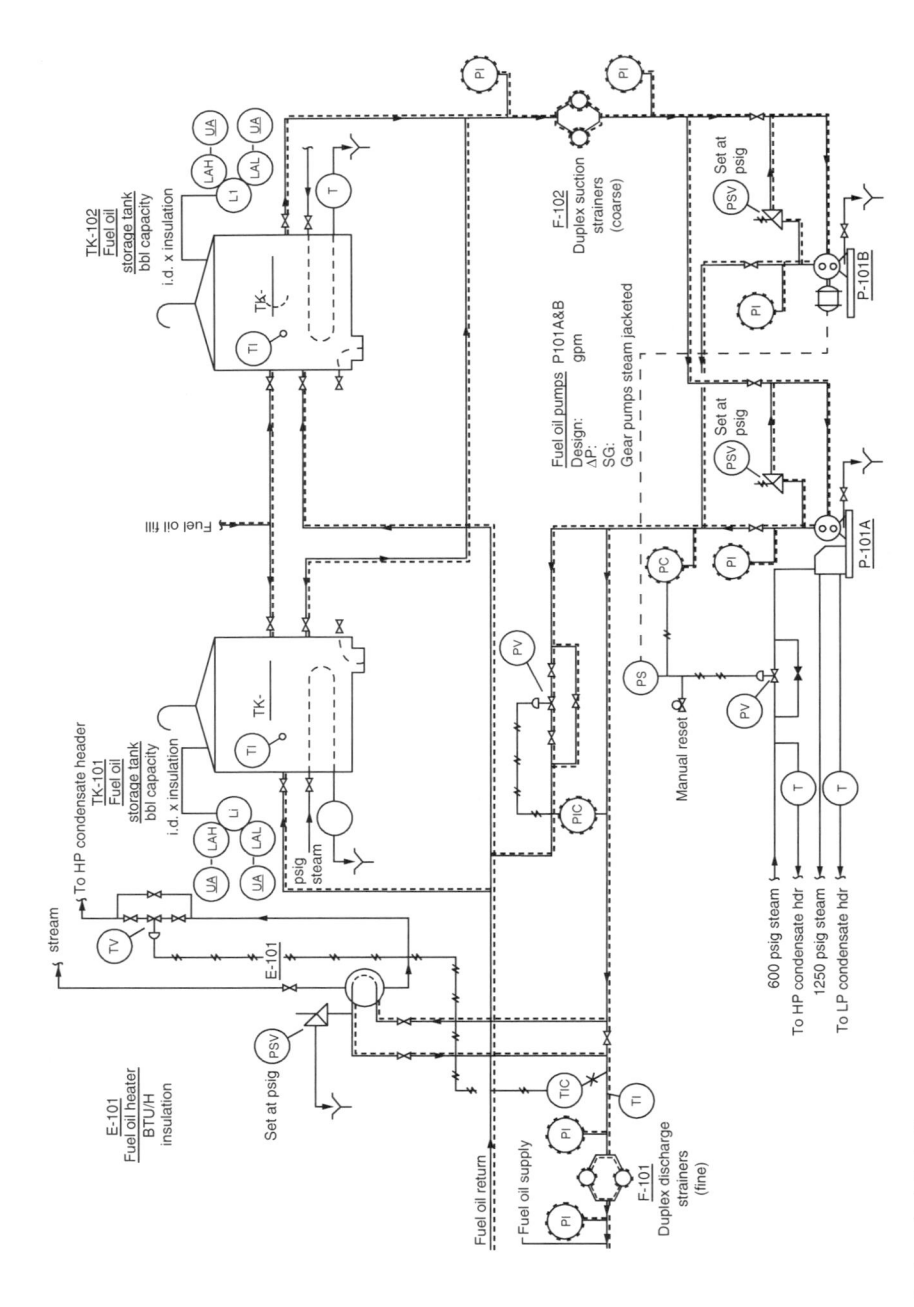

Fig. 6 A typical fuel oil system

Utilities in Petroleum Processing

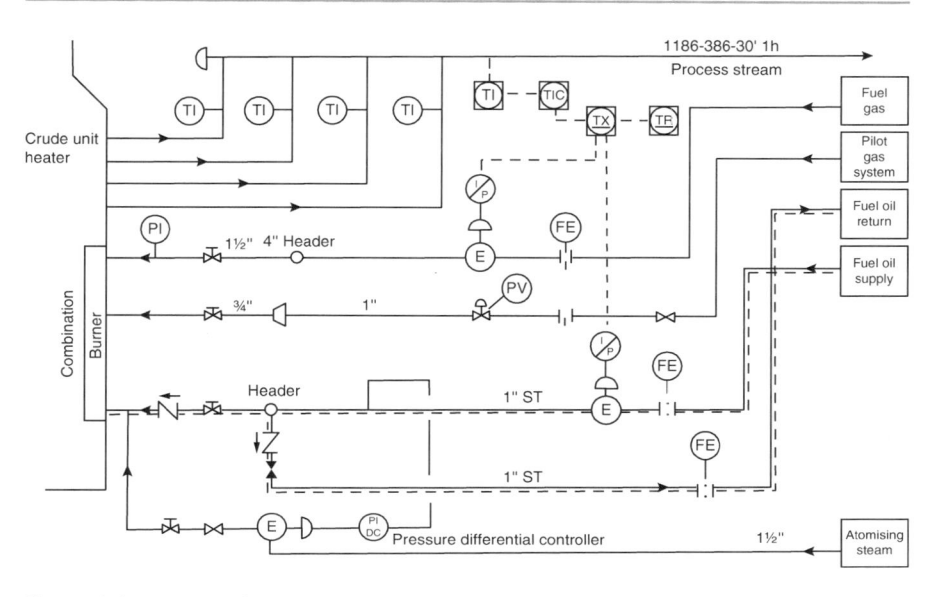

Fig. 7 A burner control system

This atomizing stream is introduced into the gun chamber and comes into contact with the oil stream just before the burner tip. The kinetic energy in the atomizing medium forces the oil into suitable droplets as it leaves the burner. Steam is normally used as the atomizing material because it is usually cheaper, more readily available, and has a more reliable source than air from a compressor.

The steam pressure for atomizing should be 15–25 psig higher than the fuel oil pressure. The quantity of steam will range from 1.5 to 5 lbs per gallon of oil. Dry steam with a superheat of about 50 °F is preferred for atomizing.

In order to control the process heater operation, oil burners require a turndown ability. That is, they must operate satisfactorily over a prescribed range of flow. In keeping with an operating range for oil flow, the atomizing medium must also have a similar operating range. Burner pressure is a critical requirement for turndown. The steam (or air) pressure at the burner should be 15 psi or 10 % (whichever is the greater) higher than oil pressure at the burner. The fuel oil supply system should be 100 psi higher than the burner requirement.

The oil burner operation, as with the fuel gas burner, is controlled by the heater's process stream outlet temperature. The temperature control valve activated by the coil outlet temperature increases or decreases the oil flow from the circulating oil stream to the burner. A proportional control valve on the atomizing steam line regulates the flow of steam to the burner in keeping with the oil flow. Figure 7 shows an example of a burner control system.

There are various methods for safety shutdown. The one shown here shuts down the oil flow on steam failure and on loss of the pilot burner. In some systems there is also an automatic changeover to gas firing on low oil pressure.

Burner Management Systems and Safe Firing of Heaters

We have already described the common shutdowns on fired heaters. Not all heaters or companies use the same matrix, but most are very similar. Some reasoning behind each of the shutdowns mentioned is worth reviewing:

- *High-high fuel gas pressure.* Fuel gas pressure that is too high causes the flame to lift off the burner tip and become unstable because the flame velocity leaving the tip exceeds the flame propagation velocity. This can result in the fuel essentially blowing the flame out. Without a shutdown, fuel would continue to enter the firebox until it found an ignition source and ignited, often explosively.
- *Low-low fuel pressure.* Similar to high-high fuel gas pressure, a low-low pressure can result in loss of flame and heat. The furnace could then fire harder to make up for the loss, introducing gas that can accumulate. While pilots help this by igniting unburned hydrocarbons continuously, flame loss is still a risk.
- *High-high process outlet temperature.* This is an indication that something has likely failed in the control system and the furnace is firing without feedback. Often high-high fuel gas pressure will shut down the furnace before high-high process outlet. HH outlet temperature may also indicate other problems, like low-low flow in the tubes. It really is a secondary monitoring for several parameters. Note that in a no-flow situation, the process outlet temperature will actually drift lower and call for more firing.
- *Loss of pilot burner* (depending on design; may also be detected as HH or LL pilot gas pressure). Pilots may have heated thermistors to tell when they are on or may use pressure sensors. Loss of pilots becomes a big problem if there is an additional loss of flame on the burner. Some people shut down a furnace on loss of pilots, while others alarm the situation. Loss of pilots is not always a problem. In a large furnace with many burners, loss of pilots may have no immediate impact, but they must be restored in short order. As noted above, some heaters do not have pilots, but have hundreds of burners that are not much larger than pilots. This shutdown would not apply to them.
- *Atomizing steam failure (oil burners).* If atomizing steam (or air) fails in an oil-fired furnace, the fuel oil will not be properly burned. Unburned oil will collect in the heater floor and flow out onto the ground, creating a mess, if not a fire hazard. Hence, it is critical that an oil-fired furnace be shut down on loss of atomization.
- *Low-low process flow to heater.* In most heaters, process flow is required to ensure the heat being fired is removed safely. Loss of flow in the heater tubes will result in damage to the tubes and may cause tube ruptures. Some heaters are designed to go to a limited temperature without flow in the heater tubes, but, in general, flow is required to operate a heater.
- *Low-low heater draft.* Loss of draft in a heater means that the firebox pressure is high enough that combustion air cannot get into the burners. The flames will go out. This is a dangerous situation with a hot firebox, fuel, and no combustion. Often explosions result if the fuel finds some air; hence, the heater is best shut down on loss of draft.

Shutdown systems for heaters are normally separate, secure systems. As shown in Fig. 5, many refiners use automatic double block and bleed (DBB) to positively stop fuel flow to the furnace. The DBB is operated by the multiple-redundant safety integrity system (SIS), after pre-alarming. Some refiners will also drive the fuel control valves closed in addition to the DBB or in lieu of DBB.

Safe heater firing needs to be practiced whenever a heater must be lit. The key steps in safe heater firing are:

1. Shut off and isolate all fuel sources (including pilot gas). Block in all burners and pilots.
2. Purge the firebox free of combustibles using snuffing steam or FD/ID flow. Verify there are no combustibles in the firebox with an LEL meter or other analyzer.
3. Unblock pilot gas and set pilot gas supply pressure.
4. Ignite all pilots, one at a time.
5. Once pilots are lit, the fuel gas control valve should be completely closed and the fuel gas supply unblocked (the low-low-pressure fuel gas trip may have to be bypassed until a couple of burners are lit).
6. Crack the fuel gas to the burners manually and light each burner, one at a time, off the burning pilot gas tips.
7. Begin controlling the fuel gas pressure manually initially as burners are added. Once enough burners are lit, the fuel gas controller can begin working.
8. Keep burners dispersed evenly throughout the firebox as they are lit to avoid impingement.
9. Maintain a high initial heater draft and bring draft down as the heater approaches a normal operation.
10. The heater can then be switched to automatic firing and draft control.

There are some preferences in how this firing sequence is executed, but it will generally follow the above pattern.

Fuel Gas Cleanup and Burner Fouling

The fuel gas mix drum also acts as a knockout drum for potential amine liquids entering the fuel gas from the various absorbers throughout a typical system (Fig. 4). This is not 100 % effective, especially in a system where plant expansion may have increased rates substantially.

Expect the fuel gas header to contain water, amines and amine salts, traces of oxygen and CO_2, chlorides, tar-like hydrocarbons, and other undesirable materials. The biggest problem presented by these materials is fouling at the burner tips, which are very small and operate at high temperatures often. The amine salts and wet environment are particularly corrosive to the fuel gas header piping.

It is advisable to provide local fuel gas knockout drums at each process unit to remove condensed or accumulated liquids, solids, and corrosion products.

Such a drum is shown in Fig. 5. In some large furnaces, the furnace may even need to have its own knockout drum.

Once the fuel gas has been cleaned and separated in the unit knockout drum, keep it clean and dry by tracing the fuel gas piping to the users. Some refiners also use stainless steel piping between the knockout drum and the furnace(s).

Some types of furnaces and burner designs expose the burner tips to extremely high temperatures. An example would be a hydrogen plant where the tips see the firebox at about 2,800 °F (1,573 °C). At these temperatures, even natural gas will crack and foul the tips. These need to be watched closely for signs of fouling and cleaned before they become problems.

Troubleshooting

A few of the more common problems related to fuels are highlighted below, along with some possible causes and remedies:

- *Frequent burner tip fouling.* Verify the knockout drums are working properly and the design flow rates are not being exceeded. Verify the absorbers are not overloaded, not foaming, not operating at excessive temperatures, and have enough lean amine flow. If the final fuel lines are traced, verify the tracing is operating. Clean the tips as required using the correct drill size or replace the tips.
- *High fuel gas volume requirements, unusual flame patterns, or sounds.* Verify the heating value of the fuel. It may have too much hydrogen. Heating values below about 500 Btu/scf can be a problem.
- *Flame impingement.* Look for burner fouling, high fuel gas pressures, not enough burners in service, and problems with air registers or draft.
- *Liquid pooling below oil-fired heater.* Loss of atomizing steam/air, burner tip problems, and leak.
- *Pilot failure to light.* Check pilot gas pressure and lineup. Check pilot gas orifice to be sure it is clear. Adjust air intake orifice on pilot. Be sure heater draft is not excessive so that it is blowing the pilot out.
- *Burner failure to light.* Be sure the pilot is burning. Check fuel gas pressure and lineup. Check fuel oil pressure, lineup, tracing, and atomization medium. Check for burner or distribution header plugging. Be sure heater draft is not excessive and that the air registers are not too far open so the flame is being blown out.

Water Systems

The major water systems found in most refineries and chemical plants are:

- Cooling water
- Treated water for BFW and process uses

- Fire water
- Potable water

We will discuss each of these systems in turn.

Note that in this discussion and in water treatment literature, hardness, alkalinity, and even individual ions are often expressed as their $CaCO_3$ equivalent. Table 1 provides a handy conversion reference for converting from the water analysis to $CaCO_3$ equivalents.

Table 1 Calcium carbonate equivalents (The Nalco Water Handbook 1979)

Substance	Mol wt.	Equiv. wt.	To convert to $CaCO_3$, multiply by	To convert $CaCO_3$ to substance, multiply by
$Al_2(SO_4)_3$ (Anhyd)	342.1	57.0	0.88	1.14
$Al(OH)_3$	78.0	26.0	1.92	0.52
Al_2O_3	101.9	17.0	2.94	0.34
$BaSO_4$	233.4	116.7	0.43	2.33
$Ca(HCO_3)_2$	162.1	81.1	0.62	1.62
$CaCO_3$	100.1	50.0	1.00	1.00
$CaCl_2$	111.0	55.5	0.90	1.11
$Ca(OH)_2$	74.1	37.1	1.35	0.74
$CaSO_4$ (Anhyd)	136.1	68.1	0.74	1.36
$Ca_3(PO_4)_2$	310.3	51.7	0.97	1.03
$Fe_2(SO_4)_3$	399.9	66.7	0.75	1.33
MgO	40.3	20.2	2.48	0.40
$Mg(HCO_3)_2$	146.3	73.2	0.68	1.46
$MgCO_3$	84.3	42.2	1.19	0.84
$MgCl_2$	95.2	47.6	1.05	0.95
$Mg(OH)_2$	58.3	29.2	1.71	0.58
$Mg_3(PO_4)_2$	262.9	43.8	1.14	0.88
$Mg_3(SO_4)_2$ (anhyd)	120.4	60.2	0.83	1.20
SiO_2	60.1	30.0	1.67	0.60
$NaHCO_3$	84.0	84.0	0.60	1.68
Na_2CO_3	106.0	53.0	0.94	1.06
$NaCl$	58.5	58.5	0.85	1.17
$NaOH$	40.0	40.0	1.25	0.80
$NaNO_3$	85.0	85.0	0.59	1.70
Na_3PO_4 (anhyd)	164.0	54.7	0.91	1.09
NaH_2PO_4	120.0	40.0	1.25	0.80
Na_2SO_4	142.1	71.0	0.70	1.42
Na_2SO_2	126.1	63.0	0.79	1.26

(*continued*)

Table 1 (continued)

Substance	Mol wt.	Equiv. wt.	To convert to CaCO$_3$, multiply by	To convert CaCO$_3$ to substance, multiply by
Al^{+3}	27.0	9.0	5.56	0.18
NH$_{4+}$	18.0	18.0	2.78	0.36
Ba^{+2}	37.4	68.7	0.73	1.37
Ca^{+2}	40.1	20.0	2.50	0.40
Cu^{+2}	63.6	31.8	1.57	0.64
H$^+$	1.0	1.0	50.0	0.02
Fe^{+3}	55.8	18.6	2.69	0.37
Fe^{+2}	55.8	27.9	1.79	0.56
Mg^{+2}	24.3	12.2	4.10	0.24
Mn^{+2}	54.9	27.5	1.82	0.55
K$^+$	39.1	39.1	1.28	0.78
Na$^+$	23.0	23.0	2.18	0.46
Sr^{+2}	87.6	43.8	1.14	0.88
Zn^{+2}	65.4	32.7	1.53	0.65
HCO$_3^-$	61.0	61.0	0.82	1.22
CO$_3^{-2}$	60.0	30.0	0.67	0.60
Cl$^-$	35.5	35.5	1.41	0.71
CrO$_4^{-2}$	116.0	58.0	0.86	1.16
F$^-$	19.0	19.0	2.63	0.38
OH$^-$	17.0	17.0	2.94	0.34
NO$_3^-$	62.0	62.0	0.81	1.24
PO$_4^{-3}$	95.0	31.7	1.58	0.63
HPO$_4^{-2}$	96.0	48.0	1.04	0.96
H$_2$PO$_4^-$	97.0	97.0	0.52	1.94
SO$_4^{-2}$	96.1	48.0	1.04	0.96
HSO$_4^-$	97.1	97.1	0.52	1.94
SO$_3^{-2}$	80.1	40.0	1.25	0.80
HSO$_3^-$	81.1	81.1	0.62	1.62
S^{-2}	32.1	16.0	3.13	0.32

A substance's equivalent weight is the molecular weight of the substance or ion divided by, either, the valence of the positive ion (or total valence of negative ions). The CaCO$_3$ equivalent is CaCO$_3$'s equivalent weight of 50.0 divided by the substance's equivalent weight. Hence sodium with valence +1 and weight of 23.0 has an equivalent weight of $23.0/1 = 23.0$. The CaCO$_3$ equivalent is $50.0/23.0 = 2.18$. Multiplying ppm sodium by 2.18 would give the CaCO$_3$ equivalent concentration.

Cooling Water

We will focus here on the normal circulating cooling water systems that cool by evaporation in cooling towers. There are special circulating, tempered or attenuated, closed-loop cooling water systems in some services. These are not addressed here.

Cooling Water System Flow Sheet and Equipment

Figure 8 is a mechanical flow sheet section showing the arrangement around a cooling tower for the collection and supply of cooling water.

The cooling water system is a circulating one. That is, there is a cold supply line with an associated warmer return line from all its users. Figure 9 shows a section of this distribution system. The typical differential between the cold supply and the warm return is 15–20 °F (8–11 °C). Cooling water from a given exchanger should generally not exceed about 140 °F (60 °C). Beyond this, excessive fouling of cooling water is observed. Chemical treatments can help, but are not completely effective.

The water returned to the cooling tower by the return header enters the top of the tower and flows down across the tower internals countercurrent to an air flow, either induced or forced by fans, passing up through the tower. The return header for a counterflow tower normally has a full line-sized standpipe extending well above the top of the tower to allow any possible gases that leak into the cooling water system to vent. These inlet standpipes must be monitored for exhausting gases – usually visually. In extreme cases of sudden, large leaks, the standpipes can prevent damage to the distributors and the tower. The open top decks of crossflow towers do not require the standpipes.

There are two common types of cooling tower used in petroleum processing facilities, as illustrated in Fig. 10:

- *Crossflow*. In these towers, the tower fill or packing is located down the sides of the tower structure. Air flows roughly horizontally though the fill as the water flows from distributors downward through the fill. There are inlet air louvers around the outside of the fill and drift eliminators inside the thick walls of fill. In the center of the tower is a plenum that collects the moist air. An ID fan pulls the air out of the plenum and exhausts it to the atmosphere through a venturi stack. These towers tend to be shorter and cheaper than counterflow towers, at the expense of efficiency.
- *Counterflow*. In counterflow towers, the air intake is located at the bottom of the fill, and the fill nearly covers the whole tower footprint. Above the fill are the drift eliminators to intercept entrained water. Above the drift eliminators are the plenum chamber and an induced draft fan exhausting to the atmosphere. These towers tend to be more expensive, but more efficient. The hyperbolic cooling towers you see around power plants are often of this type, but the hyperbolic shape eliminates the need for the fans.

A continuous blowdown of cooling water return is normally maintained for the control of solids and dissolved solids in the system. This will be discussed further along with chemicals.

The water, cooled by the air flow and evaporation, is collected in the cooling tower basin (cold water basin in Fig. 10). Makeup water (usually potable water) is

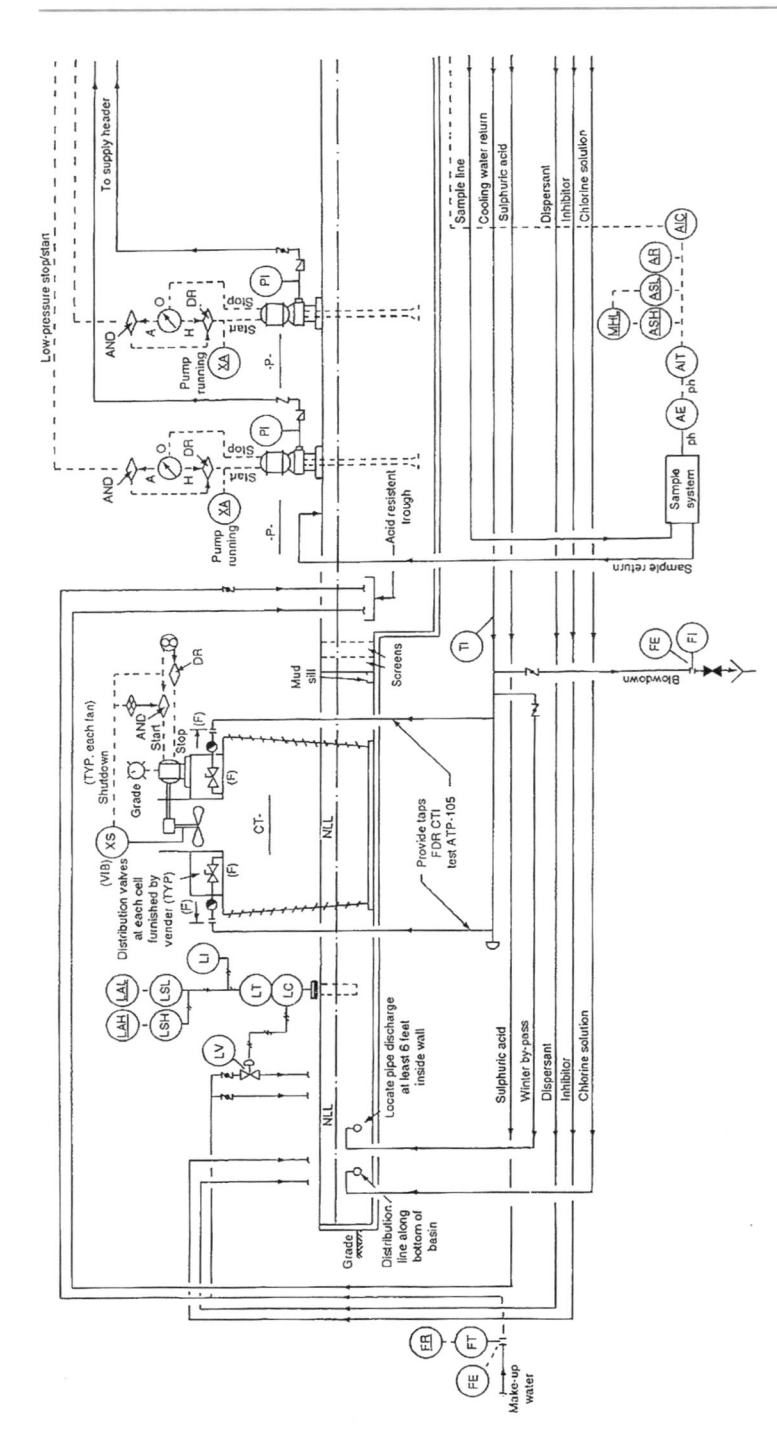

Fig. 8 Mechanical flow diagram of a water cooling system

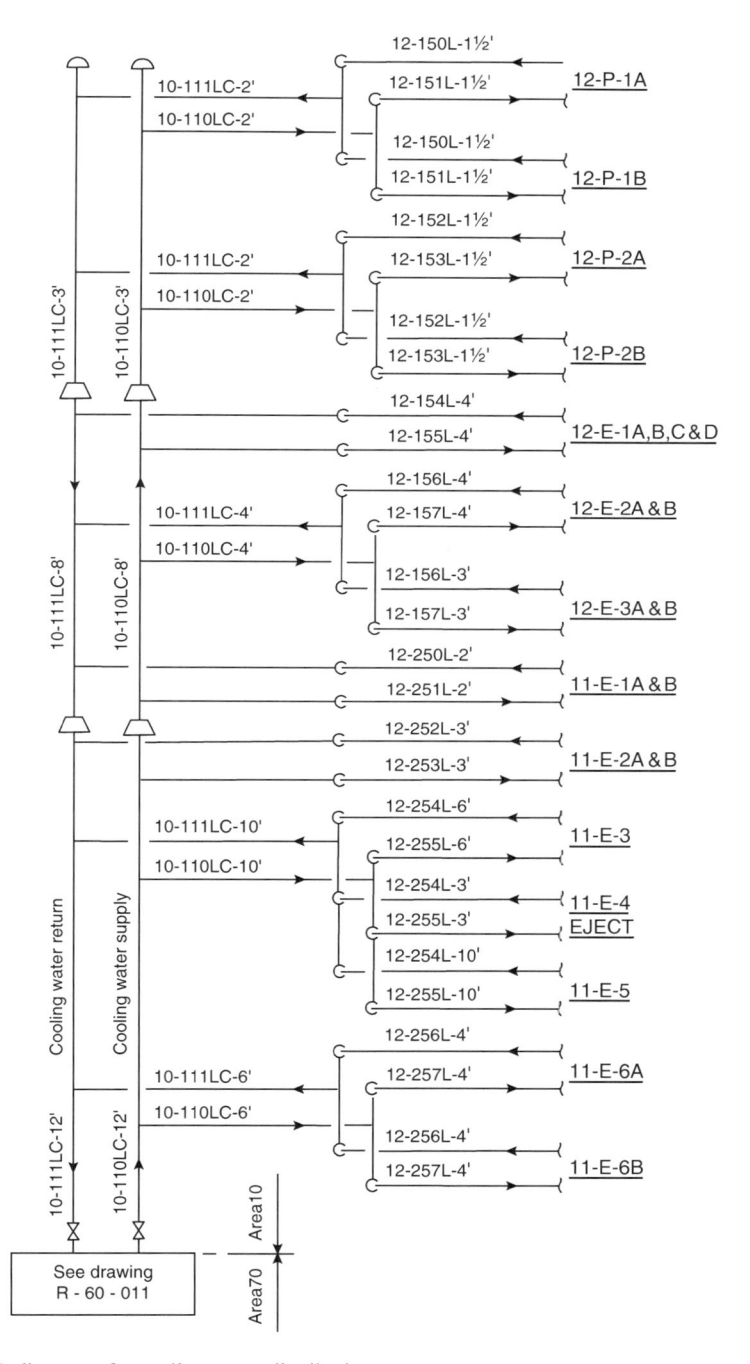

Fig. 9 A diagram of a cooling water distribution system

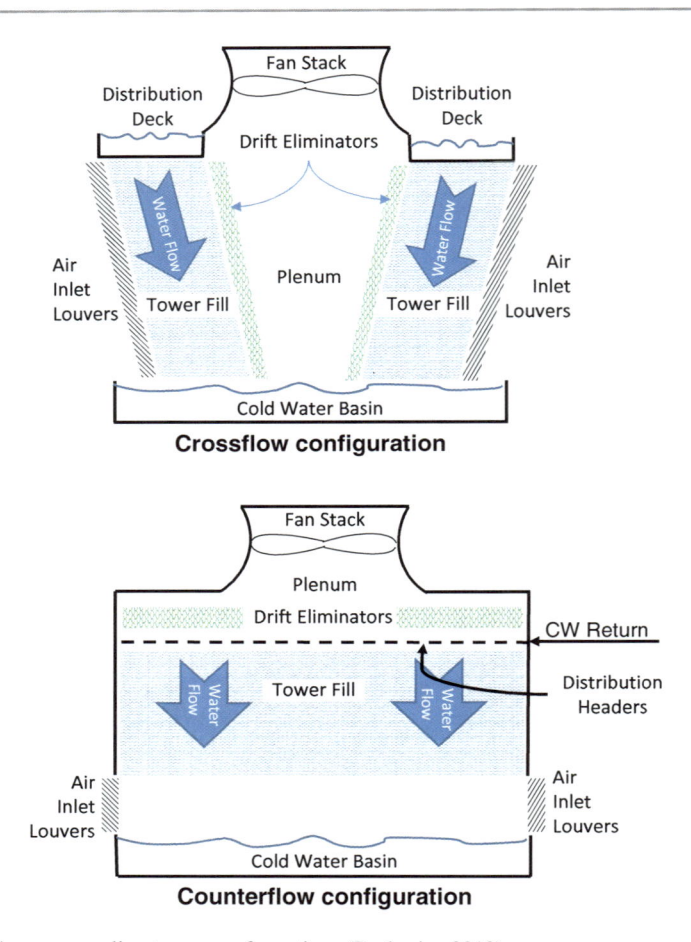

Fig. 10 Common cooling tower configurations (Puckorius 2013)

added to the basin under level control. The makeup water rate is equal to the evaporative load plus wind and drift losses plus blowdown plus leakage losses. The evaporative loss can be determined directly from the amount of heat rejected to the cooling water. Wind and drift losses are typically <0.1–0.3 % of the circulation. 0.1 % is usually a good number. Blowdown depends on makeup water quality and is discussed below. Leaks depend on the specific system and equipment condition, but should be almost negligible.

Vertical cooling water circulating pumps take suction from the cooling tower basin sump to deliver the water into the plant's distribution header. The sump design issues are discussed below. Cooling water pumps are usually high-capacity centrifugal pumps with a moderate differential head. Because of the critical nature of the water supply, the pumps are redundant. Spare pumps are commonly automatically started on low-low cooling water flow or pressure. There may be four or more pumps with each rated at around 60 % of design capacity – two in operation and two on standby. A mixture of motor and steam turbine drivers are quite common. Some facilities have two 100 % pumps with one electric and one steamer. For reliability, autostart turbine-driven pumps usually are kept with the turbine hot and may be on slow roll to ensure they can autostart when needed. The slow roll speed has to be sufficient to ensure the pump lubrication is adequate.

Cooling water pump sumps are located adjacent to the cooling tower. The entrance to each sump usually has two removable trash screens (about 0.5″ or 1.3 cm mesh) in series. There is a small sill or foot at the bottom of each screen to catch material that may fall off the screen when it is removed. The screens are removable for regular (weekly) cleaning – one at a time. After the screens, the pumps should be individually housed in bays. Design of the bays and the exact pump locations in the bays may require computerized fluid dynamic analysis (CFD) to ensure adequate pump suction. If the pumps are all in the same bay or sump or they are too close together, the flow into one pump's suction often interferes with flow into a neighboring pump. Separation of the individual pumps into bays prevents most interferences.

The cooling water supply and return lines may be quite large when central cooling water systems are employed. Lines over 5 ft. diameter (1.6 m.) are not unusual. The largest lines are often buried and are made from jointed, fiber-reinforced plastic.

The supply header pressure is kept at around 30–50 psig (2–3.5 barg), and, very often in large plants covering long distances, booster stations are installed at predetermined locations to maintain the supply header pressure. These booster stations consist of pump pits with high-capacity vertical pumps rated smaller, of course, than the main supply pumps. In some cases, inline pumps may also be used to boost pressure without a sump. The location of these booster stations is determined by a rigorous hydraulic analysis of the distribution system which also determines the header pipeline sizes. The return flow is collected from each user into the return header and flows back to the cooling tower under the users' outlet pressure.

The locations and orientations of towers can significantly affect performance. Towers are normally made from one or more cells, with each cell being like a smaller tower in itself. Every two cells share a common dividing wall. Each cell usually has its own fan. All the cells together share a common cold water basin. See Fig. 11. The orientation of the cells is normally perpendicular to the prevailing wind to avoid short-circuiting of air flow. When one or more cooling towers are present in an area, they are normally lined up end to end down the long axis and with the prevailing wind perpendicular to the long axis, again to avoid short-circuiting.

Operation of cooling towers in cold climates presents special challenges. The accumulation of ice on upper decks and air inlet louvers can result in collapse. The cold water is often too cold for many services. Approaches to dealing with these issues include reversing the cooling tower fans, installing temporary covers over top distribution decks (in crossflow towers), and shutting off fans as necessary. In extreme cases, the normal open cooling water system may be replaced by a closed-loop tempered water system with cooling supplied by air coolers instead of

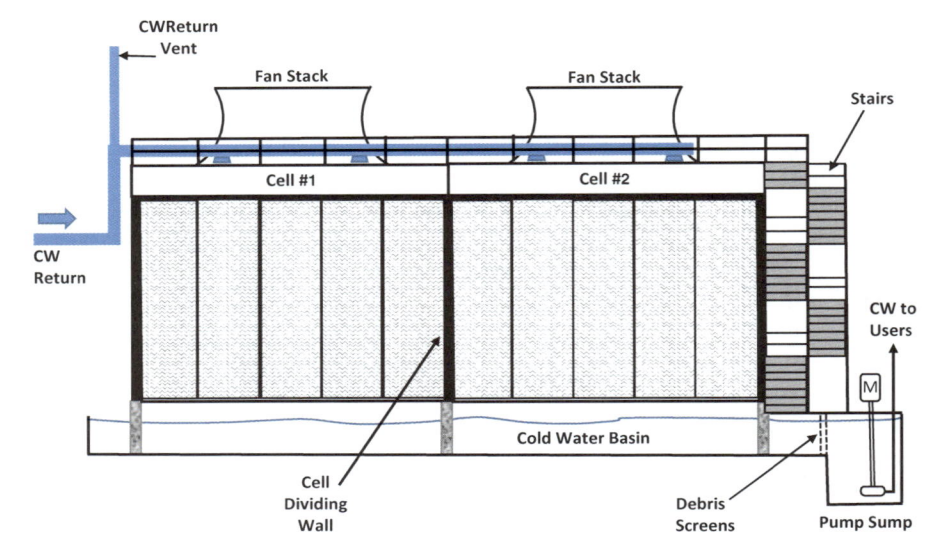

Fig. 11 A two-cell, crossflow cooling tower

evaporation. The closed-loop systems normally use glycol and different water treatments than open towers.

Cooling Water Chemistry and Treatment

The water in the cooling tower basin and in the cooling tower itself requires blowdown for hardness control and chemical treatment to prevent the buildup of algae and other undesirable contaminants. A separate small treatment plant may be used for mixing the inhibiting chemicals and injecting them into the critical sections of the system.

The blowdown rate required depends on the makeup water quality. A tower can usually operate at about 4–5 cycles of concentration. Cycles of concentration are roughly equal to ppm TDS/hardness of the circulating cooling water divided by ppm TDS/hardness of the makeup. Poor quality water with a lot of hardness will limit the cycles of concentration possible and increase the makeup requirement. High-quality water with low TDS and hardness can be cycled much higher.

The normal chemical treatments applied are:

- *Corrosion control*. At one time, chromate was the primary corrosion treatment, but environmental issues have eliminated chrome treatment programs. Today's options include orthophosphate, polyphosphates, zinc, and azole (if there are copper alloys in the system). The heavy metal treatments may be limited for environmental reasons.
- *Deposit control*. Phosphonate, carboxylate, and bio-dispersants.
- *Microbiological control*. Halogens (chlorine, bromine), sodium hypochlorite (bleach), chlorine dioxide, and organic biocides. Chlorine is by far the most

effective, but safety and environmental concerns are limiting its use. Bleach works very well. Bromine and organic biocides are fine in normal service, but have difficulty if there are any leaks into the system.
- *pH control.* To make many of the other chemicals work and to help control hardness deposits, sulfuric acid is normally injected to control pH in the required range, usually about 7.5–8.5 pH. The optimum range for pH, however, depends on the specific treatment chemicals used. The water treatment chemical supplier will set the range.

The chemicals should be injected into an actively flowing slipstream from the cooling water supply line. The stream should be reintroduced into the basin opposite the cooling water pump sumps so that the chemicals are carried across the basin and further distributed. Be careful, however, that the chemicals are not introduced into a dead area of the basin, or they will cause localized corrosion at the introduction point.

There are a number of fouling indices used to track cooling water treatment chemistry. The index definitions are presented below without much discussion or description about how to calculate them. There are several articles and a great deal of literature that explore these indices more completely.

- *Langelier Saturation Index (LSI).* This is equal to the difference between the actual pH and the pH at saturation of calcium carbonate. An LSI below zero is a non-scaling condition. An LSI above 1.5–2.5 is a scaling condition. The pH at saturation is a function of the calcium hardness, alkalinity, and temperature. Correlations are available in the references.
- *Ryznar Stability Index (RSI).* This index is given by

$$RSI = 2(pH_{\text{saturation}}) - pH^{3.5} \qquad (1)$$

where:
$pH_{\text{saturation}}$ = pH at calcium carbonate saturation
pH = actual circulating water pH
An RSI below 6.0 indicates possible scaling and RSI above 6 is progressively non-scaling. An RSI above 8 may result in corrosion of carbon steel.
- *Stiff-Davis Index* (SDI). This index can be used for high concentration water. It is defined by

$$SDI = pH - \log_{10}(I/[\text{Ca}]) - \log_{10}(I/[\text{alkalinity}]) - K \qquad (2)$$

where:
pH = actual circulating water pH
I = calculated ionic strength of the water (usually ~1)
[Ca] = calcium ion concentration, mg/l
[alkalinity] = alkalinity concentration, mg/l

$$= [\text{HCO}_3^-] + 2[\text{CO}_3^{2-}] + [\text{OH}-] \qquad (3)$$

K = constant based on ionic strength and temperature

Similar to LSI, an SDI below zero is a non-scaling condition. An SDI above zero is a scaling condition.
- *Practical (or Puckorius) Scale Index (PSI).* This index attempts to account for buffering. It is defined by

$$PSI = 2 \times pH_{saturation} - pH_{equilibrium} \tag{4}$$

where:

$$\begin{aligned} pH_{saturation} &= \text{defined above} \\ pH_{equilibrium} &= 1.465 \times \log_{10}[\text{alkalinity}] + 4.54 \end{aligned} \tag{5}$$

Interpretation of PSI values is similar to RSI.

Cooling Water System Monitoring

Monitoring a cooling water system includes both mechanical and chemistry tracking. There is no substitute for physical inspection of the tower, in addition to instrumentation and samples. Some of the factors to monitor include:

- *Mechanical.* Inspect tower visually.
  - Look at fans and drivers, condition of the fill, condition of the casing, cleanliness of the screens, corrosion, water flow patterns to the pumps, and anything else that looks unusual. Especially watch for fouling of the tower fill or packing. Plugged packing reduces the tower efficiency. Extremely plugged packings have resulted in physical collapse of towers.
  - Observe the return header standpipe for splashing or signs of leaks into the system.
  - Watch for algae buildup, as this indicates a leak. Address problems noted.
  - If the appearance of the circulating water becomes muddy or "chocolate brown," you may have a low pH.
  - Look at the air exhausts for signs of excessive drift. This may indicate problems with the drift eliminators.
- *Thermal.* Look at the data.
  - Track the difference between supply and return temperatures.
  - Track cooling water flow rates and pressures.
  - Watch the cooling water return temperatures for values over 130 °F (55 °C) which could indicate overloaded conditions or insufficient flow.
- *Chemistry.* Review and act on sample results (your's and your chemical supplier's) for the cooling water and injection system operation. Periodic samples (appropriate frequencies) should include:
  - pH
  - Hardness
  - Alkalinity
  - Total dissolved solids
  - Total suspended solids

- Scaling indices
- Residual or available chlorine (or other biocide)
- Water treatment chemical concentrations
- Blowdown biochemical oxygen demand, total organic carbon, and volatile organic carbon

Troubleshooting

A few tips are offered here on problems commonly encountered in cooling water systems:

- *Increasing algae or slim accumulation – sudden or gradual.* This indicates undertreatment of biocide for the water conditions. Check the biocide dosing system and dosage (bleach, chlorine, etc.) to be sure it is working correctly. The tower may have to have a shock treatment with a heavy biocide dose to control some cases. If there is a step change in the bio-deposit accumulations, suspect a leak. You will have to sample to narrow down the source of the leak. Allowing a leak to continue will foul the system and cause corrosion of other exchangers. Maintain a high blowdown rate until the leak is resolved.
- *Carryover or overflow of return header vent.* Indicates a leak into the system or excessive plugging of the distributors. Identify and eliminate the leak. Caution should be exercised if the vented vapors can be encountered by personnel as the leak could contain H_2S.
- *Muddy appearance of cooling water.* This normally accompanies a low pH. A control problem with the acid injection pump or valve or an overflow of the acid storage tank into the basin will cause this. The corrosion rate within the cooling water circuit will rise dramatically. Maximize your blowdown and increase corrosion inhibitor injection until the problem is resolved. This may require a few days.
- *Low cooling water flow rate or pressure.* Check the cooling water pumps for proper operation. Switch to the standby pump(s), if necessary. Check the pump inlet screens for plugging. Clean the screens. Verify that the flow and pressure instruments are correct. Check tower water level control – low level will reduce flow.

Cooling Tower Safety

- *Chemicals*
  - If the tower uses chlorine gas for biocontrol, care needs to be taken in managing the gas cylinders. Some companies locate these in sumps and cover the cylinders with water. As a minimum, they need to be in a secondary containment area.
  - All chemicals used in cooling towers need to be provided with secondary containment (dikes or pits) to control spills. The drains on the containment should be kept closed until the containment is deliberately drained, such as after a storm.

- The secondary containments must be designed to prevent contact between incompatible chemicals. For instance, acid must be separated from bleach tankage.
  - Brominators can present a problem with generation of gas sufficient to cause an explosion of the brominator vessel if the vessel is not properly managed.
  - All the chemicals used in cooling water treatment present some safety hazards. These need to be appropriately managed.
- *Legionella bacteria*
  - One safety issue that has come to light in recent years applies to tower maintenance. Cooling towers tend to grow *Legionella* bacteria in the packing and drift eliminators. *Legionella* can cause a fatal respiratory infection. Workers cleaning a tower should use air respirators suitable for this hazard until the tower is completely cleaned. No one else should enter the tower until it is cleaned.

Boiler Feed Water (BFW) Treating

All water contains impurities, no matter from what source the water originates. Sometimes, in a petroleum processing plant, water with most of these impurities can be used without treating of any kind. However, when it comes to generating steam, particularly high-pressure steam, and uses of water in the processes, these impurities become problematic. We will focus here on water for steam generation, but the discussion applies equally well to process uses.

To operate steam generators effectively, and to avoid serious damage to the unit, impurities in boiler feed water either have to be removed or be converted into compounds that can be tolerated in the system.

Types of Boiler Feed Water Impurities
In general there are three types of soluble impurities naturally present in water and which must be removed or converted in order to make the water suitable for boiler feed purposes. These are:

- *Scale-forming impurities*. These are salts of calcium, magnesium, silica, manganese, and iron.
- *Compounds that cause foaming*. These are usually soluble sodium salts.
- *Dissolved gases*. These are usually oxygen and carbon dioxide. The soluble gases must be removed to prevent corrosion.

Table 2 provides a more comprehensive listing of the various impurities, impacts, and treatment options.

Boiler Water Management
Buildup of solids in the boiler itself is managed or kept at a low level by blowdown. This is the mechanism of draining a prescribed amount of the boiling water from the

Table 2 Boiler impurity impacts and treatment options (http://energyconcepts.tripod.com/energyconcepts/water_treatment.htm)

Impurity type	Impacts	Control	Notes
Soluble gases			
H_2S	Odor, corrosion	Deaeration	From leaks or contaminated water
CO_2	Corrosion, esp. in condensate	Deaeration, neutralizing chemicals	Filming and neutralizing amines for condensate
Oxygen	Corrosion and pitting of boiler tubes	Deaeration, O_2 scavengers	
Suspended solids			
Sediment and turbidity	Sludge, scale, carryover	Clarification, filtration	5 ppm max in most applications
Organic matter	Carryover, foaming, deposits, corrosion	Clarification, filtration, chemicals	Surface and some well waters
Dissolved or colloidal solids			
Oil and grease	Foaming, boiler deposits	Coagulation, filtration	Condensate contamination
Hardness, Ca, Mg	Scale, loss of heat transfer/efficiency, tube failures	Softening plus internal treatment	HCO_3^-, $SO_4^=$, Cl^-, NO_3^- in order of increasing solubility
Sodium, alkalinity, NaOH, $NaHCO_3$, Na_2CO_3	Foaming, condensate corrosion, potential caustic embrittlement	Deaeration, ion exchange, RO	
Sulfates ($SO_4^=$)	Hard scale if Ca present	Demineralization, RO	Max 100–300 ppm as $CaCO_3$
Chlorides	Uneven boiler operation (belching), lower efficiency, deposition on turbines and superheaters, foaming	Demineralization, RO	
Iron and manganese	Boiler deposits, erosion, inhibit heat transfer	Filtration, demineralization, RO	Most common form is ferrous bicarbonate
Silica	Hard boiler scale, turbine blade deposits	Lime softening, demineralization, RO	Very hard scale. Silica is volatile at higher temperatures with water present = turbine blade deposits

boiler steam drum or mud drums at regular intervals. This amount is calculated from the analysis of the solid content of the feed water and must equal the amount brought into the system by the feed water. Figure 12 gives an example of boiler blowdown.

STEAM
Solids content – Essentially zero
Amount per day – 900 000 lb.

Feedwater
Solids content – 1000 ppm
Amount per day – 100 000 lb.
Solids removed per day – 100 lb.

Boiler water solids level – 1000 ppm.

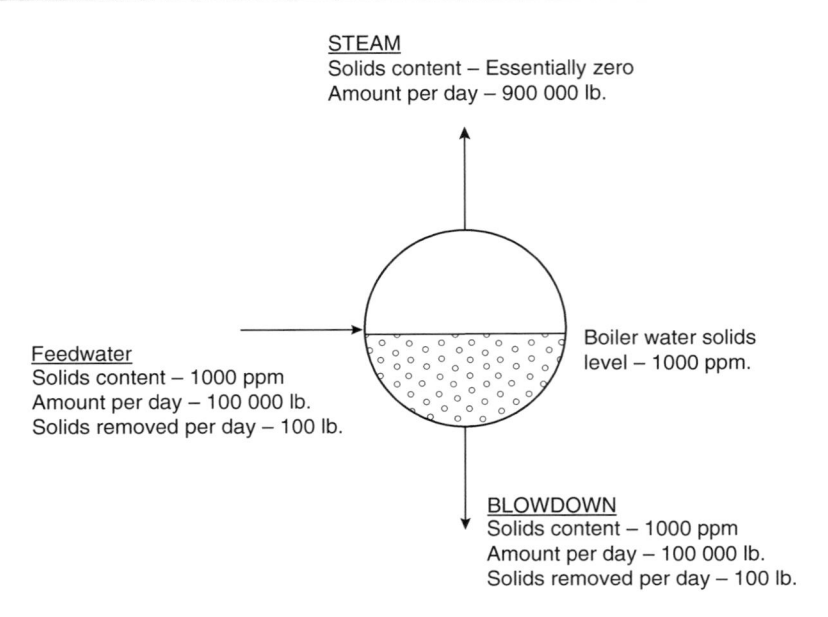

BLOWDOWN
Solids content – 1000 ppm
Amount per day – 100 000 lb.
Solids removed per day – 100 lb.

Fig. 12 How blowdown reduces the amount of solid buildup

Table 3 ABMA limits of various solids in boiler water

Boiler pressure (psig)	Total solids (ppm)	Alkalinity (ppm)	Suspended solids	Silica
0–300	3,500	700	300	125
301–450	3,000	600	250	90
451–600	2,500	500	150	50
601–750	2,000	400	100	35
751–900	1,500	300	60	20
901–1,000	1,250	250	40	8
1,001–1,500	1,000	200	20	2.5
1,501–2,000	750	150	10	1.0
Over 2,000	500	100	5	0.5

ASME has also provided some guidance for boiler feed water and boiler water along the same lines, as noted in Table 4 for BFW and boiler water. Although the table is stated for water-tube boilers, it is commonly also applied to waste heat boilers with the hot gas in the tubes (similar to a fire-tube boiler)

The American Boiler Manufacturers Association (ABMA) has developed limits for the control of various solids in boiler water. These are given in Table 3.

Other considerations regarding the limits of solids in BFW are:

- *Sludge.* This is a direct measurement of feed water hardness (calcium and magnesium salts) since virtually all hardness comes out of solution in a boiler.

- *Total dissolved solids.* These consist of sodium salts, soluble silica, and any chemicals added. Total solids do not contribute to scale formation, but excessive amounts can cause foaming.
- *Silica.* This may be the blowdown controlling factor in pre-softened water containing high silica. At elevated pressures high silica content can cause foaming and carryover.
- *Iron.* High concentration of iron in BFW can cause serious deposit problems. Where concentration is particularly high, blowdown may be based on reducing this concentration.

Preparation of Boiler Feed Water Makeup

There are three types of BFW treatment in use. There are the external type of treating and the internal processes. As the names suggest the external processes are those that treat the water before it enters the boiler. The internal treatments are those in the form of added chemicals that treat the water inside the boiler, and these were discussed under steam and condensate systems. Only the external processes are described here. A great deal of detailed information is available on each of these processes from vendors, treatment chemical suppliers, and in literature, so the descriptions here are only cursory.

The "Hot Lime" Process

This is a water softening process which uses hot lime contact to induce a precipitate of the compounds contributing to hardness (mostly Ca and Mg carbonates). The sludge formed is allowed to settle out. Very often coagulation chemicals such as alum or iron salts are used to enhance the settling and the removal of the sludge formed. In most plants that use the "lime" process, the reaction by the addition of lime and soda ash is carried out at elevated temperatures. However the reaction can be allowed to take place at ambient temperatures. The hardness of the water from the "cold" process will be about 17–35 ppm, while that from the "hot" process will be 8–17 ppm. Cleanup filters containing anthracite are often used to finish the treating process.

Ion Exchange Processes

As the name implies this process exchanges undesirable ions contained in the raw water with more desirable ones that produce acceptable BFW. For example, in the softening process, calcium and magnesium ions are exchanged for sodium ions. In dealkalization, the ions contributing to alkalinity (carbonates, bicarbonates, etc.) are removed and replaced with chloride ions. Demineralization in this process replaces all cations with hydrogen ions (H^+) and all anions with hydroxyl ions (OH^-) making pure water ($H^+ + OH^-$).

The ion exchange resins need to be regenerated after a period of operation. The operating period will differ from process to process and will depend to some extent on the amount of impurities in the water and the required purity of the treated water. Regeneration is accomplished in three steps:

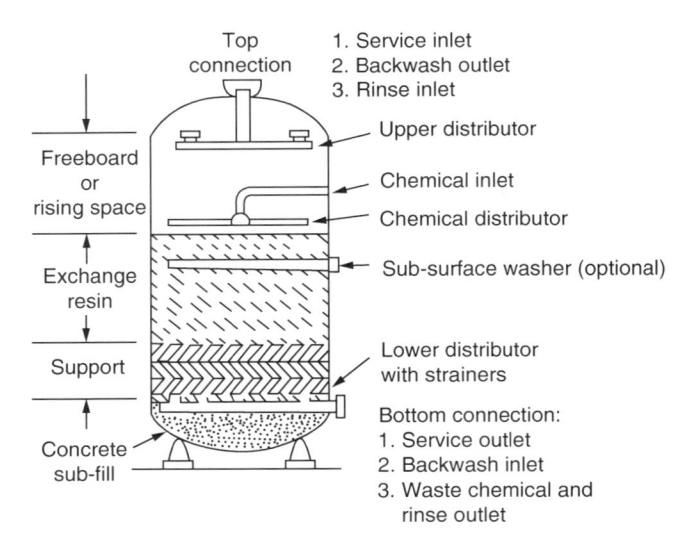

Fig. 13 A typical ion exchange vessel

- Backwashing
- Regenerating the resin bed with regenerating chemicals
- Rinsing

Figure 13 shows the internals of a typical ion exchange unit.

Under operating conditions the raw water is introduced through the top connection and distributor. The water flows through the resin bed where ion exchange takes place. The treated water is removed via the bottom connection. Under regeneration operation, raw water as backwash is introduced through the bottom connection and removed from the top connection. During its passage upward through the resin and support beds, it "fluffs" the beds and removes any waste material that has adhered to them. The backwash water is then sent to the plants wastewater disposal system.

Regenerating exchange chemical is introduced directly above the resin bed through a chemical distributor and allowed to flow downward to be removed at the bottom water outlet. The regenerating cycle is completed with the rinsing of the bed to remove any surplus regenerating chemicals. This is done by introducing a stream of raw water at the top connection and removing it from the bottom connection. This water is also disposed to waste.

Normally ion exchange units are installed in pairs (or more). When one is operating, the other is being regenerated. An automatic switch over of electronically controlled valves takes the pair of units through the correct cycles at the prescribed time intervals, without disrupting the treating process. Figure 14 shows a typical "hookup" of an ion exchange unit.

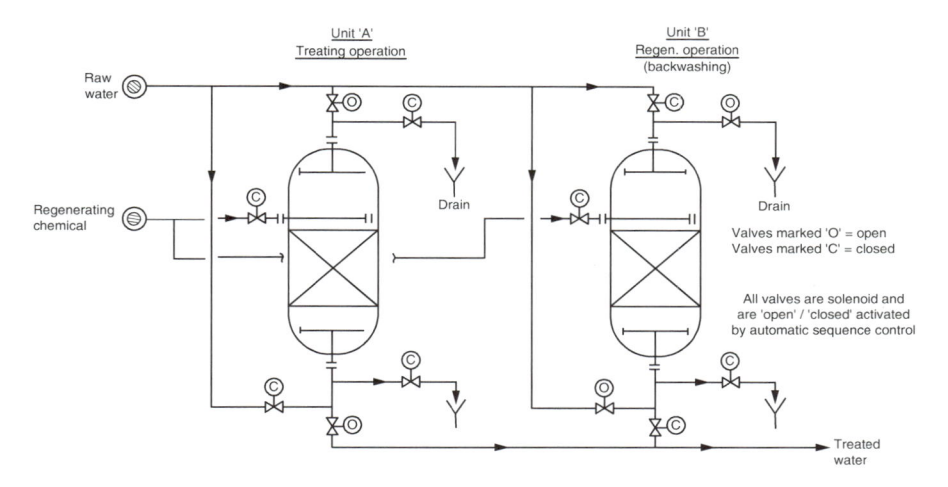

Fig. 14 An ion exchange unit hookup

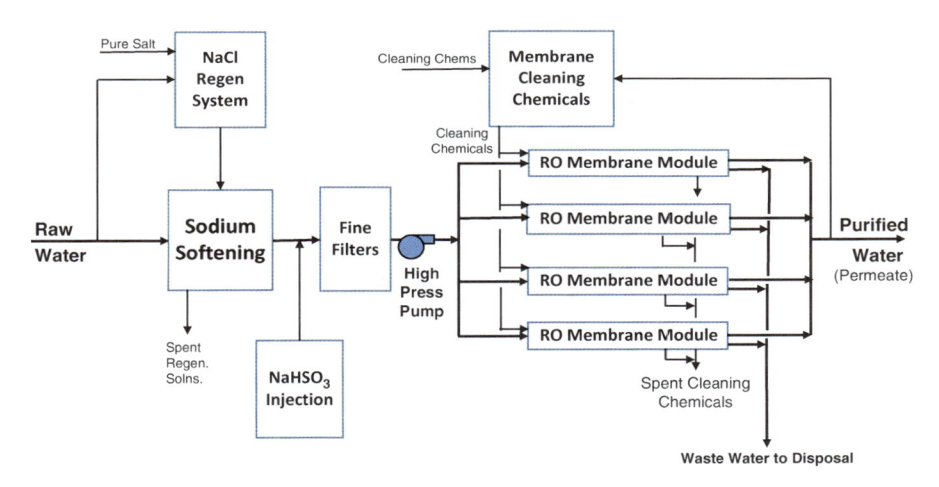

Fig. 15 Reverse osmosis system block flow diagram

Reverse Osmosis

Many newer facilities make boiler feed water by reverse osmosis (RO) using a selective membrane. The block flow diagram for an RO unit looks like Fig. 15.

In a reverse osmosis system, raw water, normally city water or other filtered and chlorinated water, is softened using sodium zeolite. Sodium replaces the calcium, magnesium, and other cations quantitatively. This is essentially the same as the first step in the other ion exchange processes. The softener resin is regenerated periodically by salt (sodium chloride) solutions. The spent regeneration solution is rejected to the sewer normally, although the final fast rinse can be used as cooling tower makeup, since it is essentially clean water. At least two 100 % softener vessels are used so one can be regenerating while the other is in service.

The softened water is filtered through a fine (micron-range) filter. Sodium bisulfite solution is injected to scavenge any residual chlorine in the softened water which can damage the RO membranes.

The water is pumped to fairly high pressure and sent to the RO membrane skids. The pumps for this use are often seal-less. The pumps may actually be integrated into and controlled by the RO skids as a self-contained unit with its own PLC.

The RO skids consist of several pressure tubes about 8–12″ diameter (0.2–0.3 m) and around 20 ft. (~6 m) long. Inside the tubes normally are wound membranes packaged in smaller modules. The softened water enters one side of the membrane. The smaller water molecules preferentially diffuse through the membrane under pressure, leaving the impurities behind. The clean water permeate leaves the modules, headed to the users, usually through a large storage tank.

The remaining water, which contains the impurities, is sent to sewer or can be used as cooling tower makeup (since it is soft).

The RO membranes must be chemically cleaned regularly (every few months) as they build up foulants, micro bio growth, and slight bits of hardness that get through. The timing for cleaning is normally determined by the increase in pressure drop through the modules and the loss of clean water recovery as a percentage of feed water. A connected, packaged cleaning system may be used, or the membrane modules can be removed from the tubes fairly quickly and sent off-site for thorough cleaning.

Modules will last several years if properly maintained. Some types are damaged by chlorine and may require replacement if the sodium bisulfite injection is not reliable. The modules also lose a little capacity each time they are cleaned because not all contaminants are normally removed. This impact can be minimized by attention to the cleaning process.

The RO modules are usually spared because of the maintenance requirements. Three 60 % modules or two 100 % modules are common configurations. Often, the RO system is owned and operated by a water treatment company, who will ensure good reliability.

Note that one thing RO systems do not like is low rates or no flow. If water is allowed to sit in the modules without movement, the membranes tend to foul. Attention to sizing of the system is necessary.

Monitoring an RO system requires watching:

- Softened water hardness (for regeneration needs)
- Sulfite injection rate
- Pre-filter pressure drop (for changeout)
- Membrane pressure drop for each skid
- Product water conductivity, pH, and hardness from each skid
- Wastewater conductivity from each skid

Utilities in Petroleum Processing

- Product water recovery vs. feed water rate from each skid
- Required cleaning frequency for each skid

A couple of RO system trouble indicators are:

- *Sudden increase in RO membrane pressure drop.* If all the skids are affected, check the softeners and the pre-filters. It is likely that the softeners have failed. If only one membrane skid is affected, it is possible that the membranes have collapsed because the pressure supplied from the pumps was too high or there may be a valve problem.
- *High pressure drop through the membranes building over time.* Indicates membrane fouling. Time to clean the membranes.
- *Sudden loss of flow.* Check the pressure pumps and lineups. The pumps sometimes fail without warning.
- *Increasing product water conductivity or TDS trend after cleaning.* Membrane damage is possible. Check the dechlorination sodium bisulfite flow for problems. It may be necessary to replace the membrane modules.

Rely on the RO system or membrane manufacturer and your water treatment chemical vendor for troubleshooting help. These systems work great normally, but troubleshooting can be difficult.

Pretreatment Process Comparison

The quality of water provided by softening, ion exchange, and reverse osmosis systems needs to be capable of meeting the requirements for the boiler feed water and drum water in Tables 3 and 4. In general, softening would apply to lower-pressure steam generation. RO systems cover the medium pressures, but can be augmented by polishing, additional stages, and ultrafiltration to supply high-pressure systems. Demineralization is for the medium- to higher-pressure generators.

Deaeration and Degasification

The deaeration and degasification process is used in almost all BFW treatment to remove dissolved gases from the water. Usually, this is just referred to generically as deaeration and we will use that convention here.

Normally treated water and returned condensate are routed to a deaerator immediately prior to entering the boiler steam drum. Figure 16 is a sketch of a typical deaerator drum layout.

The drum consists of a retention vessel surmounted by a degassing tower section (also referred to as the "stripper" or "heater" section). The degassing section contains a packed or trayed volume over which the treated water (and condensate) stream is passed. If this section is trayed, the trays are of special design with a spacing of about 2–3″ (~50–75 cm), and there may be 30–40 trays packed into the stripper section. Metallurgy is stainless steel.

Table 4 Water quality guidelines for reliable, continuous operation of water-tube boilers (ASME)

Boiler feed water limits			
Drum pressure, psig	Iron, wppm	Copper, wppm	Total hardness, wppm $CaCO_3$
0–300	0.100	0.050	0.300
301–450	0.050	0.025	0.300
451–600	0.030	0.020	0.200
601–750	0.025	0.020	0.200
751–900	0.020	0.015	0.100
901–1,000	0.020	0.015	0.050
1,001–1,500	0.010	0.010	ND
1,501–2,000	0.010	0.010	ND
Boiler water limits			
Drum pressure, psig	Silica, wppm SiO_2	Total alkalinity, wppm $CaCO_3$	Specific conductance, μmho/cm
0–300	150	350	3,500
301–450	90	300	3,000
451–600	40	250	2,500
601–750	30	200	2,000
751–900	20	150	1,500
901–1,000	8	100	1,000
1,001–1,500	2	NS	150
1,501–2,000	1	NS	100

Low-pressure steam, usually letdown saturated 50 psig (~3.4 barg) steam or 5 psig (0.3 barg) steam from condensate flash, is introduced to the bottom of the degassing section. The steam flows upward through the packed section and countercurrent to the water. This action removes or strips the dissolved gases from the water. These gases then leave with the steam from the top of the degassing section to be vented to atmosphere. In some uses, like hydrogen plants, the stripped gases may contain traces of hydrocarbons, other organic compounds, and ammonia. In these cases, the vented gas may be routed to a heater or back into the process under pressure to prevent emissions. The gas-free water free flows into the main retaining section or "tank" of the deaerator.

The treated water feed is introduced to the degassing section of the deaerator through atomizing spray/distributor. This reduces the water stream to fine droplets prior to entering the packed/trayed section. This enhances the removal of the gases in the water. Deaerators operate. Preferably, at about 2–3 psig (0.1–0.2 barg) and at this pressure, all of the CO_2 contained in the water and most of the oxygen are removed.

The BFW pumps draw suction directly from the deaerator storage section. To ensure that there is available sufficient NPSH for the pumps to operate properly, deaerator drums are installed on a structure at least 15 ft. (~5 m) above the center

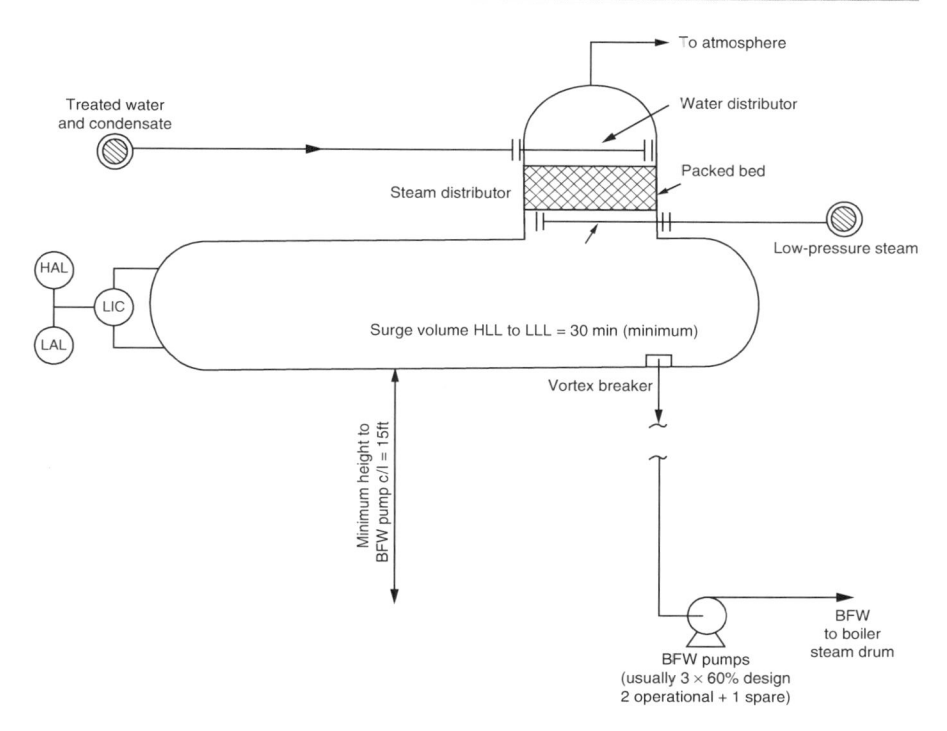

Fig. 16 A deaerator drum

line of the pump suction. Most large pumps (as BFW pumps are) usually require a relatively high NPSH when handling hot water.

The retention section of the deaerator should have as a minimum 30 min of surge between HLL and LLL. The water feed to the deaerator is normally on retention section level control. The boiler feed water from the BFW pumps will be on flow control with steam drum level or 3-element controller reset. Very often boiler feed flow has low flow alarm and at very low flow has automatic boiler shutdown device.

Chemical Treatment and Other Considerations

Boiler feed water may be treated at the point it is generated or may be treated locally in a process unit for specific situations and stresses. The chemicals needed for boiler feed water treatment are described in detail in the section "Boiler and Condensate Chemical Treatment" of this chapter. Summarizing the treatment requirements, the water needs to be treated to:

- Eliminate any traces of dissolved oxygen (down to <10 ppb)
- Prevent deposition and corrosion of the steam generation system
- Prevent corrosion of the condensate system

Condensate Recovery and Recycle

Because of the high cost in making boiler feed water, it is economically desirable to recover condensates from throughout the plant. This recovery is discussed in some detail in the section "Components of the Steam System" under the subsection "Steam Traps and Condensate Systems" of this chapter. A couple of key points in that discussion are reiterated here:

- It is necessary to monitor the condensate return streams for hydrocarbon or other contamination by process fluids. Condensate "hot well" vapors can be monitored to identify leaks. Streams with leaks should not be reused.
- Surface condensers and other systems operating at vacuum will pull in air, becoming contaminated with dissolved oxygen. Condensate from these services should not be used in processes, but may still be used for BFW after deaeration.
- Condensate chemical treatment is just about as critical as BFW treatment to avoid corrosion in the condensate headers.

Prevention of BFW Contamination

As with most utilities, it is critical to avoid contamination of BFW with process fluids. This has been mentioned under condensate reuse, but consideration needs to be given to backflow prevention wherever BFW is introduced directly into a process.

As a minimum, the process injection point should have one or more check valves and a block valve. The injection point should be capable of being blinded when not in use.

Temporary connections of BFW into a process are at particular risk. These can occur as part of the shutdown or start-up procedures or during maintenance. It is a best practice to provide backflow prevention and to remove and blind off any temporary BFW connections to a process system when they are not needed and *before* the process system is pressured up.

A little care in preventing contamination of the BFW system will avoid unexpected problems.

Plant Water

For general use in a refinery, there is normally a plant water system (AKA industrial, utility, or yard water). This system (which we will call plant water) may be supplied from city water, wells, or other sources. It is not considered potable (drinkable). Plant water is used for area wash-down, flushing equipment, and other routine uses that do not require drinking water quality or have other requirements.

The plant water is normally supplied at ambient temperature and ~100–125 psig (~7–9 barg) through a dead-ended distribution system. The water is available at numerous "utility drops" located strategically throughout the units. The utility drops are usually capable of supplying something like 50 gpm (200 l/min) of water.

In many locations, there are provisions for heating the water by direct, live steam injection manually or through a self-contained control valve.

In cold climates, the plant water headers and drops may be winterized by tracing and/or a flowing loop. Usually, the tracing will be of the "stand-off" steam type or electrical to avoid boiling the water.

Potable Water

In some, usually remote, locations the facility may need to provide its own potable or drinkable water. This water is used for drinking, but must also be provided to safety eyewashes and showers. It would be used in change room for showers, also.

Governmental authorities define the required water quality for potable water. It normally has limits on the impurities and must be chlorinated or otherwise disinfected. The potable water system piping also requires disinfection. There are requirements for sampling the potable water. The local requirements need to be determined.

Table 5 provides a sampling of some key drinking water standards. This is presented for example only. You should consult the applicable regulations and authorities for current requirements. The US standards actually include limits on at least 88 different contaminants so you can see this is just a small sampling.

A typical potable water treatment system may include several steps as noted below. These processes are not detailed here. A civil engineer would normally be consulted in design of these systems.

- Flocculation or deep bed filtration of the water to remove solids
- Reverse osmosis or softening to manage hardness, dissolved solids, and potentially hazardous inorganic compounds
- Carbon filtration or treatment for organics
- Chlorination (with a required holding time) before distribution to the plant.

The potable water system must be protected against contamination from the process or any other source. For this reason, the potable water system should *never* be connected to the process or any other utility. Where potable water is used, backflow protection (such as air gaps, preventers, or air/siphon breaks) must always be applied.

Winterization of the potable water system normally takes the form of insulation with electric or warm water tracing. Never steam-trace potable water lines.

On the other hand, the normally-no-flow lines that go to safety eyewashes or showers may get very hot in the summer or in desert locations. These potable water lines are normally run below other piping in the rack to provide shade, underground, or are provided with insulation and cooling water tracing to keep them from getting hot enough to cause injury. Insulation alone is not adequate.

Table 5 Sample of typical drinking water standards (www.lenntech.com/applications/drinking/standards)

Contaminant	US EPA (2013) Maximum	US EPA (2013) Recommended	WHO (1993) Guideline max	EU (1998) Standard max
Arsenic, mg/l	0.01	Nil	0.01	0.01
Barium, mg/l	2	2	0.3	
Benzene, mg/l	0.005	Nil		0.001
Cadmium, mg/l	0.005	0.005	0.003	0.005
Chloride, mg/l	–	250	250	250
Chlorine, mg/l (Cl_2)	(min 4 residual)	(min 4 residual)	(5 residual)	
Chromium, mg/l	0.1	0.1	0.05 (Cr^{6+})	0.05 (Cr^{6+})
Copper, mg/l	1.3	1.3	2	2
Cyanide, mg/l	0.2	0.2	0.07	0.05
Fluoride, mg/l	4	4	1.5	1.5
Iron, mg/l	–	0.3		0.02
Lead, mg/l	0.015	Nil	0.01	0.01
Manganese, mg/l	–	0.05	0.5	0.05
Nitrate, mg/l	10 (as N)	10 (as N)	50 total as N (Nitrate & Nitrite)	50 (as N)
Nitrite, mg/l	1	1		0.5 (as N)
Phenols, mg/l	–	0.001		
Selenium, mg/l	0.05	0.05	0.01	0.01
Sodium, mg/l			200	200
Sulfate, mg/l	–	250	500	250
TDS, mg/l	–	500	500	
Toluene, mg/l	1	1	0.7	
Xylenes, mg/l	10	10	0.5	
Zinc, mg/l	–	5	3	
Turbidity, NTU	Specific treatment processes req'd			
Radioactivity, pCi/l	5 (as $Ra^{226, 228}$)	Nil		
Coliform bacteria	Negative test req'd	Nil		Nil in 100 ml

Fire Water

Water for fighting fires (AKA fire water) is distributed through a secure, redundant system throughout the refinery. Fire water is sometimes used when large quantities of water are needed for flushing, area wash-down, or hydrotesting systems. These uses are only intermittent.

The fire water system should never be connected permanently to any process equipment or systems. Backflow prevention to avoid contamination of the fire water system should always be used when firewater is involved. People have been unpleasantly surprised when fighting a fire to find their firewater is contaminated with gasoline because an unwise (or prohibited) connection was made somewhere in the system.

Additional discussion of facilities and practices for fire water systems is in the chapter entitled "▶ Fire Prevention and Firefighting in Petroleum Processing."

Compressed Air Systems

All petroleum processing plants require a supply of compressed air to operate the plant and for plant maintenance. There are usually two separate systems and these are:

- Plant air system
- Instrument air system

Plant Air System

Plant air is generally supplied by a simple compressor with an after cooler. Very often, when plant air is required only for maintenance, this is furnished by a mobile compressor connected to a distribution piping system. Air for catalyst regeneration and the like is normally supplied by the regular process gas compressor on the unit.

Instrument Air System

Instrument air should always be a separate supply system; although in some refineries, instrument air is provided from the plant air system on a priority basis, with plant air distribution cut off at low pressures.

In some refineries, there are additional emergency air compressors provided locally in the event main plant instrument air fails.

Compressed air for instrument operation must be free of oil and dry for the proper function of the instruments it supplies. This is a requirement which is not necessary for most plant air usage. A reliable source of clean dry instrument air is an essential requirement for plant operation. Failure of this system means a complete shutdown of the plant.

Instrument Air System Flow Sheet

Figure 17 shows a typical, dedicated instrument air supply system.

Atmospheric air is introduced into the suction of one of two compressors via an air filter. The compressors are usually reciprocating or screw type non-lubricated. Centrifugal type compressors have been used for this service when the demand for instrument air is very high. The air compressors discharge the air at the required

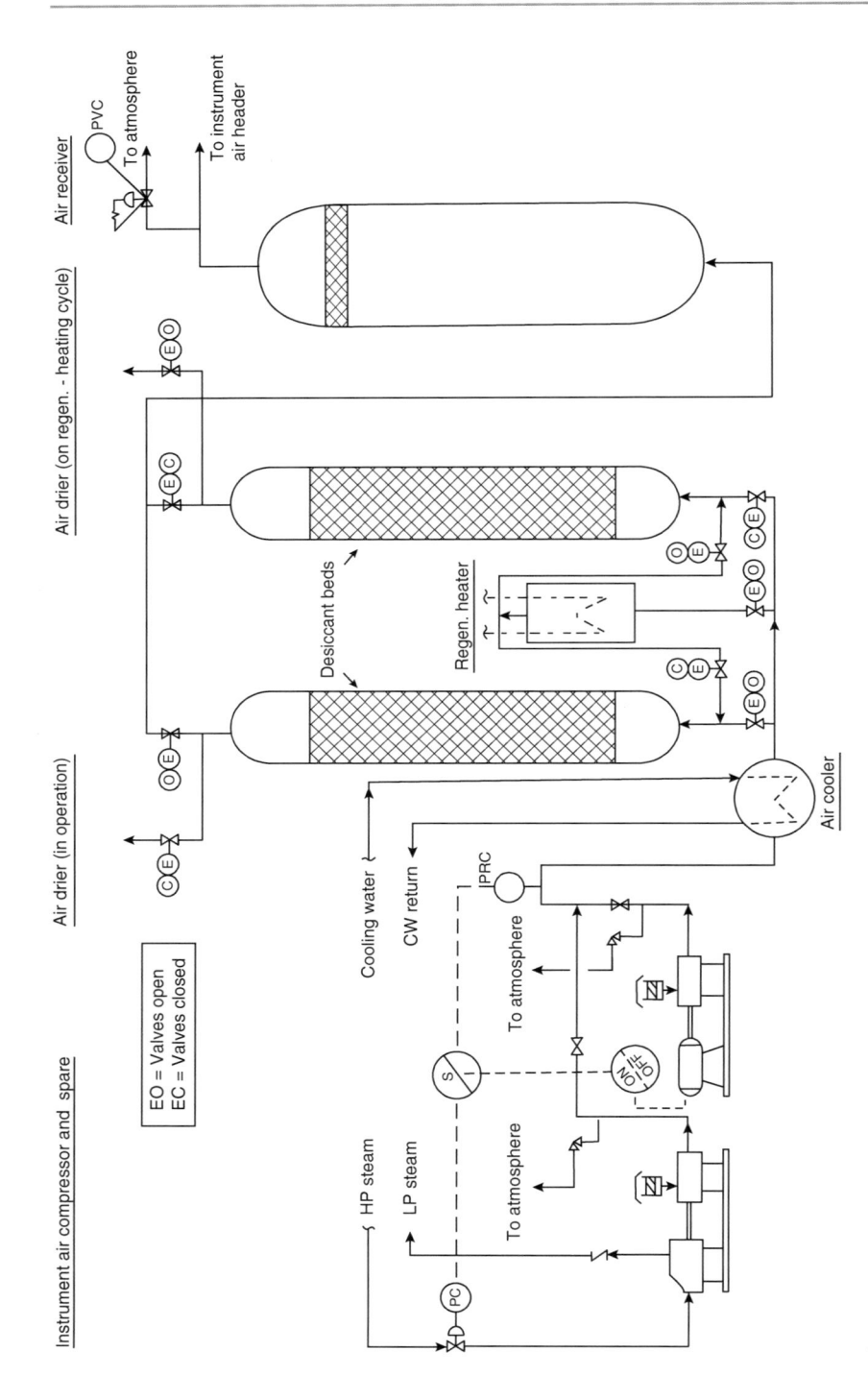

Fig. 17 A typical instrument air supply system

pressure (usually above 100 psig) into an air cooler before the air enters one of two dryers.

One of the compressors is in operation while the other is on standby. The operating compressor is usually motor driven with a discharge-pressure-operated on/off start-up switch. The standby compressor is turbine (or diesel engine) driven with an automatic start-up on low discharge pressure.

The cooled compressed air leaves the cooler to enter the dryers. There are usually at least two dryer vessels each containing a bed of desiccant material. This material is either silica gel, alumina, or, in special cases, zeolite (molecular sieve). One or more of the dryers is in operation with the compressed air flowing through to be dried and to enter the instrument air receiver. The desiccant in a spent dryer is simultaneously regenerated.

The regeneration cycle may be initiated manually, on a predetermined frequency, or by a moisture analyzer in the instrument air header.

Regeneration of a traditional desiccant bed is effected by passing a stream of heated air through the bed and venting the stream to atmosphere. This heated stream removes the water from the desiccant to restore its hygroscopic properties. At the end of this heating cycle, cooled air is reintroduced to cool down the bed to its operating temperature. When cool, the unit is ready to be switched into operation and to allow the other dryer to start its regeneration cycle. The various operating and regeneration phases are automatically controlled by a series of solenoid valves operated by a sequence timer switch control. These dryers (often including the compressor and receiver items) are packaged units supplied, skid mounted, and ready for operation.

There are also pressure swing adsorption (PSA) dryers for instrument air, normally using molecular sieves. These dryer beds are regenerated by dropping the bed pressure to atmosphere and purging the bed with dry air from the bed(s) in service. The regenerated bed is then repressured and swung back online. The sequence is controlled by a PLC.

The instrument air receiver vessel is a pressure vessel containing a crinkled wire mesh screen (CWMS) or other fine coalescing medium before the outlet nozzle. It is at high pressure protected by a pressure control valve venting to atmosphere and is also protected by a pressure safety relief valve (not shown in diagram). The air leaves the top of this vessel to enter the instrument air distribution system servicing all the units in the complex.

The instrument air must meet dryness specifications appropriate to the facility location. Very cold climates require much dryer air. Table 6 provides some typical instrument air specifications.

Table 6 Typical instrument air specifications

Specification	Typical maximum	Notes
Dew point	−40 °F (−40 °C)	Absolute max under any condition is 35 °F (2 °C)
Oil content	0.01 ppm	
Particulates	0.02 mg/m^3	100 % must be <100 μm
Supply temperature	110 °F (43 °C)	

The dry instrument air is usually distributed to the users at around 60–80 psig (say, 4–5.5 barg) by a utility header. For large, intermittent users, like pressure swing adsorption units, there may be local surge drums or additional knockouts.

Instrument Air Security and Prioritization
Instrument air is normally supplied by redundant compressors with redundant air dryers. Because instrument air is so critical, many facilities also provide emergency backup instrument air at critical locations in the plant. Approaches that have been used include:

- Standby local compressor sized to supply air to one unit. These are often non-lubricated machines with separate coalescers and knockout pots that tie into the instrument air system. They may be steam or diesel driven. These compressors must be tested regularly (weekly) to ensure they will work when required.
- Some facilities locate a large instrument air surge drum at a critical unit. The instrument air drum is held at full compressor discharge pressure (maybe 120 psig (say, 8–9 barg)). The drum is sized to contain enough air to operate the plant with no compressor for a period of time (usually something like 30 min) before the pressure decays to the point where control valves begin failing (usually around 40–60 psig (say, 2.5–4.0 barg)).
- Some locations may provide vaporized liquid nitrogen for backup to the instrument air. This approach has to be carefully managed, however, since the nitrogen will also enter control rooms and may cause suffocation, especially if the controls are heavily pneumatic. It might be acceptable for a few specific local controls where the vented nitrogen cannot reach an occupied or potentially occupied building.
- In an instrument air loss, some plants pressure their plant air into the instrument air header. This should only be temporary. Extended use of plant air as instrument air will likely foul the pneumatic instruments and control valves unless the plant air meets instrument air specs.

Connecting Instrument and Plant Air Systems
While the preferable approach to plant and instrument air is complete separation, many locations combine the systems up to the battery limit of a unit. In these cases, the entire air supply, both instrument and plant, is conditioned to meet instrument air specifications. The two services then separate. There is a valve on the plant air supply header that shuts that header off in the event of low air supply pressure – giving priority on air to the instrument system.

Instrument Air System Capacity
The required capacity for the instrument air system will be determined by the types and number of instruments consuming the air as well as the frequency with which they must act. Table 7 provides some typical air usage rates for pneumatic

Utilities in Petroleum Processing 1145

Table 7 Some instrument air consumption rates and ranges (see references at the end of the chapter)

Device	Air requirement, scfm (Nm^3/h)
Signal converter (I/P, P/I, E/P, etc.)	0.35–0.60 (0.56–1.0)
Ball valve actuator	[73 cu in/opn (0.0012 m^3/opn)]
Diaphragm control valve, w/ positioner	0.3–0.75 (0.5–1.3) – stable flow
	7–15 (11–24) – unstable flow
Piston actuated control valve, w/ positioner	5 (8.5)
Panel-mounted instrument	0.5 (0.85)
Field-mounted instrument	0.5 (0.85)

instruments as a guideline. You can estimate the air rate by counting the number of instruments of each type in the control system, applying the appropriate consumption rate, and adding the rates. Commonly, a 25–50 % contingency is applied on top of these numbers to cover unknowns, emergency situations, and an allowance for expansion.

For detailed design, vendor data on air consumption for the instrumentation selected should be used for header and capacity sizing.

Plant Air System

In many respects, a plant air system resembles the instrument air system. The compressors are usually redundant. In cold climates, both types of the air will be dried. The plant air is then distributed by headers throughout the plant. It is also one of the utilities supplied at the utility stations or utility drops for general use in the facility.

The chief differences between plant and instrument air are in the level of security applied and the requirement for drying the instrument air. Anyone can use the plant air through a utility station. The usage rates can be much larger than for instrument air. Plant air may also be connected to the process at times, whereas instrument air is normally strictly excluded from process connections.

Plant air use tends to rise, because it is uncontrolled. It is advisable to manage the plant air users, just like any other utility users to ensure adequate air supply to more important users.

Protection of Air Systems

Care must be taken to protect the air systems from contamination. Instrument air should never be connected to a process system. Plant air should only be connected to the process through check valves. Air should never be left connected into a process line that is at higher pressure. It should always be disconnected and air gapped or blinded.

Contamination of plant and instrument air has been seen as the result of strange, unauthorized, or inadequately considered connections that have resulted in air systems full of nitrogen, oil, and water. Care must be taken at all times to ensure that this does not happen to you.

Electrical Power Supply

In today's petroleum processing facilities, electrical power is arguably the most important utility, competing with steam. Failure of electrical power for even 0.5 s usually means hours, if not days, of downtime and lost production accompanied by the potential for equipment damage or worse.

We will touch only slightly on this critical system in this handbook, but this is not a reflection on its importance.

Electrical Power Distribution System

The electrical power distribution system in a facility might look something like that depicted in Fig. 18. This is a very simple version of a one-line diagram. Electrical engineers (who may cringe at this diagram) have much more detailed diagrams for the power systems, but this will suffice for the discussion here. The diagram does not show all the breakers, motor controllers, switches, etc.

Power to a facility is generally supplied from an outside utility via a nearby set of transformers and switch gear. The utility protects itself from a problem in your facility with breakers. The utility power, usually 138 KV or higher from the long distance transmission lines, is transformed down to 13.8 KV entering a large facility.

For reliability, there are normally multiple feeder lines into a facility, preferably originating from different sources or transmission lines. This provides some level of protection against power loss, but often problems on one feeder tend to propagate to the others.

Facilities with cogeneration and power recovery units have some alternate sources of power, but these units can also be knocked out by an outside problem if they are not properly protected.

The 13.8 KV feeders are transformed down to 4,160 V, usually within the facility. The 4,160 V electricity goes to the plant power distribution centers (PDCs) where it is distributed to the high voltage users, like large induction and synchronous motors. These are usually over 250 HP (~188 KW). The large motors may be over 10,000 HP (7,500 KW).

The 4,160 V power is transformed down to lower voltages in steps to supply the smaller users. Usually, there are 480 V, 240 V, and 120 V users typically in the USA. These will include smaller motors, lighting, instrumentation, electronics, and other services.

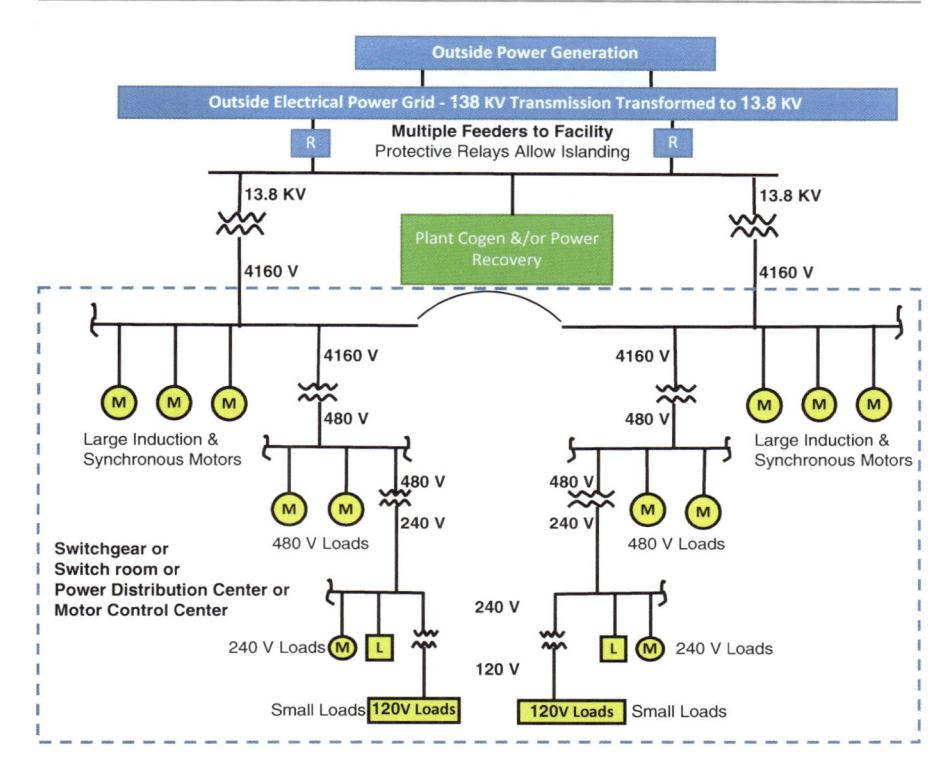

Fig. 18 A simplified electrical one-line diagram

There is usually an attempt to provide critical services with power from at least two different sources, even within a facility. For instance, two critical 480 V motors may be supplied from different 480 V lines supplied by different 4,160 V lines. Or, the redundant power supplies to the DCS may originate in different PDCs tracing back to different feeders. For control systems, power is usually automatically switched to an uninterruptable power supply (UPS, batteries) that is charged off the external power supply. The UPS normally is sized to provide continued power to the controllers for several minutes to hours.

Not really obvious in the simplified one-line diagram, the loads must be balanced among the power phases to avoid a phase imbalance and creation of high current in one phase. This is carefully considered by the electrical engineers in design and operation of the system.

Cogeneration and Power Recovery

Many facilities have surplus fuel gas available and they need steam. In these plants, cogeneration of power and steam is a good option. The large cogeneration units are available as packaged plants (many in the 50–75 MW range). The power is supplied by an aeroderivative gas turbine driving a generator, although there are engine-driven units, too. Excess heat leaving the turbine is used to raise steam, just like a conventional water-tube boiler. This system thus provides two critical utilities to the plant and allows the export of excess power to the utility grid.

Cogeneration units are normally very reliable, but they do experience shutdowns. The backup steam supply needs to be capable of taking over fast enough when the cogen trips. Sometimes, the cogen supplies so much of the plant steam demand that, when it trips, the boilers cannot fire up fast enough to supply the critical steam needs. In these cases, what starts as a cogen trip snowballs into a total utility failure and cold restart of the entire facility. To avoid this, the percentage of steam supplied by the cogen needs to be managed. It is possible to take surplus cogen steam to a separate steam turbine generator to make use of excess steam. Care needs to be taken in managing these situations.

Cogens can also be knocked down by a power disturbance in the outside utility grid. For this reason, many facilities provide protective relays that allow the plant to rapidly (automatically) isolate from the outside power grid and continue running with minimal disturbance. This is referred to as "islanding." In these cases, the cogen generator and control system may need to be designed to handle the reactive power demand it will see when it is the primary power supply to an islanded facility.

Often there are other power recovery generators within a facility. One of the most common is a turbo-expander on the FCC flue gas. These generators are much smaller than cogens. A trip of one of the smaller generators usually has little facility-wide impact, unless they create a direct ground fault as they fail. If the generator is in a pressure-letdown service, a control alternative needs to be available. Usually there is a control valve or controlled louver/damper in parallel to a letdown generator, so a rapid switchover is possible if the generator fails.

Power Versus Emergency Steam Loads

Electrical power frequently replaces steam in prime drivers of rotating equipment. This is energy efficient, but it does present some hidden problems. If the plant depends on steam turbine drivers to back up the electric motor drivers, a sudden loss of power can place a huge, instantaneous load on the steam system. This is alluded to above in the discussion of cogeneration, but the concern applies in other cases as well.

There is a limit to how fast a boiler can increase its steaming rate. As a rule of thumb, a boiler can double steam production about every minute in an emergency without tripping and assuming it has adequate back pressure.

Let's say a 250 Mlb/h (113,000 kg/h) boiler is operating at one fourth rate in parallel to a cogen that is supplying 88,000 lb/h (40,000 kg/h) of steam and there are several critical electrical services in the plant. If the cogen trips and the external electrical supply trips due to the sudden increased load (a common scenario), the boiler would pick up the cogen steam supply plus another, say, 100 Mlb/h of load from the spare steam-driven lube pumps, backup air compressors, etc., that kick on due to power failure. The new instantaneous steam demand is now about 250 Mlb/h (full boiler capacity). The boiler can increase steam rate from 63 to 125 Mlb/h within the first minute. And it gets to 250 Mlb/h after 2 min. Meanwhile the steam header pressure has sagged, and the emergency turbines are now using even more steam than they were (the nature of steam turbines). The boiler will have a hard time catching up without shedding load somewhere or reestablishment of electrical power. If the operator tries to fire the boiler faster, it will usually just trip on high steam drum level (due to the sudden increase in steam bubbles in the drum).

The point here is that you must balance the electrical and steam loads to enable transition from electricity to steam in critical services in an emergency.

Of course, the alternative is to accept that steam will fail along with the power and construct your emergency procedures for that event. That is a viable alternative and one which is often chosen. You are accepting the economic impacts of the outage in that case vs. the continued cost of higher steam generation rates.

Electrical Power Distribution Conclusion

Electrical power is one of the most critical utilities in a petroleum processing facility. It is normally supplied by an outside utility, but may be supplied or augmented by on-site generation.

Reliability of the power supply is critical. These systems normally are supplied by multiple feeders from the power company. If the facility has a cogeneration unit with sufficient capacity, it may be capable of islanding to operate off internally generated power. UPSs are normally provided for backup power to control systems.

Integration of the power and steam systems through a cogeneration unit or excessive reliance on power (with steam backup) needs to be carefully managed to avoid total utility failure in the event of a power trip. Boilers have to be capable of assuming the required plant steam load if power is lost.

Nitrogen System

General Discussion of Nitrogen Utilities

Nitrogen is a common gas provided throughout refineries for inerting, blanketing, and purging. Often nitrogen is supplied from a plant header to the continuous users, just like any other utility.

Nitrogen can be provided in several forms:

- Vaporized liquid (on-site or temporary vaporizers)
- Pressure swing adsorption
- Membranes
- Gas cylinders/tube trailers

Table 8 Typical nitrogen purity specification for purging and blanketing

Component	Limits
Nitrogen, v%	99.7 min
CO + CO_2, vppm	30–40 max
Hydrocarbon, vppm	5 max
Water, vppm	5 max
Oxygen, vppm	100 max

We will discuss the usual supply approaches briefly in the following sections. First we will look at the required purity for nitrogen within a plant.

Specifications

Nitrogen purity requirements will vary, depending on the service. Where the gas is used for hydrocarbon gas or oxygen elimination, the most common uses, the purity must be high or effectiveness is limited.

For equipment purging and blanketing, a typical purity specification for nitrogen is in Table 8:

This specification is normally satisfied by vaporized liquid nitrogen (or cylinders for small users). Purity is generally much higher than that shown. The intake source for the nitrogen plant can influence the composition. For instance, an intake near a freeway will show higher CO and CO_2 content than nitrogen recovered elsewhere. Often this is not a problem, but may become a problem if the nitrogen is used for inert vessel entry where CO is a criterion.

In some cases, a lower purity specification may be permitted, such as in gas blanketing service. This may be supplied by a PSA or membrane nitrogen plant. Still, attention has to be given to the disposition of vent gases. If the blanket gas contains some oxygen (usually the main contaminant), it may be unwise to vent it to flare or vapor recovery.

Typical Nitrogen Supply Flow Sheets

The most common nitrogen supply in a facility is provided by vaporized bulk liquid. Figure 19 shows the main parts of the LN_2 vaporizer systems commonly used. Two types of liquid supply are shown:

- *On-site liquid storage tank.* In this system, a large on-site cryogenic storage tank is installed. The tank is usually double walled and super-insulated to prevent external heating. A tiny amount of the liquid is lost through evaporation which cools the remaining liquid. Evaporative loss is usually something like 1 %. A small stream of LN_2 flows through vaporizer coils which pull heat from the air to turn the liquid to vapor. The vapor is distributed to the facility nitrogen headers at about 125–200 psig (8.5–13.8 barg) and ambient temperature.

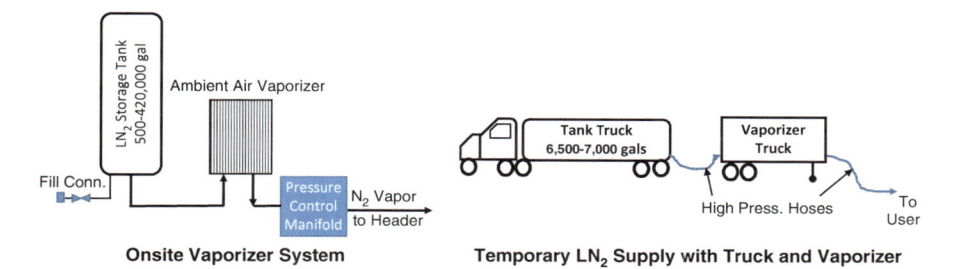

Fig. 19 Typical bulk liquid nitrogen supply options

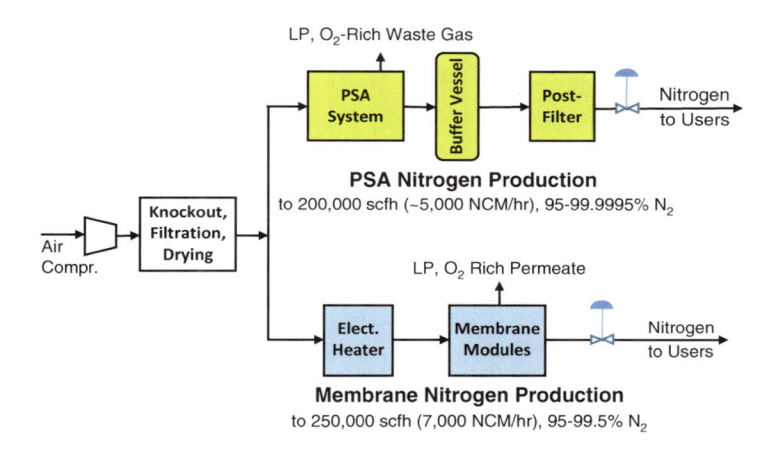

Fig. 20 On-site, non-cryogenic nitrogen generation options

There is backflow prevention to avoid contamination of the stored liquid. The tank and vaporizer are normally owned and maintained by the LN_2 vendor.
- *Temporary trailers and vaporizers.* The nitrogen vendor may supply his own LN_2 trailers and vaporizer. This is common during turnarounds when large quantities of nitrogen are needed for purging or inert entry. Full LN_2 tractor-trailer rigs are connected to a vaporizer truck. In the vaporizer truck, a propane-fired heater heats a circulating stream of, usually, oil or glycol solution, which then vaporizes the nitrogen. This system can supply liquid or gaseous nitrogen at elevated temperatures and pressures to 10,000 psi (690 barg). Care must be taken to avoid overpressuring process equipment or causing brittle fracture from unvaporized nitrogen liquid entering the system. The vendors are used to dealing with these issues.

On-site nitrogen generation is also an option by use of pressure swing adsorption and membranes. Packaged units for this are available from industrial gas suppliers. Simplified block flow sheets for these processes are shown in Fig. 20.

Nitrogen from these sources tends to be lower purity, although high purity is possible at higher cost.

Prevention of Nitrogen System Contamination

There are few worse feelings than purging a vessel with your plant nitrogen for an entire shift without success only to find your plant nitrogen supply is contaminated with hydrocarbon or hydrogen. Now you have to scramble to find a clean nitrogen supply quickly. If you are lucky, that is all you have to do.

There have also been occasions where someone has connected to the nitrogen header when they meant to tie into the plant air utility header. When the line goes into a building or enclosed space, this presents an asphyxiation hazard.

It is critical that the nitrogen utility system be protected from contamination by a foreign material and that it not be used for anything other than its intended purpose. In fairly common practice, contamination and unauthorized/unsafe use of nitrogen are prevented by specific steps:

- Nitrogen distribution header drops or outlets are separated from the normal utility stations and clearly marked with signs.
- The nitrogen drops have a specific nitrogen connection fittings on them – not the crow's feet used for air. The nitrogen fitting is often tack welded to the N_2 pipe to prevent removal.
- Special nitrogen hoses with nitrogen fittings are used which can only be connected to the nitrogen fittings on the utility drops.
- At the connection of nitrogen to each user, a special backflow prevention manifold is used. An example of such a manifold is in Fig. 21. Each company and facility will have their own standard for these manifolds.

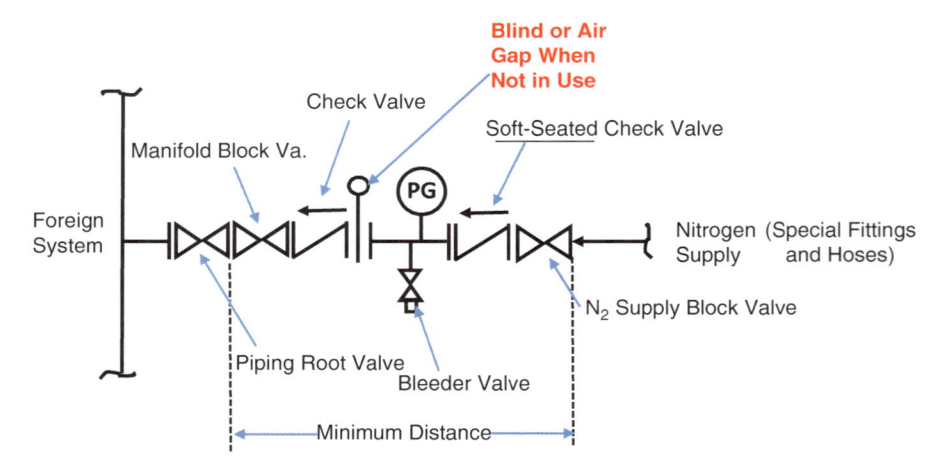

Fig. 21 Example of nitrogen backflow prevention manifold

- All nitrogen connections to the process or any other foreign systems, except permanent purge connections, are removed, blinded, or air gapped before pressure is raised on the foreign system.
- LN_2 brought into the plant should be tested for oxygen before it is used to ensure it is, in fact, nitrogen. There have been instances where liquid oxygen has accidently been unloaded into a nitrogen system – that is, about as bad as contamination can get.

Regardless of the precautions taken, it is still possible to defeat the protective measures. Constant diligence is needed to ensure that the nitrogen header provides the pure nitrogen expected.

Other Utility and Utility-Like Systems

In addition to the normally recognized utility systems, modern petroleum processing plants often have other systems that are distributed throughout a facility from a central supply location in the same manner as utilities. Among these are:

- Flushing oil
- Ammonia
- Oxygen for enrichment
- Hydrogen
- Caustic soda
- Amine solution

There are also several systems typically present in a processing facility that collect wastewaters, drainings, and other fluids from throughout the facility and convey them to a central location for disposition. Among these systems are:

- Process or oily water sewer
- Storm sewer
- Chemical sewers
- Slops or recovered oil

In this section, we will describe the functions and features of each of these systems. The discussion here will not be in the same detail as the foregoing utility discussions, but these systems are critical to operating a refinery or petrochemical plant and should not be overlooked in a comprehensive discussion.

Flushing Oil

Refineries processing heavy oils, vacuum resids, or tars with high pour points will often have a flushing oil utility available at strategic locations. The flushing oil is normally a straight-run diesel or light cycle oil. In very cold climates the flushing oil may be in the kerosene boiling range.

As an example, flushing oil may be used to dilute or cut the heavy oil and displace it from equipment when necessary. It may also be used to flush seals on heavy oil pumps when the pumpage is too heavy for a reliable seal flush.

In a flushing oil system, the oil is pressured into a dead-ended header at 50–100 psig (3.5–7 barg) and may be heated. The header may normally supply, perhaps, 100–200 gpm (370–750 l/m) – about a 2″ supply header. The users draw off the header as needed.

Flushing oil will often be tied into the process lines through a block valve and check, ready for use when needed. Some uses, such as continuous pump seal flushes, will be normally flowing, with the flushing oil going out of the pump with the pumpage.

Contamination of the flushing oil header with heavy oil by backflow is always possible, so check valves should be used at the injection points to reduce the risk. If heavy oil does get into the flushing oil system, it mostly presents a plugging problem and would have to be melted out.

Ammonia

The addition of selective catalytic reduction (SCR) and non-catalytic reduction (NCR) systems for NOx control from heaters and FCCs has resulted in some facilities providing ammonia for the reduction units as a utility throughout the plant. These systems may supply anhydrous or aqueous ammonia as a dead-ended, pressurized header throughout the facility from a central location. This consolidates the unloading and handling of the ammonia in one place, minimizing some of the handling risks.

The central ammonia facilities generally consist of a storage tank with connections for filling from a tank truck. There is a pressurizing pump that pulls from an elevated suction in the tank and provides pressure to the system – typically about 100 psig (~7 barg). Ammonia is distributed at ambient temperatures. Pumps for this service are normally positive displacement or gear pumps, with the excess flow spilled back into the storage tank on pressure control. The pumps are spared.

The ammonia is only routed to the local vaporizers or injection skids at each user from the central system. It is not provided as a utility drop.

There are a couple of idiosyncrasies about ammonia handling that can prevent a lot of problems:

- If the ammonia is for use in NOx reduction, be sure you get SCR grade. There are cheaper grades and more expensive grades. The cheaper grades have more impurities, and the expensive grades are unnecessary.
- Ammonia can have tiny amounts of oil and grease as well as other contaminants. Suction lines need to be elevated above the dead-bottom of the tanks so they do

not draw in the contaminants that settle as a sludge. It may be necessary to periodically draw the sludge out of the storage tanks.
- Aqueous ammonia is corrosive to carbon steel, despite what you may expect. This can be due to the use of undeaerated water for blending or the tendency of ammonia and deionized water to complex iron. In any event, the ammonia will have to undergo fine (submicron) filtration at each user if the distribution header is carbon steel. Once filtered, the piping should all be stainless steel. A stainless steel ammonia distribution header to the users would be nice, but probably not worth the expense.
- In vaporizers, because you are evaporating the ammonia completely, there will be an accumulation of hard deposits of iron and other materials from the ammonia. This is especially noticeable with aqueous ammonia. Watch for pressure drop in the vaporizers and plan to clean them on a regular basis. The spray-type vaporizers plug less than packed vaporizers. It is easy to set up a temporary vaporizer using bottled anhydrous ammonia and plant air mixed into the SCR supply header for emergencies. There is more discussion of this in the separate topic entitled "▶ Environmental Control and Engineering in Petroleum Processing" under SCR systems.
- Think about where you vent the relief valves from ammonia storage tanks and where you route tank drains or vents. Putting ammonia into a flare normally results in NOx emissions and/or plugging from ammonia salts. It is not unusual to route reliefs and drains to a nearby, dedicated atmospheric scrubber stack where the ammonia is absorbed in a water spray and the water is sent to the sewer.
- Avoid draining ammonia to the deck without a good-sized flowing water flush. The ammonia can easily overcome an operator.
- Secondary containment around ammonia storage facilities is required to manage spills.

So, if you have enough NOx control systems in your plant, an ammonia utility may be a viable alternative to the risks and problems with multiple delivery points and pumps.

Oxygen Enrichment Supply

Many facilities are choosing to de-bottleneck sulfur recovery units and FCCs using "oxygen enrichment." In these processes, atmospheric air that is used for combustion is augmented or enriched by injection of a controlled amount of pure or nearly pure oxygen. This is like increasing the size of the combustion air blower, without the capital cost of a new air blower. The practice is, of course, limited by temperatures in the system.

In facilities with O_2 enrichment, there may be an oxygen utility header provided. It may only be local to the unit(s) using the O_2. The oxygen is only routed to the specific users. It is not a plant-wide utility.

The options for supplying oxygen are similar to supply of nitrogen:

- Liquid oxygen with a vaporizer
- PSA oxygen (possibly the waste stream from a nitrogen generator)
- Membrane oxygen (possibly the permeate from a nitrogen generator)

Oxygen handling presents enhanced risks of fire where it may contact hydrocarbons, hydrogen, ammonia, or other combustibles. These need to be specifically managed.

Hydrogen Distribution

Because of the importance of hydroprocessing in meeting the fuel standards today, hydrogen is distributed in many refineries through one or more headers in the same manner as a utility. The systems for this are discussed in detail in the topic "▶ Hydrogen Production and Management for Petroleum Processing" in this handbook.

Caustic Soda

Sodium hydroxide or caustic soda finds uses in many units of a petroleum processing facility. In many cases it is received into a central facility and then distributed like a utility to the various users to minimize the bulk handling risks.

The caustic is usually received as ~50 w% solution, which may be sent to some users directly. At this concentration, the caustic freezes at 58 °F (14 °C). Lines carrying the 50 % caustic need to be electrically or cooling water traced to avoid freezing in many climates.

The 50 % caustic can be diluted down to 19 w%, which is a eutectic point, freezing at 18 °F (−28 °C). The 19 % caustic can then be distributed with less concern over freezing. The dilution can be done in a batch mode using cold condensate or other clean water. It is necessary to cool the solution as it is diluted to dissipate the substantial heat of solution or do the dilution slowly enough that the heat has time to dissipate. A circulating loop from the storage tank with injection of the dilution water and then flow through a cooler can be used. This system needs backflow protection to avoid contaminating the water stream. It should also ensure positive isolation of the water after use to avoid inadvertent water leakage into a tank that could create a high temperature.

Secondary containment around caustic storage facilities is required to manage spills and prevent groundwater contamination.

Caustic distribution would only go to the specific users in a facility.

Sulfuric Acid

Like caustic soda, sulfuric acid is also used at many units in a facility. Of course, a sulfuric acid alkylation unit would be a major user, but it will normally have its own dedicated supplies of acid. More commonly, acid is used in the cooling towers for pH control. In some facilities, it is supplied from a central location (often at the SA alky unit) through headers periodically to day tanks at the various users. This minimizes the hazards and risks in handling the bulk acid, which are substantial.

Secondary containment around all sulfuric acid storage and pumping facilities is required to manage spills and prevent groundwater contamination.

Also, like caustic, concentrated sulfuric acid has a high freezing point at about 50 °F. This means that sulfuric acid distribution headers need to be managed empty or traced and insulated, but the tracing temperature must be controlled to avoid rapid corrosion. Normally, carbon steel lines are fine for sulfuric acid, but, if a carbon steel sulfuric acid line is steam traced, it will only last a few minutes to a few hours before corrosion destroys it. Electric tracing to 60–70 °F and tracing with cooling water are the usual remedies.

Sulfuric acid lines also need to be protected from excessive ambient temperatures by insulation. Cooling water tracing also helps here. And, it must be protected against inadvertent dilution with water, which produces very high temperatures.

Amine Distribution

Many facilities have central amine solution regeneration, with the lean amine routed throughout the plant and rich amine returning to the central facility. The amine is routed only to the specific users and is not a normal utility drop.

Discussion of centralized amine facilities is included in the chapter entitled "▶ Refinery Gas Treating Processes" of this handbook.

Sewer Systems

Petroleum processing facilities generally have three types of sewer systems to collect aqueous fluids that are drained, purged, flushed, or are otherwise discharged and precipitation that falls on the facility. The most common systems are:

- Process or oily water sewer
- Storm sewer
- Chemical sewers
- Sanitary sewer

The following discussions provide some of the considerations in designing and operating these sewer systems.

Process or Oily Water Sewer (OWS)

Most petroleum processing plants have process or oily water sewers (OWS) as the main sewer system in the facility. These sewers accept any drained fluid from myriad sources which might be contaminated with oil or other process fluids.

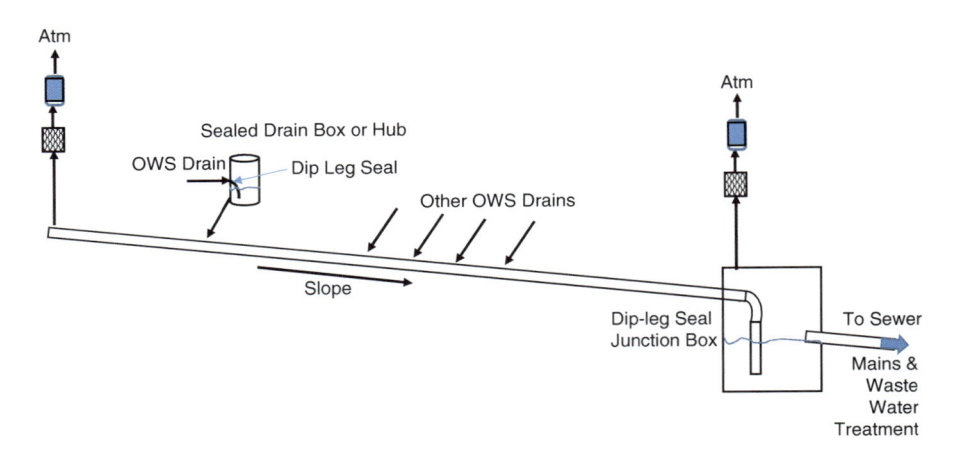

Fig. 22 Oily water sewer lateral

These sewers are located underground and convey the collected liquids to the wastewater treatment plant.

Figure 22 illustrates an oily water sewer lateral within a plant.

Some key design and operating considerations for the process or oily water sewer are:

- Drains to the sewer must be managed to avoid upsetting the wastewater treatment facilities. For instance, draining a large amount of soda ash solution to the OWS quickly will cause a pH excursion in the wastewater treatment facilities and the outfall or line to any publically owned treatment works (POTW) that receives the outfall. The same would be true if a large amount of sulfuric acid was dumped into the sewer. In the case of acid, in addition to a pH problem, the acid will cause corrosion in the treatment facilities. These would be potential permit violations.
- If a facility has a biotreatment unit, that unit is particularly upset by pH excursions, chlorine or bleach, biocides from cooling water treatment, phenols, heavy metals, and many other types of pollutants. The "bugs" get used to a regular diet of pollutants and don't usually react well to big changes in their diet. Killing the bugs requires significant recovery time.
- In many facilities, especially in the USA, the oily water sewer VOC emissions must be controlled, as seen in Fig. 22. This is normally done by providing closed, dip leg-sealed drain hubs for connections into the OWS. The hubs can be opened for inspection as needed. The dip leg-sealed junction boxes and manholes are also sealed from the atmosphere, but these must be vented to the atmosphere in order to function. To ensure the closed sewer flows, in the seal boxes and at the end of each OWS lateral, there is normally a vent to a high point through a flame arrestor and a carbon canister or drum.

- The OWS hubs are usually elevated slightly above grade so that they only collect the oily water and storm water is collected in drains located at grade. Of course, if the OWS backs up, the overflow will go to the storm sewer, which has to be designed to manage this flow. More on this is under "storm sewers."
- Drained material to the OWS can often be very hot, such as when a vessel is being steamed out or for blowdown coming from a boiler via a flash drum. The sewer needs to be able to handle the high temperatures. Typically, the design temperature is around 160 °F (70 °C) for the underground lines. In some cases, a cooler stream, like plant water or cooling water blowdown, may be flushed into the sewer along with the hot stream to reduce temperatures.
- The grade or slope in the oily water sewer must be sufficient to avoid backup. Commonly, the slope of the OWS lines is a few inches for every 10 f. of line (a few cm per meter). Slope drains generally toward the wastewater treatment facility.
- Fluids that can set up or leave deposits should never be introduced into the OWS lines. This would include things like molten sulfur.
- High vapor pressure hydrocarbons, hydrogen, H_2S, and other light combustibles or toxic gases should never be sent to the oily water sewer. They should be routed to a flash drum, and only aqueous drains from the flash drum should be allowed in OWS.
- The multiple dip leg seals or subsurface seals are important in this system as a means of preventing combustible gases and explosions from propagating through the sewer.
- The OWS is usually steel and single wall. Older OWSs may be vitreous clay tile or even redwood. Some localities may require the OWS to be double walled with leak detection.

Storm Sewer

For precipitation events, facilities have to be able to manage the large influx of water. Storm water flow rates generally dwarf any other sewer flows. It is common to have a separate storm sewer to minimize the quantity of contaminated water that the wastewater treatment plant has to deal with.

A storm sewer will generally run in parallel to the oily water sewer. The storm sewer lines are much larger than the OWS lines to handle the flows. Sizing of these gravity drain lines must include a few other runoffs, in addition to storm water.

Most notable of the added capacity requirement is the water load from firefighting in an emergency. This load can easily exceed storm water loads. Unit storm sewers may be sized to handle 5,000–12,000 gpm (~1,100–2,700 m^3/h) of fire water or as required by the capability of the firewater system. This water will contain oil, solids, debris, fire retardants, chemicals, and foam – all of which must be managed in the wastewater treatment.

Generally, the storm water flow and the firefighting load are not considered simultaneously in sizing the storm sewer. The larger load sets the sewer size.

There are advantages to keeping the OWS and storm sewer separate. If they are separated, the initial storm flow can be captured for processing through the wastewater treatment plant and, after the sewer system, has been well flushed by storm water, the remaining storm water can usually be diverted and discharged without further treatment.

Sometimes, the oily water sewer and the storm sewer are combined, however. This can be less expensive, but it means there is no difference in treatment requirements between the waters. All storm water has to be intercepted and processed just like oily water.

A storm water sewer generally consists of local open surface drains and open catch basins (with water seals) located along lateral sewer lines. There will be several catch basins, drains, and laterals in a unit.

The laterals flow to junction boxes where each line is sealed by dip leg (or subsurface entry). The junction boxes connect together into larger and larger lines ending at the wastewater treatment plant (WWTP).

In some low-point locations, there may be a lift sump where the storm waters are intercepted and pumped to the WWTP. Normally, the inlet basins and boxes at the WWTP allow the storm water to be processed through the API separator and other facilities at a controlled rate, within the capacity of the equipment. When the allowable processing rate is exceeded, the excess storm water overflows or is diverted into one or more large holding basins. It will then be brought back into the WWTP processes as capacity allows.

As noted above, some plants have permits allowing direct discharge of non-contaminated storm water after a certain amount of rain has fallen. In these plants, the potentially clean storm water will be diverted away from the wastewater facilities into a clean water basin. It is usually checked for potential contaminants before being discharged.

The multiple dip leg seals or subsurface seals are important in this system, as in the OWS, for preventing combustible gases and explosions from propagating through the sewer.

Storm sewers are generally single pipe systems, i.e., they do not usually require any double containment. The storm water holding basins, however, are often double lined with plastic.

Chemical Sewers

Chemical sewers are present in many facilities in localized areas. These are designed to accept relatively clean drain fluids containing predominantly a specific chemical that can be collected and reused or sent off to disposal. These are materials you generally do not want in the OWS and certainly not in the storm sewer.

For instance, a chemical sewer is often available within a unit for amine solutions. Level bridle drains, pump drains, equipment drains, sample purges, etc., in amine circulation systems are all collected in a holding tank or drum to be pumped back into the process periodically. The drum is usually located in a below-grade vault, so the chemical sewer can flow by gravity into the holding tank. The chemical sewer piping leading to the holding tank is often a double-walled pipe, and the vault may also be double walled. There is leak detection provided between the walls. A vertical pump pulls the solution out of the tank and pumps it back into the process. The chemical sewer can also be used to empty vessels for maintenance and save good solution.

The chemical sewer hubs are normally closed, or piping is directly connected to the chemical sewer line. The sewer line may be vented to the atmosphere at a safe location, or an equalization line may be provided from the holding tank back to the end of the sewer line. These systems may also be purged or blanketed with inert gas under slight pressure.

Care must be taken to avoid contaminating the chemical sewer. If the sewer does get contaminated, the holding tank contents may have to be sent out for disposal and the system cleaned. This would be the case if oil got into the amine sewer, for instance, since the oil would cause foaming.

Often the vault around the holding tank is open to atmosphere. The top rim is usually elevated 6–12 in. (0.15–0.3 m) above grade to avoid acting as an area catch basin, but it will still collect some rainwater. The vault usually has a drain that is normally closed. When it collects rainwater, the collected water is drained to the OWS manually, and the drain valve is immediately blocked back in.

Sanitary Sewer

The final type of sewer we will consider is the sanitary sewer. These are normally vitreous clay or plastic pipe installations that carry the sanitary wastes from occupied facilities to the sanitary sewage treatment plant. They gravity drain (with an occasional lift sump) to the treatment plant either on-site or a POTW off-site. Local municipal codes govern the sanitary sewer installation. Care must be taken to avoid contaminating the sanitary sewer. Conversely, you need to avoid sending sanitary sewage to the other sewer systems unless they are specifically designed to handle sanitary waste.

Slops and Recovered Oil

There are several locations in a plant where it is necessary to get rid of oil, emulsion, oily water, and so on, but it does not belong in the sewer. This is usually because the amount of oil involved is substantially greater than what the wastewater treatment system can or should handle. For these situations, we have hydrocarbon slops or recovered oil systems.

Uses for slops may include:

- Off-spec products
- Oil flushed before taking a sample
- Oil skimmed off wastewater and flash drums
- Oil collected in the flare drums
- Oil from vessels and other equipment being removed for maintenance
- Level bridle drains that are primarily oil
- Drains from compressors, pumps, etc.
- Emulsions from desalting or other processes
- Essentially anything that is predominantly oil

The oil streams that make up the slops are collected, usually at the source, and brought to the slops tankage area. There are commonly intermediate drums for surge, forwarding, and separation from water. The streams all end up in one or more slops or recovered oil tanks.

From the slops/recovered oil tanks, the oil is sent back to an appropriate unit in the facility. Often, fairly clean oil is sent to the crude unit and reprocessed. If the oil is heavy or has undesirable components, it may go to the delayed coker where it can be injected into the coke drum during part of the coking cycle.

Emulsions are a particular problem. They may be sent to a separate emulsion-breaking tank. An emulsion-breaking tank is usually equipped with good mixing and circulation facilities, emulsion-breaking chemical injections, and circulation through an external, steam-heated exchanger. The emulsion-breaking tank is controlled and treated to break whatever emulsions are encountered. Normally the process chemical vendor works with plant personnel to ensure the emulsion-breaking system works.

Once the emulsion is separated, the oil goes on to recovered oil tanks and the water goes to sour water. There is always a residual solid or floc that will build up in the emulsion-breaking tank until the tank is taken out of service for cleaning. Multiple suction elevations are necessary to stay above the sludge. Emulsion-breaking tanks do require more frequent cleaning than other tanks, so they are normally relatively small.

The slops rate produced within a refinery should not be more than 1 % of the total crude input. With attention to draining and sampling practices as well as controls, the slops rate can be much lower.

It is only possible to paint a picture of slops and recovered oil facilities with a broad brush as these are so dependent on the processes used in the plant and how well the processes are managed.

Summary of Practices to Prevent Utility System Contamination

Rounding out the discussion of utilities, we should reemphasize the importance of avoiding contamination of these critical systems:

- Plant water, air, flushing oil, steam, and almost any other utility should only be connected into a process or foreign system through an anti-backflow manifold that includes the equivalent of double block and bleed, with a check valve and a blinding point. In this case, foreign systems include connections to other utilities. These manifolds should be removed or blinded after use before the foreign system is pressured up above the utility supply pressure.
- Potable and city water connections can only be made to foreign systems using a code-approved backflow preventer with an atmospheric vent.
- Nitrogen connections to a foreign system require special anti-backflow precautions and managed, so the foreign system cannot enter the nitrogen under any condition.

- Air gap utility connections when not in use.
- Think before connecting a utility into a foreign system.

In spite of the precautions, utility systems will become contaminated. Monitor the systems for potential contamination and deal with it before it becomes a serious problem.

Utilities Conclusion

This concludes the discussion of utilities in petroleum processing plants. Each of the topics, covered briefly here, has been the subject of countless articles and books.

There is a tendency to ignore the utilities until you have a problem. Then the problem utility becomes the most important thing there is in the plant – until the problem is solved. The keys to reliable, trouble-free, economical utilities are:

- Good utility design practices that emphasize reliability
- Routine monitoring and early action on issues
- Educated operators and staff
- Routine preventative maintenance
- Good utility operations management
- Attention to preventing contamination

Following these keys should enable a facility to focus primarily on running the main process units, trusting that the utilities will be there to support them.

References

Steam and Condensate Systems

R.C. Andrade et al., (Drew Chemical Corp., 1983), Controlling boiler carryover. Chem. Eng. 51–53 (1983)

J. Colanniono (Colannino Consultants, 1993), Prevent boiler tube failures, Parts 1 and 2. Chem. Eng. Prog. 33–36 (Oct 1993); 73–76 (Nov 1993)

S.T. Costa, (Calgon Corp., 1994), Bulletin: Factors affecting the selection of an industrial boiler water treatment program. Bulletin. 10–327 (1994)

L. Huchler, (MarTech Systems, 2000), Basics of boiler water chemical treatment, Hydrocarb. Process. 20, 87 (2000)

B. Liptak, (Bela G. Liptak Associates, 1987), Improving boiler efficiency. Chem. Eng. 49–60 (1987)

Water Systems

B. Buecker, (CEDA, Inc., 2000), Control water chemistry in HRSGs. Chem. Eng. Prog. **9**, 55–61 (2000)

F. Caplan, Quick calculation of cooling tower blowdown and makeup. Chem. Eng. 10 (1975)
Corrosion Doctors, Item: Scaling Indices. www.corrosion-doctors.org. Accessed 9 Feb 2014
S.T. Costa, (Calgon Corp., 1994), Bulletin: Factors affecting the selection of an industrial boiler water treatment program. Calgon Bull. 10–327 (1994)
R.H.L. Howe, R.C. Howe, (Eli Lilly & Co., 1981), Combining indexes for cooling water evaluation. Chem. Eng. 157–158 (1981)
L.A. Huchler, (MarTech Systems, Inc., 1998), Select the best boiler-water chemical treatment program. Chem. Eng. Prog. **11**, 45–50 (1998)
L.A. Huchler, Can you reduce your cooling tower blowdown rate? Hydrocarb. Process. 100, 121 (2000a)
L.A. Huchler, Whatever happened to deaerator cracking? Hydrocarb. Process. 131 (2000b)
L.A. Huchler, What about Legionella in industrial cooling towers? Hydrocarb. Process. 115 (2000c)
L.A. Huchler, Basics of boiler water chemical treatment. Hydrocarb. Process. 20, 87 (2000d)
L.A. Huchler, Basic cooling water treatment. Hydrocarb. Process. 123 (2001a)
L.A. Huchler, What do scaling indices mean? Hydrocarb. Process. 167 (2001b)
L.A. Huchler, Oxidizing biocides: How about chlorine? Hydrocarb. Process. 181 (2001c)
L.A. Huchler, (MarTech Systems, Inc., 2009), Cooling towers, Part 1: Siting, selecting, and sizing. Chem. Eng. Prog. 51–54 (2009a)
L.A. Huchler, (MarTech Systems, Inc., 2009), Cooling towers, Part 2: Operating, monitoring, and maintaining. Chem. Eng. Prog. 38–41 (2009b)
F.N. Kemmer, J. McCallion (eds.), (Nalco Chemical Co., 1979), *The Nalco Water Handbook* (McGraw-Hill Book, New York, 1979)
Lenntech website, Boiler feed water, www.lenntech.com. Accessed 11 Feb 2014
Lenntech website, WHO's drinking water standards 1993 (1993). www.lenntech.com. Accessed 11 Feb 2014
Lenntech website, WHO/EU drinking water standards comparative table. www.lenntech.com. Accessed 11 Feb 2014
J.R. Macdonald, (Nalco Chemical Co., 1987), Choosing the correct cooling water program. Chem. Eng. 135–137 (1987)
H.K. Miyamoto, (MacLaren Plansearch), M.D. Silbert (Marvin Silbert & Assoc.), A new approach to the Langelier stability index. Chem. Eng. 89–92 (1986)
Multiple authors (Nalco Chemical Co. & ABB Lummus Global, 2001), Cooling towers. Chem. Eng. Prog. 29–41 (2001)
Nalco Chemical Company, Bulletin: An introduction to boiler feedwater treatment. Bulletin. 30 (1978)
Nalco Chemical Company, Technifax bulletin: Boiler system sampling. (1984)
P.R. Puckorius, (Puckorius & Assoc., Inc., 2013), Selecting the optimal cooling tower fill. Chem. Eng. Prog. 31–34 (2013)
W.J. Scott, (Hercules, Inc., 1982), Handling cooling water systems during a low pH excursion. Chem. Eng. 21–126 (1982)
S.D. Straus et al., Boiler water treatment for low and moderate-pressure plants. Power **30**, S-1–S-16 (1987)
U.S. EPA, Bulletin: National primary drinking water regulations. Bulletin EPA 816-F-09-0004, May 2009
U.S. FDA, Item: Flow diagram of a reverse osmosis system. www.fda.gov. Accessed 9 Feb 2014
Wikimedia.org, Photo: Cooling tower. www.commons.wikimedia.org. Accessed 2014
P. Zisson, (Buckman Labs. Intl., 1997), Chemically treating boilers. Chem. Process. 9 (1997)

Compressed Air Systems

B.D. Bullough, (Sebesta Blomberg & Assoc.), Piping design of instrument air distribution systems

Chere Resources, Instrument air consumption – preliminary estimates. www.chereresources.com. Accessed Feb 2014

Instrumentation Portal, Instrument air consumption calculation. (2011). http://instrumentationportal.com. Accessed Feb 2014

Electrical Power Supply

T. Brown, J.L. Cadick, (Electro Technology Laboratories, Inc., 1979), Fundamentals of electricity. Chem. Eng. 72–76 (1979)

Cleaver Brooks Div. of Aqua-Chem, Inc. Bulletin: How to read schematic wiring diagrams. (1974)

V. Ganapathy, (ABCO Industries, 1997), Efficiently generate steam from cogeneration plants. Chem. Eng. 187–190 (1997)

Nitrogen System

Air Products, *Safety Bulletin: Liquid Nitrogen*. Air Products, Safetygram 7 (2013)

Grasys, Item: Membrane nitrogen plants and packages. (2014). www.graysys.com. Accessed Feb 2014

S. Ivanova, R. Lewis (Air Products, 2012), Producing nitrogen via pressure swing adsorption. Chem. Eng. Prog. 38–42 (2012)

R.L. Lewis, D.J. Stookey, (Permea, Inc., 1992), Pressure-swing adsorbers and membrane systems: Alternative nitrogen sources. Chem. Process. 41–46 (1992)

S. Shelley, Out of thin air. Chem. Eng. 30–39 (1991)

Off-Site Facilities for Petroleum Processing

David S. J. Jones and Steven A. Treese

Contents

Introduction	1168
Storage Facilities	1168
Atmospheric Storage	1169
Pressure Storage	1175
Heated Storage Tanks	1175
A Few Other Tank Management and Design Considerations	1176
Product Blending Facilities	1178
In-Line Versus Batch Blending	1178
The In-Line Blender Operation	1179
The In-Line Blender Design	1179
Component Tankage	1180
Finished Product Tankage	1180
Road and Rail Loading Facilities	1181
Loading Rates	1181
Loading Equipment	1182
Loading Facilities Arrangement	1184
Jetty and Dock Facilities	1184
Jetty Size, Access, and Location	1185
Equipment	1186
Loading Rates	1187
Other Features	1188
Waste Hydrocarbon Disposal Facilities	1189
Blowdown and Slop/Recovered Oil	1191
The Flare	1195

David S. J. Jones: deceased.
Steven A. Treese has retired from Phillips 66.

D.S.J. Jones
Calgary, AB, Canada

S.A. Treese (✉)
Puget Sound Investments, LLC, Katy, TX, USA
e-mail: streese256@aol.com

© Springer International Publishing Switzerland 2015
S.A. Treese et al. (eds.), *Handbook of Petroleum Processing*,
DOI 10.1007/978-3-319-14529-7_18

Effluent Water Treating Facilities 1202
   Other Effluent Water Treating Systems 1204
   Biochemical Oxygen Demand Reduction 1205
   Sanitary Sewage Treatment 1205
Appendix 1: Calculating Heat Loss and Heater Size for a Tank 1205
Appendix 2: Example Calculation for Sizing a Tank Heater 1209

Abstract

In a refinery, "offsites" are the facilities outside the main refining units that support those process units. The discussion in this chapter focuses on storage tanks, product blending, loading and receiving, waste hydrocarbon disposal, and effluent water treating. A procedure and example are provided for estimating tank heat loss and heater sizing.

Keywords

Offsites • Tanks • Jetty • Effluent Water • Slops • Flare • Blending • Loading

Introduction

In most refineries the off-sites facilities are a major capital cost center, second only to the process plants themselves. Indeed in some instances, where the off-sites include one or more complex jetties, they can be the principal capital cost center.

Among the major off-sites facilities found in most refineries are:

- Storage
- Product blending
- Road and rail loading
- Jetty facilities
- Waste disposal
- Effluent water treating

Many refineries consider the flare as part of the safety systems. To a large extent this is justifiable as most relief valves exhaust to the flare as do the process and other vent systems. For the purpose of this publication, however, the flare system is considered as part of the off-sites and specifically comes under the discussion of waste disposal.

We will explore each of these types of facilities in more detail within this chapter.

Storage Facilities

The crude oil feed and the processed products are held in storage tanks of various sizes and types. These tanks are usually collected and located together in the refinery area suitably defined as "the tank farm." There will be many other tanks (usually much smaller than those in the tank farm) which will contain chemicals to be used in the processes and the slop/recovered oil or spent chemicals from the various processes and utility plants. Most refinery storage tanks fall into the following three categories:

- Atmospheric storage
- Pressure storage
- Heated storage

Atmospheric Storage

As the name implies, all atmospheric storage tanks are open to the atmosphere or are maintained at atmospheric pressure by a controlled vapor blanket. These tanks fall into two categories:

- Cone-roof tanks
- Floating-roof tanks

Cone-Roof Tanks

Among the most common atmospheric storage tanks is the cone-roof tank shown in Fig. 1. This tank is used for the storage of nontoxic liquids with fairly low volatility. In its simplest form the roof of the tank will contain a vent, open to the atmosphere, which allows the tank to "breathe" when emptying and filling. A hatch in the roof also provides access for sampling the tank contents. In oil refining this type of tank is used for the storage of gas oils, diesel, light heating oil, and the very light lube oils (e.g., spindle oil).

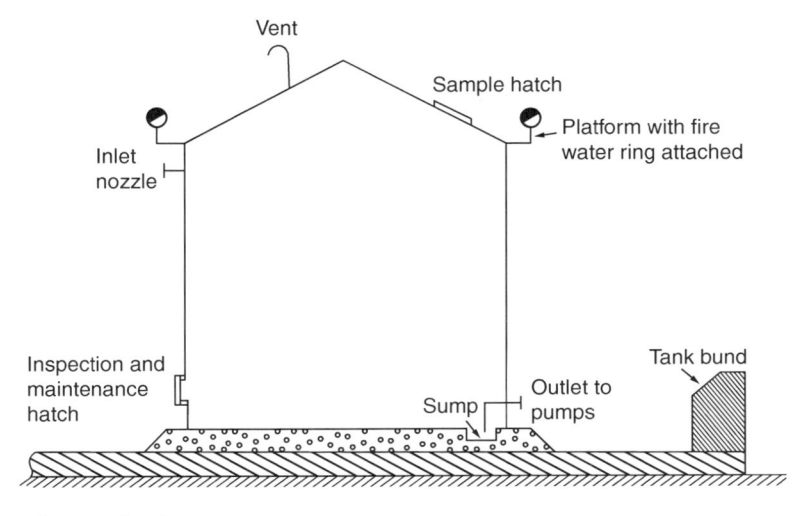

Fig. 1 Cone-roof tank

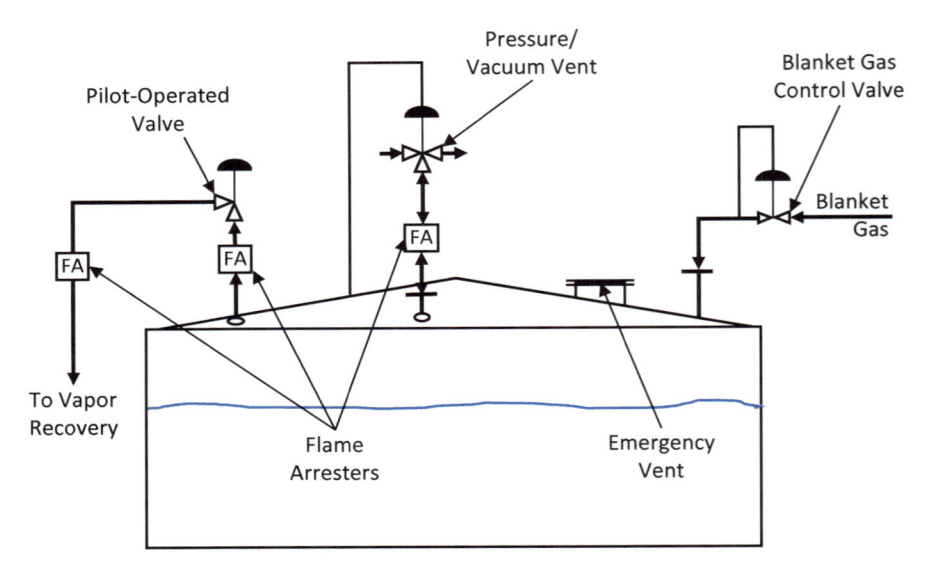

Fig. 2 Blanket gas or padding system on a tank

In keeping with a company's fire protection policy, tanks containing flammable material will be equipped with foam and fire water jets located around the base of the roof. All storage tanks containing flammable material and material that could cause environmental damage are contained within a diked area or bund. The size and volume of the bund area are fixed by regulation. Usually the bund has to contain the total contents of one of the tanks included in the area. The number of tanks per bunded area is also fixed by regulation.

Many commodities need to be protected from atmospheric oxygen exposure. This is normally done in cone-roof tanks using an inert gas blanket or padding system. The inert gas blanket is controlled tightly using a very sensitive control system that keeps the blanket pressure in about the +0.5 to +2.0 in. water range. Inert gas (usually natural gas or nitrogen) is brought in at low pressure and is vented at high pressure. The vented gases go to a vapor recovery system, usually through a small liquid ring compressor or similar device, to prevent emissions. The system uses special control valves or regulators. Blanketed tanks will also have regular pressure and vacuum vents in case the blanket system fails or the rate of change in tank inventory exceeds the blanket system capability. Figure 2 illustrates the blanket gas system concept. Note that the blanket gas inlet is opposite the outlet to help ensure a good sweep of the tank vapor space with inert gas.

Floating-Roof Tanks

Light volatile liquids may also be stored at essentially atmospheric pressure by the use of "floating-roof" tanks. A diagram of this type of storage tank is given in Fig. 3.

The roof of this tank literally floats on the surface of the liquid contents of the tank. In this way the air space above the liquid is reduced to almost zero, thereby

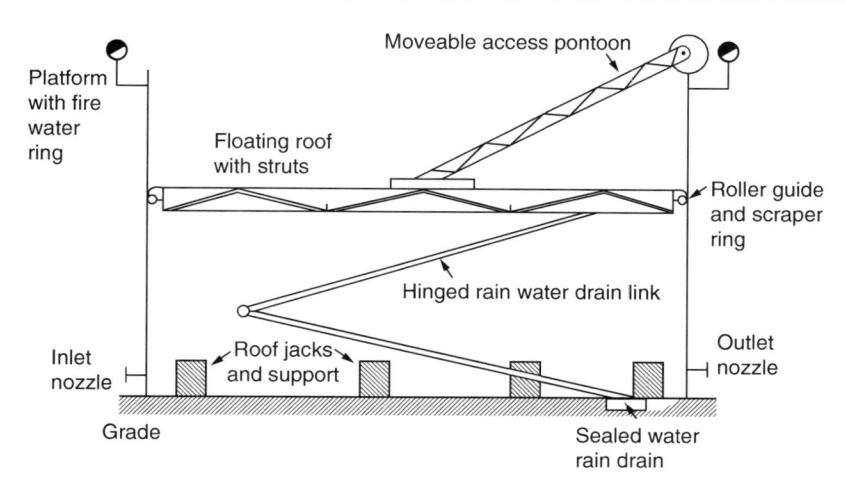

Fig. 3 Floating-roof tank

minimizing the amount of liquid vaporization that can occur. The roof is specially designed for this service and often contains a top skin and a bottom skin of steel plate, held together by steel struts. These struts also provide strength and rigidity to the roof structure. The roof moves up and down the inside of the tank wall as the liquid level rises when filling and falls when emptying. The roof movement is enhanced by guide rollers between the roof edge and the tank wall. A scraper ring or a set of shoes located around the periphery of the floating roof are pressed tightly against the tank wall to ensure a primary seal between the contents and the atmosphere. The primary seal also provides additional guide to the roof movement and stability to the roof itself.

To further reduce tank emissions, in many areas, the primary seal is supplemented by one or more secondary seals. These are tubes or wipers made of resilient material (like rubber) that press against the tank wall and seal any remaining gap from the primary seals.

In some tanks, the floating roof may only be a single roof skin. There are floats around the circumference. This is referred to as a "pan" design. The other tank design features in a pan floating roof are similar to those described above.

When the tank reaches the minimum practical level for the liquid contents, the roof structure comes to rest on a group of pillars at the bottom of the tank. These provide the roof support when the tank is empty. There is a space between the roof and the tank bottom. This space is required to house the liquid inlet and outlet nozzles for filling and emptying the tank which, of course, must always be below the roof. The space is also adequate to enable periodic tank cleaning and maintenance.

In many floating-roof tanks, the maintenance support pillars or legs move up and down with the roof. They are pinned to penetrations through the roof at a low level. The advantage of this design is that when the tank is emptied for maintenance or switching services, the roof can go almost all the way to the bottom (or at least as low as any internals). This leaves less remaining heel to deal with, effectively minimizing waste. For maintenance, then, the roof can be refloated with water to a level where the legs are reset to support the roof when the tank is entered. After maintenance, the roof is refloated with water, the pins are pulled, the roof is dropped to minimum level, the legs are re-pinned, and then the tank is filled with oil.

A drain line running inside the tank from the roof to a "below-grade" sealed drain provides the facility for draining the roof of rain water. These roof drains may be hard piping connected together by swivel joints (like hinges), or the drain may be a coiled, large-diameter flexible tubing. The flexibility in the drain line allows the line to move up and down with the roof movement.

A pontoon-type access pier from the platform around the perimeter at the top of the tank provides access to the sample hatch located at the center of the roof. This "pontoon" also moves upward and downward with the roof movement.

Automatic bleeder vents are provided on all floating-roof tanks. They vent air from under a floating roof when the tank is being filled initially from empty. After the liquid rises high enough to float the roof off its supports, the vent automatically closes. Likewise when the tank is being emptied, the vent automatically opens just before the roof lands on its support, thereby preventing the development of a vacuum under the roof.

Other accessories include rim vents, float gauges, anti-rotation devices, and manholes.

Liquids stored in this type of tank have relatively high volatilities and vapor pressures such as gasoline, kerosene, jet fuel, and the like. In oil refining the break between the use of cone-roof tank and floating-roof tank is based on "flash point" of the material. Flash point is that temperature above which the material will ignite or "flash" in the presence of air. Normally this break point is 120 °F.

Nozzle Arrangement and Location Considerations

Most tanks have multiple connections into the commodity storage section of the tank. The locations of these connections affect how the tank will operate. Typical connections include:

- Inlet lines
- Outlet lines
- Subsurface foam lines
- Water draw
- Roof drain (floating-roof tank)
- Sample lines (which may be attached to a floating pantograph)
- Low-low suction line
- Mixer
- Circulating system return
- Bayonet heater
- Maintenance manway

Some specific considerations when designing and managing these connections include:

- Inlet and outlet lines should be separated from each other by at least 10 f. (3 m) horizontally to avoid short circuiting. It is best if the lines connect to the tank on opposite sides, far apart.
- If sludge buildup is expected (and it almost always occurs), there should be high and low inlet and outlet connections. The low lines are usually a nominal distance above the tank floor, like 2–3 f. (0.7–1 m). The high lines are usually about 4 f. (1.3–1.5 m) higher.
- Many tanks use subsurface foam lines, in addition to or in lieu of the foam generators around the top of the tank. These need to be evenly spaced around the periphery. Each should enter through a backflow prevention device.
- The water draw should come from the lowest point in the tank. Usually the water draw takes suction from the bottom of the tank sump. It may exit the sump underground in a drain-dry tank.
- The roof drains for floating-roof tanks normally pass through the tank commodity to a nozzle on the wall near a sewer hub or drain sump leading to the oily water sewer. Some refiners leave these drains open in service to avoid sinking the roof in a sudden rain storm. Others are concerned about leakage of the joints or pipe leading from the roof drain (which is common), with subsequent loss of hydrocarbon to the sewer or atmosphere. Everyone follows practices based on their experience.
- Tanks often have multiple sample taps to the shell or connected to a pantograph that floats on the commodity. These are used to get top, middle, bottom, and dead-bottom samples from the tank. In the case of emulsion-breaking tanks, there are often taps every 12 in. (0.3 m) up the side of the tank to help find the oil/water interface. Tank sample taps should always be plugged when not being used. Sometimes the taps are all routed to a local drain or sink where the samples are taken. In many tanks, the samples are actually circulated from a tap back into the tank to ensure an accurate sample.
- When a tank has to switch commodities or batches frequently, it will often have a low-low suction line. This line is as close as possible to the tank floor vertically and may even originate in a sump below the floor. It allows the tank to be emptied almost completely. This "zero" NPSHA situation needs to be considered in pump design.
- As noted above, some tanks have mixers installed through the tank wall. These may be propeller-style or jet nozzles, which include venturis. The location and orientation of these is important. Propeller mixers usually enter through a packing gland, which must be maintained. Propellers often can be moved to different angles in the tank to prevent accumulation of sludge in one area. Where mechanical mixers are used, the stresses imposed on the tank wall by the mixer action will require reinforcement of the shell where the mixer nozzle is located.
- Where a circulating system is provided on a tank, either for heating or mixing, the return needs to be across the tank from the suction line to ensure proper blending action.

- Bayonet or coil heaters are installed in some tanks. These are inserted through the tank wall near grade level. It is best to encase the heater elements in an enclosure that will allow the elements to be removed for maintenance without having to empty and clean the tank. There are some tanks where steam-heating coils are installed over the tank floor. In these cases, the steam inlet and condensate outlet nozzles are usually close to each other. If a regular floor-mounted steam coil leaks, it must wait until the tank can be drained for repairs.
- Most tanks have a large maintenance manway for access when the tank is serviced. In very large tanks, this flanged manway is large enough to drive a small front-end loader through. The manway should be pointed toward the easiest access point in the tank dike, usually toward a ramp down into the dike. Sealing these manways is often difficult after use. In recent practice, some refiners have chosen to eliminate the manway and just cut an access hole in the tank when it requires servicing. The access hole is welded up after each use. Since the turnaround frequency on tanks is about 20 years, this can be an acceptable approach.

Tank Internals

Many tank internals have been described in the previous tank discussion. It is important to remember that most tanks are not just empty volumes. Tanks may contain:

- Internal mixing nozzles (sometimes they disappear in service)
- Mixing venturis
- Propeller or other mechanical mixers
- Dip lines down into the sumps
- Roof drains
- Sample pantographs
- Sample tap extensions
- Water draw lines
- Inlet line extensions
- Outlet line extensions
- Roof support legs (cone as well as floating roof)
- Drain sumps
- Gauging instruments
- Stilling wells

Once a tank is closed, you will not be able to easily examine any of these internals. It is a good practice to photograph each internal and maintain these photos on file for reference. The function, nozzle number, location (angle), and elevation of each internal should be attached to each internal photo. You will never remember what is in the tank once it is closed.

Sumps and Drain-Dry Tanks

Switching services in a tank or changing product specifications requires that as much of the previous tanks' contents be removed as possible before the new material is introduced. In most tanks this is accomplished by flushing out the tank two to three times with the new commodity while the level is kept low. This does accomplish the task, but has the disadvantages of potentially downgrading product and not completely eliminating the previous commodity. It may generate a lot of slops. The effectiveness of this approach is limited by the tank internals, the suction line elevation, the outlet pump NPHSR, and the minimum heel in the tank.

Many tanks have a drain sump located at the lowest point in the tank. This sump helps drain a tank to a low level during commodity switches, but the level is usually still limited. The biggest value for the sump is normally draining accumulated water.

Many tanks today use a "drain-dry" design. This design is limited mostly to clean services. In a drain-dry tank, there is nothing to prevent the tank from being almost completely drained by taking suction from the drain sump below grade. Only a few inches of commodity is left in the tank, instead of 2–4 f. of heel in a normal tank.

Pressure Storage

Pressure storage tanks are used to prevent, or at least minimize, the loss of the tank contents due to vaporization. These types of storage tanks can range in operating pressures from a few inches of water gauge to 250 psig. There are three major types of pressure storage. These are:

- Low-pressure tanks – these are dome-roof tanks and operate at pressures of between 3 in. water gauge and 2.5 psig.
- Medium-pressure tanks – these are hemispheroids (which operate at pressures between 2.5 and 5.0 psig) and spheroidal tanks (which operate at pressures up to 15 psig).
- High-pressure tanks – these are either horizontal "bullets" with ellipsoidal/ hemispherical heads or spherical tanks (spheres). The working pressures for these types of tanks range from 30 to 250 psig. The maximum allowable pressure is limited by tank size and code requirements. For a 1,000 bbl sphere, the maximum pressure is 215 psig. For a 30,000 bbl tank it is 50 psig. These pressure limits can be increased if the tank is stress relieved.

Although it is possible to store material in tanks with pressure in excess of 250 psig normally when such storage is required, refrigerated storage is usually a better alternative.

Heated Storage Tanks

Heated storage tanks are more common in the petroleum industry than most others. They are used to store material whose flowing properties will restrict flow at normal

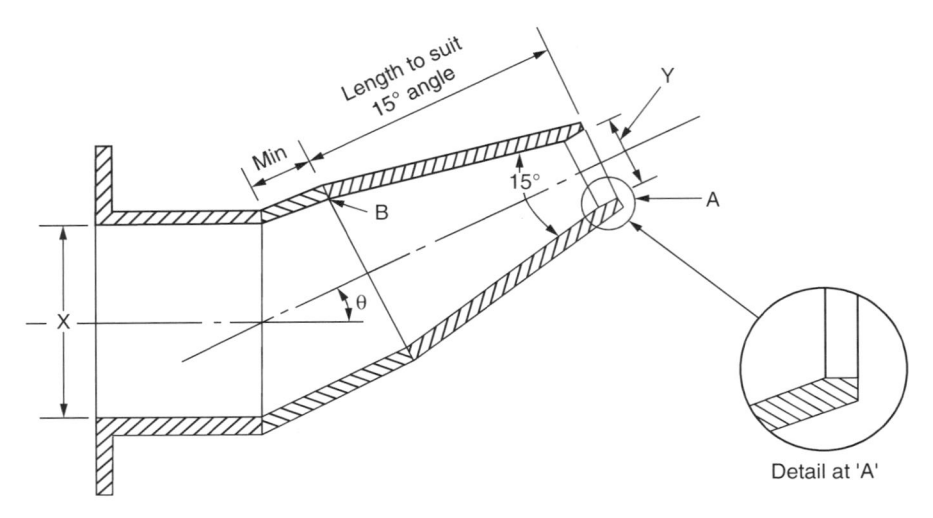

Fig. 4 Jet mixing nozzle

ambient temperatures. In the petroleum industry, products heavier than diesel oil, such as heavy gas oils, lube oil, and fuel oil, are stored in heated tanks.

Generally speaking, tanks are heated by immersed heating coils or bayonet-type immersed heaters. Steam is normally used as the heating medium because of its availability in petroleum complexes, although smaller tanks or tanks containing heat-sensitive materials may be heated electrically. You will find some tanks heated by circulation through an external heat exchanger. External tank heating is used when there is a possibility of a hazardous situation occurring if an immersed heater leaks or to enable easier heater repairs.

Very often where immersed heating is used, the tank is agitated, usually by side-located propeller agitators for large tanks. Where external circulating heating is used for tanks, the contents are mixed by means of jet mixing. Here the hot return stream is introduced into the tank via a specially designed jet nozzle as shown in Fig. 4.

Heat loss and the heater surface area to compensate for the heat loss may be calculated using the procedure outlined in Appendices 1 and 2 of this chapter.

A Few Other Tank Management and Design Considerations

Tanks and tank farms require specific management practices and may have some special design features. These include operation of the tanks as well as maintenance and record keeping. Each refiner has its own practices, but some common themes are noted here:

- When a tank is built, one of the first activities before use is strapping the tank. This means filling the tank with water that is precisely metered. The tank's

Off-Site Facilities for Petroleum Processing 1177

contained volume is noted every inch or two. The results are published as strapping tables (which may be electronic) that will be used for the life of the tank or until the strapping is repeated. More recently, strapping tanks by laser methods has become feasible. By either method, the strapping tables, with any corrections for temperature, are the most accurate gauge of contained volume in a tank.

- Each tank is normally manually gauged at least every day. In some cases, they may be gauged several times during a shift. The readings are recorded and are considered company records for regulatory purposes. Gauging usually involves going to each tank and manually activating the gauge device to get a reading. In some services, gauging may require the operator to go to the top of the tank and use the gauging hatch to drop a weighted tape into the tank through a stilling well. These manual gauges are compared to readings on the remote gauging system, if any, to verify readings and indicate when repairs are needed. A change in level of a tank that is unexpected is investigated. It may indicate a lineup error or a leak.
- The safety of the operators working around tanks needs to be considered, especially when going up to the roof of the tanks. It is common practice for an operator to let the control room know when he enters and exits a tank dike area. Many refiners also require an operator to wear fresh air and have a standby operator in fresh air when manually gauging tanks or going onto a tank roof for any reason. The low spots represented by the tank dike and the top of a floating-roof tank, as well as the vapor space in a cone-roof tank, can accumulate hydrogen sulfide or other gases in sufficient quantity to pose a hazard to personnel.
- Tanks collect water in service. The amount of water collected is usually determined by "water cutting" the tank when it is manually gauged. Water cutting is accomplished by coating the gauge tape with a special water-indicating paste and then dropping the gauge tape through the gauge hatch. When the tape is removed, the water level will be seen by color change on the tape, so the gauging gets you both oil and water levels. Some newer instruments for gauging tanks may also indicate water level.
- When the water level reaches a predetermined depth (or on a defined frequency), the water needs to be removed by "water drawing" the tank. This means that the valve from the water draw sump (at the lowest point in the tank) is opened and the drawn liquid sent to the oily water sewer until the drawn liquid is mostly oil. This is a manual operation. When water drawing a tank, the operator should stand by the draw point or in the immediate area to ensure minimal hydrocarbon is sent to the sewer.
- The tank dike is intended to accept the contents of the tank in the event of rupture. A dike has a drain sump (usually the same one where water draws go). The sump will be routed to the oily water sewer through a block valve located outside the berm or dike. Depending on the refiner, the decision needs to be made to leave the block valves open or closed. If the valves are managed normally closed, then rain water will accumulate in the dike and have to be periodically

drained. Accumulated rain water will also corrode the tank shell if left for an extended period. If the valves are managed open, a tank failure will easily overwhelm the wastewater-treating facilities. Many refiners opt to keep the valves closed and then drain the dikes after a rain at a predetermined rate. This helps keep the wastewater treatment plant to a reasonable size and provides a controllable discharge rate.
- Many tanks today have double bottoms to help prevent loss of hydrocarbon to the ground, even inside the dike. These double bottoms have provisions for leak detection between the two bottom skins. They can be retrofitted to existing tanks with a small loss of capacity.
- Tanks are normally set on sand pads. In modern tank design, it is common for the pad to also contain an impervious (plastic) membrane. A small tank leak will then be indicated by oil seepage collected on the membrane.
- The primary corrosion risk around tanks, as well as the underground piping leading to and from them, is from contact with the ground. This corrosion attacks the floor of the tank or the walls of the pipes from the outside. Tank farms may be protected by impressed current systems or buried zinc electrodes. Because of the long times between turnarounds, it is important to keep the corrosion prevention systems in good working order.
- Tanks are generally in service for many years between turnarounds. A typical service interval is 20 years. Attention to management of these facilities is important to prevent leaks and sludge buildup in the tanks.

Product Blending Facilities

Blending is combining two or more components to produce a desired end product. The term in refinery practice usually refers to process streams being combined to make a saleable product leaving the refinery. Generally these include gasolines and middle distillates (jet, kerosene, diesel, and heating oil). Other blended finished products will include various grades of fuel oil and lube oil. The blending of the process streams is accomplished either by batch blending in blending tanks or by in-line blending by in the pipeline itself.

In-Line Versus Batch Blending

In batch blending, the components are routed separately into a single receiver tank. They are mixed in this tank to meet the finished product specification. In the case of in-line blending, the component streams are routed through automatically operated flow control valves to a finished product tank. With modern computerized control technology, in-line blending is the more common form of blending process. In the case of gasolines and lube oils in particular, in-line blending is extensively the accepted method. Middle distillates and residuum blending by batch still has some advantage because there are fewer components to be handled, although the quantities involved are usually greater.

As a word of caution, in-tank, batch blending of butanes into gasoline for RVP control is a very dangerous practice. It has resulted in multiple fatalities, because the butane does not initially and quickly dissolve in the gasoline. Only in-line blending should be used for butane in gasoline.

The In-Line Blender Operation

An in-line blender is essentially a multiple stream controller with feedback. The controller itself is a computer into which the recipe for the blend is keyed. Such blending recipes have been covered in some detail in the chapter entitled "▶ Introduction to Crude Oil and Petroleum Processing" of this handbook. The controller automatically starts the pumps for the blend components and manipulates the flow control valves on the component lines to meet the required component quantities. In most cases the component lines join together to form the blended product, which is then routed to the finished product tank. A series of online analyzers, located in the blend rundown lines, monitor the finished product properties and, in turn, reset the controller adjusting the component quantities to meet the end product specifications.

The In-Line Blender Design

Most refinery companies have their own proprietary component blending recipes for their finished products. These will be in computer program form and usually use the linear programming techniques described in the chapter "▶ Petroleum Refinery Planning and Economics" of this handbook. It is this program software that is installed into the blender controller to activate the respective component systems. The blender controller acts to start the selected pumps and the control valves. It also receives data from the online analyzers located in the product rundown line.

The design of the blending system as a whole is the combined effort of the process engineer, the instrument engineer, and the computer specialist. It will be the duty of the process engineer or the blender to develop the blending recipe in terms of component percentages and quantities to meet a particular product specification. The instrument engineer will ensure that the control valves, pump starting arrangement, and the onstream analyzers meet the process requirements. A typical list of the instrument engineer's responsibilities in this regard includes:

- The control panel, panel instruments, instrument power supply, annunciators, pump switches, indicating lights, graphic display, and all panel wiring
- Turbine meters and preamplifiers
- Control valves with positioners
- Stream analyzers
- Field transmitters (flow, temperature, pressure)

Finally it will be the computer specialist's duty to translate these requirements into the software program for the controller computer.

Recent developments in online analyzers, especially NIR analyzers, are allowing these devices to keep blenders optimized continuously during the blend. The newer analyzers can accurately determine octane, cetane, and several other key specifications. They are more reliable than some of the older online methods, such as online knock engines. It is still necessary to final certify a product blend, however.

Component Tankage

Because the in-line blender permits the rapid conversion of the components to finished products, the ratio of component to finished product tankage should be quite large. For good flexibility this ratio should be 4 or 5 to 1.

The most significant process requirement for successful blending is that the properties of the individual components must not change during the blending operation. Alkylates and catalytically cracked naphtha vary little, unless feeds or unit operations are changed. Reformates and straight-run naphtha, however, have a greater variability due to changes in source crude feed quality. These are often stored in separate "running gauge" tankage, the quality of the contents of which are tested when full.

Lube oils are a particular problem in component storage, because of the tendency of the oils to "stratify" to a greater extent than other stocks. That is, these oils tend to separate in storage with the higher density and viscosity portion of the oil sinking to the tank bottom. In most cases the contents of these tanks are continually mixed using propeller-type mixers. Many companies adopt this system to all heavy component tankage such as gas oils and fuel oil product components to avoid stratification in those services, also.

When blending ethanol and many other renewable components, it is necessary to prevent exposure of the renewable components to water. These components will absorb water and then release the water in a vehicle tank later when the gas tank cools. Ethanol is particularly bad in this regard. Renewable blend stock tanks usually have cone or geodesic dome roofs, sometimes with inert purges, to prevent absorption of atmospheric moisture and collection of rain water.

Finished Product Tankage

Finished product tankage is needed even with the most efficient in-line blender. This is because of required product disposal rates and product certification. However, in many design cases, in-line blending will still only require about one half the product storage required by batch blending.

Actual requirements of product tankage will usually be dictated by the manner in which the product is to be certified and shipped from the refinery. The blend rate is usually sized to blend a day's production in a certain number of hours.

Off-Site Facilities for Petroleum Processing

The maximum rate of an economically sized blender will usually be too slow to blend into an ocean-going tanker.

Thus, product tankage and loading pumps are needed to supplement the blender. Conversely, the minimum rate of a blender will usually be too fast to blend directly into a road tanker. In some cases, however, this is done by limited volume transfer pumps taking suction directly from component tankage.

As with blend stock tankage, product tanks containing ethanol blends or other renewable blends normally require protection from water intrusion with a roof or dome.

Road and Rail Loading Facilities

The extent of product shipping facilities required in a chemical or petroleum complex depends on the size of the complex, the local market, the number of different products to be shipped, and the market to be supplied. Normally the shipping facilities installed in most plants are sufficient to cater for normal product handling and the flexibility required for seasonal demands. The capacity of these facilities will almost invariably exceed the plant's total production.

The most common method of shipping product is by road or rail in suitably designed tanker cars. In the case of large complexes located on coastal or riverside sites, shipping by barge or ships carries the bulk of the plant products. This section, however, will deal only with dispatch by road and rail. Note that many of these same considerations can be applied to receiving facilities for crudes or intermediates by truck and rail, also.

Loading Rates

Loading rates for road and rail tankers vary from as low as 150 GPM to as high as 1,000 GPM, but most terminals load at rates between 300 and 550 GPM. Road tankers have capacities from 1,300 to 6,500 gal, and one tractor can haul two 6,500 gal tankers.

The number of loading arms required for each product to be loaded varies with:

- Truck size
- Number of loading hours per day
- Number of loading days per week
- Time for positioning, hookup, and de-positioning of the truck

Figure 5 gives the number of arms or spouts required for loading a 3,500 gal truck under various conditions.

The conditions shown in Fig. 5 are for filling at a rate of 300 GPM (bottom curve) and for 500 GPM (top curve). time The loading is taken as the filling time per tank truck plus 10 min. Thus loading time is

Fig. 5 Number of loading arms for quantity shipment

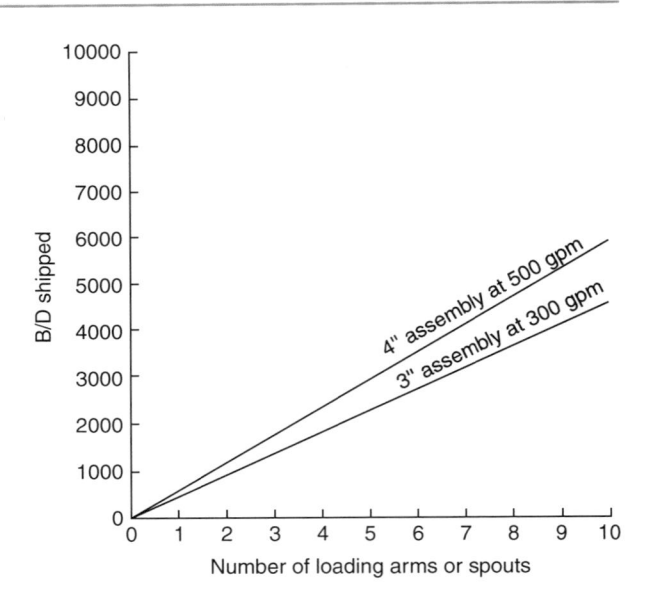

[Chart: B/D shipped vs Number of loading arms or spouts, showing lines for 4" assembly at 500 gpm and 3" assembly at 300 gpm]

$$[(\text{Tank truck capacity, gals})/\text{GPM}] + 10 \text{ min} \quad (1)$$

Tank car capacity is taken as 3,500 gal. Thus for the lower curve loading time is 22 min per tanker and for the upper curve 17 min per tanker. Assuming that a single product is loaded over 4 h in an 8-h day 5 days a week, then the number of trucks required per barrel/day is

$(\text{BPD} \times 42 \times 5 \text{ days} \times 8 \text{ h/day})/(3,500 \text{ gals/truck} \times 20 \text{ h loading per week})$
$= 0.024$ trucks per barrel/day

Then for 1,000 B/D, the number of trucks per working day $= 0.024 \times 1,000 = 24$ trucks/day of 4 h filling.

Time to load trucks at 300 GPM filling rate $= 24 \times 22$ min $= 528$ min.

To complete the loading in 4 h, the number of arms required $= 528/240 = 2.2$ arms.

Continuing with this calculation for several more shipping capacities and for filling rates of 300 and 500 GPM produces the data given in Fig. 5.

Loading Equipment

Figure 6 is a schematic drawing for a typical road or rail loading facility.

The loading pumps, which are located close to the product storage tanks, take suction from these tanks. The loading pumps are high-capacity, low head type with flat head/capacity characteristics. They operate between 35 and 45 psi differential head.

Off-Site Facilities for Petroleum Processing

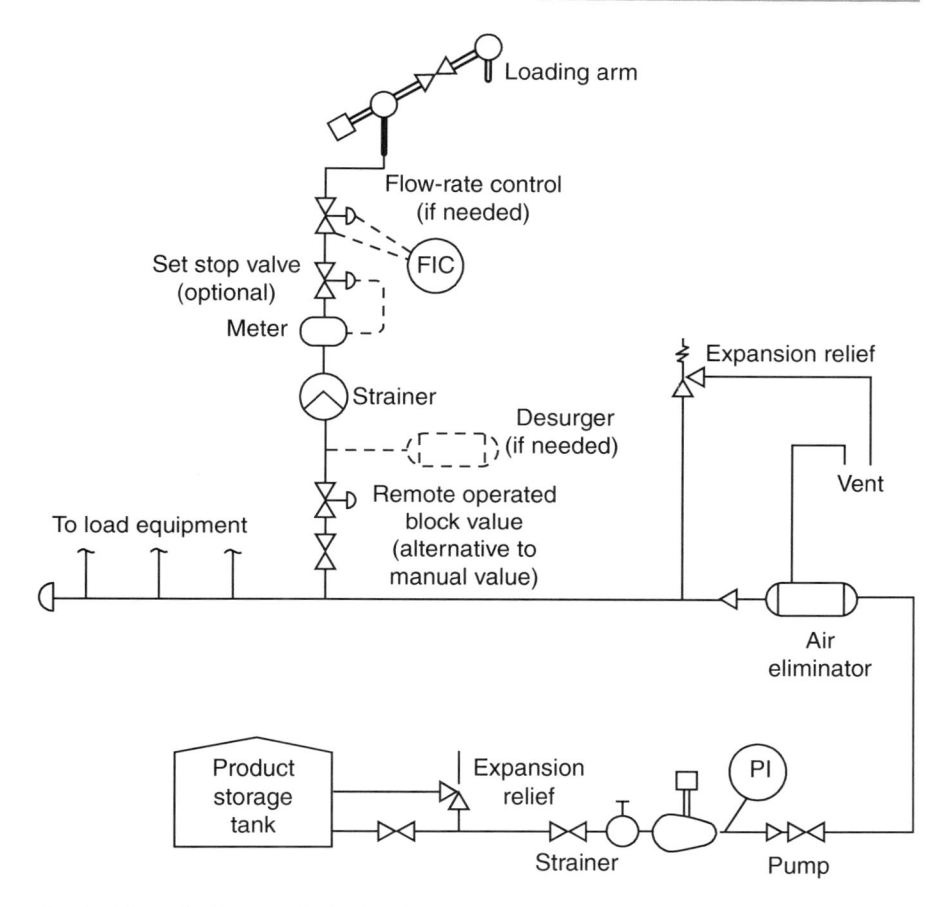

Fig. 6 Schematic diagram of a loading facility

These pumps discharge through an air eliminator drum into the loading header. Several loading arm assemblies are connected to the loading header. Each of these assemblies includes a remote-operated block valve followed by a desurger (optional) and then a strainer located before the loading meter. The product flows from the meter into a swivel-jointed loading arm and nozzle. Tank trucks and rail cars are loaded through their top hatches into which the nozzles of the loading arm fit.

Air eliminators are used to disengage air and other vapors which would interfere with the accuracy of the meters. Disengaging of the vapors is accomplished at about 3 psig. Should there not be sufficient static head at the disengaging vessel, a back pressure valve must be provided to obtain this pressure. The meters are positive displacement type, and desurgers are installed to decrease hydraulic shock resulting from quick shutoff.

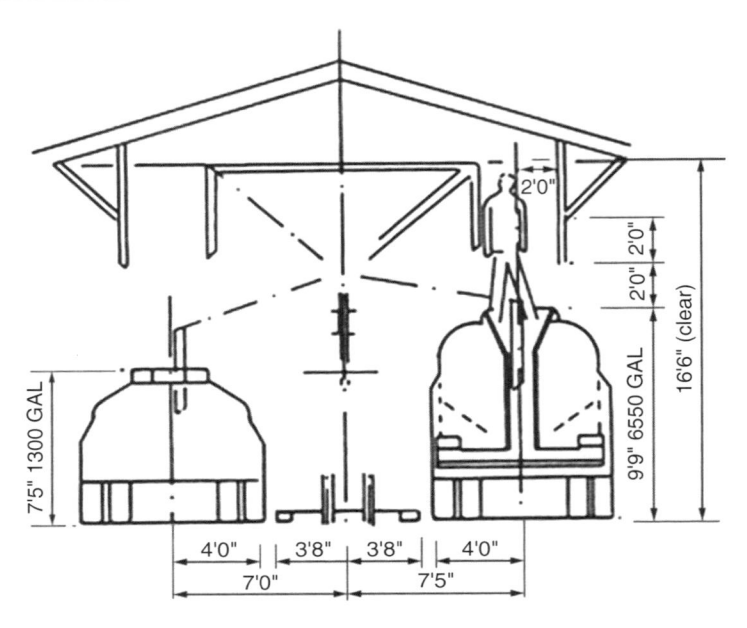

Fig. 7 Tank truck loading facility

Loading Facilities Arrangement

Figures 7 and 8 show typical arrangements of loading facilities for truck and railcar, respectively.

The dimensions shown on the diagram are applicable to one world area and may not be applicable to other localities. The equipment and its arrangement shown in the diagrams, however, are similar for most of these facilities.

In truck loading the meters and strainers are located at the loading station. The connection of the loading arm is made by an operator actually standing on the car itself. In the case of the truck loading facilities, the loading arm is operated from an adjacent platform. As in the case of truck loading, the meter and strainer, together with the on/off valve, are located near the loading site. In many newer trucks and rail cars, the connections are usually located on the bottom of the tanker. This is safer for loading personnel and less expensive. The same general facilities are used, however.

Jetty and Dock Facilities

Tankers and barges are loaded and unloaded at jetties or docks. In almost all circumstances these facilities for handling petroleum products are separate from those used for general cargo. Very often tankers, particularly the modern "supertankers," are loaded and unloaded by submarine pipelines from a deepwater

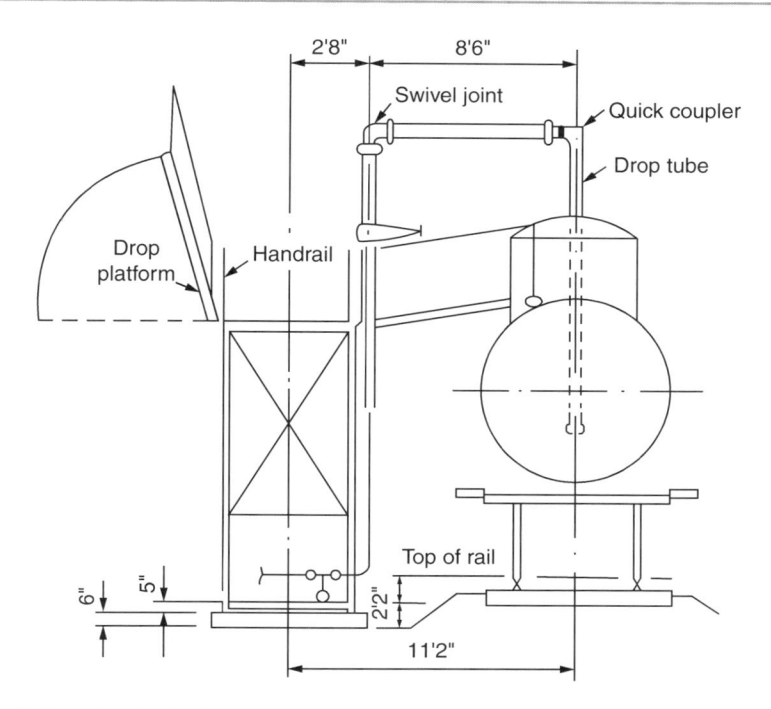

Fig. 8 Rail car loading facility

anchorage. This section of the chapter, however, deals only with onshore docking facilities.

Jetty Size, Access, and Location

Tanker sizes range from small coastal vessels of 10,000 bbl capacity to supertankers in excess of 250,000 bbl capacity. The more common tanker size is about 140,000 bbl capacity, and this size tanker is labeled a T2. This tanker is usually used for product carrying. It can carry as much as three different product parcels at the same time. The larger tankers are usually used for crude oil transportation.

Ideally the jetty size should be sufficient to accommodate several sizes of tankers and usually at least a couple of sizes at the same time. In some refineries which have jetty facilities, these usually include barge loading items also. The barge loading may, however, be located on remote docking facilities from the larger sea-going tankers.

The location of the jetty itself must consider the following:

- There is sufficient deep water to accommodate the larger crude-carrying tanker.
- It is located as close as possible to the refinery's tank farm.
- It is in an area that has a good approach road and park.

- There is sufficient room for a product/crude pipeway.
- There is sufficient waterway in which to maneuver and handle tanker docking.

A fully loaded T2 tanker has a maximum draft of 30 ft. The larger crude-carrying tankers would have a maximum draft around 45 ft. It is important to minimize rundown pipe lengths to and from the jetty loading area and the refinery tank farm. The first is a piping cost factor and the second is that often the pumping characteristics may have a negative static head during the pumping program.

A good onshore jetty approach road is mandatory for the operation of the jetty. This is required for safety reasons and the easy approach way for emergency vehicles (such as fire engines and ambulances). The approach road is also required for the transportation of the operating staff, ship's crew, and the ship's chandler vehicles. Usually this approach road is dedicated for jetty use and will be independent of any adjacent refinery road.

There must be room on shore and on the jetty itself for the loading and unloading pipeway(s). This can be extensive, depending on the refinery size and the number of products that are exported. There are several options for the location and size of this pipeway configuration. On the jetty itself, it may be carried on overhanging supports on both sides of the jetty pier, or it can be supported by an independent pipe rack adjacent to the jetty pier (much more expensive, however), and this pipe rack could be multi-tiered. On shore, the pipeway can be located along the roadway at ground level or elevated with two or more tiers. It could also be elevated and run above the roadway. This does, however, restrict access by limiting vehicle height using the roadway.

Finally the location of the jetty must allow sufficient waterway room for tankers to be berthed properly. Tankers arriving from the open sea must have room so that tugs can handle and turn the ship around to face the open sea before tying up at the jetty.

A layout plan for a typical tanker jetty is shown in Fig. 9.

Equipment

The equipment required for tanker loading includes pumps, hoses or flexible loading pipes, and handling cranes or structures. The loading pumps are located at the tank farm. These are centrifugal type with discharge pressures in excess of 100 psig. This is dependent on rundown pipe lengths, but pressure drop through the loading hoses or flexible loading pipes is within 10–25 psig. Also, and as mentioned earlier, there is often a static head loss to the deck manifold of an empty tanker.

When a rubber hose is used, it is supported by a dockside derrick plus the tanker boom. Some installations employ a combination of hose and pipe or flexible assemblies of pipe and swivel joints supported by structures. Automatic adjustment for tide and tanker draft is incorporated in this equipment. Hose and various assemblies are available in sizes from 2 to 12 in. diameter with the 8 and 10 in. most frequently used for products.

Off-Site Facilities for Petroleum Processing

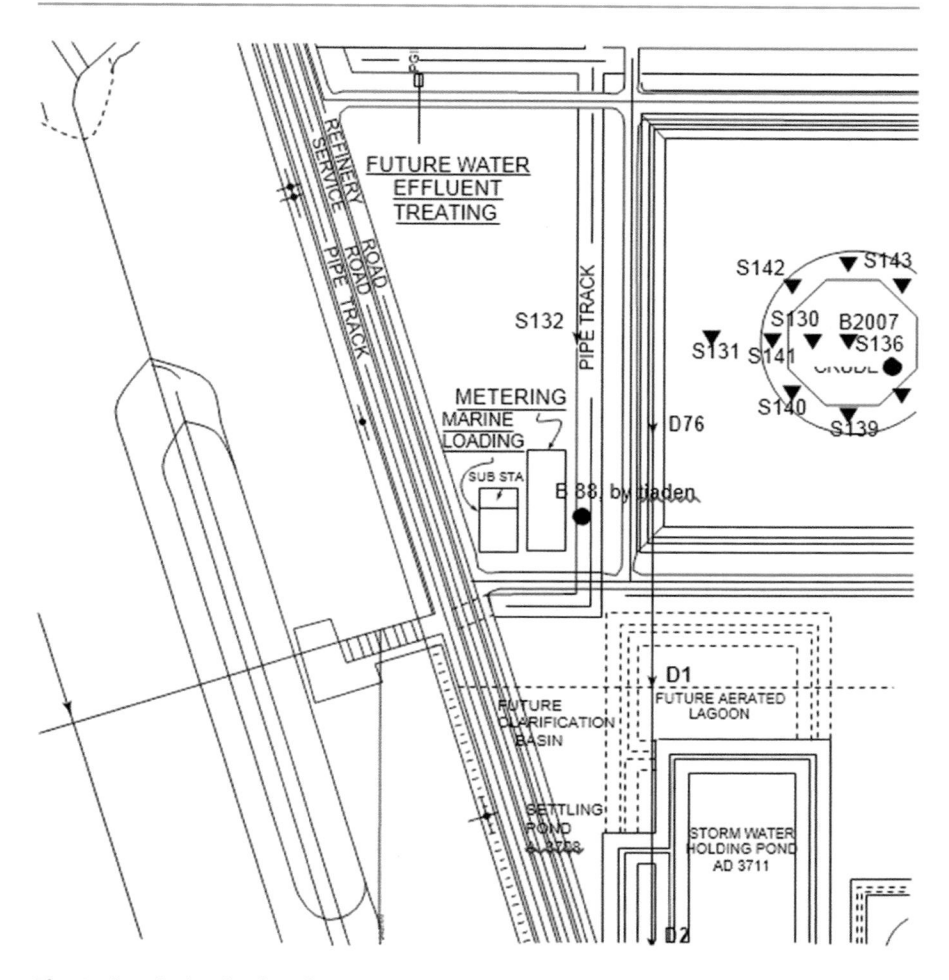

Fig. 9 A typical tanker jetty layout

Barges are usually loaded through hoses supported by dockside derricks or, in some cases, by derricks on the barges. To conserve space, barges are frequently moored two or three abreast with the loading hose being manhandled across the inboard barges to the outboard barges. This hose is accordingly limited in size to a 6 in. diameter weighing about 8 lb/ft.

Loading Rates

Tanker piping and pumping systems are designed for relatively high rates. The unloading pumps for the T2 tanker size can handle up to 8,000 bph. Supertankers unloading crude have pumps that can handle quantities of 22,000 bph or more.

A desirable product loading rate for each product is 8,000–10,000 bph. It is common practice to load two products simultaneously. Loading rates for barges are usually limited to 2,000 bph. Barges have capacities ranging from 600 to 2,000 bbl. Barges are flat-bottomed shallow draft vessels used to transport products short distances in canals, harbors, or inland waterways.

Quantities loaded aboard tankers are measured by metering storage tanks and tanker compartments. Products loaded into barges are measured by metering. Meters are located on the dock near the hose connections. These metering facilities include strainers and flow controllers, similar to truck or rail loading.

Other Features

Some of the other features that are considered in the establishing of jetty facilities are:

- Ship ballast handling
- Vented gas handling or incineration
- Tanker mooring facilities
- Slop and spill collection facilities
- Lighting and communication facilities

Ship ballast water is handled using specially allocated onshore tanks to collect the water, which will be contaminated. This contamination will be the petroleum product residue remaining in the ship's tanks after product unloading. Ships arriving at the refinery jetty under ballast are usually the smaller product tankers. These are the vessels that will load at the refinery with the finished products for shipment. The ballast water is pumped from the ship's tanks to the onshore ballast water tankage by the ship's pumps. From the ballast water tanks, the content is drained off to the refinery's effluent treating facilities. The hydrocarbon contaminants are removed from the water in the treating plant and routed to the refineries' slop/recovered oil system to finally enter the refinery processes. The treated effluent water may be drained back to the sea, unless prohibited by regulation.

As the tanker or barge is filled, gases in the holds have to be vented. In many locals, this vented gas must be incinerated or flared by regulation. A packaged incinerator, with appropriate flame arrestors and backflow prevention, is usually used. Many modern tankers contain inert gas systems which provide a blanket on the tanks using engine exhaust gases. These blanket gases will not readily support combustion, so any incinerator will require supplemental fuel gas to ensure complete combustion.

The length of the jetty's loading/unloading wharf, where the ships are moored, is sized to accommodate two or more tankers of fixed length (say two T2s). The allocated space for these vessels must conform to standard conditions usually established by the particular Port Authority Regulations. One of these regulations, which affects the length of the wharf, is that the space between moored vessels should be such that the stern and aft mooring lines of adjacent vessels, measured at an angle of 45° to the center line of the vessels, cannot overlap.

Slop and spill facilities around loading or unloading vessels at the wharf may include a temporary boom installed around the vessel during these operations. Any spillage is contained by the boom and is subsequently disposed of by the same route as the ballast water.

Jetty lighting is based on the main refinery lighting code and practice. This means that all access ways and roads will have general street lighting. Areas where personnel are employed on a 24-h basis will be floodlit between the hours of sunset and sunrise. This lighting will be supplemented by the ship's lighting facilities as required for loading/unloading activities and for ship berthing and departure. Ship to onshore communication by means of telephone, radio, and company computer systems is activated as soon as the ship has been berthed.

Waste Hydrocarbon Disposal Facilities

Although every attempt is made for economic and regulatory reasons to minimize waste hydrocarbon gases and oils, all process plants, including oil refineries, produce quantities of toxic and/or flammable material during periods of plant startup, shutdown, upset, or emergencies. A properly designed flare and slop handling system is therefore essential to the plant operation. This section describes and discusses typical disposal systems currently in use in the oil refining industry where the hydrocarbon is immiscible with water. Where the chemical is miscible in water, special separation systems must be used.

Figure 10 shows a completely integrated waste disposal system for the light ends section of an oil refinery. The system shown here consists of three separate collection systems being integrated to a flare and a slop (or recovered oil in more recent terminology) rerun system.

A fourth system is for the disposal of the oily water drainage with a connection to the flare and a separate connection for any oil-laden skimming. This later connection would be to route the skimmed oil to the refinery slop tanks.

In the three integrated systems, the first collects all the vapor effluent streams from the relief headers. The contents of this stream will be material that is normally vapor at ambient conditions. It would be the collection of the vapors from the relief valves and the vapor venting on plant shutdown or upset conditions.

The second of the three systems is the liquid hydrocarbon drainage. The material in this system is liquid under normal ambient conditions and is collected from drain headers used to empty vessels during shutdown or upset conditions.

Both the first and the second collection systems are routed to the flare knockout drum. The second (liquid system) may also be routed to the light ends slop storage drum. The liquid phase from the flare knockout drum is also routed to the slop storage drum.

The third system is the light ends feed diversion. This allows the light ends unit to be bypassed temporarily by sending the feed to the slop drum for rerunning later.

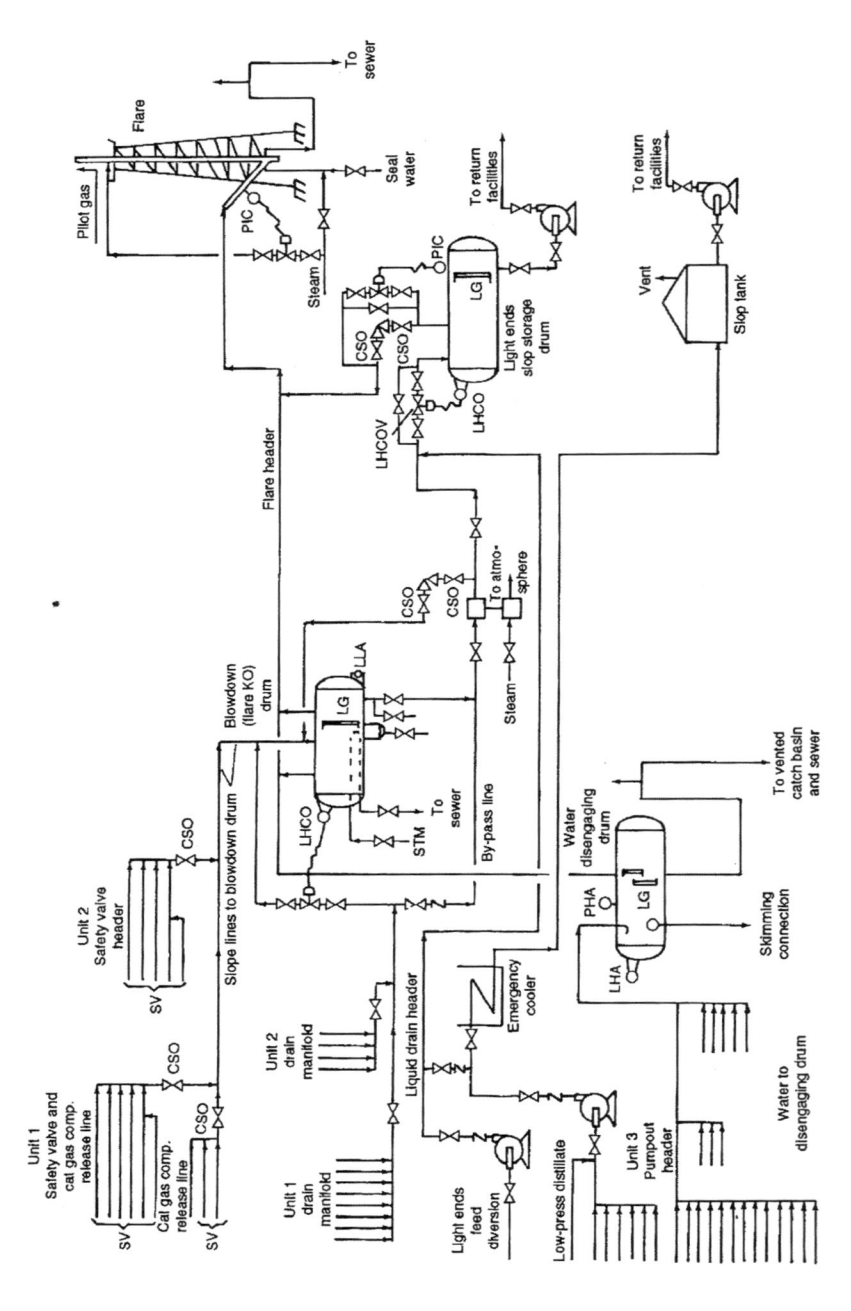

Fig. 10 An integrated waste disposal system

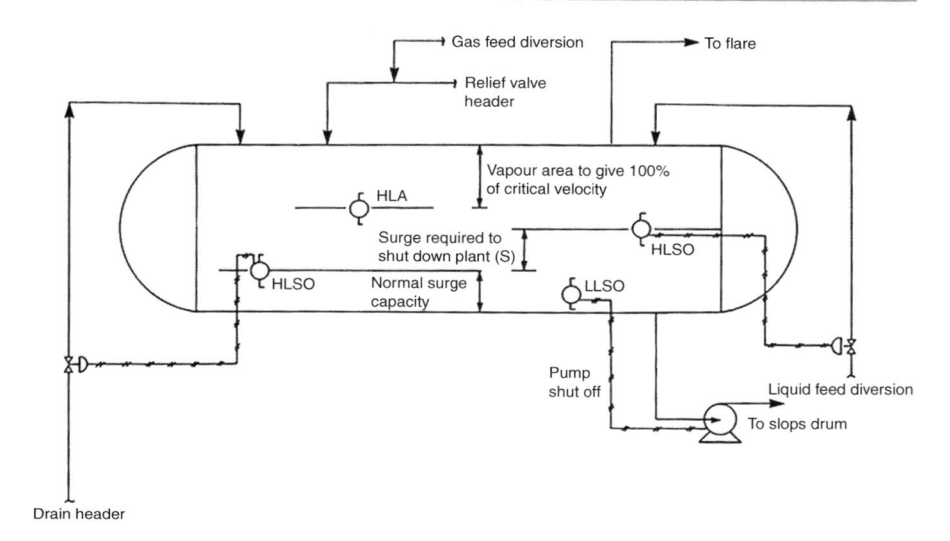

Off-Site Facilities for Petroleum Processing 1191

Fig. 11 A noncondensable blowdown drum

Further description and discussion of these disposal systems is given in the following sections:

- Blowdown and slop disposal
- Flares

Blowdown and Slop/Recovered Oil

This system generally consists of the following drums:

- Noncondensable blowdown drum
- Condensable blowdown drum
- Water disengaging drum

A typical noncondensable blowdown drum is shown in Fig. 11. These types of drums are provided for handling material normally in the vapor state and high volatility liquids. These drums receive and disengage liquid from safety valve headers and drain headers.

Blowdown drums are often referred to as flare knockout drums as the disengaged vapor is routed directly to a flare. The drum is basically a surge drum and therefore should be sized as one using the following criteria:

1. Normal liquid surge is based on the daily liquid draw-off to drain per operating day of 24 h. This includes spillage, sample point draining, etc.

2. The surge capacity between the HLSO (high level shutoff) for normal drainage and the HLSO for feed diversion should be such as to contain the total feed to a unit routed to this drum for a period sufficient to shutdown the unit producing the feed stream. Should there be more than one unit routed to the drum, then this surge capacity should be for the largest of a single feed stream.
3. The capacity between the highest HLSO and the high level alarm (HLA) should be sized to handle the largest liquid volume that can be discharged in 30 min by the relief valves constituting any single risk.
4. The drum must be sized for a vapor velocity above the HLA at a maximum of 100 % of the critical figure calculated by

$$V = 0.157[(\rho_l - \rho_v)/\rho_v]^{0.5} \tag{2}$$

where:
V = critical velocity in ft/s.
ρ_l = liquid density in lbs/cuft.
ρ_v = vapor density in lbs/cuft at drum temperature and pressure.

5. Drum pressure: The operating pressure for this drum will be about 0.5 psig or that of the water disengaging drum tied to the same flare header. In relief conditions, the pressure in the drum may be substantially higher, however. Typically, design pressure for these drums is 50 psig.

A condensable blowdown drum and system are used for collection and containment of heavier hydrocarbons with low volatility. For example, this would account for the middle and waxy distillates (kerosene, gas oils, and the like).

Figure 12 illustrates a typical blowdown drum with quench. The material entering this system is generally above ambient temperatures. Very often hot streams directly from operating units find their way into this system. To handle these materials the condensable blowdown drum is designed as a direct contact quench drum. The blowdown material leaves the unit in a drain collection system to enter the bottom section of the drum. Cooling water is introduced at the top of the drum and passes over a baffled tray section to contact the hot blowdown stream at the drum base. Any hot vapors rising from the blowdown stream are condensed in the baffle section of the drum and carried down to the bottom of the drum. Uncondensed material leaving the top of the vessel is routed to the flare. The aqueous mixture containing the condensed blowdown leaves the bottom of the drum through a seal system to enter the chemical or oily water sewer for separation and treatment.

This design works if the fluids in the drum are below about 212 °F. If the collected fluids are above 212 °F, then you absolutely do not want to introduce water into the drum. This would cause a steam explosion and likely rupture the drum. In systems where the fluids are very hot, the blowdown drum is provided with cooling by an external circulation loop. The drum may be sized to hold the largest "hot blow" volume until the collected fluid can be cooled to a temperature at which it can be sent to slop or recovered oil.

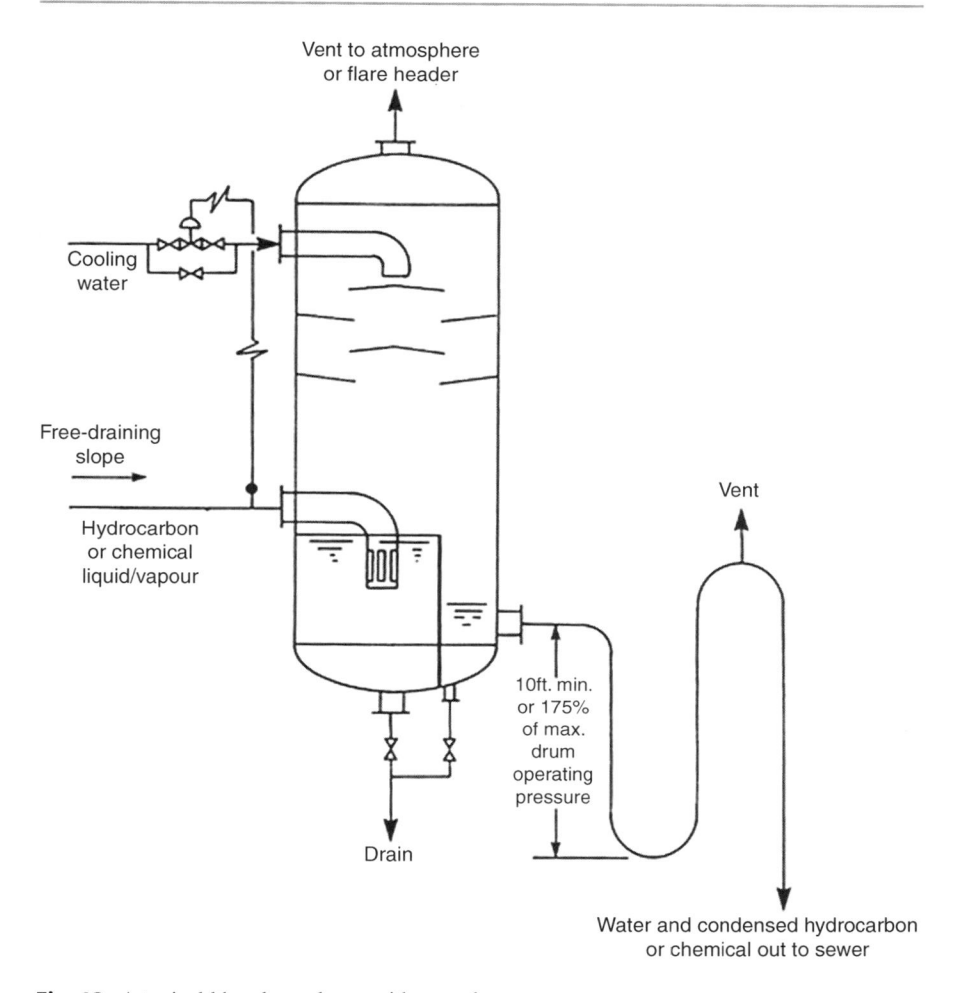

Off-Site Facilities for Petroleum Processing   1193

Fig. 12 A typical blowdown drum with quench

The following criteria are used to size a blowdown vessel with internal quench:

- The vapor load on the drum is based on the safety valve(s) constituting the largest single risk.
- The maximum operating pressure for the drum is usually 1–2 psig.
- The stack may vent to the atmosphere rather than the flare if desired. However, if vented to the atmosphere, the stack should vent at least 10 f. above the highest adjacent structure. Unfired vent stacks are strongly discouraged after multiple fatality incidents related to these stacks. In any case the vent should not release to atmosphere below 50 f. above grade. Snuffing steam should also be provided.
- The cooled effluent leaving the drum should be at 150 °F or colder. The cold water supply should be controlled either by effluent temperature or inlet

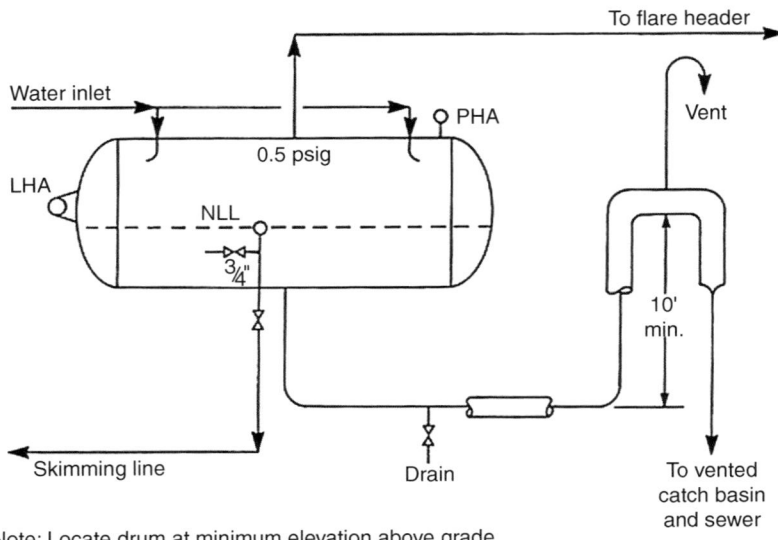

Fig. 13 A water separation drum

blowdown stream flow. There should, however, be a bypass flow of water entering the drum at all times.
- The drain system from the unit(s) to the drum should be free draining into the drum. The drum therefore should be located at a minimum height to grade. Where very waxy materials are likely to be handled, steam tracing of lines and a steam coil in the drum should be considered.

The water separation drum arrangement is shown in Fig. 13. The purpose of this drum is to remove any volatile and combustible material from certain water effluent streams before they enter the sewer or sour water system.

Thus:

- All water from distillate drums which have been in direct contact with flammable material such as light hydrocarbons is sent to the disengaging drum before disposal to the sewer. The exception is where a sour water stripper is included in the plant; then these streams are sent to this stripper.
- Cooling water drainage from coolers and condensers which may have been contaminated with flammable high volatility material is sent to the disengaging drum. So, too, are steam condensate streams which fall in the same category.

The drum is located at a minimum height above grade. It operates at about 0.5 psig and vents into the flare system. The pressure and the liquid level in the drum are maintained by the free draining of the effluent through a suitable seal. For sour water flash drums, the water would be pumped to sour water tankage.

Design criteria used for the sizing of this drum are as follows:

- The vapor load on the drum will be the result of high volatile material flashing to equilibrium conditions at the drum pressure. This design load is based on the largest amount of vapor arising from a single contingency. For exchangers this contingency will be due to a fractured tube. For liquid from a distillate drum, this will be due to a failed open control valve on the water outlet.
- The liquid seal must be such as to eliminate air from the sewer system and to allow free drainage from the drum.
- An oil or chemical skimming valve is located at the water NLL. This allows for the draw-off of the oil phase from time to time. A high interface level alarm is often included.

The Flare

Vapors collected in a closed safety system are disposed of by burning at a safe location. The facilities used for this burning are called flares. The most common of these flares used in industry today are:

- The elevated flare
- The multi-jet ground flare

The elevated flare is used where some degree of smoke abatement is required. The flare itself operates from the top of a stack, usually in excess of 150 f. high. Steam is injected into the gas stream to be burnt to complete combustion and thereby reduce the smoke emission.

The multi-jet ground flare is selected where luminosity is a problem, for example, at locations near housing sites. In this type of flare, the vapors are burned within the flare stack thus considerably reducing the luminosity. Steam is again used in this type of flare to reduce the smoke emission.

Figure 14 shows a typical arrangement of an elevated flare, and Figs. 15 and 16 show that for a multi-jet ground flare.

The Elevated Flare

This type of flare is the normal choice in the larger process industries such as the petroleum refining industry. It consists of a flare stack over 150 f. in height that contains an igniter system, pilot flame(s), and the flare pipe itself. The flare header enters the stack through a water seal at the base of the stack immediately above an anchor or concrete plinth. The water seal maintains a back pressure of around 0.5 psig on the flare header. The waste gas to be flared moves up the stack to exit at the top.

At the stack top there is an assembly of igniter and pilot gas which ensures the safe burning of the waste flare gases. A typical pilot assembly is shown in Fig. 17. It consists of three tubes all external to the stack itself and each supplied with the

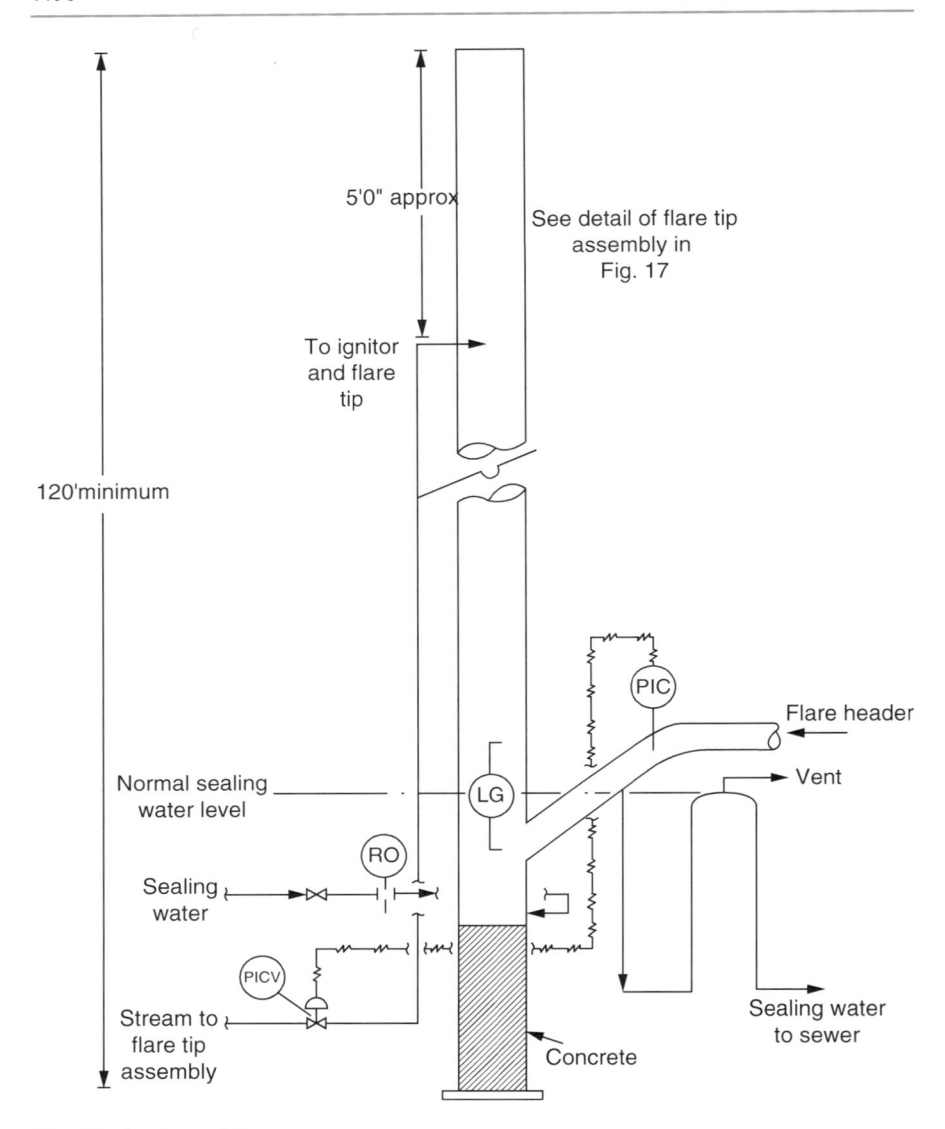

Fig. 14 An elevated flare

plant fuel gas. In some cases, the first and largest of these tubes is the igniter. Here the fuel gas supply is mixed with air (plant or instrument air supply) before passing upward through a venturi tube to an igniter chamber. A spark is induced in the igniter chamber by an electric current of 15 Amp. The chamber and the venturi tube are located near grade, and a sight glass on the igniter chamber enables the operator to check on the igniter's operation. The flame front from the igniter travels up the igniter tube to contact the waste gases that are to be flared as they exit the top of the

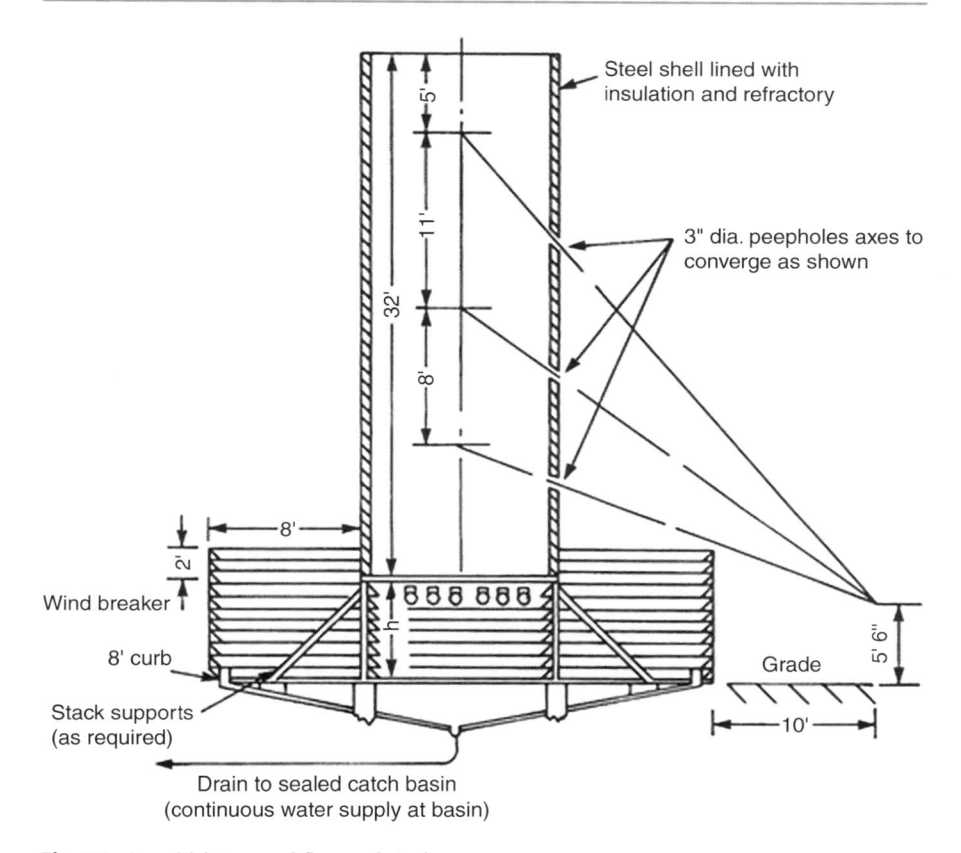

Fig. 15 A multi-jet ground flare and stack

flare stack. The same flame front ignites the "on and off" pilot burner which is the center tube of the three and initially ignites the permanent pilot burner(s) at the stack top.

The outlets of these three tubes are located at the stack top such that the prevailing wind ensures that the flame from them is blown across the stack exit.

Other configurations also exist for igniting the pilots. In one common system, there is a direct spark ignition at each pilot tip that is automatically triggered every few seconds to ensure each pilot stays lit.

Steam is often injected into the stack at some point near the top to complete combustion and eliminate or at least reduce smoke emission. The amount of steam normally used for this purpose depends on the character or composition of the waste gases. Aromatics and olefins when burnt produce a smoky flame: steam injection allows the free carbon which makes up the smoke to convert into CO and CO_2 which, of course, are invisible gases. An estimate of the amount of steam required for smoke abatement is given in Fig. 18. From a practical standpoint, steam rates above 1.0 lb steam per lb vented gas are normally unrealistic.

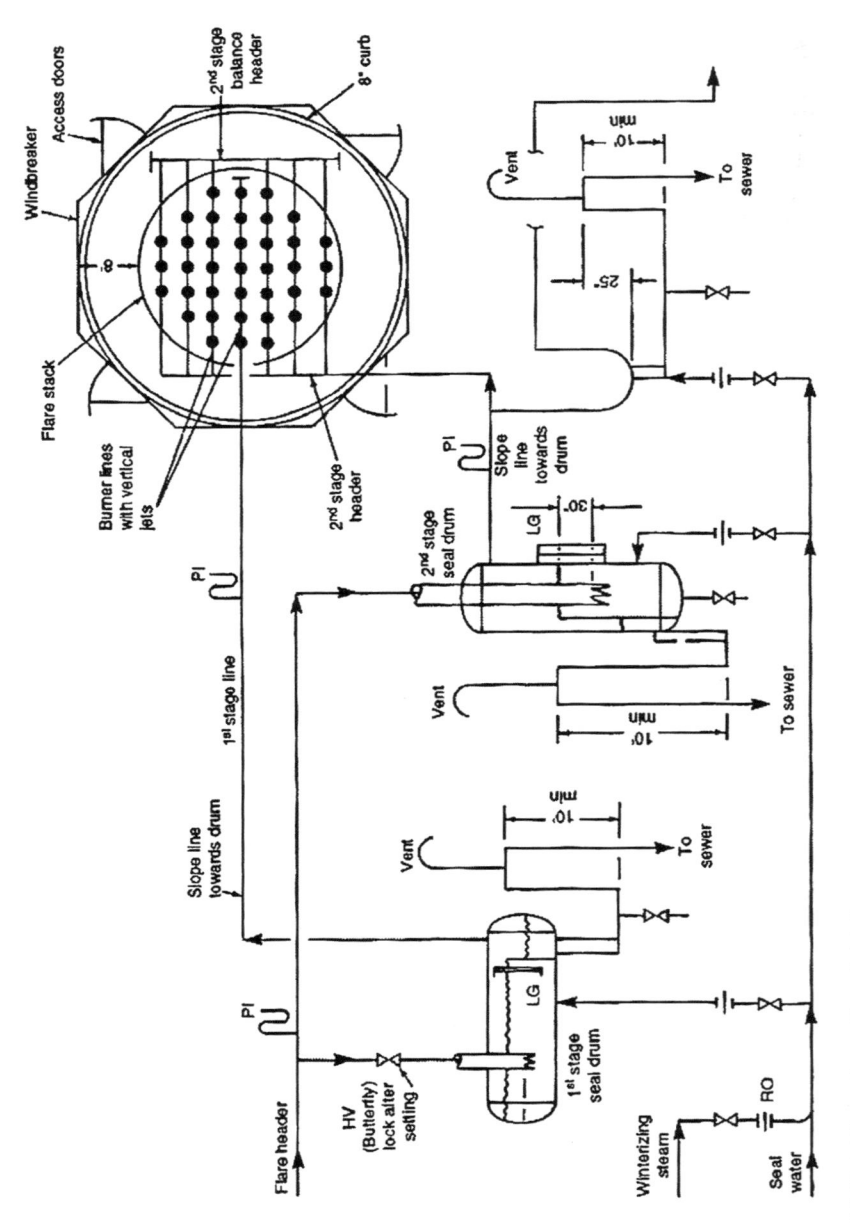

Fig. 16 Multi-jet flare plan with seal details

Off-Site Facilities for Petroleum Processing

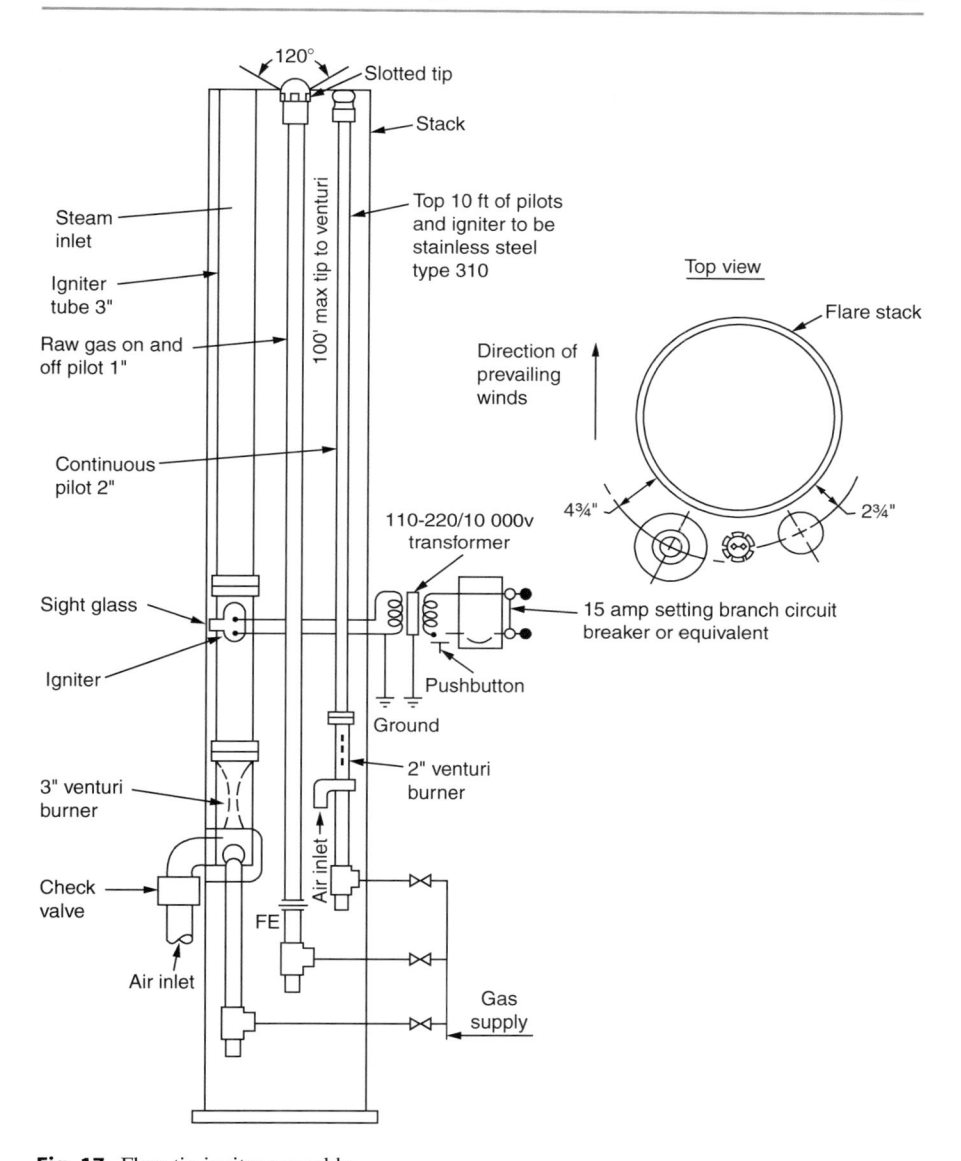

Fig. 17 Flare tip igniter assembly

A clear space around an elevated flare is required to allow for the effect of heat radiation from the flare to the ground. Flares which have a heat release of 300 million to 1 billion Btu/h should be located at least 200 f. from the plant property line or any pond, separator, tankage, or any equipment that could be ignited by a falling spark. The stack also must have a spacing of at least 500 f. from any structure or plant whose elevation is within 125 f. of the flare tip.

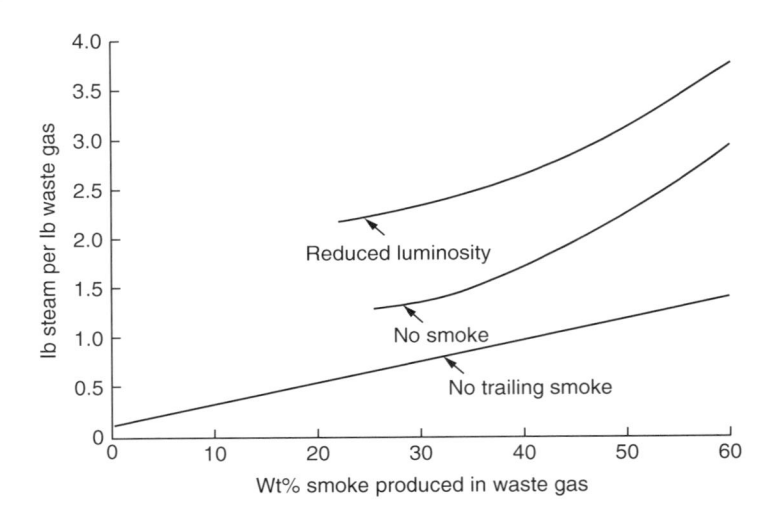

Fig. 18 Approximate amount of steam for smoke abatement

Flare tips today tend to be molecular seal types. In these proprietary tips, there is a baffle-like arrangement to trap a gas that is heavier than air, like propane, or a gas much lighter than air, like methane, in the tip to prevent air ingress back into the stack that can cause an explosion. This system is used in addition to a water seal and a purge.

The elevated flare stack is designed to maintain a gas velocity of between 100 and 160 ft/s during a major blowdown to flare. This rate is based on the maximum single emergency plus any steam added to improve the burning characteristics. Above a velocity of 160 ft/s, noise becomes a problem, and the maintenance of ignition also is dubious unless multiple ignition tubes or the intermittent spark ignition system is used. Some proprietary flare tip designs, however, do claim ability to handle satisfactorily velocities up to 400 ft/s. By API guidelines, the maximum flare tip velocity should not exceed 0.5 mach.

The Multi-jet Ground Flare

The multi-jet flare provides a completely noiseless, nonluminous flaring at a reasonable cost. At normal loads the flare is also essentially smokeless and is particularly useful where continuous flaring is required. Figures 15 and 16 show the elevation of the flare stack and the plan arrangement of a two-stage multi-jet flare, respectively.

The two-stage arrangement shown here shows the flare header being directed to one of two seal drums or to both. The first-stage seal drum operates at a back pressure of 20 in. of water at the first-stage burners at its design capacity. The second-stage burners are activated when the pressure in the flare header reaches 30 in. of water gauge.

Very often, particularly in large process complexes, the ground flare is designed to operate in conjunction with an elevated flare. The multi-jet flare takes a gas stream up to say 80 % of its rated capacity. Additional flow is then diverted to an elevated flare system. Thus, if there is need for the continuous flaring of a reasonably small quantity, the ground flare accommodates it. In an emergency or surge, the elevated flare comes into operation automatically to take the additional load.

The burners of a multi-jet flare are jet nozzles approximately 15 in. in length of 1 in. diameter stainless steel pipe. They discharge vertically from the horizontal burner lines which run across the bottom of the stack. The number of jets is based on gas velocity and is expressed by the equation:

$$N = 16.4\,V \qquad (3)$$

where:

N = number of jets.
V = flare design capacity in MMscf/day.

The jets are placed on a square or rectangular pitch of 18–24 in. A first estimate of the required pitch may be obtained from the expression:

$$P = (100 \times D^2)/(N \times C) \qquad (4)$$

where:

P = pitch in inches.
D = stack ID in feet.
N = number of jets.
C = distance between burner center line in inches.

No jet should be placed closer than 12 in. from the inside of the stack.

The inside diameter of the stack is based on the rate of heat release at design capacity. It is calculated using the following equation:

$$D = 0.826\ Q^5 \qquad (5)$$

where:

D = stack inside diameter in feet.
Q = heat release at max design in MM Btu/h.

The stack height for diameters up to 25 f. is 32 ft, and the steel shell of the stack is lined with 4" of refractory material. A windbreaker completes the construction of the stack. This is necessary to prevent high wind gusts from extinguishing the flames.

Flame holders are installed above the burners to prevent the flames "riding" up to the top of the stack. These are simply solid rods of 1" refractory material supported horizontally above each burner line. The position of these flame holders relative to the bottom of the stack is critical to the proper operation of the burners. The stack itself is elevated to allow air for combustion to enter. The minimum space between grade and the bottom of the stack is set at 6 f. or 0.3D whichever is the larger.

As for any flare, a continuous pilot burner is recommended. The proper operation of this pilot is important with respect to multi-jet-type flares because of the danger of un-burnt flammable material escaping outside the flare at ground level. A gas pilot is provided at each end of the primary burner to minimize this risk.

Effluent Water Treating Facilities

This section of the "off-site system" discussion deals with the treating of wastewater accumulated in a process complex before it leaves the complex. Over the years requirements for safeguarding the environment have demanded close control on the quality of effluents discharged from chemical and oil refining plants. This includes effluents which contain contaminants that can affect the quality of the atmosphere and those that can be injurious to plant and aquatic life. Effluent management in the oil industry has therefore acquired a position of importance and responsibility to meet these environmental control demands.

Water effluents that are discharged from the process and other units are collected for treating and removal or conversion of the injurious contaminants. In many oil refineries, imported water, in the form of ship's ballast water, is also collected on shore for treatment before discharging back to the sea. Figure 19 is a schematic of the water effluent treating system for a major European oil refinery.

Normally the water effluent treating facilities for a complex would be located at the lowest geographical point in the plant. In this way very little pumping is required to move the wastewater to and from the treating plants. The schematic in Fig. 19 is for a refinery that was sited below the sea level, so more pumps are used than would be typical.

The primary contaminant to be removed in the system shown in Fig. 19 is, of course, oil. Five separate systems are used in this refinery's treatment plant.

The first is that for handling ballast water from sea-going tankers. The second is the handling of clean water. This is included because the system bypasses all the treating processes except the last "guard" process which, in this case, are the retention ponds. The third system is also for handling non-oily water but water that would be high in certain chemicals. This system also discharges into retention or storm water ponds. The water is held in these ponds to ensure that there is no contamination. If the storm water is contaminated, it would be returned into the appropriate treating process(es) for removal of the contaminants.

The last two systems shown are for the handling of contaminated water from the refinery's paved areas, various tank and process plant drainage, and the like. These oily water systems and the ballast water stream are treated for oil removal. In the case of the ballast water, the water drained from the bottom of the holding tanks is routed through an API separator. This is a specially designed pond that reduces the

Off-Site Facilities for Petroleum Processing

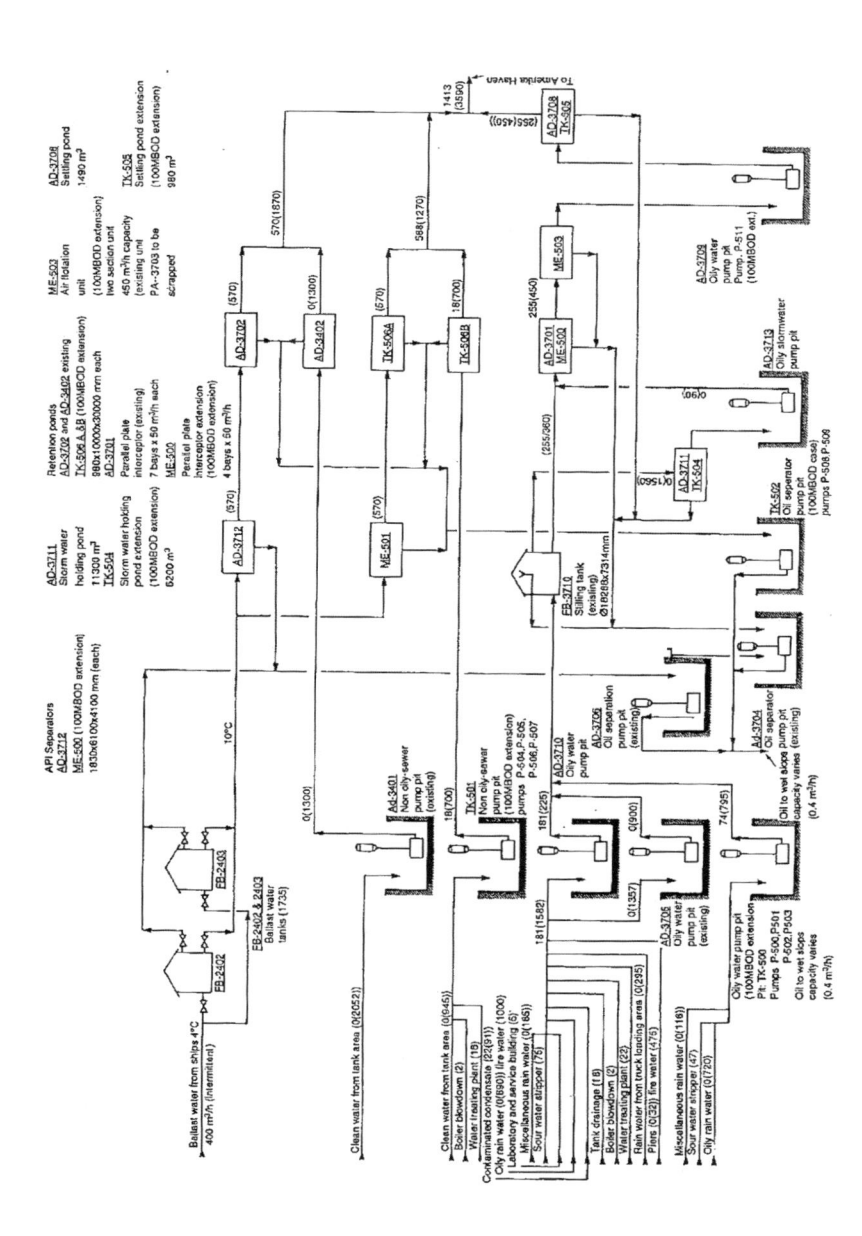

Fig. 19 A schematic of a water effluent treating system

forward velocity of the water stream to allow the separation of oil from the water by settling or gravity.

The water/oil separation for the other refinery streams takes place in a series of settling ponds. Final cleanup in this case is accomplished by the use of parallel plate interceptors and an air flotation process. The principle of the parallel plate interceptor is to force the water stream to change direction several times in rapid sequence and thus "knocking out" any oil entrained in the stream. The air flotation unit causes the contaminated water stream to be agitated so as to force the lighter oil phase to the surface where it can be removed by skimming or by baffled overflow.

There is often sludge accompanying the water. API separators and plate interceptors have provisions to collect the sludge for disposal. In many cases, the sludge is sent to the coker unit for disposal. In the US, API separator sludge is a hazardous waste by definition. Effluent and sour water treatment processes are discussed in much more detail in the handbook chapter "▶ Environmental Control and Engineering in Petroleum Processing."

Other Effluent Water Treating Systems

Most chemical plants and indeed a few oil refining plants require more complex methods for clarifying their effluent water to meet environmental requirements for its disposal. The four more common methods are:

- In-line clarification using coagulation, flocculation, and filtration
- Plain filtration
- Sedimentation
- Chemically aided sedimentation using coagulation, flocculation, and settling

Clarification is a process that removes suspended (usually organic) matter that gives the stream color and turbidity. The removal of this matter, especially in a colloidal form, requires the addition of chemicals to cause coagulation and flocculation to promote settling and separation of suspended solids. Coagulants and coagulant aids added to the influent stream chemically react with impurities to form precipitates. These, together with particles of enmeshed turbidity, are flocculated into larger masses that are then readily separated from the bulk liquid.

There are essentially three steps in the chemically aided clarification process. These are:

- Mixing of the additives
- Flocculation
- Settling

Coagulation encompasses the process of mixing the contaminated water with additives to form agglomerates, or floc. This is carried out in a series of separate compartments, with the settling basin occupying the largest volume. Coagulation is the singular most important step in the clarifying process.

Because it involves the buildup of colloidal-type particles, the chemicals and the process rate are specific to the material that is to be clarified. There are companies that specialize in the design construction and the operation of this type of effluent treating. These companies use their experience in handling the complex electrochemical kinetics associated with flocculation and coagulation principles.

Biochemical Oxygen Demand Reduction

As discussed above, the primary function of effluent treating is removal of oil and other contaminants from refinery wastewaters. In many locations, it is also necessary for the refinery to reduce the BOD or COD of the refinery effluent before discharge to a waterway or before sending the effluent to a publically owned treatment works (POTW).

This is usually accomplished by processing the water through a biological digestion system. Systems that have been used for this purpose have included:

- Trickle filters
- Activated sludge units
- Activated carbon
- Aerated ponds

The design of these systems is beyond the scope of the current Handbook.

Sanitary Sewage Treatment

In some remote refinery locations, the facility must provide its own sanitary sewage disposal. This is normally accomplished with a packaged bio-treatment unit and/or aerated holding basins. A trickle filter may also be used. We will not discuss these systems in more detail here, as they are not common.

Appendix 1: Calculating Heat Loss and Heater Size for a Tank

Heat loss and the heater surface area to compensate for the heat loss may be calculated using the following procedure:

Step 1. Establish the bulk temperature for the tank contents. Determine the ambient air temperature and the wind velocity normal for the area in which the tank is to be sited.

Step 2. Calculate the inside film resistance to heat transfer between the tank contents and the tank wall. The following simplified equation may be used for this:

$$h_c = 8.5(\Delta t/\mu)^{0.25}$$

where:
h_c = inside film resistance to wall in Btu/h sqft °F.
Δt = temperature difference between the tank contents and the wall in °F.
μ = viscosity of the tank contents at the bulk temperature in cPs.

The heat loss calculation is iterative with assumed temperatures being made for the tank wall.

Step 3. Using the assumed wall temperature made in step 2, calculate the heat loss to the atmosphere by radiation using Fig. 20. Then calculate the convection heat loss from the tank wall to the atmosphere using Fig. 21. Note the temperature difference in this case is that between the assumed wall temperature and the ambient air temperature. Correct these figures by multiplying the radiation loss by the emissivity factor given in Fig. 20. Then correct the heat loss by convection figure by the factors as described in step 4 below.

Step 4. The value of h_{co} read from Fig. 21 is corrected for wind velocity and for shape (vertical or horizontal) by multiplying by the following shape factors:

Vertical plates	1.3
Horizontal plates	2.0 (facing up)
	1.2 (facing down)

For the wind velocity correction, use:

$$F_w = F_1 + F_2$$

where:
F_w = wind correction factor.
F_1 = wind factor at 200 °F calculated from:
$F_1 = (MPH/1.47)^{0.61}$
F_2 = read from Fig. 22
Then the corrected h_{co} is
h_{co} × shape correction × F_w.

Step 5. The resistance of heat transferred from the bulk of the contents to the wall must equal the heat transferred from the wall to the atmosphere. Thus:
Heat transferred from the bulk to the wall = "a"

$$= h_c \text{ from step 2} \times \Delta t \text{ in Btu/h/sqft}$$

where Δt in this case is (bulk temp – assumed wall temp).

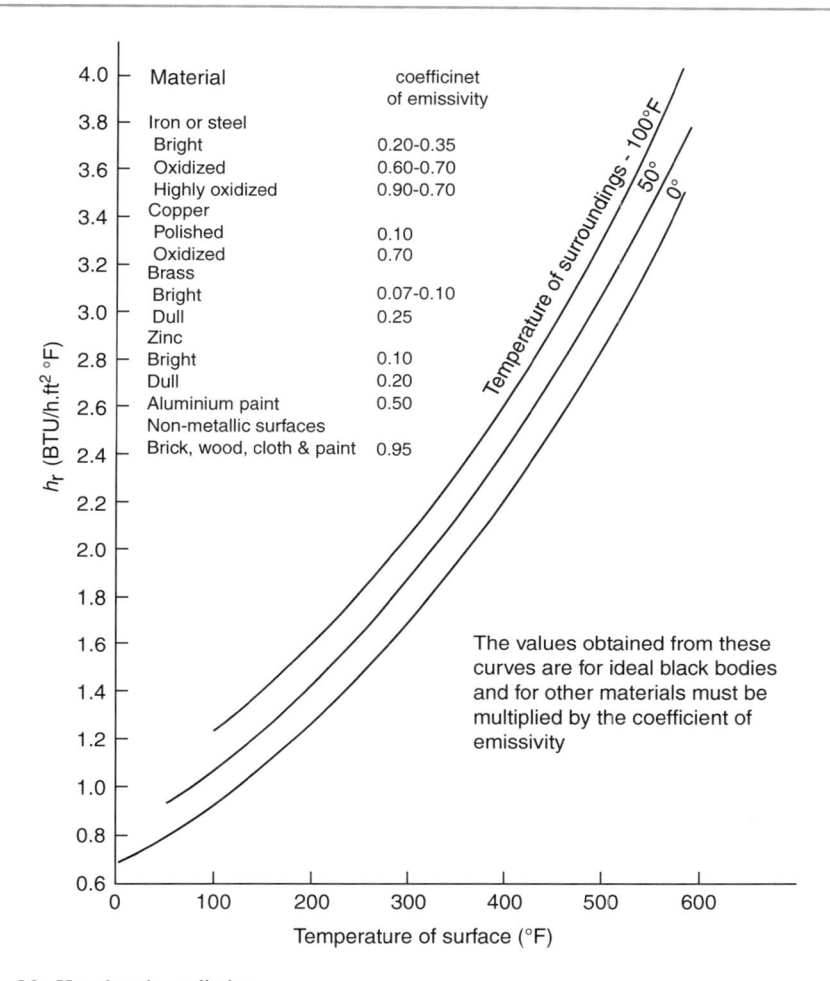

Fig. 20 Heat loss by radiation

Material	coefficient of emissivity
Iron or steel	
Bright	0.20-0.35
Oxidized	0.60-0.70
Highly oxidized	0.90-0.70
Copper	
Polished	0.10
Oxidized	0.70
Brass	
Bright	0.07-0.10
Dull	0.25
Zinc	
Bright	0.10
Dull	0.20
Aluminium paint	0.50
Non-metallic surfaces	
Brick, wood, cloth & paint	0.95

The values obtained from these curves are for ideal black bodies and for other materials must be multiplied by the coefficient of emissivity

Axes: h_r (BTU/h·ft²·°F) vs Temperature of surface (°F); curves labeled Temperature of surroundings – 100°F, 50°, 0°.

Heat transferred from the wall to the atmosphere = "b"

$$= (h_{co} + h_r) \times \Delta t \text{ in Btu/h/sqft}$$

where Δt in this case is (assumed wall temp − air temp).

Step 6. Plot the difference between the two transfer rates against the assumed wall temperature. This difference ("a" − "b") will be negative or positive, but the wall temperature that is correct will be the one in which the difference plotted = 0. Make a last check calculation using this value for the wall temperature.

Step 7. The total heat loss from the wall of the tank is the value of "a" or "b" calculated in step 6 times the surface area of the tank wall. Thus:

Fig. 21 Heat loss to the atmosphere by natural convection

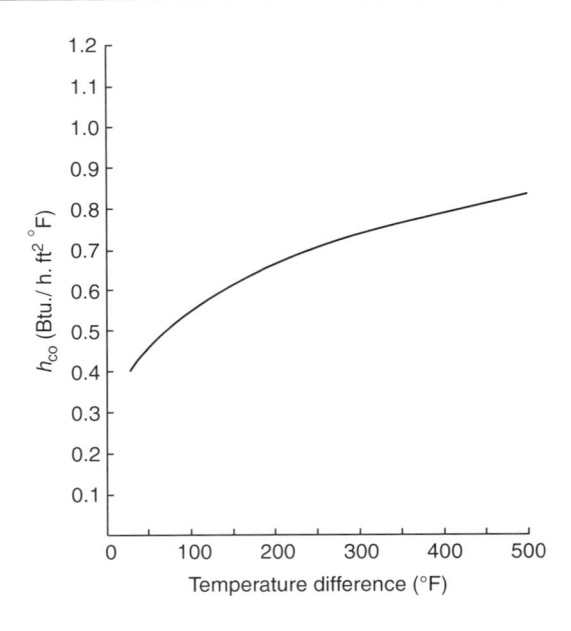

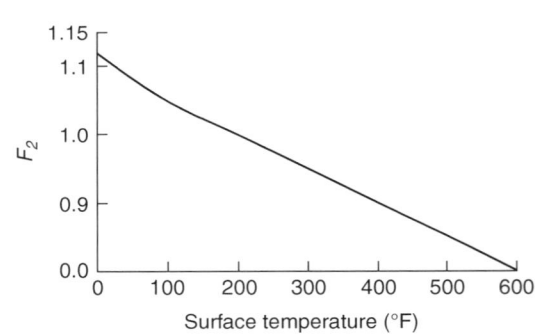

Fig. 22 Plot of "F2" versus surface temperature

$$Q_{wall} = h_c \times \Delta t \times (\pi D_{tank} \times \text{tank height}) \text{ in Btu/h}$$

Step 8. Calculate the heat loss from the roof in the same manner as that for the wall described in steps 2–7. Note the correction for shape factor in this case will be for horizontal plates facing upward and the surface area will be that for the roof.

Step 9. Calculate the heat loss through the floor of the tank by assuming the ground temperature as 50 °F and using

$$h_f = 1.5 \text{ Btu/h sqft °F}$$

Step 10. The total heat loss then is:

$$\text{Total heat loss from tank} = Q_{wall} + Q_{roof} + Q_{floor}$$

Off-Site Facilities for Petroleum Processing

Step 11. Establish the heating medium to be used. Usually this is medium pressure steam.

Calculate the resistance to heat transfer of the heating medium to the outside of the heating coil or tubes. If steam is used then take the condensing steam value for h as 0.001 Btu/h sqft °F. Take the value of steam fouling as 0.0005 and tube metal resistance as 0.0005, also.

The outside fouling factor is selected from the following:
Light hydrocarbon = 0.0013.
Medium hydrocarbon = 0.002.
Heavy hydrocarbons, such as fuel oils = 0.005.
The resistance of the steam to the tube outside = $1/(h + R)$
where $R = r_{steamfouling} + r_{tube\ metal} + r_{outside\ fouling}$

Step 12. Assume a coil outside temperature. Then, using the same type of iterative calculation as for heat loss, calculate for "a" as the heat from the steam to the coil outside surface in Btu/h/sqft. That is,

$$"a" = h \times \Delta t_i$$

Calculate for "b" as the heat from the coil outside surface to the bulk of the tank contents.

Use Fig. 23 to obtain h_o, and again "b" is $h_o \times \Delta t_o$ where the Δt_o is the temperature between the tube outside and that of the bulk tank contents. Make further assumptions for coil outside temperature until "a" = "b".

Step 13. Use "a" or "b" from step 12, which is the rate of heat transferred from the heating medium in Btu/h/sqft, and divide this into the total heat loss calculated in step 10. The answer is the surface area of the immersed heater required for maintaining the tank content's bulk temperature.

An example calculation using this technique is given in Appendix 2 of this chapter.

Appendix 2: Example Calculation for Sizing a Tank Heater

Problem We need to calculate the surface area for a heating coil which will maintain the bulk temperature of fuel oil in a cone-roof tank at a temperature of 150 °F. The ambient air temperature is an average 65 °F and the wind velocity averaged over the year is 30 MPH. The fuel oil data are as follows:

Viscosity (μ) = 36 cPs at 150 °F.
SG at 150 °F = 0.900.

The tank is to be heated with 125 psig saturated steam. The tank dimensions are 60 f. diameter by 180 f. high. It is not insulated, but is painted with nonmetallic color paint.

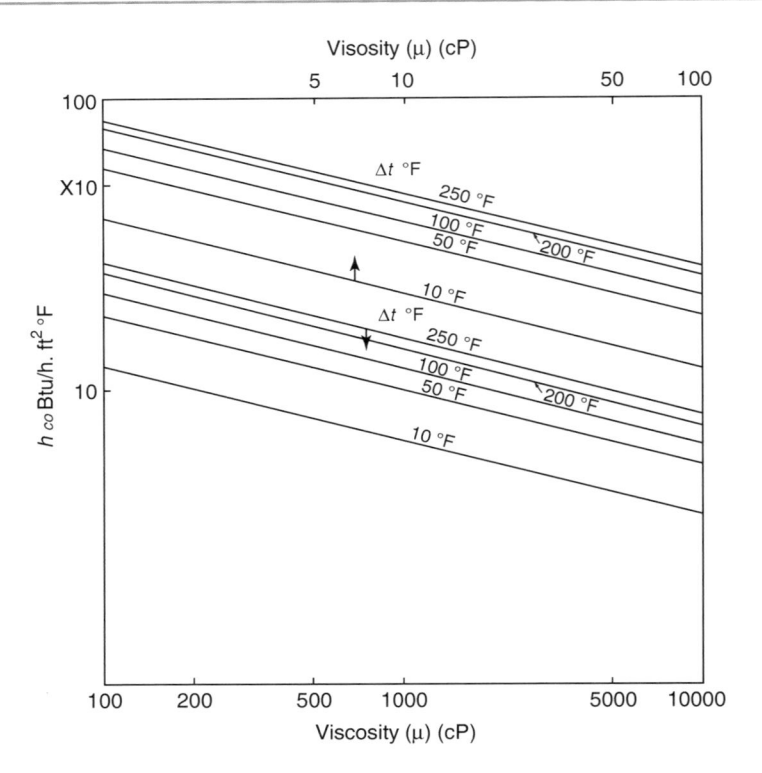

Fig. 23 Convection heat transfer coefficient

Solution (Note: figure numbers refer to Appendix 1 in this chapter)
1.0 Calculating the heat loss from the wall
1st trial. Assume wall temperature is 120 °F.

$$h_i = 8.5(\Delta t_i/\mu)^{0.25}$$

where:

$$\Delta t_i = 150 - 120 = 30 \ °F.$$

$$h_i = 8.5 \times 0.955 = 8.12 \ Btu/h.sqft°F.$$

$$"a" = 8.12 \times 30 = 243.6 \ Btu/h.sqft.$$

Δt_0 is the temperature difference between assumed wall temp and the ambient air = $120 - 65 = 55 \ °F$.
$h_{co} = 0.495 \times 1.3$ (from Fig. 21).

Wind correction factor F_w is as follows:

$$F_1 = (\text{MPH}/1.47)^{0.61}$$
$$= 6.29$$
$$F_2 = 1.04 (\text{from Fig. 22})$$
$$F_w = 6.29 + 1.04 = 7.33$$

$$h_{co} \text{ (corrected)} = 0.495 \times 1.3 \times 7.33 = 4.20 \text{ Btu/h.sqft °F}$$

Heat loss from wall due to radiation h_{ro} is found from Fig. 20 =1.18 Btu/h.sqft °F. Corrected for emissivity $h_{ro} = 1.18 \times 0.95 = 1.123$ Btu/h.sqft°F:

$$\text{"}b\text{"} = (h_{co} + h_{ro}) \times \Delta t_0$$
$$= (4.21 + 1.12) \times 55 = 293 \text{ Btu/h.sqft}$$

$$\text{"}a\text{"} - \text{"}b\text{"} = 244 - 293 = -49 \text{ Btu/h.sqft}$$

2nd trial. Assume wall temperature is 110 °F.
Carrying out the same calculation procedure as for trial 1: $\text{"}a\text{"} - \text{"}b\text{"} = +172$ Btu/h.sqft

3rd trial. Assume wall temperature is 115 °F.
Again carrying out the calculation procedure as for trial 1: $\text{"}a\text{"} - \text{"}b\text{"} = +35$
The results of the above trials are plotted linearly below:

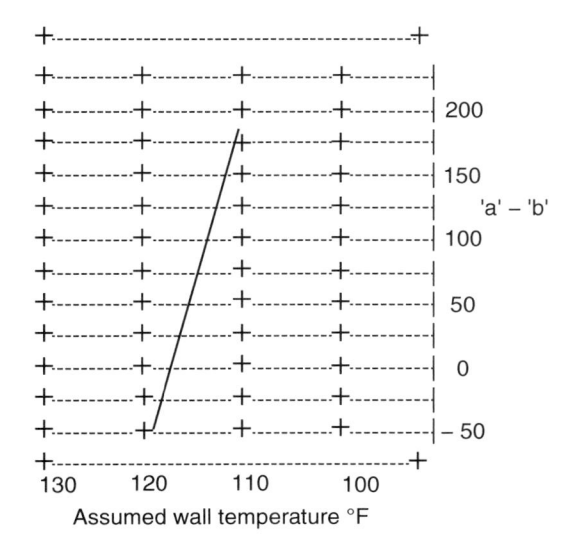

Final trial. At wall temperature of 117 °F,

$$h_{io} = 8.31 \text{ Btu/h.sqft°F}$$
$$\text{"}a\text{"} = 8.31 \times (150 - 117) = 274 \text{ Btu/h.sqft}$$
$$h_{co}(\text{corrected}) = 4.18 \text{ Btu/h.sqft°F}$$
$$h_{ro}(\text{corrected}) = 1.12 \text{ Btu/h.sqft°F}$$
$$\text{"}b\text{"} = (4.18 + 1.12) \times (117 - 65)°\text{F} = 275.7 \text{ Btu/h.sqft}$$

"a" and "b" are close enough to call the total heat loss 275.7 Btu/h.sqft:

$$\text{Surface area of wall} = \text{circumference} \times \text{height} = \pi D \times 180 \text{ ft} = 33,929 \text{ sqft}$$

$$\text{Total heat loss through wall} = 275.7 \times 33,929 = 9.35 \text{ MMBtu/h}$$

2.0 Calculating heat loss through roof
Trial 1. Assume roof temperature is 116 °F:

$$h_i = 8.38 \text{ Btu/h.sqft°F}$$

$$\text{"}a\text{"} = 8.38 \times (150 - 116)°\text{F} = 284.9 \text{ Btu/h.sqft}$$

$$h_{co} \text{ (corrected)} = (0.470 \times 2.0) \times 1.04 \times 6.29 = 6.35 \text{ Btu/h.sqft°F}$$

(Note: the number read from Fig. 21 is multiplied by 2.0 in this case as the roof is an upward facing plate.)

$$h_{ro} \text{ (corrected)} = 1.165 \times 0.95 = 1.11 \text{ Btu/h.sqft°F}$$
$$\text{"}b\text{"} = (6.35 + 1.11) \times (116 - 65)°\text{F} = 380 \text{ Btu/h.sqft}$$
$$\text{"}a\text{"} - \text{"}b\text{"} = -95 \text{ Btu/h.sqft}$$

Trial 2. Assume a wall temperature of 110 °F.
"a" − "b" in this case = +23 which is within acceptable limits.
The heat loss is taken as an average of "a" and "b" = 338 Btu/h.sqft
Total roof heat loss = roof area × 338 = 2,827 sqft × 338 = 0.956 MMBtu/h
3.0 Calculating the heat loss through the floor
Assume the ground temperature is 50 °F, and the heat transfer coefficient is 1.5 Btu/h.sqft°F:

$$\text{Then heat loss} = 1.5 \times (150 - 50) \times 2,827 \text{ sqft} = 0.424 \text{ MMBtu/h}$$

4.0 Total heat loss from the tank
Heat loss from the walls = 9.350 MMBtu/h.
Heat loss from the roof = 0.956 MMBtu/h.
Heat loss from the floor = 0.424 MMBtu/h.
Total heat loss = 10.730 MMBtu/h.
5.0 Calculating the tank heater coil surface area required
The heating medium is saturated 125 psig steam.

Temperature of the steam = 354 °F.
Steam side calculations:

Approx. resistance of steam, $h_s = 0.001$ h.sqft°F/Btu
Fouling factor on steam side, $r_1 = 0.0005$ h.sqft°F/Btu
Tube metal resistance $r_2 = 0.0005$ h.sqft°F/Btu
Outside fouling factor $r_3 = 0.005$ h.sqft°F/Btu

$$\text{Steam side heat transfer coefficient} = 1/(0.001 + 0.0005 + 0.0005 + 0.005)$$
$$= 143\,\text{Btu/h.sqft°F}.$$

Oil side heat transfer coefficient is obtained from Fig. 23.
1st trial. Assume a tube wall temperature of 310 °F.
For steam side "a" = 143 × (354 − 310) = 6,292 Btu/h.sqft.
For oil side ho = 31 Btu/h.sqft°F (Fig. 23),

$$\text{"}b\text{"} = 31 \times (310 - 150) = 4,960 \text{ Btu/h.sqft}$$

$$\text{"}a\text{"} - \text{"}b\text{"} = +1,332$$

2nd trial. Assume a tube wall temperature of 320 °F.
"a" in this case was calculated to be 4,862.
"b" was calculated to be 5,355.
"a" − "b" = −493
Plotted on a linear curve, the tube wall temperature to give "a" = "b" was 317 °F.
Final trial. At a tube wall temperature of 317 °F,

$$\text{Steam side} = 143 \times (354 - 317)$$
$$\text{"}a\text{"} = 5,291\,\text{Btu/h.sqft}.$$

Oil side ho =31.2 (From Fig. 23):

$$\text{"}b\text{"} = 31.2 \times (317 - 150) = 5,210\,\text{Btu/h.sqft}$$

"a" − "b" = +81 which is acceptable.
Make the rate of heat transfer the average of the calculated rates:

$$U = (5,291 + 5,210)/2 = 5,251 \text{ Btu/h.sqft}.$$

Coil surface area required is:

$$A = 10.73 \text{ MM Btu/h}/\,5,251\,\text{Btu/h.sqft} = 2,043\,\text{sqft}.$$

Environmental Control and Engineering in Petroleum Processing

David S. J. Jones and Steven A. Treese

Contents

Introduction	1216
Air Emissions	1218
Introduction	1218
Impacts and Sources of Refinery Emissions	1219
Controlling Air Emissions	1220
Water and Aqueous Effluents	1250
Pollutants in Aqueous Waste Streams	1250
Treating Refinery Aqueous Wastes	1256
Solid Wastes	1271
Solid Waste Disposal Regulations	1272
Some Definitions	1275
Solid Waste Management Techniques and Practices	1276
Noise Pollution	1278
Noise Problems and Typical In-Plant/Community Noise Standards	1278
Fundamentals of Acoustics and Noise Control	1280
Coping with Noise in the Design Phase	1284
A Typical Community/In-Plant Noise Program	1286
Environmental Discussion Conclusion	1289
Appendix 1 Partial Pressures of H_2S and NH_3 Over Aqueous Solutions of H_2S and NH_3	1289
Appendix 2 Example of the Design of a Sour Water Stripper with No Reflux	1289
Specifications	1289
The Design	1289

David S. J. Jones: deceased.
Steven A. Treese has retired from Phillips 66.

D.S.J. Jones
Calgary, AB, Canada

S.A. Treese (✉)
Puget Sound Investments LLC, Katy, TX, USA
e-mail: streese256@aol.com

Appendix 3 Example Design of an API Separator 1301
    Specification 1301
    The Design 1301
References 1302

Abstract
Operating a petroleum processing facility cleanly is a requirement for doing business. The environmental regulations that affect refineries and other facilities have become increasingly restrictive and the trend is expected to continue. This chapter explores the range of relevant regulations, processes, and practices that apply to modern refineries and other petroleum processing facilities for environmental management. Topics include air emissions, water effluents, solid wastes, and noise. Detailed design examples are provided for a sour water stripper and an oil/water separator.

Keywords
Refinery • Environment • Emissions • Effluent • Hazardous waste • Noise

Introduction

During the last half of the 1900s, control of contaminants from processing crude oil became an increasingly important driver in the design and operation of refineries and other petroleum processing facilities. The topic "▶ Petroleum Products and a Refinery Configuration" in this handbook discusses some of the changes that have taken place in the specifications for petroleum products to meet legislative environmental control. These products have been singled out because they impact more prominently on the daily lifestyle and health of the general public, particularly in the more advanced industrial countries. The product specifications continue to ratchet downward with respect to sulfur. At the same time, the requirements are becoming more stringent for fuel quality.

Similar changes and legislative controls have been adopted on the levels of emissions and streams leaving the petroleum processing facilities in the air, water, and solid wastes. These changes have become more restrictive as greater knowledge of their effects on the environment and health has developed.

This chapter deals with many of the measures adopted in the design and operation of processes to meet these environmental protection measures. We will focus on four areas of direct environmental controls in facilities, with several sub-areas:

- Air – sulfur, volatile organic carbon (VOC), nitrogen oxides, carbon monoxide, particulate matter, carbon dioxide, ozone, and other emissions
- Water effluents – oxygen demand, oil and grease, sulfides, ammonia compounds, phenols, caustic, benzene, selenium, metals, and other potential contaminants

Environmental Control and Engineering in Petroleum Processing

- Solid wastes – sludge, catalysts, waste oils, and characteristic hazardous wastes
- Noise levels – internal, fenceline

Table 1 highlights some of the environmental concerns in air, water, solids, and noise from refineries, providing a map to potential areas of concern in each type of process. This chapter will explore each of these areas in more detail.

Table 1 Some key areas of environmental concern for oil refining process unit streams

Process area	Air	Water	Solids	Noise
Atmospheric & vacuum distillation	Flue gases, fugitives	Sour water (NH_3, H_2S, amines), desalter water (oil), spent caustic	Sludges, exchanger cleaning	Rotating equipment, fluid flow
Coking, thermal cracking	Flue gases, fugitives	Sour water, coke cutting water, process waste water	Sludges, exchanger cleaning, coke	Rotating equipment, fluid flow
Fluid catalytic cracking, unsaturated gas plant	Flue gases, fugitives, regenerator gas	Sour water (NH_3, H_2S, phenols), spent caustic, process waste water	Spent catalyst, exchanger cleaning	Rotating equipment, fluid flow, catalyst flow
Hydroprocessing units	Flue gases, fugitives	Sour water (NH_3, H_2S, oil), process waste water	Spent catalyst, exchanger cleaning	Rotating equipment, fluid flow, HP relief, venting
Hydrogen plants	Flue gases, deaerator/degasifier vent		Spent catalyst, exchanger cleaning	Rotating equipment, fluid flow, hydrogen vents
Saturated gas plant, alkylation		Spent caustic and acid, process waste water	Exchanger cleaning	Rotating equipment, fluid flow
Catalytic reforming	Flue gases	Process waste water	Spent catalyst, exchanger cleaning	Rotating equipment, fluid flow
Sulfur plant, sour water stripping	Flue gases, incinerator, sulfur pit or tank vent (SO_X, H_2S)	Sour water, tail gas unit fluid	Sludges, sulfur, exchanger cleaning	Rotating equipment, fluid flow
Waste water treating	Fugitive vapors	Refinery effluent, storm water	Sludges	Rotating equipment, fluid flow
Tankage areas	Fugitive vapors	Tank dike drains, tank water draws	Tank sludges	Rotating equipment, fluid flow

(*continued*)

Table 1 (continued)

Process area	Air	Water	Solids	Noise
Flare	Combustion gases	Sour water, oily water		Flare noise, emergency vents, rotating equipment
General refinery concerns	Fugitive vapors, cooling towers	Storm water, oily water drainage	Trash (hazardous and non-hazardous)	

Air Emissions

Introduction

Air emissions limits today are largely driven by government legislation and the resulting regulations that have been promulgated. Some of the key air pollution regulations with which petroleum processing plants must contend include those listed below, among many others.

- US Clean Air Act (CAA) and amendments – this act, initially passed in 1970, contained a number of innovative provisions to reduce air pollution. With amendments, it is still a primary driver behind air emissions control design and practices in the United States today. Some of the important provisions and regulations resulting from the act include:
  - Roles for both the federal and state governments
  - Environmental Protection Agency (EPA) sets limits on air pollutants anywhere in the United States – states can regulate levels lower, which is particularly notable in California.
  - EPA reviews and approves State Implementation Plans (SIPs).
  - CAA includes provisions for interstate and international pollution limits.
  - Prevention of Significant Deterioration (PSD) – classified areas in the United States into attainment (meeting all standards) or non-attainment (do not meet some standards) with different permitting for facilities in each type of area.
  - Established a permitting system for emitters.
  - Enforcement provisions – including both civil and criminal penalties.
  - Requires public participation in permitting.
  - Market incentives for reduction of pollutants.
  - New source performance standards (NSPS) for facilities – best available control technology (BACT).
  - Air toxics rules – Maximum Achievable Control Technology (MACT) and Residual Risk and Technology Reviews (RTR).

- European Union (EU) regulations and directives
  - EU climate package – targets 20 % reduction in greenhouse gases (GHGs) by 2020 (vs. 1990) along with 20 % increase in renewables use and 20 % reduction in total energy use
  - Standards for higher fuel quality and diesel production
  - Industrial Emissions Directive (IED) – limits SO_2, NOx, particulates, and VOCs from industrial sources – integrated approach, best available technologies (BATs), and local conditions by local authorities
  - Thematic Strategy on Air Pollution (TSAP) – recommends emission limit values (ELVs) be consistent with local impacts
  - Energy Taxation Directive (ETD) – still proposed (2014) – potential taxes on energy content and CO_2 emissions
  - Renewable Energy Directive – affects product specifications
  - Fuel Quality Directive – affects product specifications, GHG emissions
- International Maritime Organization (IMO) legislation
  - International Convention on the Prevention of Pollution from Ships (MARPOL 73/78 or MARPOL Convention) – regulates maximum sulfur in fuel oil and emissions for ships – mostly product quality impacts, but more sulfur to deal with in the refinery.
- State regulatory decisions
  - California Air Resources Board and local air districts – these have been leaders in ratcheting down allowable emissions in refineries. Often, US regulations follow the actions of the California authorities within a few years.
  - Fuel oils – there is a drive to reduce the sulfur in fuel oils used primarily on the US East Coast. Individual states are adopting ultra-low sulfur fuel oil requirements ahead of the EPA deadlines. This is resulting in more sulfur to deal with in refineries.
- Asia
  - The focus has been on SO_2 reduction primarily – especially driven by severe conditions in some areas.
  - China has limits on allowable SO_2 based on an area classification, e.g., protected areas like scenic spots, historical sites, and protected habitats have a more severe limit than other areas – Regional Air Quality (RAQ) regulations (took effect in 2014) – SOx, NOx, particulates.
  - India Environment (Protection) Act – initially 1986 – SOx, NOx, VOCs, etc.

Impacts and Sources of Refinery Emissions

Table 1 provided an initial indication of some of the air emissions sources and the pollutants of concern. Here we will look more closely at the most common air emissions that are regulated, with additional attention to where these emissions may arise in the refinery. Specifically, we will talk about:

- Sulfur (SOx, H_2S, organic sulfur compounds)
- Nitrogen oxides (NOx)
- Carbon oxides (CO, CO_2)
- Particulate matter (esp. PM-10)
- Volatile organic carbon (VOC)
- Ozone
- Ammonia
- Methanol
- Lead
- Mercury

Table 2 provides a summary of some of the refinery sources, health effects, environmental effects, and property damage aspects of each of these pollutants.

Controlling Air Emissions

We will now examine the most common approaches to controlling the types of air emissions listed in Table 2.

Sulfur Compounds

Sulfur Oxides (Primarily SO_2)

For most combustion sources of sulfur oxides, the primary control approach is elimination of sulfur compounds from the fuel. Refinery gas treatment is discussed extensively in the chapter entitled "▶ Refinery Gas Treating Processes" of this handbook. We will not repeat that material except to note that the normal treatment approach is scrubbing of sour refinery gases with an amine (or other) solution to absorb H_2S (the primary sulfur compound) creating a low sulfur fuel gas. The amine is regenerated to release the H_2S, which is routed to the sulfur plant for sulfur recovery (to be discussed later).

Amine absorbers do not remove all sulfur compounds from the gases. COS and mercaptans will usually still be present. Normally, the levels of sulfur remaining are low enough that the gas can be used as fuel, but sometimes the level is still too high after removal of H_2S. In these cases, the options are:

- Caustic treatment (e.g., Merox or Merifining) to remove the mercaptans
- Physical solvent, sorbent treatment or hydrotreating to remove COS

Caustic Treating

A typical caustic treating system is illustrated in Fig. 1. In the caustic treater, the vapor containing mercaptans and other sulfur compounds is contacted with a circulating caustic stream containing a proprietary catalyst. The caustic absorbs the sulfur compounds selectively. Any residual H_2S will also be absorbed. Absorption occurs by reaction of mercaptan with caustic in solution:

Table 2 Sources and impacts of air emissions from a refinery

Air pollutant	Vapor sources	Health effects (in air)	Environmental effects (in air)	Property damage (in air)
Sulfur compounds				
Sulfur dioxide (primarily SO_2)	Fired heaters and other combustion sources burning the sulfur compounds fuels, sulfur recovery plant exhaust and vents, some processes (e.g., Edeleanu), combustion of pyrophoric iron sulfide in foulants. FCC regenerator	Short term: respiratory problems (irritation, bronchoconstriction, asthma symptoms). Long term: similar respiratory symptoms, decreased lung function	Reacts with other compounds in the air to form particulates and aerosols (smog). Increased respiratory problems in general public. Primary source of acid rain	Corrosion of stone and structures from sulfurous and sulfuric acids in low concentrations
Hydrogen sulfide (H_2S)	Sewer vents, tank seals (especially sour water storage tanks), process releases or leaks, fugitive emissions from sour streams, and maintenance activities	Highly toxic. Short term: maximum 10 ppm (8 h shift) with the max exposure of 15 ppm (15 mins). Short exposure above 500 ppm may be fatal. See the chapter "▸ Hazardous Materials in Petroleum Processing." Long term: damage to lungs, upper respiratory tract, eyes, and the central nervous system	Severe odor issue near a facility, H_2S would be oxidized to sulfur oxides and have similar environmental impacts to SO_2	See SO_2
Organic sulfur (R-SH)	Fugitive emissions from piping, valves and tanks; leaks	Using ethyl mercaptan as a guide. Short term: respiratory or skin irritation, cough, loss of sense of taste, nausea, vomiting, organ damage, unconsciousness, and coma. Long term: central nervous system damage in addition	Severe odor issue near a facility; RSH would be oxidized to sulfur oxides and have similar environmental impacts to SO_2	See SO_2

(continued)

Table 2 (continued)

Air pollutant	Vapor sources	Health effects (in air)	Environmental effects (in air)	Property damage (in air)
Nitrogen oxides (NOx, primarily NO_2)	Burning of gasoline, fuel gas, fuel oil, natural gas, coal, etc. Vehicles are an important source of NO_2 from refined product use	Lung damage, illnesses of breathing passages and lungs (respiratory system)	Ingredient of acid rain (acid aerosols), which can damage trees and lakes. Acid aerosols can reduce visibility	Corrosion of stone and structures from nitrous and nitric acids in low concentrations
Carbon oxides				
Carbon monoxide (CO)	Incomplete burning of vehicle fuels, fuel gas, fuel oil, natural gas, coal, etc. FCC regenerator	Reduces ability of blood to carry oxygen. May exacerbate heart or circulatory problems and respiratory damage	Oxidation to CO_2 with similar effects	No particular issues noted as CO
Carbon dioxide (CO_2)	Burning of vehicle fuels, fuel gas, fuel oil, natural gas, coal, etc. SMR hydrogen plants. Incinerators. FCC regenerator	Short term (acute): asphyxiant, respiratory distress, headache, dizziness, circulatory problems, irritation, excitation, vomiting, unconsciousness, fatality. Long term: low-level exposure: no significant impacts. Rapid unconsciousness >10 %	Primary greenhouse gas. Factor in acid rain. Positive impact on plant growth at normal concentrations of a few hundred ppm in air	Similar to SO_2 in very high concentrations
Particulate matter (especially PM-10) – dust, smoke, soot	Fuel combustion, industrial activities, catalyst handling, construction activities, unpaid roads, activities from outside the refinery. FCC regenerator	Nose and throat irritation, lung damage, bronchitis. Early death with long-term, heavy exposure. Hazards of chemicals in dust may present hazards: e.g., silica and nickel in catalyst dusts	Particulate haze reduces visibility. Chemicals in some dusts may have environmental impacts	Ashes, soots, smokes, and dusts can dirty and discolor structures and other property, including clothes and furniture. Some dusts may promote corrosion

(*continued*)

Table 2 (continued)

Air pollutant	Vapor sources	Health effects (in air)	Environmental effects (in air)	Property damage (in air)
Volatile organic carbon (VOC)	Incomplete burning of fuel, solvents, light hydrocarbon fugitive emissions and leaks (e.g., benzene, toluene, xylene, naphtha, etc.), paints, glues, vehicles, chemical fugitives or leaks, flares, and vents	In addition to ozone (smog) effects, it can cause serious health problems such as cancer and other effects. Many are hazardous air pollutants, which can cause varying serious illnesses (e.g., benzene)	In addition to ozone (smog) effects, some VOCs such as formaldehyde and ethylene may harm plants	Some solvents and organics can damage paint, elastomers, and fabrics
Ozone (O_3)	Chemical reaction of other pollutants normally: VOCs and NOx. Some cooling water or wastewater treatments use ozone	Breathing problems, reduced lung function, asthma, irritates eyes, stuffy nose, reduced resistance to colds and other infections, may speed up aging of lung tissue	Can damage plants and trees; smog can cause reduced visibility	Damages elastomers, rubber, paints, fabrics, etc.
Ammonia (NH_3)	NOx reduction system leaks, sour water leaks and tanks, hydrogen plant degasifier, other fugitive emissions	Irritant and corrosive to skin, eyes, respiratory tract, and mucous membranes. High concentrations may cause chemical burns. Low concentrations are an irritant	Potential oxidation to NO_2. In normally expected concentrations, NH_3 is not a major environmental issue. Could be a wastewater issue	Not significantly corrosive as a vapor in normally expected concentrations
Methanol (CH_3OH)	Leaks and fugitive emissions, hydrogen plant degasifier	Toxic in high concentrations. Primarily an irritant. Sensitizer on long-term skin contact and toxic to some organs. May be absorbed through skin	Reacts with metals, acids, and oxidizers. Oxidizes in the atmosphere to less hazardous materials. Half-life = 3–30 days	Not significantly corrosive as a vapor in normally expected concentrations

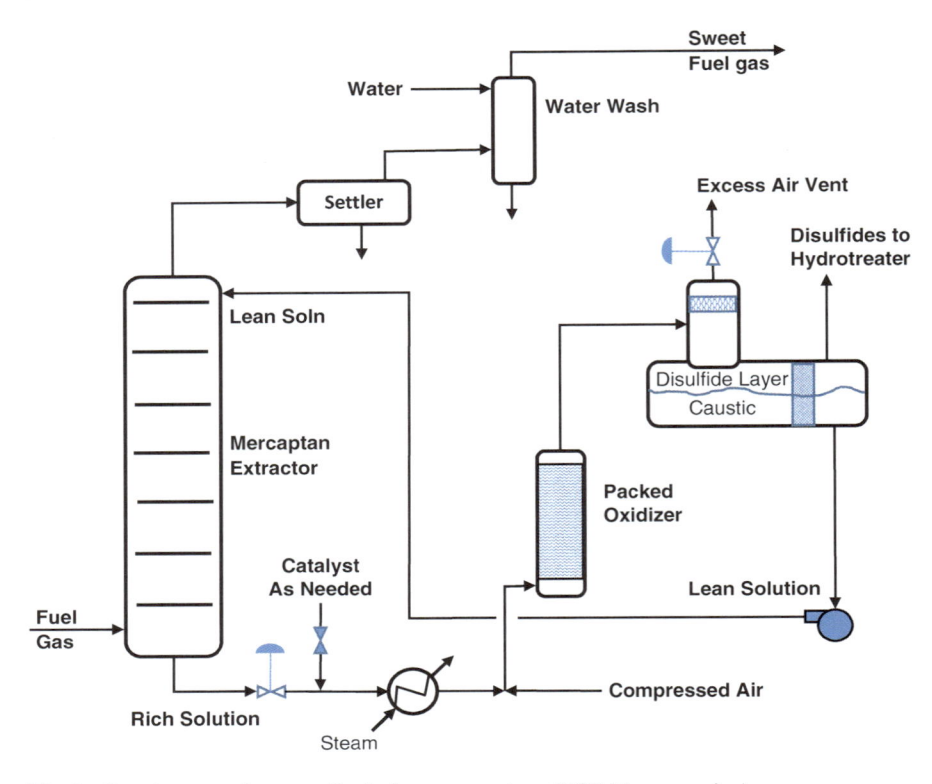

Fig. 1 Caustic removal process for fuel gas mercaptans (UOP Merox version)

$$2\,\text{RSH}(v) + 2\,\text{NaOH}(aq) \rightarrow 2\,\text{NaSR}(aq) + 2\,\text{H}_2\text{O}(l) \tag{1}$$

The rich caustic stream is sent to a regenerator where the NaSR is then oxidized with a slight excess of air to a disulfide:

$$4\,\text{NaSR}(aq) + \text{O}_2(aq) + 2\,\text{H}_2\text{O} \rightarrow 2\,\text{RSSR}(aq) + 4\,\text{NaOH}(aq) \tag{2}$$

The disulfides (RSSR) are finally separated and sent to a hydrotreater for removal as H_2S.

The remaining mercaptans are at trace levels. These processes do not significantly remove COS.

COS Removal by Solvent or Hydrotreating

Usually, COS is present in fuel gas at low enough levels that there is no issue after H_2S and mercaptans are removed, but if there is still too much sulfur in the fuel gas, then selective solvents, absorbents, or direct hydrotreating can be used.

Some solvents that remove H_2S also remove COS. Most notable among these are physical solvents like Sulfinol. MEA, DEA, and DGA also remove some COS and other sulfides, but these solutions are degraded by the COS. A Sulfinol system, as an example, uses a concentrated mixed solvent of diisopropanolamine (DIPA) or MDEA, sulfolane (tetrahydrothiophene dioxide), and water in a scrubber/regenerator arrangement like an amine absorption system.

There are a number of direct absorbents available for COS removal. Processes using these materials come in regenerable and non-regenerable flowsheets. In some processes, all the sulfur is actually hydrotreated first to H_2S and then absorbed. Processes are offered by several vendors, including Johnson Matthey, Sud-Chemie, BASF, UOP, Alcoa, and others.

Finally, direct hydrotreating of the fuel gas followed by amine absorption is an option. In the hydrotreating step, all sulfur is converted to H_2S before absorption – similar to some of the solid sorbent options.

Fuel Oil Sulfur Control

When fuel oil is burned, the sulfur in the oil will be converted to SOx. Control of SOx emissions from fuel oil combustion can be analogous to controlling SOx from fuel gases through removal from the fuel before firing. In this case, the sulfur would have to be removed by severe hydrotreating of the fuel oil to the desired level. The cost of this often means that it is more economical to convert to fuel gas firing instead. Hence, where environmental emissions limits are tight, you will seldom find fuel oil firing of heaters. If you do need to fire low sulfur fuel oil, the handbook chapter entitled "▶ Hydrotreating in Petroleum Processing" discusses sulfur removal.

Another option for fuel oil firing of large furnaces or the FCC stack would be flue gas scrubbing. Added benefits of scrubbers are that they can help control NOx and particulate emissions. A simplified flowsheet for a flue gas scrubbing system is illustrated in Fig. 2.

In the flue gas scrubber, flue gases are contacted with a circulating slurry of limestone, caustic, or soda ash in a tower or venturi scrubbers. There is continuous makeup of chemical into the circulating stream and continuous blowdown of spent chemical. This type of system does create a significant spent slurry stream that requires disposal.

Management of Other SOx Sources

Sulfur pit vents and sulfur storage tank vents will contain SO_2 vapors. Options for managing the emissions from these sources include:

- Combining the vapors with the sulfur plant vent and routing through the stack incinerator, if the emissions limits allow.
- Routing to the front-end reaction furnace with the process air.
- Routing to a sodium bicarbonate scrubber.
- Routing to the tail gas unit ahead of the hydrogenation reactor.

Since the pit vapors will also contain H_2S, sulfur vapor, and other compounds, those materials will influence the choice.

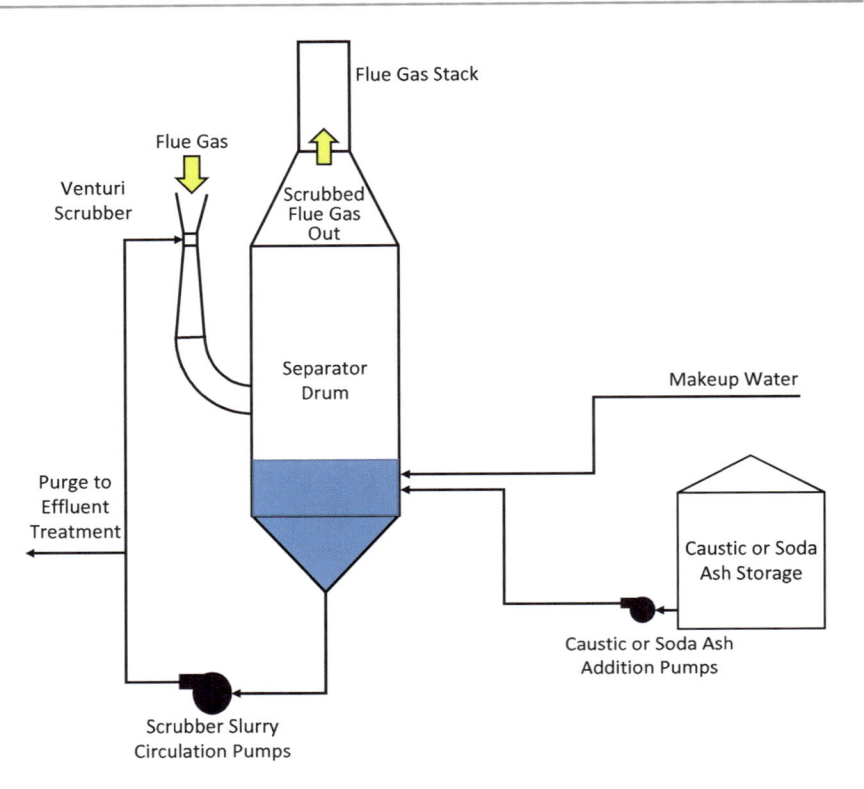

Fig. 2 A flue gas SOx scrubbing system

There are few processes that use SO_2 deliberately, so we will not address those here other than to note that fugitive SO_2 from those processes would be managed by use of proper design and operating practices to eliminate fugitives.

Iron sulfide is found throughout refineries as a product of corrosion. The FeS is normally pyrophoric, that is, it will spontaneously burn when it is dried out and exposed to air. The combustion products will include SO_2. The amount of SO_2 generated can be impressive. If the iron sulfide is in tank or other sludges, these materials can be deliberately oxidized by treating them with peroxide or another oxidizer that will make them into stable sulfates for disposal. Other solids and semisolids that can generate SO_2 if they begin burning can be kept water-wet to prevent the initiation of combustion until they can be properly disposed of.

Hydrogen Sulfide (H_2S)

The most pervasive sulfur form in a refinery after processing is normally hydrogen sulfide. Primary emissions of hydrogen sulfide are controlled by recovery of the H_2S and routing it to a sulfur recovery unit or sulfur plant. We will discuss the sulfur plant in a moment.

Because H_2S is almost everywhere in a plant, it can be released from multiple locations, among these are:

- Sewer vents
- Tank vapors and tank seals, especially on sour water tanks
- Process releases during venting or sampling
- Leaks from packings, flanges, and other pipe elements
- Compressor and pump seals
- Maintenance activities, such as blinding

Control methods for H_2S from sewer vents are discussed in the topic entitled "▶ Utilities in Petroleum Processing." If H_2S is an issue in these vents, the vapors can be routed through a canister on their way to the vent location.

For tank vapors, there are a couple of options, depending on the type of tank:

- Cone roof tanks – these can be blanketed and the blanket gas recovered and or incinerated. Some tanks have internal floating roofs, and then the floating roof options apply.
- Floating roof tanks – these use double seals to minimize fugitive emissions. In sour water tanks, the annulus can be filled with diesel or kerosene to further suppress H_2S fugitive emissions.
- Generally, it is best not to put an H_2S sour stream into an open, unblanketed, or single seal tank.

Managing process releases is primarily application of good operating practices, i.e., never open an H_2S sour line to atmosphere if there is any other choice. Always route the sour fluids to flare, sour water, a flash tank, or another enclosed system. It is safer to burn H_2S to SO_2 before emission than to let the H_2S get out.

When sampling a sour stream, in addition to personnel protection, including fresh air breathing apparatus, the venting should be minimized. If a sample requires purging for a period, design the sample purge as a closed-loop system with the sample pulled from the closed fast loop, so minimal material is vented.

Leaks from piping, valves, fitting, flanges, and other joints are managed by attention to proper fit-up and leak testing before use. It is a good practice to investigate any little whiff of H_2S that is found and eliminate the leak at the source. Control valves are often a problem. Special packing designs are available to minimize these leak points combined with routine "sniffing" of the potential leak points with a detector system (not your nose!).

Options for compressor and pump seals to prevent loss of sour process vapors continue to develop. Some of the options today are:

- Double mechanical seals, including barrier fluids
- Sealless pumps
- Purged recip compressor packing with nitrogen barrier
- Closed/controlled recip compressor distance pieces

Eventually you have to open up piping and equipment for maintenance. Many maintenance activities present opportunities for H_2S releases. Of particular concern are those activities that could expose someone to H_2S, such as initially setting isolation blinds. Every effort is usually made to eliminate all H_2S from systems, but it can still find its way to unexpected locations from which it can be released to the atmosphere. Good shutdown and maintenance practices should minimize H_2S fugitive emissions.

So this brings us to our primary tool in reducing H_2S emissions: the sulfur recovery plant.

Claus Process

The primary process used today for recovering sulfur is the Claus process.

Claus process chemistry depends on the overall Claus reaction, which makes elemental sulfur from H_2S:

$$10\,H_2S + 5\,O_2 \rightarrow 2\,H_2S + SO_2 + 7/2\,S_2 + 8\,H_2O \qquad (3)$$

The process is diagrammed in Fig. 3.

Conceptually, the reaction is carried out in several steps:

1. Partial combustion of the H_2S to produce a 2:1 molar ratio of H_2S to SO_2. This reaction is highly exothermic. It is conducted in a high temperature, low-pressure (~21 psig/1.5 barg) reaction furnace with a stoichiometric amount of air:

$$6\,H_2S(g) + 3\,O_2(g) \rightarrow 2\,SO_2(g) + 2\,H_2O(g) + 4\,H_2S(g)$$
$$(\Delta H = -4147\,kJ/mol) \qquad (4)$$

2. The H_2S and SO_2 now react to form sulfur via the Claus reaction. By the end of the reaction furnace, about two-thirds of the original H_2S has been converted to sulfur. The temperature is ~1,830 °F (~1,000 °C) at the reaction furnace outlet:

$$2\,H_2S(g) + SO_2(g) \rightarrow 3\,S(g) + 2\,H_2O(g)$$
$$(\Delta H = -1166\,kJ/mol) \qquad (5)$$

3. The reaction furnace fluids are cooled in a waste heat boiler and steam generator to condense the sulfur formed. The sulfur is removed from the steam generator as liquid through a steam-jacketed seal leg or sulfur trap.
4. The remaining gases are reheated by a high pressure, superheated steam exchanger or an in-line fired reheater and then pass through a catalyst bed that is primarily alumina. The pressure is still low and the temperature is ~580 °F (~305 °C). The alumina catalyzes the Claus reaction, converting about two-thirds of the remaining H_2S/SO_2 to sulfur.
5. The reactor effluent is again cooled. The sulfur is condensed and removed. Overall, about 89 % of the original H_2S has been recovered as liquid sulfur.

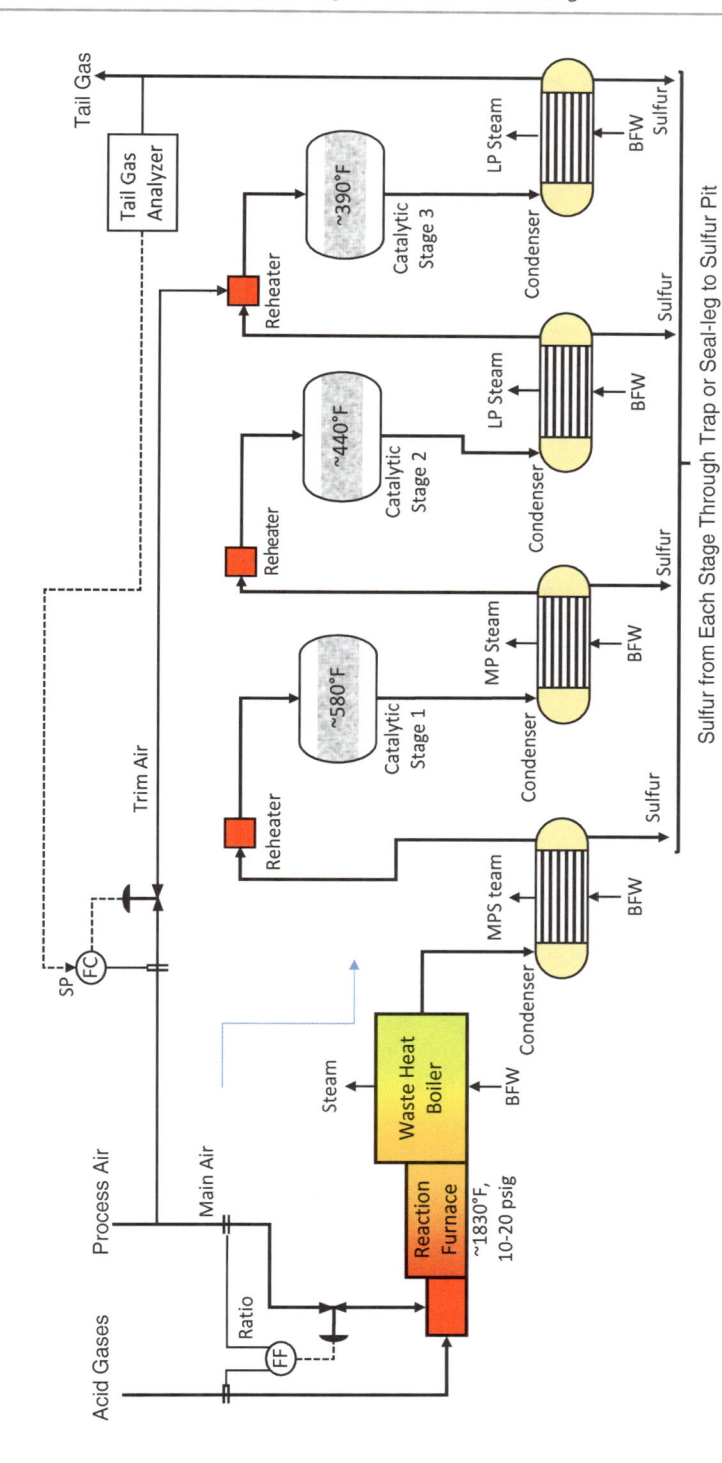

Environmental Control and Engineering in Petroleum Processing

Fig. 3 A three-stage Claus plant flowsheet

6. Steps 4 and 5 are usually repeated twice more, at progressively lower temperatures – 437 & 392 °F (225 & 200 °C). The overall recovery at the end of this "3-stage Claus" configuration would be about 98–99 %. The sulfur condenser for each subsequent stage operates at slightly lower temperatures, but still in a range to keep sulfur molten.
7. The condensed, molten sulfur from the Claus stages is collected, usually in an enclosed sulfur pit, and maintained at about 285 °F (140 °C) using steam coils. The vapor space of the pit may be vented at a high point through a steam-jacketed vent or it may be swept by air or inert gas. Disposal options were discussed under SO_2 emissions. If a pit is allowed to accumulate gases, it can catch fire and must be extinguished with snuffing steam.
8. All piping along the Claus process path is maintained at about 285 °F (140 °C) – not much hotter or colder – to keep sulfur from setting up.

More than three Claus catalytic stages are of marginal value, although some plants use four stages. Usually, if recovery has to be higher, the process moves into a tail gas unit, which we will cover in the next section.

Ammonia-Burning Claus Plants

Claus plants have one other use in a refinery. The sour water collected from all the refinery units is normally steam stripped free of H_2S and ammonia in the sour water strippers (SWS). This process is detailed in section "Water and Aqueous Effluents." For the moment, it is important to know that the strippers produce a stream containing very high concentrations of ammonia and H_2S, along with water vapor, at about 170 °F and at a pressure high enough to go to the Claus furnace, maybe 30 psig (~2 barg).

The SWS gas can be introduced into the reaction furnace of the Claus unit. This is normally done by staging the injection of the SWS gas with the H_2S from the amine regenerators. The design of these systems is proprietary, but you end up with an ammonia-burning Claus plant that can deal with the ammonia as well as the sulfur.

The ammonia is destroyed in the reaction furnace. The required temperature is 2,250 °F (1,230 °C). The resulting ammonia remaining will be less than 30 ppm, which will prevent downstream deposition of ammonium salts in the cooler catalyst beds:

$$2\,NH_3(g) + 3/2\,O_2(g) \rightarrow N_2(g) + 3\,H_2O(g)(endothermic) \qquad (6)$$

Monitoring the Claus Process

Now, what do you have to watch out for in a Claus plant?

- The feed streams to a Claus plant are extremely toxic and would be fatal with just one whiff. The acid gas from the amine regenerators will be around 85 % H_2S. The SWS gas has both H_2S and ammonia in high concentration. Personnel

safety and leaks are always concerns. Generally, sulfur plants actually have the least odor in the area because the operators understand the risks and remain diligent.
- To optimize a Claus plant, control of the H_2S/SO_2 ratio is absolutely critical. There are H_2S/SO_2 analyzers available. Many units analyze the tail gas leaving the Claus unit (or in the tail gas unit) and use that result to fine-tune the air/feed ratio.
- The reaction furnace must be kept in the correct temperature range for both the Claus reaction and ammonia burning. If it is too low, conversion will be inadequate and catalyst beds and exchangers will plug. If the furnace is too hot, there will be refractory or waste heat boiler damage.
- Claus plant recoveries are affected by atmospheric pressure. Plants at high elevations will not recover quite as much sulfur as those at lower elevations because the sulfur vapor pressure has an impact on the performance of the condensers.
- Claus plants have limited turndown capability. Some plants have supplementary natural gas firing for startup or "hot-standby," but this is hard to control and generates a lot of water. There is a high probability of an air/fuel ratio error that will plug the catalyst beds with soot.
- Air/feed ratio is critical. It must be kept exactly at stoichiometric for the Claus reaction. Slight differences often mean plugged beds or equipment damage.
- Any significant amount of hydrocarbon in the feed is usually a major problem in a Claus plant. The hydrocarbons will consume a large amount of air and rob the H_2S and NH_3 of their combustion air. The air systems on Claus plants normally cannot deliver enough air to support combustion of very much hydrocarbon, so the hydrocarbons will form soot instead and plug the catalyst beds.
- So, from a monitoring standpoint:
  - Track feed rates, compositions, and conditions.
  - Calculate recovery periodically.
  - Trend the air/feed ratio.
  - Track the residual ppm sulfur in the tail gas.
  - Watch your steam generation systems carefully.
  - Watch the reaction furnace temperatures and firing pattern – keep it as close to optimum as possible.
  - Ensure the catalytic stage temperatures are in the best ranges.

Claus Tail Gas Treating Processes

The Claus tail gas after processing can often be sent to an incinerator and emitted to the atmosphere in many locations. After 3-Claus catalytic stages, the recovery would be close to 99 % of the sulfur. The balance is burned to SO_2 before discharge.

In many countries and locales, however, 99 % recovery of sulfur is still insufficient. Recovery needs to be at least 99.9 % or less than 10 vppm total sulfur compounds in the stack. The need for recovering more of the sulfur has resulted in the creation of tail gas treating technologies.

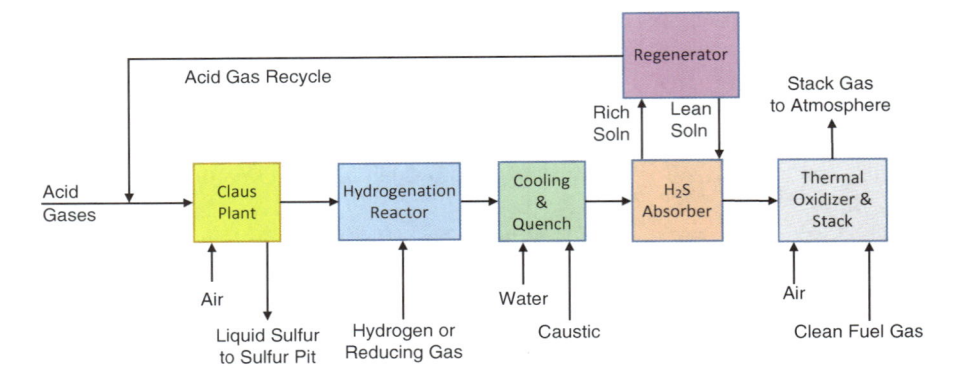

Fig. 4 Typical amine-based Claus tail gas process block flowsheet

Two approaches have been commonly taken:

- Hydrogenation of all sulfur compounds to H_2S and removal by an alkaline solution – the most common approach
- Combustion of all sulfur compounds to SO_2 and removal by an alkaline solution

A typical block flow for the most common hydrogenation + absorption approach using an amine solvent is Fig. 4. Various flavors of this process are offered on a proprietary basis by several licensors.

The gas leaving the Claus plant has a 2:1 molar ratio of $H_2S:SO_2$ with a few traces of other sulfur species and elemental sulfur vapor. The solvent system is intended to work on H_2S, so the first step in the tail gas unit is hydrogenation of all sulfur compounds to H_2S. The hydrogen for this step can be supplied by an external hydrogen source (hydrogen plant) or by an in-line reducing gas generator burning sub-stoichiometrically.

The tail gas is heated to about 570 °F (300 °C) and enters the hydrogenation reactor. The reactor contains a cobalt-molybdenum catalyst which promotes the reduction of all oxidized sulfur compounds to H_2S. The actual temperature in the catalyst bed can be different, depending on the catalyst choice and catalyst age. The primary reaction is

$$SO_2 + 3H_2 \rightarrow H_2S + 2H_2O \qquad (7)$$

The hydrogenation reactor effluent is cooled in an exchanger followed by a quench tower. In some processes, the quench tower has a caustic scrubbing section to absorb any trace SO_2 that remains, since this can degrade the absorber solution. The gas leaving the quench and cooling system will be around 100–130 °F ($\sim$80–90 °C).

The cool gases enter a trayed or packed absorber where the H_2S is scrubbed out of the gases using a lean solution. Absorber solutions that have been used include the same amines used for fuel gas scrubbing (MEA, DEA, MDEA, UCARSOL, etc.) or chemical solutions like Stretford (vanadium).

The clean absorber gas can have less than 10 ppm total sulfur compounds remaining. It is sent to a final incinerator to convert all the H_2S to SO_2 for a safer emission at the final stack. Over 99.9 % of the original sulfur is now removed.

The rich absorber solution is sent to a regenerator system. For amines and most other absorbents, the rich solution is simply steam stripped of the H_2S it absorbed, like in any amine absorber system. The H_2S released is routed back to the front end of the Claus unit to join the incoming acid gas feed.

In the rare case of a Stretford tail gas plant, the vanadium in the solution has been reduced by absorbing H_2S. It is oxidized back to its higher oxidation state using air in a large, stirred, multistage oxidizer tank. The absorbed H_2S is converted to a sulfur froth at the same time, which is skimmed off the top of the oxidizer tank.

Alternative Flowsheets
As noted above, another approach to tail gas cleanup is complete oxidation of the sulfur compounds remaining to SO_2 and subsequent absorption in caustic solution and oxidation to sulfate. These processes are also offered by licensors on a proprietary basis. The final sulfur removal in this case would be in the form of a sodium sulfate solution, which is relatively innocuous.

Tail Gas Plant Monitoring and Operation
Monitoring and operation of the tail gas system would commonly include:

- Analyzing (normally continuously) and trending the stack gas sulfur content. Problems would be immediately investigated.
- Absorber solution chemistry analysis and trending, including lean and rich solution loading, alkalinity and pH (in some cases), and the accumulation of degradation salt products.
- Reclamation or salt removal from the absorber solution when needed. There are firms and processes available who perform this service.
- Chemistry within the quench tower circulating solution – pH, alkalinity, available caustic.

Organic Sulfur Compounds
There are a number of other organic sulfur compounds handled in a petroleum processing facility. These include the mercaptans in fuel gas along with chemicals used to start up hydroprocessing and catalytic reforming units. The former have been covered adequately already. We will comment here about the sulfur chemicals used in processes.

The common sulfur chemicals found in refining processes include:

- Dimethyl disulfide (DMDS) – hydroprocessing activation, reformer startup
- Di-tert nonyl Polysulfide (DNPS) – hydroprocessing activation
- Di-tert butyl Polysulfide (TBPS) – hydroprocessing activation

These chemicals tend to be respiratory irritants in air and are skin sensitizers, but their use is limited to local areas of the refinery. From experience, note that DMDS is a very good solvent for shoes! Management is normally accomplished by ensuring no leaks occur and neutralizing any spills immediately. A 5 % bleach solution is adequate to control moderate spills of these very stinky chemicals.

In all, the potential organic sulfur fugitives and emissions, diligence in handling and addressing leaks immediately will render them non-problems.

Nitrogen Oxides (NOx, NO_2)

NOx emissions from a petroleum processing facility almost exclusively come from combustion sources, so control is focused on those sources. There are four areas that are commonly addressed:

- Elimination of nitrogen compounds, like tramp amines and ammonia, from fuel gas
- Burner design to lower NOx generation in the furnaces
- Control of excess air in the furnaces
- Selective catalytic and non-catalytic reduction of NOx that is generated (SCR and SNCR, respectively)

Fuel Gas Tramp Nitrogen Compounds

Fuel gas will have traces of amine compounds that are carried over from the amine scrubbers along with ammonia that may result from amine degradation or other sources. If these tramp nitrogen compounds reach the burners, the nitrogen will be at least partially converted to NOx.

In addition, the wet amines in the fuel gas headers promote corrosion and formation of salts, which plug burner tips and orifices. The plugged tips result in undesirable flame patterns that co-opt the low-NOx characteristics of the burners.

Control of the tramp nitrogen compounds is addressed by:

- Ensuring good knockout of amines at the scrubbers and in the fuel gas mix drum to keep the fuel gas header dry (See the utilities chapter of this handbook).
- Avoiding foaming or overloading of the amine scrubbers
- Good local knockout and other measures to keep foulants from reaching the burners within a process unit as described in the chapter entitled "▶ Utilities in Petroleum Processing"
- Attention to amine degradation and addressing accumulations of degradation products as described in the chapter "▶ Refinery Gas Treating Processes" of this handbook.
- Cleaning burner tips when fouled

The best strategy is to keep the tramp nitrogen compounds and the resulting foulants from reaching the burners.

Table 3 Impact of burner modifications on NOx production (reference: Baukal 2008)

Modification type	Approx. NOx, ppm at 3 % excess air	Est NOx emission, lb/MMBtu fired
Conventional burner	120	0.14
Staged air	70	0.08
Staged fuel	40	0.05
Internal flue gas recirculation	15	0.02
Ultra-lean premix	10	0.01

Burner Design

Emphasis on reducing NOx emission over the past several years has resulted in tremendous advancements in understanding combustion and how NOx is made within the burners of a furnace. This understanding has led to new, cleaner burner designs – primarily for gas burners. The details of the resulting burner designs are proprietary, but the general understanding has become well known.

Some key factors affecting NOx generation in a burner include:

- NOx production increases exponentially with combustion temperature.
- NOx is increased by high excess air up to a point.
- Air preheat increases NOx production.

Burner manufacturers have developed several approaches to address these issues, resulting in low-NOx and ultra-low-NOx burner designs. Burner features to reduce NOx include:

- Staging the air addition so that the initial fuel burn is in a rich environment, with the rest of the air added later in the burner.
- Staging the fuel to burn lean initially and then adding the balance of the fuel later in the burner.
- Flue gas recirculation to moderate temperatures. This can be done internal or external to the burner.
- Water or steam injection to reduce flame temperatures.
- Ultra-lean premix of fuel and air ahead of the burner.

The value of these burner modifications is seen in the NOx reduction at 3 % excess air in Table 3 with data from a John Zink paper.

Controlling Excess Air in the Furnace

Excess air has a significant impact on NOx emissions. If we look at the excess air in a furnace, the NOx emissions increase roughly linearly up to the stoichiometric air/fuel ratio – i.e., while the combustion is lean. Above this, the NOx generation continues to increase until the amount of air present cools the flame enough to reduce the NOx. This typically would occur with, perhaps, 25 % excess air (or ~5 % excess O_2). At large amounts of excess air, the NOx production drops off, but so does the furnace efficiency.

NOx control from limiting excess air must be balanced with CO emissions control, which is favored by higher excess air values. Usually the compromise means that the heaters run about 2–3 % excess oxygen (10–15 % excess air).

Selective Catalytic and Non-catalytic NOx Reduction

Ironically, the introduction of ammonia (or another easily reacted nitrogen compound like urea) into the furnace at just the right location in just the right way reduces the amount of NOx emitted. The equation that this approach promotes overall is typically

$$4\,NO + 4\,NH_3 + O_2 \rightarrow 4\,N_2 + 6\,H_2O \tag{8}$$

In the non-catalytic application, ammonia (anhydrous or aqueous) is injected into flue gas at high temperature. The injection occurs through a grid in the flue gas that distributes the ammonia evenly into the large flue duct. Turbulent mixing of the ammonia with the flue gas creates the conditions for the reduction reactions. Temperature for this process is in the range 1,400–2,000 °F (760–1,090 °C). This is one of the approaches that has been used for boiler and FCC NOx control. This approach can generally reduce NOx by about 40 % from an uncontrolled level. Stack NOx, relative to the above table, would be about 30–80 ppm (0.033–0.085 lb NOx/MMBtu fired).

When the emissions requirements are very low (such as those for hydrogen plants, large boilers, crude and vacuum heaters, cogeneration units, and some FCCs), selective catalytic reduction is used. In an SCR system, vaporized ammonia or urea mixed with air is injected into a grid similar to SNCR, but the injection location is selected to be within a tight temperature range. There is a short turbulent mixing zone followed by a low-pressure drop, modular grid, or baskets containing NOx reduction catalyst. This system can result in 90–97 % reduction of NO_X from the uncontrolled level. Stack NOx would be down to the 4–10 ppm range (0.006–0.015 lb/MMBtu fired).

SCR catalysts are generally available in base metal and noble metal versions. There are also zeolite catalysts for high temperatures. The base metal catalysts use a ceramic substrate, like titania with vanadium, molybdenum, or tungsten oxides as the active metal. Noble metal catalysts typically use Pt or Pd. The operating temperature for base metal catalysts is typically around 600–750 °F (315–400 °C). Precious metal catalysts are effective at lower temperatures, say <570 F (<300 °C). Zeolites work at higher temperatures, up to 1,160 °F (627 °C).

The base metal catalysts tend to be the most common in refinery applications since they are cheaper than the alternatives and fit the normal temperature windows available in flue gas ducts.

Most SCR systems cannot be started until the flue gas reaches the required temperature range. This protects against accumulation of ammonium nitrate salts

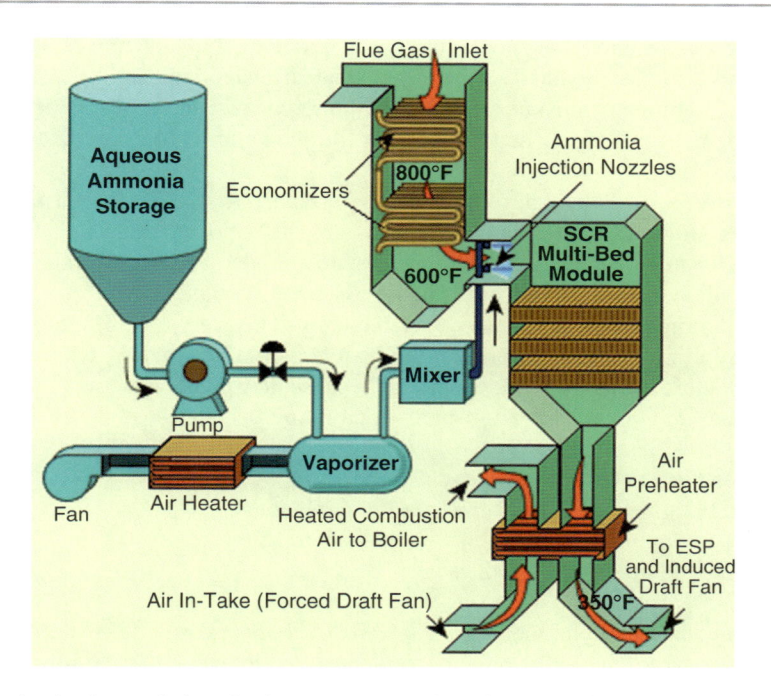

Fig. 5 A selective catalytic reduction system example (reference: United States Department of Energy)

on the exchangers downstream of the SCR. Ammonium nitrate accumulation creates the possibility of an explosion in the duct or, at the least, a significant loss of heat transfer in the downstream convection coils.

A typical SCR installation is shown in Fig. 5.

In operation of an SCR or SNCR system, the primary area of concern is the ammonia vaporizer and injection skid. If the skid fails, then NOx reduction stops, usually resulting in an immediate permit violation. The process flow in the vaporizer is simple. Atmospheric air from a fan is preheated to about 1,000 °F (537 °C) in an electric heater. Ammonia is sprayed into the hot air stream. A packed vaporizer may be used to ensure complete vaporization of the ammonia. The air/ammonia ratio in the vaporizer is carefully controlled to stay outside the explosive limits for ammonia in air. The air/ammonia mixture is sent to the ammonia injection grid. The grid may have several local meters and adjusting valves available to ensure even distribution to the grid.

The vaporizer systems have a few areas of concern:

- You are essentially mixing very hot air with a fuel in an enclosed system. An explosion is possible in the vaporizer if the air/ammonia ratio becomes explosive. These systems usually monitor the ratio and shut down if the ratio approaches the explosive range.

- Anhydrous ammonia is not a particular problem, but aqueous ammonia is corrosive to carbon steel and tends to plug the vaporizer and other systems. Aqueous ammonia must be filtered at a submicron level ahead of any metering or vaporization. All piping and tubing after filtration and before vaporization must be stainless steel.
- Packed vaporizers tend to plug. Spray-type vaporizers are more robust.
- It is advisable to have 100 % redundancy of the vaporizers and the air fans to allow for maintenance. A temporary system can also be built using anhydrous ammonia bottles and plant air, if necessary for short outages.
- The ammonia rates that must be metered are usually very tiny. The flow meters tend to plug with debris. Consider this in selection of the meters. Ultrasonic and mass meters tend to be more successful because they do not have small passages. Turbines and orifices are not usually good in this service.

NOx Emissions Monitoring and Operations

Many facilities have tight NOx emissions requirements imposed on them by permit. These require monitoring of the factors relevant to NOx. Among the more common monitoring and operating practices are:

- Continuous emissions monitoring (CEM) of the stack may be required, especially for large sources like hydrogen plants or boilers. The CEM information on NOx and CO should be watched and trended. Action should be taken on undesirable tends. Remember that you are balancing NOx against CO.
- Burner flame patterns should be examined at least every shift on all burners. Patterns that show stray flames, plugged tips, impingement, and other problems should be corrected immediately. If there is a pattern of tip fouling in a furnace, the fuel supply systems need to be reviewed for possible foulants and poor knockout facilities.
- Furnace excess oxygen needs to be kept within the correct range. Running high excess oxygen will increase NOx production and lower furnace efficiency.
- Ensure the ammonia vaporizer skid is working properly if you have an SCR or SNCR. Watch the blower/fan discharge flow and the preheated air temperature. Watch the ammonia/air ratio. Periodically check the flows to the injection grid.
- Trend the ammonia rate normalized for firing rate – this will increase as SCR catalyst ages. Your permit will often have a maximum ppm ammonia in the flue gas. When this is approached, you need to plan to replace the SCR catalyst. The usual limit is around 10 vppm. Test the flue gas periodically for the residual ammonia concentration.
- Watch your flue gas temperatures to ensure that fouling does not force the SCR into a temperature range outside optimum.

So, in short:

1. Keep your fuel gas clean.
2. Operate your burners correctly.

Environmental Control and Engineering in Petroleum Processing

3. Keep your burners clean.
4. Keep your excess oxygen under control.
5. Make sure your SCR vaporizer is working.
6. Make sure your SCR catalyst is working properly at the right temperature.

Carbon Oxides (CO, CO_2)

Emissions of carbon monoxide have been regulated for many years. The emissions of carbon dioxide are only now beginning to come under regulation as part of the concerns over global warming. In this section, we will explore management of emissions for both of these gases.

Carbon Monoxide

Carbon monoxide is formed primarily by incomplete combustion of hydrocarbon fuels. In a petroleum processing facility, CO will be emitted from fired heat in several applications.

- Process heaters
- FCC regenerator
- Boilers
- Cogeneration units
- Incinerators and combustors
- Flares

The objective is complete combustion of fuel in the fired services. This will ensure efficient use of fuel while keeping CO emissions low. Some heater permits will also have CO emissions limits, including requirements for a CEM.

The flue gas from a combustion source can be analyzed using an Orsat apparatus. The Orsat provides readings for oxygen, CO, and CO_2 in a flue gas. It has largely been replaced by simpler, online, and portable analytical instruments.

Control of CO emissions parallels, and sometimes conflicts, with NOx control in these heaters. The primary control methods for CO are:

- Maintaining correct excess oxygen levels at the burners.
  - Avoid tramp air entry.
  - Keep your registers and draft under control.
  - Keep your oxygen analyzer working.
- Keeping burners clean
  - Clean up your fuel gas.
  - Clean tips as needed.
- Minimize flaring to reduce CO from the flare
- In FCCs:
  - Ensure good combustion air distribution to the regenerator.
  - If you have a CO boiler, ensure it is working properly.

Carbon Dioxide (CO_2)

In a petroleum processing plant, carbon dioxide is produced from several sources, including:

- Fired heaters, boilers, incinerators, and combustors
- Cogeneration
- Flare
- Hydrogen plant process
- FCC regenerator

Control and minimization of CO_2 emissions from these sources include:

- Fired services:
  - Maintain excess air in the required range (no tramp air).
  - Keep coils and fins in heaters clean (inside and outside).
  - Use air preheat to reduce firing rate (may increase NOx).
  - Do not heat more than necessary (e.g., do not over-treat products).
  - Maximize use of process preheat within control limits.
- Flare:
  - Minimize flaring.
  - Recover make gases for fuel use.
- Hydrogen plants:
  - Minimize excess on-purpose hydrogen production.
  - Minimize hydrogen losses to fuel gas.
  - Recover CO_2 for sale as a product.
- FCC:
  - Ensure good combustion air and catalyst distribution in the regenerator.
  - Maintain correct combustion conditions.
  - Watch heat balance and conversion.

As concerns over greenhouse gases increase, CO_2 is one of the largest concerns worldwide. Facilities are being driven to reduce CO_2 emissions and, in some cases, may also be held responsible for the emissions that result from use of their products. All you can do as a petroleum processor is to minimize your generation of CO_2 through good practices and efficiency. In more extreme cases, you may have to recover CO_2 for disposal or sequestration.

Particulate Matter (PM, PM-10)

Particulates can come from several sources within a refinery. Figure 6 illustrates the size ranges for various types of particulates. Of particular concern when discussing particular matter are particles smaller than 10 μm (PM-10). These particles can be inhaled and stay in the lungs, causing respiratory damage. They are specifically regulated in the United States.

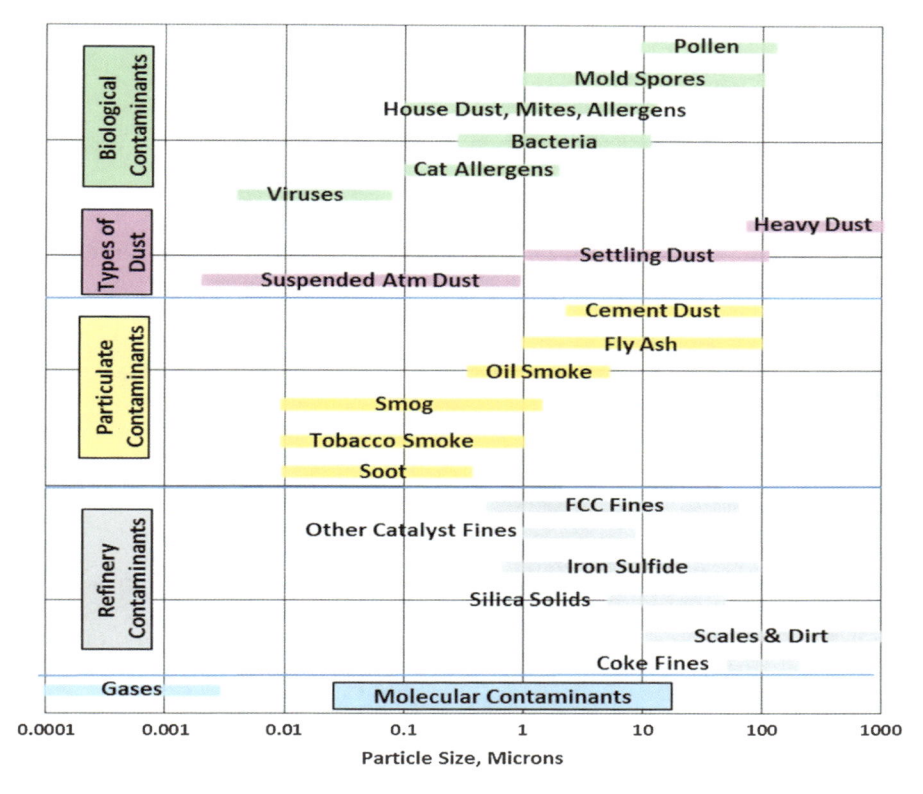

Fig. 6 Typical particulate matter size ranges

Among the more common sources are:

- Incomplete fuel combustion (<10 μm)
- FCC (0.5–80 μm)
- Catalyst and sorbent handling (1–10 μm)
- Construction activities (<10–100 μm)
- Maintenance activities (<10–100 μm)
- Decoking/soot blowing (<10 μm)
- Unpaved roads and operating areas (1–1,000 μm)

We will talk about these sources and some possible particulate control measures.

Incomplete Combustion

In burning various fuels, incomplete combustion will generate soot that can be emitted from the stack. Maintaining normal excess oxygen levels in the fired services ensures levels of soot in the stacks are controlled. In some large stacks, opacity monitoring may be used to ensure minimal PM emissions. Permits sometimes require PM emissions be tested periodically using an apparatus that takes isokinetic samples across the entire stack.

Fluid Catalytic Cracking (FCC)

In a refinery, an FCC is the biggest potential generator of PM emissions. The emission of the particulates can arise from an operating problem or from a breakdown of protection equipment, like the cyclone separators located in the critical areas of the process or the electrostatic precipitator.

There is always some carryover of small particulates from the regenerator. In most modern units, these catalyst fines are collected by cyclone separators followed by bag houses or electrostatic precipitators. The efficiency of these devices is limited. Some FCCs are driven to wet scrubbing of the FCC flue gas. This can catch the PM as well as SOx and NOx in the flue. The collected fines and solution then become an easier waste disposal issue. Failure of these systems may require shutdown for repair. Regular rapping of an electrostatic precipitator is used to keep it functioning.

A breakdown in the catalyst fines collection systems (like a precipitator ground fault), as well as certain startup and upset conditions (like a reversal), will result in large amounts of PM emitted to the atmosphere. One of the most common problems in fluid bed processes are air surges that cause a disruption of the fluidized bed. In the case of an FCC, this fluidized bed is the catalyst bed being regenerated in the regenerator. This disruption results in loss of the catalyst to the atmosphere from the regenerator exit stack. The incorporation of a CO boiler or an electrostatic precipitator does help minimize this emission, but they can be overwhelmed. Usually air surges are minimized by good flow control systems with some anti-surge system. Control of the startup and upset emissions primarily depends on moving as quickly as possible through the condition. In extreme cases, a unit may be forced to shut down instead of continuing to emit PM.

Catalyst and Sorbent Handling

Many refining catalysts and sorbents used in petroleum processing are solid extrudates, spheres, or pellets. These include hydrotreating, hydrocracking, reforming, isomerization, and Claus alumina catalysts as well as PSA adsorbents. The catalysts themselves are relatively small in size (1/20–1/8 in. nominal diameter) but are large enough not to become fugitive PM emissions.

However, the dust produced by attrition in handling the catalysts is small enough in size to become airborne. This is especially an issue during turnarounds and catalyst changes. Maximum attrition loss for catalysts is normally a specification.

FCC and other types of fluidized catalysts are small by their nature and they are abraded in service to generate dust. They easily can become fugitive PM emissions as the catalyst fines are moved around. This occurs where the fines are collected at the electrostatic precipitator or where fresh or equilibrium catalyst is transferred into the unit.

Catalyst dust may carry additional chemical health hazards, as well as PM-10 hazards. Nickel-containing catalysts are potential carcinogens because of the nickel content. FCC and other zeolite catalysts contain silica.

Control of PM from catalyst and sorbent handling can be accomplished in several ways, depending on the situation:

- Spills of fines or catalyst can be kept wet to suppress dust.
- Wet down and clean up spills immediately.
- Don't allow vehicles to drive over spilled catalyst or sorbent and crush it to powder.
- Many catalysts are handled in an inert or enclosed system using vacuum or pneumatics to move the material. These systems use cyclones and bag filters to prevent atmospheric dust.
- Personnel protection should prevent exposure of employees to the catalyst and sorbent dusts during handling. These measures would include HEPA filters as a minimum up to use of fresh air equipment.

Construction Activities

Most construction activity is going to involve earthmoving. The fugitive dust for moving soil around is sometimes difficult to control. Many construction permits require some sort of dust control.

In a petroleum processing facility, the soil may also be contaminated by chemicals from past activities when people weren't so environmentally conscious. Lead, caustic, acid, and other soil contamination may be encountered as construction activities progress. Often this material has to be completely removed for hazardous waste disposal. This generates even more potential for dust.

In some cases, construction will require removal or revamp of old equipment that may have been painted with lead-based paints or insulated with asbestos. Specific regulations may address the management of these materials. In the United States, they must generally be removed and handled by personnel or companies specifically trained and equipped for the activity. The area is usually tented and personnel wear fresh air equipment. They may need to go through decontamination after working in the area.

Aside from the specific hazardous dusts, keeping disturbed soil damp, including spraying of dust suppressants, is the normal approach to managing the PM fugitive emissions from construction. Common dust suppressants (from the Michigan Department of Natural Resources and Environment guidelines) are:

- Water
- Calcium or magnesium chloride
- Lignosulfonate products
- Emulsified asphalt or resin stabilizers
- Vegetable by-products

Use of dust suppressants is not without other environmental consequences. The impacts on groundwater need to be considered.

Maintenance Activities
The normal activities inherent in maintenance can generate fugitive PM. We have already talked about a few of these: catalyst and sorbent handling, dirt movements, and removal of lead paint or asbestos insulation.

A few other maintenance activities that should be mentioned are:

- Abrasive blasting
- Exchanger and tank cleaning
- Grinding and welding

Abrasive blasting of small parts may be done in a controlled system in the shop. These are not particularly a problem. For larger outdoor areas, PM can usually be controlled by temporarily confining the dust in the blast area using a tent. The grit itself tends to not be entrained in air, but the rust, paint, and other material being blasted off can become airborne. If dust becomes excessive from these operations, some additional thought should be put into how they are being done and how to better contain the dust. A wet spray may be needed.

Like abrasive blasting, exchanger and tank cleaning can generate fugitive PM as deposits, sludges, and other residues are removed and dry. The usual approaches to dust control include keeping the materials wet, cleaning up spills promptly, and not driving through the material and generating dust in the first place.

Grinding and welding can generate some PM. Some of the PM may contain hazardous forms of metals. The small quantity generated is not normally a general environmental problem but may present a personnel hazard in the immediate area. Attention to good industrial hygiene practices in these activities should be adequate to manage fugitive dust.

Decoking and Soot Blowing
These could be classified as maintenance activities, but they are specifically listed here. Decoking is the removal of coke and other deposits on the process (inside) of fired heater tubes. Soot blowing is the removal of carbon or soot deposits on the outside of heater tubes. Both activities can generate fugitive PM, as well as other emissions.

At one time, steam-air decoking was the common approach to the removal of process-side tube deposits. In this approach, steam was run through the tubes at high velocity to spall the solids off the tube walls. The spalled solids were collected in a knockout drum vented to atmosphere. Dust from the drum could escape. After spalling, air was introduced with the steam to burn out remaining deposits, generating more potential for PM emissions.

Today, spalling may be done online with the spalled material retained in the process. This is especially common in cokers where the spalled material is collected along with the coke.

Another approach that has become common in process-side tube cleaning is "pigging" where fairly large balls or other-shaped devices are pushed through the tube to physically abrade the deposits off the tubes. The removed material is then collected in a system that controls the PM emissions.

In soot blowing, the soot collected on the firebox sides of furnace tubes is literally blown off the tubes using air or, more commonly, steam. The soot thus removed leaves the furnace with the flue gas. This soot will also contain heavy metals present in the fuel and deposited on the tubes. Management of PM from soot blowing depends on the removal of the soot from the flue gas. Wet scrubbing, filtration, or other flue gas treatments on a furnace will catch the soot. Since soot is primarily the result of fuel oil burning, firing a furnace on fuel gas and eliminating the fuel oil would also eliminate soot blowing PM emissions. Soot blowing is not allowed in many jurisdictions.

Unpaved Roads and Operating Areas

In many facilities, the roads outside the main operating areas, and many operating areas themselves, are not paved. The exposed soil can become fugitive dust as wind or vehicle traffic picks it up.

Use of dust suppressants (see the list above) and keeping the exposed soil damp can greatly reduce fugitive dusts.

In tank dikes, sleeper ways, and below pipe racks, it is also possible to chip and seal or oil the areas. In the chip-and-seal approach, gravel is laid down in a layer a couple of inches thick and then sprayed with tar or asphalt to seal it. This forms a hard crust over the ground that resists dust generation. Oiling an exposed area just means spraying it with tar or asphalt to seal the ground directly. Local environmental practices for air or groundwater protection may limit the chip-and-seal or oiling options. The seal must be renewed about every 10 years by re-spraying the area with tar.

Of course, exposed areas can always be paved to eliminate PM generation.

Volatile Organic Carbon (VOC)

The major sources of atmospheric pollution by VOC emission in refinery include:

- Relief and vent valves open to atmosphere
- Control valve, flange, seal, and packing leaks
- Storage tanks
- Rail and road tanker filling facilities
- Ship loading and unloading facilities (jetty area)
- Water effluent treatment facilities and sewers

Attention to control of these sources continues to push toward more monitoring, more controls, and better operating practices. The following discussion will address some of the key VOC emissions issues.

Relief and Vent Valves

Relief and vent valves normally discharge to a closed relief header which is routed to flare system. However, in some cases, because of the location of the relief and/or vent valve and the quantity of the discharge material, the valve(s) are open to atmosphere. A typical example is crude distillation unit relief. These are located at the top of the distillation towers, about 200 ft. (60+ m) above grade. Steam injection facilities may be installed at the valve discharge to facilitate atomization and dispersion. Nevertheless when these valves do open, there will be some considerable VOC emission to atmosphere. Upset events that would cause these valves to discharge are extremely rare and do not usually warrant the extensive costs of increased flare header and flare design that would be necessary to cater for these valve discharges. Still, many refineries are making the necessary changes to include these valves in the flare systems.

A multiple fatality explosion event at Texas City in the United States has prompted several refiners to review the use of unfired vent stacks and US regulators to more tightly review these VOC sources. Most refiners are eliminating any unfired vents. The trend is to route all hydrocarbon or VOC vents to the relief and recovery system or flare.

Control Valve, Flange, Seal, and Packing Leaks

Leaks from control (and other valves), pump seals, packing, and flanged joints are related entirely to the refinery's maintenance policy and program.

In the United States and other locations, there are required leak monitoring programs that involved "sniffing" all the flanges, packings, seals, and other potential leak points on a regular basis, usually quarterly, with a leak detection instrument. The sniffing may be done using a combustible gas detector or may even use a portable mass spectrometer. Fugitive VOCs identified above specific levels require repair of the identified leak point within specific, defined time frames and proof that the repair was made. The repair may be as simple as tightening a packing follower or may involve clamping and injecting an exchanger channel-to-shell flange.

Seal leaks of VOCs from rotating equipment are often controlled through the use of double or tandem seals, often with barrier fluids, with appropriate maintenance programs for these seal systems. In some parts of the United States, a leak of as little as a drop or less per minute of a VOC from a rotating equipment seal may require repair actions that may include shutdown of the equipment.

Storage Tank Emissions

Perhaps the biggest sources of VOCs from a refinery and its operation are the storage tanks. Emission occurs during the tank filling, particularly in the case of fixed roof tanks and LPG bullets and spheres, or by evaporation of tank contents.

The emissions from floating roof tanks are generally low, and the function of a properly installed roof and seals should minimize the emission from these sources. Highly volatile liquids should not normally be contained in floating roof tanks, but there are occasions when high vapor pressure material may find its way into a tank. These upsets need to be dealt with promptly when they occur to avoid damage to the roof as well as the VOC emission. The seals need to be inspected and repaired regularly. These emissions controls are discussed further in the off-sites section of this handbook: "▶ Off-Site Facilities for Petroleum Processing."

Many of today's refineries have installed closed inert gas blanketing for fixed roof tanks containing high vapor pressure hydrocarbons using nitrogen or natural gas. Blanketing is sometimes required in intermediate tanks whose contents are feed to processes which need to avoid oxygen contamination or contact of the feed. Extension of this inert gas to all fixed roof tanks as a blanket can prevent the emission of VOC during their movements. These systems are also discussed in the off-sites section of this handbook.

Tanks containing low vapor pressure materials (say <2–3 psi RVP) are not generally a VOC emission problem, so do not usually have inert gas blankets.

The storage of LPG also has potential for emitting VOCs to the atmosphere. One method used to combat this is to float the make/break valves of the spheres and bullets on the refinery fuel gas main header.

In highly regulated areas, the VOC controls may extend to smaller storage tanks or sumps. Often VOCs from the small tanks or sumps can be managed by connection to carbon canisters through a flame arrestor.

Rail and Road Tanker Filling Facilities

Spills from truck and railcar loading facilities have always been a problem. Most spills occur when filling the vessels. Most loading facilities today have automatic shutoff on the loading arm nozzles in the event of failure, very similar to those commonly used in public filling stations. These use a level-sensitive device to close the filling valve at a prescribed filling level in the tank or on excessive flow rate. The other source of VOC emission is a poorly designed slops/recovered oil system that allows the draining of bottoms from road tankers and railcars. Such a system should allow for steam-out facilities on vessels carrying heavy petroleum cuts. Vents on this system should be routed to the flare header.

Ship Loading and Unloading Facilities (Jetty Area)

The jetty along with ship loading and unloading is also a potential source of VOC emissions. Again, as in the case of the tank farm and product loading facilities, the most vulnerable product in this respect is the handling of the light hydrocarbon streams. As the ships or barges are filled, the vented vapors will contain VOCs. The vapors should be routed to an incineration system through a flame arrestor to ensure the displaced vapors are combusted to eliminate VOCs.

In the loading of LPG, a flash recycle system has been found to be effective. In this system, the LPG flashes as it first enters the ship's empty tank at close to atmospheric pressure and temperature. This flashed vapor is routed, under pressure control, to a compressor and cooler system, where it is liquefied and returned to the LPG feed stream. The system continues to operate throughout the loading activity, acting as the loading relief system.

Water Effluent Treatment Facilities and Sewers

The sewer and wastewaters collected from a petroleum processing facility will contain hydrocarbons, some of which will flash in the system to generate VOCs.

This is especially notable if any light ends or naphtha is drained or spills into the sewer.

The methods for control of fugitive VOCs from sewer and wastewater treating systems are addressed in detail in the utilities and off-sites discussions of this handbook ("▶ Utilities in Petroleum Processing" and "▶ Off-Site Facilities for Petroleum Processing." The approaches generally involve sealing sewers and sumps with vents through flame arrestors to hydrocarbon collection systems, such as canisters. Oil water basins may be covered by membranes, with VOC vapors collected and recovered. API separators are often covered for vapor collection or may have floating roof covers with skimming and seals similar to a floating roof sour water tank.

A dissolved air floatation unit (or DAF) may be a particularly difficult source of VOC emissions because the process involves deliberately bubbling air through an oily water. Ideally, you want to separate any volatile hydrocarbons from wastewater before it is routed to a DAF.

Aerobic digestion in a trickle bed, activated sludge, or other bio unit has the potential for creating VOC fugitives. Light hydrocarbons should be eliminated from the wastewater before it is routed to a "bug" unit. The bio unit is usually the last processing unit in the water effluent treatment system.

There is further discussion of effluent water treatment later in this chapter.

Ozone

Ozone is not normally directly emitted by the refinery. It is primarily the product of the atmospheric reactions of other air pollutants. NOx catalyzes the formation of ozone in the lower atmosphere through the reaction path:

$$O_2 + UV\ light(< 240\,nm) \longleftrightarrow 2O \tag{9}$$

$$O_2 + O + NO_x \longleftrightarrow O_3 + NO_x \tag{10}$$

VOCs can also contribute to ozone generation.

Control of ozone from a refinery generally comes down to control of the compounds that contribute to its formation, which we have already addressed.

One additional, but uncommon, ozone source in a refinery can be disinfection of water using ozone directly. It is a strong disinfectant and can replace chlorine or bleach, but the cost is usually higher to generate ozone and it offers no residual disinfectant action. When used, ozone is generated and used immediately. Fugitive ozone does not survive long in the air, so it is not really an emissions concern, except in the immediate area of use.

Ammonia (NH$_3$)

Ammonia can usually be emitted as a fugitive gas:

- Where ammonia is handled
- As a residual from NOx reduction equipment

- Through sour water leaks
- From the hydrogen plant degasifier vent

In the air, ammonia is relatively quickly removed by formation of ammonium salts and by rain, with some pH implications.

Ammonia supply systems for SCRs, SNCRs, and other uses in a refinery offer the chance to leakage and emissions of ammonia. Attention to maintenance and operations of these systems will prevent any significant releases. Methods to manage ammonia releases from a storage and distribution system are detailed in the section of this handbook concerning "▶ Utilities in Petroleum Processing."

When used in NOx reduction processes, a slight excess of ammonia is normally added. The residual should normally be less than 10 vppm in the flue gas. Still, 10 ppm in a large flue gas stream can represent a significant amount of ammonia. Control of this ammonia emission centers around control of the NOx removal process so the minimum amount of ammonia is used. Refer to the discussion of NOx reduction above.

Sour water contains ammonia in significant concentrations. The ammonia is tied up as a bisulfide, normally. As a result, the ammonia emitted is usually low. The most significant risk of ammonia emission from sour water would be in storage tanks where good tank sealing practices, as discussed in the chapter on "▶ Off-Site Facilities for Petroleum Processing," would minimize losses to atmosphere. If the ammonia reaches the sulfur recovery plant, it will be destroyed in the Claus reaction furnace or an oxidizer, so it is not emitted.

A hydrogen plant makes ammonia in trace quantities. This is especially true in plants running a low steam/carbon ratio. The ammonia ends up in the shift condensate and is emitted when the condensate is stripped in the degasifier/deaerator. The amount emitted is generally below regulatory interest, but newer plants running very low steam/carbon ratios can make a significant amount of ammonia. In these units, prevention of ammonia emissions entails either sending the degasifier off-gas to the hydrogen plant SMR furnace to be combusted or the degasifier may be pressurized with the off-gas plus steam sent to the SMR process side.

For most petroleum processing facilities, ammonia would not be a major problem emission.

Methanol (CH_3OH)

Methanol often gets lumped with VOC emissions, so most control methods for methanol are similar to those of VOCs.

One exception is in hydrogen plants where methanol can be made in significant quantities in low temperature shift reactors. In these cases, the methanol ends up in the shift condensate, which is routed to the degasifier for reuse. The methods noted above for control of fugitive ammonia from hydrogen plants also effectively capture and destroy methanol to prevent its emission. It is also possible to use special low methanol shift catalysts to reduce the process make of methanol.

As with ammonia, methanol is generally not an emissions problem for most refiners.

Water and Aqueous Effluents

In this section, we will discuss the aqueous or water-based refinery streams and their dispositions. The basic aqueous waste stream origins were highlighted initially in Table 1 for reference. Now we will explore these wastes and their management in greater depth.

The discussion is divided into a description of the pollutants in aqueous wastes, the processes that generate the wastes, and the treatment of the various aqueous waste streams. Recycle opportunities are highlighted where these are feasible as a means of waste reduction.

Pollutants in Aqueous Waste Streams

The potential pollutants in refinery aqueous streams are conveniently listed in greater detail, with some typical rates and pollutants, in Table 4, from a US EPA report.

A number of laws and regulations populate the regulatory landscape controlling the allowable contaminants in water effluents and permit requirements. A sampling of the key, applicable regulations includes:

- US laws and amendments:
  - EPA regulations:
    Clean Water Act (CWA)
      Permit requirements, including National Pollutant Discharghe Elimination System (NPDES) permits
      Storm Water Pollution Prevention Plans (SWPPP)
      Spill Prevention, Control, and Countermeasures (SPCC) plans
    Clean Air Act (CAA) and amendments (CAAA, 1990)
      Title V permitting
      New Source Performance Standards (NSPS)
      National Emission Standards for Hazardous Air Pollutants (NESHAP, especially benzene, relates to water)
    Resource Conservation and Recovery Act (RCRA)
    Emergency Planning and Community Right-To-Know Act (EPCRA)
      Notifications of local emergency agencies
      Release notifications
      MSDS and inventory reporting requirements
      Requirement for annual toxic release inventory reporting
    Safe Drinking Water Act (SDWA)
      Regulation of drinking water
      Regulation of underground water injection
  - Numerous state and regional laws and regulations, over and above state regulations

Table 4 Some of the aqueous wastewater streams and the potential pollutants from a refinery (Source: "Leavitt et al. (2004))

Process	Wastewater description (possible pollutants)	Wastewater flow rate (gallon/barrel of crude petroleum)	Percentage of total wastewater flow rate
Distillation	Sour water (hydrogen sulfide, ammonia, suspended solids chlorides, mercaptans, and phenol)	26.0	44 %
Fluid catalytic crackling	Sour water (hydrogen, sulfide, ammonia, suspended solids, oil phenols, and cyanides)	15.0	26 %
Catalytic reforming	Sour water (hydrogen, sulfide, ammonia, suspended solids, mercaptans, oil)[a]	6.0	10 %
Alkylation	Spent potassium hydroxide stream (hydrofluoric acid)	2.6	4 %
Crude desalting	Desalting wastewater (salts, metals, solids, hydrogen sulfide, ammonia, and phenol)	2.1	4 %
Thermal crackling/ visbreaking	Sour water (hydrogen, Sulfide, ammonia, suspended solids, dissolved solids, and phenol)	2.0	3 %
Catalytic hydrocracking	Sour water (hydrogen sulfide, ammonia, solid suspended solids)	2.0	2 %
Coking[b]	Sour water (hydrogen, sulfide, ammonia) and suspended solids	1.0	2 %
Isomerization	Sour water (hydrogen sulfide and ammonia) and caustic, wash water (calcium chloride or other chloride salts)	1.0	2 %
Additive production: ethers, manufacture	Pretreatment wash water (nitrogen contaminants)	<1.0	
Catalytic hydrotreating	Sour water (hydrogen, sulfide, ammonia, suspended solids, and phenol)	1.0	2 %
Chemical treating: sweetening/ Merox process		[c]	
Sulfur removal/ Claus process	Sour water (hydrogen sulfide and ammonia)	<1.0	
Lubricating oil manufacture	Stream stripping wastewater (oil and solvents) and solvent recovery wastewater (oil and propane)	<1.0	
Total		58.7	100 %

US DOE. *Water Use in Industries of the Future: Petroleum Industry*. July 2003
[a]Additional pollutants identified in EPA's *Industry Sector Notebook: Petroleum Refining*, September 1995
[b]Fluid coking produces little or no effluents
[c]Little or no wastewater generated

- Several European Union directives are applicable to wastewater; members of the Union implement these individually. Among the requirements are:
  - Registration, Evaluation, Authorisation and Restriction of Chemicals (REACH) – registration and evaluations
  - European Pollutant Release and Transfer Register (E-PRTR) – reporting
  - Industrial Emissions Directive (IED) – reduction requirements
  - Water Framework Directive (WFD) – obligation of member countries to achieve water quality and reduce effluents
  - Groundwater Directive (GWD) – similar to WFD applicable to groundwater
  - Environmental Quality Standards Directive (EQSD) – specific limits for priority substances, obligations to inventory and reduce losses
  - QA/QC Directive – monitoring standards
  - Marine Strategy Framework Directive (MSFD) – similar to WFD for marine environment
- Similar regulatory structures apply throughout most of the world. Some regions are less restrictive.

In general, the regulations target the most undesirable types of pollutants in aqueous waste streams:

- Those that deplete the dissolved oxygen content of the waterways into which they discharge
- Those contaminants that are toxic to almost all forms of life, such as arsenic, cyanide, mercury, selenium, and the like
- Those contaminants that impart undesirable tastes and odors to streams and other waterways into which they discharge

A description and discussion on these contaminants are expressed in the following paragraphs.

Impacts of Contaminants on Oxygen Balance in Natural Waterways

Natural waterways have a complex and delicate oxygen balance. The water contains an amount of dissolved oxygen which has an equilibrium of between 14 ppm in winter and 7 ppm in summer. Aquatic life (fish and tadpoles) continually consumes oxygen from the water, but aquatic plants produce oxygen naturally maintaining the balance required to sustain fish life.

Any oxidizable contaminant introduced into the natural waterways consumes the dissolved oxygen to be oxidized. The dissolved oxygen will be depleted below the saturation point and will be replenished only by re-aeration. The relative rate of replenishment to its saturation level will depend on a time factor related to the stream distance or the stream flow rate.

Oxygen depletion occurs by the introduction of one or more of the following contaminants into the waterway:

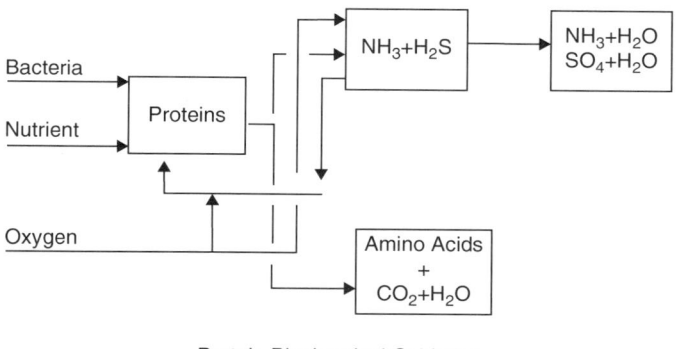

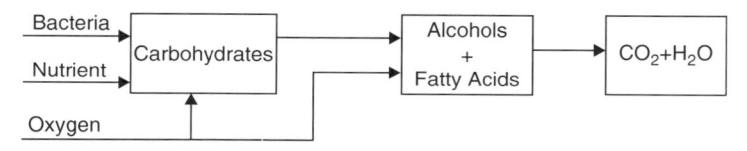

Fig. 7 Typical BOD chemistry

- Natural pollution by surface runoff rainwater, or melting snow, in the form of soluble salts leached from the earth – consider paved and parking areas
- Natural pollution caused by decay of organic plants from swamps or other sources
- Human and animal life excretion – consider sanitary sewer systems
- Chemical pollution from reducing agents in industrial plant wastes. Such as sulfides, nitrites, ferrous salts, etc.
- Biochemical pollution from such industrial wastes as phenols, hydrocarbons, carbohydrates, and the like

The degree of oxygen depletion from the pollution sources described above may be catalogued by the following terms:

BOD – biological oxygen demand
COD – chemical oxygen demand
IOD – immediate oxygen demand

Biological Oxygen Demand (BOD)
Since all natural waterways will contain bacteria and nutrients, almost any waste compound introduced into the waterway will initiate biochemical reactions. These reactions will consume some of the dissolved oxygen in the water. This is illustrated in Fig. 7.

Fig. 8 Lab determination of 5-day BOD versus ultimate BOD

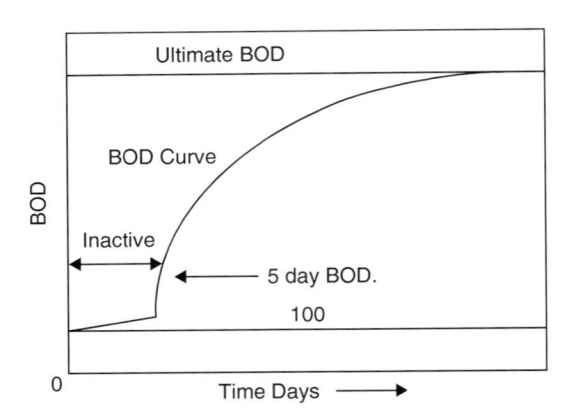

The depletion of oxygen due to biological pollution is not very rapid. It follows the laws of a first-order reaction. Because of this, the effect of BOD is measured in the laboratory on a 5-day basis and has been universally adopted as the measure of pollution effect.

The "ultimate" BOD is a measure of the total oxygen consumed when the biological reaction proceeds to near completion. The "5 day" BOD is believed to be approximately the ultimate. Figure 8 illustrates how the "5 day" BOD relates to the ultimate BOD.

In summary BOD measures organic wastes that are biologically oxidizable.

Chemical Oxygen Demand (COD)

The COD is a measure of the oxygen depletion due to organic and inorganic wastes which are chemically oxidizable. There are several laboratory methods accepted to measure the oxygen depletion effect of this pollution. The two most widely accepted are the "4 h permanganate" method or the "2 h dichromate" method. Although there is no generalized correlation between BOD and COD, usually the COD will be larger than the BOD. The following table illustrates how different wastes exhibit a different relationship between COD and BOD (Table 5).

Immediate Oxygen Demand (IOD)

Oxygen consumption by reducing chemicals, such as sulfides and nitrates, is typified by the following equations:

$$S^{2-} + 2O_2 \rightarrow SO_4^{2-}$$
$$NO_2^- + \tfrac{1}{2}O_2 \rightarrow NO_3^-$$

(11, 12)

These types of inorganic oxidation reactions are very rapid and create what is measured in the laboratory as immediate oxygen demand (IOD). If waste contaminants contain these inorganic oxidizers, the "5 day" BOD test will include the consumption of the oxygen due to IOD also. A separate test to determine IOD must be made and this result subtracted from the "5 day" BOD to arrive at the true BOD result.

Table 5 Relationship between BOD and COD

Source	Pollutants	BOD "5 day"	COD "2 h dichromate"
Brewery	Carbohydrates, proteins	550	–
Coal gas	Phenols, cyanides, thiocyanates, thiosulphates	6,500	16,400
Laundry	Carbohydrates, soaps	1,600	2,700
Pulp mill	Carbohydrates, lignins, sulfates	25,000	76,000
Domestic sewage	Solids, oil and grease, proteins, carbohydrates	350	300
Petroleum refinery (sour water)	Phenols, hydrocarbons, sulfides	850	1,500
Petroleum refinery	Phenols, sulfides, hydrocarbons, mercaptans, chlorides	125	2,600

Toxic Pollutants Common to Oil Refining

Toxic pollutants that are of greatest concern in untreated refinery aqueous wastes include:

Hydrocarbons, Oil, and Grease

Heavy oil and other hydrocarbons are the most problematic pollution to be found in refinery water effluent. All refineries exercise the most stringent methods to control and remove these undesirable pollutants. Indeed, in many cases, the treated effluent streams leaving the refinery may well be purer than incoming potable water used in the processes.

Aromatics are a particular problem and are specifically regulated to very low ppb levels in effluents. Benzene is specifically targeted by both water and air regulations.

Phenols

These chemicals are often formed in refinery processes such as catalytic and thermal crackers. They are highly toxic to aquatic life in concentrations of 1–10 ppm. Apart from their toxicity, phenols also impart an unpleasant taste and odor to drinking water in the range of 50–100 ppb. In concentrations of 200 ppm and higher, these chemicals can also deactivate biological water treatment plants such as trickle filters and activated sludge units. This could include a downstream publicly owned treatment works (POTW) that received refinery effluent water.

Caustic Soda and Derivatives

Solutions containing sodium hydroxide (NaOH) and related chemicals (like soda ash) are used in a number of refinery processes and maintenance activities. Inevitably some of these chemicals enter the refinery's wastewater system. Most of these contaminants are toxic to living organisms, even in low concentrations. The spent caustic (compounds leaving the process) such as sodium sulfide can be even more injurious.

Aqueous Solutions of Ammonium Salts
The most common of these are ammonium sulfide (NH_4SH) and ammonium chloride (NH_4Cl). Both these salts are present in effluent water from the crude distillation unit overhead accumulator; from the cracking processes, and from hydroprocessing units (hydrocrackers and hydrotreaters). Other ammonium salts are also present in hydrocracking and deep oil hydrotreating sour waters and from processing of some alternative feeds and renewable stocks.

Acids in Aqueous Effluents
The most common of the acids in wastewater are from the alkylation processes, which use either hydrofluoric acid or sulfuric acid. In the reforming and isomerization processes, hydrochloric acid is also used as a catalyst promoter. In some older processes (e.g., Edeleanu), sulfur dioxide is used to remove aromatics. Sulfuric acid is also used at cooling towers and other locations as an additive, where spills may reach the water treatment system. The most common acidic effluent form is dilute sulfuric acid.

Ketones, Furfural, and Urea
These compounds are used in the refining of lube oils. MEK and urea are used in some dewaxing processes. Furfural is used in some extraction processes for finished lube oil stock. All of these compounds are toxic.

Selenium and Metals
The recent years have seen an increase in the regulations to control toxic metals in effluents. Specific emphasis has been placed on selenium in refining because of its bioaccumulation effects. Selenium occurs naturally in crude oil at very low concentrations (0.03–1.4 ppm). When oil, especially distillate, is refined, the sour water produced in processes like hydrotreating contains enough selenium to require separate treatment to reduce the selenium to ppb levels in the refinery effluent.

Treating Refinery Aqueous Wastes

The treatment options for aqueous wastes from oil refineries fall into three general categories:

- In-plant treatment – these are onsite processes, usually sour water strippers, spent caustic oxidizers, and spent caustic neutralizers
- The API separator, or similar oil/water separating device
- Secondary treatment, which includes chemical coagulation, activated sludge processes, trickle filters, air flotation, and aerators

Most energy refineries contain the first two of the above categories. Some refineries also have secondary treatment, especially if they discharge effluent directly to a stream or body of water.

We will explore some specific treatments in more detail in this section, including:

- Sour water strippers
- Spent caustic disposal
  - Neutralization
  - Oxidation
- Oil/water separation
  - API separator
  - Corrugated plate interceptors
- Storm surge ponds
- Other processes
  - Oxidation ponds
  - Air floatation/dissolved air floatation (DAF)

Sour Water Strippers

Sour water stripping is one of the most common treating processes. Its purpose is to remove the pollutant gases included in the sour waters produced in the facility. The primary sour water pollutants are ammonia, ammonium salts, and hydrogen sulfide. The sour water stripper is often located in the sulfur recovery area of the refinery, because the stripper off-gas is usually routed to the sulfur recovery unit for disposal.

Feed to the stripper may include effluent water from the crude unit overhead condensers, water phase from the desalters, condensed water from the vacuum unit's hot well, and all the water condensate and wash waters from the hydroprocessing units.

A sour water stripper is usually a trayed tower (with about 20 trays) with no reflux or a similar tower with an overhead reflux stream. Figure 9 shows a simple stripper tower with no reflux and Fig. 10 shows a stripper with an overhead condenser and reflux. In many refineries, there are two or more sour water strippers (SWS) that feed different water streams:

- Sour water containing phenols (AKA phenolic sour water)
- Sour water without phenols (AKA non-phenolic sour water)
- Dedicated hydroprocessing unit wash water stripper

Separation of the streams helps minimize waste, especially a dedicated hydroprocessing wash water stripper. With a separate hydroprocessing wash water stripper, the stripped water can usually be reused in the units if it meets certain quality limits and has not collected nonvolatile salts.

These types of sour water strippers lend themselves to tray-by-tray mass and heat transfer calculations for design. The sour water feed is introduced on the top tray of the tower while steam, usually at a rate of 0.5–1.5 lbs/gal of feed, is introduced below the bottom tray. The steam is usually generated by reboiling the bottoms from the tower.

Fig. 9 Sour water stripper with no reflux

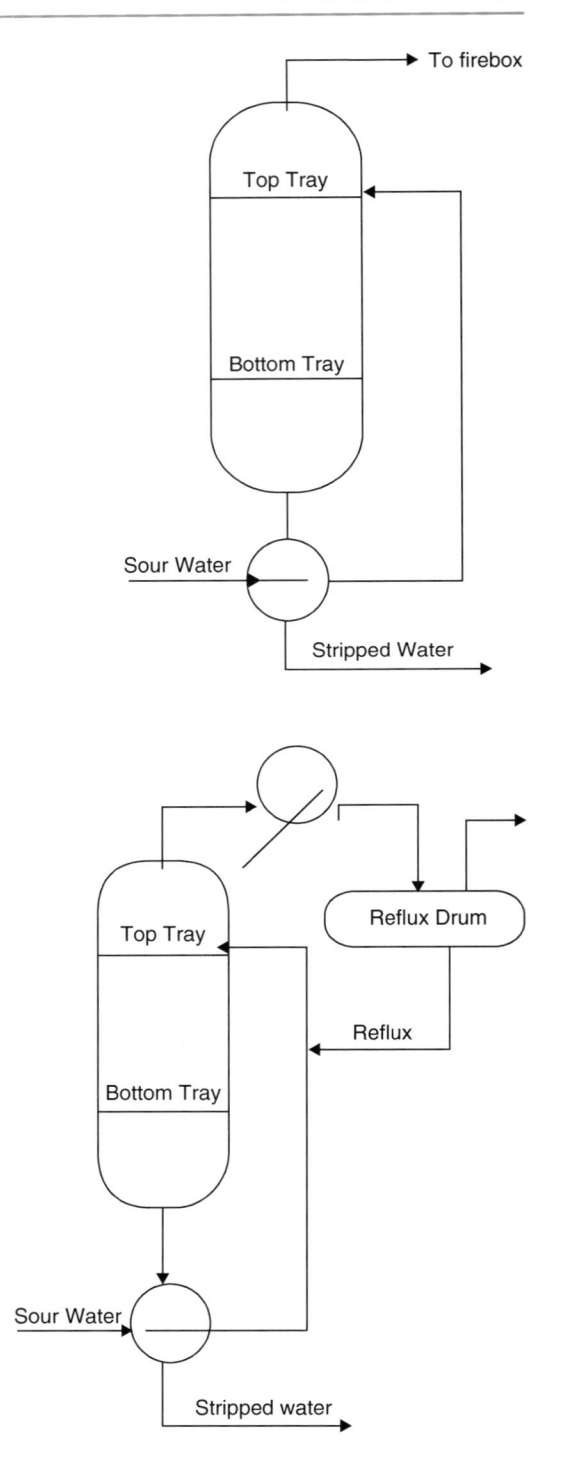

Fig. 10 Sour water stripper with reflux

In the case of a tower with reflux, the reflux enters the tower with the fresh feed. The design of both towers takes advantage of the equilibrium relationships of NH_3 and H_2S in aqueous solutions. A series of graphs giving these relationships is included in Appendix 1 of this chapter as Figs. 20, 21, 22, 23, 24, 25, 26, 27, 28, and 29.

An example of the design for a sour water stripper with no reflux is given in Appendix 2 of this chapter.

Selenium Removal from Stripped Sour Water

Selenium in a refinery normally ends up in the sour water. The selenium in the wastewater may be present in many forms: e.g., selenide, selenite, selenate, and selenocyanate. These are potential environmental hazards. Normally, selenium is removed from oil refinery wastewater or stripped sour water by proprietary chemical treatments that promote precipitation and/or adsorption followed by separation techniques. Some biological agents are also useful in facilitating selenium removal.

Spent Caustic Disposal

Another major aqueous effluents from oil refining are the spent caustic streams from hydrogen sulfide and phenol removal. Refiners usually have the following options in the disposal of these streams. In order of preference, these can be listed as follows:

- Phenolic spent caustic:
  - Disposal by sales
  - Disposal by dumping at sea
  - Neutralizing with acid
  - Neutralizing with flue gas
- Sulfidic spent caustic:
  - Disposal by sales
  - Oxidation with air and steam
  - Neutralization with acid and stripping
  - Neutralization with flue gas and stripping

Both the phenolic and sulfidic spent caustics may have a commercial value in industrial areas and where transportation costs make sales an attractive economic option. In the case of the phenolic spent caustic, processors recover, separate, and purify various cresylic acid fractions for commercial use and sales. In the case of the sulfidic spent caustic, this can be sold as sodium sulfide with some additional processing.

Neutralizing Phenolic Spent Caustic

As listed above, phenolic spent caustic can be neutralized using acid or flue gas. When neutralized the mixture separates into two liquid phases. The upper phase contains the acid oils, while the lower phase is an aqueous solution of sodium sulfate or sodium carbonate. The neutralization using either acid or flue gas can be accomplished in either batch or continuous operations.

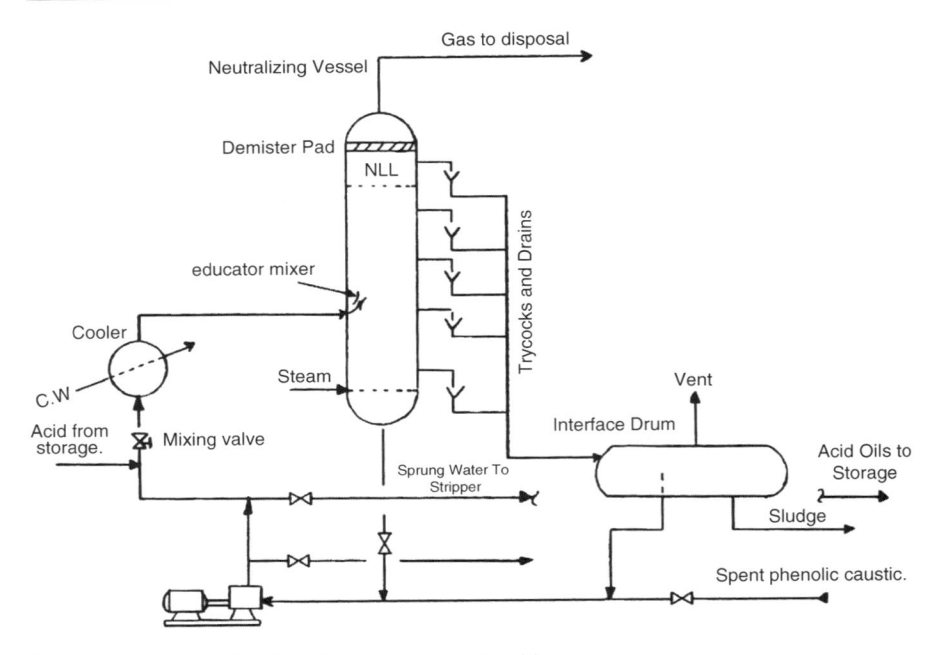

Fig. 11 Batch neutralization of spent caustic with acid

Batch neutralization, using acid, is considered to be the preferable route in this case. Flue gas neutralization, although practical and commercially used in many installations, is more complex in design than the acid application and does not lend itself readily to batch operation. It may require, in some instances, large piping, switch valves, and a blower.

A diagram of a phenolic caustic batch neutralizer is shown in Fig. 11.

The process is fairly simple involving the following steps:

1. Charging the spent caustic batch into the neutralizer vessel.
2. Addition and thorough mixing of the acid neutralizer into the batch. This has to be accomplished slowly and carefully to avoid lowering the pH of the batch too far. A pH of between 4 and 5 will be required to free the phenols completely.
3. A settling period is necessary to allow the sprung acids to separate from the sprung water.
4. Pumping out and disposal of the sprung water.
5. Sprung acid is steam stripped to remove any residual H_2S and light mercaptans.
6. Pump out the interface material which will be included in the next spent caustic batch.
7. Finally pumping out of the sprung acid.

The neutralization step is exothermic, giving out around 125 Btu/lb of sprung acid. As the objective of the process is to produce a phenol-free sprung water for disposal, the system temperature should be kept as low as possible until the sprung water is removed. For example, the solubility of phenol in pure water at 120 °F is about 11 w%, while its solubility at 100 °F is 8.5 w%. For this reason, steam stripping should not begin until the sprung water has been removed. Routing the sprung water to the sour water stripper ensures the removal of any entrained H_2S in that stream.

Mixing of the acid/spent caustic is usually achieved using a mixing valve at the inlet of the neutralizer. An educator or a jet nozzle inlet to the vessel itself would ensure complete mixing if this is required.

Spent Caustic Oxidation

Spent caustic cannot be steam stripped to remove the sulfides contained in it due to the H_2S removal process in which the caustic was used. This is because sodium sulfide does not hydrolyze even when heated. Acids could be used to neutralize the spent caustic, which would release gaseous H_2S. This could be a costly procedure and could cause a potential air pollution problem, although some sour water strippers are designed to dispose of caustic in this manner using waste acid.

The alkaline sulfide can be economically oxidized to form thiosulfates and sulfates. This is the process most commonly used in refineries where only sulfides are the pollutants in the spent caustic and the release of gaseous H_2S is a problem. The oxidation process can also be used in the presence of ammonium sulfide in the stream. The process has not been applied to wastes containing a higher percentage of phenols, because the phenols interfere with the oxidation process.

The oxidation options are described in the following, more detailed discussion.

Oxidation of Sulfides to Thiosulfates

The oxidation of Na_2S or $(NH_4)_2S$ to the corresponding thiosulfates may be expressed in ionic terms as follows:

$$2S^{2-} \quad + \quad 2O_2 \quad + \quad H_2O \quad \rightarrow \quad S_2O_3^{2-} \quad + \quad 2OH^-$$

	$2S^{2-}$	$2O_2$	H_2O	$S_2O_3^{2-}$	$2OH^-$	
lb	2(32)	2(32)	18	112	2(17)	(13)
H_f	2(−10)	2(0)	+68	+154	2(+55)	

where H_f is the theoretical heat of formation in kcals/gm-mole in dilute solution. The plus sign denotes heat evolved, while the minus sign denotes heat absorbed.

Similarly the oxidation of NaSH or NH_4SH to the corresponding thiosulfate can be expressed as follows:

$$2SH^- \quad + \quad 2O_2 \quad \rightarrow \quad S_2O_3^{2-} \quad + \quad H_2O$$

	$2SH^-$	$2O_2$	$S_2O_3^{2-}$	H_2O	
lb	2(33)	2(32)	112	18	(14)
H_f	2(+3)	2(0)	+154	+68	

In both cases, the theoretical oxygen required is 1 lb/lb of S, and the theoretical air required would be 4.33 lbs/lb of S. The theoretical heat of reaction in both cases is 216 kcals/gm-mole of S_2O_3 or 6,100 Btu evolved per lb of sulfur.

Figure 12 shows a typical oxidizing unit.

Briefly, the oxidizer unit consists of an oxidizing tower and an overhead separator. The sour water feed enters the oxidizing tower, after preheating by exchange with the hot oxidizer overhead effluent. Heating steam and air are injected into the sour water feed stream before entering the oxidizer tower. The steam flow is controlled by the temperature of the inlet mixture, while the air is flow controlled to meet the reactor condition premise.

The oxidizing tower itself is divided into three sections, with mixing nozzles connecting the inlet to each section. The reactants (air and the sour water feed) enter the base of the oxidizer, also through a mixing nozzle, and flow upwards through the sections of the tower. The oxidized effluent leaves the top of the tower to be cooled and partially condensed by exchange with the sour water feed. The cooled effluent enters a separator, operating under pressure control. The vent gases leave through the pressure controller to flare or other suitable disposal equipment (such as a heater firebox).

The oil phase of the effluent is skimmed off as a side stream, while the water phase, after settling, leaves the bottom of the separator as the oxidized effluent.

High temperature and low pressures tend to vaporize some of the water in the top section of the oxidizer. This results in more sulfides being stripped out into the vapor phase. Thus, it is recommended that the oxidizer top pressure be at least 25–40 psi above the vapor pressure of the water at the oxidizer top temperature. Typical design criteria for these oxidizing units are given in Table 6.

Oxidation of Mercaptans

If caustic soda has been used to remove H_2S, there will almost certainly be mercaptans present in the sour stream. Caustic soda will also react with and remove these from a process stream. In treating these light streams from sour crudes, the mercaptan content of the spent caustic stream could be as high as 3,000 ppm of sodium mercaptide sulfide. The greater part of the mercaptides will be oxidized to hydrocarbon disulfides according to the equation

$$2\,RSNa + \tfrac{1}{2}O_2 + H_2O \rightarrow RSSR + 2\,NaOH. \tag{15}$$

Almost all the mercaptides will be oxidized in a concurrent flow reactor. However, some of the mercaptides will be partially hydrolyzed, and it can be assumed that there will be a certain amount of free mercaptan present in accordance with the equilibrium:

$$RSNa + H_2O \leftrightarrow RSH + NaOH. \tag{16}$$

The free mercaptan formed by this reaction will be stripped out of solution and leave the reactor in the overhead vapor phase. Now, if the separator is operating at 62 psia and 125 °F, the vapor pressure of water at 125 °F is 2 psia, and then the partial pressure acting to condense the overhead vapor is 60 psia. Methyl mercaptan may condense under these conditions, but the heavier mercaptans and the disulfide

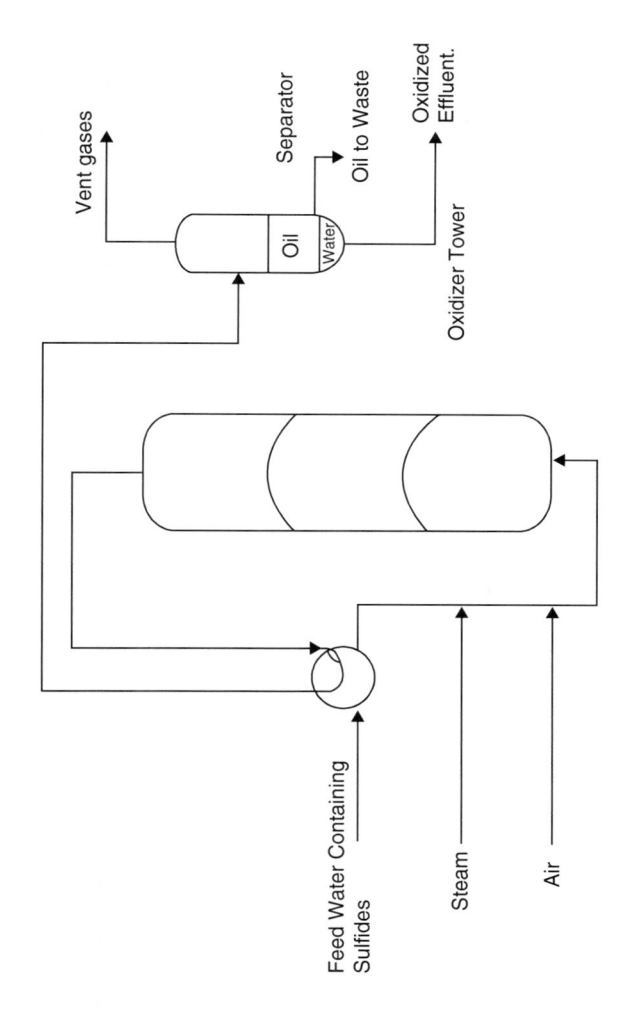

Fig. 12 Schematic of an oxidizer plant

Table 6 Typical design criteria for sulfide oxidizers

Temperature at bottom, °F	165–225
Pressure at bottom, psia	75–100
Inlet air (lbs/gal)/(1,000 ppm S)	0.05–0.075
Inlet air (lb/lb S)	1.4–2.1
Superficial air (ft/s)	0.08–0.1
Sulfide oxidation rate (lbs/h)/cuft	0.35

certainly will condense. The disulfides are immiscible in aqueous alkaline effluent and will form the upper oil layer, to be skimmed off for disposal. The mercaptans that are condensed will form mercaptides with the NaOH present. This will really occur in the feed/effluent exchanger when the reformed mercaptides are still in contact with the reactor air supply. Most of the mercaptides, if not all, will be oxidized to disulfides again.

The theoretical oxygen requirement for oxidizing mercaptides to disulfides is 0.25 lb per pound of mercaptide sulfur, and the theoretical air requirement is 1.09 lb per pound of sulfur. For convenience, the following value of the theoretical amount in this case can be taken as

$$1\,\text{Theory} = 0.0091 (\text{lb air/gal})/(1,000\,\text{ppm RSH} - \text{S}) \tag{17}$$

Oxidation of Sulfide to Sulfate

The oxidation of sulfides to thiosulfides only reduces the ultimate oxygen demand of the sulfides from two pounds of oxygen per pound of sulfides to one pound per pound of sulfide. The oxidation to thiosulfate will remove the short term or immediate oxygen demand. Thiosulfate limits are not usually stipulated in water pollution regulations, while sulfides are always stipulated.

In those cases where thiosulfates are a regulated limit, it will be necessary to convert the thiosulfate to the sulfate. The chemistry relating to this may be written as follows:

$$S^{2-} + 2O_2 \rightarrow SO_4^{2-} \tag{18}$$

The above equation applies equally to aqueous ammonium sulfide and to sodium sulfides.

The theoretical oxygen requirement is 2 lb per pound of S, and the theoretical air requirement is 8.66 lb per pound of S. The theoretical heat of reaction is 12,700 Btu evolved/lb of S. That is, one theory of air is

$$0.072 (\text{lb air/gal})/(1,000\,\text{ppm S}) \tag{19}$$

Table 7 compares those units oxidizing to thiosulfates to those oxidizing to sulfates. Although many units processing the sulfides to sulfates do not use a catalyst, they convert the sulfide to a mixture of thiosulfates and sulfate.

Table 7 Comparison of sulfide oxidation processes

	Oxidizing to thiosulfate only	Partial oxidizing to sulfate	Total oxidizing to sulfate
Anions produced			
SO_4^{2-} %	Nil	46	100
$S_2O_3^{2-}$ %	100	54	Nil
% Sulfide oxidized to			
SO_4^{2-}	Nil	34	100
$S_2O_3^{2-}$	100	66	Nil
Sulfide oxidation rate (lbs/h)/cuft	0.33	0.033	0.035
Air rate in theories			
Oxidation to $S_2O_3^{2-}$	2.05	1.55	6.53
Oxidation to actual anions	2.05	1.17	3.27
Air/water flow	Concurrent	Countercurrent	Countercurrent
Catalyst used	None	None	$CuCl_2$

To convert the sulfide completely to the sulfate, a copper chloride catalyst is used to accelerate the reaction rate. The reaction rate of this conversion is quite slow without the catalyst.

Oil/Water Separation

Most aqueous effluents from a refinery will contain oil. This oil content has to be reduced to less than about 10 ppm before it can be deposited into a river, lake, or ocean. Publicly owned treatment works (POTWs) will have similar requirements for what they will accept for processing to avoid causing problems in their systems.

The oil contamination sources are from multiple areas, including process water rundown, paved area drainage, storm catch pots, tanker ballast pump-out, and tank farm diked areas. All the water from these sources is treated in oil separation processes. The most important of these oil/water separation processes is the API separator.

There are also proprietary corrugated plate interceptors available, which operate on similar principles to the API separator. These contain slanted packs of multiple, parallel, corrugated metal plates, which enhance water separation.

Here we will primarily focus on the common API separator in the discussion.

The API Oil/Water Separator

The design of an API separator is based on the difference in specific gravity of oil and water in accordance with the general laws of settling. The applicable laws are represented by the following equations:

$$V = 8.3 \times 10^5 \times \frac{d^2 \Delta S}{\eta} \quad \text{(Stokes' Law)}$$

$$V = 1.04 \times 10^4 \times \frac{d^{1.14} \times \Delta S^{0.71}}{S_c^{0.29} \times \mu^{0.43}} \quad \text{(Intermediate Law)} \quad (19, 20, 21)$$

$$V = 2.05 \times 10^3 \times \left[\frac{d \times \Delta S}{S_c}\right]^{1/2} \quad \text{(Newton's Law)}$$

where:

V = settling rate in inches per min
d = droplet diameter in inches
ΔS = specific gravity differential between the two phases
η, μ = viscosity of the continuous phase (water in this case) in centipoise
S_c = continuous phase specific gravity

When the Reynold's number is <2.0, use Stokes' law. When the Re number is between 2 and 500, use the intermediate law. When the Re number is >500, use Newton's law:

$$\text{The Re number in this case is given as} \frac{10.7 \times d V S_c}{\eta} \quad (22)$$

As a guide, the droplet size of the oil in water can be taken (as a minimum) as 0.008 in., when the oil specific gravity is 0.850 or lighter, and as 0.005 in., when the gravity of oil is greater than 0.850.

The above laws for settling in oil refining processes usually apply to the hydrocarbon being the continuous phase with water (the heavier phase) being the one that settles out. In the case of the API separator, the reverse is true. That is, the continuous phase is water and the lighter oil phase is the one that is separated to be disposed as the product skimmed from the surface. In the design of this system, Stokes' law is used to reflect the rate at which the oil rises to the surface through the water. This modified Stokes' law may be written as

$$V_r = 6.69 \times 10^4 \frac{d^2 \Delta S}{\eta} \quad (23)$$

where:

V_r = rising rate of the oil phase in feet/min
d = droplet diameter in inches
ΔS = difference in specific gravity of the phases
η = viscosity of the continuous phase (water) in centipoises

An example of the application of this equation in the design of an API separator is given in Appendix 3 of this current chapter.

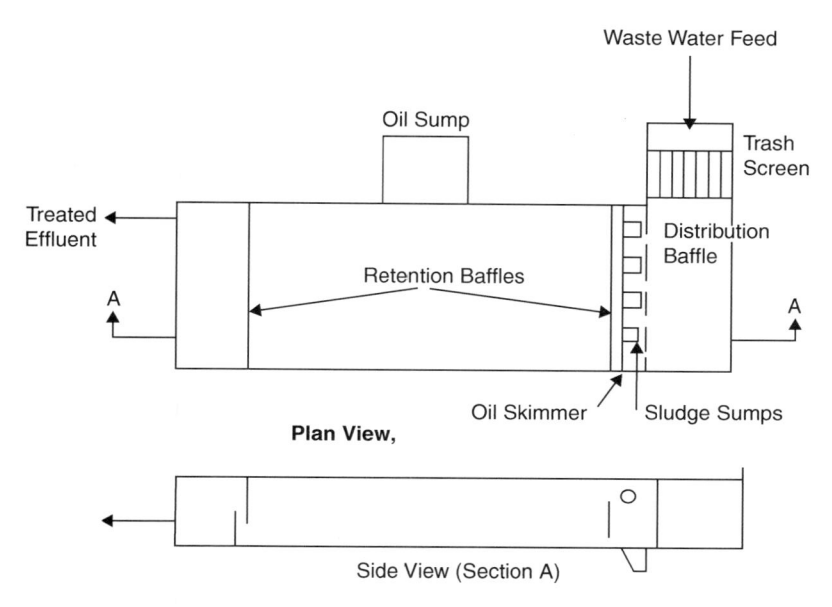

Fig. 13 Schematic diagram of a typical API separator

The oil phase from the API separator is removed using specially designed skim pipes through an oil sump.

A simplified diagram of a typical API separator is given in Fig. 13.

Ancillary equipment used in the design and operation of the API separator are described in the following items.

Oil Skim Pipes

As a good part of the oil will separate from the water on leaving the inlet flume, two oil retention baffles with skim pipes are provided. One is located ahead of the flow baffle and the other ahead of the overflow weir. The distance between these baffles will be the function "L" as calculated in Appendix 3. The API manual recommends 10 in. diameter skim pipes where the horizontal run is less than 40 ft. Larger diameter skim pipes will be installed in greater horizontal runs.

The Oil Sump

The oil recovery sump is located at one edge, usually midway along the horizontal run. The size of the sump is based on the oil content of the inlet stream to the separator plus four times that amount as water delivered by the skim pipes. For example, if the inlet stream contains 200 ppm by volume of oil and the water outlet contains 50 ppm, then the oil removal from the separator will be 150 ppm or 0.015 v%, which, in the case of the specification given in Appendix 3, will equate to

$$0.00015 \times 600 \times 1,440 = 130 \text{gals/day}. \tag{24}$$

Use the API recommendation to add to this figure four times the 130 gal per day for the amount of water contained with the skimmed oil delivered by the skim pipes. Then the total liquid to the sump will be 650 gal/day.

For ease of operation a sump size is often based on a predicted weekly accumulation which amounts to 4,550 gal in this example. Assume the sump cross-sectional area is 144 sqft (12 × 12 ft.) then the liquid depth to NLL should be 4,550/(7.48 × 144) = 4.2 ft. and the HLL can be set at 5 ft. To ensure adequate gravity head for the oil removed by the skim pipes, the bottom of the sump will be 8 ft. below the NLL of the separator.

The Sump Pump

The sump pump will be a standard vertical pump with a capacity to pump out the entire sump in about 1 h or less. Normally refineries pump the sump oil to the "Bottom Settling Tanks" of the refinery. These are tanks used to store the bottoms of crude oil tanks which usually contain high water and sediment content not suitable for the normal process plants. The contents are allowed to sit for some time and are then returned as feed to the normal refinery processes via recovered oil or slops. The bottom settling (water) from these tanks is routed back to the API separator.

Oil Retention Baffles

API recommends that oil retention baffles have a depth below the level in the separator that is 60–80 % of the liquid level. Further the baffle should extend 1 f. above the normal liquid level of the separator.

Distribution Baffles

An important feature in the design of the API separator is the means of distributing the influent evenly. API recommends either "V"-shaped (V-notched) baffles or pipes with multiple holes installed at the inlet of the separator to be used for this purpose. It is also recommended that the open area between slots be 3–7 % of the chamber cross-sectional area.

Sludge Sumps

Sludge sumps are located immediately downstream of the distribution baffles. API recommends that these sumps should be at least 30 in. deep – meaning, at least 30 in. below the floor of the separator. The sludge pump suction is to extend to within 6 in. of the sump bottom and the suction line would be 3 in. diameter. Normally the sludge bottom will be four square inches in cross-sectional area (or 2 in. square). API also recommends that these sumps slope a minimum of 1.7 in. vertically to 1.0 in. horizontal. In the example design in Appendix 3, if we chose 18 × 18 in. sumps sloping to 2 × 2 in. bottoms, we would have a slope of 1.9:1, which would meet the API recommendation. Such a layout could accommodate six sumps.

Trash Rack

API recommends that a trash rack, made up of series of parallel, 3/8 in. rods on 1.5 in. centers, be installed as a trash trap at the separator inlet.

Flight Scrapers

Many API separators have flight scrapers. These are baffles on a moving chain that follow a circuit: moving along the bottom of the API separator bay scraping sludge accumulations toward the sludge sumps then moving backwards down the bay at the top level pushing the oil into the first set of skimmers. The scrapers themselves are usually about 2 in. by 8 in. slats made of heart redwood or fiber-reinforced plastic spaced about 2–3 f. apart.

Storm Surge Ponds

These ponds are installed to provide storage for maximum rainfall conditions. There are several forms of these surge ponds, some requiring pumps, some located upstream of the API separator, and some downstream of the separator.

In most cases the storm drain system is directed to the storm surge pond. Thus, in a storm, the excess rainfall is held in this pond and fed to the API separator over a period of time, at a rate that will not exceed the separator's capability to handle the water effectively. In this way, the refinery ensures that any oily water will not bypass the separator under the worst condition.

The surge pond(s) must be sized to handle the maximum rainfall expected. Normally, the size is based on a 100-year precipitation event, but some facilities design for more. Understanding the precipitation event definitions is critical in selecting the sizing basis. The definitions depend heavily on application of probability theory. The pond(s) must be capable of accepting flow from all the catch basins and open culverts that form part of the refinery drainage system.

Surge ponds are constructed with a shallow depth over a large surface area. The ponds are usually clay-lined to limit loss to groundwater. In some areas, they may be lined or double lined with thick, impermeable (e.g., UV-stabilized polyethylene) membranes to prevent groundwater contamination. They may have a pump-out sump or other provisions for water transfer.

Surge pond elevations are usually only slightly above the API separator, but near the low point of the refinery. Thus, even when the pond is full of water, it will not "back up" into the process area drains. In some cases, where it is not possible to locate the ponds at the lowest elevation, people have installed intermediate lift sumps with multiple vertical lift pumps on uninterruptable power or steam to move water into the ponds at high rate in a large precipitation event.

Refer to the topic "Utilities in Petroleum Processing" in this handbook for more discussion of sewers and how they interact with storm water basins.

Other Refinery Water Effluent Treatment Processes

Oxidation Ponds

Oxidation ponds are usually used as a secondary effluent cleanup after the API separator. There are three types of oxidation ponds:

- *Aerobic* – where the oxidation of the water utilizes oxygen from the atmosphere plus oxygen produced by photosynthesis
- *Anaerobic* – where oxidation of the wastes does not utilize oxygen
- *Aerated* – where oxidation of the wastes utilizes oxygen introduced from the atmosphere by mechanical aeration

Looking at the more common aerobic oxidation ponds, consider a shallow pond containing bacteria and algae. The bacteria will utilize oxygen to biochemically oxidize the incoming water. In so doing, they will produce H_2O, CO_2, and, perhaps, NH_3. The algae will use sunlight plus the H_2O and CO_2 to produce oxygen. Oxygen in turn produces additional bacteria and algae growth. Aerobic oxidation ponds will often contain aerators to ensure an oxygen supply and keep the ponds from "going anaerobic," which tends to be more odorous.

The aerobic cycle is shown in Fig. 14.

Air Flotation and Dissolved Air Flotation (DAF)

The purpose of the air flotation process is the clarification of wastewater by the removal of suspended solids and oil. This is achieved by dissolving air in wastewater under pressure and then releasing it at atmospheric pressure. The released air forms bubbles which adhere to the solid matter and oil in the wastewater. The bubbles cause the adhered matter to float up to the surface of the water as a froth.

The dissolved air in the water also achieves a reduction in the oxygen demand of the effluent stream. Figure 15 illustrates the principle elements of a typical air flotation process.

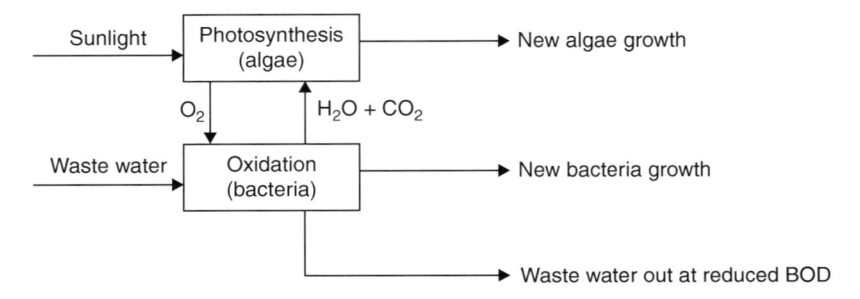

Fig. 14 Aerobic cycle in oxidation ponds

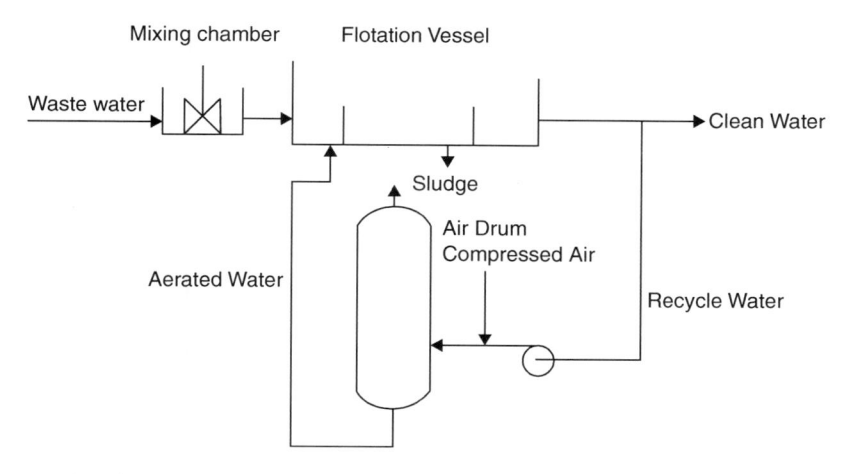

Fig. 15 A typical air flotation process

Table 8 Design criteria for a typical recycle air flotation unit

Recycle rate	50 % of raw feed rate
Air drum pressure	35–55 psig
Air drum retention time	2 min of recycle flow (from liquid level to drum bottom)
Flotation tank retention	15–20 min of total flow (raw feed plus recycle)
Flotation tank rise rate	3.0 gal/min of total flow per sqft of liquid surface
pH	7.5–8.5
Flocculating chemical	25 ppm of alum in total flow
Air rate	0.25–0.5 scf/100 gal of total flow
Flotation tank liquid depth	6–8 f. (to meet requirements above)

The process shown is the preferred recycle type. The process can also be designed for a once-through operation. The typical design criteria for the recycle process are given in Table 8.

Referring to Fig. 15, the flocculating chemical and pH control chemical are mixed in the mixing tank with the raw water feed before the feed enters the flotation vessel.

The processes described in this section have been those found in oil refining most often. Indeed, of the processes described above, most refineries only use the API separator and the surge ponds to meet the oil/water separation required.

Solid Wastes

We have discussed air emissions and water emissions management from petroleum processing facilities. Now the discussion turns to solid wastes generated by a refinery.

There are several types of solids generated in processing petroleum. Not all are "solids" in the normal sense of the word. Some are semisolids or near liquids, but we will treat them here as solids for the discussion.

Some typical solid wastes generated in a refinery are listed in Table 9 along with some of their characteristics and disposal options. Some are considered and defined as hazardous, while others are relatively innocuous. All must be dealt with appropriately.

The table lists some of the options for disposal of the different types of wastes. The actual disposition, however, will be heavily influenced by the applicable regulations from many different governing authorities. An acceptable disposal method in one jurisdiction may not be acceptable in another location.

Solid Waste Disposal Regulations

As with all environmental issues, there is a complex web of cross-connected regulations that define what you can and cannot do with a solid waste in a particular locale. Some of the applicable regulations in the United States and the European Union are listed below:

- United States
  - Resource Conservation and Recovery Act (1970).
    Lists specific wastes defined as hazardous (F & K wastes, for instance)
    Established "characteristic" hazardous waste definitions
    Universal wastes defined – e.g., batteries, pesticides, mercury-containing equipment, and light bulbs/lamps
  - Comprehensive Environmental Response, Compensation, and Liability Act (CERCLA) (1980).
  - Hazardous and Solid Waste Amendments (HSWA) (1984).
  - Federal Facility Compliance Act (1992).
  - Land Disposal Program Flexibility Act (1996).
  - Used Oil Recycling Act (1980).
  - Solid Waste Disposal Act Amendments (1980).
  - Superfund Amendments and Reauthorization Act (SARA) (1986) – CERCLA reauthorization and strengthening.
  - Ocean Dumping Ban Act (1988) – bans ocean dumping of wastes.
  - RCRA Cleanup Reforms I & II (1999, 2001).
  - Used Oil Management Standards (2003).
  - Emergency Planning and Community Right-To-Know Act (EPCRA) (1980).
  - RCRA Expanded Public Participation Rule (1996).
  - Pollution Prevention Act (1990).
  - Hazardous Waste Combustors; Revised Standards; Final Rule – Part 1 (1998).
  - Numerous state and local regulations also govern solid wastes, hazardous and non-hazardous. Many encourage reclamation and recycle over disposal in a landfill.

Table 9 A survey of refinery "solid" wastes

Type of waste	Source(s)	Hazardous? (RCRA designation)	Typical disposition options
Defined hazardous wastes under US regulations (RCRA)			
Primary oil/water/solids separator sludges	Miscellaneous sources	Yes (F037)	Coker injection, combustion, wet oxidation
Secondary oil/water/solids sludges	Miscellaneous sources	Yes (F038)	Coker injection, combustion
Dissolved air flotation (DAF) float	Water effluent treatment DAF unit	Yes (K048)	Coker injection, combustion
Slop oil	Refinery-wide sources	Yes (K049)	Recycle, coker injection, combustion
Heat exchanger bundle cleaning sludge	Maintenance or within a unit	Yes (K050)	Landfill, encapsulation, combustion
API separator sludge	Water effluent treatment	Yes (K051)	Coker injection, combustion
Tank bottoms (leaded)	Leaded gasoline tanks	Yes (K052)	Combustion, wet oxidation
Crude oil storage tank sediment	Crude tanks	Yes (K169)	Coker injection, combustion
Clarified slurry oil tank sediment/filter solids	FCC or tankage	Yes (K170)	Combustion
Spent hydrotreating catalyst	Hydrotreaters, hydrocrackers	Yes (K171)	Regeneration and reuse, reclamation, landfill
Spent hydrorefining catalyst	Hydrotreaters, hydrocrackers, other units	Yes (K172)	Regeneration and reuse, reclamation, landfill
Other types of "solid" refinery wastes			
Oily sludges	Tank bottoms, biotreatment, interceptors, wastewater treatment, contaminated soils, desalters	Probable (some classified above)	Coker injection, combustion, wet oxidation
Oily solids	Contaminated soils, oil spill debris, filter clay acid, tar, rags, filters, packing, lagging, activated carbon	Possible	Combustion, landfill
Drums and containers	Miscellaneous sources	Possible	Recycle, landfill
Spent catalysts (non-listed)	FCC, cat poly, reformers, etc.	Not normally	Reclamation, landfill

(*continued*)

Table 9 (continued)

Type of waste	Source(s)	Hazardous? (RCRA designation)	Typical disposition options
Non-oily materials	Resins, BFW sludges, dessicants, adsorbents, neutral alky sludges, flue gas desulfurization wastes	Not normally	Reclamation, landfill
Radioactive wastes	Catalysts (NORM), lab, level instruments	Possible	Reclamation, landfill, encapsulation
Scales	Leaded and unleaded deposits, rust	Possible	Landfill, encapsulation
Construction and demolition debris	Scrap metal, concrete, asphalt, soil, asbestos, mineral fibers, plastic, wood	Some (asbestos, mineral fibers)	Landfill, encapsulation
Lab wastes	Laboratory	Possible	Reclamation, landfill
Pyrophoric wastes	Iron sulfide-containing scales, certain spent catalysts	Probable	Reclamation, combustion, landfill
Mixed wastes	Domestic refuse, vegetation	Not normally	Landfill, composting
Waste oils – not listed above	Spent lube oil, cutting oil, transformer oil, recovered oils, engine oil, slops	Possible	Reclamation, combustion

- European Union
  - Council Directive 75/439/EEC – disposal of waste
  - Council Directive 75/442/EEC – wastes
  - Council Directive 91/689/EEC – hazardous waste
  - Council Regulation (EEC) No 259/93 – shipments of waste within, into, and out of the European Community
  - Council Directive 94/67/EC – incineration of hazardous waste
  - Council Decision 97/640/EC – control of transboundary movements of hazardous wastes and their disposal
  - Council Directive 1999/31/EC – landfill of waste
  - Council Decision 2000/33/EC – criteria and procedures for acceptance of waste at landfills
  - Commission Decision 2000/532/EC – lists of wastes and categories, incineration
  - Regulation (EC) No 2150/2002 – waste statistics

The regulatory thrusts have moved from controlling disposal toward encouraging waste reduction, reuse, and reclamation over simple disposal.

Some Definitions

We have used several terms without definition so far. Here we will review a couple of the important definitions and considerations in solid waste management.

Hazardous Waste

A hazardous waste is normally defined as a spent material which may pose a substantial risk to public health or the environment. In the United States, the EPA defines some refinery wastes specifically as hazardous. These are highlighted by a designation like the F or K numbers in Table 9 and are called "listed hazardous wastes."

Other wastes are deemed hazardous by their characteristics, as determined by specific tests:

- Ignitability (i.e., flammable) – in air – may be pyrophoric or self-heating
- Reactivity – with other materials or air
- Corrosivity – acids or caustics
- Toxicity – to animals or other life

Jurisdictions outside the United States have generally taken similar approaches for waste classification.

Non-hazardous Waste

The definition of non-hazardous solid wastes in the United States is included in RCRA, Subtitle D. The definition actually includes some hazardous wastes that were exempted from the regulations (e.g., small quantity household hazardous wastes). Oil and gas exploration and production wastes are also exempted. Subtitle D mostly applies to things like garbage, appliances, scrap metal, construction materials, some POTW sludges, and drinking water treatment plant sludges.

Waste Minimization and Reduction

This is the practice of reducing the amount of net waste produced from a facility through changes to the process, raw materials, or disposition of spent materials. It may mean:

- Redesign and operation of a process to reduce by-products or wastes – e.g., minimizing offgrade product through good operation practices
- Reuse of waste from one process in another process, can be internal or external – e.g., cascading slightly spent caustic to other users or reprocessing offgrade products
- Recycling of waste – e.g., sending API separator sludge to the coker or sending slops/recovered oil back to the crude unit as feed
- Reclamation of usable materials from a facility waste – e.g., reclaiming Co, Ni, and Mo from spent catalysts

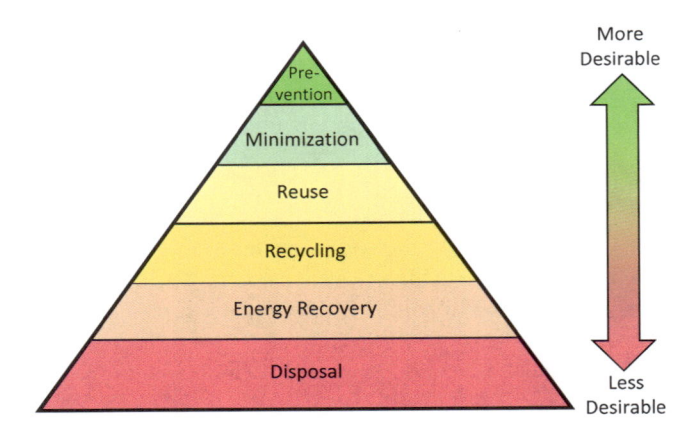

Fig. 16 The waste minimization pyramid concept

- Recovery of heat from sludges and waste oils – e.g., incinerating sludges to generate steam

The concept of waste minimization is illustrated by the pyramid in Fig. 16. The least desirable disposition is by disposal in a landfill for most solid wastes.

Solid Waste Management Techniques and Practices

Many techniques and practices have been developed for management and minimization of solid wastes. We will only list a few of the common approaches here, along with some comments and examples.

Waste Minimization Techniques (from CONCAWE Report No. 6/03)
- Reduction at the source:
  - Process selection and design – e.g., consider another technology for trace sulfur removal instead of caustic treating; cascade caustic solutions among units
  - Process/equipment modifications – e.g., closed-loop sampling systems, mechanical seals on pumps, selection of chemicals for better operations
  - Alternative treatments – e.g., regeneration and reuse of hydrotreating catalysts, hydrotreat lube oils instead of acid treating
- Recycle and reuse of wastes:
  - Internal to facility – e.g., re-running off-spec products or recovered oil, reuse of spent caustic in less demanding applications (like crude unit corrosion control)
  - External to facility – e.g., catalyst metals reclamation, FCC catalyst use as road aggregate, cresylic acid recovery from spent caustic

- Use materials efficiently:
  - Good catalyst activation and conditioning to extend life
  - Not over-treating products
  - Not using more chemical than optimum
- Good housekeeping:
  - Avoid/limit spills – e.g., proper handling of materials to eliminate leaks and losses, tank farm dike design to contain spills, quality spill prevention and containment plan.
  - Limit the need for cleaning – e.g., use antifoulants in service to minimize exchanger cleaning; spray-neutralizing with soda ash rather than flooding reactors; dewatering sludges before disposal (avoid "treating" them, however).
  - Prevent solids from entering the sewer system – e.g., keep dirt, plants, and trash from accumulating in facilities, like in sleeper ways; control the pH of drainings to sewer; segregate storm and process sewers.
- Waste handling:
  - Segregate wastes – e.g., do not mix hazardous and non-hazardous wastes (it all becomes hazardous).
  - Pretreat if possible to reduce volume or render non-hazardous.
- Minimize caustic wastes:
  - Use alternatives of high efficiency caustic contacting.
  - Reuse caustic in refinery.
  - Sell caustic to others for reuse.
  - Manage through the effluent system.
  - Several alternatives are discussed in this chapter under wastewater disposal.
- Do not put a non-hazardous material in a hazardous material storage area, lest it get contaminated or inadvertently disposed of at a higher classification than needed. As a corollary, do not classify a material as more hazardous than it is "just to be safe." Ensure a material is correctly classified with the information you have at the time. It can be reclassified later.

Storage Facilities

The wastes, especially hazardous wastes, awaiting disposal require adequate, environmentally acceptable storage facilities. Here are a few storage considerations:

- Specific permitting is usually required for a facility to be allowed to treat and/or store hazardous wastes beyond a specific time frame. In the United States, hazardous wastes must be out of a facility within 90 days of its generation, unless the facility has a treatment and long-term storage permit. Other countries have similar storage time limits. Most refineries do not have these permits. There are provisions for small, satellite hazardous waste accumulation locations in a facility for up to 1 year – such as for waste lube oil accumulation drums.

- It is a good practice to ensure all containers, whether holding hazardous or non-hazardous materials, are labeled with the contents and hazards. In the United States, an unlabeled container is an immediate violation. Labeling is easy enough to do, so there is no sense in incurring this penalty. Ensure all containers are labeled with material, generation date, hazards (as you know them), and other pertinent data. You may need to change these labels based on any test results.
- The hazardous waste storage area normally must be segregated from other storage, with storm and other runoff collected and managed separately from general plant runoff. If the runoff is not contaminated by anything in storage, it can usually be pumped into the plant sewer system, but if it becomes contaminated, it will have to be disposed of separately.
- Often, the haz-waste storage area must be double contained, with an impermeable membrane under it to prevent groundwater contamination.
- Materials in the haz-waste storage area should be in drums, flow-bins, bags, or other sealed containers. If the materials are pyrophoric or self-heating, they should be sealed under an inert gas blanket and monitored for temperature rises.
- The haz-waste area should have adequate firefighting with hazmat handling PPE and materials immediately available to manage any problems.
- Keep all the wastes in the haz-waste area separated by type and specific waste.

Disposal Practices
- Ensure your selected disposal location is licensed and qualified to dispose of the materials you send them. Often facilities are vetted by a company group or team specifically charged with ensuring the quality of disposal facilities to be used. The facilities should be periodically re-reviewed to maintain their qualifications.
- The vetting practices should include hazardous waste testing labs and transportation companies. We depend on these vendors a great deal. Transportation of a solid waste actually presents the greatest risk of something going wrong of all the handling steps for wastes.
- When a hazardous material is sent for reclamation or disposal, be sure you get confirmation in writing that the required disposition has, in fact, occurred. In the regulatory structures, you retain liability for a waste until it is properly disposed of. Many companies have been surprised when they find out that wastes they thought had been destroyed were just stored (or worse, dumped) by someone. The company ends up paying for the cleanup and disposal twice (at the least).

Noise Pollution

Noise Problems and Typical In-Plant/Community Noise Standards

Noise has been widely recognized as a major industrial/environmental problem in most processing plants because of the risk of hearing loss involved when workers are exposed to high noise levels. Occasionally, the noise levels affect the general public: recent focus has been placed on flaring and the resulting noise in communities, as well as the resulting emissions.

The high noise levels in process plants can be attributed to a great number of sources. Major noise sources are compressors, fans, pumps, motors, furnaces, control valves, steam and gas turbines, and piping systems. The noise-generating mechanism for each piece of equipment is complex, but, in most cases, the noise levels can be reduced to desired limits through the implementation of proper noise control measures.

With increasing awareness of the noise problem and its effect on the general public, regulations on noise standards have been adapted in many countries throughout the world. In the United States, the Occupational Safety and Health Act (OSHA) of 1970 (29 CFR 1910.95) and the Noise Control Act of 1972 and later amendments have served as basic guidelines for noise control requirements. OSHA contains maximum permissible sound pressure levels for each daily time of exposure. These guidelines are presented in Table 10 and serve as a basis for in-plant noise criteria for any process plant constructed in the United States.

Conversely, community noise criteria are more variable and depend on a number of factors, including local ordinances, existing noise levels, and the site of the plant with respect to the community. Some typical community noise limits are shown in Table 11.

The art of acoustics and noise control is beyond the scope of this book. Only the most important concepts necessary for an analysis of process plant noise will be considered. The classification of different areas of the community in terms of environmental noise zones is usually determined by the regulatory authorities, based upon the assessment of community noise survey data.

In the United States, the Noise Control Act of 1972, administered by the Environmental Protection Agency, was intended to establish federal noise emission standards. This act serves a broader scope by coordinating all noise control efforts. It places the primary responsibility for noise control on the states. Please refer to current and local regulations.

Table 10 OSHA noise exposure limits

Duration per day, hr	Sound level, DBA
8	90
6	92
4	95
3	97
2	100
1.5	102
1	105
0.5	110
0.5 or less	115

Table 11 Some typical community noise limits

		Noise levels, DBA		
		Noise zone classification		
Receiving land use category	Time period	Rural suburban	Suburban	Urban
One and two family residential	10 pm–7 am	40	45	50
	7 am–10 pm	50	55	60
Multiple dwelling residential	10 pm–7 am	45	50	55
	7 am–10 pm	50	55	60
Limited commercial some dwellings	10 pm–7 am		55	
	7 am–10 pm		60	
Commercial	10 pm–7 am		60	
	7 am–10 pm		65	
Light industrial, heavy industrial	Any time		70	
			75	

Fundamentals of Acoustics and Noise Control

Several factors contribute to this problem:

1. *Sound pressure level:*
   Sound is a fluctuation in the pressure of the atmosphere at a given point. Sound pressure level is expressed as a ratio of the particular sound pressure and a reference sound pressure:

$$\text{SPL} = 10 \log \frac{p^2}{P_{\text{ref}}^2} \quad (25)$$

   where:
   SPL = sound pressure level in dB (decibels)
   p_2 = mean-square amplitude of the pressure variation
   $P_{\text{ref}} = 2 \times 10^{-5} \text{ N/m}^2$

2. *Sound power level:*
   Sound power level is defined as the ratio of the particular sound power and the reference power:

$$\text{PWL} = 10 \log \frac{W}{W_{\text{ref}}} \quad (26)$$

   where:
   W = sound power (rate of acoustic energy flow) in acoustic watts
   $W_{\text{ref}} = 10^{-12} \text{ W}$
   The relationship between SPL and PWL is

$$\text{SPL} = \text{PWL} + K \quad (27)$$

Environmental Control and Engineering in Petroleum Processing 1281

where:
K = a constant dependent upon geometry and other aspects of the situation
3. *Wavelength:*
Consideration of the wavelength is important to noise control. It is defined as

$$\lambda = \frac{C}{f} \qquad (28)$$

where:
λ = wavelength in feet
C = speed of sound in feet per second
f = frequency, cycles per second (Hz)
4. *Octave band:*
An octave refers to a doubling of frequency. Generally, the audible frequency range consists of ten preferred octave bands with following center frequencies: 31.5, 63, 125, 250, 500, 1,000, 2,000, 4,000, 8,000, and 16,000.
5. *A-weighted sound pressure level (dBA):*
Most noise regulations set the maximum allowable noise limits based on the use of the "A" weighting network which provides a popular means of rating noise. This network is designed to account for the response of the human ear. The other "B" and "C" weighting networks are no longer in common use. Table 12 shows the A-weighting network band corrections for each octave band to convert the sound pressure level (dB) to an A-weighted sound pressure level (dBA).
6. *Adding decibels:*
The noise levels expressed in decibels cannot be added arithmetically, but the addition should be performed on the basis of energy addition. Therefore, the combined dB level is determined by

$$dB_{total} = 10 \log \sum_{i}^{N} 10 \, dB_i/10 \qquad (29)$$

where:

Table 12 A-weighting network band correction

Octave band (Hz)	Band correction (dB)
63	−25
125	−16
250	−9
500	−3
1,000	−0
2,000	+1
4,000	+1
8,000	−1

dB_{total} = the combined dB level
dB_i = the individual dB level
N = the total number of dB levels

7. *Sound fields:*

   A sound field is a description of the relationship between the PWL of the source and the SPL at different points in the surrounding space.

   (a) *Idealized sound field:*

   The sound field for an idealized sound sources which can be considered as a very small, uniformly pulsating sphere is given by the following equation:

   $$SPL = PWL - 20\log r + K \text{ dB} \tag{30}$$

   where:

   r = distance from source to measurement point
   K = a constant

   (b) *Non-idealized sound fields:*

   i. *Outdoor:*

   The sound field of a directional source radiating over a plane (hemispherical radiation) is given by

   $$SPL = PWL + 10\log \frac{Q}{r^2} + 2.5 \text{ dB} \tag{31}$$

   where:

   Q = directivity factor in the direction of interest
   r = distance from source in feet

   ii. *Enclosed space:*

   If source is radiating in an enclosed space, the field equation becomes

   $$SPL = PWL + 10\log \left(\frac{Q}{4r^2} + \frac{4}{R}\right) + 10.5 \text{ dB} \tag{32}$$

   where:

   R = room constant in square feet

8. *Directivity:*

   The directivity factor may be defined as the ratio of the mean-square sound pressure at a given distance in a particular direction to the value which would exist if the source were non-directional. The directivity index is defined by

   $$DI = 10\log Q \tag{33}$$

   where:

   Q = directivity factor in the direction of interest

   Some typical directivity indices are shown in Table 13.

9. *Sound propagation:*

Table 13 Some typical directivity indices

Location	Q	DI
Near a single plane surface	2	3 dB
Near the intersection of 2 plane surfaces	4	6 dB
Near a corner formed by 3 plane surfaces	8	9 dB

The propagation of sound waves can be affected by a number of factors. The factors important to noise control consist of sound absorption, transmission loss, barriers, atmospheric, and terrain effects.

(a) *Sound absorption:*

Sound waves traveling in an enclosed space are affected by the absorptive quality of the incident surface. The amount of absorption is expressed as

$$R = \frac{S \cdot ab}{1 - ab} \quad (34)$$

where:
R = room constant in sqft.
S = total interior surface area in square feet
ab = average absorption coefficient

(b) *Transmission loss:*

The sound isolating capability of a wall is defined as

$$TL = 10 \log \frac{1}{\tau} \text{ dB} \quad (35)$$

where:
τ = transmission coefficient (ratio of transmitted sound intensity to incident sound intensity)

(c) *Barriers:*

Appreciable sound attenuation can often be obtained by interposing a barrier or acoustical shield between the source and receiver. The sound attenuation of a barrier is given approximately by

$$B = 10 \log \frac{20 H^2}{\lambda r} \quad (36)$$

where:
B = reduction of the sound pressure level at a given frequency
H = effective barrier height
r = distance from source to barrier
λ = wavelength of sound at the frequency being considered

(d) *Atmospheric and terrain effects:*

Table 14 Molecular effects T = 72 °F; RH = 50 %	Frequency (Hz)	Attenuation (dB per 1,000 ft)
	500	1
	1,000	2
	2,000	3
	4,000	8
	8,000	15

The propagation of sound outdoors at long distances may be significantly influenced by atmospheric and terrain effects. Sound propagated through the atmosphere is subject to small energy losses due to molecular effects. This loss is dependent upon air temperature and relative humidity. The molecular effects for 72 °F and 50 % RH are shown in Table 14.

The other effects resulting in noise reduction include attenuation due to substantial vegetation, the effects of uneven terrain and tall buildings, and the effects of wind and temperature gradients. These effects are more complex and, in most cases, can be neglected in a process plant noise analysis.

Coping with Noise in the Design Phase

Because of local, state, and/or federal noise regulations which establish the maximum noise reception limits, permission to build or expand any significant industrial facility may be dependent on predicting that the reception limits set by the controlling agency will not be exceeded.

Consequently, noise control engineering must be commenced early in the design stage. A typical noise control program adopted by some major engineering companies is shown in Fig. 17:

(a) *Plant noise design targets*:
Design criteria should be developed early in the design stage, with considerations given to federal, state, and local laws, company standards, proximity and type of adjacent communities, as well as anticipated community growth patterns.
(b) *Development of mechanical equipment and control valve noise criteria*:
Criteria for individual items of equipment will be developed to meet plant noise design objectives and will be made part of inquiry specifications for all noise-generating equipment.
(c) *Preparation of preliminary noise contour map*:
Noise reception levels will be predicted using the engineering company's in-house estimate data bank and noise prediction computer programs.
(d) *Preparation of noise control budget*:
A budget is prepared which identifies funds necessary to implement noise control measures.
(e) *Plot plan assistance:*

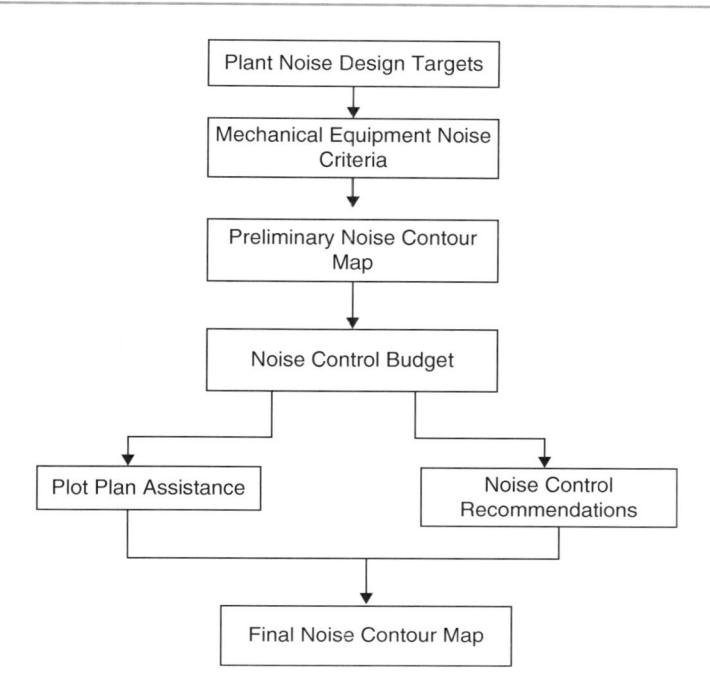

Fig. 17 A typical noise control program flowchart

Equipment location can be optimized based on projected noise levels which will minimize the need of attenuation treatment.

(f) *Noise control recommendations:*
Recommendations for noise attenuation are prepared based on the contribution of all major equipment to the composite noise level of a given plant area. It may be necessary to apply special coatings, insulations, or acoustical lagging (e.g., lead or heavy PVC) to meet the required noise levels.

(g) *Preparation of final noise contour map:*
After final noise level data have been obtained, the plot plan has been finalized, and any noise control measures have been implemented, a noise contour map can be prepared. The noise contour map identifies that the plant noise design objectives have been met. This contour map is normally prepared using the company's own modeling program. Alternatively a proprietary program can be leased from an appropriate software company.

An effective noise control program requires an analysis in the early stage of plant design, when no equipment has been purchased. The most efficient and economical approach to noise control is to include noise control features as an integral part of equipment design through equipment specifications. An optimal plot plan arrangement can also minimize the need for attenuation treatment by strategically locating noisy equipment or positioning process areas or known noise sources at maximum distances from sensitive areas.

The task of predicting noise levels to be used in the design phase can be overwhelming without the aid of the computer. A typical project may involve hundreds of thousands of noise sources, and attempting to do the noise level predictions by hand is almost impossible. Normally a computer program is used to predict the community/in-plant noise levels. The features of this type of program are discussed in the next section.

A Typical Community/In-Plant Noise Program

The noise pollution cycle is one of emission, propagation, and reception. A computer program is usually developed to simulate the noise propagation from several types of noise sources with different configurations. Here we will describe the necessary capabilities of such a program and its application to plant design.

1. *Capabilities:*
   The computer program calculates sound pressure levels generated by single or multiple noise sources at specified grid points or special receptors. This program should utilize a simple algorithm to simulate the propagation of four different source models. These models include point, line, discrete points on a line, and plane sources.
2. *The mathematical model:*
   The basic equation used in the program is

$$\mathrm{SPL} = \mathrm{PWL} + 10\log[F(\mathrm{R})] + DI + K - AE \tag{37}$$

where:
SPL = sound pressure at any receptor in dB
PWL = source sound power level in dB
$10 \log[F(R)]$ = distance attenuation factor for various types of source which will be defined in the following paragraphs in dB
DI = directivity index in dB
K = characteristic resistance of air in dB
AE = total excess attenuation factor (molecular absorption, ground absorption, screening effect, barrier effect, etc.) in dB
- *Distance attenuation $F(R)$:*
  (A) *For a point source:*

$$F(R) = \frac{1}{R^2} \tag{38}$$

where:

Fig. 18 Continuous line source

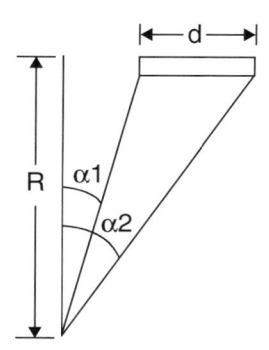

R = distance between source and receptor

(B) *For a continuous line source:*

$$F(R) = \frac{\alpha_2 - \alpha_1}{R_{Od}} \qquad (39)$$

All terms are defined in Fig. 18.

(C) *For discrete sources on a line:*

$$F(R) = \sum_{n=1}^{N} \frac{1}{R_n^2} \qquad (40)$$

where:

N = the number of sources on the line
R = the distance between receptor and each source

(D) *For a plane source:*

$$F(R) = 1/A \int_{-L/2}^{L/2} \int_{-L/2}^{L/2} R \, dxdy \qquad (41)$$

where:

R = distance from the receptor to a differential area $dxdy$ on the plane
All terms are defined on Fig. 19.

- *Excess attenuation A:*
  Excess attenuation due to ground and molecular absorption can be entered as input data. When these data are not entered, the default values shown in Table 15 should be used. In-plant shielding corrections may be included in the ground absorption correction. Corrections due to the effect of wind, temperature gradients, rain, sleet, and barriers can be added later.

3. *Input data requirements:*
   The requirements for each noise source are basically the source sound power spectrum, the source location, and the desired noise level prediction model. Additionally, excess attenuation factors including molecular and ground absorption data can be input if available.

Fig. 19 Plane source

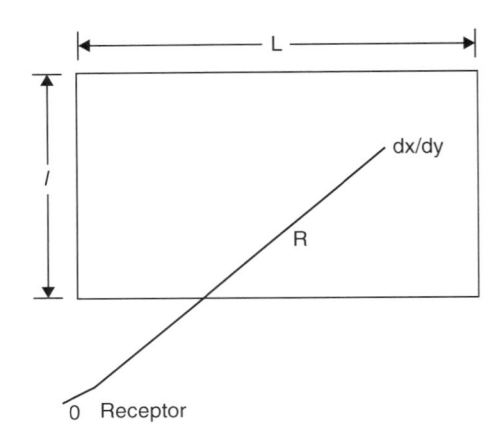

Table 15 Default values for excess attenuation

Frequency, Hz	Ground absorption, dB	Molecular absorption, dB/1,000 ft
63	0	0
125		0
250		0.3
500		1.0
1,000		2.0
2,000		3.0
4,000		6.0
8,000		11.0

4. *Output:*
   In the program described here, all the input data would be summarized and the calculation results will be available electronically or in hard copy. The calculated octave band sound pressure level, overall sound pressure levels, and A-weighted sound pressure levels will be tabulated for all the special receptors specified. The calculated A-weighted sound pressure levels for all grid points will also be listed. A contour map can be produced.
5. *Applications:*
   The community/in-plant noise program described here can be used to perform the following applications:
   - Estimating the net noise impact due to a prospective industrial activity
   - Checking plant design for compliance with noise regulations
   - Establishing noise emission levels and plant design necessary to comply with the applicable environmental regulations

As with other computer programs, the accuracy of computer-calculated noise levels depends on the accuracy of the input data as well as the application of the model selected for a particular noise source.

Environmental Discussion Conclusion

Complete compliance with environmental regulations today is part of the right to operate a petroleum processing facility. Some refiners have found themselves on the wrong side of the regulations, sometimes resulting in shutdown until they comply. Some regulations contain provisions under which criminal prosecution is possible.

Good, diligent, comprehensive design and operating practices along with awareness of the requirements will keep a plant operating cleanly and efficiently.

Appendix 1 Partial Pressures of H_2S and NH_3 Over Aqueous Solutions of H_2S and NH_3

This appendix includes Figs. 20, 21, 22, 23, 24, 25, 26, 27, 28, and 29 for partial pressures of H_2S and ammonia over aqueous solutions of H_2S and NH_3. The figures are in the next few pages. Refer to Appendix 2 for the interpretation and use of these charts.

Appendix 2 Example of the Design of a Sour Water Stripper with No Reflux

Specifications

Feed: The tower is to be designed to handle 200 gpm (at 100 °F) of sour water containing 10,000 ppm of H_2S and 7,500 ppm of NH_3 (by weight).
Unit: The unit shall be a trayed column using sieve trays (efficiency of 0.5) with no reflux.
Steam rate: Refinery 50 psig saturated steam shall be used at a rate of 1.3 lbs/gpm of feed.
Tower pressure and temperature: The vapors leaving the tower top shall have sufficient pressure to enter a "rat-tail" burner in a nearby heater. The tower top pressure shall be 20 psia. The feed entering the top tray shall be preheated to a temperature of 200 °F. The total tower pressure drop shall be 2 psi.
Stripped water specification: The tower shall be designed and operated to remove 99.0 % of the H_2S in the feed and 95 % of the NH_3.

The Design

Assume the stripping will be accomplished using four theoretical trays. Then at an efficiency of 0.5, the number of actual trays will be eight. The pressure drop per tray will be 0.25 (i.e., 2 psi/8 trays).

Calculate Feed Mass Per Hour

Water at 100 °F has a specific volume of 0.1207 gal/lb.
Then 200 gpm of water $= \frac{200 \times 60}{0.1207} = 99,420 \text{lbs/hr}$.

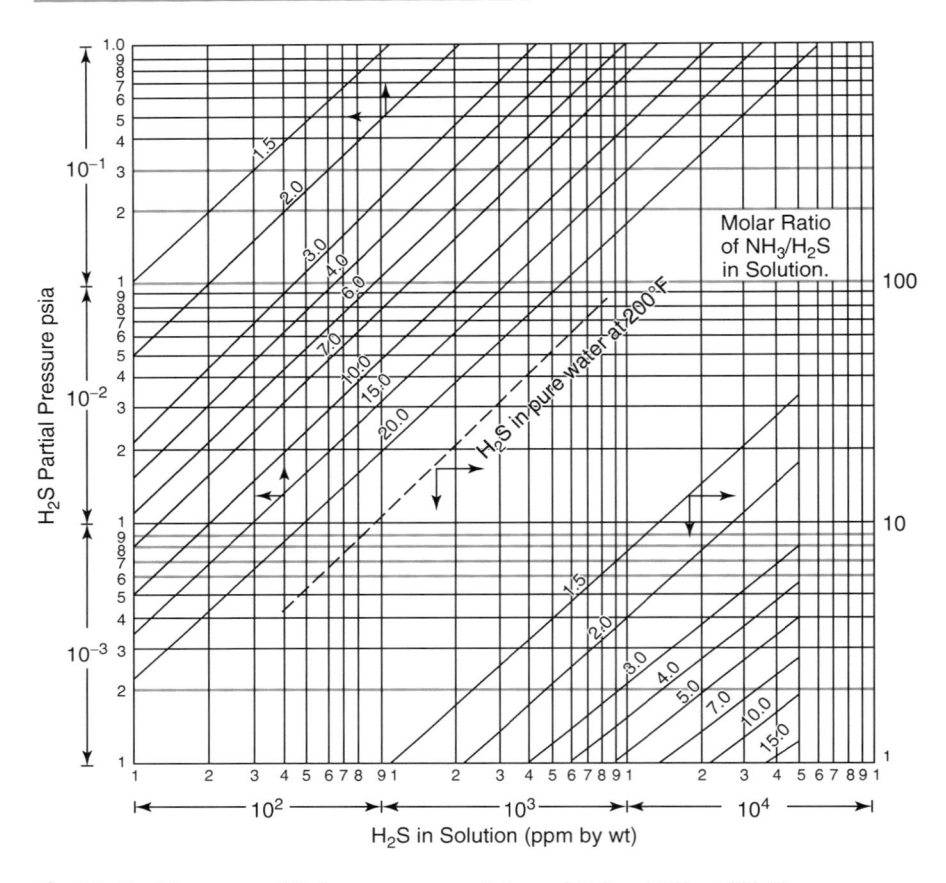

Fig. 20 Partial pressure of H_2S over aqueous solutions of H_2S and NH_3 at 200 °F

The feed will be as follows:

	lbs/h	Moles/h	
Water	99,420	5,523.3	
NH_3	745.65	43.86	(7,500 ppm by wt)
H_2S	994.2	29.24	(10,000 ppm by wt)

Calculate Stripping Steam

Tower bottom pressure will be 20 psia + (8 × 0.25) = 22 psia.
 From steam tables, tower bottom temperature will be water at 22 psia = 233 °F.
 Feed temperature = 200 °F.
 Then steam used for heating = 99,420(33/924) = 3,550 lbs/h.
 And steam used for stripping = 15,550–3,550 = 12,000 lbs/h.

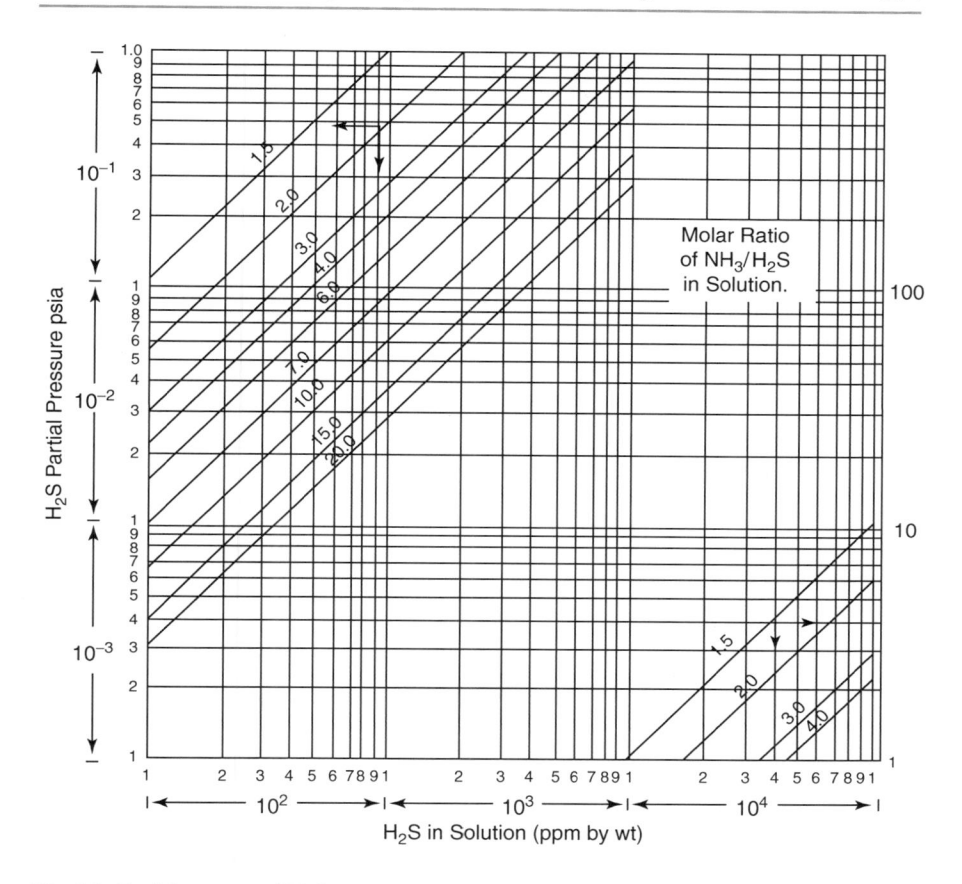

Fig. 21 Partial pressure of H_2S over aqueous solutions of H_2S and NH_3 at 210 °F

Calculate Stripped Water Quantity and Composition

Stripped water shall contain feed water plus condensate. $99,420 + 3,550 = 102,970$ lbs/h.

NH_3 in stripped water shall be 5 % of total $= 37.28$ lbs/h $= 2.19$ moles/h $= 361$ ppm by wt.

H_2S in stripped water shall be 1 % of total $= 9.94$ lbs/h $= 0.29$ moles/h $= 100$ ppm by wt.

Calculate the Overhead Vapor Partial Pressures

Overhead vapor leaving the tower V_o will be as follows:

	Moles/h	lbs/h	PP psia
NH_3	41.67	708	1.13
H_2S	28.95	984	0.78
Steam	667	12,000	18.09
Total	737.62	13,692	20.00

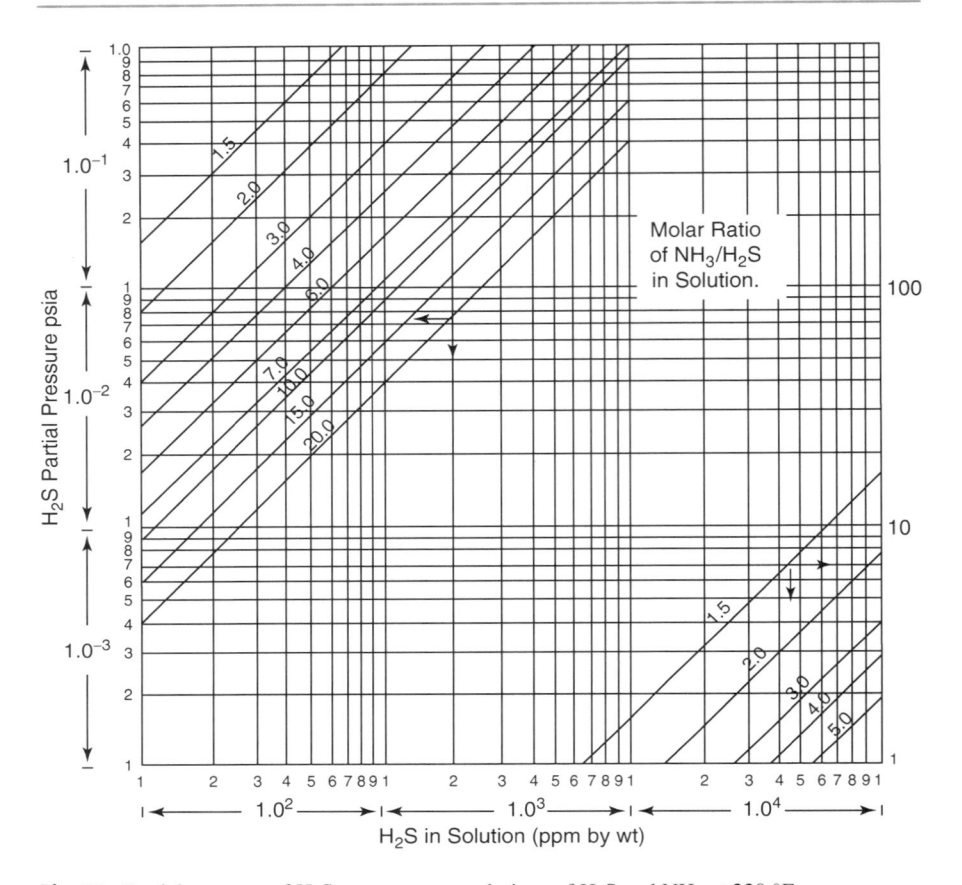

Fig. 22 Partial pressure of H_2S over aqueous solutions of H_2S and NH_3 at 220 °F

Temperature of top theoretical tray is steam at a saturation pressure of 18.09 psia = 223 °F.

Top Tray Calculation

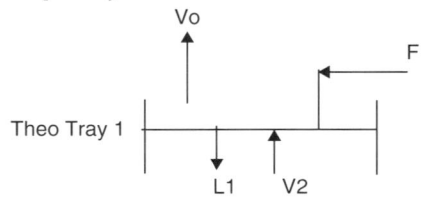

Assume a ratio of NH_3/H_2S as 3.8 moles, that is, $(17/34) \times 3.8 = 1.9$ by wt.

From Appendix 1, Fig. 28, NH_3 at a partial pressure of 1.13 will be 4,600 ppm by wt and H_2S ppm by wt will be $4,600/1.9 = 2,421$ ppm.

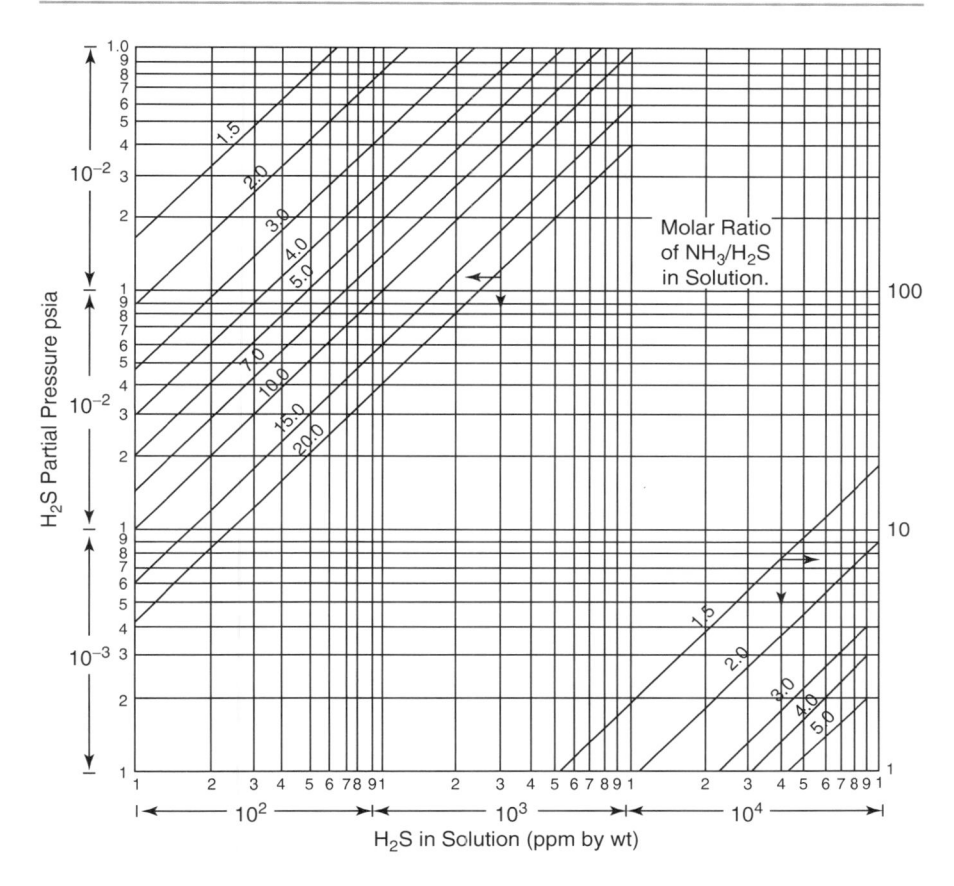

Fig. 23 Partial pressure of H_2S over aqueous solutions of H_2S and NH_3 at 225 °F

From Fig. 23 partial pressure of $H_2S = 0.75$ which is acceptably close to 0.78, which was established.

Note: Should the partial pressure of H_2S be substantially different to that for V_o, then a different ratio of the two components would have to be chosen and the calculation repeated.

Calculate Liquid from Top Tray L_1

	ppm	lbs/h	Moles/h
NH_3	4,600	457	26.9
H_2S	2,421	241	7.08

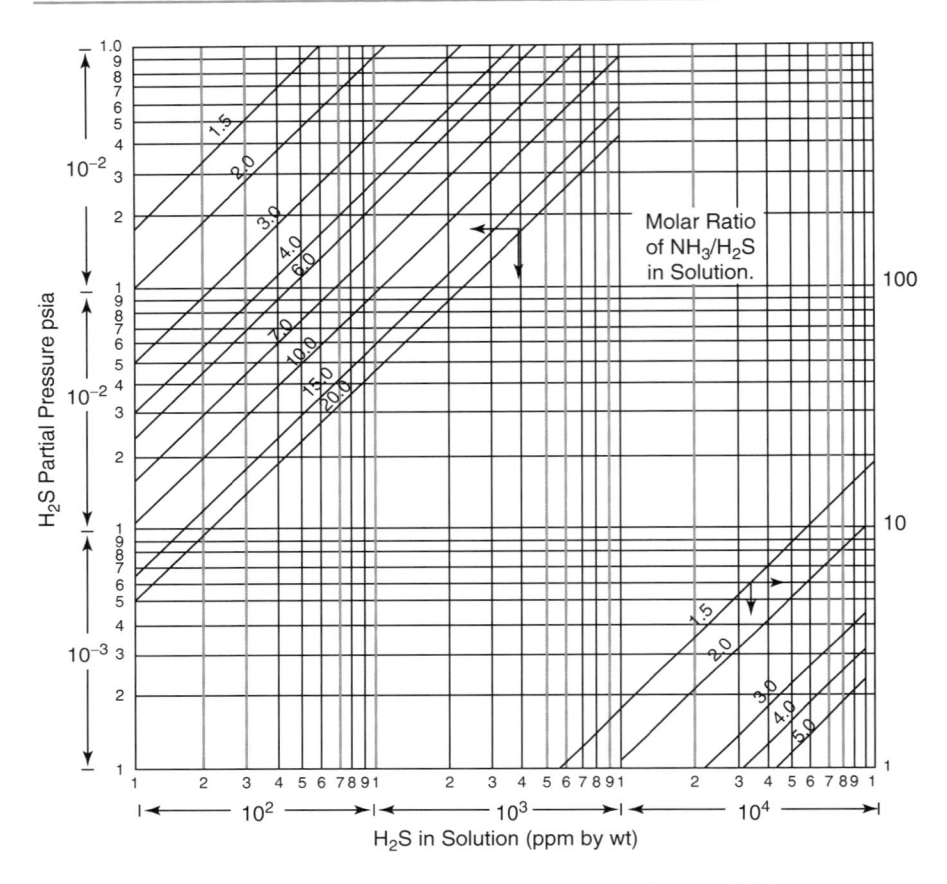

Fig. 24 Partial pressure of H_2S over aqueous solutions of H_2S and NH_3 at 230 °F

Calculate Vapor from Theo Tray 2 V_2

Moles vapor for NH_3 and H_2S will be those moles in $V_o + L_1 - F$

Pressure on Theo 2 will be $(1/0.5) = 2$ actual trays at 0.25 psi pressure drop = 20 psia + 0.5 = 20.5 psia.

	Moles/h	PP psia	Notes
NH_3	24.71	0.73	
H_2S	6.79	0.2	
Steam	667	19.57	
Total	698.5	20.5	Tray temp (from steam tables = 226 °F)

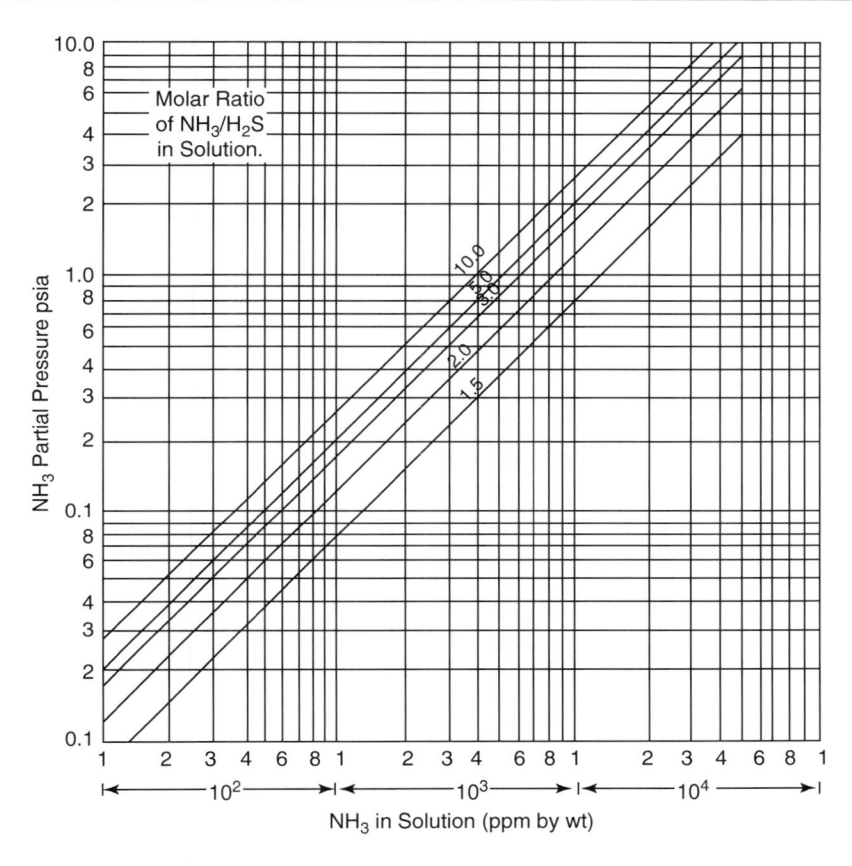

Fig. 25 Partial pressure of NH_3 over aqueous solutions of H_2S and NH_3 at 200 °F

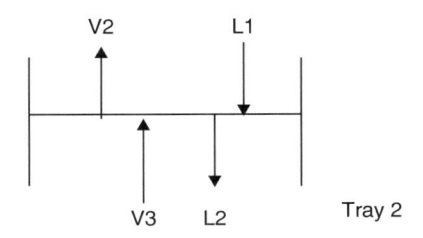

Calculate Balance Over Theoretical Tray 2
Assume NH_3/H_2S ratio is 5.0 molar and 2.5 ppm by weight.
  NH_3 at a PP of 0.73 and 5.0 molar ratio = 2,400 ppm by wt (from Appendix 1, Fig. 29).
  H_2S ppm is 2,400/2.5 = 960.
  From Fig. 24, H_2S ppm = 0.19, which is a satisfactory match.

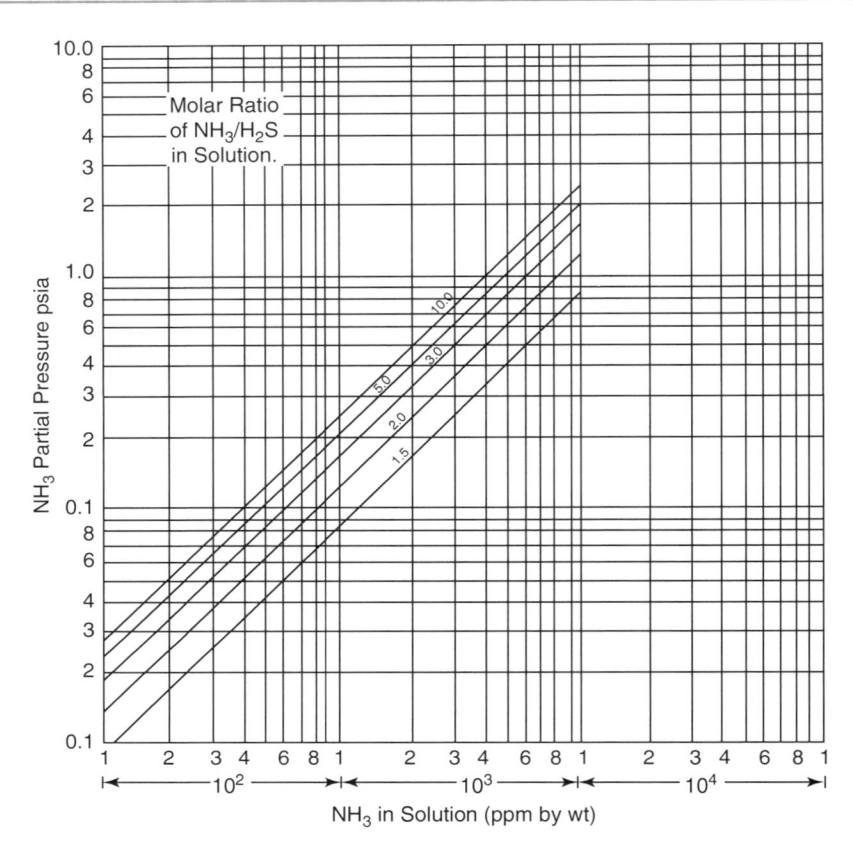

Fig. 26 Partial pressure of NH_3 over aqueous solutions of H_2S and NH_3 at 210 °F

Liquid from Tray 2. L_2

	ppm	lbs/h	Moles/h
NH_3	2,400	278.4	16.37
H_2S	960	95.5	2.81

Vapor from Tray 3 V_3

Total tray pressure = 21 psia and temperature is 231 °F:
 Vapor from tray $3 = V_2 + L_2 - L_1$

	Moles/h	PP psia
NH_3	14.18	0.44
H_2S	2.52	0.08

(*continued*)

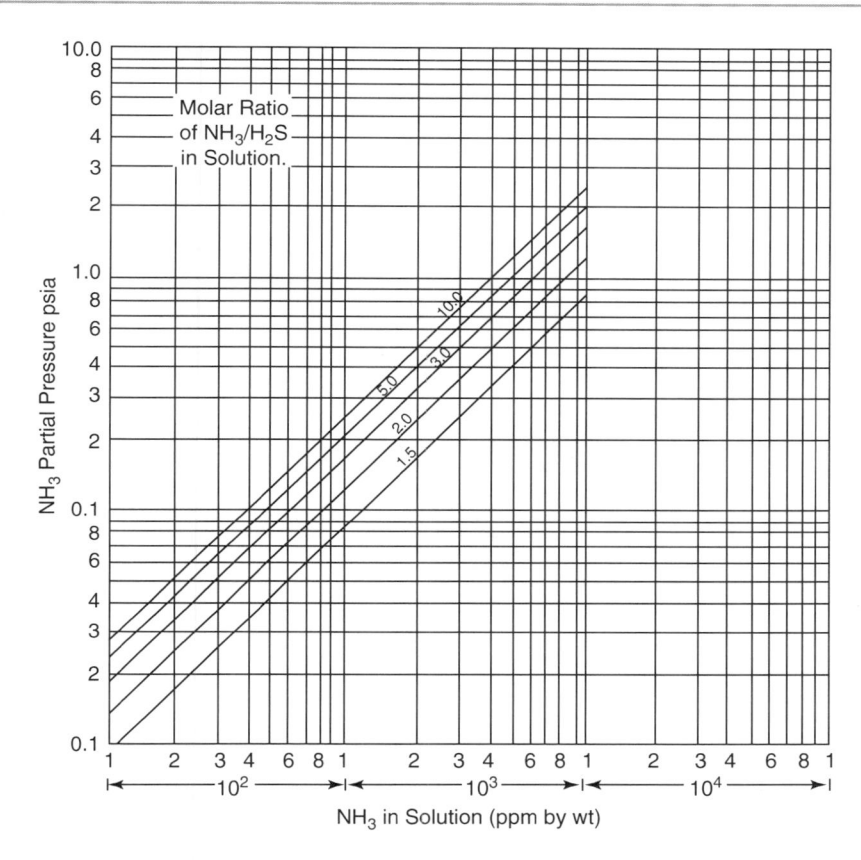

Fig. 27 Partial pressure of NH_3 over aqueous solutions of H_2S and NH_3 at 220 °F

	Moles/h	PP psia
Steam	667	20.52
Total	683.7	21.00

Calculate Balance Over Theoretical Tray 3

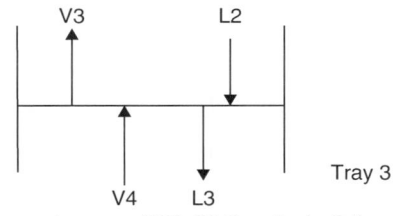

Assume NH_3/H_2S ratio is 6.5 molar and 3.25 by weight.

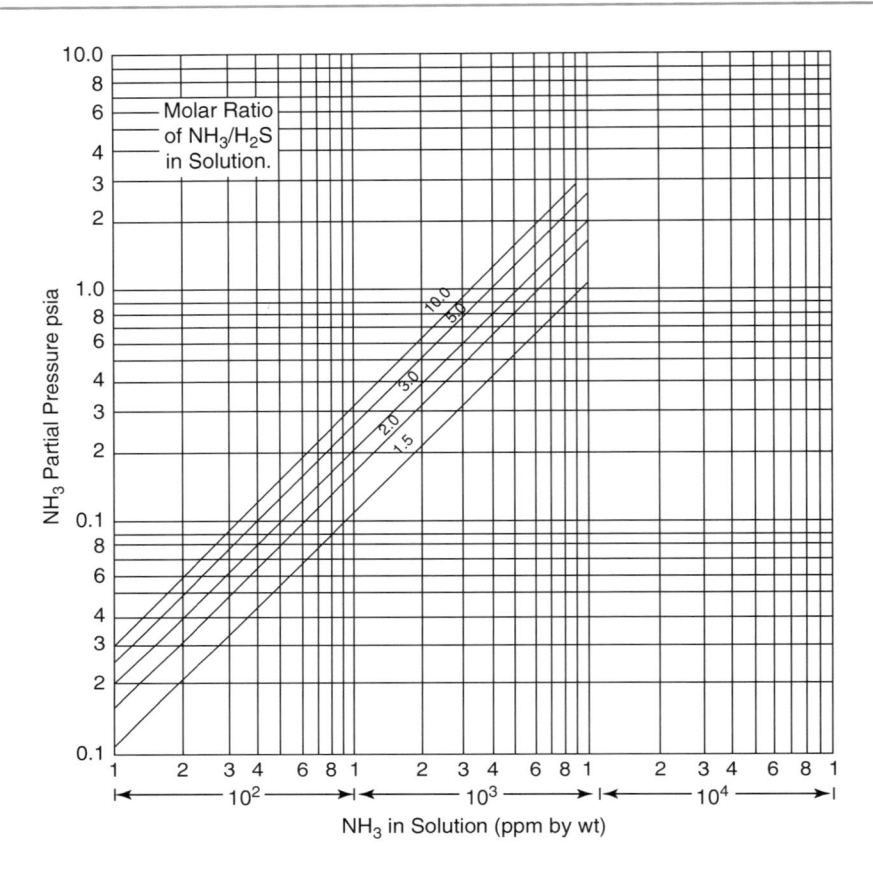

Fig. 28 Partial pressure of NH_3 over aqueous solutions of H_2S and NH_3 at 225 °F

NH_3 ppm from Fig. 29 = 1,600 ppm by weight.
H_2S is 493 ppm from Fig. 25. PP of H_2S is 0.085, which is a satisfactory match.

Liquid from Theoretical Tray 3, L_3

	ppm	lbs/h	moles/h
NH_3	1,600	159.1	9.35
H_2S	493	49	1.3

Vapor from Theoretical Tray 4, V_4

Tray pressure 21.5 psia temperature 232 °F

$$V_4 = V_3 + L_3 - L_2$$

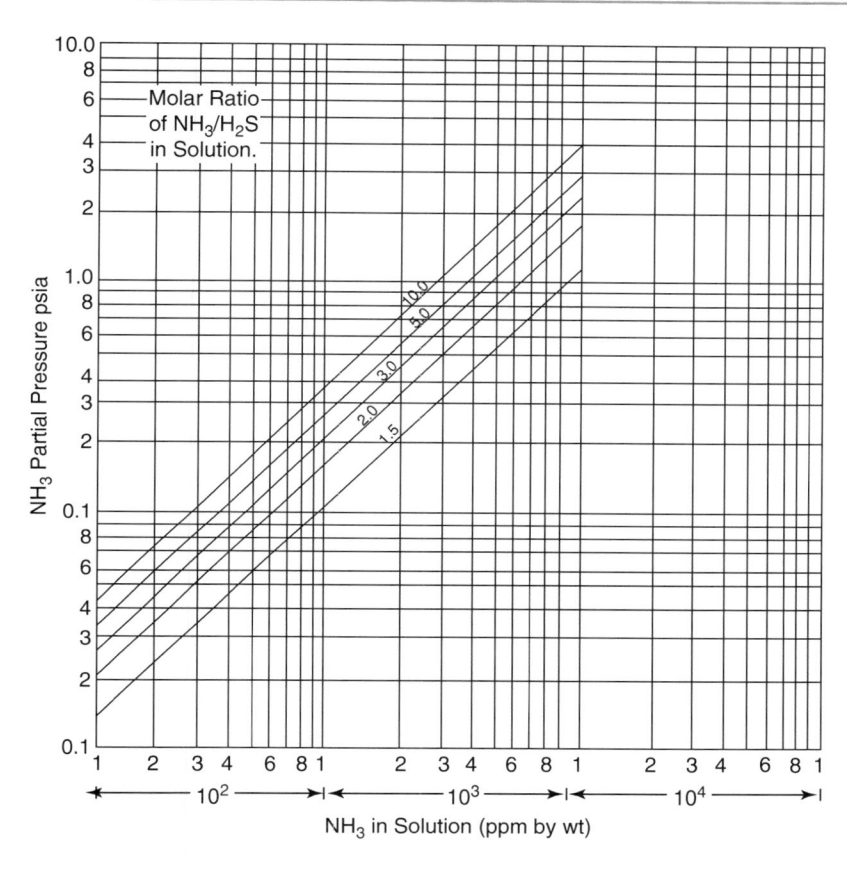

Environmental Control and Engineering in Petroleum Processing

Fig. 29 Partial pressure of NH_3 over aqueous solutions of H_2S and NH_3 at 230 °F

	Moles/h	PP psia
NH_3	7.16	0.23
H_2S	1.01	0.032
Steam	667	21.24
Total	675.17	21.5

Calculate Balance Over Theoretical Tray 4

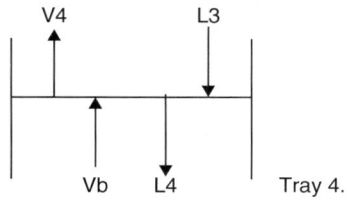

Tray 4.

Assume NH_3/H_2S ratio is 6.6 molar and 3.3 by weight.
NH_3 ppm from Fig. 29 is 610.
H_2S ppm is 185. PP psia of H_2S from Fig. 25 is 0.032, which is a satisfactory match.

Liquid from Theoretical Tray 4, L_4

	ppm	lbs/h	Moles/h
NH_3	610	60.6	3.57
H_2S	184.8	18.4	0.54

Vapor from Tower Bottom V_b

Pressure 22 psia temperature 233 °F
$$V_b = V_4 + L_4 - L_3$$

	Moles/h	PP psia
NH_3	1.38	0.045
H_2S	0.25	0.008
Steam	667	21.947
Total	668.83	22.0

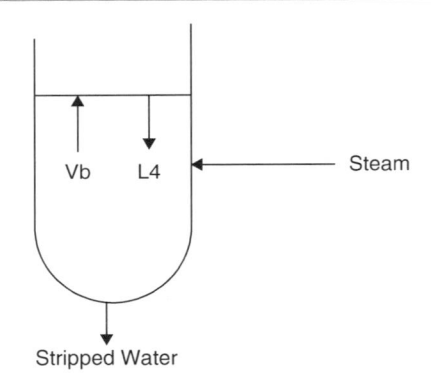

Calculate the Bottom of the Tower

Assume NH_3/H_2S ratio is 7.0 molar and 3.5 by weight.
NH_3 ppm from Fig. 29 is 180.
H_2S ppm is 51.4. PP psia of H_2S from Fig. 25 is 0.008 (extrapolated), which is a satisfactory match.
Contaminants in stripped liquid product:

	ppm	lbs/h
NH_3	180	17.9
H_2S	51	5.1

Conclusion

The tower will handle 200 gpm of sour water and remove 97.6 % weight of NH_3 and 99.49 % weight of H_2S using eight actual trays and a steam rate of 1.3 lbs/h per gpm of feed. Feed will be preheated to 200 °F by heat exchange with tower bottoms before entering on the top tray of the tower.

Appendix 3 Example Design of an API Separator

Specification

We are asked to design an oil/water separator to handle the normal quoted rainfall and process waste from a 4,500 BPSD hydro-skimming refinery. The quantity of inflow to the separator is estimated to be 600 gpm. The normal rundown temperature for this stream is taken as 100 °F. The design of the separator shall be in accordance with the appropriate section of the *API Manual* – sixth edition. The following data shall be used in the design:

Specific gravity of water	0.995
Specific gravity of the oil	0.890
Viscosity of water	0.7 cP
Diameter of the oil globules	0.006 in

The oil shall be removed from the separator by means of oil skimming pipes and an oil sump designed to meet the oil influent content of 400 ppm by volume.

The Design

The rising rate of the oil is calculated from the equation

$$V_r = 6.69 \times 10^4 \times \frac{d^2 \Delta S}{\mu}$$

where:

V_r = rising rate of oil in ft/min.
d = diameter of oil globule in inches = 0.006 inches
ΔS = difference in the SGs of oil and water phases = 0.105
μ = viscosity of the water phase in centipoise = 0.7

Then

$$V_r = \frac{6.69 \times 0.36 \times 0.105}{0.7}$$
$$= 0.361 \text{ft/min}.$$

V_h = the horizontal velocity of the effluent = $15 \times V_r$ but not to exceed 3 ft/min; so use 3 ft/min for the design.

Volumetric rate of flow = $Q = \frac{600}{7.48} = 80 \, \text{cuft/min}$
Minimum cross sectional area of flow = $\frac{80}{3} = 26.7 \, \text{sqft}$ say 27 sqft.
API recommends the following limits:

Depth – 3–6 ft.
Width – 6–20 ft
Ratio of depth to width – 0.3–0.5
Width of section = 27/3 = 9 f. which meets the API recommended depth to width ratio.

Calculating the effective separation length of unit:

The API manual gives the following factors to include in the determination of the unit's length.
Turbulence factor $F_t = 1.22$
Short circuit factor $F_s = 1.2$
The length is then calculated by the expression

$$L = (F_t + F_s)(V_H/V_r) \times \text{depth.}$$
$$= 2.42 \times 8.31 \times 3$$
$$= 60 \, \text{ft.}$$

This 60 f. is the length between the oil retention baffles (see Fig. 13).

Recommendations regarding the types and sizes of the internal equipment are found in the appropriate section of the API manual. A brief description of these is given in the text of this chapter.

References

J.S. Alshammari et al., Solid waste management in petroleum refineries. Am J Environ Sci **4**(4), 353–361 (2008). Suez Canal University, 2008

M.R. Beychok, *Aqueous Wastes from Petroleum and Petrochemical Plants* (Wiley, New York, 1967)

M. Beychok, Image: Claus sulfur recovery.png; Wikimedia Commons, 2 Mar 2012. Accessed Feb 2014

C.E. Baukal (John Zink Co. LLC, 2008), Article: NOx 101: a primer on controlling this highly regulated pollutant, Process Heating reprint, Feb 2008. Accessed Feb 2014

CAI-Asia, Fact sheet: Sulfur dioxide (SO_2) standards in Asia, Clean Air Initiative for Asian Cities, CAI-Asia Fact Sheet No. 4, Aug 2010, http://cleanairinitiative.org/portal/sites/default/files/documents/4_SO2_Standards_in_Asia_Factsheet_26_Aug_2010.pdf. Accessed Feb 2014

D.A. Dando, D.E. Martin (CONCAWE, 2003), Report 6/03: A guide for reduction and disposal of waste from oil refineries and marketing installations, Brussels, Nov 2003, https://www.concawe.eu/content/default.asp?PageID=569. Accessed Feb 2014

K. den Haan, Presentation to Japan Petroleum Energy Center conference: Water in EU: legislation & the refining industry's response (CONCAWE, 2011), http://www.pecj.or.jp/japanese/overseas/conference/pdf/conference11-12.pdf. Accessed Feb 2014

W.L. Echt (Union Carbide Corp.) and C.J. Wendt (The Ralph M. Parsons Co.), Paper: Reduce sulfur emissions from Claus sulfur recovery unit tail gas treaters, presentation to AIChE National Meeting, 1993

Fuelling Europe's Future website, Item: Fuelling Europe's future: impact of EU legislation (2013), http://www.fuellingeuropesfuture.eu/en/refining-in-europe/economics-of-refining/demand-trends. Accessed Feb 2014

GPSA, *Book: Engineering Data Book,* 10th edn. (Gas Processors Suppliers Association, Tulsa, 1987)

T. Huntsman (ed.), S. Watson, et al., Article: Study compares COS-removal processes (Pearl Development Co., 2003), www.ogj.com, 101, 36. 22 Sep 2003

India Environment Portal, Article: Effluent and emission standards for oil refineries: guidelines 8 May 2008, www.indiaenvironmentportal.org. Accessed Feb 2014

IPIECA Operations Best Practice Series, Best practice entitled: Petroleum refining water/wastewater use and management (2010), http://www.ipieca.org/publication/petroleum-refining-waterwastewater-use-and-management. Accessed Feb 2014

IPIECA, Publication: Refinery air emissions management: Guidance for the oil and gas industry, IPIECA (2012), http://www.ipieca.org/publication/refinery-air-emissions-management. Accessed Feb 2014

M. Leavitt et al. (EPA, 2004), Technical support document for the 2004 effluent guidelines program plan, Section 7, U.S. Environmental Protection Agency, Office of Water, EPA-821-R-04-014, Aug 2004. Accessed Feb 2014

Matheson Tri-Gas, Inc., MSDS: ethyl mercaptan (2009), Accessed Feb 2014

Michigan Department of Resources and Environment, Guidance document: Guidelines for selecting dust suppressants to control dust and prevent soil erosion, Feb 2010. Accessed Feb 2014

M. Mihet et al., Low temperature hydrogen selective catalytic reduction of NO on Pd/Al_2O_3. Rev Roum Chim **56**(6), 659–665 (2011). Academia Română, 2011

W.L. Nelson, *Petroleum Refinery Engineering*, 4th edn. (McGraw-Hill Book Co., New York, 1969)

P.K. Niccum et al. (KBR, 2002), Paper: FCC flue gas emission control options, Paper AM-02-27, NPRA 2002 Annual Meeting. 17–19 Mar 2002

Praxair, Inc. MSDS: carbon dioxide, Jul 2007. Accessed Feb 2014

M. Quinlan, A. Hati (KBR, 2010), Article: Processing NH_3 acid gas in a sulphur recovery unit, Apr 2010, www.digitalrefining.com Accessed Feb 2014

M. Rameshni, Paper: Selection criteria for Claus tail gas treating processes (WorleyParsons, Approx. 2008), http://www.worleyparsons.com/CSG/Hydrocarbons/SpecialtyCapabilities/Documents/Selection_Criteria_for_Claus_Tail_Gas_Treating_Processes.pdf. Accessed Feb 2014

Sciencelab.com, MSDS: methyl alcohol, 21 May 2013. Accessed Feb 2014

D. Seligsohn, Article: China's new regional air quality regulations: a win-win for local air quality and the climate (China FAQs, 2013), www.chinafaqs.org, 14 Nov 2013. Accessed Feb 2014

Shell Global Solutions, Fact sheet: Sulfinol-M (2012). Accessed Feb 2014

Tanner Industries, Inc., MSDS: aqua ammonia, Feb 2006. Accessed Feb 2014

The Lubrizol Corp., MSDS: Sulfrzol 54. 4 Jan 2013

U.S. EPA, Reference document (Draft): Hazardous waste listings: a user-friendly reference document, U.S. Environmental Protection Agency, Mar 2008, http://www.epa.gov/osw/hazard/refdocs.htm. Accessed Feb 2014

U.S. EPA, Public outreach presentation: Addressing air emissions from the petroleum refinery sector: risk and technology review and new source performance standard rulemaking, U.S. Environmental Protection Agency (2011), http://www.epa.gov/apti/video/10182011Webinar/101811webinar.pdf. Accessed Feb 2014

U.S. EPA Fact sheet: Sulfur dioxide: health, U.S. Environmental Protection Agency, 28 Jun 2013, http://www.epa.gov/airquality/sulfurdioxide/health.html. Accessed Feb 2014

Wikimedia Commons, Image: Airborne particulate size chart.jpg, 30 Sep 2010. Accessed Feb 2014

Wikipedia and Wikimedia images. Multiple items accessed Feb 2014

Wikipedia, Article: Selective catalytic reduction, 6 Jan 2014. Accessed Feb 2014

Wikipedia, Article: Selective non-catalytic reduction, 15 Mar 2013. Accessed Feb 2014

Wikipedia, Article: Merox, www.wikipedia.org, 15 Oct 2013. Accessed Feb 2014

WorleyParsons, White paper: Options for handling vent gases in sulfur plants, http://www.colteng.com/CSG/Hydrocarbons/SpecialtyCapabilities/Documents/Options_for_Handling_Vent_Gases_in_Sulfur.pdf. Accessed Feb 2014

Part IV

Safety Systems

Safety Systems for Petroleum Processing

David S. J. Jones and Steven A. Treese

Contents

Petroleum Processing Safety	1308
Personal Protective Equipment and Systems	1309
Process Safety Management	1311
Elements of PSM	1311
Foundational Needs for PSM	1315
Pressure Safety	1315
Determination of Relief Cases	1315
Definitions	1317
Types of Pressure Relief Valves	1321
Relief Capacity	1322
Sizing of Required Orifice Areas	1326
Temperature Safety	1332
Thermal Relief	1333
Managing Runaway Reactions	1335
Oxygen-Deficient Environments	1338
Hazards and Potential Exposures	1339
Managing Oxygen-Deficient Environments	1340
Confined Space Entry	1341
Hazards of Confined Space Entry	1342
Managing Confined Spaces	1343
Facility Siting Hazards and Considerations	1344
Material Failures and Prevention Programs	1345

David S. J. Jones: deceased.
Steven A. Treese has retired from Phillips 66.

D.S.J. Jones
Calgary, AB, Canada

S.A. Treese (✉)
Puget Sound Investments LLC, Katy, TX, USA
e-mail: streese256@aol.com

© Springer International Publishing Switzerland 2015
S.A. Treese et al. (eds.), *Handbook of Petroleum Processing*,
DOI 10.1007/978-3-319-14529-7_20

Appendix: Example Calculation for Sizing a Relief Valve 1347
    Problem 1347
    Solution 1347
References 1348
    PSM and Related Programs 1348
    Oxygen Deficient Environments 1349
    Confined Space Entry 1349
    Facility Siting Hazards and Consideration 1349

Abstract

The processing of petroleum can be an inherently hazardous activity. The processes we use are handling a flammable material. They use strong, often hazardous, chemicals and employ high pressures and temperatures to convert the oil into finished products. All these factors present safety risks for both personnel and those living near a process facility. Still, the industry experiences very few incidents while processing millions of barrels of oil each day. Refineries place a high value on safety. This chapter discusses four key areas which contribute toward safety performance: personal protective equipment and systems, Process Safety Management (PSM), pressure safety, and temperature safety. A relief valve sizing example is presented.

Keywords

Refinery • Safety • PPE • Process safety management • PSM • Relief • Runaway • Facility siting

Petroleum Processing Safety

The processing of petroleum can be an inherently hazardous activity. The processes we use are handling a flammable material. They use strong, often hazardous, chemicals and employ high pressures and temperatures to convert the oil into finished products. All these factors present safety risks for both personnel and those living near a process facility.

In spite of the possible risks, very few significant incidents occur as the industry processes millions of barrels of oil each day. It is, in fact, statistically one of the safest industries in which to work in. The total personnel incident rate (recordable incidents per 100,000 man-hours) was 0.9 in 2012 for refining and 0.4 for petrochemicals. Few industries approach these low values. Most that do are light industries, like retail sales and electronic product manufacturing. All industries together had an incident rate of 3.7 in 2012.

This is not to say that incidents do not happen. Because of the materials present, extreme conditions, and huge volumes involved in petroleum processing, when an incident does occur, it is usually large and garners a lot of media and regulatory attention.

Safety Systems for Petroleum Processing

The petroleum industry's excellent safety history is not by "accident." A great deal of focus is placed on plant safety from design through operations to final shutdown. The focus is on ensuring that petroleum processing facilities are safe places to work and that the public is protected from harm resulting from any incidents in the facilities.

In this chapter, we will focus on several areas of safety:

- Personal protective equipment and systems
- Process safety management
- Pressure safety
- Temperature safety

Additional safety discussions are included in many areas of this handbook. These include the topics entitled:

"▶ Process Controls in Petroleum Processing"
"▶ Environmental Control and Engineering in Petroleum Processing"
"▶ Hazardous Materials in Petroleum Processing"
"▶ Fire Prevention and Firefighting in Petroleum Processing"

We will start with a general discussion of personnel protection practices.

Personal Protective Equipment and Systems

One of the primary and immediate focuses of safety in a facility is personnel safety. You must be able to protect your people from harm.

The facility design, equipment codes and standards, process design, control systems, and many other practices all work together to provide basic layers of protection for people working in a plant. These, however, can only go so far in protecting the workers. The use of personal protective equipment (PPE) provides an additional layer of protection.

In most facilities, the following are the minimum PPE required for a person entering a process area:

- Hard hat
- Flame-retardant outer clothing (like Nomex®) – long sleeves
- Safety glasses, usually with side shields
- Work boots, usually steel or safety toe

There are specific, applicable regulations that define the performance requirements for each of these PPE items. Companies often go above and beyond the regulations in their local specifications. Hard hats, in the USA, for instance, are specified by ANSI Z89.1-2009. Safety glasses normally must meet the OSHA specification ANSI Z87.1-2010.

In addition, specific PPE may be required in certain areas of the plant or during certain activities. Examples include:

- Hearing protection (earplug and/or muffs) in high-noise areas
- Gloves for general work or impermeable gloves for chemicals
- Goggles and face shield in chemical handling areas
- Personal monitors for H_2S, CO, and other hazardous substances potentially in an area
- Acid suits in acid handling areas
- Full PVC or neoprene suit with supplied air for HF acid
- Fresh air breathing apparatus for potential exposure to hazardous atmospheres or O_2 deficiency
- Respirators of various kinds (dust, chemical, etc.) appropriate to a potential hazard
- Fire suits for intense fires or high radiant heat areas
- Turnout gear for firefighting
- Man-down monitors
- Radios

Plants also contain a number of systems designed to protect workers from harm in the event of a problem. Included in these systems are the following:

- Safety showers and eyewashes (see the topic "▶ Utilities in Petroleum Processing")
- Firefighting systems (see the topic "▶ Fire Prevention and Firefighting in Petroleum Processing") – including deluges and water curtains
- Blast-resistant buildings and shelters
- Emergency shutdown buttons at safe distances from equipment
- Alarm notification systems
- Area public address systems
- Emergency breathing air stations and SCBAs throughout units
- Unit log in/log out system for anyone entering an operating area
- Installed area monitors with alarms for hazardous atmospheres like H_2S or CO
- Operator training
- Fire and emergency response plans and systems

This heavy focus on personnel protection is reflected in the low industry incident rates.

The tracking of personnel incident rates is important to comply with regulations; but what makes the difference between just tracking and developing improvements in personnel safety are (1) analyzing the data for trends and (2) taking action on undesirable trends. You can get ahead of the game by gathering data on "near misses," where an incident came close to occurring or could have happened under slightly different conditions. There will normally be many related near misses for each type of incident. Identifying and discussing those near misses with your workers can help build awareness and influence attitudes toward safety that will prevent future incidents. You have to have the information and get it to the workers to be effective, however.

Entire books can be written (and have been written) on improving personnel safety. You must always be mindful and proactive in protecting the workers to operate safely.

Process Safety Management

For many years, the focus of safety in industry has been on the protection of workers from generally recognized hazards and the programs focused on how the employee interacted with the facility and processes. This emphasis worked up to a point, greatly improving personal safety. In the processing industry, however, the actual chemical and physical processes employed can also create significant safety risks. Under a traditional safety program approach, the process received limited attention.

This situation changed in the 1980s and 1990s, after a series of major process safety incidents resulted in multiple fatalities, both inside and outside processing plants. Recognizing the risks, there was new emphasis placed on process safety management (PSM), in addition to the traditional "people"-based approaches to safety.

Specific requirements for PSM have been incorporated into legislation and regulation throughout the world. Here, we will discuss the US version of PSM as it applies to refineries in a general way. We will go into a little more depth on some specific aspects of the regulations where necessary.

Elements of PSM

The key elements of the US federal regulations on process safety management are in the Occupational Safety and Health Administration (OSHA) 1910.119 along with several guidelines OSHA has developed. There are similar regulations promulgated in many other countries, in addition to local (state, province, etc.) governments.

There are 14 key elements to a required PSM program. The remarks below indicate the general nature of the PSM elements. Comments and considerations in complying with the PSM program elements are offered, but these are not part of the regulations. This is not intended as a legal document, so reviewing and complying with the applicable regulations is left to the reader.

1. *Process safety information* – This is a compilation of written information about:
    (a) Highly hazardous chemicals used in the process – toxicity, PEL, physical data, reactivity, corrosivity, thermal and chemical stability, and hazards of mixing with other substances

(b) Process technology – up-to-date block flow diagram or PFD, process chemistry and properties, maximum intended inventory, safe upper and lower operating conditions, and consequences of deviations (including safety and health consequences)
(c) Process equipment – materials of construction, P&ID or MFD, electrical classification, relief system design and design basis, ventilation design, codes and standards used, compliance with recognized and generally accepted good engineering practices (RAGAGEP), heat and material balances, and dedicated safety systems

2. *Process hazard analysis (PHA)* – This essentially requires completion of a PHA using one of the approved techniques, for each process unit. The PHA must be periodically reviewed and updated based on new learnings. A PHA is a detailed study of each process or unit to identify potential process hazards and evaluate the adequacy of safeguards to prevent incidents and releases. It is conducted by a team with the appropriate engineering and operating knowledge of the process being reviewed. A PHA must address:
   (a) Process hazards
   (b) Previous incidents – to benefit from learnings
   (c) Engineering and administrative controls
   (d) Consequences of failure of the controls
   (e) Facility siting – much more emphasis following an industry multiple fatality incident
   (f) Human factor considerations
   (g) Evaluation of effects of failures on personnel and off-site

   The results of a PHA are normally contained in a report summarizing all things that were evaluated and the conclusions. These reports can be hundreds of pages long and very detailed. Problems found and recommendations made must be resolved in a timely manner. Usually the time frame depends on the potential severity of the risk. The results must be communicated to employees and contractors.

   Methods employed for PHAs include:
   - Hazard and operability reviews (HazOp)
   - Layers of protection analysis (LOPA)
   - What-ifs
   - Checklists
   - Failure modes and effects analysis (FMEA)

   Each company defines its own approach to meeting the PHA requirement, including the risk matrix or allowable risk profile for a PHA. OSHA offers some guidelines. API has also provided guidance in API Recommended Practice 750.

3. *Operating procedures* – Procedures must address specific hazards of the process, along with defining the operating limits for each phase of unit operations (start-up, shutdown, normal operations, etc.). The procedures should indicate the required PPE, precautions, and emergency procedures in the event of a problem. There must be procedures for the safety systems. These all need to be

in writing. The detail in the operating procedures varies from company to company, even site to site within a company.

Sometimes writing good procedures to cover all these requirements and ensure an operator will be thinking about the right things at the right times becomes difficult. By nature, procedures are a series of sequential steps. In practice, plants are started up, operated, and shutdown by multiple parallel steps. The challenge is to coordinate among several parallel steps – easier said than done.

4. *Training* – There is no substitute for the operator or staff knowing what they are doing and how to do it. This is where training plays a critical role in PSM. The training must include the procedures, equipment, operating limits, and, especially, safety systems. It is usually periodically refreshed. Each process change requires retraining on the changes. This must all be documented. Practices to comply with the regulatory requirements vary in accomplishing the objectives.
5. *Contractors* – This element of PSM requires that visitors and contractors to a facility be made aware of the process hazards and the work procedures that apply in that plant. The extent of training may vary, depending on what the contractor or visitor is doing in the plant. Someone who will not be physically working on the equipment does not need to know how to get a hot work or vessel entry permit, for instance; but they still need to know the process hazards that may be around when they enter a unit, what to do if they hear an alarm siren, and that they have to check in at the control room before entering. Conversely, a welder entering a unit will need much more complete training, including how to get a hot work permit and what to do if an alarm sounds.
6. *Mechanical integrity* – This element of PSM focuses on maintaining the process equipment in good operating condition. It would include the maintenance and inspection programs, as well as spare parts. Safety systems may require additional, periodic proof-testing to verify if they will function when needed.
7. *Hot work* – The regulations require that there be specific, appropriate hot work procedures before welding or any potentially sparking-type operation is performed in a process area. Many refiners extend this to other permitting procedures.
8. *Management of change (MOC)* – Facilities are seldom static collections of equipment and process fluids. There are ongoing improvements and changes for many reasons. The MOC provisions require that each change in a unit, which is not a direct like-in-kind replacement, be evaluated for the potential process hazards it might cause. Essentially, the change has to undergo a form of PHA, the procedures and process safety information must be updated, the impacts addressed, the operators trained, and all the other myriad details of the change fully integrated into the PSM system.

The definition of a change is something people struggle with. Many refiners say any change is a "change" for purposes of MOC. Others would classify more changes as "like in kind" by not including changes where, say, replacement equipment satisfies the same process intent as the original equipment.

Each company must reach its own consensus on what is and is not a change for purposes of MOC.

Within a company, the MOC process itself can be very onerous and time consuming. Consideration to streamlining the process, without compromising the value and validity of process safety, is usually worthwhile.

9. *Incident investigation* – There is a requirement that serious incidents and near misses be investigated in a timely manner. To make the best use of this element, investigations should be documented and the lessons taught should be conveyed to plant personnel to improve their knowledge and understanding. The objective here is to prevent future incidents of the same type. In some very serious incidents, a company may be required to make the learnings known to other companies in the industry. This is important in maintaining a safe industry overall – not everyone should have to make the same mistake to learn the lessons.

The incidents will usually have corrective actions, with time deadlines for follow-up. These must be documented and completed.

10. *Compliance audits* – There must be a program of regular auditing of compliance with the PSM elements. The audits include PSI information, operating procedure certifications, current PHAs, etc. Quality of the PSM program, materials, and compliance is audited. Companies perform these audits internally for their own edification and improvements more frequently than OSHA audits. There will be an OSHA audit scheduled on a less frequent basis. Penalties can be assessed by OSHA as a result of gaps in the PSM program identified in their audits.

11. *Trade secrets* – It is recognized that some of the process information required to operate a plant safely is proprietary (i.e., a trade secret). The fact something is in this category cannot be used as a reason for not sharing the information with operators, engineers, maintenance personnel, and contractors who could be affected by it and need to know the risks. Employees and contractors are usually bound by confidentiality agreements to not divulge such information to outside companies or competitors. Usually, there are ways to provide the information needed for safe activities while protecting the trade secret owner's intellectual property (IP). Involving the IP owner in how the information is disclosed may be helpful.

12. *Employee participation* – It is a requirement of the PSM regulations that the employees be involved in the PSM program. Usually, participation by the operators and maintenance personnel, as well as the technical staff, is valuable in PHAs, incident investigations, near miss reporting, PSI verification, PSSRs, and other areas. This also helps with the training requirements.

13. *Pre-start-up safety review (PSSR)* – A PSSR is required for new facilities or "changes." The PSSR verifies if the facilities are constructed correctly, the procedures to operate them are in place, and the required training related to the facilities has been completed. Basically, the PSSR confirms the changed plant is ready to operate safely. A PSSR is normally conducted by a team which includes technical and operations personnel. The team physically examines the

new or changed facilities, essentially punch-listing them. They ensure all aspects of PSM that must be completed before start-up are, in fact, done. Only after agreement of the PSSR team can the changed facilities be used.
14. *Emergency planning and response* – Using PHAs and the related information as a guide, this element requires that a plant have an appropriate emergency plan and response plan in place. These plans are often integrated and/or coordinated with local public emergency services groups. Often, if there are other processing facilities in an area, there are mutual aid agreements among the different facilities. The important thing is that there must be a plan to address process emergencies and you must be ready and capable of implementing the plan.

Foundational Needs for PSM

Now, to ensure that the PSM elements work and that you actually get value from it (not just meet a regulation), there are two foundational elements that need to be in place but may not necessarily be required by regulation:

- There must be a management commitment to develop, operate, and maintain the PSM system. Failure to have a demonstrated commitment from management in this area can result in gaps that can lead to citations.
- There must be a system available for appropriately sharing and managing the documentation developed by the PSM system. Current, correct information must be easy to get when needed.

Pressure Safety

A major requirement in the design and engineering of a plant or system is to ensure safe equipment operation. Much of the effort in this respect is directed to determining the pressure limits of equipment and to protect that equipment from dangerous overpressure conditions. Pressure relief valves are normally used for this protection service, although under certain conditions, bursting disks (rupture disks) may be used. This section describes the various types of relief valves and a procedure for calculating the correct orifice size of the valve and the valve selection. There is also software available (including some from relief valve vendors) to perform these calculations.

Determination of Relief Cases

The cost of providing facilities to relieve all possible emergencies simultaneously can be high. Every emergency arises from a specific scenario. The simultaneous occurrence of two or more emergencies or contingencies is not common; however, problems, such as utility failures, must be considered on a plant- or area-wide basis and may create simultaneous relief requirements that set relief system sizes and back pressures. Electrical power failures are notorious for causing large relief loads.

Unless a failure can cascade into multiple types of failure, an emergency which can arise from two or more types of failure (e.g., the simultaneous failure of a control valve and cooling water) is not usually considered when sizing safety equipment.

Each unit or piece of pressure equipment must be studied individually and every contingency must be evaluated. The safety equipment for an individual unit is then sized to handle the largest load resulting from the controlling contingency. If a certain emergency would involve more than one unit, then all must be considered as an entity. The equipment judged to be involved in any one emergency is termed "single risk." The single risk which results in the largest load on the safety facilities in any system is termed "largest single risk" and forms the design basis for the equipment.

Note that the emergency which results in the largest single risk on the overall basis may be different from the emergency(ies) which form the basis for individual pieces of equipment.

Several contingencies must be evaluated when considering the relief load for an individual piece of equipment. The minimum scenarios to consider (as applicable) are the following:

- Fire (external)
- Blocked discharge:
  - On the process side flow(s)
  - Include cases for liquid, vapor, or mixed phase
- Electrical power failure:
  - May be local or refinery-wide event
  - May be total or partial
  - Increases superimposed back pressure on a relief device
  - May cascade into instrument air, steam, and other utility failures
  - Often sets the relief sizing and flare size
  - May be necessary to look hard at timing of different reliefs
  - May require emergency depressuring of some units (e.g., hydrocrackers), which would impose a large back pressure
- Coolant failure:
  - Especially critical on towers
- Reflux failure (towers)
- Other failures to consider:
  - Mal-operation
  - Temperature runaway – depressuring
  - Mechanical failure of equipment (e.g., reboiler steam valve fails to open)
  - Abnormal heat input
  - Split exchanger tube
  - Automatic control failure

Safety Systems for Petroleum Processing 1317

- Instrument air failure
- Valve left open
- Steam-out
- Thermal (blocked-in line or equipment)
- Any other type of failure that can create a relief condition

Be sure you consider *all* the credible failure scenarios in sizing the relief systems and devices.

Definitions

The terms used and the descriptions given in this item are based on data given in two API publications. These are API RP-520 and RP-521. References are also made to Part 1 ANSI Proposed Standard, "Terminology for Pressure Relief Devices," and to ASME PTC 25.2. These publications are the safety standards commonly in current use. The following definition of terms used in the design of safety systems helps to understand the design and criteria of safety systems.

Accumulation	Accumulation is the pressure increase over the maximum allowable working pressure of the vessel during discharge through the pressure relief device expressed as a percent of that pressure or in psi or bar.
Atmospheric discharge	Atmospheric discharge is the release of vapors and gases from pressure relief and de-pressurizing devices to the atmosphere.
Back pressure	Back pressure is the pressure existing at the outlet of the pressure relief device due to pressure in the discharge system at the time the device must relieve. This is sometimes imposed by relief or venting of other devices simultaneously.
Balanced safety relief valves	A balanced safety relief valve incorporates means for minimizing the effect of back pressure on the performance characteristics – opening pressure, closing pressure, lift, and relieving capacity.
Blowdown	As used here, blowdown is the difference between the set pressure and the resealing pressure of a pressure

Burst pressure	relief valve, expressed as a percent of the set pressure or in psi or bar. Burst pressure is the value of inlet static pressure at which a rupture disk device functions.
Conventional safety relief valve	A conventional safety relief valve is a closed bonnet pressure relief valve that has the bonnet vented to the discharge side of the valve. The performance characteristics (opening pressure, closing pressure, lift, and relieving capacity) are directly affected by changes of the back pressure on the valve.
Design pressure	Design pressure is the pressure used in the design of a vessel to determine the minimum permissible thickness or physical characteristics of the different parts of a vessel.
Flare	A flare is a means for safe disposal of waste gases by combustion. With an *elevated flare*, the combustion is carried out at the top of a pipe or stack where the burner and igniter are located. A *ground flare* is similarly equipped except that combustion is carried out at or near ground level. A *burn pit* differs from a flare in that it is normally designed to handle both liquids and vapors. Flare systems are described and discussed more fully later in the handbook chapter entitled "▶ Off-Site Facilities for Petroleum Processing".
Lift	Lift is the actual travel of the disk away from closed position when the valve is relieving.
Overpressure	Overpressure is the pressure increase over the set pressure of the primary relieving device; it would be termed *accumulation* when the relieving device is set at the maximum allowable working pressure of the vessel. *Note*: When the set pressure of the

	first (primary) pressure relief valve to open is less than the maximum allowable working pressure of the vessel, the overpressure may be greater than 10 % of the set pressure of that valve.
Pilot-operated pressure relief valve	A pilot-operated pressure relief valve is one that has the major flow device combined with and controlled by a self-actuated auxiliary pressure relief valve. This type of valve does not utilize an external source of energy. These are usually used where the operating pressures are very high or the set pressure is close to operating pressure (within about 5 %).
Pressure relief valve	Pressure relief valve (PRV) is a generic term applied to relief valves (RV), safety valves (SV), pressure safety valves (PSV), or safety relief valves (SRV).
Relieving conditions	Relieving conditions pertain to pressure relief device inlet pressure and temperature at a specific overpressure. The relieving pressure is equal to the valve set pressure (or rupture disk burst pressure) plus the overpressure. The temperature of the flowing fluid at relieving conditions may be higher or lower than the operating temperature.
Rupture disk (RD or PSE)	A relief device that consists of a thin metal plate or disk designed to burst or fail when a specific pressure differential is imposed on the disk. Once the disk has failed, it must be replaced. It will *not* stop flowing after the overpressure condition is relieved. These are sometimes used to protect a relief valve from corrosive or otherwise compromising process fluids, in addition to providing full flow relief in a large overpressure event (e.g., failure of a high-pressure hydrogen stream into a

Set pressure	cooling water exchanger). See below under types of devices. Set pressure, in psig or barg, is the inlet pressure at which the pressure relief valve is adjusted to open under service conditions. In a safety or safety relief valve in gas, vapor, or steam service, the set pressure is the inlet pressure at which the valve pops under service conditions. In a relief or safety relief valve in liquid service, the set pressure is the inlet pressure at which the valve starts to discharge under service conditions.
Superimposed back pressure	Superimposed back pressure is static pressure existing at the outlet of a pressure relief device at the time the device is required to operate. It is the result of pressure in the discharge system from other sources. This type of back pressure may be constant or variable; it may govern whether a conventional or balanced-type pressure relief valve should be used in specific applications.
Vapor depressing (or depressuring) system	A vapor depressing (or depressuring) system is a protective arrangement of valves and piping intended to provide rapid reduction of pressure in equipment by release of vapors. Actuation of the system may be automatic or manual. These are often found in hydrocrackers and some hydrotreaters where the hydrogen must be rapidly eliminated to control a runaway temperature.
Vent stack	A vent stack is the elevated vertical termination of a disposal system which discharges vapors into the atmosphere without combustion or conversion of the relieved fluid. Following a multiple fatality incident in the 2000s, this type of stack is largely being eliminated.

Types of Pressure Relief Valves

The following describes the types of relief valves commonly used in industry. These have been approved according to the ASME VIII "Boiler and Pressure Vessel" Code.

Conventional Safety Relief Valves
In a conventional safety relief valve, the inlet pressure to the valve is directly opposed by a spring closing the valve. The back pressure on the outlet of the valve changes the inlet pressure at which the valve will open. A diagram of a conventional relief valve is shown below as Fig. 1.

Balanced Safety Relief Valves
Balanced safety valves are those in which the back pressure has very little or no influence on the set pressure. The most widely used means of balancing a safety relief valve is through the use of a bellows. In the balanced bellows valve, the effective area of the bellows is the same as the nozzle seat area. Back pressure is prevented from acting on the top side of the disk. Thus the valve opens at the same inlet pressure, even though the back pressure may vary. A diagram of a balanced safety relief valve is shown as Fig. 2.

Pilot-Operated Safety Relief Valves
A pilot-operated safety relief valve is a device consisting of two principal parts: a main valve and a pilot valve. Inlet pressure is directed to the top of the main valve piston with more area exposed to pressure on the top of the piston than on the bottom. Pressure, not a spring, holds the main valve closed. At the set pressure, the pilot opens, reducing the pressure on top of the piston, allowing the main valve to go fully open.

Resilient-Seated Safety Relief Valves
When metal-to-metal seated conventional or bellows-type safety relief valves are used where the operating pressure is close to the set pressure, some leakage can be expected through the seats of the valve (refer to API Standard 527, "Commercial Seat Tightness of Safety Relief Valves with Metal-to-Metal Seats").

Resilient-seated safety relief valves, with either an O-ring seat seal or a plastic seat, such as Teflon, provide seat tightness. Limitations of temperature and chemical compatibility of the resilient material must be considered when using these valves. They often do not reseat after a relief event. They also do not work well in high-pressure differential applications.

Rupture Disk (RD, PSE)
A rupture disk consists of a thin metal diaphragm held between flanges. The disk is designed to rupture and relieve pressure within tolerances established by the ASME Code.

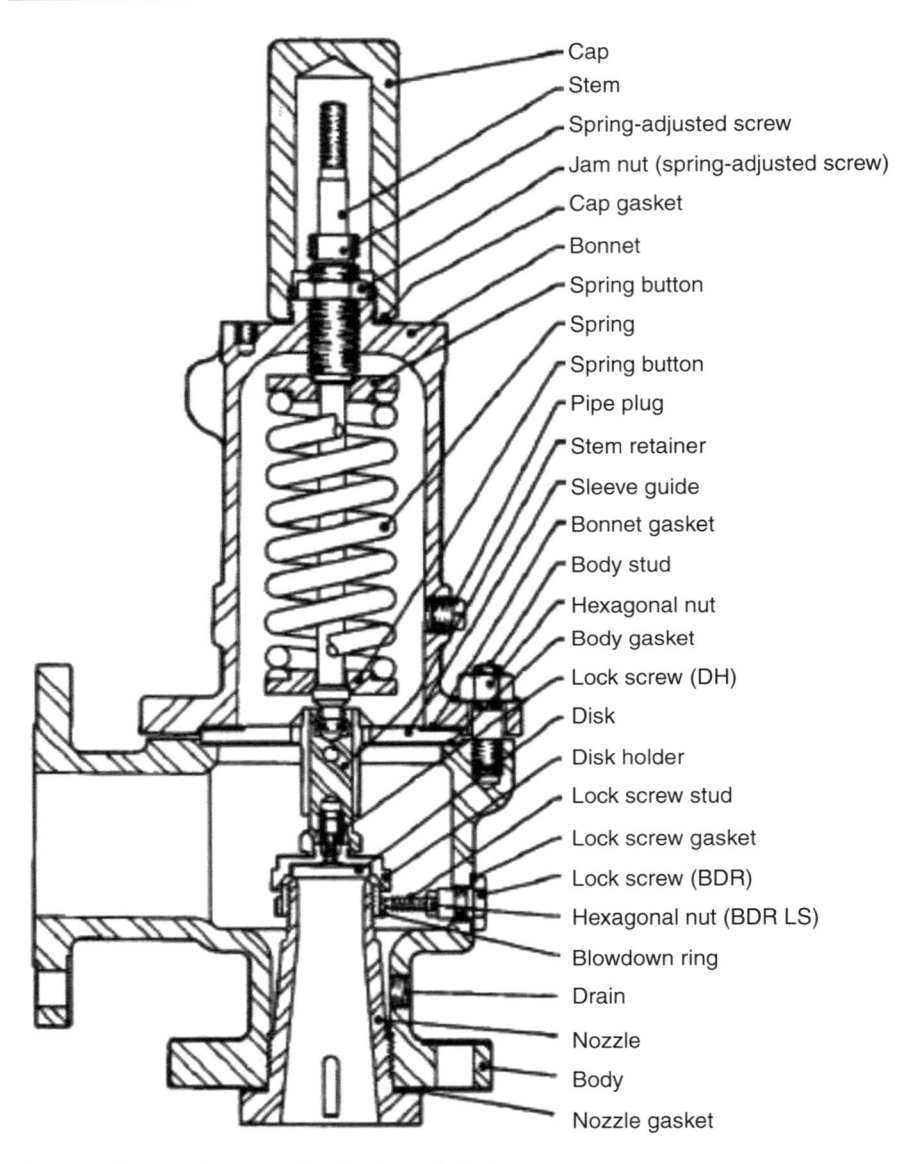

Fig. 1 A diagram of a conventional safety relief valve

Relief Capacity

The maximum amount of material to be released during the largest single risk emergency determines the size of the safety relief valves in any given system. Any calculation to determine valve sizing must therefore be preceded by a calculation or some determination of the maximum amount. Among the most common sizing

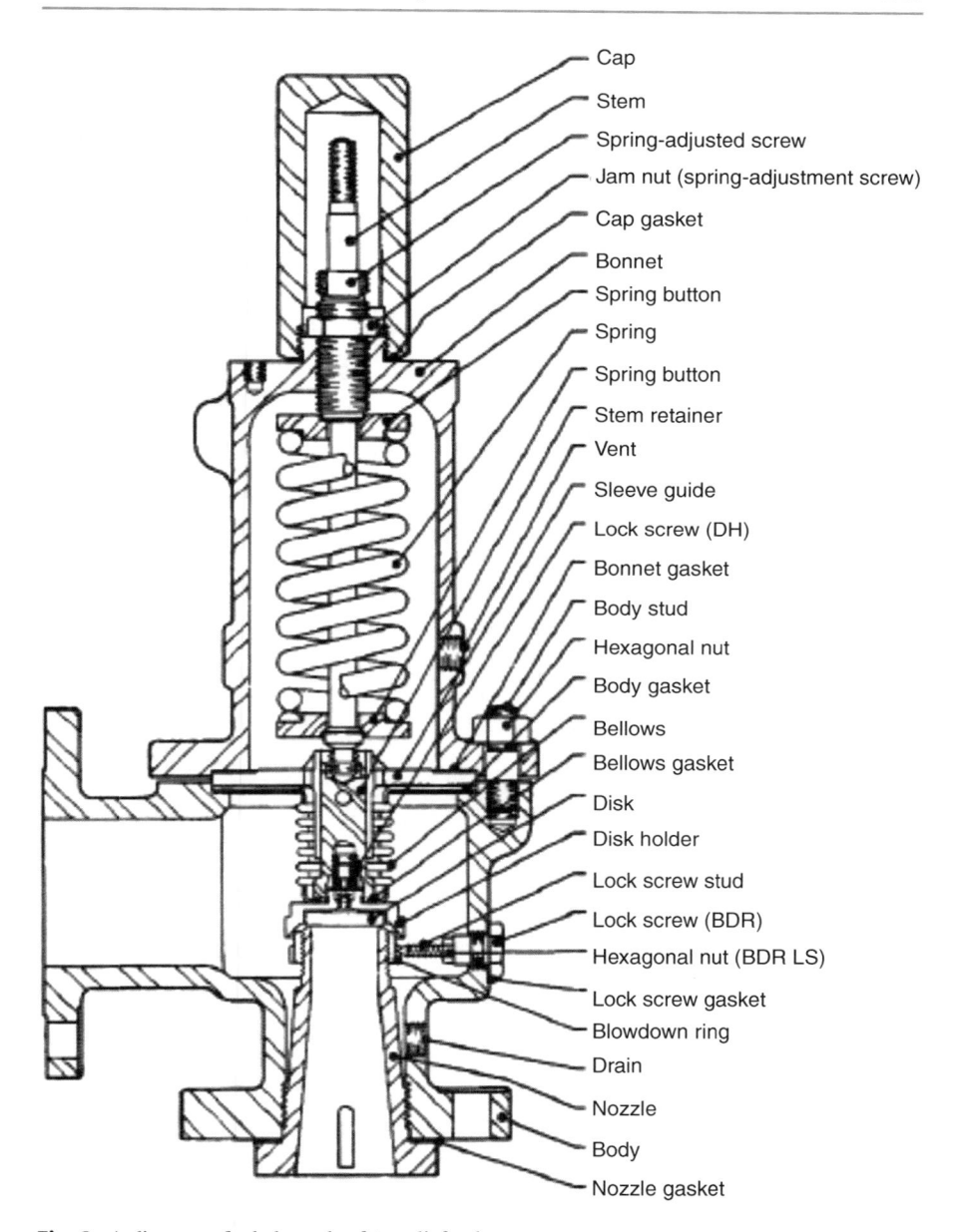

Fig. 2 A diagram of a balanced safety relief valve

criteria are the event of fire and its effect on the contents of exposed vessels. There are also other criteria which can determine maximum release that are attributable to operational failure or one of the other scenarios listed in section "Determination of Relief Cases."

Capacity Due to Fire

The exact method of making this calculation must be established from the appropriate codes which apply, API RP-520, Part I, API Standard 2510, and NFPA No. 58, or local codes which may apply. Each of the listed codes or standards approaches the problem in a slightly different manner.

Liquid systems – Many of the systems that are encountered will contain liquids or liquids in equilibrium with vapor. Fire relief capacity in this situation is calculated on the basis of heat energy from the fire translated in terms of vapor generated in the boiling liquid.

API RP-520, Part I, applies to refineries and process plants. It expresses requirements in terms of heat input:

$$Q = 21,000 F A^{0.82} \tag{1}$$

where:

Q = Btu/h heat input.
A = area in sq ft of wetted surface of the vessel up to 25 f. above grade.
Wetted surface is calculated at the maximum fill level. Grade is the ground level under the vessel.
F = fireproof factor due to insulation becoming 1.0 for a bare vessel (see Fig. 3).

The amount of vapor generated with this is then calculated from the latent heat of the material at the relieving pressure of the valve by the following equation. For fire relief only this may be calculated at 120 % of maximum allowable working pressure. All other conditions must be calculated at 110 % of maximum allowable working pressure; thus,

$$Q/H_t = W \tag{2}$$

where:

Q = Btu/h heat input to the vessel.
H_t = Btu/lb latent heat of the material being relieved.
W = lb/h of vapor to be relieved by the relief valve.

The latent heat of pure and some mixed paraffin hydrocarbon materials may be estimated using the data given in API RP-520. A more accurate latent heat evaluation for mixed hydrocarbons will be found by utilizing vapor-liquid equilibrium K data and making a flash calculation. Mixed hydrocarbons will fractionate, beginning with the lowest temperature boiling mixture and progress to the highest temperature mixture; therefore, consideration must be given to the condition which will cause the largest vapor generation requirements from the heat input of a fire.

Fig. 3 Insulation factors for fire relief loads

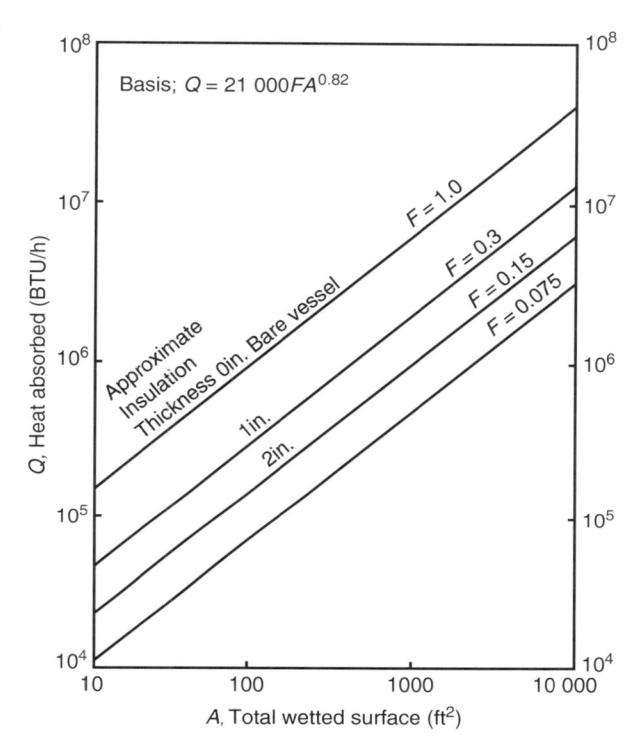

[Graph: Q, Heat absorbed (BTU/h) vs A, Total wetted surface (ft²); Basis: $Q = 21\,000 F A^{0.82}$; curves for $F = 1.0$ (Bare vessel, 0 in.), $F = 0.3$ (1 in.), $F = 0.15$ (2 in.), $F = 0.075$; labeled "Approximate Insulation Thickness"]

Latent heat will approach a minimum value near critical conditions; however, the effect does not go to zero. An arbitrary minimum value that may be used is 50 Btu/lb.

Other Capacity Scenarios

Section "Determination of Relief Cases" listed a few of the other scenarios that must be considered when sizing relief valves and systems. The other capacity scenarios, briefly, are the following:

- Blocked discharge
- Electrical power failure (partial and total)
- Coolant/cooling water failure
- Reflux failure (towers)
- Other failures in the process

Considerable effort may be required to define the other capacity scenarios. It is common that the capacity required for one of these scenarios greatly exceeds the fire relief case and will set the relief valve capacity requirement.

Overall Relief and Recovery System Loads

All the various vents and relief valves are tied into the relief and recovery system, usually connected to the flare. It is often necessary to perform a flare system or relief study to determine the superimposed back pressure on the relief devices before the final sizing can be completed. For this study, the entire flare or relief system must be hydraulically modeled with the various scenarios for relief load each calculated.

The modeling must be done carefully because the pressure drops are very low and small changes in the pressure drop can have significant implications on sizes and design flows. You may find that the back pressure in some scenarios exceeds the allowable working pressure for a vessel or tower – meaning it could not relieve under that scenario. If it does not need to relieve anyway, then there is little consequence (unless you consider double jeopardy). If the device would need to relieve, then alternatives and other controls need to be considered to reduce back pressure/backflows to an acceptable level.

Having a flare system model is a good practice and is very helpful in understanding options in plant emergency scenarios.

Relief Valve Load Specification

It is normal practice for the relief valve load calculations and bases to be summarized on a "relief valve load calculation data sheet." There is normally one sheet for each relief valve service. The sheet typically includes source (e.g., vessel) data, process data, each relieving condition (including quantity, relieving fluid properties, phases, required valve area), sketches, and other data required for the valve.

The load calculation sheet, with the supporting calculations, satisfies one of the PSM compliance requirements.

Sizing of Required Orifice Areas

The safety relief valve manufacturers have standard orifice designations for area and the valve body sizes which contain these orifices (API Standard 526, "Flanged Steel Safety Relief Valves for Use in Petroleum Refineries"). The standard orifices available, by letter designation and area, are in Table 1.

Sizing for Gas or Vapor Relief and Sonic or Critical Flow

Safety relief valves in gas or vapor service may be sized by use of one of these equations:

$$A = \frac{W\sqrt{T}\sqrt{Z}}{CKP_iK_b\sqrt{M}} \quad (3)$$

$$A = \frac{V\sqrt{T}\sqrt{M}\sqrt{Z}}{6.32\, CKP_iK_b} \quad (4)$$

Safety Systems for Petroleum Processing

Table 1 Standard relief valve orifice sizes

D orifice	0.110 in.^{2}
E orifice	0.196 in.^{2}
F orifice	0.307 in.^{2}
G orifice	0.503 in.^{2}
H orifice	0.785 in.^{2}
J orifice	1.287 in.^{2}
K orifice	1.838 in.^{2}
L orifice	2.853 in.^{2}
M orifice	3.60 in.^{2}
N orifice	4.34 in.^{2}
P orifice	6.38 in.^{2}
Q orifice	11.05 in.^{2}
R orifice	16.0 in.^{2}
T orifice	26.0 in.^{2}

Note: However, many small safety relief valves are manufactured with orifice areas smaller than "D," and many pilot-operated types contain orifice areas larger than "T"

$$A = \frac{V\sqrt{T}\sqrt{G}\sqrt{Z}}{1.175\, CKP_i K_b} \tag{5}$$

where:

$W =$ flow through valve, lb/hr.
$V =$ flow through valve, scfm.
$C =$ coefficient determined by the ratio of the specific heats of the gas or vapor at standard conditions.
$K =$ coefficient of discharge, obtainable from the valve manufacturer (usually 0.6–0.7).
$A =$ effective discharge area of the valve, inch^2.
$P_i =$ upstream pressure, psia. This is set pressure plus overpressure plus the atmospheric pressure.
$K_b =$ capacity correction factor due to back pressure. This can be obtained from Fig. 4 for conventional valves or pilot-operated valves and from Fig. 5 for balanced bellows valves.
$M =$ molecular weight of gas or vapor.
$T =$ absolute temperature of the inlet vapor in °R (°F +460).
$Z =$ compressibility factor for the deviation of the actual gas from a perfect gas.
$G =$ specific gravity of gas referred to air, 1.00 at 60 °F and 14.7 psia.

Sizing for Liquid Relief

Conventional and balanced bellows safety relief valves in liquid service may be sized by use of the following equation. Pilot-operated relief valves should be used

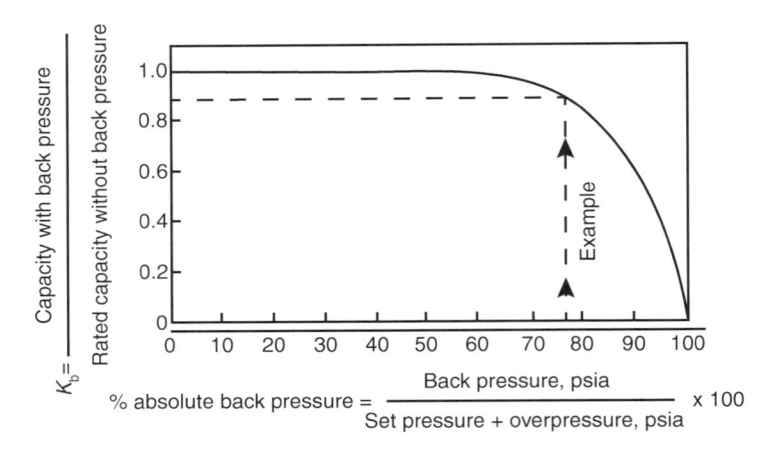

Fig. 4 Back pressure sizing factor K_b for conventional valves

$$\% \text{ absolute back pressure} = \frac{\text{Back pressure, psia}}{\text{Set pressure} + \text{overpressure, psia}} \times 100$$

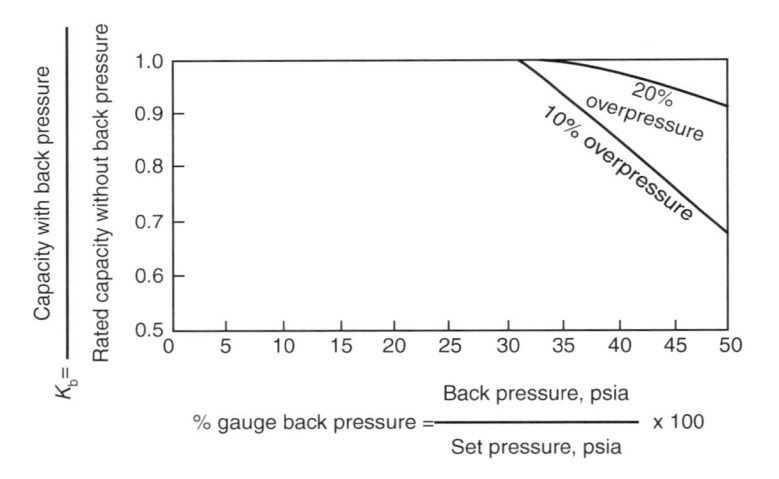

Fig. 5 Back pressure sizing factor K_b for balanced valves

$$\% \text{ gauge back pressure} = \frac{\text{Back pressure, psia}}{\text{Set pressure, psia}} \times 100$$

in liquid service only after determining from the manufacturer that they are suitable for the service:

$$A = \frac{\text{gpm}\sqrt{G}}{38K\left(K_p K_w K_v\right)\left(1.25P - P_b\right)^{1/2}} \quad (6)$$

Note: A coefficient of discharge of 0.62 is normally used for K.
where:

gpm = flow rate at the selected percentage of overpressure, in US gals.
A = effective discharge area, inch^2.

Safety Systems for Petroleum Processing 1329

Fig. 6 Capacity correction factor due to overpressure K_p

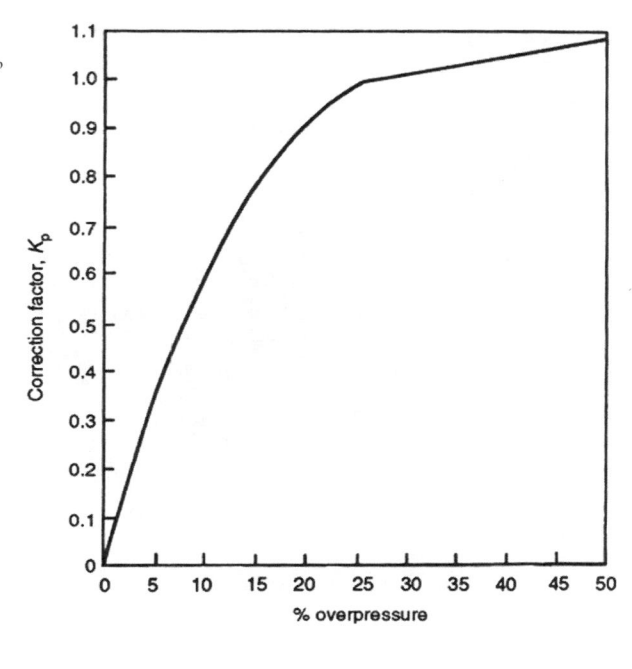

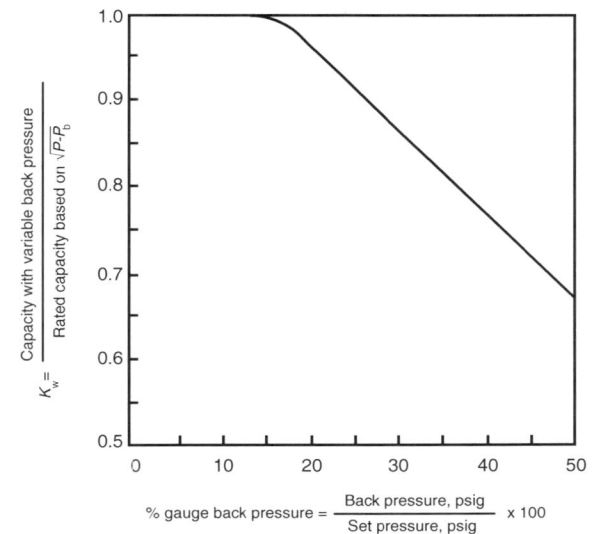

Fig. 7 Capacity correction factor K_w

$K_w = \dfrac{\text{Capacity with variable back pressure}}{\text{Rated capacity based on }\sqrt{P-P_b}}$

% gauge back pressure = $\dfrac{\text{Back pressure, psig}}{\text{Set pressure, psig}} \times 100$

K_p = capacity correction factor due to overpressure, for 25 % overpressure. $K_p = 1.00$. The factor for other percentages of overpressure can be obtained from Fig. 6.

K_w = capacity correction factor due to back pressure and is required only when balanced bellows valves are used. K_w can be obtained from Fig. 7.

K_v = capacity correction factor due to viscosity. For most applications, viscosity may not be significant, in which case $K_v = 1.00$. When viscous liquid is being relieved, see method of determining K_v as described below.
P = set pressure at which relief valve is to begin opening, psig.
P_b = back pressure, psig.
G = specific gravity of the liquid at flowing temperature referred to water, 1.00 at 70 °F.

When a relief valve is sized for viscous liquid service, it is suggested that it be sized first as for nonviscous-type application in order to obtain a preliminary required discharge area, A. From manufacturers' standard orifice sizes, the next larger orifice size should be used in determining the Reynolds number, R, from either of these relationships:

$$R = \frac{\text{gpm}(2,800G)}{\mu\sqrt{A}} \tag{7}$$

$$R = \frac{12,700\,\text{gpm}}{U\sqrt{A}} \tag{8}$$

where:

gpm = flow rate at the flowing temperature, in US gpm.
G = specific gravity of the liquid at the flowing temperature referred to water, 1.00 at 70 °F.
μ = absolute viscosity at the flowing temperature, in centipoise.
A = effective discharge area, in sq in. (from manufacturers' standard orifice areas).
U = viscosity at the flowing temperature, in Saybolt universal seconds.

After the value of R is determined, the factor K_v is obtained from Fig. 8. Factor K_v is applied to correct the "preliminary required discharge area." If the corrected area exceeds the "chosen standard orifice area," the above calculations should be repeated using the next larger standard orifice size.

Sizing for Mixed-Phase Relief

When a safety relief valve must relieve both liquid and gas or vapor, it may be sized by the following steps:

(a) Determine the volume of gas or vapor and the volume of liquid that must be relieved.
(b) Calculate the orifice area required to relieve the gas or vapor as previously outlined.
(c) Calculate the orifice area required to relieve the liquid as previously outlined.
(d) The total required orifice area is the sum of the areas calculated for liquid and vapor.

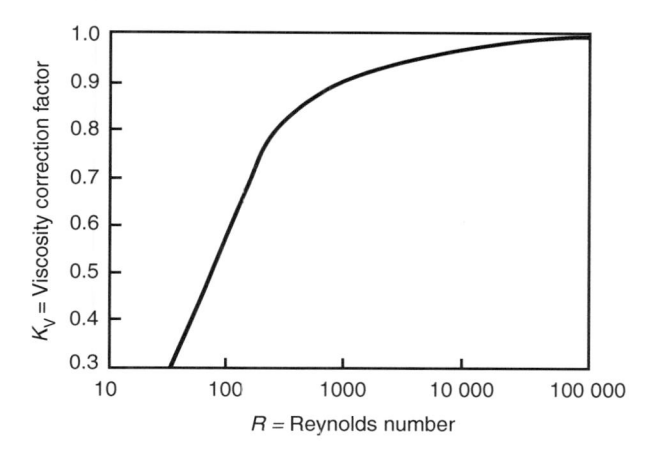

Fig. 8 Capacity correction factor due to viscosity K_v

Sizing for Flashing Liquids

The conventional method is to determine the percent flashing from a Mollier diagram or from the enthalpy values. Then consider the liquid portion and vapor portion separately as in mixed-phase flow. A method to calculate the percent flashing is shown in the following equation:

$$\% \text{ Flash} = \frac{h_f(1) - h_f(2)}{h_{fg}(2)} \times 100 \tag{9}$$

where:

$h_f(1)$ = enthalpy in Btu/lb of saturated liquid at upstream temperature.
$h_f(2)$ = enthalpy in Btu/lb of saturated liquid at downstream pressure.
$h_{fg}(2)$ = enthalpy of evaporation in Btu/lb at downstream pressure.

Sizing for Gas or Vapor on Low-Pressure Subsonic Flow

When the set pressure of a safety relief valve is very low, such as near atmospheric pressure, the K_b values obtainable from Fig. 5 are not accurate. Safety relief valve orifice areas for this low-pressure range may be calculated:

$$A = \frac{Q_v \sqrt{GT}}{863 KF \sqrt{(P_1 - P_2)P_2}} \tag{10}$$

where:

A = effective discharge area of the valve, inch^2.
Q_v = flow through valve, scfm.

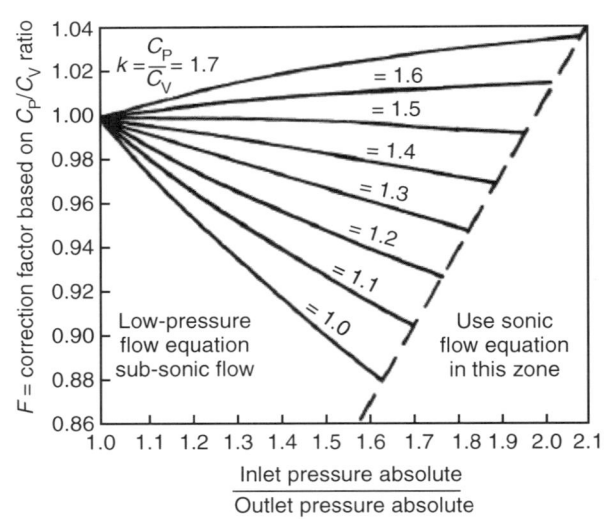

Fig. 9 Flow correction factor "F" based on specific heats

[Graph: F = correction factor based on C_p/C_v ratio vs. Inlet pressure absolute / Outlet pressure absolute; curves for $k = C_p/C_v$ = 1.0, 1.1, 1.2, 1.3, 1.4, 1.5, 1.6, 1.7. Left region labeled "Low-pressure flow equation sub-sonic flow"; right region labeled "Use sonic flow equation in this zone".]

G = specific gravity of gas referred to air, 1.00 at 60 °F and 14.7 psia.
T = absolute temperature of the inlet vapor, °R = °F +460.
K = coefficient of discharge, available from the valve manufacturer.
F = correction faction based on the ratio of specific heats. This can be obtained from Fig. 9.
P_1 = upstream pressure in psia = set pressure plus overpressure plus atmospheric pressure.
P_2 = downstream pressure at the valve outlet in psia.

An example calculation covering the sizing of a relief valve is given in Appendix of this chapter.

Temperature Safety

Most petroleum processes involve elevated temperatures. These are normally well controlled, but there are safety risks posed by temperature. Of course, high temperatures always present physical hazards from simple heat that can cause burns. Many processes operate at temperatures which, if someone were exposed to them, would immediately cause severe burns. We will take the hazards of burns from operating conditions and fluids as a given.

Our focus here will be on two cases in which temperature can lead essentially to overpressuring equipment with a resulting loss of containment:

- Thermal relief
- Reaction runaway

Thermal Relief

When a hydrocarbon or another fluid is trapped in a fixed equipment volume, usually between block valves in a pipeline, and that fluid is heated, the pressure within the trapped fluid increases as the fluid expands. At the same time, the piping also expands with temperature, but not necessarily at the same rate as the fluid. For fluids like LPGs, the expansion of the fluid can be substantial and can include vaporization.

If the fluid expands more than the space in which it is contained, the pressure can exceed the yield strength of the equipment and the equipment will rupture. This releases the fluid to the atmosphere – an uncontrolled release.

For equipment, this situation is addressed as the fire load when sizing the relief valve load. For a blocked-in pipeline, the heat source can be the sun, heat tracing, or nearby hot process equipment; but the case can be much worse if the heat source is a pool fire or other nearby fires. Piping has also ruptured due to hydrogen evolution from corrosion in the blocked-in line. Release of additional combustible materials from pipe rupture can just make a fire worse.

To avoid rupturing a long, blocked-in pipeline, it is customary to provide thermal relief valves (TRVs). These are also called thermal safety valves (TSVs). They are normally small, conventional relief valves. They may be vented to grade or, in many cases, they can be designed to jump around one of the block valves at either end of the line to contain the fluids (these have to be managed when isolating equipment). In many facilities, there are more of these TSVs in the plant, especially in off-site areas, than there are any other relief valve services.

The set points for the TSVs are normally much higher than the operating pressures of the lines they protect. They are set at the maximum pressure allowed in the pipe (considering temperature).

There are two primary codes and standards applicable to this situation:

- API Recommended Practice 521, section 3.14
- ASME B31.3, paragraph 301.2.2

Some practical exceptions are usually made if the contained fluid is not a dangerous liquid (loosely defined):

- Short lines generally do not require TSVs.
- Small-diameter pipes do not require TSVs.
- Pipes that are blocked in with a contained hot fluid will subsequently cool.

To determine the need for thermal relief in a line, the following analysis applies:

- Calculate the net volume of liquid expansion beyond the pipe volume using the equation:

$$\Delta V = (V_t - V_{t0}) - \Delta V_p - \Delta V_{Pr} \tag{12}$$

Table 2 Some typical coefficients of cubic expansion for liquids(Brahmbhatt 1984)

Liquid	Range, °C	a, 10^{-3}	b, 10^{-6}	c, 10^{-8}
Benzene	11–81	1.17626	1.27755	0.80646
Toluene	0–100	1.028	1.779	0
Methanol	−38– + 70	1.18557	1.56493	0.91113
Ethyl acetate	−36– + 72	1.2585	2.95688	0.14922
Sulfuric acid	0–60	0.5758	−0.864	0
n-Pentane	−190– + 30	1.50697	3.435	0.975

Coefficients for tr =0 °C
Example for benzene: $V_t = V_{tr}(1 + 1.17626 \times 10^{-3} t + 1.27755 \times 10^{-6} t^2 + 0.80646 \times 10^{-8} t^3)$

where:
V_t = volume of liquid at the final temperature.
V_{t0} = volume of fluid at the initial temperature.
ΔV_p = pipe volume expansion due to temperature.
ΔV_{Pr} = increase in pipe volume from increasing pressure.
- Now we use the cubic form of the equation to determine the volumetric expansion of the fluid:

$$V_t = V_{tr}(1 + at + bt^2 + ct^3) \quad (13)$$

where:
V_{tr} = base volume at a reference temperature.
a, b, c = cubic volumetric expansion coefficients (see Table 2 examples).
 The easiest way to do this is to calculate the relative fluid volume at the initial temperature and the relative volume at the final temperature. The change in volume is the ratio of final to initial relative volumes times the total initial volume.
- The pipe expansion is given by

$$\Delta V_p = \pi/4 \left[(D+CD)^2(L+CL) - D^2 L\right] \quad (14)$$

where:
D = pipe ID in feet (initial condition).
C = linear coefficient of expansion at final temperature (Table 3).
L = total pipe length including all fittings in feet (initial condition).
- We will assume that the ΔV_{Pr} is negligible. This is a relatively safe assumption.
- To determine the relief need, compare the change in volume of the liquid to the change in volume of the pipe, both due to thermal expansion. If the liquid volume change exceeds the pipe volume change, it is normally advisable to include a TSV.

The most common TSVs are ¾ × 1 in. with a special 0.06 square inch orifice. This size of TSV can be used for most fluids up to total piping lengths of

Table 3 Typical linear coefficients of thermal expansion for metals

Temp, °F	Carbon steel, × 10^{-6}	18-8 stainless, × 10^{-6}	25 Cr-20 Ni, × 10^{-6}	Wrought iron, × 10^{-6}
50	−0.14	−0.21	−0.16	−0.16
70	0	0	0	0
100	0.23	0.34	0.28	0.26
125	0.42	0.62	0.51	0.48
200	0.99	1.46	1.21	1.14
250	1.40	2.03	1.70	1.60
300	1.82	2.61	2.18	2.06
400	2.70	3.80	3.20	3.01
500	3.62	5.01	4.24	3.99
700	5.67	7.50	6.44	6.06
900	7.81	10.12	8.78	8.26

Example: Carbon steel at 100 °F is expanded by a factor of 0.23×10^{-6} compared to carbon steel at 70 °F – i.e., 100 f. of carbon steel at 70 °F will be $100 \times 1.00000023 = 100.000023$ f. long at 100 °F

10,000 f. of 4 in. piping. For longer pipes, larger pipes, and fluids with large expansion coefficients (like liquefied gases), the appropriate TSV size would have to be determined using the fluid properties and the heat input rate. The TSV is set at the design pressure of the pipe.

A couple of final thoughts on TSVs:

- It is easy to forget about the TSVs on piping and equipment, especially in remote areas. The TSVs need to be tracked and routinely serviced just like any other relief valves in order to serve their purpose when needed.
- TSV discharges need to be monitored to ensure there is no leakage. If a TSV discharges to the atmosphere, monitoring is easy. If a TSV jumps around a block valve, monitoring is harder.
- When isolating a line, do not forget to include to isolate the TSV if it jumps around a block valve.
- Provide root valves and bleeders for TSVs to enable servicing them online, like any other relief valve.

Managing Runaway Reactions

Many processes in petroleum refining are exothermic, that is, they evolve heat. Combustion is an obvious example; but here we are talking about processes like:

- Hydroprocessing (e.g. hydrocracking, hydrotreating, isomerization)
- Acid/base reactions – e.g., amine absorption
- Methanation
- Alkylation unit acid runaway

The processes depend on various mechanisms to control the heat of reaction:

- Hydrocrackers, hydrotreaters, and isomerization units depend on the continuous flow of oil and hydrogen through the reactors along with injected cold quench gas to move the heat away.
- Polymerization reactors may depend on coolant-jacketed or exchanger-style reactors or may have large recycle cooling streams with quenches.
- Amine absorption and other acid/base reactions depend on dilution of the heat in a flowing mass or a circulating cooling system.
- Methanators depend on flow and short residence times to carry heat away.
- An alkylation unit depends on maintaining proper chemistry to avoid acid runaway conditions.

Failures of the control mechanisms can lead to very high temperatures in the units. Hydrocrackers, for example, can exceed 1,900 °F in a runaway and those temperatures can be reached in only a few minutes or seconds.

Loss of containment can result when the reaction temperature exceeds the allowable metal temperature with the equipment still at pressure (i.e., you exceed the ultimate yield stress of the material under the process condition).

Industry, including process licensors, has developed specific methods to manage the affected processes in runaway temperature conditions. We will briefly discuss some of the techniques that work. Working with your process licensor can help add more information on runaway control methods.

Hydroprocessing Units

If temperatures in a hydroprocessing unit exceed about 850 °F (~454 °C), the unit can enter a hydrocracking runaway condition in which catalytic and thermal cracking will drive temperatures up rapidly. Once started, the reaction is self-sustaining. A hydrocracking catalyst does *not* have to be present for this to occur.

The reactor seldom fails in a runaway because the metal mass slows its heat up. The normal failure point is at the reactor outlet line or the first effluent exchanger. Fatalities have resulted from these failures.

Control of a true hydrocracking runaway requires immediate and aggressive depressuring of the unit to the flare. The depressuring point is usually the cold high-pressure separator; although if there is an amine absorber, depressuring may be downstream of it. Depressuring rate would typically be somewhere between 100 and 300 psi/min to start (this decays as pressure drops). The unit should be at flare pressure within about 15 min. It should not be repressured until cooled, possibly using nitrogen gas. Licensors have specific guidelines for these procedures and the required equipment.

Not all units can get into a runaway, so analysis is needed to determine the requirements. Some specific scenarios to consider when determining the needs for a depressuring system are the following:

- Does the unit have a hydrocracking catalyst?
- Does the unit operate near 850 °F at any point in its run?

Safety Systems for Petroleum Processing   1337

- If flow stops (either treat or recycle gas or a single quench flow), will the bed get too hot?
- Can a sudden change in feedstock (like a lot more cracked feed) drive temperatures quickly to runaway conditions?

The depressuring system is normally manually activated for hydrotreaters. It may be automatically activated under defined conditions for hydrocrackers, with manual activation also available.

For units where runaway is only a remote possibility, the unit should still be capable of depressuring in a reasonable amount of time in case of a fire or other release scenarios. This is part of the API Recommended Practices.

Temperature coverage can be a challenge in these units. You only know temperatures at a few points in a reactor and, possibly, on the reactor shell. Maldistribution (from, e.g., internal damage, previous runaway, or bad catalyst loading) can result in undetected hot spots. If a hot spot happens to be near a reactor wall, you can miss it completely. Careful consideration of the numbers and placement of temperature measuring points is required, both in the reactors and on the reactor shells. Licensors have guidelines for temperature monitoring locations, and many companies have their own, more complete guidelines or requirements.

If an automatic depressuring system is prescribed, it normally will be a safety integrity, integrated, or instrumented system (SIS), with appropriate security, redundancy, maintenance, and testing requirements.

Polymerization Reactions

Polymerization reactions in petroleum processing can be highly exothermic. Loss of temperature control in these processes can occur if the coolant is lost – such as loss of boiler feed water in an exchanger-style reactor.

Control approaches may include:

- Stopping feed for continuous processes.
- Ensuring relief capacity is adequate for generated gases and polymer.
- Alternate coolant sources or types.
- In some reactions, an inhibitor injection may be used to kill the reaction.
- For some batch polymerizations, the reactor contents may be dumped to a special relief system designed to contain and manage the reacting mass with a separate flare.

Acid/Base Reactions

For many acid/base reactions, temperature is controlled by a large circulating stream or a large base volume of cool material, with injection of acid or base into the much larger mass. If temperatures get too high, they can be controlled just by stopping feed of one or more of the reactants.

An interesting case is presented in amine and other absorbers that depend on an acid/base reaction. The reaction generates a significant amount of heat.

The heat is moderated by using a much larger circulating mass of the absorbent solution. This acts as a flywheel to control temperature. If the solution rate gets too low, the temperatures in the absorber can rise to the point where the solution actually begins regenerating in the absorber tower. This results in a phenomenon where the absorber accumulates hot liquid in the trays and periodically blows the hot liquid overhead. The process repeats on some frequency. The remedy is increasing the absorbent rate to ensure the rich solution loading is at the correct level. This is not normally a safety problem, but it can become one if not controlled.

Methanation
Methanation is a highly exothermic reaction that is part of older hydrogen plants and other units where carbon oxides in hydrogen may poison a process. A methanator can get into a runaway condition if the upstream carbon oxide removal process fails or the incoming CO and CO_2 exceed the tolerable levels (usually only a few %). These reactors generally have good thermocouple coverage. Because they are in a vapor phase, maldistribution is normally not a problem.

The reactors have normally relatively thinner walls than hydroprocessing units, so they can heat up extremely quickly. Fast action is needed to avoid loss of containment.

If a runaway condition is detected in a methanator by any high reactor or outlet temperature, the reactor is automatically shut down by a feed block valve and the incoming gas is sent to flare until the problem is resolved. The auto-shutdown system will generally be an SIS-type design with appropriate security, redundancy, maintenance, and testing requirements. There are provisions for manually activating the shutdown also.

Alkylation Unit Acid Runaway
An acid runaway in an alkylation unit can result when the acid strength drops too low. The more dilute acid promotes rapid, exothermic polymerization of olefins over the reaction of an olefin and isobutane as desired. The polymers tend to dissolve in the acid, further diluting it and pushing more polymerization.

From a simple standpoint, an acid runaway is controlled by stopping the olefinic feed, but leaving the paraffinic feed in. Fresh acid is brought in to restore the acid strength.

Vendors have recommended approaches in much more detail for managing acid runaway.

Oxygen-Deficient Environments

The following discussion is presented for information only. Governmental regulations, company/contractor procedures, and legal considerations control all aspects of dealing with oxygen-deficient environments at a given facility.

Hazards and Potential Exposures

Many materials handled in petroleum processing facilities are hazardous, and some are immediately dangerous to life or health (IDLH). Examples of IDLH materials include some of the more obvious compounds, like H_2S and HF acid. These types of materials have well-known hazards and defined approaches to dealing with them.

Other potential IDLH exposures are not so obvious, like nitrogen or CO_2 atmospheres. In these cases, one of the highest risks is oxygen deficiency. According to the US Chemical Safety Board, nitrogen asphyxiation resulted in 80 deaths between 1992 and 2002. Lack of oxygen in your lungs can quickly lead to death, unless you can get to oxygen quickly. A person will lose consciousness after only one or two breaths of nitrogen. There is no warning.

Even low oxygen levels can be an issue. Below 16 % of O_2, the brain makes you breathe faster and deeper, which can make the situation worse. This can affect someone simply walking near where a large nitrogen vent is located. Exposure to 4–6 % of O_2 for about 40 s will result in coma. A person must get oxygen within minutes to survive. There is still residual risk of cardiac arrest.

Carbon dioxide has the added hazard of being toxic in concentrations above about 3 % and fatal on short exposures (5 min) over 9 %.

So, where could these types of atmospheres be found in a refinery? Some of the possible exposure routes include:

- Nitrogen:
    - Loading and unloading of pyrophoric or self-heating catalysts (both inside a reactor and outside during handling).
    - Vessels or containers kept under inert gas blanket or purge to prevent air intrusion.
    - Gas-blanketed tanks or drums.
    - Air-freeing procedures for equipment prior to start-up.
    - Gas-freeing procedures for equipment during shutdown.
    - Equipment protected from oxygen contact to avoid polythionic acid stress corrosion cracking or other materials damage mechanisms caused by air exposure.
    - Backup instrument air systems.
    - Bad breathing air (incorrectly mixed or using nitrogen bottle instead of air).
    - Leakage of nitrogen from an inerted pipe or system into a system that is not supposed to be inert through leaks or improper blinding or valve lineup. An example would be nitrogen accumulation in a firebox from a leaking furnace tube that is being held under nitrogen.
    - Blinding or blanking piping and equipment under nitrogen purge.
    - Some welding procedures.
- Carbon dioxide:
    - Gas-freeing tanks or other equipment for start-up or maintenance
    - Blinding or blanking piping and equipment under carbon dioxide blanket

- Hydrogen plant CO_2 product lines
- Flue gases and ducting
- Sewer boxes and manholes
- Some welding procedures

Since these gases are colorless, there is little warning they may be present. Carbon dioxide at least has some warning once encountered – it has a bad taste in your mouth, even at low concentrations.

Managing Oxygen-Deficient Environments

The remarks in this section are meant to indicate only some of the types of practices that have been employed to manage oxygen-deficient environments and prevent worker exposure to IDLH conditions. The comments are not all-inclusive. Each facility should establish its own requirements. Most companies and contractors already have their own programs and requirements for use and control of these types of situations.

Some of the good practices for managing oxygen-deficient environments (focusing mostly on nitrogen) include:

- Implementation of a warning system and exclusion zones around any area where inert is being used for purging or inerting.
  - Persons within the "hot zone" closest to the vent location must usually be in fresh air equipment.
  - Signs are posted indicating nitrogen or CO_2 in use and potential for inadequate oxygen environment.
  - The area is usually taped or barricaded for a defined distance around the vent point.
- Testing before entry is permitted into equipment, followed by continuous monitoring of confined or enclosed space atmosphere during entry:
  - Testing would include oxygen, toxics, and combustibles/% LEL.
- Special procedures are used for inert vessel or reactor entry when work must be performed within an inert atmosphere:
  - Normally, an effort is made to avoid inert entry, but it is sometimes used – such as in pyrophoric catalyst dumping or for metallurgy protection.
  - Procedures usually include much more detailed testing, permitting, and equipment requirements and require higher approval levels than a normal entry.
  - Fresh air equipment with backup and emergency egress air supplies would be used.
  - Often the workers use hard helmets with supplied air that they cannot remove themselves while in a reactor or IDLH environment.
  - Entry is normally done by contract personnel who do this type of work on a regular basis.

Safety Systems for Petroleum Processing

- Monitoring of the vapor space is continuous during entry – normally for oxygen (which you want near zero in this case), H_2S, LEL, CO, and temperature. Entry is immediately terminated if any of the parameters goes out of the allowable range.
- For equipment that is supposed to be full of air (not inerted), fresh air ventilation is normally maintained before and during any entry or other work.
- Adequate ventilation is required anywhere inert gas may accumulate. For instance, if a plant uses vaporized nitrogen to back up the instrument air system, the nitrogen should not be allowed in any operator shelter or control room without ventilation. It is actually probably better to avoid this situation by addressing instrument air reliability rather than use nitrogen.
- Ensure you have a rescue system and plan for a "man down" in an inert environment. Unfortunately, a large number of the fatalities resulting from oxygen-deficient environments are would-be rescuers that enter without the proper air supply.
- If fresh breathing air is being used:
  - Test the breathing air (even if it is "certified" breathing air) to ensure it has sufficient oxygen. Some breathing air is made by blending vaporized liquid nitrogen and oxygen. Sometimes, this ends up being 100 % nitrogen instead due to a control problem. Often every air bottle is tested for oxygen.
  - Alternately, some companies insist on the breathing air in the bottles being compressed atmospheric air – no blending. In this case, the air intake location is carefully reviewed to ensure the incoming air is not contaminated.
  - Air connections should not be capable of being connected to nitrogen or CO_2 supplies. The standard connectors for the gases are different. Sometimes, there is "creativity" in making connections, however.
  - Ensure breathing air supply is uninterruptable with backup supply and egress bottles.
- Everyone who may be potentially exposed to an oxygen-deficient atmosphere, both employee and contractors, should be trained in the hazards, procedures, and equipment used at the specific facility.

People work safely around, and sometimes in, inert and oxygen-deficient environments every day. Understanding the hazards and using good management practices can prevent harm to people who must deal with this type of IDLH environment.

Confined Space Entry

The following discussion is presented for information only. Governmental regulations, company/contractor procedures, and legal considerations control all aspects of confined space entry at a given facility.

Hazards of Confined Space Entry

We have talked a little about confined space entry in discussing oxygen-deficient environments above. Here we will review some of the other hazards when someone enters an enclosed or confined space to work.

From the US Occupational Safety and Health Administration (OSHA) definition of a confined space [1910.146(b)]:

"Confined space" means a space that:

1. Is large enough and so configured that an employee can bodily enter and perform assigned work
2. Has limited or restricted means for entry or exit (e.g., tanks, vessels, silos, storage bins, hoppers, vaults, and pits are spaces that may have limited means of entry)
3. Is not designed for continuous employee occupancy

Examples of confined spaces in a petroleum processing facility would normally include:

- Reactors and absorbers/adsorbers
- Vessels
- Tanks
- Sewer boxes
- Heat exchanger shells
- Fireboxes
- Convection section ducting
- Very large piping

Some facilities may include equipment such as cooling tower cells and other relatively open structures in their definition of confined space because of specific hazards.

Confined spaces are of concern to us because they may present hazards from:

- A hazardous atmosphere:
  - Flammable (e.g., hydrocarbon vapors evolved from sludge)
  - Dusty (e.g., catalyst handling)
  - Oxygen deficient (e.g., inert reactor entry)
  - Toxic or hazardous gas (e.g., H_2S or NH_3)
  - Any other IDLH condition
- Possible engulfment risk from materials (e.g., catalyst falling in on a worker)
- Possible risk of being trapped or asphyxiated by configuration (e.g., inward sloping walls like an FCC cyclone)
- Other recognized, serious hazards

Safety Systems for Petroleum Processing

For confined spaces which do, or could, present one or more of these hazards, a permit is usually required to enter the space. In the OSHA definitions, these would be "permit-required confined spaces (permit spaces)."

There are other types of confined spaces for which some facilities may not require a permit. Many facilities just default to defining essentially all confined spaces as "permit spaces," with only some permit conditions differing.

Managing Confined Spaces

Facilities generally have a specific safety program to deal with confined spaces. There are legal/regulatory requirements for these programs. In general, programs governing confined spaces will have provisions that include:

- Identification of any opening into a confined space (e.g., signs and/or tape over all possible entry routes into the confined space).
- If the confined space is inert or may present some other atmospheric hazard, warning signs indicate the hazards. There may also be barriers or exclusion areas around any opening.
- A permit is required for anyone entering the confined space to work. The permit conditions will usually address things such as:
  - Allowable activities within the space for the specific permit.
  - Verification of complete isolation (blinds and blanks) of the confined space from other systems or potential energy sources.
  - Testing of the atmosphere inside the space to verify it meets the required conditions –usually flammability (LEL), oxygen (inert or air), CO, H_2S, and temperature.
  - Requirements for continuous monitoring of the space during entry.
  - Requirements for appropriate PPE.
  - Requirements for a "hole-watch" and/or standby personnel, including what equipment they must have available.
  - Verification of training of the entrant, hole-watch, and other personnel involved.
  - Conditions which would terminate or cancel the permit.
  - A time limit on the permit validity. Usually permits have to be renewed periodically, like each shift or each day. And entry must start within a specific time frame or the permit is canceled.
  - Requirement for continuous presence of someone at the entry point until entry is complete. This ensures the space is never left unattended, even over lunches or breaks.
  - Requirement to sign in and sign out of the space, with entry and exit through the same "hole." This is to ensure everyone is accounted for.
  - A rescue plan if needed. The hole-watch may be required to have a radio or another emergency signal device available.

- Additional permits and plans may also be required, for example:
  - Hot work permit for welding work inside equipment
  - Inert entry permit for catalyst or other work requiring exclusion of air during entry

Entry into confined spaces and work within the spaces are safely performed on a routine basis. The risks, hazards, requirements, and contingencies have to be thoroughly considered to protect the entrants from harm.

Facility Siting Hazards and Considerations

On 23 March 2005, an explosion occurred at BP's refinery in Texas City, Texas, in the USA that resulted in 15 fatalities and over 170 injuries. One of the refinery units was starting up after turnaround when a problem caused the release of hydrocarbon to an unfired vent stack. The hydrocarbon vapors found an ignition source and exploded. There is a great deal more detail in the BP and U.S. Chemical Safety Board (CSB) reports on the incident, but of interest to us here is a statement from the BP Executive Summary report:

> "The severity of the incident was increased by the presence of many people congregated in and around temporary trailers which were inappropriately sited too close to the source of relief."

Consideration of facility siting has always been an element of process safety management, but this incident elevated the importance of considering where facilities, both permanent and temporary, are located relative to process equipment. It prompted reevaluation and actions on the part of many facilities which could have similar risks.

Hazards relevant to facility siting can include injuries to workers in occupied structures from blast, fire, and release of toxic materials. In some cases, there may be off-site consequences.

Operator shelters are generally located close to the units they operate. These structures are often occupied. Consideration of their locations is appropriate relative to the different hazards presented by the units. The response time required to reach the unit when necessary is normally balanced against the hazards and risks. Some facilities "harden" the operator shelters to resist blasts and provide the capability for positive pressurization and sheltering in place to protect workers in the event of a toxic release. Alarm systems are provided in most plants to notify workers of releases or other problems and initiate egress in emergencies. Self-contained breathing apparatuses (SCBAs) are usually located throughout a unit and at operator shelters. The shelters usually have minimal windows, to prevent flying glass in the event of an incident.

Control rooms, increasingly, are centralized and located away from the process units. These "central control rooms" are often blast resistant, pressurized, and capable of isolating themselves for sheltering in place.

Administration, maintenance, firehouses, and other buildings are usually located far enough away from the process units and hazards that they are not a major risk. However, the locations of these buildings relative to the possible pressure waves or releases of toxics have been considered by some refiners. In some cases, some of these building are not in good locations. Some have been replaced with buildings further from the potential hazards.

The considerations for siting of structures have grown, with the BP incident, to also include temporary facilities. It used to be common to locate project trailers, turnaround trailers, lunch tents, and other temporary facilities close to where the work was to be done. This can expose workers to risks similar to the operator shelters, but these facilities are not designed for blast or atmospheric hazards.

Today, the temporary facility locations are considered much more carefully. Some rental companies have developed hardened, blast-resistant temporary "buildings" (really converted shipping containers), for use in process facilities when the office has to be near the work. When the office does not have to be in the plant, temporary trailers, lunch tents, and other facilities are being located away from the units in a safe distance.

As an additional protective measure, many facilities have a policy of evacuating all unnecessary personnel from any unit that is in start-up or shutdown. While this slows work, it minimizes the potential exposure of workers to a problem.

Practices and options for both permanent and temporary facility siting continue to develop and improve.

Material Failures and Prevention Programs

Process safety management regulations require a facility to maintain mechanical integrity (MI) of the equipment. Facilities should have a well-documented inspection and repair program to ensure the equipment and piping are well maintained to prevent losses of containment and process safety incidents. A primary MI concern is materials damage and prevention. The MI program is usually based on the risks and corrosion rates of the various process services and materials used (e.g., risk-based inspection).

API 510 (Pressure Vessel Inspection Code: In-Service Inspection, Rating, Repair, and Alteration) defines the US refining industry requirements for mechanical integrity. API 579 (Fitness for Service) provides methods and requirements for evaluation of equipment to continue operating in a facility. Refer to these publications in creating or evaluating your own mechanical integrity programs.

API Recommended Practice 571 (RP-571) provides a detailed discussion of the materials damage mechanisms and methods of control for each major type of process unit in a refinery. Many of these can be extrapolated to other types of petrochemical facilities. We will not propose to repeat the information in RP-571. Table 4 provides a summary of the materials damage mechanisms discussed in RP-571 for reference. Please refer to that publication for details.

Table 4 Refinery materials damage mechanisms (Reference: API Recommended Practice RP-571)

Damage Mechanisms	Crude/Vacuum	Delayed Coker	Fluid Catalytic Cracking	FCC Light Ends Recovery	Continuous Catalytic Reforming (CCR)	Catalytic Reforming (Cyclic, Semi-regen)	Hydroprocessing (Treating, Cracking)	Alkylation, Sulfuric Acid	Alkylation, HF	Amine Treating	Sulfur Recovery	Sour Water Stripper	Isomerization	Hydrogen Plant
Mechanical & Metallurgical Failure														
Graphitization			X											
Softening (Spheroidization)			X											
Temper Embrittlement			X			X	X							
Strain Aging														
885°F Embrittlement	X													
Sigma Phase / Chi Embrittlement			X											
Brittle Fracture														
Creep / Stress Rupture	X	X	X		X	X	X				X			X
Thermal Fatigue		X	X											
Short-term Overheating - Stress Rupture	X	X	X											
Steam Blanketing											X			
Dissimilar Metal Weld (DMW) Cracking	X	X					X							
Thermal Shock														
Erosion / Erosion-Corrosion	X	X	X											
Cavitation														
Mechanical Fatigue														
Vibration-Induced Fatigue														
Refractory Degradation		X	X											
Reheat Cracking														
Uniform or Localized Loss of Thickness														
Galvanic Corrosion										X				
Atmospheric Corrosion														
Corrosion Under Insulation (CUI)		X								X				
Cooling Water Corrosion					X									
Boiler Water / Condensate Corrosion			X									X		
CO2 Corrosion	X													
Flue Gas Dew Point Corrosion											X			
Microbiologically Induced Corrosion (MIC)		X												
Soil Corrosion														
Caustic Corrosion	X						X							
Dealloying														
Graphitic Corrosion														
Amine Corrosion										X				
Ammonium Bisulfide Corrosion (Alkaline Sour Water)		X	X	X			X			X		X		
Ammonium Chloride Corrosion	X	X	X	X		X	X							
Hydrochloric Acid (HCl) Corrosion	X	X		X		X	X						X	
High Temperature H2/H2S Corrosion							X							
Hydrofluoric Acid (HF) Corrosion									X					
Naphthenic Acid Corrosion (NAC)	X	X	X											
Phenol (Carbolic Acid) Corrosion														
Phosphoric Acid Corrosion														
Sour Water Corrosion (Acidic)														
Sulfuric Acid Corrosion								X						
High Temperature Corrosion [400°F, 204°C]														
Oxidation	X	X	X		X	X	X				X			X
Sulfidation	X	X	X				X				X			
Carburization						X								
Metal Dusting						X								X
Fuel Ash Corrosion	X													
Nitriding														
Decarburization		X												
Environment-Assisted Cracking														
Chloride Stress Corrosion Cracking (Cl⁻ SCC)	X						X						X	
Corrosion Fatigue														
Caustic Stress Corrosion Cracking (Caustic Embrittlement)	X						X			X				
Ammonia Stress Corrosion Cracking	X			X										
Liquid Metal Embrittlement (LME)	X													
Hydrogen Embrittlement (HE)							X							
Polythionic Acid Stress Corrosion Cracking (PASCC)		X				X	X							
Amine Stress Corrosion Cracking										X				
Wet H2S Damage (Blistering, HIC, SOHIC, SCC)	X	X	X	X		X	X			X		X		
Hydrogen Stress Cracking - HF									X					
Carbonate Stress Corrosion Cracking			X	X										
Other Mechanisms														
High Temperature Hydrogen Attack (HTHA)						X	X						X	X
Titanium Hydriding														

Appendix: Example Calculation for Sizing a Relief Valve

Problem

A vessel containing naphtha C_5–C_8 range is uninsulated and is not fireproofed. The vessel is vertical and has a skirt 15' in length. Dimensions of the vessel are I/D 6'0", T-T 20'0", and liquid height to HLL 16'0". Calculate the valve size for fire condition relief. Set pressure is 120 psig.

Solution

Latent heat of naphtha at 200 °F is 136 Btu/lb = H_L
$Q = 21,000\ FA^{0.82}$
A = Wetted area and is calculated as follows:

$$\text{Liquid height above grade} = 15 + 16\ \text{ft} = 31\ \text{ft}$$

Therefore wetted surface of vessel need only be taken to 25 f. above grade which is 25–15 = 10 f. of vessel height:

$$\text{Wetted surface} = \pi D \times 10\ \text{ft for walls}$$
$$= 188.5\ \text{sq ft}$$
$$\text{plus } 28.3\ \text{sq ft for bottom}$$
$$= 216.8\ \text{sq ft}$$
$$Q = 21,000 \times 1.0 \times (216.8)^{0.82}$$
$$1.729 \times 10^6\ \text{Btu/h}$$
$$Q/H_L = \frac{1.729 \times 10^6}{136} = 12,713\ \text{lbs/h} = W$$
$$A = \frac{W\sqrt{T}\sqrt{Z}}{CKP_1 K_b \sqrt{M}}$$

where:

A = effective discharge area in sq in.
W = flow through valve in lbs/h, 12,713.
T = absolute temp of inlet vapor, 460 + 200 = 660 R (Bubble point of C_5-C_8 at, say, 10 psig).
Z = 0.98 (nC_5).
C = 356.06 (based on $C_A/C_V = 1.4$).
K = 0.65 (typical coefficient of discharge).

$K_b = 0.9$
$M = 100$ (use C_7).
$P_1 =$ set pressure of valve, 134.7 psia.

$$A = \frac{12,713 \times 25.7 \times 0.99}{356 \times 0.65 \times 134.7 \times 0.9 \times 1} = 1.153 \text{ in.}^2$$
$$\text{nearest orifice size} = \text{'}J\text{' at } 1.287 \text{ in.}^2$$

References

PSM and Related Programs

American Petroleum Institute, Management of Process Hazards, *API Recommended Practice 750* (1990, Since Withdrawn)

J. Atherton, F. Gil, (AIChE CCPS, 2008), *Incidents that Define Process Safety* (American Institute of Chemical Engineers, Center for Chemical Process Safety/Wiley, New York/Hoboken,, 2008)

Center for Chemical Process Safety (CCPS), American Institute of Chemical Engineers, New York, several references

J. B. Christian, (Osram Sylvania Inc., 1997), Combine fault and event trees for safety analysis, Chem. Eng. Progress. 72–75 (1997)

M.M.R. Eastman, J.R. Sawers (Knowledge Technologies, 1998), Learning to document management of change. Chem. Process. 15–22 (1998)

K.A. Ford, A.E. Summers, (Triconex, 1998), Are your instrumented safety systems up to standard? Chem. Eng. Progress 55–58 (1998)

R.W. Garland (Eastman Chem. Co., 2012), An engineer's guide to management of change. Chem. Eng. Progress 49–53 (2012)

L. Goodman (Fluor Daniel, 1996), Speed your hazard analysis with the focused what if? Chem. Eng. Progress 75–79 (1996)

W.K. Goddard (Solutia Inc., 2007), Article: Use LOPA to determine protective system requirements, Chem. Eng. Progress, Feb, pp. 47–51 (2007)

R.K. Goyal, (BAPCO, 1993), FMEA, the alternative process hazard method. Hydrocarb. Process. 95–99 (1993)

R.K. Goyal (BAPCO, 1994 and 1995), Understand quantitative risk assessment. Hydrocarb. Process. 105–108 (Dec 1994), 95–99 (Jan 1995)

ISA, *Standard: Application of Safety Instrumented Systems for the Process Industries*, ANSI/ISA-SP 84.01-1996 (ISA, Research Triangle Park, 1996)

R.E. Knowlton, (Chemetrics, 1987), *An Introduction to Hazard and Operability Studies: The Guide Word Approach*. (Chemetrics Int'l, Vancouver, 1987)

H. Ozog, H.R. Forgione, (ioMosaic Corp., 2012), OSHA's new chemical NEP: how to make the grade and earn extra credit, Chem. Eng. Progress 27–34 (2012)

K.E. Smith, D.K. Whittle (EQE Int'l, Inc., 2001), Six steps to effectively update and revalidate PHAs. Chem. Eng. Progress 70–77 (2001)

R.P. Stickles, G.A. Melhem, (Arthur D. Little, Inc., 1998), How much safety is enough?, Hydrocarb.. Process. 50–54 (1998)

Wikipedia, several articles on www.wikipedia.org. Accessed Feb 2014

Oxygen Deficient Environments

J. Cable, NIOSH report details dangers of carbon dioxide in confined spaces. Occup. Hazards. (30 Dec 2004)

National Institute for Occupational Safety and Health (NIOSH), *Guidance Document: Criteria for a Recommended standard, Occupational Exposure to Carbon Dioxide* (National Institute for Occupational Safety and Health, Menlo Park, 1976)

U.S. Bureau of Land Management, Health risk evaluation for carbon dioxide (CO_2). (2005)

U.S. Chemical Safety and Hazard Investigation Board, Bulletin: Hazards of nitrogen asphyxiation, Safety Bulletin No. 2003-10-B, June 2003 (2003)

Confined Space Entry

U.S. Occupational Safety and Health Administration, *Standard: Permit-Required Confined Spaces*, Standard No. 1910.146. www.osha.gov

Facility Siting Hazards and Consideration

BP USA, *Public report: Fatal Accident Investigation Report, Isomerization Unit Explosion*, Final Report, Texas City. (BP, Texas, 2005)

S.R. Brahmbhatt, (MG Ind., 1984), Are liquid thermal relief valves needed? Chem. Eng. 70 (1984)

Hazardous Materials in Petroleum Processing

David S. J. Jones and Steven A. Treese

Contents

Introduction to Handling Hazardous Materials	1353
Amines for Gas Treating	1354
Discussion and Hazards of Amines	1354
Safe Handling of Amines	1356
Equipment and Piping for Amines	1356
Ammonia (NH_3)	1357
Discussion of Ammonia	1357
Properties of Ammonia	1357
Ammonia Hazards	1357
Clothing and Personal Protective Equipment for Ammonia	1358
Release Measures	1359
Storage and Handling	1359
Benzene	1359
General Discussion of Benzene	1359
Key Properties of Benzene	1360
Hazards of Benzene	1360
Personnel Protection from Benzene	1362
Storage and Handling of Benzene	1362
Carbon Monoxide (CO)	1363
Discussion of CO	1363
Chemical and Physical Properties of CO	1363
Hazards of CO	1364
Protective Equipment for CO	1365
Handling of CO Streams	1365

David S. J. Jones: deceased.
Steven A. Treese has retired from Phillips 66.

D.S.J. Jones
Calgary, AB, Canada

S.A. Treese (✉)
Puget Sound Investments LLC, Katy, TX, USA
e-mail: streese256@aol.com

© Springer International Publishing Switzerland 2015
S.A. Treese et al. (eds.), *Handbook of Petroleum Processing*,
DOI 10.1007/978-3-319-14529-7_21

Catalysts and Sorbents 1366
    General Discussion of Catalysts and Sorbents 1366
    Hazardous Catalyst and Sorbent Properties 1366
    Catalyst and Sorbent Handling 1366
Caustic Soda (Sodium Hydroxide, NaOH) 1370
    Discussion and Hazards of Caustic Soda 1370
    Handling Hazards for Caustic Soda 1370
    First Aid and Personal Safety for Caustic Soda 1372
    Protective Clothing for Caustic Soda 1373
    Materials of Construction for Caustic Soda 1373
Furfural 1374
    Discussion and Hazards of Furfural 1374
    Hazards Associated with Handling Furfural 1374
    First Aid and Personal Safety with Furfural 1375
    Protective Clothing and Equipment for Furfural Handling 1375
    Fire Prevention and Firefighting for Furfural 1376
    Materials of Construction and Storage for Furfural 1377
Hydrofluoric Acid (HF, AHF) 1377
    Discussion and Hazards of HF 1377
    Safe Handling Practices for HF 1378
    Personal Protective Equipment Requirements for HF 1379
    Unloading and Transfer of AHF 1379
    Equipment for HF Service 1380
    Personal Safety When Handling HF 1382
Hydrogen 1383
    General Discussion of Hydrogen 1383
    Physical and Chemical Properties of Hydrogen 1383
    Hazards of Hydrogen 1384
    Protective Equipment for Hydrogen Handling 1384
    Storage and Handling of Bulk Hydrogen 1384
Hydrogen Sulfide (H_2S) 1385
    Discussion and Hazards of H_2S 1385
    Hazards and Toxicity of Hydrogen Sulfide 1385
    Protective Clothing and Personal Safety for H_2S 1387
    Materials of Construction in H_2S Service 1387
Methyl Ethyl Ketone (MEK) 1388
    Discussion of MEK 1388
    Hazards Associated with MEK 1388
    First Aid and Personal Protection 1389
    Clothing and Protective Equipment 1390
    Fire Prevention and Fighting 1390
    Storage and Handling 1391
Nickel Carbonyl ($Ni(CO)_4$) 1391
    General Discussion of Nickel Carbonyl 1391
    Chemical and Physical Properties of Nickel Carbonyl 1392
    Hazards of Nickel Carbonyl 1393
    Protective Equipment for Nickel Carbonyl 1394
    Management of Nickel Carbonyl Hazards 1394
Nitrogen (N_2, LN_2) 1395
    General Discussion of Nitrogen 1395
    Chemical and Physical Properties of Nitrogen 1396
    Hazards of Nitrogen 1396
    Protective Equipment for Nitrogen Handling 1397

Hazardous Materials in Petroleum Processing 1353

```
Storage and Handling of Nitrogen ..................................................... 1397
Sulfiding Chemicals ................................................................. 1397
    General Discussion of Sulfiding Chemicals .......................................... 1397
    Chemical and Physical Properties of Sulfiding Agents ................................ 1398
    Hazards of Sulfiding Agents ...................................................... 1398
    Protective Equipment for Sulfiding Chemicals Handling .............................. 1399
    Storage and Handling of Sulfiding Chemicals ....................................... 1402
Sulfuric Acid (H₂SO₄) .............................................................. 1403
    General Discussion of Sulfuric Acid .............................................. 1403
    Properties of Concentrated Sulfuric Acid .......................................... 1403
    Hazards of Sulfuric Acid ........................................................ 1403
PPE for Handling Sulfuric Acid ..................................................... 1404
Storage and Handling of Sulfuric Acid ................................................ 1405
Hazardous Materials Handling Summary ............................................... 1405
```

Abstract

Although the fire hazard is always a primary concern in the refining of petroleum, there are other hazards present that need to be addressed. Among these are the handling of some of the chemicals that are used or generated in the refining processes. This chapter provides general information on the hazards and handling of several chemicals. The information is not intended to substitute for regulatory, local, or company standards or medical expertise. Included here, are discussions of: amines, ammonia, benzene, carbon monoxide, catalysts and sorbents, caustic soda, furfural, hydrofluoric acid, hydrogen, hydrogen sulfide, methyl ethyl ketone, nickel carbonyl, nitrogen, sulfiding chemicals, and sulfuric acid.

Keywords

Refinery · Hazardous materials · Materials handling · Safety · Toxicity · Exposure limits

Introduction to Handling Hazardous Materials

Disclaimer: Please note that the discussions presented in this chapter are for informational purposes only. They are not meant to be specifications and are not all-encompassing. We defer to industrial hygiene, medical, and other experts and to your company to evaluate and determine the appropriate, current methods for handling these hazardous materials and for emergency responses and procedures relevant to each chemical. Many resources (e.g., MSDSs, chemical suppliers, licensors) are available to assist in these efforts.

From the very nature of crude oil, its refining and processing can create potentially hazardous situations. The most commonly recognized hazard we expect is, of course, the danger of fire. Considerable effort is made to prevent fires and, if a fire does occur, to combat and restrain it in the most effective manner. Fire prevention and firefighting are discussed in detail in the chapter "▶ Fire Prevention and Firefighting in Petroleum Processing" of this handbook.

Although the fire hazard is always a primary concern in the refining of petroleum, there are other hazards that are present and always need to be addressed. Among these are the handling of toxic and dangerous chemicals that are used in the refining processes. For many of the chemicals used in refining, there is also a danger to life or health from exposure to toxic materials that are produced in some of the refining processes. Perhaps the most notable is hydrogen sulfide, which is common to all modern refineries.

This chapter deals with the nature and handling of some of the common hazardous materials used or produced in refining. The specific materials addressed are:

- Amines
- Ammonia
- Benzene
- Carbon monoxide
- Catalysts and sorbents
- Caustic soda (AKA sodium hydroxide, NaOH)
- Furfural
- Hydrofluoric acid (HF, AHF)
- Hydrogen
- Hydrogen sulfide
- Methyl ethyl ketone
- Nickel carbonyl
- Nitrogen
- Sulfiding chemicals
- Sulfuric acid (H_2SO_4)

Some of these materials are discussed in other chapters, but we have tried to pull the relevant materials all together here.

In all cases, it is strongly recommended that you consult the material safety data sheet (MSDS) for each material for the most up-to-date information about its hazards and handling.

Amines for Gas Treating

Discussion and Hazards of Amines

Amine solvents are used in petroleum and natural gas refining to remove hydrogen sulfide and carbon dioxide from the various streams. In petroleum refining, the monoethanolamine (MEA) compound of the homologue used to be common in the treating processes. Diethanolamine (DEA) and methyl diethanolamine (MDEA) have become much more common. Proprietary compounds of amine such as Sulfinol and ADIP follow in usage. All of these amines, however, are similar with respect to their hazards, handling, and health effects.

Table 1 Properties of amines used in petroleum refining

Type	MEA	DEA	MDEA	DGA	ADIP	Sulfinol
Mole weight	61.1	105.1	119.16	105.14	133.19	120.17
Boiling Pt °F (°C)	338.5 (170)	515.1 (268)	477 (247)	405.5 (208)	479.7 (249)	545 (285)
Boiling range 5–95 % °F (°C)	336–341 (169–172)	232–237 (111–114)	477 (247)	205–230 (96–110)	–	–
Freezing Pt °F (°C)	50.5 (10.2)	77.2 (25.1)	−9.3 (−23)	9.5 (−12.5)	107.6 (42)	81.7 (27.6)
S.G. @ 77 °F (25°C)	1.0113	1.0881	1.0418 (20 °C)	1.0572	0.99	1.256
Viscosity @ 77 °F (25 °C)	18.95 cP	351.9 cP	36.8 cSt (100 °F [38 °C])	40 cP	870 cP	12.1 cP
Viscosity @ 140 °F (60 °C)	5.03 cP	–		6.8 cP	86 cP	4.9 cP
Flash Pt °F (°C)	200 (93.3)	295 (146)	265 (129)	260 (127)	255 (124)	350 (177)
Fire Pt °F (°C)	205 (96)	330 (166)	–	285 (141)	275 (135)	380 (193)
Typical solvent concentration, wt%	15–25	25–35	50	50–70	NA	Varies

For the purpose of this work, only MEA will be considered. However, as a point of reference, Table 1 provides the physical properties of those amines used in petroleum refining (see also the chapter entitled "▶ Refinery Gas Treating Processes").

MEA is corrosive and a combustible liquid and requires special handling and personnel protection considerations.

All amines are injurious to personnel. As a guide, the effects of exposure to MEA are as follows:

- *Target organs:* Kidneys, central nervous system, and liver.
- *Potential health effects*
  - *The eyes:* MEA causes severe eye irritation and burning.
  - *The skin:* May be absorbed through the skin in harmful amounts. Causes moderate skin irritation.
  - *Ingestion:* Causes gastrointestinal tract burns.
  - *Inhalation:* Inhalation of high concentrations may cause central nervous system effects. This is characterized by headaches, dizziness, unconsciousness, and coma. Also causes respiratory tract irritation.
  - *Chronic:* May cause liver and kidney damage.

Safe Handling of Amines

MEA is transported by road or rail tanker in its concentrated form. It is transferred in the normal way to an onsite storage bullet or tank, which is blanketed by an inert gas. MEA is degraded on exposure to air. The use of this, and other ethanolamines, in the refinery processes is in a dilute form. This dilution and its onsite storage are very often in a suitably constructed pit, usually in the proximity of the user plant. In some cases, a cone roof tank may be used for onsite storage. In all cases, though, the product must be kept free from exposure to air by inert gas blanketing. The dilution of MEA for use in the refinery process is between 15 and 25 wt%. The water for this dilution is usually treated boiler feed water or condensate, which are essentially free of impurities.

Vessels and piping in the process in which the amine is used should be of a suitable grade of carbon steel. The amine process temperatures should not exceed 300 °F (149 °C) anywhere in the process.

Personnel likely to be exposed to the amine should wear protective clothing, including eye protection (goggles and/or face shield in addition to safety glasses). As in the case of handling AHF, the minimum protection for operating and maintenance personnel should be:

- Coveralls, with sleeves to the wrist
- Chemical safety goggles
- Hard hat
- Gauntlets (polyvinyl chloride)
- Standard safety footwear

In addition to the standard, protective clothing listed, certain operating and maintenance work requires the use of a respirator. Such an instance would be in the changing of the amine filter (see the chapter entitled "▶ Refinery Gas Treating Processes") cartridge. Although the filter will have been steam cleaned prior to opening the filter vessel, respiratory protection is essential until no presence of amine or H_2S is verified. The verification is established by gas testing and the special processes to determine the absence of the sulfide (lead acetate test).

In some services, such as hydrogen plants, where MEA is used, arsenic is sometimes used as a corrosion inhibitor. This can increase the hazards of the amine and the equipment during servicing. Procedures to operate and decontaminate equipment for maintenance must ensure the operators and maintenance hands do not come in contact with the arsenic compounds.

Equipment and Piping for Amines

As is the case with all alkaline substances, amines cause stress corrosion. Consequently, all vessels and piping (welds) are stress relieved. Valves and piping are normally carbon steel, as are pumps and heat exchanger tubes.

Ammonia (NH$_3$)

Discussion of Ammonia

In petroleum processing facilities, ammonia may be used for NOx control in combustion flue gases and for refrigeration. In some facilities it is used to control corrosion in tower overheads and to suppress catalyst activity during hydrocracker startup.

Ammonia is produced in a refinery by hydroprocessing, when nitrogen compounds are removed from petroleum products. It is a major constituent of sour water stripper gases. When economic, some facilities recover ammonia as a by-product.

Ammonia itself may be present as a colorless, pressurized liquid or a colorless gas with a characteristic, pungent smell. Under typical ambient conditions, it is a gas. It is common to handle ammonia in an anhydrous liquid form under pressure or as an aqueous solution (5–30 % ammonia, AKA aqua ammonia) at ambient conditions. A typical aqua ammonia concentration is 19 wt% (for SCR-grade ammonia). See the separate topic "▶ Utilities in Petroleum Processing" for a discussion of ammonia systems and uses in the refinery.

Properties of Ammonia

Table 2 summarizes some of the key properties of anhydrous and aqua ammonia grades.

Ammonia Hazards

Ammonia, in any form, is an irritant and corrosive to the skin, eyes, respiratory tract, and mucous membranes. It can cause severe chemical burns to eyes, lungs, and skin. Any existing skin or respiratory illnesses are aggravated by exposure to ammonia. Ammonia is not recognized as a carcinogen. Evaporating anhydrous ammonia can produce frostbite if skin contacts equipment where it is being vaporized.

When mixed with air, ammonia vapor is flammable between 15 and 28 vol%. Because the autoignition temperature is high, a fairly hot ignition source is needed to start combustion. Welding and similar activities are hot enough, so care should be taken to ensure equipment that has contained ammonia is completely clear before welding. Ensure there are no vapors in the area if welding near an ammonia storage tank or a line.

Ammonia does not normally corrode steel or iron but can react rapidly with copper, brass, zinc, and many alloys, especially those containing copper. Experience has shown that aqua ammonia will corrode carbon steel slowly, however. Only steel and ductile iron are normally used in ammonia service. For aqua ammonia, stainless steel should be used ahead of any vaporizers.

The exposure limits for ammonia (as vapor) are summarized in Table 3.

Table 2 Some key ammonia properties

Concentration	Anhydrous	29.4 wt% (26° Baumé)	19 wt% (~20.5° Baumé) SCR grade
Specific gravity	Vapor: 0.59 (Air = 1)	0.8974 (60/60 °F, 15/15 °C)	0.9294 (60/60 °F, 15/15 °C)
	Liq: 681.9 kg/m^3 (−33.3 °C), 42.6 lb/ft^3 (−28 °F)		
Boiling point	−28.01 °F (−33.34 °C)	85 °F (29 °C)	123 °F (~51 °C)
Freezing point	−107.9 °F (−77.7 °C)	−110 °F (−79 °C)	−30 °F (−34 °C)
Solubility in water	47 wt% @ 32 °F (0 °C)		
	31 wt% @ 77 °F (25 °C)		
	28 wt% @ 122 °F (50 °C)		
Solution pH		13.5	~12.9
Viscosity	0.276 cP (−40 °F,−40 °C)		
Autoignition temperature	1204 °F (651 °C)		
Explosive limits	15–28 vol%		

Table 3 Ammonia exposure limits and toxicity

Agency	Limit type	ppm	mg/m^3	Basis
US OSHA	PEL	50	35	8 h TWA
US NIOSH	STEL	35	27	15 min
	REL	25	18	10 h TWA
	IDLH	300		
ACGIH	TLV	25	18	8 h TWA
	STEL	35	27	15 min

Toxicity: LD50 (Oral, Rat) =350 mg/kg

Clothing and Personal Protective Equipment for Ammonia

Proper ammonia handling PPE includes safety goggles and/or face shield, rubber gloves, in addition to the normal flame-retardant clothing, safety glasses, safety-toe shoes, and hardhat. Do not wear contact lenses when handling ammonia. It is also a good idea to wear ammonium hydroxide impervious clothing and rubber boots when loading or unloading.

Adequate ventilation should be provided in any ammonia handling area. There should be a safety shower and eyewash near any ammonia handling area.

If there is a possibility that the inhalation exposure limits may be approached, then respiratory protection approved by NIOSH/MSHA (or equivalent agencies) should be used. For escape in emergencies, approved respiratory protection that consists of a full-face mask, and canisters or SCBA are needed. Regulations 29 1910.134 and ANSI: Z88.2 apply. Only a positive-pressure SCBA should be used for entry into an ammonia-contaminated area if the atmosphere could contain over 300 vppm ammonia (IDLH).

For hazardous materials response to an ammonia spill, Level A and/or Level B gear should be used, including a positive-pressure SCBA.

Release Measures

In the event of a significant ammonia release, the following steps are recommended:

- Initiate the facility emergency response system.
- Evacuate the immediate area. This evacuation area may need to be very large if the spill is large.
- Eliminate all possible ignition sources in the area.
- Shutoff or isolate the leak, if possible. This will either stop or minimize the amount.
- Control ammonia vapors with water spray around the spill. The runoff water will be contaminated and may require treatment.
- Response personnel should have SCBAs with escape bottles. They should be in chemical-resistant suits with resistant gloves for major spills.

Storage and Handling

Anhydrous ammonia is shipped in pressurized tank trailers and tank wagons. Aqua ammonia is normally shipped in 6,700 gal ($\sim$26,000 l or $\sim$25 MT) tank trucks (at low or ambient pressure) or in drums.

Refer to the topics entitled "▶ Utilities in Petroleum Processing" and "▶ Environmental Control and Engineering in Petroleum Processing" for discussions of the types of bulk storage facilities used for ammonia. For drum storage of aqua ammonia, the drums should be kept cool and out of the weather. The ammonia in the drums will be near its boiling point at about 85 °F (29 °C).

Sometimes anhydrous ammonia is supplied in pressurized cylinders. These should also be stored in a covered location out of the sun.

Proper PPE as discussed in section "Clothing and Personal Protective Equipment for Ammonia" should be used when handling cylinders or transferring ammonia. Gloves are especially important in handling cylinders and drums. Cylinders and drums should never be dragged.

Proper safety showers and eye washes should be available near any handling area.

Benzene

General Discussion of Benzene

Benzene is found in many processes and petroleum products within a refinery and related facilities. In some cases, the product may be benzene that is deliberately recovered. In other streams, such as gasoline or reformate, the benzene is present as part of the hydrocarbon mixture.

Some benzene finds its way into waste water effluents, where it is governed by both water and air regulations (NESHAP).

We will briefly discuss benzene in a petroleum processing setting along with handling hazards. Many of the hazards and protective measures for benzene are also applicable to other aromatics (like toluene and xylene) which are a little less hazardous but still need some of the same precautions.

Key Properties of Benzene

Benzene is the simplest aromatic compound found in petroleum. It occurs naturally but is also made in refining. Some of the key chemical and physical properties of benzene are listed in Table 4.

Hazards of Benzene

There are two primary areas where benzene presents hazards in handling:

- Health hazards
- Fire hazard

The fire hazard can be perceived from the key properties table. Benzene can cause a flash fire.

From a health standpoint, the exposure limits and toxicity of benzene are summarized in Table 5.

Table 4 Some key properties of benzene

Molecular formula	C_6H_6
Molecular weight	78.11
Appearance	Clear, colorless, stable liquid
Odor	Sweet aromatic, gasoline-like
Odor threshold	61 ppm
Specific gravity, 20/20 °C (68/68 °F)	0.8765
Boiling point	176.18 °F (80.1 °C)
Vapor pressure	99 mmHg (77 °F, 25 °C)
Freeze/melting point	41.9 °F (5.5 °C)
Solubility in water	1.79 g/l (59 °F, 15 °C)
Flash point	11.9 °F (−11.15 °C) CC
Autoignition temperature	1097 °F (591.65 °C)
Vapor density	2.7 (Air =1)
Flammable limits in air, vol%	LEL: 1.3; UEL: 7.1
Extremely flammable, especially near open flames	

Table 5 Exposure limits and toxicity of benzene

Agency	Type	Limit	Notes
ACGIH (2012)	TLV	0.5 ppm	8 h TWA
	STEL	2.5 ppm	15 min
NIOSH (2008)	REL	0.1 ppm	10 h TWA
	STEL	1 ppm	15 min
OSHA (2006)	PEL	1 ppm	8 h TWA
	STEL	5 ppm	15 min
Toxicity data			
LC50 (rate)	Inhalation	10,000 ppm	7 h
LD50 (rat)	Ingestion	930–1,800 mg/kg	

Known carcinogen (leukemia). Suspected mutagen.

Looking at the health hazards and effects from an exposure route standpoint:

- Inhalation: Irritating to mucous membranes and upper respiratory tract. May be fatal in high concentrations. Can cause central nervous system depression, headaches, dizziness, fatigue, and excitation followed by depression.
- Eyes: Irritant.
- Skin: Irritating. Harmful if absorbed through the skin. Skin inflammation characterized by itching, scaling, reddening, or blistering. Prolonged or repeated skin contact can defat the skin and lead to irritation, cracking, and/or dermatitis.
- Ingestion: Aspiration hazard. Can enter lungs and cause damage. May be fatal if swallowed. Can cause dizziness, headaches, breathing difficulties, diarrhea, vomiting, and possible pneumonia.
- Chronic effects: Following low-level exposure, adverse effect on hematological (blood-forming) system and myelodysplastic syndrome (disease that affects bone marrow and blood).
- Benzene is a known carcinogen.

Summarizing, symptoms of over-exposure to benzene include:

- Dizziness
- Excitation
- Pallor followed by flushing
- Weakness
- Headache
- Breathlessness
- Chest constriction
- Irritation of the eyes, skin, nose, and respiratory system
- Nausea
- Staggered gait
- Fatigue

Benzene is extremely reactive or incompatible with oxidizers, nitric acid, and many fluorides and perchlorates.

The HIMA (USA) rating for benzene is 2-3-0 (health-fire-reactivity). The US NFPA rating is also 2-3-0.

Personnel Protection from Benzene

The primary PPE worn by refinery personnel in the plant will provide some level of protection from limited fire hazards. It does not protect from the chemical and other hazards, however.

Additional PPE for handling high benzene content streams or for spills include:

- Respiratory protection appropriate to the exposure risk
- Impervious gloves
- Chemical splash goggles (possibly face shield)
- Full impervious suit for large quantities or concentrations
- Rubber boots
- SCBA or other positive fresh air equipment

As for most materials discussed in this chapter, an eyewash and safety shower should be immediately available near locations where high concentrations of benzene are handled.

Storage and Handling of Benzene

Benzene is not corrosive, so handling is normally in carbon steel equipment – tanks, lines, vessels, etc. Metallurgy may be determined more by the other materials mixed with the benzene.

When transferring benzene, the safety precautions required are essentially the same as those for gasoline, including grounding or bonding all equipment involved in the transfer electrically.

For releases of benzene:

- Small spills:
  - Absorb in absorbent material.
  - Put absorbent in metal container.
  - Dispose as hazardous waste.
- Large spills:
  - Flammable liquid risk.
  - Don the appropriate PPE.
  - Contain spill if possible and stop flow.
  - Evacuate area.

- Eliminate possible ignition sources from area – remember the vapors can spread through low-lying areas.
- Ventilate area.
- Absorb benzene in dry earth, clay, sand, or other noncombustible material.
- Do not allow runoff to enter drains or sewers.
- Dike area for containment.
- There will be notifications necessary.

A fire or explosion involving benzene will evolve large amounts CO and CO_2 vapors as well as being very hot. Fresh air equipment must be worn when fighting these fires. Cooling water sprays may be used for equipment containing benzene. Minimize and contain runoff which may contain benzene for subsequent processing. Do not let it enter the wastewater treatment system, drains, or sewers.

Most facilities never handle benzene as a pure substance. Because of this, the normal concerns are with flammability of the product containing the benzene. Higher concentrations of benzene would trigger more diligence toward preventing personnel exposure.

Carbon Monoxide (CO)

Discussion of CO

Large quantities of carbon monoxide are present in all combustion gases (stacks, flares, incinerators, etc.), in partial-burn FCC regenerator gases, and as a by-product of manufacturing hydrogen. It may also be created by smoldering hydrocarbon solids, like coke or coke-coated catalysts. In a plant with cryogenic separation of refinery make gases, CO may be encountered as both gas and liquid. Care is normally taken to ensure the CO is exhausted at a safe location from elevated stacks or, in the case of hydrogen plants, destroyed by further reaction or completion of combustion. Solids are normally kept wet to avoid smoldering.

It is still possible for personnel to encounter CO in a facility. Atmospheric conditions may cause stack gases to drift or settle in locations where someone may be exposed – such as a nearby tower or structure. Sampling of flue gas or hydrogen plant gases can expose someone to CO. Burner problems in furnaces can result in CO outside the firebox and stack.

Precautions and response plans need to consider the possible presence of CO. Here we will discuss the hazards and basic protection methods available.

Chemical and Physical Properties of CO

Table 6 lists the relevant key physical and chemical properties of CO. Note that CO will normally be a colorless, odorless, tasteless gas at conditions encountered in a refinery. The exception would be in cryogenic units, where it may be a liquid, and in

Table 6 Some key properties of CO gas

Molecular weight	28.01
Specific gravity	Vapor: 0.97 (Air = 1)
Boiling point	−311.8 °F (−191 °C)
Freezing point	−337 °F (−205 °C)
Solubility in water	27.6 mg/l (77 °F/25 °C)
Autoignition temperature	1121 °F (605 °C)
Explosive limits	12.5–74.2 vol%

flue gases, where it is mixed with other gases that do have a warning odor (like SO_2). We will focus here on CO as a gas, however.

Things to note in the properties of CO are that it has essentially the same molecular weight as nitrogen, so is likely to be found along with nitrogen. It also has a wide explosive limit, meaning that it will support combustion fairly readily.

Hazards of CO

CO is a flammable gas and may cause a flash fire. It may be fatal if inhaled in sufficient concentrations. See Table 7 below for exposure limits.

Specific organs at risk from CO include the blood, lungs, nervous system, heart, cardiovascular system, and central nervous system. Frostbite can also occur as a result of rapidly expanding gas or vaporizing liquid, but this is unlikely in a refinery setting.

The primary route of exposure that is most frequently encountered is inhalation. Once in the lungs, CO bonds to the hemoglobin in the blood and reduces the blood's ability to carry oxygen. This is often hard to identify because blood saturated with CO is still red, while the victim is effectively suffocating. Studies indicate that 667 ppm of CO in the atmosphere can bind 50 % of the hemoglobin in the blood. This loss of oxygen capacity may result in seizure, coma, and fatality. Half-life of CO in the blood is about 5 h, so there is a cumulative effect of exposure.

Symptoms of CO poisoning include:

- Headache
- Nausea
- Vomiting
- Dizziness
- Fatigue
- Feeling of weakness
- Confusion
- Disorientation
- Visual disturbances
- Seizures
- Cherry-red blood and appearance (not necessarily reliable)

Most victims of CO poisoning believe they have food poisoning.

Table 7 Exposure limits for CO

Agency	Limit type	ppm	mg/m^3	Basis
US OSHA	PEL (2010)	50	55	8 h TWA
US NIOSH	STEL	200	229	
	REL	35	40	10 h TWA
	IDLH	1,200		
ACGIH (USA)	TLV (2010)	25	29	8 h TWA

LC50 (Rat) by inhalation: 6,600 ppm (30 min); 3,760 ppm (1 h); 1,807 ppm (4 h)

Protective Equipment for CO

The first level of protection against CO exposure is adequate ventilation in locations where CO may accumulate or be encountered. Stack elevations are normally specified to be a minimum height above and a minimum distance from any nearby structure to ensure dissipation of stack gases. These distances are specified by each company.

The normal PPE worn by a refinery worker will not protect against CO inhalation hazards. If exposure to CO is anticipated or possible the following may be used:

- Personal CO monitor with visual, vibration, and audible alarms. In the event of an alarm, the worker should immediately evacuate the area to a safe location.
- Air-purifying respirator up to the maximum CO concentration for the specific respirator. It is advisable to consider this only for escape. These are not for an oxygen-deficient atmosphere.
- Supplied-air respirator with an escape bottle (SCBA or airline respirator) for any significant exposure risk (spills, fires, etc.).

Handling of CO Streams

The PPE measures for handling streams which may contain CO are listed above.

Aside from small amounts of calibration gases, CO is normally a product of the refining processes. There is normally no reason to receive any bulk CO in a facility. The primary handling focus is on getting rid of the CO produced.

CO in stack gases is dispersed in the atmosphere. Operation of the fired heaters, CO boilers, and other services minimize the amount of CO emitted.

CO from processes, such as hydrogen production, is either burned as fuel gas in the SMR heater or reduced to methane (methanated) in the case of a conventional hydrogen plant.

When exposure to any hazardous level of CO is possible, fresh air respirators should be worn. This would include sampling of hydrogen plant intermediate streams, for instance.

Catalysts and Sorbents

General Discussion of Catalysts and Sorbents

Petroleum processing facilities employ a wide range of catalysts and sorbents in the various units. Many of these are designated as hazardous wastes when they are spent. Others may be classified as hazardous wastes based on their characteristics. In some cases, the fresh catalyst or sorbent may present a hazard that should be recognized. Here we will focus on some of the more common materials that present hazards and the methods employed to manage the hazards:

- Fresh and spent hydroprocessing catalysts
- Fresh and spent hydrogen plant nickel catalysts
- PSA adsorbents
- Mole sieve or zeolite adsorbents

Hazardous Catalyst and Sorbent Properties

Table 8 lists some of the physical properties and hazards of concern for the materials of interest.

Many of these materials are self-heating or pyrophoric. They may combust, generating hazardous gases. Others are hazardous because they heat up on or may release adsorbed hazardous vapors when exposed to moisture, such as in the air. Methods to manage these factors are well established.

The topic titled "▶ Environmental Control and Engineering in Petroleum Processing" of this handbook discusses the management of spent catalysts and characteristic hazardous wastes in detail.

Health effects of the above materials by various exposure routes are summarized in Table 9.

Catalyst and Sorbent Handling

In addition to the normal petroleum facility safety gear (hardhat, safety glasses, flame-retardant clothing, safety boots, gloves), the following PPE are also common when handling the catalysts and sorbents:

- Handling in air:
    - Dust respirator or HEPA half-face mask is adequate normally.
    - If hazardous gases may be released, an appropriate air-purifying or supplied-air respirator may be used.
    - In the case of nickel catalysts, refer to the discussion of nickel carbonyl in this chapter.

Hazardous Materials in Petroleum Processing 1367

Table 8 Some common catalyst and sorbent properties and hazards

Material	Hydroprocessing catalysts	Hydrogen plant catalysts	PSA adsorbents	Molecular sieves or zeolites
Service	Hydrotreating, hydrocracking	Reforming, methanation	Hydrogen recovery, other	Air drying. O_2 or N_2 generators
Composition				
Fresh	Oxides of Ni, Mo, and Co on alumina base: May contain silica. Sulfided forms, also	NiO and reduced nickel	Activated carbon. Mole sieves (silica)	Mole sieves, zeolite
Spent	Sulfides of the metals. Coke. FeS. Heavy metal or As accumulations. Adsorbed aromatics	NiO and reduced nickel. Adsorbed CO and other gases. Coke	Activated carbon. Mole sieves (silica). Residual adsorbed gases like CO and H_2S	Mole sieves, zeolite, contaminants picked up in service
Typical density range	25–75 lb/cft (400–1,200 kg/m^3)	~40–100 lb/cft (640–1,600 kg/m^3)	30–45 lb/cft (480–720 kg/m^3)	30–60 lb/cft (480–960 kg/m^3)
Hazardous properties				
Fresh	Metal dusts. Some may be self-heating	Ni dust	Heat of adsorption for water. Silica dust	Heat of adsorption for water. Silica dust
Spent	Self-heating. Potentially pyrophoric. Desorption of gases	Ni dust. CO. Ni(CO)$_4$. Other hazards presented by gases	Heat of adsorption for water. Release of adsorbed CO, H_2S, and other gases. Silica dust	Heat of adsorption for water. Silica and other dust (coke, metals)
Waste designation	Most are K172 (U.S. RCRA)			
Comments	Refer to the topic on "Environmental control and engineering" for discussion of hazardous waste characteristics for these materials			
	Materials may be tested and reclassified based on actual waste characteristics as defined by US EPA			

- Inert handling:
  - Positive-pressure, supplied-air respirator – the inert atmosphere overrides other risks generally.
  - Refer to the inert entry safety discussion in the separate topic "▶ Safety Systems for Petroleum Processing."
- Suspected hazardous contaminants (e.g., arsenic):
  - Wear protective clothing appropriate for the hazard.
  - Provide decontamination for persons leaving the handling area.
  - Use positive-pressure, supplied-air, full-face respirator.

Table 9 Health effects of typical catalysts and sorbents

Material	Hydroprocessing catalysts	Hydrogen plant catalysts	PSA adsorbents	Molecular sieves or zeolites
Fresh material				
Inhalation	Excessive dust. Respiratory tract irritation. Cough. Difficulty breathing	High levels of dust irritating to resp. tract. Slow to clear	Irritation of respiratory tract. Potential lung damage (e.g., silicosis)	Irritation of respiratory tract. Potential lung damage (e.g., silicosis)
Ingestion	Irritation of mouth and throat. Discomfort. Malaise, nausea, diarrhea	Low oral toxicity. Irritation of the GI tract	Burns to moist body tissue from heat of adsorption	Burns to moist body tissue from heat of adsorption
Eyes	Irritant	Dust causes irritation to eyes	Irritant. Dust heats up in contact with eyes	Irritant. Dust heats up in contact with eyes
Skin	Skin sensitizer. Irritation	Unlikely to be absorbed	Irritant. Dust heats up in contact with skin. Burns if prolonged, moist contact	Irritant. Dust heats up in contact with skin. Burns if prolonged, moist contact
Chronic	Carcinogenic (Ni, Co)	Carcinogenic (Ni). Liver damage	May cause damage to lungs. May contain silica, which is a carcinogen	May cause damage to lungs. May contain silica, which is a carcinogen
Symptoms	Coughing, difficulty breathing. Delayed onset of flu-like symptoms	Allergic skin reactions. Redness, inflammation, itching, burning. Asthma	Irritation, burns to wet surfaces, respiratory difficulties from heavy dust	Irritation, burns to wet surfaces, respiratory difficulties from heavy dust
Hazard ratings (H/F/R)	HMIS: 3/1/0 NFPA: 3/1/0	HMIS: 2/1/0	NFPA: 1/0/1	NFPA: 1/0/1
Spent material				
Inhalation	Respiratory tract burns. Allergic respiratory response. H_2S may be evolved	High levels of dust irritating to resp. tract. Slow to clear. Additional hazards may be present from adsorbed gas release	Irritation of respiratory tract. Potential lung damage (e.g., silicosis). Additional hazards may be present from adsorbed gas release	Irritation of respiratory tract. Potential lung damage (e.g., silicosis). Additional hazards may be present from adsorbed gas release

Ingestion	May be fatal. Burns to mucous membranes, throat, esophagus, and stomach	Low oral toxicity. Irritation of the GI tract	Burns to moist body tissue from heat of adsorption	Burns to moist body tissue from heat of adsorption
Eyes	Causes eye burns	Dust causes irritation to eyes	Irritant. Dust heats up in contact with eyes	Irritant. Dust heats up in contact with eyes
Skin	Skin burns. Allergic skin reaction. May be fatal if materials absorbed through skin	Unlikely to be absorbed	Irritant. Dust heats up in contact with skin. Burns if prolonged, moist contact	Irritant. Dust heats up in contact with skin. Burns if prolonged, moist contact
Chronic	Allergic respiratory and skin reactions. As and Ni may increase cancer risk. Co and Ni are carcinogens	Carcinogenic (Ni). Liver damage	May cause damage to lungs. May contain silica, which is a carcinogen	May cause damage to lungs. May contain silica, which is a carcinogen
Symptoms	Burns to skin, eyes, and mucous membranes. Aggravates preexisting conditions. See H$_2$S discussion	Allergic skin reactions. Redness, inflammation, itching, burning. Asthma	Irritation, burns to wet surfaces, respiratory difficulties from heavy dust	Irritation, burns to wet surfaces, respiratory difficulties from heavy dust
Hazard ratings (H/F/R)	HMIS: 3/1/0 NFPA: 3/1/0	Normally HMIS: 2/1/0	Normally NFPA: 1/0/1	Normally NFPA: 1/0/1
Environmental effects	Toxic to aquatic life. Long effect	Limited information. Product is insoluble, so limited aquatic exposure	Limited information. Products not particularly hazardous once wetted	Limited information. Products not particularly hazardous once wetted

To avoid release of hazardous gases, keep the solids dry or coated with an oil film. It is also possible to fully saturate the materials with enough water to manage the heat of adsorption, but this will generate gases that will require control. Dust management is normally the key control measure to prevent exposure or release of the materials.

If the materials are potentially self-heating, pyrophoric, or otherwise present a hazard if exposed to air, they should be handled in purged and/or sealed systems and stored in purged and sealed containers. Refer to the discussion of inert handling of solids.

Caustic Soda (Sodium Hydroxide, NaOH)

Discussion and Hazards of Caustic Soda

Caustic soda solution is used in oil refining mostly for the absorption of hydrogen sulfide or light mercaptans from light petroleum products from LPG through the kerosene cut. Very often the compound is delivered to the refinery in a strong aqueous solution to be further diluted on site to the strength required by a specific process. Sodium hydroxide in a solution is a white, odorless, nonvolatile liquid. It will not burn, but it is highly reactive. It can react violently with water and numerous commonly encountered materials, generating enough heat to ignite nearby combustible materials. Contact with many organic and inorganic chemicals may cause fire or explosion. Reaction with some metals (like aluminum and zinc) releases flammable hydrogen gas.

Sodium hydroxide is produced mainly in three forms:

- ~50 % aqueous solution (most commonly used)
- ~73 % aqueous solution
- Anhydrous sodium hydroxide in the form of solid cakes, flakes, or beads

The major impurities include sodium chloride, sodium carbonate, sodium sulfate, sodium chlorate, potassium, and heavy metals such as iron and nickel. The following discussion reviews the information relevant to solutions.

Chemical data are in Table 10.

Handling Hazards for Caustic Soda

Skin
Sodium hydroxide is extremely corrosive and is capable of causing severe burns with deep ulceration and permanent scarring. It can penetrate to deeper layers of the skin, and corrosion will continue until removed. The severity of injury depends on the concentration of the solution and the duration of exposure. Burns may not be immediately painful; onset of pain may be delayed minutes to hours. Several human

Table 10 Physical and chemical properties of caustic soda

Molecular weight	40.00
Melting point	12 °C (53.6 °F) (50 % soln.; freezing point)
	62 °C (143.6 °F) (70–73 % solution)
Boiling point	140 °C (284 °F) (50 % solution)
Specific gravity	1.53 (50 % solution)
	2.0 at 15.5 °C (70–73 % solution)
Solubility in water	Soluble in all proportions
Solubility in other liquids	Soluble in all proportions in ethanol, methanol, and glycerol
pH values	12 (0.05 % solution)
	13 (0.5 % solution)
	14 (5 % solution)

studies and case reports describe the corrosive effects of sodium hydroxide. A 4 % solution of sodium hydroxide, applied to a volunteer's arm for 15–180 min, caused damage which progressed from destruction of cells of the hard outer layer of the skin within 15 min to total destruction of all layers of the skin in 60 min. Solutions as weak as 0.12 % have damaged healthy skin within 1 h. Sodium hydroxide dissolved the hair and caused reversible baldness and scalp burns when a concentrated solution (pH 13.5) dripped onto a worker's head and treatment was delayed for several hours.

Owing to its corrosive nature, repeated or prolonged skin contact with dust and weak solutions would be expected to cause drying, cracking, and inflammation of the skin (dermatitis).

Eyes

Sodium hydroxide is extremely corrosive to the eye tissues. The severity of injury increases with the concentration of the solution, the duration of exposure, and the speed of penetration into the eye. Damage can range from severe irritation and mild scarring to blistering, disintegration, ulceration, severe scarring, and clouding. Conditions which affect vision such as glaucoma and cataracts are possible late developments. In severe cases, there is progressive ulceration and clouding of eye tissue which may lead to permanent blindness.

Inhalation

A worker, exposed for 2 h per day over 20 years to mists from boiling a solution of sodium hydroxide in two large containers in a small room with inadequate ventilation, developed severe obstructive airway disease. It was concluded that the massive and prolonged exposure induced irritation and burns to the respiratory system eventually leading to the disease. It was noted that chronic exposure had not previously been reported, probably since the strong and immediate irritation would normally deter workers from further exposure. Actual exposures to sodium hydroxide aerosols were not measured, and it could not definitely exclude late onset asthma as a cause of the man's condition.

A report of workers exposed to sodium hydroxide aerosol for at least 16 months was confounded by the presence of high concentrations of Stoddard solvent and other solvent vapors, as well as other chemicals.

There was no trend of increased mortality in relation to duration (up to 30 years) or intensity of exposure (0.5–1.5 mg/m^3) among 291 workers exposed to sodium hydroxide dust during the production of flakes or beads of concentrated sodium hydroxide from chlorine cell effluent. This study was limited by the small population size.

Ingestion
There are no reported cases of industrial workers ingesting sodium hydroxide solutions. Non-occupational ingestion has produced severe corrosive burns to the esophageal tissue, which has in some cases progressed to stricture formation. Should ingestion occur, severe pain; burning of the mouth, throat, and esophagus; vomiting; diarrhea; collapse; and possible death may result.

Long-Term Effects
Sodium hydroxide has been implicated as a cause of cancer of the esophagus in individuals who have ingested it. The cancer may develop 12–42 years after the ingestion incident. Similar cancers have been observed at the sites of severe thermal burns. These cancers may be due to tissue destruction and scar formation rather than the sodium hydroxide itself.

A case-control study reported an association between renal cancer and history of employment in the cell maintenance area of chlorine production. The major exposures in this work were presumed to be to asbestos and sodium hydroxide. An association was made between renal cancer and sodium hydroxide exposure. This study was limited by factors such as small numbers of exposed workers, multiple exposures, and reliance on work histories and is not considered sufficiently reliable.

First Aid and Personal Safety for Caustic Soda

Contact with the Skin
Avoid direct contact with this chemical. Wear chemical-resistant protective clothing, if necessary. As quickly as possible, remove contaminated clothing, shoes, and leather goods (e.g., watchbands, belts). Flush contaminated area with lukewarm, gently flowing water for at least 60 min, by the clock. DO NOT INTERRUPT FLUSHING. If necessary, keep emergency vehicle waiting. Transport victim to an emergency care facility immediately. Discard contaminated clothing, shoes, and leather goods.

Contact with the Eyes
Avoid direct contact. Wear chemical-resistant gloves, if necessary. Quickly and gently blot or brush away excess chemical. Immediately flush the contaminated eye (s) with lukewarm, gently flowing water for at least 60 min, by the clock, while

holding the eyelid(s) open. Neutral saline solution may be used as soon as it is available. DO NOT INTERRUPT FLUSHING. If necessary, keep emergency vehicle waiting. Take care not to rinse contaminated water into the unaffected eye or onto the face. Quickly transport victim to an emergency care facility.

Suffering from Inhalation

Remove to fresh air. If not breathing, give artificial respiration. If breathing is difficult, give oxygen. Get medical attention immediately.

Ingestion

DO NOT INDUCE VOMITING! Give large quantities of water or milk if available. Never give anything by mouth to an unconscious person. Get medical attention immediately.

Protective Clothing for Caustic Soda

Personal Respirators (NIOSH Approved)

If the exposure limit is exceeded and engineering controls are not feasible, a half-face piece particulate respirator (NIOSH type N95 or better filters) may be worn for up to 10 times the exposure limit or the maximum use concentration specified by the appropriate regulatory agency or respirator supplier, whichever is lowest. A full-face-piece particulate respirator (NIOSH type N100 filters) may be worn up to 50 times the exposure limit or the maximum use concentration specified by the appropriate regulatory agency or respirator supplier, whichever is lowest. If oil particles (e.g., lubricants, cutting fluids, glycerin, etc.) are present, use a NIOSH type R or P filter. For emergencies or instances where the exposure levels are not known, use a full-face-piece positive-pressure, air-supplied respirator. WARNING: Air-purifying respirators do not protect workers in oxygen-deficient atmospheres.

Skin Protection

Wear impervious protective clothing, including boots, gloves, lab coat, apron, or coveralls, as appropriate, to prevent skin contact.

Eye Protection

Use chemical safety goggles and/or a full-face shield where splashing is possible. Maintain eye wash fountain and quick-drench facilities in work area.

Materials of Construction for Caustic Soda

Carbon steel can be used throughout at moderate temperatures. At temperatures in excess of 350 °F (177 °C), nickel or nickel alloy is recommended. All carbon steel piping, flanges, welds, and vessel must be stress relieved. Caustic soda solution also attacks glass and dissolves it to some extent. As caustic soda is used in cleaning process plants during commissioning, sight glasses and level gauges are removed and replaced with silica glasses for this cleaning phase of the commissioning. Plastic or plastic lined-pipe can also be used for handling caustic soda at low temperatures. Consult plastic manufacturers for compatibility information.

Furfural

Discussion and Hazards of Furfural

Furfural is used in petroleum refining for the production of lube oils. It is a solvent in the extraction of undesirable compounds such as naphthenes and aromatics from lube oil stocks to improve the color of the lube oil product (see the separate topic "▶ Non-energy Refineries in Petroleum Processing"). Furfural or furfuraldehyde, C_4H_3OCHO, is a viscous, colorless liquid that has a pleasant aromatic odor, which upon exposure to air turns dark brown or black. It boils at about 160 °C (320 °F). It is commonly used as a solvent. It is soluble in ethanol and ether and somewhat soluble in water.

Furfural is the aldehyde of pyromucic acid. It has properties similar to those of benzaldehyde. A derivative of furan, it is prepared commercially by dehydration of pentose sugars obtained from cornstalks and corncobs, husks of oat and peanut, and other waste products. It is used in the manufacture of pesticides, phenolfurfural resins, and tetrahydrofuran. Tetrahydrofuran is used as a commercial solvent and is converted in starting materials for the preparation of nylon.

Chemical properties of furfural are in Table 11.

Hazards Associated with Handling Furfural

Inhalation
Causes irritation to the mucous membranes and upper respiratory tract. Symptoms may include sore throat, labored breathing, and headache. Higher concentrations act on the central nervous system and may cause lung congestion. Inhalation may be fatal.

Table 11 Physical properties of furfural

Appearance	Colorless to yellowish liquid
Odor	Almond odor
Solubility	8 g/100 g water @ 70 °F (21 °C)
Specific gravity	1.16 @ 77 °F (25 °C)
% Volatiles by volume @ 70 °F (21 °C)	100
Boiling point	324 °F (162 °C)
Melting point	−38 °F (−39 °C)
Vapor density (Air = 1)	3.3
Vapor pressure (mm Hg)	1 @ 64 ° F (18 °C)

Ingestion
Highly toxic. May cause gastrointestinal disorders. Can cause nerve depression and severe headache. May be fatal. Other effects are not well known.

Skin Contact
Irritant to skin. May cause dermatitis and possibly eczema, allergic sensitization, and photosensitization. May be absorbed through the skin with possible systemic effects.

Eye Contact
Vapors irritate the eyes, causing tearing, itching, and redness. Splashes may cause severe irritation or eye damage.

Chronic Exposure
Can cause numbness of the tongue, loss of sense of taste, and headache. Other effects are not well known.

Aggravation of Preexisting Conditions
Persons with preexisting skin disorders or eye problems or impaired liver, kidney or respiratory function may be more susceptible to the effects of the substance.

First Aid and Personal Safety with Furfural

Inhalation
Remove to fresh air. If not breathing, give artificial respiration. If breathing is difficult, give oxygen. Get medical attention immediately.

Ingestion
If swallowed, give large quantities of water to drink and get medical attention immediately. Never give anything by mouth to an unconscious person.

Skin Contact
Immediately flush skin with plenty of soap and water for at least 15 min while removing contaminated clothing and shoes. Get medical attention, immediately. Wash clothing before reuse. Thoroughly clean shoes before reuse.

Eye Contact
Immediately flush eyes with plenty of water for at least 15 min, lifting lower and upper eyelids occasionally. Get medical attention immediately.

Protective Clothing and Equipment for Furfural Handling

Airborne Exposure Limits
The recommended airborne exposure limit is between 2 and 5 ppm.

Ventilation System

A system of local and/or general exhaust is recommended to keep employee exposures below the airborne exposure limits. Local exhaust ventilation is generally preferred because it can control the emissions of the contaminant at its source, preventing dispersion of it into the general work area. Please refer to the ACGIH document, *Industrial Ventilation, A Manual of Recommended Practices*, most recent edition, for details.

Personal Respirators (NIOSH Approved)

If the exposure limit is exceeded, a full-face-piece respirator with organic vapor cartridge may be worn up to 50 times the exposure limit or the maximum use concentration specified by the appropriate regulatory agency or respirator supplier, whichever is lowest. For emergencies or instances where the exposure levels are not known, use a full-face-piece positive-pressure, air-supplied respirator. WARNING: Air-purifying respirators do not protect workers in oxygen-deficient atmospheres.

Skin Protection

Wear impervious protective clothing, including boots, gloves, lab coat, apron, or coveralls, as appropriate, to prevent skin contact.

Eye Protection

Use chemical safety goggles and/or a full-face shield where splashing is possible. Maintain eye wash fountain and quick-drench facilities in work area.

Fire Prevention and Firefighting for Furfural

Fire

The relevant combustion properties of furfural are in Table 12.

Explosion

Above the flash point, vapor-air mixtures are explosive within flammable limits noted above. Reacts violently with oxidants. Reacts violently with strong acids and bases causing fire and explosion hazards. Sealed containers may rupture when heated. Sensitive to static discharge.

Fire Extinguishing Media

Water spray, dry chemical, alcohol foam, or carbon dioxide can be used as fire extinguishing media. Water spray may be used to keep fire-exposed containers cool.

Table 12 Furfural combustion properties

Flash point	140 °F (60 °C) Pensky-Martens
Autoignition temperature	601 °F (316 °C)
Flammable limits in air % by volume	LEL: 2.1; UEL: 19.3
Flammable liquid and vapor!	

Water may be used to flush spills away from exposures and to dilute spills to nonflammable mixtures.

Special Information

In the event of a fire, wear full protective clothing and NIOSH-approved self-contained breathing apparatus with full-face piece operated in the pressure demand or other positive-pressure mode.

Materials of Construction and Storage for Furfural

There are no special materials of construction required for furfural. A suitable grade of carbon steel is adequate. Storage and handling of furfural must exclude air. Furfural polymerizes readily on exposure to air. In refinery practices, startup of furfural extraction plants usually requires that the equipment which handles furfural be first filled with either the lube oil feed or a suitable middle distillate to eliminate air.

Hydrofluoric Acid (HF, AHF)

Discussion and Hazards of HF

Because of anhydrous hydrofluoric acid's (AHFs) highly toxic and corrosive nature, this section highlights its characteristics and the safe handling of the acid.

Anhydrous hydrofluoric acid is a colorless, mobile liquid which boils at 67 °F (19.4 °C) at atmospheric pressure and therefore requires pressure containers. The acid is also hygroscopic; therefore, its vapor combines with the moisture of air to form "fumes." This tendency to fume provides users with a built-in detector of leaks in AHF storage and transfer equipment. On the other hand, care is needed to avoid accidental spillage of water into tanks containing AHF. Dilution is accompanied by a high release of heat. The physical properties of AHF are given in Table 13.

AHF vapor, even at very low concentrations in air, has a sharp penetrating odor that is an effective deterrent to willful overexposure by operating personnel. Both the vapor and liquid forms of AHF cause severe and painful burns on contact with the skin, eyes, or mucous membranes.

Hydrofluoric acid is very corrosive. It attacks glass, concrete, and some metals – especially cast iron and alloys which contain silica (e.g., Bessemer steels). The acid

Table 13 The physical properties of anhydrous hydrofluoric acid (AHF)

Boiling point at 1 atm, °F (°C)	66.9 (19.4)
Freezing point, °F (°C)	−117.4 (−83)
Specific gravity at 32 °F (0 °C)	1.00
Weight per gallon at 32 °F (0 °C), lb (MT/m^3)	8.35 (1.0)
Viscosity at 32 °F (0 °C), cP	0.31

also attacks such organic materials as leather, natural rubber, and wood, but does not promote their combustion.

Although AHF is nonflammable, its corrosive action on metals, particularly in the presence of moisture, can result in hydrogen forming in containers and piping to create a fire and explosion hazard. Potential sources of ignition (sparks and flames) should be excluded from areas around equipment containing hydrofluoric acid.

Despite its corrosive nature, AHF can be handled with relative safety if the hazards are recognized and the necessary precautions taken. The next few sections describe certain procedures for the safe handling of large bulk quantities of AHF.

Safe Handling Practices for HF

The safe handling of AHF requires that well-designed equipment be properly operated and maintained by well-trained, adequately protected, responsible personnel.

Tanks and other containers of AHF should be protected from heat and the direct rays of the sun. Storage-area temperatures should preferably remain below 100 °F (38 °C). If they reach or exceed 125 °F (51.7 °C), means for cooling the containers must be applied.

Acid transfer lines between the unloading station and the storage tank should be designed to free drain toward the storage tank. Thermal relief valves should be installed in those sections of acid transfer lines where acid may be entrapped between two closed valves in the line, because expansion of the liquid might create excessive pressure and rupture the line. The relief will need to discharge toward the tank, not to atmosphere.

No open fires, open lights, or matches should be allowed in or around acid containers or lines. The possibility of acid acting on metal to produce hydrogen gas is ever present. Only non-sparking tools and spark-proof electrical equipment should be used in the AHF storage and handling areas.

Safety showers should be readily accessible at the unloading station, in the storage area, and at other locations where acid is handled. The showers should be capable of supplying volume flows of 30 GPM (114 l/min) through quick-opening valves in 2 in. (50 mm) water lines. Handles at hip level should actuate the valves which, with a 0.25 in. (6.4 mm) weep hole directly above the valve, should be positioned below the frost line and surrounded by crushed rock or gravel to provide drainage.

A water hydrant and hose should also be available in the unloading area to flush away spilled acid. Good drainage should be provided and also a supply of dry soda ash, ground limestone, or hydrated builders lime. Accidental spills of acid on walkways or equipment should be washed off immediately with large volumes of water and, if necessary, neutralized with one of the agents mentioned.

Personal Protective Equipment Requirements for HF

Personal protective equipment is not a substitute for good, safe working conditions. Its purpose is to protect the wearer in the event of an accident – major or minor. The extent of protection needed depends upon the degree of potential exposure attending the particular job at hand. Protective equipment should not be worn or carried beyond the operating area. It should be thoroughly washed with sodium bicarbonate solution immediately after each use.

The minimum protection required for operating and maintenance personnel includes the following items:

- Coveralls with sleeves to the wrists
- Face shield or chemical safety goggles
- Hard hat
- Poly(vinyl chloride) – or neoprene-dipped gauntlets
- Poly(vinyl chloride) – or neoprene-soled rubber shoes

When taking acid samples, opening equipment which may contain hydrofluoric acid, or performing similar hazardous duties, operators should wear the following:

- Poly(vinyl chloride) or neoprene overalls
- Poly(vinyl chloride) or neoprene boots
- Lightweight poly(vinyl chloride) or neoprene gloves under poly(vinyl chloride) – or neoprene-dipped gauntlets
- Poly(vinyl chloride) or neoprene jumper
- Airline hood

Air should be applied to the hood until the absence of fumes in the work area has been fully established.

Unloading and Transfer of AHF

AHF is shipped in rail tank cars having capacities ranging from approximately 5,400 to 25,000 gal ($\sim$20–94 MT) and in road tank trucks of approximately 5,250 gal ($\sim$20 MT) AHF capacity. Compressed dry gas (air, hydrocarbon, or nitrogen) is the preferred means for transferring bulk quantities of AHF, but a centrifugal, rotary, or positive-pressure pump can be used, if necessary.

The unloading of AHF tank cars or tank trucks, with transfer of the acid to plant storage, consists of five steps:

1. Spotting the tank car or tank truck at the unloading station
2. Connecting the plant compressed-gas (or vapor) and AHF-unloading lines to the proper valves on the carrier tank

3. Transferring the AHF from the carrier tank to the storage tank
4. Disconnecting the plant compressed-gas (or vapor) and AHF-unloading lines from the carrier tank valves
5. Releasing the tank car or tank truck for return to the shipper

Equipment for HF Service

Mild steel is satisfactory for storing and handling AHF at temperatures up to 150 °F (~65 °C) maximum. Type 300 stainless steels are useful up to 200 °F (~93 °C). "Monel" nickel-copper alloy and "Hastelloy" C nickel steel are suitable for higher temperatures. TEFLON TFE fluorocarbon resin is completely resistant to all concentrations of hydrofluoric acid at temperatures up to 500 °F (260 °C).

Steel should not be used for movable parts because the corrosion-product film will cause movable parts to "freeze." Cast iron, type 400 stainless steel, and hardened steels are unsatisfactory for AHF handling. Copper is velocity sensitive. Stressed Monel may stress crack if exposed to moist vapors or aerated acid containing water. Welds in Monel corrode rapidly.

The selection of construction materials used for AHF equipment depends very much on such corrosion-affecting variables as moisture, temperature, aeration, fluid velocity, and impurities. Each storage and handling situation requires separate study to evaluate these factors before selecting materials which must meet the requirements of the installation.

Additional information on metallurgy and materials for HF service can be found in API Recommended Practice 571 and is available from process licensors.

Storage Tank

The capacity of the storage system should be approximately 1.5 times the maximum quantity normally ordered to insure against running out of acid between receipts of shipments. As a rule, too large a storage system is preferable to too small a system. The additional investment required for the larger installation is not great. The larger installation permits further expansion, less precise scheduling of shipments, and larger inventories when desired.

The horizontal cylindrical storage tank should be manufactured according to the current ASME Code for Unfired Pressure Vessels or other equivalent codes which meet state or local mandatory requirements. It is further recommended that the wall thickness of the tank be at least 1/8 in. (3.2 mm) in excess of the ASME Code requirements. The tank should be double-welded, butt-joint construction, the welds to be slag-free (conforming to ASME Code, Section 8) and ground smooth inside to facilitate inspection. X-ray inspection of welds is recommended.

The storage tank should be suitably supported above ground level. Structural steel supports or concrete saddles (protected with an acid-resistant paint) are satisfactory. Secondary containment is normally required. An emergency water wall system may also be required.

Safety devices for relieving abnormal internal tank pressures should be obtained from qualified manufacturers who are familiar with AHF. The maximum working pressure of the storage system should not exceed two thirds the rated relief or bursting pressure of the safety devices. A dual relief system is recommended which has a two-way valve and rupture disks ahead of the relief valves and also a separate rupture-disk line in case of relief valve failure.

Piping

All pipe lines should be installed so that they drain toward the storage tank or toward the point of consumption. This will prevent the accumulation of acid in low points, thereby eliminating possible safety hazards when repairs are necessary. Relief valves should be installed in the various sections of the lines in case acid becomes confined between two closed valves in the line. All flanges in the lines should preferably be coated with an acid-indicating paint, such as Mobil #220-Y-7 hydrofluoric acid-detecting paint, which changes in color from orange to yellow in the presence of AHF liquid or vapor.

The line from the unloading station to the storage tank should be equipped with a gate valve so acid flow can be stopped at any time. The line should also be securely anchored to the storage tank as considerable vibration may occur, especially when unloading by means of compressed gas.

Extra heavy (Schedule 80) or, better, triple extra-heavy black seamless or welded steel pipe, which is free from non-metallic inclusions, is satisfactory.

Fittings

Larger lines (2 in. [50 mm] and larger) should preferably be welded to conform to ASME Code, Section 8. Alternatively, properly gasketed forged steel flanges can be used.

On smaller lines, extra-heavy forged steel, screw-type unions with steel-to-steel seats can be used for pipe joints. Graphite-and-oil is satisfactory as lubricant.

Gaskets

Gaskets made of TEFLON TFE fluorocarbon resin are recommended.

Valves

Jamesbury "Double-Seal" ball valves have given excellent service to AHF manufacturing operations. The valve seats are preferably of TEFLON TFE; the balls and bodies of 316 stainless steel, Durimet 20 austenitic stainless steel alloy, or equivalent. Gate valves should be of the O, S, and Y flange type, with a ring of TEFLON TFE or "Kel-F" fluorocarbon thermoplastic material on the plug seat and packing of either TEFLON TFE or Kel-F.

Globe valves can be of Monel nickel-copper alloy or have a forged steel body and trim of Monel.

Good service has been reported for Hills-McCanna diaphragm valves with body of Durimet 20 or equivalent, or Monel alloy; diaphragm of polyethylene, Kel-F, or a laminate of neoprene and TEFLON TFE; and a wheel closure.

Plug valves of Monel with a sleeve of TEFLON TFE have been found satisfactory. Check valves should be of the forged ball and body type – made of Monel metal.

Pressure Gauges

Pressure gauges should be constructed of 316 stainless steel or Monel metal Bourdon tubes. The bottom connection of the gauge should be 1/2 inch. The case should have a "blow-out" back.

Pumps

Centrifugal, rotary, or positive-pressure types of pumps are satisfactory. The 300 series stainless steels; Durimet 20 or equivalent, Hastelloy C, Monel alloys; nickel, bronze, and acid bronze have been recommended as construction materials.

Level Gauges

AHF Manufacturers suggest the use of a magnetic-type level gauge, such as a Fischer & Porter Model 13 C 2265 W Liquid "Levelrator" with donut-type float.

An alternative means for monitoring storage-tank content is to set the tank on load cells or strain gauges.

Filters

Where the critical nature of the process has warranted, cartridge-type filters in the storage tank-to-process line have been recommended. Two such filters are normally mounted in parallel to permit replacing the cartridge in one line while diverting the acid flow through the other. Construction materials used in fabricating the filters are the same or similar to those described above for other auxiliary equipment.

Polypropylene can be used as filter material for AHF alone if the liquid temperature remains below 200 °F (93 °C).

When a fouled filter is removed for replacement, it should be promptly flushed with water, neutralized with a solution of soda ash, and rinsed before discard.

Personal Safety When Handling HF

Liquid AHF causes immediate and serious burns to any part of the body on contact.

Dilute solutions of hydrofluoric acid often do not cause an immediate burning sensation where they came in contact with skin. Several hours may pass before the solution penetrates the skin sufficiently to cause redness or a burning sensation.

Wearing clothing which may have absorbed small amounts of hydrofluoric acid (such as leather shoes or gloves) can result in painful delayed effects similar to those caused by dilute acid solutions.

Hydrofluoric acid vapor causes skin irritation and inflammation of the mucous membranes; the burns become apparent a few hours after exposure. Inhaling the vapor in high concentrations may cause lung damage (pulmonary edema).

In the United States, the American Conference of Governmental Industrial Hygienists recommends a threshold limit value of 3 parts (by volume) AHF vapor (hydrogen fluoride) per million parts air. This value refers to a time weighted concentration for a seven- or 8-h workday and 40-h work week.

The 3 vppm figure is based on both experimental and occupational evidence; however, nosebleeds and sinus troubles have reportedly occurred among metal workers exposed to even lower concentrations of a fluoride or fluorine in air. Therefore, for protection against acute irritation, 3 ppm should be considered a ceiling limit.

Anyone who knows or even suspects he has come in contact with hydrofluoric acid should immediately seek first aid.

In the event of an accident, the medical response should be called immediately; however, all plant supervisors should be aware of first aid procedures for HF burns. All affected persons should be referred to a physician even when the injury seems slight.

Hydrogen

General Discussion of Hydrogen

Hydrogen is present throughout refining and petroleum processes today. It is:

- Generated in reformers and hydrogen plants
- Recovered from fuel and refinery make gases
- Consumed in hydrotreaters, hydrocrackers, isomerization units, benzene saturators, and sulfur plants

It is critical to meeting modern product specifications.

Most hydrogen is produced within a facility and used within the facility. In some cases, hydrogen may be received in bulk from a supplier for specific purposes, such as naphtha reformer startup or calibration gases. Hydrogen may also be supplied in bulk by pipeline from an outside facility.

The primary hazards presented by hydrogen are fire and explosion. When released to the atmosphere, hydrogen quickly rises and dissipates, but local concentrations may present risks.

High velocity releases of hydrogen (as in a vent) may generate a static electrical charge that can create a spark to ignite the vent.

Physical and Chemical Properties of Hydrogen

Table 14 summarizes the key properties of hydrogen of concern here.

Hydrogen gas is extremely flammable and easily ignites, even from hot equipment. It is colorless, odorless, and tasteless.

Hydrogen has the unusual property of a reverse Joule-Thompson effect – it will heat up when it rapidly expands.

Table 14 Key properties of hydrogen gas

Molecular weight	2.02
Specific gravity	Vapor: 0.0696 (air =1)
Boiling point	−423.4 °F (−253 °C)
Freezing point	−434.5 °F (−259.15 °C)
Solubility in water	27.6 mg/l (77 °F/25 °C)
Autoignition temperature	932–1060 °F (500—571 °C)
Explosive limits	4—76 vol%

Hazards of Hydrogen

The most obvious and well-known hazard of hydrogen is its flammability. It burns with a clear flame that is hard to detect. Personnel have been known to walk into a hydrogen flame without seeing it.

The heat generated from the reverse Joule-Thompson effect of hydrogen, along with the static electricity generated in a leak, can result in fires or explosions from hydrogen leaking from pressures over about 1,000 psig (69 barg).

Sometimes hydrogen is received as compressed gas (2,000 psig [138 barg] plus). In these cases, additional hazard is presented by the high pressure.

Hydrogen will burn aggressively with any oxidizer.

Hydrogen is not particularly toxic but may cause asphyxiation in high enough concentrations, e.g., near a hydrogen vent. The rapid dissipation of vented hydrogen normally limits asphyxiation risk.

Protective Equipment for Hydrogen Handling

The common refinery PPE (hardhat, safety glasses, flame-retardant clothing, and safety shoes) provide some protection against a hydrogen flash fire. They will not provide protection from a significant fire or explosion, an oxygen deficiency, or physical hazard from high pressures involving hydrogen. Major events involving hydrogen require SCBA and flash suits. There should be an SCBA nearby when handling bulk hydrogen.

It is best to focus on preventing a hydrogen fire from affecting surrounding equipment until the hydrogen can be stopped at the source.

Ventilation of areas where hydrogen may accumulate is the best way to protect personnel.

Storage and Handling of Bulk Hydrogen

Hydrogen is normally generated and used within a facility or may be received by pipeline over the fence. As such, no more handling is involved than for the hydrocarbon streams. Precautions and PPE used for handling hydrocarbon gases are adequate for hydrogen in general.

When hydrogen is received in bulk, normally pressurized gas in tube trailers or in cylinders, the normal precautions that apply to compressed gases are adequate.

Hydrogen Sulfide (H_2S)

Discussion and Hazards of H_2S

Hydrogen sulfide in a refinery is usually formed during the desulfurizing processes used to sweeten product streams. Hydrogen sulfide (H_2S) is a colorless, extremely poisonous gas that has a very disagreeable odor, much like that of rotten eggs, in low concentrations. In high concentrations that are IDLH, it deadens the sense of smell.

It is slightly soluble in water and is soluble in carbon disulfide. Dissolved in water, it forms a very weak dibasic acid. Hydrogen sulfide is flammable, and in excess air, it burns to form sulfur dioxide and water. Where less than stoichiometric oxygen is present, it forms elemental sulfur and water.

It may be made by reacting hydrogen gas with molten sulfur or with sulfur vapors or by treating a metal sulfide (e.g., ferrous sulfide, FeS) with an acid. Hydrogen sulfide reacts with most metal ions to form sulfides; the sulfides of some metals are insoluble in water and have characteristic colors that help to identify the metal during chemical analysis.

Hydrogen sulfide also reacts directly with silver metal, forming a dull, gray-black tarnish of silver sulfide (Ag_2S). One method of detecting small concentrations of hydrogen sulfide is to expose it to a filter paper impregnated with lead acetate. The paper turns black (due to the precipitation of lead sulfide). The degree of H_2S concentration is measured by the shade of "blackness" of the lead acetate paper compared with standard colors.

The relevant properties of hydrogen sulfide are given in Table 15.

Hazards and Toxicity of Hydrogen Sulfide

Table 16 indicates the toxicity of hydrogen sulfide.

The following discussion uses the present US OSHA limits as a basis.

At 1 ppm, most people can smell the gas. A strong smell does not necessarily mean a high concentration, and a slight smell does not mean a low concentration. A person could work in a 10 ppm concentration of H_2S for 8 h. If the concentration exceeds 10 ppm for a short period of time, then the time must be reduced.

A concentration of 15 ppm can be tolerated for a period of time not exceeding 15 min. There can be no more than four exposures of 15 ppm in an 8 h shift with 1 h between exposures. If the concentration of H_2S exceeds 20 ppm, a worker must wear approved breathing apparatus. If the concentration is not known, a worker must wear breathing apparatus until the concentration is determined.

Table 15 Properties of hydrogen sulfide

Chemical formula	H_2S
Relative density	1.189 (air = 1.0)
Autoignition temperature	500 °F (260 °C)
Flammability	Very flammable
Lower explosive limit	4.3 vol% in air
Upper explosive limit	46 vol% in air
Color	Colorless, invisible
Odor	Strong rotten egg at low concentrations; cannot smell at IDLH concentrations
Vapor pressure	17.7 atm at 20 °C (68 °F)
Boiling point	−77 °F (−61 °F)
Melting point	−122 °F (−86 °F)
Reactivity	Dangerous with acids and oxidizers
Solubility	In water, hydrocarbons, alcohol

Quoted from National Safety Council Data Sheet 1-284-67

Table 16 Toxicity of hydrogen sulfide

ppm	Percent	Comments
1	0.0001	Most people can smell the gas
10	0.001	Occupational exposure limit. Maximum continuous exposure for 8 h
15	0.0015	Occupational exposure for 15 min
20	0.002	Ceiling occupational exposure limit. This level of exposure cannot be exceeded at any time without respiratory protection
100	0.01	Dulls sense of smell. Causes burning sensation in the eyes and throat
500	0.05	Attacks the respiratory center of brain; causes loss of reasoning and balance
700	0.07	Victim quickly loses consciousness; breathing will stop, and death will result if not rescued promptly
1,000	0.1	Unconscious immediately; permanent brain damage or death occurs if victim is not rescued and resuscitated immediately

If exposed to a concentration of 100 ppm (1/100 of 1 %), the sense of smell will be lost or become ineffective within 2–15 min. The H_2S might cause a burning sensation to the eyes, throat, and lungs and could cause headache or nausea.

A 200 ppm concentration will cause immediate loss of smell and a burning sensation in the eyes, throat, nose, and lungs. (The hydrogen sulfide combines with alkali in body fluids to form caustic sodium sulfide.)

At a concentration of 500 ppm, the victim will appear to be intoxicated, and will lose his sense of balance and reasoning. In this state, the victim may attempt to continue with the job he was doing when he encountered the gas. For this reason, a person *must* know the people he works with and be able to detect any unusual behavior of a coworker. Obviously, persons under the influence of alcohol, or any other mind-altering drugs, should never be allowed in an area which may contain sour gas. A victim must be watched very closely and may require resuscitation. A victim should be taken for medical attention and not allowed to return to work for at least 8 h.

At 700 ppm, the victim will be rendered unconscious very quickly and may develop seizures similar to those caused by epilepsy. Loss of bladder and bowel control can be expected. Breathing will stop, and death will result, if not rescued and resuscitated promptly. At a concentration of 1,000 ppm (1/10 of 1 %), the victim will be rendered unconscious immediately. THE VICTIM WILL NOT BEGIN BREATHING VOLUNTARILY IF BROUGHT TO FRESH AIR. ARTIFICIAL RESUSCITATION MUST BE COMMENCED WITHIN THREE MINUTES OF EXPOSURE TO THIS LEVEL OF HYDROGEN SULFIDE!

Protective Clothing and Personal Safety for H_2S

The appropriate personal protective equipment when working around gaseous H_2S, or in areas where H_2S may be present and is typically the same as the minimum required in a facility:

- Hardhat
- Flame-retardant clothing
- Safety glasses
- Safety-toe shoes

In addition, many (if not most) refiners now require workers to wear personal H_2S monitors, which provide an audible and visual alarm at a specified concentration of H_2S. These personal monitors complement area H_2S monitors in many facilities.

In areas where H_2S exposure is possible, appropriate supplied-air respirators must be readily available, if not actually worn. If exposure to H_2S is expected, such as during sampling of sour gas streams or when setting blinds on a sour line, procedures generally call for the use of fresh air equipment.

In confined areas, such as enclosed compressor or pump houses which handle sour gas or liquids, constant monitoring for H_2S concentration in the atmosphere by automatic area air analyzers with alarms is common. Failing this, a routine analysis using a lead acetate paper should be made. In addition, all such buildings should always be properly vented using an exhaust fan system.

Materials of Construction in H_2S Service

Atmospheres containing hydrogen sulfide and completely or almost free of oxygen give rise to rapid corrosion of unalloyed steel by forming a sulfide film on its surface. The corrosion rate in hydrogen/hydrogen sulfide increases as the content of hydrogen sulfide increases up to about 5 vol.%, while increases beyond that point generally only have a slight effect on the corrosion rate. Steels alloyed with chromium and aluminum have improved resistance to hydrogen sulfide, while nickel has no deleterious effect.

Moist and aqueous solutions of hydrogen sulfide cause some minor pitting in unalloyed steel, and there is a risk of stress corrosion. This pitting corrosion is about 1 mm/year. This rate can be considerably reduced to about 0.1 mm/year using an alloyed steel of 18 % chrome and 9 % nickel. Vessels and piping should all be stress relieved.

The API provides detailed guidance on materials of construction for various types of H_2S services.

Methyl Ethyl Ketone (MEK)

Discussion of MEK

Methyl ethyl ketone (MEK) is used in oil refining for the removal of wax from lube oil stock (see the separate chapter "▶ Non-energy Refineries in Petroleum Processing"). Methyl ethyl ketone is a colorless liquid with a sweet/sharp, fragrant, acetone-like odor. It is extremely flammable in both the liquid and vapor phase. The vapor is heavier than air and may spread long distances, and distant ignition and flashback are possible. MEK is highly volatile. Its basic properties are in Table 17.

Hazards Associated with MEK

Inhalation
Causes irritation to the nose and throat. Concentrations above 200 ppm may cause headache, dizziness, nausea, shortness of breath, and vomiting. Higher concentrations

Table 17 Properties of methyl ethyl ketone

Appearance	Clear, colorless, stable liquid
Purity, % minimum	99.5
Water content, % maximum	0.30
Acidity, % maximum (as acetic acid)	0.003
Color, Pt-Co maximum	10
Specific gravity, 20/20 °C (68/68 °F)	0.805–0.807
Nonvolatile matter (g/100 ml), maximum	0.002
Boiling point	176 °F (80 °C)
Vapor pressure	3 in Hg @ 68 °F (76 mmHg @ 20 °C)
Flash point	−9 °C (16 °F) CC
Autoignition temperature	404 °C (759 °F)
Flammable limits in air, vol%	LEL: 1.4; UEL: 11.4
Extremely flammable	

may cause central nervous system depression and unconsciousness. The airborne exposure limits are:

- Permissible exposure limit (PEL): 200 ppm (TWA)
- Threshold limit value (TLV): 200 ppm (TWA), 300 ppm (STEL)

Ingestion
May produce abdominal pain, nausea. Aspiration into lungs can produce severe lung damage and is a medical emergency. Other symptoms expected to parallel inhalation.

Skin Contact
Causes irritation to skin. Symptoms include redness, itching, and pain. May be absorbed through the skin with possible systemic effects.

Eye Contact
Vapors are irritating to the eyes. Splashes can produce painful irritation and eye damage.

Chronic Exposure
Prolonged skin contact may defat the skin and produce dermatitis. Chronic exposure may cause central nervous system effects.

Aggravation of Preexisting Conditions
Persons with preexisting skin disorders or eye problems or impaired respiratory function may be more susceptible to the effects of the substance.

First Aid and Personal Protection

Inhalation
Remove to fresh air. If not breathing, give artificial respiration. If breathing is difficult, give oxygen. Get medical attention.

Ingestion
Aspiration hazard. If swallowed, vomiting may occur spontaneously, but DO NOT INDUCE. If vomiting occurs, keep head below hips to prevent aspiration into lungs. Never give anything by mouth to an unconscious person. Call a physician immediately.

Skin Contact
Immediately flush skin with plenty of soap and water for at least 15 min while removing contaminated clothing and shoes. Get medical attention. Wash clothing before reuse. Thoroughly clean shoes before reuse.

Eye Contact
Immediately flush eyes with plenty of water for at least 15 min, lifting upper and lower eyelids occasionally. Get medical attention.

Clothing and Protective Equipment

Ventilation System
A system of local and/or general exhaust is recommended to keep employee exposures below the airborne exposure limits. Local exhaust ventilation is generally preferred because it can control the emissions of the contaminant at its source, preventing dispersion of it into the general work area.

Personal Respirators
If the exposure limit is exceeded and engineering controls are not feasible, a full-face-piece respirator with organic vapor cartridge may be worn up to 50 times the exposure limit or the maximum use concentration specified by the appropriate regulatory agency or respirator supplier, whichever is lowest. For emergencies or instances where the exposure levels are not known, use a full-face-piece positive-pressure, air-supplied respirator. WARNING: Air-purifying respirators do not protect workers in oxygen-deficient atmospheres.

Skin Protection
Wear impervious protective clothing, including boots, gloves, lab coat, apron, or coveralls, as appropriate, to prevent skin contact. Butyl rubber is a suitable material for personal protective equipment.

Eye Protection
Use chemical safety goggles and/or a full-face shield where splashing is possible. Maintain eye wash fountain and quick-drench facilities in work area.

Fire Prevention and Fighting

Fire
MEK is extremely flammable. Refer to Table 17 for flammability data.

Explosion
Above the flash point, vapor-air mixtures are explosive within flammable limits noted in Table 17. Vapors can flow along surfaces to distant ignition sources and flash back. Contact with strong oxidizers may cause fire. Sealed containers may rupture when heated. They are sensitive to static discharge.

Fire Extinguishing Media
Dry chemical, foam, or carbon dioxide are suitable extinguishing agents. Water spray may be used to keep fire-exposed containers cool, dilute spills to nonflammable mixtures, protect personnel attempting to stop leaks, and disperse vapors.

Special Information
In the event of a fire, wear full protective clothing and NIOSH-approved self-contained breathing apparatus with full-face piece operated in the pressure demand or other positive-pressure mode. This highly flammable liquid must be kept from sparks, open flame, hot surfaces, and all sources of heat and ignition.

Accidental Release Measures
Ventilate the area of any leak or spill. Remove all sources of ignition. Wear appropriate personal protective equipment as noted above. Isolate the hazard area. Keep unnecessary and unprotected personnel from entering. Contain and recover liquid when possible. Use non-sparking tools and equipment. Collect liquid in an appropriate container or absorb with an inert material (e.g., vermiculite, dry sand, earth) and place in a chemical waste container. Do not use combustible materials, such as sawdust. Do not flush to sewer! If a leak or spill has not ignited, use water spray to disperse the vapors, to protect personnel attempting to stop leak, and to flush spills away from exposures.

Storage and Handling

MEK is usually delivered to a refinery by road or rail truck. It may be stored in small bullets or a cone roof tank under an inert gas blanket. The materials of construction normally include appropriate grades of carbon steel.

Nickel Carbonyl (Ni(CO)$_4$)

General Discussion of Nickel Carbonyl

Nickel carbonyl is one of the most toxic substances which may be encountered in a petroleum processing facility anywhere CO is present along with nickel in a reduced state. From experience, the nickel does not need to be present as metal directly; Ni(CO)$_4$ has been detected when spent NiMo hydrotreating catalyst has been exposed to air and smoldered.

It is stable at ambient pressure below ~400 °F (~204 °C).

Nickel carbonyl is formed by the reaction of CO with nickel via the equation:

$$\text{Ni(s)} + 4\text{CO(g)} \longleftrightarrow \text{Ni(CO)}_4\text{(g)} \tag{1}$$

While most literature on Ni(CO)$_4$ discusses it as a liquid, it would be most likely encountered as a gas in the refining environment. It has been found in hydrogen plant vapors from the reforming furnace (Ni catalyst and CO present) and is likely to be present in methanator reactor effluent (also CO and Ni). As noted above, it has been detected in gases from smoldering spent NiMo hydrotreating catalysts. It can also be adsorbed on NiMo catalysts and released on exposure to air.

There should be no reason to handle nickel carbonyl in bulk. Your primary concern is the possible presence of it as a trace gas and managing it as a hazardous material within the process equipment, especially during maintenance activities.

Chemical and Physical Properties of Nickel Carbonyl

Some of the key chemical and physical properties of nickel carbonyl are listed in Table 18.

Note that it is a heavy vapor that may accumulate in low elevations.

The formation of stable nickel carbonyl depends strongly on the partial pressure of CO, assuming nickel is present. The equilibrium equation is:

$$P_{Ni(CO)_4} = [P_{CO}]^4 \times 10^{[8546/T - 21.64]} \tag{2}$$

where:

$P_{Ni(CO)_4}$ = partial pressure of nickel carbonyl, atmospheres
P_{CO} = partial pressure of CO, atmospheres
T = absolute temperature, °K

Once formed, the dissociation back to nickel and CO is not instantaneous but may require several minutes. The formation of $Ni(CO)_4$ is hindered by the oxide

Table 18 Some key properties of nickel carbonyl

Chemical formula	$Ni(CO)_4$
Molecular weight	170.73
Relative density	5.9 (air = 1.0)
Autoignition temperature	140 °F (60 °C)
Flammability	Very flammable
Explosive limits	2–34 vol% in air
Color	Colorless, invisible
Odor	Musty, like brick dust (don't try to find out, however) (threshold = 1–3 ppm)
Boiling point	109 °F (43 °C)
Melting point	1.0 °F (−17.2 °C)
Reactivity	Dangerous with oxidizers and bases
Solubility in water	0.018 g/100 ml (50 °F, 10 °C)
Solubility in organics	Soluble

coating of Ni catalyst and oxygen. Formation of $Ni(CO)_4$ is accelerated by ammonia and H_2S.

Hazards of Nickel Carbonyl

This is one of the most toxic materials encountered in refining (or any industry for that matter). Table 19 lists the applicable exposure limits.

Via the various exposure routes, the toxic effects are:

- Inhalation:
  - Short-term: Cough, fever, nausea, vomiting, diarrhea, chest pain, difficulty breathing, irregular heartbeat, headache, dizziness, disorientation, bluish skin color, blood disorders, liver enlargement, convulsions, and death
  - Long term: Reproductive effects and cancer
- Skin contact:
  - Short-term: Irritation, allergic reactions, rash, itching. It may be absorbed through the skin and produce effects similar to inhalation.
  - Long term: Irritation and allergic reactions. May experience reactions similar to inhalation.
- Eye contact:
  - Short-term: irritation.
  - Long term: No information available.
- Ingestion:
  - Short-term: Cough, fever, nausea, vomiting, diarrhea, chest pain, difficulty breathing, irregular heartbeat, headache, dizziness, disorientation, and bluish skin color
  - Long term: No information available

In addition to the health impacts, nickel carbonyl in sufficient quantities is highly flammable and can be detonated by shock, friction, or moderate heating. You should not normally encounter it in any concentration that would cause these hazards, however.

NFPA rating for nickel carbonyl is 4 (health) – 3 (flammability) – 3 (reactivity).

Table 19 Exposure limits and toxicity of nickel carbonyl

Agency	Limit type	pp*b*	mg/m^3	Basis
US OSHA	PEL	1	0.007	8 h TWA
US NIOSH	REL	1	0.007	10 h TWA
	IDLH	(~3 ppm)		
ACGIH (USA)	TLV	1	0.007	10 h TWA

LC50 (Rat) by inhalation: 35 ppm (30 min)
Known human carcinogen

Protective Equipment for Nickel Carbonyl

If exposure to significant amounts of nickel carbonyl may be possible, the normal precautions (in addition to the normal hardhat, safety glasses, flame-retardant clothing, and safety shoes) are:

- Provision of adequate ventilation, which may include explosion-resistant fans.
- Eye protection: splash-resistant chemical goggles with face shield. Emergency eyewash in immediate area.
- Clothing: chemical-resistant clothing if liquid may be present.
- Gloves: chemical-resistant gloves if liquid may be present.
- Respirator: Full-face, self-contained, or supplied-air respirator operating on pressure demand (or other positive-pressure mode) with an escape bottle. A full-face, air-purifying respirator with a suitable canister is only usable for emergency escape.

If there is any possibility of IDLH atmosphere with nickel carbonyl, a supplied-air breathing apparatus is required. Do not take any chances.

Note that these precautions apply for personnel at catalyst dump points (such as at the top of flow bins) as well as those working in enclosed spaces with catalysts.

Management of Nickel Carbonyl Hazards

The primary focus of nickel carbonyl management in a refinery is prevention of its formation.

In hydroprocessing units, where nickel catalysts and deposits are common, it is normal practice to cool reactors to about 450–500 °F (232–260 °C) and then test the circulating gases for CO. Different companies and units use different limits, but the CO in the gas must normally be less than 10–100 ppm before cooling further. This ensures the nickel carbonyl content will not exceed 1 ppb as the catalyst is cooled. Purging with nitrogen further reduces the risk of formation. You have to be careful to avoid introducing any more CO after the test through the hydrogen or nitrogen used. After testing, it is normally best to avoid bringing in anything except vaporized, cryogenic nitrogen.

If the CO test shows a CO level above the allowable, the reactor should be held at temperature (or re-heated above 450 °F). The system should be purged and any possible CO (or oxygen or CO_2) sources eliminated until the CO level drops to the allowable range. Once a CO source is stopped in a circulating system with hydrogen, the remaining CO will be methanated and disappear from the system.

In hydrogen production, CO and reduced nickel catalysts are part of the process. To eliminate the possibility of nickel carbonyl formation in hydrogen plants on shutdown, the feedstock should be pulled at high temperature. No oxygen, CO_2, or other possible CO formation sources should be permitted to reenter the system during cooling. The system should be thoroughly purged with nitrogen (CO-free) before cooling below 400 °F (204 °C).

Precautions during catalyst handling include:

- Continuous monitoring of the inert atmosphere for CO and oxygen. Stop and purge if these are detected above allowable limits.
- Use fresh air equipment at the dump point where catalyst smoldering may occur.
- Keep catalyst spills wet to prevent smoldering.
- Dump potentially self-heating catalysts into inert-purge containers or flood and dump wet.

CO testing by detection tubes is normally employed to verify no CO is present. Follow the detector tube manufacturer's instructions. For Dräger tubes:

- Hydrogen gas atmosphere: use tube 8/a for CO with solid sodium hydroxide and carbon pre-tubes.
- Nitrogen or natural gas atmosphere: use tube 5/c for CO with the NaOH and carbon pre-tubes.
- Olefins present in the gas will interfere with the analysis.

While detector tubes are available for nickel carbonyl, the detector range is 0.1 ppm (100 ppb), which is too high to help in this case.

Nickel carbonyl itself can be analyzed using wet chemical or adsorption/desorption methods. US NIOSH has procedures available. This tests are too slow for most practical monitoring, however.

Nitrogen (N_2, LN_2)

General Discussion of Nitrogen

The uses and hazards of nitrogen are addressed somewhat under the topics "▶ Utilities in Petroleum Processing" and "▶ Safety Systems for Petroleum Processing" of this handbook. We will explore the hazards of handling these materials a little more quantitatively here.

As background and refresher, nitrogen is used extensively in most refineries today. Some of the common uses are:

- Purging of equipment and piping to eliminate air or hydrocarbons/hydrogen
- Purging of compressor seals
- Storage tank or vessel blanketing to prevent air contact
- As emergency backup pneumatic gas
- Chilling of some analytical instruments

In normal practice, the nitrogen is received as a refrigerated liquid. The liquid is stored in a double-walled, insulted cryogenic storage tank and is vaporized as needed to supply facility needs. For large, intermittent uses, nitrogen trailers and fired vaporizers are used. Some facilities generate their own nitrogen cryogenically or, more commonly, using a PSA or membrane system. This is discussed in more detail in the chapter on "▶ Utilities in Petroleum Processing."

Chemical and Physical Properties of Nitrogen

Table 20 lists the key properties of nitrogen relevant to the current discussion.

Nitrogen may be found in a refinery as a refrigerated liquid or as a vapor. The most common exposure risk is from inhalation of excessive nitrogen vapors.

Note that the cryogenic liquid that is normally handled in bulk is extremely cold.

Nitrogen will not support combustion at any condition normally encountered in a refinery.

Hazards of Nitrogen

Nitrogen, as used in a petroleum processing facility, presents the following hazards:

- Oxygen depletion in the atmosphere
- Risk of burns or frostbite from cryogenic liquid contact
- Physical injury from high pressure gas

The most important of these hazards is normally asphyxiation from an oxygen-deficient atmosphere. Since nitrogen is colorless and odorless, it provides no warning of oxygen deficiency.

Otherwise, nitrogen is not toxic and causes no other issues. These hazards are discussed more fully in the handbook topic "▶ Safety Systems for Petroleum Processing."

Symptoms of oxygen deficiency from a nitrogen atmosphere would include:

- Headache
- Dizziness

Table 20 Some key properties of nitrogen

Molecular weight	28.02
Specific gravity	Vapor: 0.967 (air = 1)
	Liquid: 50.46 lb/ft^3 (808.3 kg/m^3)
Boiling point	−320 °F (−196 °C)
Freezing point	−346 °F (−210 °C)
Autoignition temperature	Will not support combustion
Explosive limits	Not applicable

- Fatigue
- Nausea
- Euphoria
- Or simply unconsciousness without warning (within seconds at low oxygen levels)

Most often, unconsciousness is so rapid there is no time for other symptoms to manifest.

In working with the cryogenic liquid, any exposure of personnel to the liquid (as in skin) will result in immediate/instantaneous, severe burns.

Protective Equipment for Nitrogen Handling

The best protection against nitrogen creating an oxygen deficiency is adequate ventilation and exclusion of all personnel from areas near inerted equipment. Anyone that must approach inerted equipment must wear fresh air breathing apparatus (SCBA or supplied air).

In addition to the normal refinery PPE, areas where there may be a nitrogen deficiency are normally monitored by fixed or personal monitors for oxygen with audible, vibration, and visual alarms. Any alarm would require immediate evacuation.

Refer to the topic "▶ Safety Systems for Petroleum Processing" for additional discussion.

When handling cryogenic liquid nitrogen, chemical-resistant, insulated gloves, and face shield should be worn.

Large spills of liquid will require fresh air breathing apparatus.

Storage and Handling of Nitrogen

Refer to "▶ Utilities in Petroleum Processing" for a detailed discussion of nitrogen handling systems and to "▶ Safety Systems for Petroleum Processing" for a discussion of managing potentially oxygen-deficient atmospheres.

Fresh air equipment (e.g., SCBAs) should be immediately available where liquid nitrogen is handled and in areas where oxygen deficiency may occur.

Sulfiding Chemicals

General Discussion of Sulfiding Chemicals

Most hydroprocessing catalysts are not very active as received. The most active forms of these materials are the sulfides or the catalytic metals. To create the active forms, the catalysts are activated or "sulfided" using H_2S or one of several other sulfiding agents in a circulating hydrogen stream. Today, H_2S is not commonly used.

At one time carbon disulfide (CS_2) would have been included here; but it is rarely used today because of the extreme fire hazard it represents.

In another application, sulfiding chemicals are used to temporarily suppress cracking activity in naphtha reformers during startup.

Aside from H_2S, the most common sulfiding agents include:

- Dimethyl sulfide (DMS)
- Dimethyl disulfide (DMDS)
- Di-tertiary butyl polysulfide (TBPS)
- Di-tertiary nonyl polysulfide (TNPS)

There are proprietary versions of some of these chemicals, such as Sulfrzol® 54, but the hazards are similar.

These chemicals present a range of hazards from relatively low hazard to suspected carcinogens. We will explore the hazards and their management here.

Chemical and Physical Properties of Sulfiding Agents

Table 21 provides some of the key, relevant properties for various sulfiding agents.

Most of these chemicals have an unpleasant smell. Often the smell is associated with residual traces of the mercaptans left over from manufacturing.

Note that the polysulfides decompose on heating before they boil. The decomposition products include mercaptans and H_2S. Under some conditions, the polysulfides may form solid deposits and plug lines, especially if boiled at high temperatures without the presence of other liquid hydrocarbons.

Some of these compounds are excellent solvents for many common materials, especially shoes.

Hazards of Sulfiding Agents

The odors of these materials provide an early warning of potential exposure. Spills of the heavier liquids, however, just look like water.

Note that some people react much more strongly to even the slightest odor of some of these compounds. Reactions may include nausea and vomiting.

All the compounds present a fire hazard, in addition to health hazards. DMS is especially volatile and tends to load up the treat gas circulation with methane. It is not used much anymore. DMDS, TBPS, and TNPS (or their branded counterparts) are most common.

Table 22 summarizes the key hazards and exposure limits for these materials.

All of these materials are flammable. They are all toxic to fish and long lasting in an aquatic environment, so they should not be allowed into storm drains or effluents without treatment.

The products of combustion will include CO, SO_2, and H_2S if the materials burn.

Hazardous Materials in Petroleum Processing

Table 21 Some key properties of sulfiding agents

Chemical	Dimethyl sulfide (DMS)	Dimethyl disulfide (DMDS)	Di-t-butyl polysulfide (TBPS)	Di-t-nonyl polysulfide (TNPS)
Formula	$(CH_3)_2S$	$(CH_3)_2S_2$	$C_8H_{18}S_x$ (x = avg ~2.8)	$C_{18}H_{38}S_x$ (x = avg 5.0)
Molecular weight	62.13	94.2	~204 avg	414 avg
Color (liq)	Clear	Pale yellow	Yellow	Yellow to yellow-orange
Odor	Stench	Strong garlic-like odor	Slightly acrid	Mildly unpleasant
Odor threshold	<1 ppm	~8–10 ppb		
Sulfur, wt%	51.5	67.9	~44 avg	35–39
Specific gravity	Gas: 2.1 (air=1) Liq: 0.85 (68 °F, 20 °C)	Gas: 3.25 (air=1) Liq: 1.06 (39 °F, 4 °C)	Liq: 1.07 (68 °F, 20 °C)	Liq: 1.05 (68 °F, 20 °C)
Boiling point	99 °F (38 °C)	228 °F (109 °C)	342–356 °F (172–180 °C) (decomposes)	407–507 °F (208–264 °C) (decomposes)
Freezing point	−145 °F (−98.3 °C)	−121 °F (−85 °C)	37 °F (3 °C)	<−4 °F (<−20 °C)
Flash point	−54 °F (−48 °C)	59 °F (15 °C)	217 °F (103 °C) CC	277–291 °F (136–144 °C) PMCC
Solubility in water	2.0 wt% (68 °F/20 °C)	~0.25 wt% (68 °F/20 °C)	Insoluble	Insoluble
Autoignition temperature	403 °F (206 °C)	Not available	437 °F (225 °C)	464 °F (240 °C)
Explosive limits	2.2–19.7 vol%	1.1–16 vol%	Not available	Not available
Viscosity (liq)	0.29 cP (68 °F, 20 °C)	0.62 cP (68 °F, 20 °C)	10 cP (68 °F, 20 °C)	24–34 cP (122 °F, 50 °C)

Protective Equipment for Sulfiding Chemicals Handling

In addition to the normal petroleum processing facility PPE (hardhat, flame-retardant clothing, safety glasses, safety shoes), the recommended personal protective equipment when handling these materials includes:

- Eye/face protection: Goggles, face shield. Eye wash station should be available nearby.

Table 22 Hazards and exposure limits of common sulfiding agents

Chemical	Dimethyl sulfide (DMS)	Dimethyl disulfide (DMDS)	Di-t-butyl polysulfide (TBPS)	Di-t-nonyl polysulfide (TNPS)
Exposure effects				
Inhalation	Headache, memory loss, confusion, convulsions, unconsciousness; ACGIH TLV = 10 ppm (8 h)	Nausea, headache, or dizziness, drowsiness, unconsciousness; ACGIH limit 0.5 ppm TWA	Unlikely route of exposure	Unlikely route of exposure
Eyes	Irritation, inflammation	Irritation, may be irreversible	Almost no irritation	Slight irritation
Skin	Stinging, reddening, removal of skin oils	Irritation, redness, rash, removal of oils, may also see central nervous system effects like inhalation	Allergic reaction, redness, rash	Slight irritation
Ingestion	Irritation of the mouth, throat, stomach; nausea, vomiting	Irritation of the mouth, throat, stomach; nausea, vomiting	Aspiration hazard to lungs, increased breathing and heart rate, coughing, respiratory distress	Slightly toxic
Chronic	None identified	Sensitizer	Sensitizer	No effects expected
Preexisting conditions	Aggravates respiratory diseases	No information	No information	No information
Carcinogenic	Not listed	Suspected	No information	No information
Toxicity				
Inhalation	LD50 = 40,250 ppm (102 mg/l, rat)	Slightly toxic, LC50 805 ppm (rat, 4 h)	Not available	LC50 >50 mg/l (4 h, rat)
Ingestion	LD50 = 3,700 mg/kg (mouse)	Moderately toxic, LD50 = 290–500 mg/kg (rats)	Slightly toxic, LD50 >2,000 mg/kg (rat)	Slightly toxic, LD50 = 19,550 mg/kg (rat)

Dermal	LD50 = 10,200 mg/kg (rat)	Slightly toxic, LD50 >2,000 mg/kg	Slightly toxic, LD50 >2,000 mg/kg (rat)	Slightly toxic, LD50 >2,000 mg/kg (rabbit)
Skin irritation	Mild (rabbit), LD50 >5 g/kg	Slight (rabbit)	Slight (rabbit)	Slight
Eye irritation	Severe (rabbit)	May cause irreversible eye damage	Slight (rabbit)	Slight
Fire				
Flammability	High, vapors may ignite from remote sources, possible explosion	High, vapors may ignite from remote sources, possible explosion	Moderate	Moderate
Hazardous combustion products	Sulfur dioxide, H$_2$S, CO	Sulfur dioxide, H$_2$S, CO	Sulfur dioxide, H$_2$S, CO	Sulfur dioxide, H$_2$S, CO
Hazard ratings (health/flammability/reactivity)				
HMIS	2/4/0	3/3/1	1/2/0	Not available
NFPA	1/4/0	2/4/0	2/1/0	0/1/0

References: Various MSDSs and product data sheets from manufacturers and distributors. These are the best data available, but there are conflicts among some of the data sources on effects, exposures, and ratings

- Respiration: if strong vapor or mist present, use appropriate organic vapor cartridge respirator, SCBA, or supplied-air breathing apparatus. For vessel entry or high levels of potential exposure, use SCBA or supplied air only.
- Skin/clothing: Butyl or nitrile rubber gloves (not leather). In some cases, aprons, rubber boots, or a rubber suit may be needed. A safety shower should be available in the immediate area.
- Other: Do not smoke in areas where these chemicals may be present. Wash hands and other potentially exposed areas of skin thoroughly with soap and water before eating, smoking, or using toilet facilities.

In the event of a fire involving the chemicals, wear SCBA, protective clothing (which may include a rubber suit), and gloves.

Have bleach (2–5 %) or LAB solution available in a sprayer to manage odors from small spills. See below.

Storage and Handling of Sulfiding Chemicals

These chemicals can be handled like other hydrocarbons of similar vapor pressure. They are stable under normal conditions, but the vapors can ignite readily.

Carbon and stainless steels are suitable for vessels and piping. When making up threaded piping, normal pipe compound is not suitable. Makeup temporary threaded piping with Teflon® (Rect-R-Seal™ or equal). Pressure test all temporary piping before allowing any sulfiding chemical into the lines. It is a big, stinky mess if the lines leak once they contain the sulfiding chemicals.

The sulfiding chemicals are normally brought into a facility by truck or iso-container. Some may also be handled in barrels or tote-bins, depending on the quantity required. Small quantities may be kept onsite for activities like naphtha reformer startup, but large quantities are seldom stored onsite. They are normally used directly from the tank truck or iso-container.

The trucks are staffed by experienced and trained personnel (supplied by the chemical vendor or trucking company) around the clock until the sulfiding activity is completed. Sulfiding chemicals are normally pressured out of the trucks or iso-containers by nitrogen through metering and an injection pump into the process.

In the event of small spills, spray the chemical with a *dilute* solution (~2–5 %) of bleach to neutralize odors. This converts the chemicals to dimethyl sulfoxide, which may be treated in a wastewater treatment plant. "Liquid Alive Bacteria (LAB)" solution is also effective when used in accordance with the manufacturer's instructions. Concentrated bleach or solid bleach should *not* be used as it may react violently or cause a fire.

Absorb small spills in a sorbent. Collect the sorbent in a metal container and dispose of as a hazardous waste.

In the event of a large spill or fire:

- Get people out of the area.
- Contain the spill if possible. Avoid disposal or runoff. Runoff will be an ecological problem. Get your environmental department involved.

- Eliminate all ignition sources – remember the vapors of these chemicals are heavy and can follow low spots or sleeperways for a long distance to an ignition source.
- Wear impervious gloves, skin protection (including a rubber suit).
- If the material ignites or explodes:
  - Combustion and thermal decomposition products will be toxic (CO, SO_2, H_2S), so full-face, positive-pressure fresh air equipment is required.
  - Extinguishing media: dry chemicals, CO_2, alcohol-resistant foam, and other foams. Water spray may be used also to cool and protect equipment, but contain runoff. High volume water jets are *not* suitable.
  - Keep surrounding equipment cool with water spray.

These materials are routinely handled in petroleum facilities without incident. Their odor reminds people to use care in handling them.

Sulfuric Acid (H_2SO_4)

General Discussion of Sulfuric Acid

Sulfuric acid is found in several locations in refineries or other processing facilities. It is used for pH control in cooling towers and other services. It is the primary catalyst in some alkylation processes. Some refineries even have sulfuric acid plants associated with them to dispose of sulfur removed by other processes.

Sulfuric acid is safely handled in bulk regularly in facilities, but it can be very hazardous if not handled properly. The most common form of H_2SO_4 is concentrated acid. Dilute acid is corrosive to steel, whereas concentrated acid is not very corrosive to steel at most ambient conditions. Hence, we like to handle the acid as a concentrate.

Properties of Concentrated Sulfuric Acid

Table 23 presents the relevant properties of concentrated acid, the most common used in refining. More dilute grades may be encountered.

Hazards of Sulfuric Acid

Sulfuric acid presents a number of hazards. Table 24 summarizes the exposure limits and toxicity data.

Health effects described by exposure route include:

- Inhalation: May cause severe irritation of respiratory tract with sore throat, coughing, shortness of breath, and delayed lung edema. Causes chemical burns to respiratory tract. May cause inflammation. Destructive to mucous membranes. May cause headache, vomiting, nausea, pulmonary edema. Corrosive and toxic. May be fatal.
- Eyes: Severe eye burns. May be irreversible.

Table 23 Some key properties of concentrated sulfuric acid

Chemical formula	H_2SO_4
Molecular weight	98.079
Appearance	Clear, colorless, odorless liquid
Purity, %	98
pH, 1N solution	0.3
Specific gravity	1.841 (~1.84 kg/l)
Boiling point	554 °F (290 °C)
Vapor pressure	<0.00005 in Hg @ 68 °F (<0.0012 mmHg @ 20 °C)
Freezing point	50.6 °F (10.3 °C)
Vapor density	1.2 kg/m^3
Decomposition temperature	644 °F (340 °C)
Viscosity	26.7 cP (68 °F, 20 °C)

Table 24 Exposure and toxic effects of concentrated sulfuric acid

Agency	Limit type	mg/m^3	Basis
US OSHA	Limit	1	8 h TWA
US NIOSH	REL	1	10 h TWA
ACGIH (USA)	TLV	1	10 h TWA
	STEL	3	
Toxicity			
Inhalation	LC50 (rat)	51 mg/m^3	2 h
Oral	LD50 (rat)	2140 mg/kg	
Ecology	Harmful to aquatic life in very low concentrations		
Workers exposed to sulfuric acid mist showed increase in laryngeal, nasal, sinus, and lung cancer			

- Skin: Skin burns. Defatting dermatitis with prolonged exposure. May be fatal.
- Ingestion: May cause severe and permanent damage to digestive tract, burns in mouth, pharynx, and gastrointestinal tract. May cause nausea, vomiting, and abdominal pain. Corrosive and toxic. May be fatal.
- Chronic exposure: Nosebleeds, nasal congestion, erosion of teeth, perforation of nasal septum, chest pain, and bronchitis. Eye exposure may cause conjunctivitis. May cause death. Corrosive to body tissues.

PPE for Handling Sulfuric Acid

The following minimum PPE are recommended beyond the typical refinery worker's PPE when handling sulfuric acid:

- Eye protection: Splash-proof chemical safety goggles and face shield.
- Skin protection: Neoprene or polyethylene gloves.
- Clothing: Apron or other impervious clothing (e.g., rain coat), rubber boots, other clothing to prevent skin contact.

- Respiratory protection: Use a US NIOSH-approved (or equal) respirator when necessary.
- Ventilation: Use only with adequate ventilation. In a lab, hand the acid in a fume hood.
- Other: Eye wash and safety shower should be immediately available in any area where sulfuric acid is handled.

Storage and Handling of Sulfuric Acid

For concentrated sulfuric acid, carbon steel tanks and piping are normally adequate. Temperature must be kept low in carbon steel to prevent corrosion. If higher temperatures or dilute acid are handled, higher alloys are required. Consult with a metallurgist on the proper alloys for your application.

The acid is normally stored in atmospheric vessel vented through a desiccant cartridge to prevent moisture from the air entering the tank. Sulfuric acid is a desiccant in its own right, so avoiding any moisture is critical. The storage tanks are typically within chemical containment dikes (epoxy-coated concrete). The dike drain is managed closed to ensure any leaks are not sent to the sewer uncontrolled.

There should always be an eyewash and safety shower immediately adjacent to the acid storage area.

Acid is generally unloaded and moved by nitrogen or air pressure among users. The final users will generally use pumps for controlled injection.

Sulfuric acid will not burn but can decompose to yield poisonous sulfur oxide vapors if the storage area becomes involved in a fire. Positive-pressure, supplied-air respiratory equipment is required is fighting a fire near an acid storage tank. Use water spray on the outside of the tank to keep it cool.

Do not allow water to be pumped into an acid storage tank. The heat of solution will cause foaming and possible tank failure.

Safe handling of concentrated sulfuric acid is common in a refinery, in spite of the hazards. The key points to remember are:

- Keep the acid contained and cool.
- Do not allow any personnel contact with the acid.
- Keep water away from the concentrated acid until you are ready for use, then add acid into a large excess of water.

Hazardous Materials Handling Summary

Table 25 pulls together some of the most important information from the foregoing sections of this chapter for convenience. For more details, consult the individual sections and the MSDS sheets from your supplier or company for the specific materials of interest.

Table 25 Summary of key hazardous materials handling information[a]

Material	Category							Ammonia		
	Amines for gas treating						Sulfinol	Anhydrous	26 1Bé	20.5 Bé
	Monoethanolamine	Diethanolamine	Methyldiethanolamine	Diglycolamine	Diisopropanolamine					
Abbreviation/alias	MEA	DEA	MDEA	DGA	DIPA				Aqua ammonia	Aqua ammonia
Chemical formula	HOC$_2$H$_4$NH$_2$	(HOC$_2$H$_4$)$_2$NH	(HOC$_2$H$_4$)$_2$NCH$_3$	H(OC$_2$H$_4$)$_2$NH$_2$	(HOC$_3$H$_6$)$_2$NH		Mixture	NH$_3$	NH$_3$	NH$_3$
Molecular weight	61.1	105.1	119.16	105.14	133.19		120.17	17.0305	–	–
Physical states	Pure liquid, as received	Pure liquid, as received	Pure liquid, as received	Pure liquid, as received	Pure liquid, as received		Aq. solution	Liq, vap	Aq. liq, vap	Aq. liq, vap
Color	Colorless	Colorless to yellow	Colorless to Lt. yellow	Colorless	Colorless		Colorless	Colorless	Colorless	Colorless
Odor	Ammoniacal	Ammoniacal	Amine-like	Mild amine	Ammonia-like		Fishy hydrocarbon	Strong acrid	Strong acrid	Strong acrid
Vapor Sp. Gr. (air = 1)	2.1	3.6	4.1	3.6	4.6		4.6	0.59	0.59	0.59
Liquid Sp. Gr. (nominal)	1.01	1.09	1.04	1.06	0.99		1.26	0.68	0.90	0.93
Boiling point, °F/°C	338.5/170	515.1/268	477/247	405.5/208	479.7/249		545/285	−28/−33	85/29	123/51
Freeze point, °F/°C	50.5/10.2	77.2/25.1	−9/−23	9.5/−12.5	107.6/42		81.7/27.6	−108/−78	−110/−79	−30/−34
Flash point, °F/°C	200/93.3	295/146	265/129	260/127	255/124		350/177	Not available	Not available	Not available
Autoignition, °F/°C	Not available	Not available	509/265	698/370	545/285		Not Available	1,204/651	See anhydrous	See anhydrous
Decomposition, °F/°C	Not available	Not available	Not available	Not available	Not available		Not available	Not available	Not available	Not available

Hazardous Materials in Petroleum Processing

Flammability limits (air)	Not available	Not available	0.9–8.4	2.6–11.7	1.1–8.5	Not available	15–28 %	See anhydrous	See anhydrous
Exposure limits (always verify current values)	TLV=3 ppm STEL=6 ppm	TLV=2 mg/m³	Not available	Not available	US PEL=5 ppm, UK STEL=15 ppm	Not available	TLV= 25 ppm STEL= 35 ppm IDLH = 300 ppm	See anhydrous	See anhydrous
Primary hazard routes	Skin, inhalation	Skin, inhalation	Skin, inhalation	Skin, inhalation	Skin, inhalation	Skin, inhalation	Inhalation, skin, eyes	Inhalation, skin, eyes	Inhalation, skin, eyes
Exposure type									
Fire/explosion	Remote	Remote	Remote	Remote	Remote	Remote	X	Unlikely	Unlikely
Toxicity	X	X	X	X	X	X	X	X	X
Ecological impacts									
Aqueous env.	Toxic	Medium toxic	Slightly toxic	Slightly toxic	Not Available	Not Available	Highly toxic	Moderately toxic	Moderately toxic
Biodegradable	Yes	Yes	Yes	Yes	Not available	Not available	Yes	Yes	Yes
Firefighting media	Water spray/fog, foam, dry chemical, CO_2	Water spray/fog, foam, dry chemical, CO_2	Water spray/fog, foam, dry chemical, CO_2	Water spray/ fog, foam, dry chemical, CO_2	Water spray/fog, foam, dry chemical, CO_2	Water spray/ fog, alc. foam, dry chemical, CO_2	Water spray/ fog, dry chemical, CO_2	Water spray/ fog, dry chemical, CO_2	Water spray/ fog, dry chemical, CO_2
Possible added PPE[1]									
Handling area (normal)	Goggles, face shield, impermeable clothing (gloves, coveralls, boots)	Goggles, face shield, impermeable clothing (gloves, coveralls, boots)	Goggles, face shield, impermeable clothing (gloves, coveralls, boots)	Goggles, face shield, impermeable clothing (gloves, coveralls, boots)	Goggles, face shield, impermeable clothing (gloves, coveralls, boots)	Goggles, face shield, impermeable clothing (gloves, coveralls, boots)	Goggles, face shield, impermeable clothing (gloves, coveralls, boots), vapor respirator or supplied air	Goggles, face shield, impermeable clothing (gloves, coveralls, boots), vapor respirator or supplied air	Goggles, face shield, impermeable clothing (gloves, coveralls, boots), vapor respirator or supplied air
Fire or large spill (additional)	SCBA or =, eye protection, protective clothes	SCBA or =, eye protection, protective clothes	SCBA or =, eye protection, protective clothes	SCBA or =, eye protection, protective clothes	SCBA or =, eye protection, protective clothes	SCBA or =, eye protection, protective clothes	SCBA or =, eye protection, protective clothes	SCBA or =, eye protection, protective clothes	SCBA or =, eye protection, protective clothes

(*continued*)

Table 25 (continued)

a

Category	Amines for gas treating						Ammonia		
Material	Monoethanolamine	Diethanolamine	Methyldiethanolamine	Diglycolamine	Diisopropanolamine	Sulfinol	Anhydrous	26 °Bé	20.5 °Bé
Abbreviation/alias	MEA	DEA	MDEA	DGA	DIPA	–	–	Aqua ammonia	Aqua ammonia
Hazard ratings (health – flammability – reactivity)	NFPA 3-1-0	NFPA 2-1-0	HIMS 1-1-0	HIMS 3-1-0	NFPA 2-1-0	NFPA expect 2-10 based on ingredients	NFPA 3-1-0	NFPA 3-1-0	NFPA 3-1-0
References MSDS SOURCES	Dow	Dow	Acros Organics, Dow, Huntsman	Huntsman, Univar	BASF, Hazel Mercantile	Burlington Resources, Shell	Tanner Industries, Air Products	Tanner Industries, Air Products	Tanner Industries, Air Products
ARTICLES; BULLETINS; OTHER				Huntsman			Tanner Industries, Wikipedia	Tanner Industries, Wikipedia	Tanner Industries, Wikipedia

b

Category	Catalysts and sorbents								
Material	Benzene	Carbon monoxide	Hydro processing	Hydrogen plant	PSA adsorbents	Mole sieves, zeolites	Caustic soda (50%)	Furfural	Hydrofluoric acid
Abbreviation/ alias	Bz	–	K172	–	–	–	–	–	AHF, HF
Chemical formula	C_6H_6	CO	Co, Ni, Mo, sulfides and oxides; alumina; silicates; coke; with contam	Ni and NiO on ceramic support	Molecular sieves, activated carbon, activated alumina	Molecular sieves, zeolites	NaOH	C_4H_3OCHO	HF
Molecular weight	78.11	28.01	N/A	N/A	N/A	N/A	40.00	96	20.01

Hazardous Materials in Petroleum Processing

Physical states	Liq, vap	Vap	Solid to oily solid	Solid	Solid	Solid	Aq, liq, solid	Liq, vap	Liq, Vap
Color	Colorless	Colorless	Varies	Gray	White to black depends on product	White to tan	Colorless to white	Lt. yellow to brown	Colorless
Odor	Sweet aromatic	Odorless	Hydrocarbon	None	None	None	None	Almond	
Vapor Sp. Gr. (air = 1)	2.7	0.97	N/A	N/A	N/A	N/A	N/A	3.3	0.69
Liquid Sp. Gr. (nominal)	0.88	–	N/A	N/A	N/A	N/A	1.53	1.16	1.00
Boiling point, °F/°C	176/80	−312/−191	N/A	N/A	N/A	N/A	284–140	324/162	67/19
Freeze point, °F/°C	42/6	−337/−205	N/A	N/A	N/A	N/A	54/12	−38/−39	−117/−83
Flash point, °F/°C	12/−11	Not available	Varies	N/A	N/A	N/A	N/A	140/60	N/A
Autoignition, °F/°C	1097/592	1121/605	Not available	N/A	N/A	N/A	N/A	601/316	N/A
Decomposition, °F/°C	Not available	N/A	Not available (may liberate CO, H_2S, other gases)	Not available (may liberate CO, $Ni(CO)_4$, other gases)	Not available (may liberate CO or other gases if wet)	Not available (may liberate CO, H_2S, or other gases if wet)	N/A	Not available – unstable in light and air	Not available – decomposition products include halogens
Flammability limits (air)	1.3–7.1	12.5–74.2	N/A	N/A	N/A	N/A	N/A	2.1–19.3	N/A
Exposure limits (always verify current values)	TLV=0.5–1 ppm STEL=2.5–5 ppm	TLV=25 ppm STEL=200 ppm	N/A	Refer to nickel hazards	Refer to silica hazards	Refer to silica hazards	2 mg/m³ mist or dust	TLV=2 ppm STEL=5 ppm	TLV=0.5 ppm skin, STEL=6 ppm
Primary hazard routes	Inhalation, skin	Inhalation	Inhalation (dust, vapors), skin	Inhalation (dust, vapors)	Inhalation (dust, vapors)	Inhalation (dust, vapors)	Skin, eyes, inhaled mist or dust	Inhalation, skin	Inhalation, skin
Exposure type									
Fire/explosion	X	X	Smoldering	N/A	X (act. carbon)	N/A	N/A	X	N/A

(*continued*)

Table 25 (continued)

b

Category		Catalysts and sorbents						
Material	Benzene	Carbon monoxide	Hydro processing	Hydrogen plant	PSA adsorbents	Mole sieves, zeolites	Caustic soda (50%)	Furfural	Hydrofluoric acid
Abbreviation/ alias	Bz	–	K172	–	–	–	–	–	AHF, HF
Toxicity	X	X	X	Ni dust, carcinogen	Low	Low	X	X	High
Ecological impacts									
Aqueous env.	Not major	N/A	Slightly toxic, potential groundwater contamination	Possible hazard	Not available	Not available	Toxic, avoid contamination of effluent	Toxic	Not available
Biodegradable	Yes	N/A	No	No	Not available	N/A	No	Yes	No
Firefighting media	Water spray/fog, alc. foam, dry chemical, CO_2, Halon	Water spray/ fog, foam, dry chemical	Water spray/fog, foam, dry chemical, CO_2	N/A	Water	N/A	N/A	Water spray/fog, foam, dry chemical, CO_2	N/A
Possible added PPE[1]									
Handling area (normal)	Goggles, face shield, impermeable clothing (gloves, coveralls, boots), vapor respirator or supplied air	SCBA or ≡, personal CO monitor	Dust respirator, goggles, face shield, gloves, chemical-resistant clothing	Dust respirator, goggles, gloves	Dust respirator, goggles, gloves	Dust respirator, goggles, gloves	Goggles, face shield, impermeable clothing (gloves, coveralls, boots)	Goggles, face shield, impermeable clothing (gloves, coveralls, boots), vapor respirator or supplied air	Goggles, face shield, impervious acid suit, boots and gloves, SCBA
Fire or large spill (additional)	SCBA or ≡, eye protection, protective clothes	SCBA or ≡	SCBA or ≡	N/A	SCBA or ≡	N/A	N/A	SCBA or ≡, eye protection, protective clothes	SCBA
Hazard ratings (health – flammability – reactivity)	HIMS 2-3-0, NFPA 2-3-0	HIMS 2-4-0, NFPA 3-4-0	Varies	HIMS 2-1-0, NFPA 2-1-0	Not Available	NFPA 1-0-1	NFPA 3-0-1	NFPA 3-2-0	NFPA 4-0-1 (acid)

Hazardous Materials in Petroleum Processing

References											
MSDS SOURCES	Total		Airgas	CRI/Criterion, Haldor Topsoe, Porocel, Valero	Johnson Matthey		Delta Adsorbents		Orica Chemicals	Int'l Furan Chemicals	Airgas
ARTICLES; BULLETINS; OTHER	Wikipedia			Wikipedia							Wikipedia
c											
Category						Sulfiding chemicals					
Material	Hydrogen	Hydrogen sulfide	Methyl ethyl ketone	Nickel carbonyl	Nitrogen	Dimethyl sulfide	Dimethyl disulfide	Di-t-butyl polysulfide	Di-t-nonyl polysulfide	Sulfuric acid (Conc.)
Abbreviation/alias	—	H2S	MEK	—	N_2, LN_2	DMS	DMDS	TBPS	TNPS	SA
Chemical formula	H_2	H_2S	$CH_3C(O)CH_2CH_3$	$Ni(CO)_4$	N_2	H_3CSCH_3	H_3CSSCH_3	$C_8H_{18}S_x$	$C_{18}H_{38}S_x$	H_2SO_4
Molecular weight	2.02	34.08	72	170.73	28.02	62.13	94.2	~204	~414	98.08
Physical states	Vap	Vap	Liq, Vap	Vap (Liq unlikely)	Cryogenic Liq, Vap	Liq, Vap	Liq, Vap	Liq	Liq	Liq
Color	Colorless	Colorless	Colorless	Colorless	Colorless	Clear	Pale Yellow	Yellow	Yellow	Colorless
Odor	Odorless	Rotten eggs	Sweet alcohol	Musty/wet brick	Odorless	Stench	Strong garlic	Slightly acrid	Mildly unpleasant	Odorless to slight sulfur odor
Vapor Sp. Gr. (air = 1)	0.0696	1.19	2.4	5.9	0.97	2.1	3.25	—	—	$(1.2\ kg/m^3)$
Liquid Sp. Gr. (nominal)	—	—	0.81	—	0.81	0.85	1.06	1.07	1.05	1.84
Boiling point, °F/°C	−423/−253	−77/−61	176/80	109/43	−320/−196	99/38	228/109	342−356/172−180	407−507/208−264	554/290
Freeze point, °F/°C	−435/−259	−122/−86	−123/−87	1/−17	−346/−210	−145/−98	−121/−85	37/3	≤4/≤20	50.6/10.3
Flash point, °F/°C	—	—	16/−9	4/20	N/A	−54/−48	59/15	217/103	277−291/136−144	N/A
Autoignition, °F/°C	932−1,000/500−571	500/260	759/404	140/60	N/A	403/206	Not available	437/225	464/240	N/A
Decomposition, °F/°C	N/A	N/A	Not available	~300/~149	N/A	Yes (H_2S, mercaptans)	Yes (H_2S, mercaptans)	Yes (H_2S, mercaptans)	Yes (H_2S, mercaptans,	644/340 (SO_x)

(*continued*)

Table 25 (continued)

Category	Hydrogen	Hydrogen sulfide	Methyl ethyl ketone	Nickel carbonyl	Nitrogen	Sulfiding chemicals					Sulfuric acid (Conc.)
						Dimethyl sulfide	Dimethyl disulfide	Di-t-butyl polysulfide	Di-t-nonyl polysulfide		
Material											
Abbreviation/alias	–	H2S	MEK	–	N_2, LN_2	DMS	DMDS	TBPS	TNPS		SA
Flammability limits (air)	4–76	4.3–45.5	1.4–11.4	2–34	N/A	2.2–19.7	1.1–16	elemental sulfur)	elemental sulfur)		N/A
Exposure limits (always verify current values)	N/A	TLV=1 ppm STEL=5 ppm IDLH=100 ppm	TLV=200 ppm STEL=300 ppm	TLV=1 **ppb** IDLH=3 ppm carcinogen	O_2 depletion hazard	TLV=10 ppm	TLV=0.5 ppm	Not available	Not available		TLV=1 mg/m^3 STEL=3 mg/m^3
Primary hazard routes	Inhalation	Inhalation	Inhalation, skin	Inhalation, skin (absorption)	Inhalation, skin (cryogenic liquid)	Inhalation, skin (absorption)	Inhalation, skin (absorption)	Skin	Skin		Skin, eyes, inhalation
Exposure type											
Fire/explosion	High	X	X (CO evolved)	Unlikely	N/A	X (CO, SO_2, H_2S)	X (CO, SO_2, H_2S)	X (CO, SO_2, H_2S)	X (CO, SO_2, H_2S)		N/A
Toxicity	N/A	High	Moderate	Extreme (CO, Ni)	O_2 depletion hazard	High	Moderate	Slight	Slight		High, tissue damage
Ecological impacts											
Aqueous env.	N/A	Toxic	Moderate to low toxicity	Toxic, but short lived	N/A	Moderate (evaporates so short-lived effect)	Toxic, long lived	Very toxic	Insufficient data		Harmful
Biodegradable	N/A	Yes	Not available	Not available	N/A	Evaporates	No	Not available	Insufficient data		No, but reacts
Firefighting media	Water, steam	Water spray/fog, foam, dry chemical	Water spray/fog, alcohol-resistant foam, dry chemical, CO_2	Water spray/fog, alcohol-resistant foam, dry chemical, CO_2	N/A	Alcohol foam, foam, dry chemical, CO_2	Water spray/fog, alcohol-resistant foam, dry chemical, CO_2	Water spray/fog, foam, dry chemical, CO_2	Water spray/fog, foam, dry chemical, CO_2		N/A

Hazardous Materials in Petroleum Processing

Possible added PPE[1]										
Handling area (normal)	Not normally required	Personal monitor for H_2S	Organic respirator, goggles, face shield, gloves, chemical-resistant clothing	SCBA or equal, goggles, face shield, gloves, chemical-resistant clothing	SCBA or equal, personal O_2 monitor, cryogenic gloves when handling liquid	Organic respirator, goggles, face shield, gloves, chemical-resistant clothing (including boots)	Organic respirator, goggles, face shield, gloves, chemical-resistant clothing (including boots)	Goggles, face shield, impermeable clothing (gloves, coveralls, boots)	Goggles, face shield, impermeable clothing (gloves, coveralls, boots)	Goggles, face shield, impervious clothes and gloves, SCBA or respirator may be needed
Fire or large spill (additional)	SCBA or =, do not enter leak area	SCBA or =, do not enter leak area	SCBA or =	SCBA or =	N/A	SCBA or =	SCBA or =	SCBA or =	SCBA or =	SCBA or =, acid suit
Hazard ratings (health – flammability – reactivity)	HIMS 0-4-0, NFPA 0-4-0	HIMS 4-4-0, NFPA 4-4-0	NFPA 1-3-0	NFPA 4-3-3	HIMS 0-0-0, NFPA 0-0-0 (simple asphyxiant)	NFPA 4-3-0	HIMS 1-0-0, NFPA 1-0-0	NFPA 2-1-0	HIMS 1-1-0, NFPA 1-1-0	NFPA 3-0-2 (no water)
References										
MSDS SOURCES	Airgas	Airgas	Fisher Scientific	Matheson TriGas	Air Products, Airgas	Gaylord Chemical	Airgas, Arkema, Chevron Phillips	Arkema, Benntag, Chevron Phillips	Arkema, Chevron Phillips	Seastar
ARTICLES; BULLETINS; OTHER	Wikipedia			NJ Dept. of H&SS, Wikipedia, ChevronTexaco	Wikipedia					Wikipedia

Fire Prevention and Firefighting in Petroleum Processing

David S. J. Jones and Steven A. Treese

Contents

Introduction	1416
Design Specifications	1417
Duty Specifications and Design Premises	1417
Mechanical Specifications	1417
Electrical and Instrument Specifications	1418
Piping and Layout Specifications	1418
Other Design Specifications	1418
Fire Prevention Through Equipment Design and Operation	1418
Fired Heaters	1418
Pressure Vessels	1419
Heat Exchangers and Coolers	1420
Rotating Equipment	1420
Tanks and the Tank Farm	1421
Jetty and Onshore Loading Stations	1422
Fire Main and Installed Firefighting Equipment	1423
Fire Main Description and Control	1423
Firefighting Equipment and Layout	1424
Inspection and Maintenance Program	1426
Fire Foam and Foam Systems	1426
Introduction	1426
Potential Firefighting Foam Applications	1427
Types of Firefighting Foam	1428
Environmental Impact and Toxicity	1429
Shelf Lives of Foam Concentrates	1429

David S. J. Jones: deceased.
Steven A. Treese has retired from Phillips 66.

D.S.J. Jones
Calgary, AB, Canada

S.A. Treese (✉)
Puget Sound Investments LLC, Katy, TX, USA
e-mail: streese256@aol.com

© Springer International Publishing Switzerland 2015
S.A. Treese et al. (eds.), *Handbook of Petroleum Processing*,
DOI 10.1007/978-3-319-14529-7_22

Fire Extinguishers 1429
  Types and Applications of Fire Extinguishers 1430
  Locations of Fire Extinguishers 1431
  Fire Extinguisher Inspection and Testing Program 1431
  Concluding Comments 1432
Mobile Firefighting Equipment 1433
Fire Protection and Firefighting Plans 1433
  Incident Command System (ICS) 1434
  Notification Systems and Procedures 1434
Conclusion 1435
References 1436

Abstract

Because refineries handle large quantities of highly combustible materials, fire and explosion are constant risks that must be managed. Refineries are designed to minimize the risks through proper specifications and fire prevention measures by design. Should a fire occur, however, a facility has extensive firefighting capabilities, including fixed and portable firefighting equipment, extinguishers, and an Incident Command System (ICS). These systems help in response to other incidents, such as releases and medical emergencies as well. This chapter addresses these issues, from design through operations.

Keywords

Refinery • Firefighting • Fire • Fireproofing • Extinguisher • Fire foam • Incident Command System • ICS • Prevention

Introduction

Fire prevention and protection are of paramount importance in the operation of any hydrocarbon facility. These are more important, perhaps, in oil refining than any other related facility because of the relative size of most refineries compared with petrochemical or chemical facilities. Refineries process and store huge quantities of combustible materials.

Refinery fire prevention and protection begin at the early stages of the refinery design and engineering. This chapter begins with the petroleum processing company's development of the design and engineering specifications. These instruct the design, equipment, and construction contractors in the details of the standards that are to be implemented in the building of the facility. These specifications are critical in ensuring the facility is well constructed in the first place, to minimize the risks of release, accompanied by fire or explosion.

The discussion begins with a list of the types of items usually contained in the design specifications and continues with more details of those items to illustrate

Fire Prevention and Firefighting in Petroleum Processing

how they pertain to fire prevention and firefighting. We will then discuss additional considerations in facility equipment, layout, operation, and management of the firefighting systems.

Refer also to the chapters on "▶ Safety Systems for Petroleum Processing" and "▶ Utilities in Petroleum Processing" for additional discussion of fire protection systems and emergency response.

Design Specifications

Companies each have their own flavor of design specifications in terms of format and details for the various types of equipment. Each company has preferences about how they want things done. The design specifications define all these items for the contractors.

Most design specifications will contain the following subject as a minimum.

Duty Specifications and Design Premises

These are of interest to the design process and mechanical engineers. These specifications state exactly what facilities are to be built and the duty of each item in terms of throughput and in some cases composition of the various streams. They will also give specific details of the local meteorological data and the parameters to be considered in economic decisions.

The duty specifications will include the products required and the composition of the products. They will also give full details of utilities required or available for the process. In cases of "grassroots" facilities, the duty specification will cover off-sites such as tankage, blending, loading, unloading facilities, etc.

Mechanical Specifications

These specifications define with the requirements and standards (including the codes to be used) the client wants for the equipment that will be installed. The resulting narrative specifications that the mechanical engineer will produce based on this will form part of the package to be used for procuring these items of equipment.

Included in the equipment defined by the mechanical specifications are:

- Pumps
- Compressors and turbines
- Heat exchangers (including air coolers, double pipe, etc.)
- Fired heaters
- Other miscellaneous equipment such as ejectors, blenders, and the like

Electrical and Instrument Specifications

These documents set the standards to be used for all electrical equipment and the bulk materials associated with the equipment. They will deal with the "Area Classification Codes." These set the parameters for equipment in terms of fire proofing (e.g., whether the item is to be spark proof) that will be located in the various areas of the refinery.

For instrumentation they specify the types of instruments to be used and the basic control and measuring systems to be used. The depth of the instrument standards and the resulting specifications can be extensive.

Piping and Layout Specifications

These are usually the largest section of the design specifications. They will detail the piping codes to be used and the material break points. They will proceed to establish the criteria for equipment and tankage layout, including:

- Maintenance access
- Fire prevention (e.g., distance of fired heaters from other equipment, fire mains, hydrant locations, monitor locations, etc.)
- Tank area layout and size of tank bunds
- Extent of fire protection piping for plant units and tank farm
- Underground piping corrosion protection

Other Design Specifications

These include detailed requirements for vessels, civil, and structural equipment and associated materials. They also include requirements for equipment spacing, especially as it relates to fires.

Fire Prevention Through Equipment Design and Operation

Fired Heaters

Design of fired heaters must incorporate snuffing steam facilities for the fire box for most styles of heater. Exceptions would normally be made for down-fired heaters.

With respect to the maintenance and operation of the heaters, the following points should be included in the operating procedures:

- Implementation of a formal, regularly scheduled preventive maintenance program for burners and the cleanup of any refractory debris to prevent flame impingement.

- Introduction of a mechanism to monitor and record, more accurately and thoroughly, tube skin temperature to prevent hot spots.
- Implementation of a thorough and complete procedure to deal with hot spots once detected and reported. This includes the recording of descriptions and locations of any suspected hot spot detected in a visual inspection.
- Introduction of an accurate method to determine the impact on tube life once a hot spot is detected.
- Training of the refinery personnel to recognize and respond to the changes in metallurgical properties and characteristics demonstrated by a different material in the furnaces that are subjected to high temperature and pressure, with documentation updated to reflect this information.

Pressure Vessels

These include towers, horizontal and vertical process vessels, reactors, etc. All of these must be protected properly with pressure-relieving devices (see the chapters/topics in this Handbook entitled "▶ Off-Site Facilities for Petroleum Processing" and "▶ Safety Systems for Petroleum Processing" for discussion of pressure relief facilities).

In the design of vessels, the correct wall thickness is based on the API and ASME codes, covering the design temperatures and pressures. Fireproofing of vertical vessel skirts and horizontal vessel supports must be specified. In the operation and maintenance of pressure vessels, the following items must be included as standard procedures:

- Vessel alarms must be recognized and acted upon.
- Action must be taken immediately when excessively high temperatures and/or high or low liquid levels are observed.
- In the case of a fire in the unit or an adjacent unit, the towers should be shut down according to the plant's emergency procedures.
- In the event of an emergency shutdown, a vessel may need to be purged free of hydrocarbons either by steam or inert gas. Be careful to avoid creating another problem during this procedure.
- No entry or work must be done on a vessel before it is certified as "gas free" and isolated.
- All vessels must be inspected before recommissioning after a shutdown. The inspection must ensure that all flanges are secure and the correct gaskets have been properly installed. Ensure all the internals are in place and correctly installed.
- Instruments and relief systems on the vessels must be checked and recalibrated as necessary on every scheduled or unscheduled shutdown of the plant. Safety instrumented or integrity systems (SISs) should be thoroughly tested and serviced to maintain their critical functions.
- Relief devices are usually serviced, tested, and reset during each turnaround.

- Vessels must be pressure tested before commissioning and after any work has been completed on them which could affect the vessel pressure containment (e.g., welding on the wall).
- All drains and vents on a vessel must be checked during scheduled plant turnaround or unscheduled shutdown before recommissioning the vessel to ensure they are open. And, they must all be blinded or plugged after the equipment is closed back up and placed in operation.

Heat Exchangers and Coolers

Shell and tube heat exchangers are designed and fabricated to ASME and TEMA codes. Their shell design follows closely to that of vessels with respect to design temperatures and pressures. The thickness of tubes and tube sheets are also calculated using the ASME code. In the case of air coolers, the tube sheets and tubes are also calculated to ASME codes as are the inlet and outlet manifolds. Air cooler tubes are usually finned to enhance the heat exchanger properties of the units.

The operation and maintenance of shell and tube exchangers parallel those conditions stated for vessels with respect to fire (or explosion) prevention. In the case of air coolers, however, some different rules and procedures apply. For example, most air coolers in a refinery are installed above an elevated pipe rack. The pipe rack often runs through the center of a process unit, which means there will be equipment on both sides of the rack. It is common for the pipe rack structure to be encased in a fireproof concrete (i.e., fireproofed) to prevent collapse of the rack in a pool fire. This fireproofing should extend upwards to cover the air cooler structures as well, since the fans will pull a flame up into themselves during a fire. Firewater and/or foam sprays should be considered above the tube bundles to snuff out any outbreak of tube leaks and subsequent fires.

Rotating Equipment

This group of equipment includes all pumps and compressors.

Near fired heaters, the compressor units, which may include gas-driven turbine or turbo expanders, are the items most vulnerable to fire hazards. The reason in this case is the high pressures of the gas they handle. This is more so in those units handling hydrogen (e.g., hydrotreaters, hydrocrackers, isomerization units, reformers, etc.).

All reciprocating compressors must have pressure relief facilities on the compressor discharge. Most also have an emergency shutdown system in the case of high discharge pressure, among other conditions.

Centrifugal compressors, and their drivers, require more sophisticated emergency shutdown procedures, however. Here, critical conditions can arise due to poor suction conditions, as well as high-pressure discharge. All centrifugal compressors therefore require protection against "surge" (or "pumping") conditions.

As an example, this condition can occur when the compressor suction pressure is so low that the compressor cannot pull any of the feed gas into its impellers. Severe vibration of the unit follows. In extreme cases, the vibration can cause considerable structural damage to the compressor and even loss of containment resulting in an explosion and fire. Most modern centrifugal compressors have anti-surge systems, which, if properly maintained, will shut the compressor down long before any severe damage. These systems may be SISs.

Pumps, even those handling high-pressure liquefied petroleum gases, are a little less hazardous than compressors, but seal leaks remain a concern. Overheating due to defective seals or packing is the more common hazard that can cause a fire around a pump. Pump cavitation due to NPSH problems presents a second hazard that can cause a loss of containment.

In most present-day refineries, rotary equipment is installed in the open or in an area which has only a roof as a cover. This helps a great deal to fight any rotary equipment fire. Usually the area containing rotary equipment will have both firewater and foam facilities installed. Close monitoring of the rotary equipment during operation and regular maintenance of the items are essential for the prevention of fires from these sources.

Tanks and the Tank Farm

Tank fires are probably some of the worst types of fires in a refinery. This is because tanks hold a high inventory of hydrocarbons. The one thing that lowers the degree of hazard in the tank farm is that there is usually no direct source of ignition. There are no fired heaters or large compressors present in tank farms as a rule. Fires can occur, however, from accidents remote from the area such as a fire on a jetty or a ship nearby. There was a tank farm fire in New York during February 2003 that was caused by an explosion aboard a barge unloading gasoline at a nearby jetty. Large fuel oil tanks on shore were set ablaze by hot debris from the explosion. An explosion and fire in a process unit could have a similar impact if debris landed in the tank farm.

Another source of fire is an explosion in a product loading station while loading a light product such as gasoline. These explosions are usually caused by static electricity. Most petroleum companies have strict procedures in place to prevent such occurrences, but accidents do happen from time to time. Constant attention to their prevention is warranted.

Middle distillates and fuel oil tanks present the most difficult fires to fight and extinguish. The large inventory of these tanks and their low volatility, but high heating value, can present problems. It may take a long time to extinguish a fire in these services, even using the most up-to-date foam extinguishing techniques, once the fire has really established itself. In a large fire, the major effort is often directed at keeping adjacent tanks cool using an extensive water spray.

Fires on tanks storing lighter liquid products are a little easier to combat. Foam can be used to smother the oil inventory in these tanks. As these light products are stored in floating-roof tanks, it provides the means to build a depth of foam on the roof itself or directly on the surface of the liquid if the roof has been damaged by an earlier explosion.

Firewater sprays are installed on LPG spheres and bullets. In almost every case, the initial fire in LPG tankage causes an explosion. The material remaining can be snuffed out by foam. Again, the main focus of the firefighting effort is to activate firewater sprays on adjacent tanks to keep them cool.

Refineries today have firewater and foam spray rings on storage tanks (see the topic "▶ Off-Site Facilities for Petroleum Processing"). Water on its own should only be used to keep the tank cool from adjacent fire sources. The danger of using water alone to extinguish a fire is that it may cause the fire to spread by the hydrocarbon continuing to burn on the surface of the water stream flowing from the tank area.

In a fire, as noted above, it may be safest to let the fire burn itself out, with cooling of the nearby tanks to keep the fire from getting larger. In these cases, and if it can be done safely, as much of the contents of the burning tank(s) may be transferred to other tanks to minimize the amount burned. This is not always feasible.

The design of storage tanks and their installation must comply with the local fire regulations. In general, crude oil and petroleum products with a flash point of <130 °F (54 °C) may be stored in floating-roof tanks or, in the case of LPG, in pressurized bullets or spheres. Middle distillates and heavier are usually stored in cone roof tanks. Storage tanks, both floating roof and cone roof, are normally installed on sand or concrete pads (which will be piled if required). The space between tanks should not be less than 10 ft. (or according to local regulations). The tank area should be diked to hold at least 110 % of the largest single tank inventory. This diked or bunded area should contain adequate drainage leading to the API separator or holding pond, with the capability to block in the drain if necessary. LPG storage generally does not require a diked area; but both bullets and spheres should be installed with proper fireproofed support structures.

Jetty and Onshore Loading Stations

One of the biggest fire hazards in loading rail cars and ships is fire by spark caused by static electricity. The proper grounding, bonding, or earthing of the equipment being loaded is essential in preventing this fire hazard. There is also usually a specified wait time before the grounding connections are removed to allow static charges to dissipate.

These concerns apply to the unloading of hydrocarbons, such as crude oil feed or intermediate product streams transferred from other facilities or depots. In small refinery installations, the jetty facilities and the onshore loading islands are protected by an extension of the fire main and foam systems. In larger refineries, the jetty may have its own fire main using seawater as the main firewater medium.

Fire Main and Installed Firefighting Equipment

Fire Main Description and Control

Fire mains are usually designed as a pressurized circulating water system with an atmospheric break tank. A schematic sketch of such a system is given in Fig. 1. There may be more than one independent main in a given facility. The fire main is maintained at a pressure of around 100–150 psig (~7–10 barg) by a circulating pump or a smaller, parallel jockey pump. The pump is automatically switched on or off to maintain the fire main line pressure. When monitors, hydrants, and any other large firewater users need to deliver water to fire sources, to foam systems, or for any other application, the pressurizing pump will stay on and additional pumps normally automatically start to deliver the required firewater flow. The pumps are usually electric motor driven, with one or more spares that are usually driven by a diesel motor or a steam turbine. The spare pump(s) are automatically activated on electrical failure and the inability of the normal fire circulating pump to maintain line pressure.

The main and the main pumps are sized based on the area and types of facilities to be covered. Some companies size the pumps assuming "X" monitors and "Y" fire hoses are in use simultaneously. It is not unusual to be capable of supplying 8–15,000 gpm (1,800–3,400 MT/h) from the system.

Once a firewater pump is autostarted, it should remain running until it is intentionally shut down by the operators or firefighting crew when it is no longer needed. This means that main firewater pumps do not ever have shutdowns for high temperature bearings or low lube oil pressure. They are designed to run to destruction, if necessary. This is critical in an emergency. You do not want to ever lose firewater when you are in the middle of a fire, especially if you are on the hose!

Makeup water to the break tank is maintained under break tank level control from the refinery's fresh (or salt water) main or from an outside utility supply. Some plants have large quantities of firewater stored in basins, but the same guidelines apply. Once high firewater use begins, a large capacity makeup water supply will be needed.

The break tank is often really a very large storage tank, representing the available water supply if no makeup can be brought in. It is common to design the break tank to hold 6–8 h of firewater at full pump rates. This is especially important at remote locations, where additional water supply is limited.

In smaller refinery facilities and smaller systems, such as a jetty fire main, the system may be "dead ended." That is, the firewater pump would be used only to

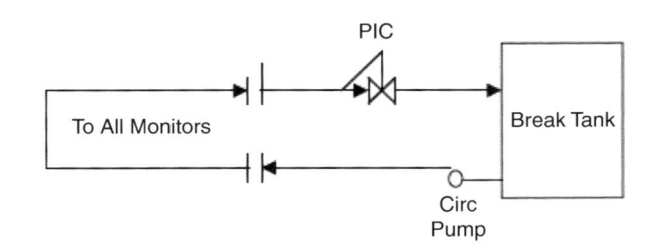

Fig. 1 Schematic fire main system

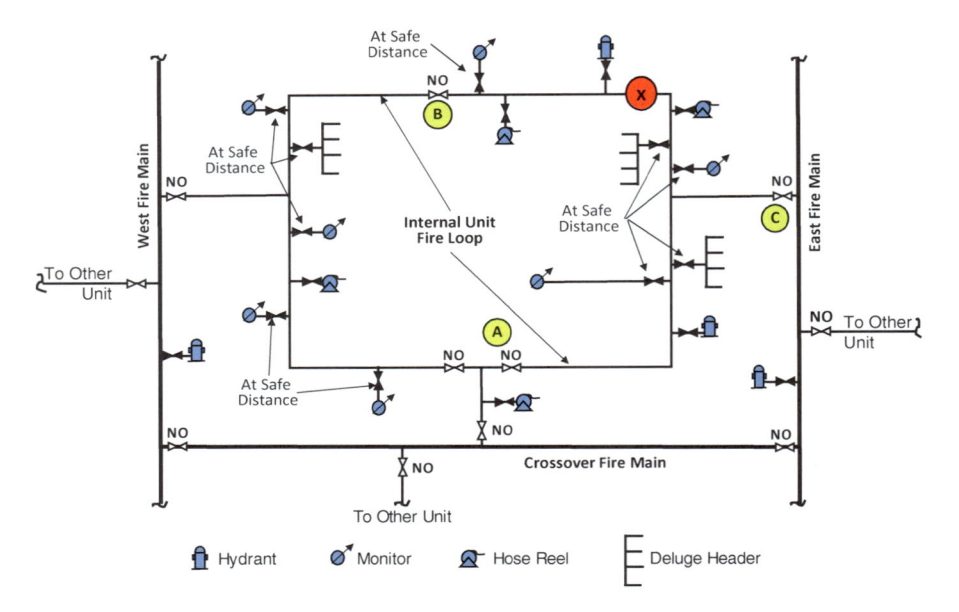

Fig. 2 Example fire loop arrangement

maintain the fire main pressure. The water would not be circulated to a break tank. A water source tank or basin would be used as the firewater source reservoir. In the case of the jetty fire main, usually seawater is used and this would be delivered from a suitable seabed pit by a vertical pump. Again an electric motor driver would be used for the normal pump with usually a steam turbine driver for the spare pump.

Firewater piping is frequently underground for freeze protection and to prevent blast or fire damage. Many of the system valves are also located underground and designed to automatically drain.

Firefighting Equipment and Layout

Generally, a firewater system is designed in loops with isolation valves for parts of the loop. The intent is to avoid any situation where damage to one part of the loop takes down the whole supply. The concept is illustrated in Fig. 2. As illustrated, the unit can pull firewater from either the east or west main or the crossover main. Within the unit, the supply lines go to different parts of a single loop. There are block valves that split the internal loop. These are normally left open. If the east internal loop is damaged at the point indicated by the X in the red circle, for instance, Blocks A, B, and C can be closed to isolate the damaged line and still get water from the west part of the internal loop. A great deal of thought must go

Fire Prevention and Firefighting in Petroleum Processing 1425

into how the mains and submains are connected and routed to ensure water supply when needed.

Aside from the mains and supply loops, there are several types of firefighting equipment normally connected to the firewater system. The most common devices are as follows:

- *Hydrants* provide locations to connect hoses for direct use on a fire or for pumper truck suction. They are usually located at defined intervals along the mains and in the loops. Some hydrants can deliver over 1,500 gpm (~5,700 l/min). Normal capacities are 500–1,000 gpm (1,800–3,800 l/min), however.
- *Monitors* are used to provide continuous water streams on equipment. There are both permanent and portable types of monitors. They can be used directly on the fire or can be used to cool neighboring equipment and prevent a larger problem. They can be set and then left to do their job, thereby reducing exposure of personnel to the fire. A monitor can be selected to deliver a stream, cone, or fog on the fire – many can perform multiple functions. They can also be setup to be activated and controlled remotely and can be elevated on poles to better reach equipment. The block valves for monitors are normally underground to prevent blast and freeze damage. A monitor can typically deliver about 350 gpm (1,300 l/min). Their reach is usually limited to about 150 f. (~50 m) and as high as the system pressure allows. The locations of monitors need to consider the equipment to be protected as well as all the interferences around that equipment. Monitor block valves are normally located a safe distance, like 50–100 f. (~15–30 m.), from the equipment the monitor protects.
- *Hose reels and hose stations* normally have about 50–100 f. (~18–36 m) of hose with a variable nozzle on the end. The hose (AKA booster hose), normally about 1–1½ in. (2.5–4 cm) diameter, is spooled on a hose reel. The capacity of these hoses is limited to what one person can manage – usually less than 100 gpm (~380 l/min). These hoses are sometimes also used for area washdown; hence, they have to be within the ability of one person to handle. They are useful primarily just for small fires.
- *Deluge systems* may be positioned around specific equipment to provide fire or other protection in an emergency. You may find them around hot pump seal areas or around large air coolers containing hydrogen or around HF acid handling facilities as water curtains to control an acid release. These systems vary in size considerably. Many are automatically activated in an emergency. Like monitors, the block valves for deluge systems are located a safe distance from the equipment being protected.
- *Foam generators* are devices that aspirate foam chemicals at a rate of 3–6 % into a flowing water stream along with air to create a foam. Often these are permanently installed on foam lines from the fire mains. They may or may not have a foam chemical supply connected to them permanently. Often the foam is brought in for use when needed because it has a limited shelf life. Installed

foam generators are often found in the tankage or offsite area. Capacity and number of foam generators are a function of the size of equipment being protected. Hand foam generators can deliver 60–120 gpm (~220–440 l/min, on a hose reel) and monitors can push 600–2,000 gpm (~2,200–7,600 l/min) of water. The water is mixed in the nozzle with chemical to make 3 % or 6 % foam. The actual foam volume is much higher (50:1 to 200:1 expansion).

The locations of the fixed fire equipment and the portable equipment, like exchangers, are coordinated to provide the best possible coverage for a facility. The layout and protection systems are normally worked out with a fire protection engineer.

Inspection and Maintenance Program

Firewater systems require routine inspection and maintenance. Aside from inspection for corrosion, the systems are normally tested on a regular interval for operation (about quarterly). The hydrants are opened and flushed. Hose is unspooled, inspected, flushed, and respooled. Monitors may be activated and the nozzles exercised. Any problems that are noted are repaired.

Foam generators are usually just inspected and may be cleaned. You don't usually generate foam with them on any routine basis.

The deluge and water curtain systems may be part of the safety instrumented systems (SISs) and so may be included in a required testing program. Activation of the systems for functional testing may be carried out on some defined schedule when it won't hurt the operation. Repairs are made as indicated by the tests. Hot pump deluges are usually only tested when the pump is cold to avoid creating a problem from thermal stresses.

With the other activities, the firewater pumps will be cycled on and off. Firewater standby pumps are normally run at least monthly for a few hours. The pressure switches that control the autostart sequences are checked in the process.

These types of inspection and maintenance activities in the firewater systems are critical to ensure the systems are available and will perform their functions when needed.

Fire Foam and Foam Systems

Introduction

Firefighting foam is simply a stable mass of small, air-filled bubbles with a lower density than oil, gasoline, or water. The foam is made up of three ingredients:

- Water
- Foam concentrate
- Air

The water is mixed with the concentrate (proportioned) to form a foam solution. This solution is then mixed with air (aspirated) to produce the foam which is very fluid and flows readily over liquid surfaces.

Balanced pressure proportioning is the most common method used for foam system applications. The foam concentrate pressure is balanced with the water pressure at the proportioner inlets allowing the proper amount of foam concentrate to be metered into the water stream. With an aspirating discharge device, foam solution passes through an orifice, past air inlets, and into an expansion area to produce an expanded foam. With non-aspirating devices, foam solution passes through the orifice and discharge outlet where it mixes with air en route to the fire.

Potential Firefighting Foam Applications

Firefighting foam is used in a variety of applications to (1) extinguish flammable and combustible liquid fires, (2) control the release of flammable vapors, and (3) cool fuels and sources of ignition. Typical foam applications include:

- Loading racks
- Refineries
- Pumping stations
- Power plants
- Airports
- Heliports
- Marine applications
- Manufacturing plants
- Storage tanks
- Chemical plants
- Flammable liquid storage
- Offshore platforms
- Aircraft hangars
- Crash rescue vehicles
- Mining facilities
- Warehouses
- Hazardous material spill control.

Firefighting foam agents suppress fire by separating the fuel from the air (oxygen). Depending upon the type of foam agent, this is done in several ways:

- Foam blankets the fuel surface, smothering the fire and separating the flames from the fuel surface.
- The fuel is cooled by the water content of the foam.
- The foam blanket suppresses the release of flammable vapors that can mix with air.

Types of Firefighting Foam

A film-forming foam (AFFF) agent forms an aqueous film on the surface of a hydrocarbon fuel. An alcohol-resistant concentrate (ARC) will form a polymeric membrane on a polar solvent fuel. Essentially there are six general types of foam agents in two classes:

Class B fire foams
Film-forming foams
   Film-forming foams (AFFF) are based on combinations of fluorochemical surfactants, hydrocarbon surfactants, and solvents. These agents require a very low energy input to produce a high-quality foam. Consequently, they can be applied through a wide variety of foam delivery systems. This versatility makes AFFF an obvious choice for airports, refineries, manufacturing plants, municipal fire departments, and any other operation involving the transportation, processing, or handling of flammable liquids. These foam-forming agents are used as 1 % and 3 % freeze-protected concentrates.

Alcohol-resistant concentrates
   Alcohol-resistant concentrates are based on AFFF chemistry to which a polymer has been added. ARCs are the most versatile of the foam agents in that they are effective on fires involving polar solvents like methanol as well as hydrocarbon fuels like gasoline. When used on a polar solvent-type fuel, the ARC concentrate forms a polymeric membrane which prevents destruction of the foam blanket. When used on hydrocarbon fuels, the ARC produces the same rugged aqueous film as a standard AFFF agent. ARCs provide fast flame knockdown and good burn back resistance when used on both types of fuels.

Protein foam concentrates
   Protein foam concentrates are recommended for the extinguishment of fires involving hydrocarbons. They are based on hydrolyzed protein, stabilizers, and preservatives. Protein foams produce a stable mechanical foam with good expansion properties and excellent burn back resistance characteristics. Protein foam concentrates are used in 3 % and 6 % concentrations.

Fluoroprotein foam concentrates
   Fluoroprotein foam concentrates are based on hydrolyzed protein, stabilizers, preservatives, and synthetic fluorocarbon surfactants. When compared to protein foams, fluoroproteins provide better control and extinguishment, greater fluidity, and superior resistance to fuel contamination. Fluoroprotein foams are useful for hydrocarbon vapor suppression and have been recognized as very effective fire-suppressing agents for subsurface injection into hydrocarbon fuel storage tanks. Fluoroprotein foam is used as in a 3 % solution.

High-expansion foam concentrates
   High-expansion foam concentrates are based on combination of hydrocarbon surfactants and solvents. They are used with foam generators for applying foam to large areas in total flooding and three-dimensional applications such as warehouses, ship cargo holds, and mine shafts. They are especially useful on

fuels such as liquefied natural gas (cryogenic fuels) for vapor dispersion and control. In certain concentrations, high-expansion foams are effective on hydrocarbon spill fires of most types and in confined areas.

Class A fire foams

Class A foams are typically formulated from a combination of specialty hydrocarbon surfactants, stabilizers, inhibitors, and solvents. They reduce the surface tension of water for improved wetting and penetrating characteristics and create a clinging foam blanket that suppresses combustible vapors while cooling the fuel. Class A foams can be applied using a variety of proportioning/discharge devices and have proven to be highly effective for use in structural and forest firefighting, coal mines, tire, and rubber manufacturing, lumber mills, coal bunkers, paper warehouses, and other hazards involving ordinary combustible materials. Please note that Class B foams are acceptable for use on Class A fires; however, they are not designed for use on Class A fires as such. The foam concentrate percentage refers to the amount of concentrate that is proportioned or premixed with water to give the resulting foam solutions.

Environmental Impact and Toxicity

Most foam concentrates are formulated to maximize performance and minimize environmental impact and human exposure hazards. All concentrates are readily biodegradable – both in the natural environment and in sewage treatment facilities. However, all foam agents should be metered into the facility to prevent overloading the plant due to foam formation. They are not considered skin irritants; however, prolonged contact may cause some dryness of the skin. For this reason, it is normally recommended that areas of the skin which have come in contact with the foam concentrate be flushed with fresh water.

Shelf Lives of Foam Concentrates

"Shelf life" is the length of time over which foam concentrates remain stable without significant changes in performance characteristics. Many AFFF, high-expansion and Class A foam concentrates – if stored in accordance with recommended guidelines – have a normal shelf life of 20–25 years. Other foam agents – those which are not totally synthetic – have a normal shelf life of 7–10 years.

Fire Extinguishers

While the larger fires make use of the firewater system and related equipment, smaller fires are often kept from becoming larger fires through prompt action using fire extinguishers. A fire extinguisher can make short work of many seal, trash, and other relatively small fires to prevent a larger problem.

Table 1 Classes of fire extinguishers (source: Farm Bureau Insurance Company of Michigan)

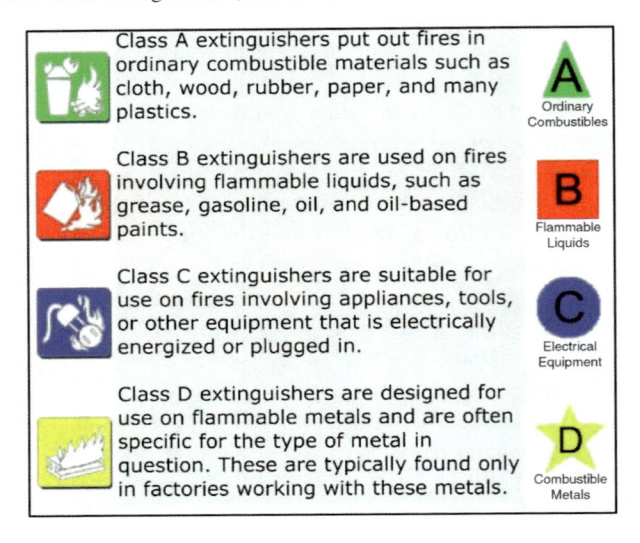

Class A extinguishers put out fires in ordinary combustible materials such as cloth, wood, rubber, paper, and many plastics.	**A** Ordinary Combustibles
Class B extinguishers are used on fires involving flammable liquids, such as grease, gasoline, oil, and oil-based paints.	**B** Flammable Liquids
Class C extinguishers are suitable for use on fires involving appliances, tools, or other equipment that is electrically energized or plugged in.	**C** Electrical Equipment
Class D extinguishers are designed for use on flammable metals and are often specific for the type of metal in question. These are typically found only in factories working with these metals.	**D** Combustible Metals

Types and Applications of Fire Extinguishers

Table 1 lists the primary classes of fire extinguishers that are recognized.

The types of fire extinguishing agents for the different types of fires are listed in Table 2.

If you correlate the charts together, in general:

- Water and water-based agents are Class A.
- Class B would include foam, dry powder, and CO_2.
- Class C would include dry powder and CO_2.
- Class D agents are usually solids, like sand.

Within a refinery or petroleum processing facility, you will normally only find Classes ABC, BC, or C extinguishers. Class D is rare. You may see Class A extinguishers inside buildings. You will see Class C inside electrical switchgear buildings.

In special cases, you will also find fire suppression agents like Halon (or one of the newer substitutes) inside buildings. Halon is essentially an ABC agent and acts by eliminating air from an enclosed space – which means no personnel can be in the enclosed space when a Halon system is activated. It is not poisonous but is designed to create an oxygen-deficient atmosphere. There is normally a time delay with visual and audible warnings between the system being triggered and actuation of the Halon to give personnel time to evacuate. While Halon is a severe ozone-depleting gas, it does not damage electronics or electrical equipment. Substitutes have been developed that are less damaging to ozone and still effective. Halon also is available in a handheld extinguisher rather than a room system.

Table 2 Types of fire extinguishing agents (Texas A&M University (2012))

Extinguisher type	Type of fire				
	Solids (wood, paper, cloth, etc.)	Flammable liquids	Flammable gases	Electrical equipment	Cooking oils and fats
Water	Yes	No	No	No	No
Foam	Yes	Yes	No	No	Yes
Dry powder	Yes	Yes	Yes	Yes	No
Carbon dioxide (CO_2)	No	Yes	No	Yes	Yes

The extinguisher size required for a particular type of hazard is specified in the United States by the National Fire Protection Association (NFPA, 2002) based on the type of hazard and distance of the exchanger from the fire.

Locations of Fire Extinguishers

Fire extinguishers are normally distributed throughout a petroleum processing facility to allow fast attack on small fires. The extinguishers are located about 50–75 ft. (15–25 m) apart, normally on pipe rack columns. They are also often located near places where they may be used, such as near hot pumps, but still far enough away to enable access in an emergency.

Extinguishers must be easy to remove from whatever bracket they are attached to – usually just a hook. The extinguisher locations should be clearly marked, usually in red. The extinguisher types within an area are appropriate to the fire risks in that area, with the normal types of extinguishers being BC and ABC dry chemicals.

Inside a control room or operator shelter, there will normally be dry chemical extinguishers and CO_2. In some cases Halon may be provided.

In power distribution centers, you do not find powder chemical agent, but may find CO_2 and Halon (or equivalent) systems or extinguishers.

Fire extinguishers in the field are exposed to the elements and may deteriorate. Extinguishers in the field should be covered for protection. The covers must be easy to remove when needed and should be replaced when they deteriorate.

Some facilities have large, cart-mounted extinguishers. These should be easy to move and should not be blocked by other materials, debris, or trash.

Fire Extinguisher Inspection and Testing Program

The availability of fire extinguishers brings with it the need to make sure the extinguishers will work when required. From the US NFPA standards, there are four inspection intervals required:

Monthly inspection

This inspection includes:
- Correctly installed at the designated location
- No obstructions to access
- Legible operating instructions on nameplate
- Safety seals and tamper indicators not broken or missing
- Fullness checked by weighing or lifting
- Examination for obvious damage, leakage, or clogged nozzle
- Pressure gauge reads in "good" range
- Condition of wheel cart and wheels good (wheeled units)
- HMIS label in place

Most facilities log the inspection. There is supposed to be a tag on each extinguisher noting the date of the last inspection and who did the inspection. The field extinguisher covers should be replaced when they are damaged or no longer protect the extinguisher adequately.

Annual inspection

A more thorough annual inspection of extinguishers is required. This inspection includes:
- Complete servicing of the extinguisher
- Recharging
- Must be done by trained personnel

Six-year maintenance

This is required for "stored-pressure" extinguishers, like dry chemicals or halogens (e.g., Halon). The extinguisher is emptied and thoroughly examined. The extinguisher must be hydrotested every 12 years, so every other 6-year inspection must include a hydrotest.

Hydrostatic testing

There is a specific hydrotesting interval defined for types of exchangers:
- Every 5 years for stored-pressure water, water mist, loaded stream, antifreeze, wetting agent, foam, dry chemicals with stainless shells, carbon dioxide, and wet chemicals
- Every 12 years for dry chemicals with stored pressure, halogenated agents, and dry powder with cartridge pressurization

There are hydrotest record requirements also.

A facility needs to define a program to ensure all the extinguisher inspection and testing requirements are satisfied. Often, the operators are asked to inspect extinguishers weekly in addition to the required inspections as a means of identifying problems early.

Concluding Comments

A couple of final words on extinguishers:

Fire Prevention and Firefighting in Petroleum Processing 1433

- Any time an extinguisher gets used, immediately replace it with an unused extinguisher and get the used one serviced completely. You do not want any of your personnel to try to use an extinguisher, only to find that it is not operable.
- It should be a standard practice that whoever finds a fire must call in the emergency first before they decide to try to use an extinguisher. They will need backup.
- Only personnel trained in the use of an extinguisher should try using it. Improper use can make the situation worse. We will not go into detail on actual extinguisher use here. Many companies, and your own safety or firefighting organization, should be able to provide adequate training. There is also no substitute for actual, hands-on training.

Mobile Firefighting Equipment

In addition to the fixed equipment and extinguishers, there is also usually an array of portable firefighting equipment in a petroleum facility. We will not go into detail on this equipment, but this critical equipment can include:

- Fire pumper trucks
- Fast attack fire trucks – with tanks and monitors or hoses, possibly foam
- Portable monitors – can be hooked to hoses and set to cool equipment
- Foam pumpers
- Emergency medical technical support truck
- Large monitor trucks – some throw 12–15,000 gpm (~45,000–57,000 l/min)

Some of this may be owned by the facility. Some may be from the local fire department. Some may come through mutual aid agreements with nearby facilities.

A facility far-removed from other plants will tend to be more self-sufficient in mobile firefighting equipment and may have more complete integration with the local community firefighting capability.

Fire Protection and Firefighting Plans

We have focused so far on the equipment used for refinery and petroleum processing firefighting. The use of the various parts of the facility fire protection systems is only marginally effective unless there are specific fire protection and firefighting plans in place to manage fire emergencies when they arise.

The emergency management system for most facilities covers other emergencies, in addition to fire. This enables consistent and trained response to all types of emergencies. Here we will describe the standard emergency management system employed commonly in the United States. There may be some local flavors of this, but the basics apply.

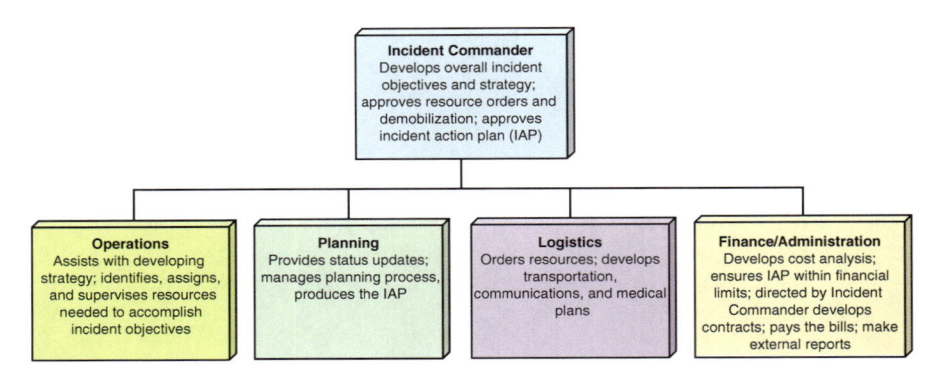

```
┌─────────────────────────────┐
│     Incident Commander      │
│ Develops overall incident   │
│   objectives and strategy;  │
│ approves resource orders and│
│  demobilization; approves   │
│ incident action plan (IAP)  │
└─────────────────────────────┘
```

Operations	Planning	Logistics	Finance/Administration
Assists with developing strategy; identifies, assigns, and supervises resources needed to accomplish incident objectives	Provides status updates; manages planning process, produces the IAP	Orders resources; develops transportation, communications, and medical plans	Develops cost analysis; ensures IAP within financial limits; directed by Incident Commander develops contracts; pays the bills; make external reports

Fig. 3 Simplified incident command system

Incident Command System (ICS)

The most common incident command system (ICS) in petroleum processing facilities looks something like that illustrated in Fig. 3. The ICS is a standardized organization and responsibilities that represents best practices. It is a key feature of the US National Incident Management System (NIMS). This system can be used for fires, releases, medical emergencies, or any other type of situation where fast response is required.

In this system, one person is designated as the incident commander (IC). This is normally a specific person, on-site initially, who has the knowledge, has the training, and is given the authority to develop an attack plan and successfully manage the incident response. If a public emergency response organization (like the city fire department or the Federal government) becomes involved, usually they will assume incident command once they are on-site.

Supporting the IC are organizations to provide:

- Operations support – help decide what to do and execute the plan
- Planning section – manages the planning
- Logistics section – gets the resources needed to execute the plan
- Finance/administration section – tracks costs, keeps plan within financial limits, pays the bills, makes external reports

This system works well with fire and explosion responses, in addition to other emergencies.

Notification Systems and Procedures

One thing we have not mentioned so far is notification of plant personnel when a fire or other emergency has occurred.

All facilities have an emergency notification system (ENS) in the event of a fire, release, or other situation that could place people in danger. The ENS normally is activated by someone activating an alarm or by calling an emergency number that is staffed 100 % of the time.

The nature of the emergency and location is reported. Additional information is provided (usually logged) and then the person receiving the call initiates the alarm, which initiates the incident command system. In medical cases, the general alarm may not be sounded, but the ICS would be activated. In some cases, an alarm will be local to one unit, with an appropriate, less extensive, response. For those with radios and in units with public address systems, the nature of the emergency may be announced in more detail.

Plant alarms are usually audible and may also include visual alarms. The audible alarms may differ for different types of emergencies. Usually there are at least three types of audible notifications, with each having a different sound:

- Fire alarm (for fires, explosions, or other situations that are not otherwise covered)
- Evacuation alarm (for releases primarily)
- All clear

All personnel on site, including contractors and visitors, are trained in how to respond if they hear the alarm. Nonessential personnel, visitors, and contractors are normally cleared off the site or out of the units – avoiding the emergency area. In some cases, however, it may be necessary to "shelter in place" (i.e., go to a building with a specifically designed air handling system to enable isolation). All active permits for maintenance activities are normally canceled by the alarm.

The appropriate first responders will go to the emergency location with the incident commander to assess the problem and begin an attack. Meanwhile, at a safe location, the rest of the ICS assembles to begin their support work.

When the emergency is under control or the condition is eliminated, the "all clear" is called and the all clear notification is sounded.

Periodic practice drills help ensure emergency response runs smoothly so that energy can be focused on the incident.

Conclusion

A well-prepared petroleum processing facility will have:

- Adequately designed, constructed, operated, and maintained equipment
- Well-designed and maintained firefighting system
- Adequate and maintained fire extinguishers
- Appropriate mobile firefighting equipment
- Appropriate and functional alarm system
- Emergency response plan staffed by trained personnel

With these elements in place, a facility should (1) be able to avoid having a fire or explosion and (2) be able to deal with a fire, explosion, or any other emergency if it does occur.

References

National Fire Protection Association, *Guidelines: Fire Extinguishers*, NFPA 10, 2002 edition, IFC 2008. Quincy, MA, USA, (2002)

Texas A&M University, *Fire Extinguishers: Are You Prepared?* (Texas A&M Engineering Extension, Bryan, TX, USA, 2012)

Part V
Reference

Process Equipment for Petroleum Processing

David S. J. Jones and Steven A. Treese

Contents

Introduction	1441
Vessels and Towers	1441
Fractionators, Trays, and Packings	1442
Drums and Drum Design	1470
Specifying Pressure Vessels	1476
Pumps	1485
Pump Selection	1489
Selection Characteristics	1490
Evaluating Pump Performance	1494
Specifying a Centrifugal Pump	1497
Centrifugal Pump Seals	1502
Pump Drivers and Utilities	1505
Reacceleration Requirement	1509
The Principle of the Turbine Driver	1509
Cooling Water Requirements for Hot Pumps	1510
Compressors	1513
Types of Compressors and Selection	1513
Calculating Horsepower of Centrifugal Compressors	1515
Centrifugal Compressor Surge Control, Performance Curves, and Seals	1524
Specifying a Centrifugal Compressor	1527
Example Calculation	1530
Calculating Reciprocating Compressor Horsepower	1534
Reciprocating Compressor Controls and Intercooling	1538
Specifying a Reciprocating Compressor	1540
Compressor Drivers, Utilities, and Ancillary Equipment	1550

David S. J. Jones: deceased.
Steven A. Treese has retired from Phillips 66.

D.S.J. Jones
Calgary, AB, Canada

S.A. Treese (✉)
Puget Sound Investments LLC, Katy, TX, USA
e-mail: streese256@aol.com

Heat Exchangers 1557
   Types and Selection of Heat Exchangers 1557
   General Design Considerations 1559
   Choice of Tube Side Versus Shell Side 1561
   Estimating Shell and Tube Surface Area and Pressure Drop 1562
   Air Coolers and Condensers 1573
   Condensers 1579
   Reboilers 1584
Fired Heaters 1594
   Types of Fired Heaters 1594
   Codes and Standards 1597
   Thermal Rating 1599
   Heater Efficiency 1601
   Burners 1604
   Refractories, Stacks, and Stack Emissions 1606
   Specifying a Fired Heater 1611
Pressure Drop Calculations 1617
   Incompressible Flow Pressure Drop in Piping 1617
   Compressible Flow Pressure Drop in Piping 1622
   Fittings and Piping Elements 1625
   Overall Pressure Drop, Including Elevation and Velocity Changes 1625
   Two-Phase (Liquid + Vapor) Pressure Drop and Flow Regimes in Piping 1629
   General Comments on Pressure Drops in Beds of Solids 1635
   Single-Phase Pressure Drop in Beds 1635
   Two-Phase Pressure Drop in Beds 1639
   Estimating Overall Pressure Drop in Beds 1641
   Moving Solids Bed Pressure Drops 1643
Appendices 1643
   Appendix 1: Chord Height, Area, and Length for Distillation Trays and
   Circular Cross-Sections 1643
   Appendix 2: Valve Tray Design Details 1646
   Appendix 3: Pressure-Temperature Curves for Hydrocarbon Equilibria 1648
   Appendix 4: ASTM Gaps and Overlaps 1651
   Appendix 5: Values of Coefficient C for Various Materials 1652
   Appendix 6: LMTD Correction Factors 1653
   Appendix 7: Common Overall Heat Transfer Coefficients – U_O 1654
   Appendix 8: Standard Heat Exchanger Tube Data 1655
   Appendix 9: Line Friction Loss for Viscous Fluids 1655
   Appendix 10: Heats of Combustion for Fuel Oils 1674
   Appendix 11: Heats of Combustion for Fuel Gases 1675
   Appendix 12: Resistances of Valves and Fittings 1676
   Appendix 13: Flow Pressure Drop for gas Streams 1677
   Appendix 14: Example Hydraulic Analysis of a Process System 1679
References 1684

Abstract

This chapter deals with the equipment normally found in the petroleum refining industry. Many of the items that will be described and discussed here are also common to other process industries. Knowledge of these equipment items is essential for good refinery design, operation, and troubleshooting when necessary. The equipment and process design calculations described here include vessels, fractionators, pumps, compressors, heat exchangers, fired heaters, and piping and packed bed pressure drops. Several examples of these calculations are provided.

Keywords
Process equipment • Refinery • Petroleum • Vessel • Tower • Pump • Compressor • Exchanger • Heater piping • Design • Fractionation

Introduction

This chapter deals with the equipment normally found in the petroleum refining industry. Indeed, many of the items that will be described and discussed here are also common to many other process industries. Knowledge of these equipment items is essential for good refinery design, operation, and troubleshooting when necessary. The equipment and design methods described here fall into six categories and will be presented in the following main sections:

- Vessels (columns, towers, fractionators, drums, separators, accumulators, and surge vessels)
- Pumps (variable head-capacity, positive displacement)
- Compressors (centrifugal, axial, reciprocating, rotary)
- Heat exchangers (shell and tube, double-pipe, air coolers, box coolers, direct-contact exchangers)
- Fired heaters (horizontal, vertical)
- Pressure drop calculations (piping, solid beds, single- and two-phase flows)

These sections will include a description of the various equipment types, an in-depth discussion, and review of design features. They will also provide examples of the data sheets usually forwarded to manufacturers for the items required. These data sheets are included as part of the "Mechanical Catalogues" or "Record Data Books" for each unit and supported by narrative specifications, which give details of metallurgy and fabrication codes. These documents are provided by the equipment supplier and are part of the information dossier on each item. Included also are such items as installation details, start-up procedures, routine maintenance procedures, and the like. In most refineries today, the catalogues are maintained on the computer or microfilm. There are also paper copies of the data books maintained.

Vessels and Towers

This section addresses the pressure vessels that are common to most refineries. These include:

- Columns and towers
- Knockout drums and separators
- Accumulators and surge vessels

Storage tanks have been dealt with in the chapter on "off-sites" in this handbook.

Fractionators, Trays, and Packings

Trayed Towers
Columns normally constitute the major cost in any chemical process configuration. Consequently it is necessary to exercise utmost care in handling this item of equipment. This extends to the actual design of the vessel or evaluating a design offered by others. Normally columns are used in a process for fractionation, extraction, or absorption as unit operations. Columns contain internals which may be trays, or packing. Both types of columns will also contain suitable inlet dispersion nozzles, outlet nozzles, instrument nozzles, and access facilities (such as manways or handholes). This item deals primarily with the trayed towers.

Tray Types
There are three types of trays in common use today:

- Bubble cap
- Sieve
- Valve

Bubble Cap Trays
This type of tray was in wide use up until the mid to late 1950s. Its predominance was displaced by the cheaper sieve and valve trays. The bubble cap tray consists of a series of risers on the tray which are capped by a serrated metal dome. Figure 1 shows two types of caps. One is used in normal fractionation service while the other is designed for vacuum distillation service. Vapor rises up through the risers into the bubble cap. It is then forced down through the serrated edge or, in some cases, slots at the bottom of the cap. A liquid level is maintained on the tray to be above the slots or serrations of the cap. The vapor therefore is forced out in fine bubbles into this liquid phase thereby mixing with the liquid. Mass and heat transfer between vapor and liquid is enhanced by this mixing action to effect the fractionation mechanism.

Capacity. Moderately high with high efficiency.
Efficiency. Very efficient over a wide capacity range.
Entrainment. Much higher than perforated-type trays due to the "jet" action that accompanies the bubbling.
Flexibility. Has the highest flexibility both for vapor and liquid rates. Liquid heads are maintained by weirs.

Process Equipment for Petroleum Processing

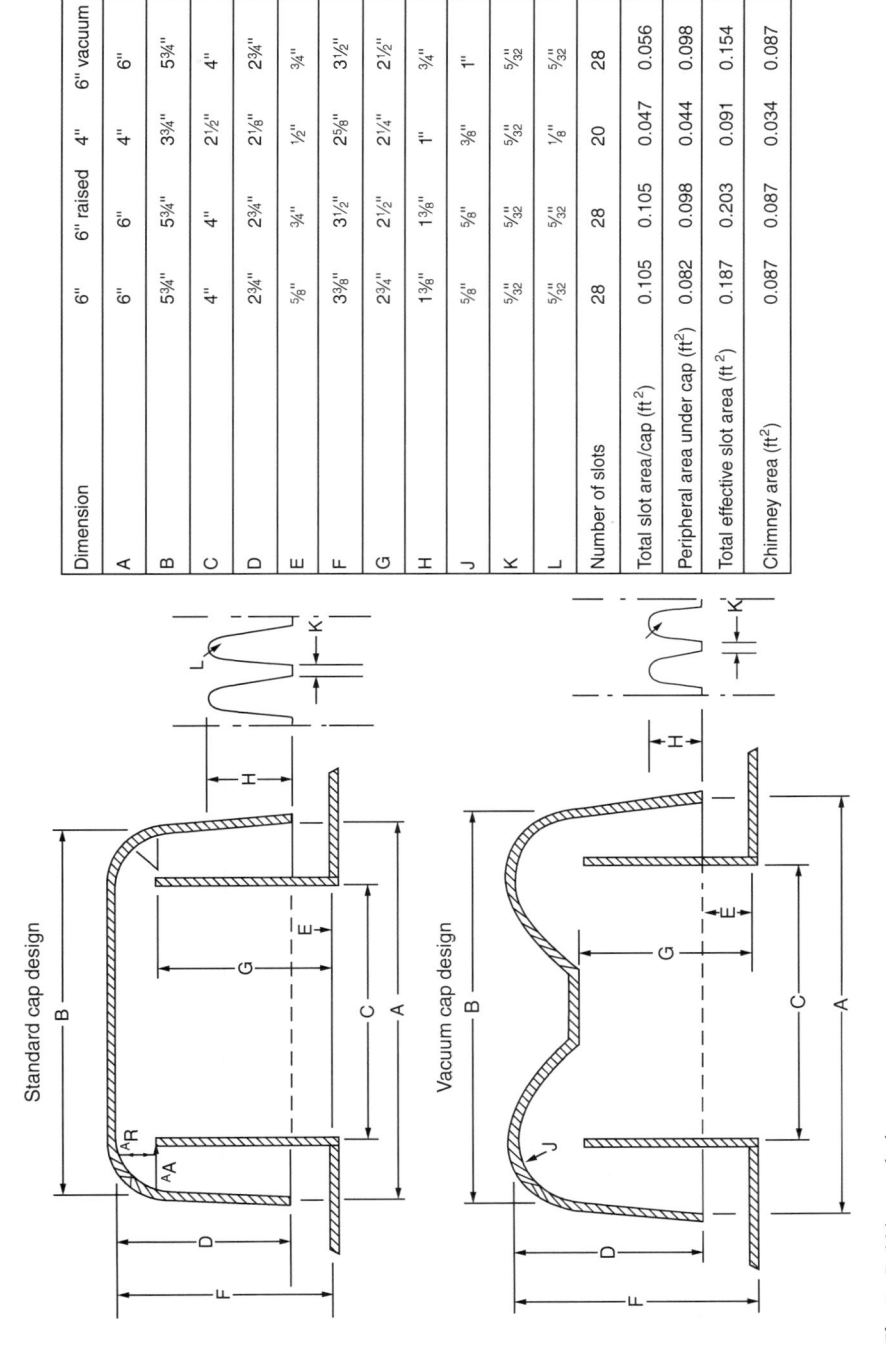

Dimension	6"	6" raised	4"	6" vacuum
A	6"	6"	4"	6"
B	5¾"	5¾"	3¾"	5¾"
C	4"	4"	2½"	4"
D	2¾"	2¾"	2⅛"	2¾"
E	⅝"	¾"	½"	¾"
F	3⅜"	3½"	2⅝"	3½"
G	2¾"	2½"	2¼"	2½"
H	1⅜"	1⅜"	1"	¾"
J	⅝"	⅝"	⅜"	1"
K	5/32"	5/32"	5/32"	5/32"
L	5/32"	5/32"	⅛"	5/32"
Number of slots	28	28	20	28
Total slot area/cap (ft^2)	0.105	0.105	0.047	0.056
Peripheral area under cap (ft^2)	0.082	0.098	0.044	0.098
Total effective slot area (ft^2)	0.187	0.203	0.091	0.154
Chimney area (ft^2)	0.087	0.087	0.034	0.087

Fig. 1 Bubble cap design

Application. May be used for all services except for those conditions where coking or polymer formation occurs. In this case baffle or disc and donut or grid-type trays should be used.

Note: Because of the relatively high liquid level required by this type of tray, it incurs a higher pressure drop than most other types of trays. This is a critical factor in tray selection for vacuum units.

Tray spacing. Usually 18–24″. For vacuum service, this should be about 30–36″.

Sieve Trays

This is the simplest of the various types of trays. It consists of holes suitably arranged and punched out of a metal plate. The vapor from the tray below rises through the holes to mix with the liquid flowing across the tray. Fairly uniform mixing of the liquid and vapor occurs and allows for the heat/mass transfer of the fractionation mechanism to occur. The liquid flows across a weir at one end of the tray through a downcomer to the tray below.

Capacity. As high as or higher than bubble cap trays at design vapor/liquid rates. Performance drops off rapidly at rates below 60 % of design.

Efficiency. High efficiency at design rates to about 120 % of design. The efficiency falls off rapidly at around 50–60 % of design rate. This is due to "weeping" which is the liquid leaking from the tray through the sieve holes.

Entrainment. Only about one-third of that for bubble cap trays.

Flexibility. Not suitable for trays operating at variable loads.

Application. In most mass transfer operations where high capacities in vapor and liquid rates are required. Handles suspended solids and other fouling media well.

Tray spacing. Requires less tray spacing than bubble cap. Usually spacing is rarely less than 15″ although some services can operate at 10″ and 12″. In vacuum service, a spacing of 20–30″ is acceptable.

Valve Trays

These trays have downcomers to handle the liquid traffic and holes with floating caps that handle the vapor traffic. The holes may be round or rectangular and the caps over the holes are moveable within the limits of the length of the "legs" which fit into the holes. Figures 2 and 3 show the type of valves and valve trays offered by Glitsch as their "Ballast" trays.

Valve trays are by far the most common type of tray used in the chemical industry today. The tray has good efficiency and a much better flexibility in terms of turndown than the sieve or bubble cap trays. Its only disadvantage over the sieve tray is that it is slightly more expensive and cannot handle excessive fouling as well as the sieve tray. The remainder of this item will now be dedicated to the sizing and analysis of the valve tray tower.

Trayed Tower Sizing

The height of a trayed tower is determined by the number of trays it contains, the liquid surge level at the bottom of the tower, and the tray spacing. The number of

Process Equipment for Petroleum Processing

trays is a function of the thermodynamic mechanism for the fractionation or absorption duty required to be performed. This is described in the chapters on atmospheric and vacuum distillation and light ends fractionation of this handbook.

The diameter of the tower is based on allowable vapor and liquid flow in the tower and the type of tray. The next section deals with determining the tower diameter using valve trays.

a

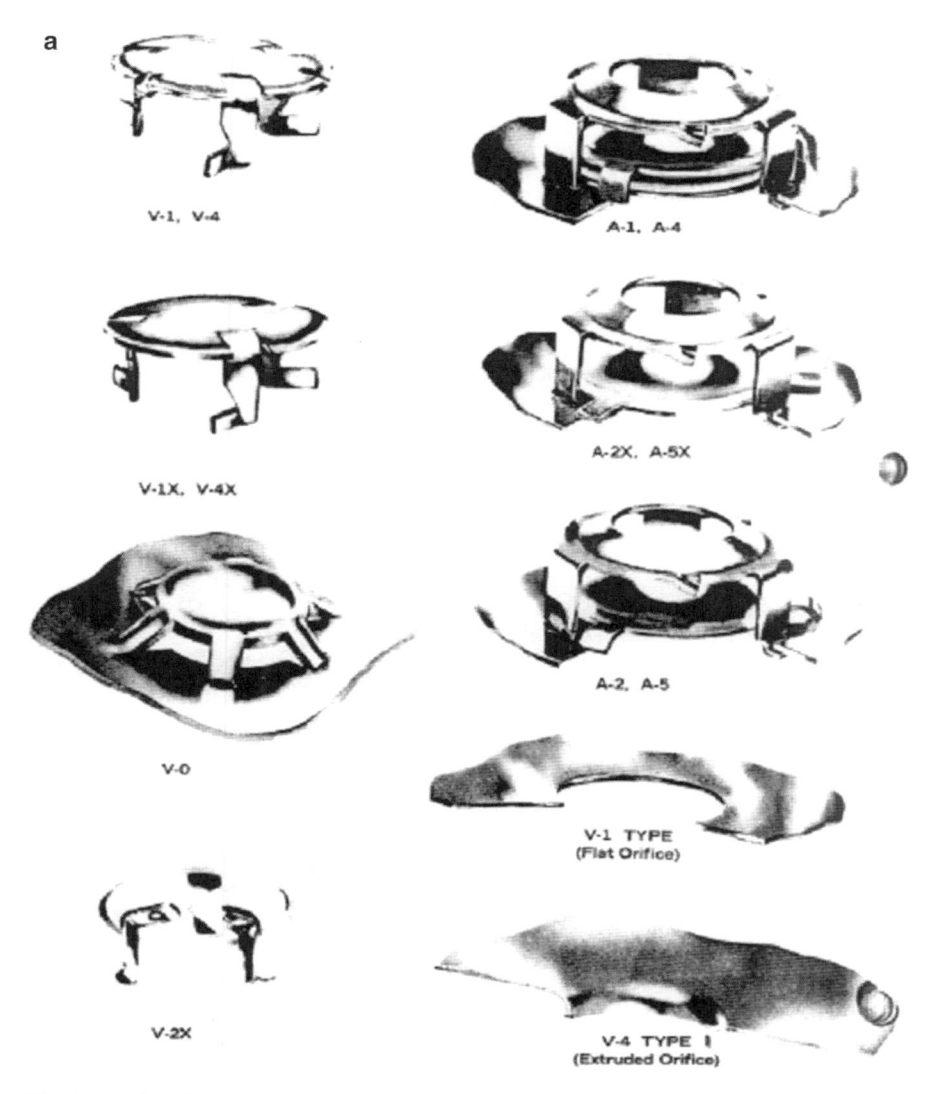

V-1, V-4

A-1, A-4

V-1X, V-4X

A-2X, A-5X

V-0

A-2, A-5

V-1 TYPE
(Flat Orifice)

V-2X

V-4 TYPE I
(Extruded Orifice)

Fig. 2 (continued)

b

Description of Ballast® Units

The various types of Ballast units are shown on the page 881. A description of each unit follows:

V-0 A non-moving unit similar in appearance to the V-1 in a fully open position. It is used in services where only moderate flexibility is required and minimum cost is desired.

V-1 A general purpose standard size unit, used in all services. The legs are formed integrally with the valve for deck thicknesses up to ⅝".

V-2 The V-2 unit is similar to the V-1 unit except the legs are welded-on in order to create a more leak-resistant unit. The welded legs permit fabrication of Ballast units for any deck thickness or size. Large size units are frequently used for replacement of bubble caps.

V-3 A general purpose unit similar to the V-2 unit except the leg is radial from the cap center.

V-4 This signifies a venturi-shaped orifice opening in the tray floor which is designed to reduce substantially the parasitic pressure drop at the entry and reversal areas. A standard Ballast unit is used in this opening normally, although a V-2 or V-3 unit can be used for special services. The maximum deck thickness permissible with this opening is 10 gage.

V-5 A combination of V-0 and V-1 units. It normally is used where moderate flexibility is required and a low cost is essential.

A-1 The original Ballast tray with a lightweight orifice cover which can close completely. It has a separate Ballast plate to give the two-stage effect and a cage or travel stop to hold the Ballast platc and orifice cover in proper relationsip.

A-2 The same as A-i, except the orifice cover is omitted.

A-4 An A-1 unit combined with a venturi-shaped orifice opening in order to reduce the pressure drop.

A-5 An A-2 unit combined with a venturi-shaped opening.

The diameter of the standard size of the V-series of Ballast units is 1⅞". The V-2 and V-3 units are available in sizes up to 6".

Photographs of several Ballast trays are shown on page 8 and 9.

Fig. 2 (**a, b**) Valve unit types – Glitsch Ballast

The "Quickie" Method

This method is good enough for a reasonable estimate of a tower diameter which can be used for a budget-type cost estimate or initial plant layout studies. The steps used for this calculation are as follows:

Step 1. Establish the liquid and vapor flows for the critical trays in the section of the tower that will give the maximum values. These are obtained by heat balances (see the chapters on "▶ Atmospheric and Vacuum Crude Distillation Units in Petroleum Refineries" and "▶ Distillation of the "Light Ends" from Crude Oil in Petroleum Processing" of this handbook). The critical trays are:

Process Equipment for Petroleum Processing

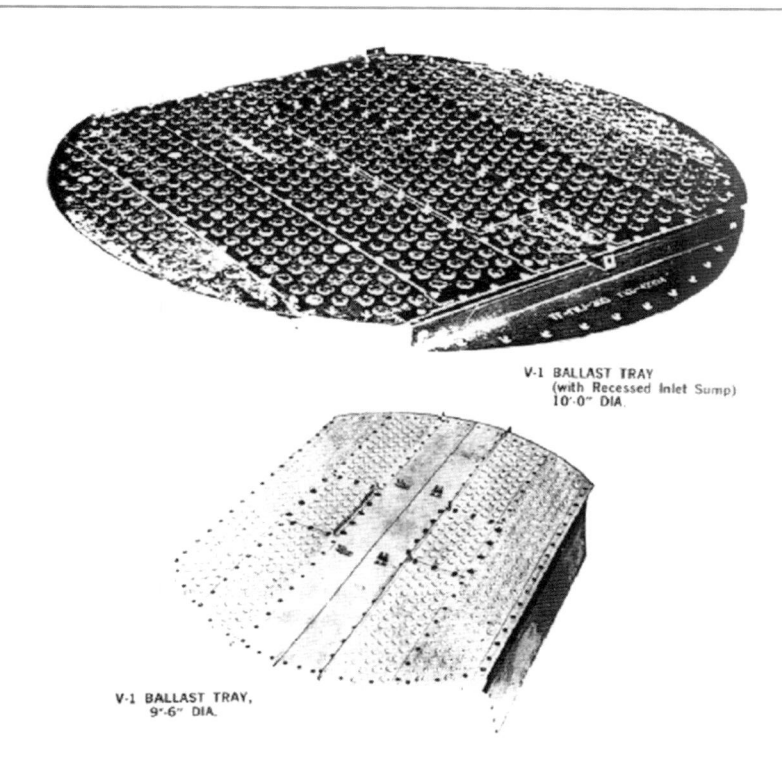

Fig. 3 Valve trays – Glitsch Ballast

The top tray
Any side stream draw-off tray
Any intermediate reflux draw-off tray
The bottom tray

Step 2. Calculate the actual cuft/h at tray conditions of the vapor. Then using the total mass per hour of the vapor, calculate its vapor density in lbs/cuft.

Step 3. From the heat balance, determine the density of the liquid on the tray at tray temperature in lbs/cuft.

Step 4. Select a tray spacing. Start with a 24″ space. Read from Fig. 4 a value for "K" on the flood line, using the equation:

$$G_f = K\sqrt{\rho_v \times (\rho_l - \rho_v)} \tag{1}$$

where:

G_f = mass vapor velocity in lbs/h sqft at flood.
K = the constant read from the flood curve in Fig. 4.
ρ_v = density of vapor at tray conditions in lbs/cuft.
ρ_L = density of liquid at tray conditions in lbs/cuft.

Step 5. Multiply G_f by 0.82 to give mass velocity at 82 % of flood which is the normal recommended design figure. Divide the actual vapor rate in lbs/h by the vapor mass velocity to give the area of the tray. Calculate tray diameter from this area.

Fig. 4 Tray spacing versus "K" factor

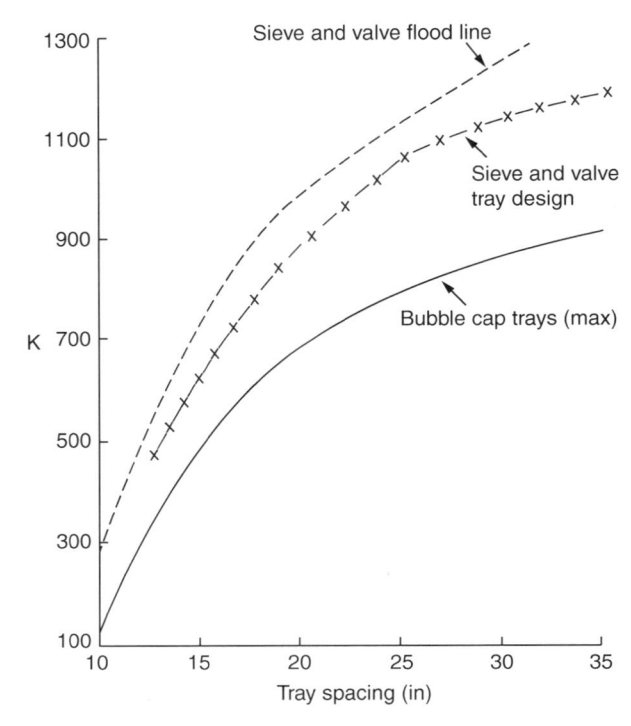

An example calculation now follows.

Example Calculation
Calculate the diameter of the tower to handle the liquid and vapor loads as follows:

Vapor to tray	Liquid from tray
lbs/h = 47,700	GPH at 60 = 119.7
moles/h = 929.7	Hot GPH = 153.0
ACFS = 7.83	Hot CFS = 0.339
lbs/cuft ρ_v = 1.69	lbs/h = 33,273
Temp °F = 167	lbs/cuft ρ_L = 27.3
Pressure PSIA = 220	Temp °F = 162

Tray spacing is set at 24″ and the trays are valve type.
From Fig. 4, "K" = 1,110 at flood.

$$\rho_v = 1.69 \, \text{lbs/cuft at tray conditions.}$$

To calculate ρ of vapor at tray conditions, use

$$\rho = \frac{\text{wt/h} \times 520 \times \text{pressure (psia)}}{378 \times 14.7 \times \text{moles/h} \times \text{Temp °R}}$$

Process Equipment for Petroleum Processing

where:

Press is tray pressure.
Temp °R is tray temperature in °F + 460.

$$\rho_L = 27.3 \text{ lbs/cuft.}$$

$$\text{Then } G_f = 1,110\sqrt{1.69 \times (27.3 - 1.69)}$$

$$= 7,302 \text{ lbs/h sqft}$$

$$\text{Area of tray at 82\% of flood} = \frac{47,700}{0.82 \times 7,302} \text{ sqft}$$

$$= 7.97 \text{ sqft}$$

$$= 3.18 \text{ ft. say } 39''.$$

The tray dimensions and configuration for design purposes are subject to a much more rigorous examination. This is normally undertaken by the tray manufacturer from data supplied by the purchaser's process engineer. However, this process engineer needs to be able to check the manufacturer's offer before committing to purchase. The following calculation procedure offers a rigorous method for this purpose which establishes tray size and geometry. This calculation is based on a method developed by Glitsch Inc., a major manufacturer of valve and other types of trays and packing. Computer simulation software is also available and used for design checks. Software has largely replaced hand calculations.

The Rigorous Method

A rigorous method used in the design of valve trays is described by the following calculation steps:

Step 1. Establish the liquid and vapor flows as described earlier for the "quickie" method.

Step 2. Calculate the downcomer design velocity V_{dc} using the following equations (or by Fig. 5):

(a)
$$V_{dc} = 250 \times \text{system factor.} \tag{2}$$

(b)
$$V_{dc} = 41 \times \sqrt{(\rho_L - \rho_v)} \times \text{system factor.} \tag{3}$$

(c)
$$V_{dc} = 7.5 \times \sqrt{TS} \times \sqrt{(\rho_L - \rho_v)} \times \text{system factor} \tag{4}$$

where TS = tray spacing, in inches.
Use the lowest value for the design velocity in gpm/sqft.
Downcomer system factors are given in Table 1.

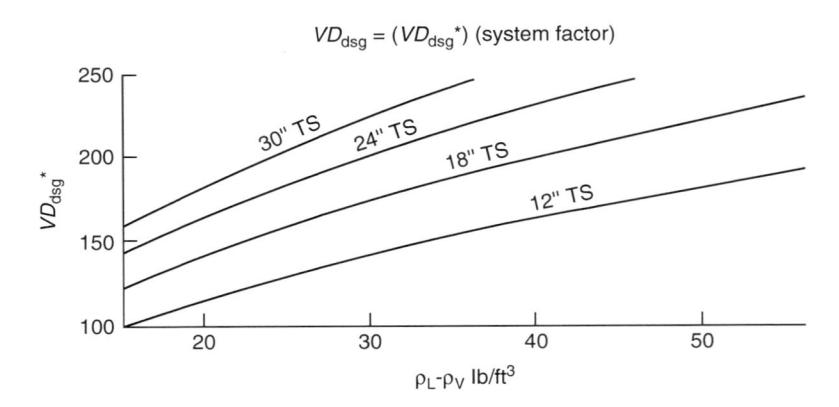

Fig. 5 Downcomer design velocity

$$VD_{dsg} = (VD_{dsg}^*)\text{ (system factor)}$$

(x-axis: $\rho_L - \rho_v$ lb/ft³; y-axis: VD_{dsg}^*; curves labeled 30″ TS, 24″ TS, 18″ TS, 12″ TS)

Table 1 Downcomer system factors

Service	System factor
Non-foaming, regular system	1.0
Fluorine systems	0.9
Moderate foaming (amine units)	0.85
Heavy foaming (glycol, amine)	0.73
Severe foaming (MEK units)	0.6
Foam stable systems (caustic regen)	0.3

Step 3. Calculate the vapor capacity factor (CAF) using Fig. 6.

$$\text{CAF} = \text{CAF}_o \times \text{system factor.} \tag{5}$$

System factors used for this equation are given in Table 2.

Step 4. Calculate the vapor load using the equation:

$$V_1 = \text{CFS}\sqrt{\rho_v/(\rho_L - \rho_v)} \tag{6}$$

where CFS = actual vapor flow in cuft/s.

Step 5. Establish tower diameter using Fig. 7. Tray spacing is usually 18″, 24″, or 30″ for normal towers operating at above atmospheric pressures. Large vacuum towers may have tray spacing 30–36″. Note this diameter may be increased if other criteria of tray design are not met.

Step 6. Calculate the approximate flow path length (FPL) based on tower diameter from step 5 using the equation:

$$\text{FPL} = 9 \times \text{DT}/\text{NP} \tag{7}$$

Process Equipment for Petroleum Processing

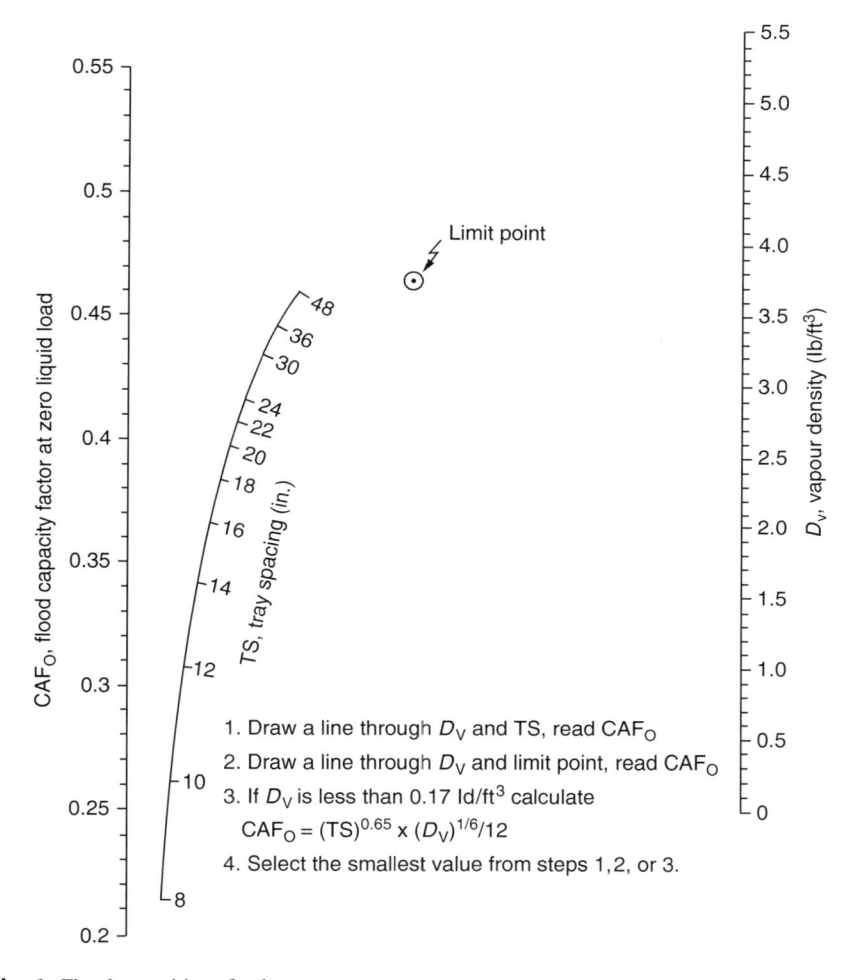

Fig. 6 Flood capacities of valve trays

Chart instructions:
1. Draw a line through D_V and TS, read CAF_O
2. Draw a line through D_V and limit point, read CAF_O
3. If D_V is less than 0.17 ld/ft^3 calculate
   $CAF_O = (TS)^{0.65} \times (D_V)^{1/6}/12$
4. Select the smallest value from steps 1, 2, or 3.

Axes: CAF_O, flood capacity factor at zero liquid load (0.2–0.55); TS, tray spacing (in.) (8–48); D_V, vapour density (lb/ft^3) (0–5.5). Limit point marked on chart.

Table 2 Vapor system factors

Service	System factor
Non-foaming, regular	1.0
Fluorine systems	0.9
Moderate foaming	0.85
Heavy foaming	0.73
Severe foaming	0.6
Foam stable system	0.3–0.6

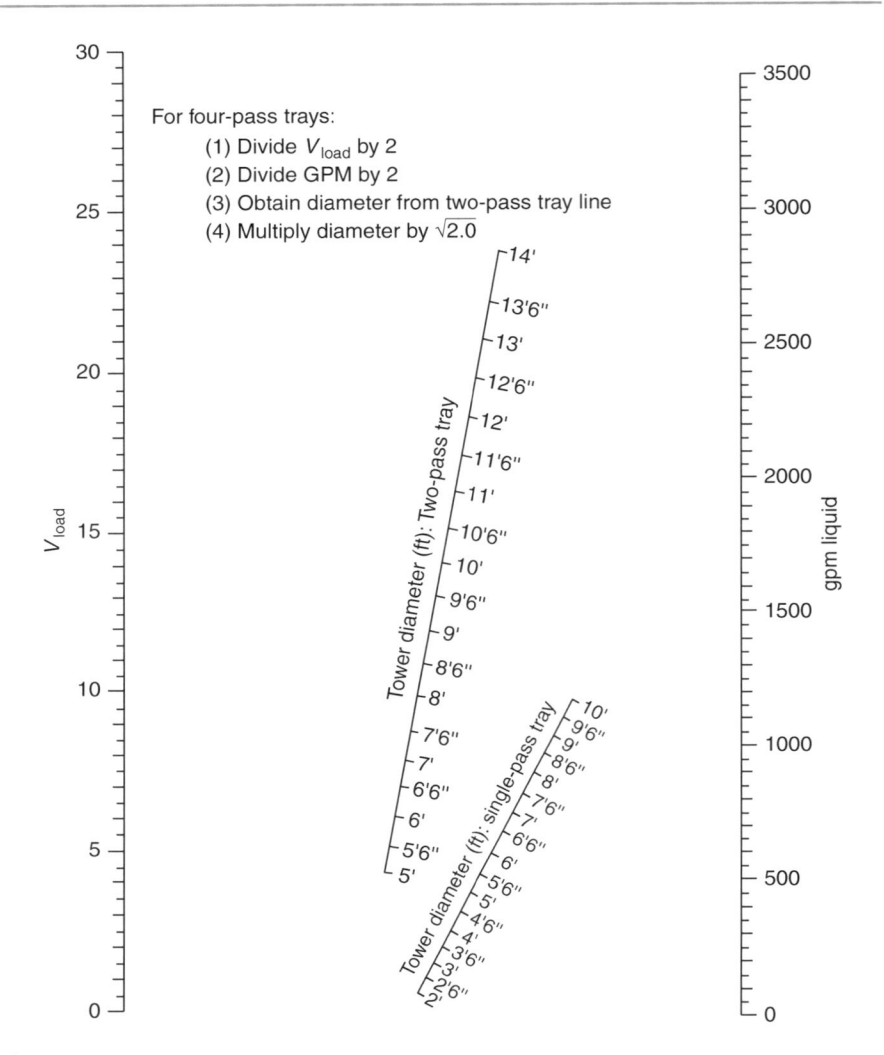

Fig. 7 Tray diameter versus vapor loads

where:

FPL = flow path length in inches.

DT = tower diameter from step 5 in ft.

NP = number of passes. For small towers with moderate liquid flows, this will be 1. For larger towers, this will depend on liquid velocities in the downcomer. The highest number of passes is usually 4.

Step 7. Calculate the *minimum* active area (AA_m) using the expression

$$AA_m = \frac{V_1 + (L \times FPL/13,000)}{CAF \times FF} \tag{8}$$

Process Equipment for Petroleum Processing 1453

where:
AA_m = minimum active area in sqft.
V_1 = vapor load in CFS.
L = liquid flow in actual gpm.
FPL = flow path length in inches.
CAF = capacity factor from step 3.
FF = Flood Factor usually 80–82 %.

Step 8. Calculate *minimum* downcomer area (AD_m) using the equation

$$AD_m = \frac{L}{V_{dc} \times 0.8} \qquad (9)$$

where:
AD_m = minimum downcomer area in sqft.
L = actual liquid flow in gpm.
V_{dc} = design downcomer velocity from step 2.

Note: The downcomer liquid velocity using the calculated minimum downcomer area should be around 0.3–0.4 ft/s.

Step 9. Calculate the *minimum* tower cross-sectional area using the following equations:

$$AT_m = AA_m + 2AD_m \qquad (10)$$

or

$$AT_m = \frac{V_1}{0.78 \times CAF \times 0.8} \qquad (11)$$

where
AT_m = minimum tower cross-sectional area in sqft.

Step 10. Calculate actual downcomer area using the following equation:

$$AD_c = \frac{AT \times AD_m}{AT_m} \qquad (12)$$

where:
AD_c = actual downcomer area in sqft.
AT = tower area in sqft from the diameter calculated in step 5.

Step 11. Determine downcomer width (H_1) (from Appendix 1 of this chapter) for side downcomers. For multi-pass trays, use the following equation with the width factors given in Table 3:

$$H_i = WF \times \frac{AD}{DT} \qquad (13)$$

Table 3 Allocation of downcomer area and width factors

		Fraction of total downcomer area		
No of passes	AD1	AD2	AD3	AD4
1	0.5 % each	1.0	–	–
2	0.34 % each	–	0.66	–
3	0.25 % each	0.5	0.5 each	–
4	0.2	–	0.4	0.4
		Width factors (WF)		
Passes	H4	H5	H7	
1	12.0	–	–	
2	–	8.63	–	
3	6.0	6.78 each pass	–	
4	–	5.66	5.5	

where:
H_i = width of individual downcomers in inches.
WF = width factor from Table 3.
AD = total downcomer area in sqft.
DT = actual tower diameter in ft.
See Fig. 8 for allocation of downcomers in multi-pass trays.

Step 12. Calculate the actual FPL from the equation:

$$FPL = \frac{12 \times DT - (2H_1 + H_3 + 2H_5 + 2H_7)}{NP} \quad (14)$$

where:
H_{1-7} are individual downcomer widths in inches (see Fig. 8).
NP = number of passes.

Step 13. Calculate actual active area (AA) using values for H calculated in step 12 and Appendix 1 of this chapter to establish inlet areas of multi-pass tray downcomers.

$$AA = AT - (2AD_1 + AD_3 + 2AD_5 + 2AD_7)$$

where:
AA = actual active area in sqft.
AT = actual tower area in sqft.
AD_{1-7} are individual downcomer areas in sqft for multi-pass trays corresponding to H_{1-7}.

Step 14. From the data now developed, calculate the actual percent of flood or flood factor (FF). The following expression is used for this:

$$\% Flood = \frac{V_1 + (L \times FPL/13,000) \times 100}{AA \times CAF}. \quad (15)$$

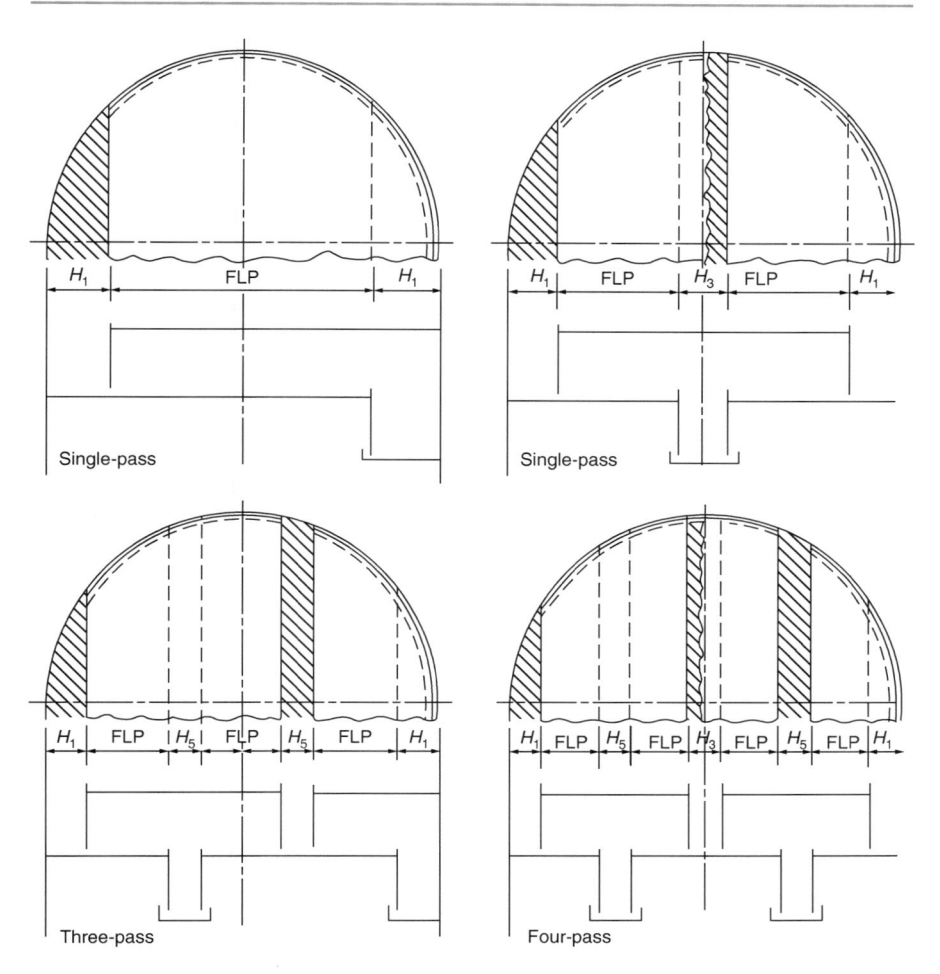

Process Equipment for Petroleum Processing 1455

Fig. 8 Types of trays

Step 15. Calculate vapor hole velocity V_h assume 12–14 units (holes) per sqft of AA. Then:

$$V_h = \frac{\text{CFS} \times 78.5}{\text{NU}} \qquad (16)$$

where:
 V_h = hole velocity in ft/s.
 CFS = actual cuft/s of the vapor.
 NU = total number of units.

Step 16. Calculate dry tray pressure drops from:

$$\text{Valves partly open}: \Delta P_D = 1.35\, t_m \rho_m / \rho_1 + K_1(V_h)(\rho_v / \rho_1) \qquad (17)$$

Table 4 Pressure drop coefficients (Glitch Ballast-type trays)

Type of unit	K1	← K$_2$ for deck thickness (in.) →			
		0.074	0/104	0.134	0.25
V1	0.2	1.05	0.92	0.82	0.58
V4	0.1	0.68	0.68	0.68	–

← Thickness →		← Densities of valve material →	
Gauge	tm (ins)	Metal	Density (lbs/cuft)
20	0.37	CS	400
18	0.5	SS	500
16	0.6	Ni	553
14	0.074	Monel	550
12	0.104	Titanium	283
10	0.134	Hastelloy	560
		Aluminum	168
		Copper	560
		Lead	708

where:
ΔP_D = dry tray valve pressure drop in inches liquid.
t_m = valve thickness in inches (see Table 4).
ρ_m = valve metal density in lbs/cuft (see Table 4).
K_1 = pressure drop coefficient (see Table 4).

$$\text{Valves fully open}: \Delta P_D = K_2(V_h)^2 \rho_v / \rho_l \qquad (18)$$

where K_2 = pressure drop coefficient (see Table 4).
Step 17. Calculate total tray pressure drop.

$$\Delta P = \Delta P_D + 0.4(L/L_{wi})^{0.87} + 0.4 H_w \qquad (19)$$

where:
ΔP = total tray pressure drop in inches of liquid.
L_{wi} = weir length in inches (from Appendix 1).
H_w = weir height in inches (usually 1–2).
Step 18. Calculate height of liquid in downcomer.
First calculate the head loss under the downcomer H_{UD}, where $H_{UD} = 0.65 (V_{UD})^2$. V_{UD} is calculated from the liquid velocity in CFS or gpm/450 divided by the area under the downcomer. Use weir length times weir height for the area. This velocity should be around 0.3–0.6 ft/s for most normal towers.
Then:

$$H_{dc} = H_w + 0.4(L/L_{wi})^{0.67} + (\Delta P + H_{UD})(\rho_l/(\rho_l - \rho_v)). \qquad (20)$$

Process Equipment for Petroleum Processing 1457

where:

H_{dc} = height of liquid in downcomer in inches.
For normal design, this should not exceed 50 % of tray spacing.

An example calculation now follows:

Example calculation

In this example, the same liquid and vapor flows and data will be used as in the "quickie calculation." The objective of this calculation will be to determine the tower diameter, tray pressure drop and configuration, and the percent flood for the design flow rates given.

Calculating the downcomer design velocity V_{dc}:
System factor in this case is 1.

$$V_{dc} = 250 \times 1.0 = 250.$$
$$\text{Or } V_{dc} = 41 \times (\rho_l - \rho_v) \times 1.0 \quad (21)$$
$$= 41 \times 5.06 \times 1.0 = 207.$$

$$\text{Or } V_{dc} = 7.5 \times \text{TS} \times (\rho_l - \rho_v) \times 1.0$$
$$= 7.5 \times 4.9 \times 5.06 \times 1.0 = 186 \text{ gpm/sqft (TS = tray spacing = 24''}).$$
$$\text{Or } V_{dc} = 188 \text{ from Fig. 5, use } V_{dc} = 186 \text{ gpm/sqft}. \quad (22)$$

Vapor capacity factor CAF:
System factor in this case is 1.
From Fig. 6, CAF = 0.43.

Actual vapor load V_1:

$$V_1 = \text{CFS} \sqrt{\rho_v/(\rho_l - \rho_v)}$$
$$= 7.83 \sqrt{1.69/(27.3 - 1.69)} \quad (23)$$
$$= 2.01 \text{ cft/s}.$$

Approximate tower diameter AT:
Using Fig. 7:
$V_1 = 2.01$ cft/s.
TS = 24''.
$L = 153$ gpm.
Tower ID = 3.25 ft = 39''.
Area = 8.30 sqft.

Calculate approximate flow path length FPL:

$$\text{FPL} = \frac{9 \times 39}{\text{No. of passes}} \times 12 = 29.25 \text{ in.}$$

Calculate minimum active area AA_m:

$$AA_m = \frac{V_1 + (L \times FPL/13{,}000)}{CAF \times FF}$$
$$= \frac{2.01 + (153 \times 29.25/13{,}000)}{0.43 \times 0.8} \text{ (using 80\% flood)} \quad (24)$$
$$= 3.01 \text{ sqft.}$$

Calculating minimum downcomer area AD_m:

$$AD_m = \frac{L}{V_{dc} \times 0.8}$$
$$= \frac{153}{186 \times 0.8} = 1.028 \text{ sqft.} \quad (25)$$

Calculating the minimum tower cross-sectional area:

$$\text{Either } AA_m + 2AD_m = 3.01 + 2.056$$
$$= 5.066 \text{ sqft.}$$
$$\text{or } \frac{V_1}{0.78 \times CAF \times 0.8} = 7.49 \text{ sqft} \quad (26, 27)$$

Use the larger which is 7.49 sqft.
Min diameter therefore is 3.09 f. say 37".
Calculating actual downcomer area AD_c:

$$AD_c = \frac{AT \times AD_m}{AT_m} = \frac{8.3 \times 1.028}{7.49}$$
$$= 1.14 \text{ sqft.}$$

Downcomer area/tower area ("A" in Appendix 1) $= AD/AT = 1.14/8.30 = 0.137$.
$$(28)$$

From Appendix 1 "R" $= 0.197$, then

$$H = 0.197 \times 3.25 = 0.633 \text{ ft} = 7.6'' \text{ DC width.}$$

Recalculating flow path length FPL:

$$FPL = 12 \times D_t - (2H)$$
$$= 12 \times 3.25 - (2 \times 7.6) \quad (29)$$
$$= 23.8 \text{ in.}$$

Recalculating active area based on actual downcomer area:

$$\begin{aligned} AA &= AT - (2AD) \\ &= 8.3 - 2.28 \\ &= 6.02 \text{ sqft, which is greater than min allowed.} \end{aligned} \tag{30}$$

Checking percent of flood:

$$\begin{aligned} \% \text{ flood} &= \frac{V_1 + (L \times \text{FPL}/13{,}000)}{AA \times \text{CAF}} \times 100 \\ &= \frac{2.01 + (153 \times 23.8/13{,}000)}{6.02 \times 0.43} \times 100 \\ &= 88.0\% \quad \text{which is a little high for design, but is acceptable.} \end{aligned} \tag{31}$$

Check downcomer velocity:

$$\text{CFS of liquid} = 0.34 \text{ cfs.}$$
$$\text{Area of down comer} = 1.14 \text{ sqft.}$$
$$\text{Velocity of liquid in downcomer} = \frac{0.34}{1.14}$$
$$= 0.3 \text{ ft/s.}$$

Calculating pressure drops and downcomer liquid height
- *Dry tray pressure drop*

Partially open valves:

$$\begin{aligned} \Delta P_D &= 1.35\, t_m \rho_m / \rho_1 + K_1 (V_h)^2 (\rho_v / \rho_1). \\ V_h &= \frac{7.83 \times 78.5}{72} \quad \text{(assumes 12 units per sqft of AA)}. \\ &= 8.5 \text{ ft/s.} \\ \Delta P_D &= 1.35 \times 0.74'' \times (490/27.3) + (0.2 \times 72.25 \times 0.062) \\ &= 2.69'' \text{ liquid.} \end{aligned} \tag{32}$$

Fully open valves:

$$\begin{aligned} \Delta P_D &= K_2 (V_h)^2 \cdot (\rho_v / \rho_1) \\ &= 0.92 \times 72.25 \times 0.062 \\ &= 4.12'' \text{ liquid. This will be used.} \end{aligned} \tag{33}$$

- *Total tray pressure drop:*

$$\Delta P = \Delta P_D + 0.4 \, (L/L_{wi})^{0.67} + (0.4 \times H_w). \tag{34}$$

H_w (weir height) is fixed at 2".

L_{wi} (downcomer length) is calculated from Appendix 1 of this chapter as 30.9".

$$\Delta P = 4.12 + 0.4\,(153/30.9)^{0.67} + 0.89''$$
$$= 6.09'' \text{ of liquid.}$$

Height of liquid in downcomer:

$$H_{dc} = H_w + 0.4\,(L/L_{wi})^{0.67} + (\Delta P + H_{UD})(\rho_1/(\rho_1 - \rho_v)).$$
$$H_{dc} = 2 + 0.4 \times 1.16 + (6.09 + 0.405)(27.3/25.61) \tag{35}$$
$$= 10.08'' \text{ liquid. This is } 42.0\% \text{ of tray spacing which is acceptable.}$$

(H_{UD} was calculated using a downcomer outlet area of $L_{wi} \times 2''$ giving a velocity of 0.339 CFS divided by 0.429 sqft which is 0.79 ft/s. H_{UD} is then 0.65 $(0.79)^2 = 0.405$).

Calculating the actual number of valves for tray layout. With truss lines parallel to liquid flow:

$$\text{Rows} = \left[\frac{\text{FPL} - 8.5}{0.5 \times \text{base}} + 1\right]\text{NP} \tag{36}$$

where:

Base = spacing of units usually 3.0", 3.5", 4.0", 4.5", or 6.0".

$$\text{Units/row} = \frac{\text{WFP}}{5.75 \times \text{NP}} - (0.8 \times \text{number of beams}) + 1. \tag{37}$$

With truss lines perpendicular to liquid flow

$$\text{Rows} = \left[\frac{\text{FPL} - (1.75 \times N_o \text{ trusses} - 6.0)}{2.5}\right]\text{NP}.$$
$$\text{Units/row} = \frac{\text{WFP}}{\text{base} \times \text{NP}} - (2 \times N_o \text{ major beams}) + 1. \tag{38, 39}$$

where:

$$\text{WFP} = \text{width of flow path in inches}$$
$$= \text{AA} \times 144/\text{FPL}.$$

Using a base pitch of 3.5", the number of rows on the trays with trusses parallel to flow was calculated to be 9.7. Units per row were then calculated to be 8.73. This gives a total number of valves over the active area as 84.7. Thus, the number of valves per sqft of AA is 14. The assumption of 12 in the calculation gives a more stringent design; therefore, the assumption is acceptable.

Calculation summary:

> Tower diameter = 3.25 ft or 39″.
> Downcomer area (ea.) = 1.14 sqft (single pass).
> Active area = 6.02 sqft.
> Percent of flood = 88%.
> Tray spacing = 24″.
> Downcomer backup = 42.0% of tray spacing.
> Number of valves = 85.
> Number of rows = 10.
> Valve pitch = 3.5″.

Packed Towers and Packed Tower Sizing

Although trayed towers are generally the first choice for fractionation and absorption applications, there are a number of instances where packed towers are preferable. For example, on small-diameter towers (below 3 f. diameter), packed towers are generally cheaper and more practical for maintenance, fabrication, and installation. At the other end of the spectrum, packing in the form of grids and large stacked packed beds have superseded trays in vacuum distillation towers whose diameter range up to 30 f. in some cases. This is because packing offers a much lower pressure drop than trays.

The packing in the tower itself may be stacked in beds on a random basis or in a defined structured basis. For towers up to 10–15 ft, the packing is usually dumped or random packed. Above this tower size and depending on its application, the packing may be installed on a defined stacked or structured manner. For practical reasons and to avoid crushing the packing at the bottom of the bed, the packing is installed in beds. As a rule of thumb, packed beds should be around 15 f. in height. About 20 f. should be a maximum for most packed sections.

Properties of good packing are as follows:

- Should have high surface area per unit volume.
- The shape of the packing should be such as to give a high percentage of area in active contact with the liquid and the gas or in the two liquid phases in the case of extractors.
- The packing should have favorable liquid distribution qualities.
- Should have low weight but high unit strength.
- Should have low pressure drop but high coefficients of mass transfer.

Some data on the various common packing available commercially are given in Tables 5, 6, and 7. Figure 9 shows a sectional layout of a typical packed tower. Note this tower has bed supports designed for gas distribution and includes intermediate weir liquid distributors between some of the beds.

Table 5 Physical properties of some common packing

Packing type	Size (in.)	Wall thickness (in.)	OD and length	Approx. no./per ft^3	Approx. wt/per ft^3	Approx. surface area (ft^2 ft^3)	% void volume
Raschig rings (ceramic)	1/4	1/32	1/4	88,000	46	240	73
	5/16	1/16	5/16	40,000	56	145	64
	3/8	1/16	3/8	24,000	52	155	68
	1/2	1/32	1/3	10,600	54	111	63
	1/2	1/16	1/2	10,600	48	114	74
	5/8	3/32	3/8	5,600	48	100	68
	3/4	3/32	3/4	3,140	44	80	73
	1	1/3	1	1,350	40	58	73
	1-1/4	3/16	1-1/2	680	43	45	74
	1-1/2	1/4	1-2/3	375	46	35	68
	1-1/2	3/16	1-1/2	385	42	38	71
	2	1/4	2	162	38	28	74
	2	3/16	2	164	35	29	78
	3	3/8	3	48	40	19	74
Raschig rings (metal)	1/4	1/32	1/4	88,000	150	236	69
	5/16	1/32	5/6	45,000	120	190	75
	5/16	1/16	5/16	43,000	198	176	60
	1/2	1/32	4/2	11,800	77	128	84
	1/2	1/16	1/2	11,000	132	118	73
	19/32	1/32	19/32	7,300	66	112	86
	19/32	1/32	19/32	7,000	120	106	75
	3/4	1/32	3/4	3,400	55	84	88
	3/4	1/16	3/4	3,190	100	72	78
	1	1/32	1	1,440	40	63	92
	1	1/16	1	1,345	73	57	85
	1-1/4	1/16	1-1/2	725	62	49	87
	1-1/2	1/16	1-1/2	420	50	41	90
	2	1/16	2	180	38	31	92
	3	1/16	3	53	25	20	95
Raschig rings (carbon)	1/4	1/16	1/4	85,000	46	212	55
	1/2	1/16	1/2	10,600	27	114	74
	3/4	1/8	3/4	3,140	34	75	67
	1	1/3	1	1,325	27	57	74
	1-1/4	1/16	1-1/4	678	31	45	69
	1-1/2	1/2	1-1/2	392	34	37	67
	2	1/5	2	166	27	28	74
	3	5/16	3	49	33	19	78
Berl saddles (ceramic)	1/4	–	–	113,000	56	274	60
	1/2	–	–	16,200	54	142	63
	3/4	–	–	5,000	48	82	66

(*continued*)

Table 5 (continued)

Packing type	Size (in.)	Wall thickness (in.)	OD and length	Approx. no./per ft^3	Approx. wt/per ft^3	Approx. surface area (ft^2 ft^3)	% void volume
	1	–	–	2,200	45	76	69
	1-1/2	–	–	580	38	44	75
	2	–	–	250	40	32	72
Intalox saddles (ceramic)	1/4	–	–	117,500	54	300	75
	1/2	–	–	20,700	47	190	78
	3/4	–	–	6,500	44	102	77
	1	–	–	2,385	42	78	77
	1-1/2	–	–	709	37	60	81
	2	–	–	265	38	36	79

Table 6 Coefficients for use in the HETP equation

Packing type	Packing size (in.)	K$_1$	K$_2$	K$_3$
Raschig rings	0.375	2.10	−0.37	1.24
	0.500	0.853	−0.24	1.24
	1.000	0.57	−0.10	1.24
	2.000	0.52	0.00	1.24
Saddles	0.500	5.62	−0.45	1.11
	1.000	0.76	−0.14	1.11
	2.000	0.56	−0.02	1.11

Table 7 Recommended packing sizes

	← Tower diameter ft →				
Packing type	1.0	2.0	3.0	4.0	+ 5.0
Raschig rings	0.5	0.75	1.00	1.5	3.00
Berl saddles	0.75	1.50	2.00	2.00	3.00
Intalox saddles	0.75	1.50	2.00	2.00	3.00
Pall rings	1.00	1.50	2.00	2.00	3.00

Other salient points concerning packed towers are as follows:

1. Reflux ratios, flow quantities, and number of theoretical trays or transfer units are calculated in the same manner as for trayed columns.
2. Internal liquid distributors are required in packed towers to ensure good distribution of the liquid over the beds throughout the tower.

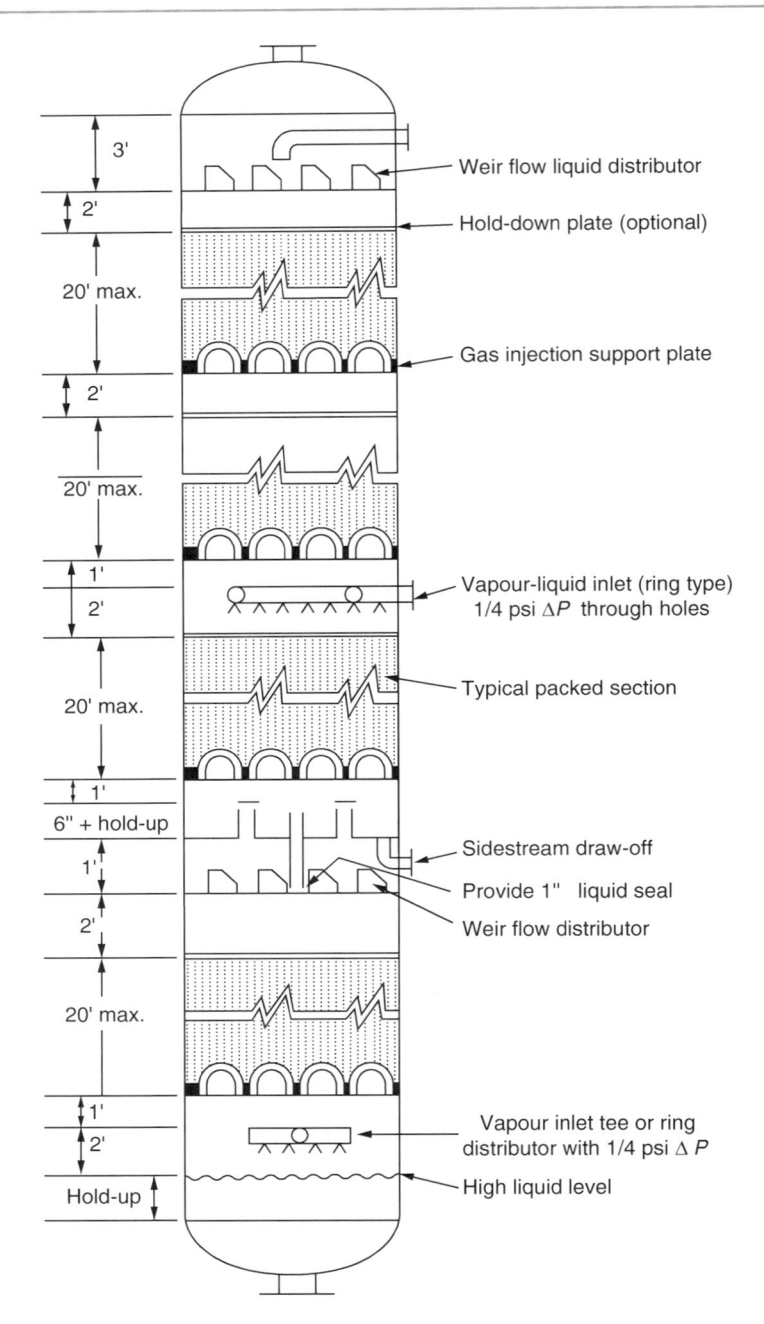

Fig. 9 A typical packed tower

Process Equipment for Petroleum Processing   1465

3. The packed beds are supported by grids. These are specially designed to ensure good flows of the liquid and the gas phases.
4. Every care must be taken in the design of the packed tower that the packing is always properly "wetted" by the liquid phase. Packing manufacturers usually quote a minimum wetting rate for their packing. This is usually around 2.0–2.5 gpm of liquid per sqft of tower cross section. Most companies prefer this minimum to be around 3.0–3.5 gpm/sqft (Tables 5 and 6).

Sizing a Packed Tower

The height of the tower is determined by the methods used to calculate the number of theoretical trays required to perform a specific separation. These have been discussed earlier in the introductory chapter to this handbook. Normally, this calculation is done by process simulation software. A figure equivalent to the height of a theoretical tray is then calculated to determine the height of packing required. This is used as the basis to determine the overall height of the tower by adding in the space required for distributors, support trays, and the like.

The diameter of the tower is calculated using a method which allows for good mass and heat transfer while minimizing entrainment. The same principle of tower flooding is applicable to packed towers as for trayed towers. A calculation procedure for determining a packed tower diameter and the height of packed beds now follows:

Step 1. From examination of the flows of vapor and liquid in the tower, determine the critical section of the tower where the loads are greatest. Usually this is at the bottom of an absorption unit and either the top or bottom of a fractionator.

Step 2. Determine the conditions of temperature and pressure at the critical tower section. This is usually accomplished by bubble and dew point calculations as described in the introductory chapter to this handbook. That is, bubble point of the bottoms liquid (either in a fractionator or an absorber) determines the bottom of the tower conditions, and dew point calculation of the overhead vapor determines the tower top conditions.

Step 3. Establish the liquid and vapor stream compositions at the critical tray conditions. See the chapter on "▶ Atmospheric and Vacuum Crude Distillation Units in Petroleum Refineries" of this handbook for determining vapor/liquid streams in absorption and fractionation towers. Calculate the properties of these streams such as densities, mass/unit time, moles/unit time, viscosity, etc. at the conditions of the critical tower section. You can use process simulation software for this, too. Next select a packing type and size. Use Table 7 for this.

Step 4. Commence the tower sizing by calculating the diameter. First calculate a value for

$$(L/G)\sqrt{(\rho_v/\rho_l)} \tag{40}$$

where:
L = mass liquid load in lbs/s sqft.
G = mass vapor load in lbs/s sqft.

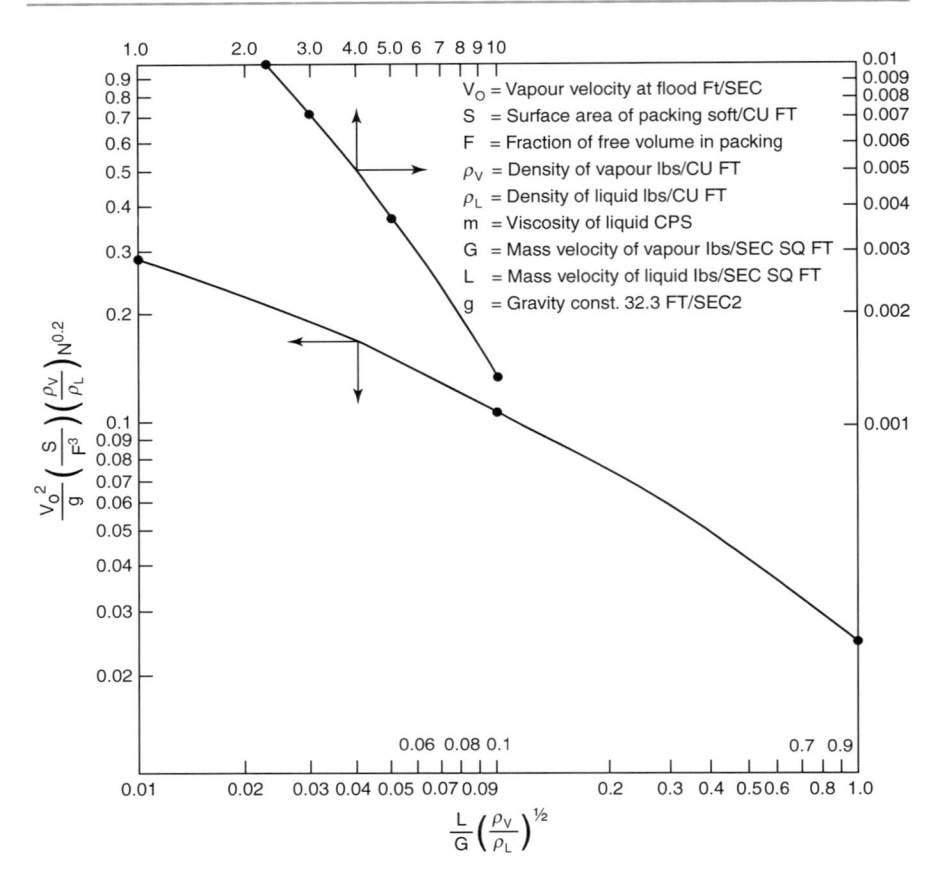

```
                    1.0         2.0    3.0 4.0 5.0 6 7 8 9 10                                           0.01
              0.9                                                                                       0.009
              0.8                                                                                       0.008
              0.7                            V₀ = Vapour velocity at flood Ft/SEC                       0.007
              0.6                            S  = Surface area of packing soft/CU FT                    0.006
              0.5                            F  = Fraction of free volume in packing                    0.005
                                             ρᵥ = Density of vapour lbs/CU FT
              0.4                            ρ_L = Density of liquid lbs/CU FT                          0.004
                                             m  = Viscosity of liquid CPS
              0.3                            G  = Mass velocity of vapour lbs/SEC SQ FT                 0.003
                                             L  = Mass velocity of liquid lbs/SEC SQ FT
              0.2                            g  = Gravity const. 32.3 FT/SEC2                           0.002


              0.1                                                                                       0.001
              0.09
              0.08
              0.07
              0.06
              0.05

              0.04

              0.03

              0.02
                                      0.06 0.08 0.1                               0.7 0.9
              0.01       0.02    0.03 0.04 0.05 0.07 0.09           0.2    0.3 0.4 0.5 0.6  0.8  1.0
```

Fig. 10 Packed tower flooding criteria

ρ_v = density of vapor in lbs/cuft at tower conditions.
ρ_1 = density of liquid in lbs/cuft at tower conditions.
Then, using Fig. 10, read off a value for the equation:

$$(L/G)\sqrt{(\rho_v/\rho_1)} = \frac{V^2 \cdot S \cdot \rho_v \cdot \mu^{0.2}}{g(F)^3 \rho_1} \qquad (41)$$

where:
  V = the vapor velocity at flood in ft/s.
  g = 32.2 ft/s/s.
  S = surface area of packing in sqft/cuft of packing (see Table 5).
  F = fraction of void (see Table 5).
  μ = viscosity of liquid in cPs.
Step 5. Solve the equation from step 4 to give a value for V. This is the superficial velocity of the vapor at flood. Designing for 80 % of flood, multiply V by 0.8.

Process Equipment for Petroleum Processing

Step 6. Divide the total cuft/s of the vapor flowing in the tower by 0.8 V to give the tower cross-sectional area in sqft. Calculate the tower diameter from this area.

Step 7. The next part of the calculation is to determine the height of the tower. The number of theoretical trays has been determined by either the fractionation or absorption calculation described in the handbook chapter entitled "▶ Atmospheric and Vacuum Crude Distillation Units in Petroleum Refineries" or by computer simulation. It is now required to establish either the actual number of trays for a trayed tower or the height of packing in the case of a packed tower. This calculation deals with the second of these.

The next step sets out to establish the HETP which is the height equivalent to a theoretical tray.

Step 8. The HETP is calculated from the following equation:

$$\text{HETP} = K_1 \cdot G_h^{K_2} \cdot D^{K_3} \cdot \frac{62.4 \times \alpha \times \mu^2}{\rho_1} \qquad (42)$$

where:

$K_{1,2,3}$ = factors from Table 6.
D = tower diameter in inches.
α = relative volatility of the more volatile component in the liquid phase (see step 9 below).
μ = viscosity in cPs.
G_h = mass velocity of the vapor in lbs/h sqft.

Step 9. To determine the relative volatility α, select a key light component that is in the lean liquid and the wet gas. The relative volatility is the equilibrium constant of the lightest significant component in the rich liquid divided by the equilibrium constant of the light key component. Solve for a value of HETP and multiply this by the number of theoretical stages to give the total packed height. Software can also perform this calculation.

Step 10. Determine the number of beds to accommodate the packed height. Allow space between the beds for vapor/liquid redistribution and holdup plates. Use Fig. 9 as a guide for this. The tower height will be the sum of beds, internal distributors packing support trays, liquid holdup, and vapor disengaging space. An example calculation now follows.

Example calculation

In this example, the number of theoretical trays for an absorption unit has been fixed as 4. The compositions of the "wet" gas and the lean liquid have been given and used to determine the composition and quantities of the rich liquid and the lean gas. The quantities to be used in the following calculation are as follows:

Rich liquid leaving the bottom of the absorber = 452.66 moles/h.
Wet gas entering the bottom of the tower = 1,018.35 moles/h.

Their respective compositions and conditions are as follows:
Wet gas

	Mole frac	Mole wt	Weight
H_2	0.467	2.0	0.93
C_1	0.190	16.0	3.40
C_2	0.059	30.0	1.77
H_2S	0.242	34.0	8.24
C_3	0.604	44.0	1.32
iC_4	0.006	58.0	0.35
nC_4	0.006	58.0	0.35
Total	1.000	16.0	16.00

Temperature = 95 °F; pressure = 175 psia.

$$\text{Cuft} = \frac{378 \times 14.7 \times 555}{175 \times 520} = 33.89.$$

$$\rho_v = 16/33.89 = 0.473 \text{ lbs/cuft}.$$

Rich liquid

	Mole frac	Mole wt	wt fact	lbs/Gal	vol fact
H_2	0.004	2.0	0.002	–	–
C_1	0.013	16.0	0.208	2.5	0.083
C_2	0.016	30.0	0.480	2.97	0.162
H_2S	0.092	34.0	3.128	6.56	0.477
C_3	0.028	44.0	1.232	4.23	0.291
iC_4	0.011	58.0	0.638	4.68	0.136
nC_4	0.013	58.0	0.754	4.86	0.155
C_9	0.823	128.0	105.344	6.02	17.499
Total	1.000		111.786	5.95	18.803

Temperature = 95 °F
SG at 60°F = 0.715
SG at 95°F = 0.696
ρ_1 at 95°F = 43.3 lbs/cuft

$$(L/G)\sqrt{(\rho_v/\rho_1)} = \frac{V^2 \cdot S \cdot \rho_v \cdot \mu^{0.3}}{gF^3\rho_1} \tag{43}$$

where:
    V = the vapor velocity at flood in ft/s.
    g = 32.2 ft/s.
    S = surface area of packing in sqft/cuft of packing (see Table 5).
      = 36 sqft/cuft.
    F = fraction of void (see Table 5) = 0.79.

μ = viscosity of liquid in cps. = 0.56 cPs.

$$(L/G)\sqrt{(\rho_v/\rho_1)} = \frac{50,670}{16,325} \times \left[\frac{0.473}{43.3}\right]^{1/2} = 0.324.$$

From Fig. 10, y value is = 0.055 for x = 0.324.
Then:

$$\frac{V^2 \times S \times \rho_v \times \mu}{g \times F^3 \times \rho_1} = 0.055.$$

$$V^2 = \frac{0.055}{0.022} = 2.5.$$

$$V = 1.58 \text{ ft/s}.$$

at 80% of flood $V = 1.58 \times 0.8$
$= 1.26$ ft/s.

Total vapor flow = 16,325 lbs/h
$= \frac{16,325}{0.473} = 34,514$ cuft/h.
$= 9.59$ cuft/s

Cross − sectional area $= \frac{9.59}{1.26} = 7.6$ sqft.

Tower diameter = 3.1 ft say 3.25 ft or 39″.

To calculate HETP, use 2″ Berl saddles.

$$\text{HETP} = K_1 \cdot G_H^{K_2} \cdot D^{K_3} \cdot \frac{62.4 \times \alpha \times \mu_1}{\rho_1} \tag{44}$$

where:
$K_1 = 0.56.$
$K_2 = -0.2.$
$K_3 = 1.11.$
α = relative volatility (neglect H_2 and C_1 in liquid composition as noncondensable).
 The key component is C_3 then is KC_2/KC_3, which is 4.6/1.0 = 4.6.
$G_H = 16,325/7.6 = 2,148$ lbs/h·sqft.
D = tower diam = 39″.

$$\text{HETP} = \frac{0.56 \times 39 \times 62.4 \times 4.6 \times 0.56}{(2,148)^{0.02} \times 43.3}$$
$= 25.83$ ft per theoretical tray.

The number of theoretical trays was fixed at four for this separation. Then using four theoretical trays, the total height of packing = 4 × 25.83 ft = 103 ft call it 100 ft. Five packed beds each 20 f. would satisfy the required duty. Using Fig. 9 as a guide, the tower height is developed as follows:

Bottom tan to HLL (holdup). Liquid is fed to a stripping column; therefore, let the holdup time be 3 min to NLL. Then NLL = 6.9 f. say 7.0 f. and HLL = 10 ft.

HLL to vapor inlet distributor. This will be set at 2.0 ft.

Distributor to bottom bed packing support. This will be set at 1.0 ft.

Bottom packed bed support to top of packed bed. Packed height is 20 ft.

Top of bottom bed to bottom of next bed. Set this at 3.0 f. to allow for a liquid weir-type distributor.

Height to top of top bed.
Packed height which is 4 × 20 ft = 80.
4 distributors ⇒ 12 ft.
Total = 92 ft.

Top of top bed to top tan (tangent). Make this 5 f. to allow for liquid distribution tray and liquid inlet pipe.

Total height tan to tan = 130 ft.

Drums and Drum Design

Drums may be horizontal vessels or vertical. Generally drums do not contain complex internals such as fractionating trays or packing as in the case of towers. They are used however for removing material from a bulk material stream and often use simple baffle plates or wire mesh to maximize efficiency in achieving this. Drums are used in a process principally for:

- Removing liquid droplets from a gas stream (knockout pot)
- Separating vapor and liquid streams
- Separating a light from a heavy liquid stream (separators and decanters)
- Surge drums to provide suitable liquid holdup time within a process
- To reduce pulsation in the case of reciprocating compressors

Drums are also used as small intermediate storage vessels in a process.

Vapor Disengaging Drums

One of the most common examples of the use of a drum for the disengaging of vapor from a liquid stream is the steam drum of a boiler or a waste heat steam generator. Here the water is circulated through a heater where it is raised to its boiling point temperature and then routed to a disengaging drum. Steam is flashed off in this drum to be separated from the liquid by its superficial velocity across the area above the water level in the drum. The steam is then routed to a superheater and thus to the steam main. The performance of the steam superheater depends on receiving fairly "dry" saturated steam, that is, steam containing little or no water droplets. The separation mechanism of the steam drum is therefore critical. The design of a vapor disengaging drum depends on the velocity of the vapor and the area of disengagement. This is expressed by the equation:

$$V_c = 0.157\sqrt{\frac{\rho_1 - \rho_v}{\rho_v}} \tag{45}$$

where:

V_c = critical velocity of vapor in ft/s.
ρ_1 = density of liquid phase in lbs/cuft.
ρ_v = density of vapor phase in lbs/cuft.

The area used for calculating the linear velocity of the vapor is:

- The vertical cross-sectional area above the high liquid level in a horizontal drum
- The horizontal area of the drum in the case of vertical drums

The allowable vapor velocity may exceed the critical, and normally design velocities will vary between 80 % and 170 % of critical. Severe entrainment occurs however above 250 % of critical. Table 8 gives the recommended design velocities for the various services. The minimum vapor space above the liquid level in a horizontal drum should not be less than 20 % of drum diameter or 12″, whichever is greater.

Crinkled wire mesh screens (*CWMS*) screens are effective entrainment separators and are often used in separator drums for that purpose. When installed they improve the separation efficiency so vapor velocities much above critical can be tolerated. They are also a safeguard in processes where even moderate liquid entrainment cannot be tolerated.

CWMS are now readily available as packages that include support plates and installation fixtures. Normally for drums larger than 3 f. in diameter 6″, thick open mesh-type screen is used.

Liquid Separation Drums

The design of a drum to perform this duty is based on one of the following laws of settling:

Stoke's law

$$V = 8.3 \times 10^5 \times \frac{d^2 \Delta S}{\mu} \tag{46}$$

when the Re number is <2.0.
Intermediate law

$$V = 1.04 \times 10^4 \times \frac{d^{1.14} \Delta S^{0.71}}{S_c^{0.29} \times \mu^{0.43}} \tag{47}$$

when the Re number is 2–500.
Newton's law

$$V = 2.05 \times 10^3 \times \left[\frac{d \Delta S}{S_c}\right]^{1/2} \tag{48}$$

Table 8 Some typical drum applications

Service	Liquid surge and distillate	Settling drums	Compressor suction Cent	Compressor suction Recip	Fuel gas KO drums	Steam drums	Water disengaging drums
Allowable vapor velocity without CWMS, % V_c	170		80	80	170	–	170
Allowable vap velocity with one CWMS, % V_c			150	120		100	
Allowable vap velocity with two CWMS, % V_c			–	–		150	
Liquid holdup set by:	Water settling	Settling requirements	10 min liquid spill		Should be at least volume of a 20 f. slug of condensate	1/3 the heater and steam piping volume	50 in. per minimum settling rate for hydrocarbon vapors from water
	Minimum instrument	Minimum instrument	When taking suction from absorbers		Following an absorber – 5 min on total lean oil circulation		Minimum height to low level 1.5 ft
	Controlling Process	Controlling process	For refrigerators – 5 min based on largest cooling unit				
	Inventory requirement						
Normal drum position	Horizontal	Horizontal	Vertical		Vertical	Vertical	Horizontal
Type of nozzle inlet	90° bend	90° bend	Tee dist		Flush	Tee Dist or proprietary	90° bend
Outlet vapor	Flush	–	Flush		Flush	Flush	Flush
Outlet liquid	Flush	Flush	Flush		Flush	Flush	Flush

Process Equipment for Petroleum Processing

Table 9 Typical droplet sizes for liquid-liquid separation calculations

Lighter phase	Heavy phase	Minimum droplet size
0.850 SG and lighter	Water	0.008 in.
Heavier than 0.850	Water	0.005 in.

when the Re number is >500
where:
$$\text{Re number} = \frac{10.7 \times d \cdot V \cdot S_c}{\mu}. \qquad (49)$$

V = settling rate in inches per minute.
d = droplet diameter in inches.
S = droplet specific gravity.
S_c = continuous phase specific gravity.
ΔS = specific gravity differential between the two phases.
μ = viscosity of the continuous phase in cPs.

Table 9 may be used as a guide to estimating droplet size.

The holdup time required for settling is the vertical distance in the drum allocated to settling divided by the settling rate. Some typical applications of drums for this service are given in Table 8.

Settling baffles are often used to reduce the holdup time and the height of the liquid level.

Surge drums

This type of drum, the calculation of holdup time, and surge control have been described in section "Overall Pressure Drop, Including Elevation and Velocity Changes" of the handbook chapter entitled "▶ Process Controls in Petroleum Processing."

Pulsation drums or pots

This type of drum will be described in some detail in part 3 of this chapter in the section on reciprocating compressors.

An example calculation of drum sizing now follows.

Example Calculation

It is required to provide the dimensions and process data for the design of a reflux drum receiving the hydrocarbon distillate, water, and uncondensed hydrocarbon vapor from a distillation column. Details of flow and drum conditions are as follows:

Vapor: 12,000 lbs/h, 40 mole wt, 300 moles/h.
Distillate product: 76,650 lbs/h, SG at $100°F=0.682$.
Reflux liquid: 61,318 lbs/h, SG at $100°F=0.682$.
Water: 17,381 lbs/h.
Temperature of drum: $100°F$.
Pressure of drum: 30 psia.

The drum is to be a horizontal vessel located on a structure 45 f. above grade. The liquid product is to feed another fractionating unit and therefore requires a holdup time of 15 min between LLL and HLL. The vapor leaving the drum is to be routed to fuel gas via a compressor; therefore, complete disengaging of liquid droplets is required. Complete separation of water from the oil is required. However, as the water is routed to a desalter separator from the drum, separation of oil from the water is not critical.

In all probability, the surge volume required by the product will be the determining feature of this design. Setting the liquid levels in the drum will depend on the settling of the water from the hydrocarbon phase. The design will be checked for satisfactory vapor disengaging.

The Design
Step 1 – Calculating the surge volume for the distillate product

$$\text{Holdup time} = 15 \text{ min.}$$
$$\text{Product rate} = \frac{76,650 \text{ lbs/h}}{0.682 \times 62.2} = 1,807 \text{ cuft/h.}$$
$$\text{Holdup volume} = \frac{1,807 \times 15}{60} = 452 \text{ cuft.}$$

Then the volume of liquid between HLL and LLL is 452 cuft. Let this be 60 % of the total drum volume. Then drum volume $452/0.60 = 753$ cuft.

Using a length-to-diameter ratio (L/D) of 3, diameter and length are calculated as follows:

$$753 \text{ cuft} = \frac{\pi \cdot D^2}{4} \times 3D.$$

$$D = \sqrt[3]{\frac{753 \times 4}{3\pi}} = 6.8 \text{ ft make it 7.0 ft.} \tag{50}$$

$$L = 3 \times 7.0 \text{ ft} = 21.0 \text{ ft.}$$

Step 2 – Calculating water settling rate
Using "intermediate law" then:

$$V = 1.04 \times 10^4 \times \frac{d^{1.14} \Delta S^{0.71}}{S_c^{0.29} \times \mu^{0.43}}.$$

$V =$ settling rate in ins/min.
$d =$ droplet size in ins $= 0.008''$.
$S_c =$ SG of continuous phase $= 0.682$.
$S_w =$ SG of water $= 0.993$.
$\Delta S = 0.311$.

$$V = 1.04 \times 10^4 \times \frac{0.004 \times 0.44}{0.895 \times 0.78}$$
$$= 29.2 \text{ in./min.}$$

(51)

Check Re number:

$$\text{Re} = \frac{10.7 \times 0.008 \times 29.2 \times 0.682}{0.56}$$
$$= 3.0 \text{ so use of "intermediate law" is correct.}$$

Step 3 – Setting the distance between bottom tan and LLL

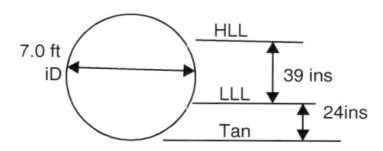

Sufficient distance or surge should be allowed below LLL to provide a LLL alarm at a point about 10 % below LLL and bottom tangent. The remaining surge should be sufficient to provide the operator with some time to take emergency action (such as shutting down pumps).

Let LLL be 2 f. above bottom tan. Then the surge volume in this section is as follows:

$R = 2/7 = 0.286$. From Appendix 1 at 0.237, area of section $= 0.237 \times 38.48 = 9.1$ sqft and volume $= 21 \times 9.1 = 191$ cuft.

$$\text{Total flow rate} = \text{product} + \text{reflux} + \text{water}$$
$$= 3,531.4 \text{ cuft/h} = 58.86 \text{ cuft/min.}$$

Minutes of holdup below LLL $= 3.25$ min.

By the same calculation, holdup after alarm $= 2.9$ min, which is satisfactory.

Step 4 – Checking settling time for the water

At the LLL, a distance of 2 f. from tan.
Residence time for liquid below LLL $= 3.25$ min.
Minimum settling time required:

$$= \frac{\text{Vert distance to bottom of drum}}{\text{Settling rate}} \qquad (52)$$
$$= \frac{24''}{29.2 \text{ in./min}}$$
$$= 0.82 \text{ min,}$$

which is adequate.

Step 5 – Calculating height of HLL above LLL

$$\text{Total volume to HLL} = 191 + 452 = 643 \text{ cuft}$$
$$\text{Area above HLL} = \frac{808 \text{ cuft} - 643 \text{ cuft}}{21 \text{ ft}}$$
$$= 7.58 \text{ sqft.}$$

Using table in Appendix 1 of this chapter:

$$\frac{A_D}{A_s} = \frac{7.58}{38.48} = 0.197 \quad R = 0.251.$$

$$r = 0.251 \times 7.0 = 1.76 \text{ ft.}$$

$$\text{Height of HLL above LLL} = 7 - (1.76 + 2.0)$$
$$= 3.25 \text{ ft. } (39 \text{ in.}).$$

Step 6 – Checking the vapor disengaging space

$$V_c = 0.157\sqrt{\frac{\rho_l - \rho_v}{\rho_v}} \quad (53)$$

where:
 V_c = critical velocity of vapor in ft/s.
 ρ_l = density of liquid phase in lbs/cuft = 42.42.
 ρ_v = density of vapor phase in lbs/cuft = 0.216.
 V_c = 2.28 ft/s.
 Actual velocity of vapor is as follows:

Cross − sectional area of vapor space above HLL = 7.58 sqft

$$\text{Vapor linear velocity} = \frac{59,840 \text{ cuft/h}}{7.58 \times 3,600} = 2.19 \text{ ft/s}$$

which is 96 % of critical.
The drum design meets all necessary criteria.

Specifying Pressure Vessels

The process engineer's responsibility extends to defining the basic design requirements for all vessels. These data include:

- The overall vessel dimensions
- The type of material to be used in its fabrication
- The design and operating conditions of temperature and pressure
- The need for insulation for process reasons
- Corrosion allowance and the need for stress relieving to meet process conditions
- Process data for internals such as trays, packing, etc.
- Skirt height above grade
- Nozzle sizes, ratings, and location (not orientation)

Typical process data sheets used for specifying columns and horizontal drums are given in Figs. 11 and 12 (with their attachments in Fig. 13), respectively. These data sheets have been completed to reflect the examples calculated in this chapter. The following paragraphs describe and discuss the contents of these data sheets.

Process Equipment for Petroleum Processing

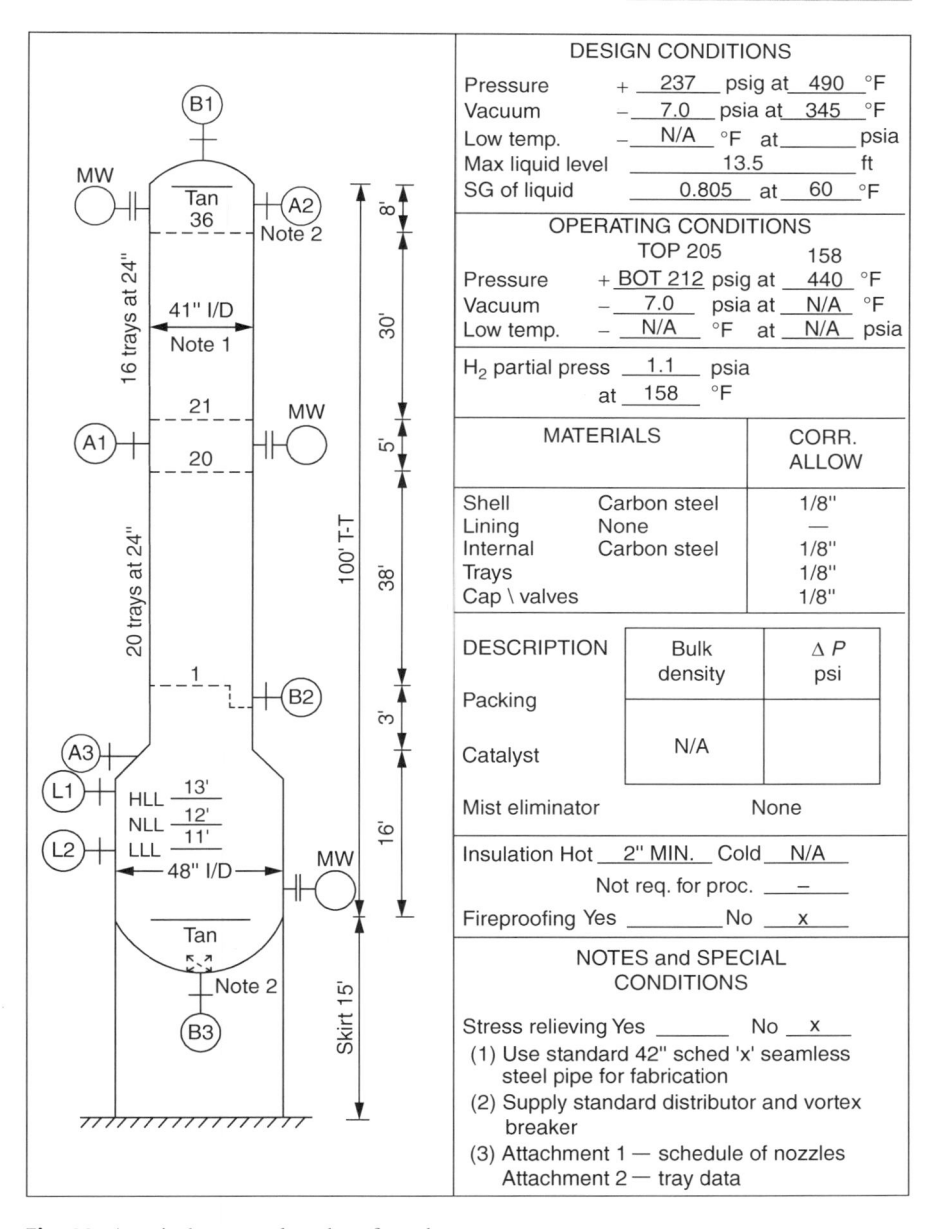

DESIGN CONDITIONS				
Pressure	+ 237	psig at	490	°F
Vacuum	− 7.0	psia at	345	°F
Low temp.	− N/A	°F at		psia
Max liquid level		13.5		ft
SG of liquid		0.805	at 60	°F

OPERATING CONDITIONS				
		TOP 205	158	
Pressure	+ BOT 212	psig at	440	°F
Vacuum	− 7.0	psia at	N/A	°F
Low temp.	− N/A	°F at	N/A	psia
H$_2$ partial press	1.1	psia		
at	158	°F		

MATERIALS		CORR. ALLOW
Shell	Carbon steel	1/8"
Lining	None	—
Internal	Carbon steel	1/8"
Trays		1/8"
Cap \ valves		1/8"

DESCRIPTION	Bulk density	ΔP psi
Packing		
Catalyst	N/A	

Mist eliminator   None

Insulation Hot   2" MIN.   Cold   N/A
Not req. for proc.   —
Fireproofing Yes _____ No x

NOTES and SPECIAL CONDITIONS

Stress relieving Yes _____ No x
(1) Use standard 42" sched 'x' seamless steel pipe for fabrication
(2) Supply standard distributor and vortex breaker
(3) Attachment 1 — schedule of nozzles
    Attachment 2 — tray data

Fig. 11 A typical process data sheet for columns

The Vessel Sketch

This particular vessel (Fig. 11) is a light ends' fractionator and has a single tray diameter (i.e., it is not swaged). The tower contains 36 valve trays on a 24″ tray spacing and has a calculated diameter of 39″ for the trayed section.

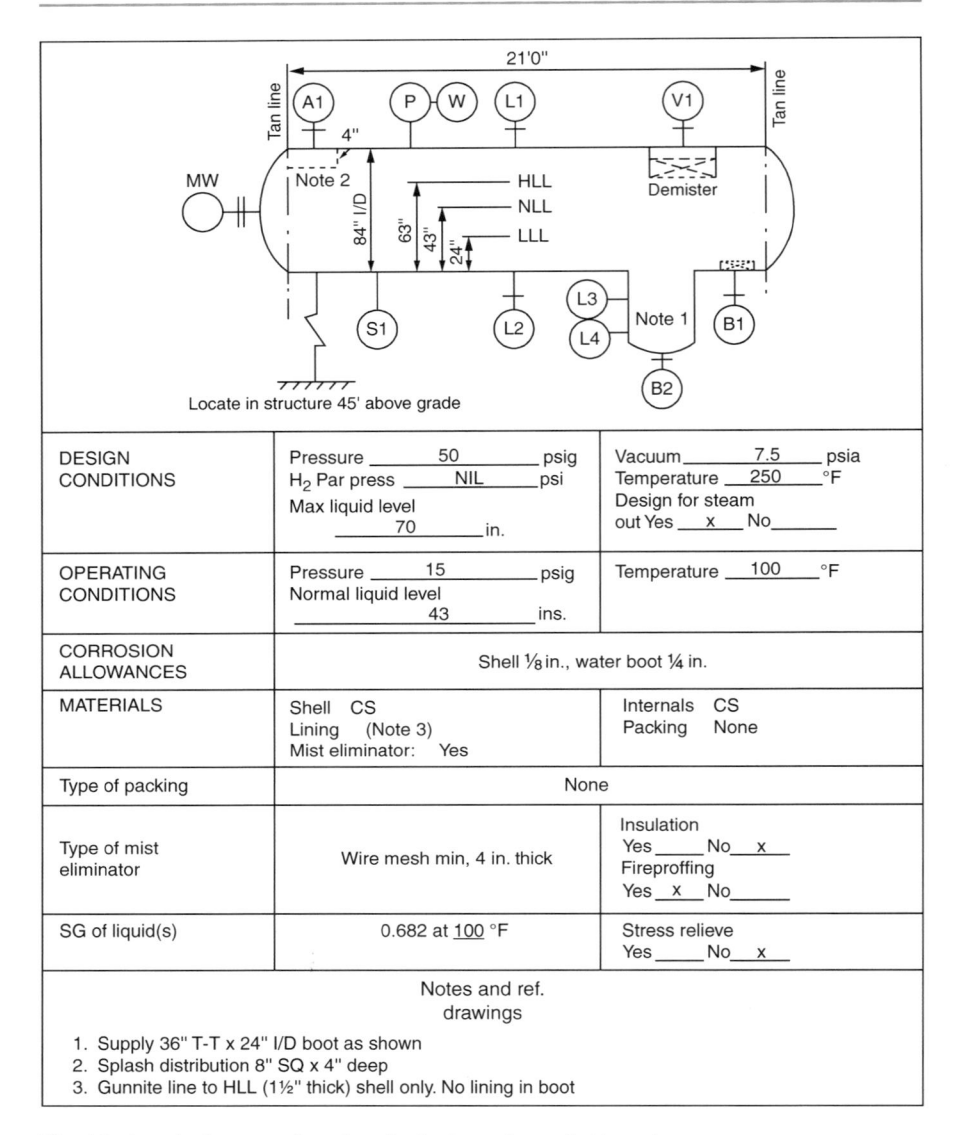

Fig. 12 A typical process data sheet for horizontal vessels (drums)

DESIGN CONDITIONS	Pressure ___50___ psig H$_2$ Par press ___NIL___ psi Max liquid level ___70___ in.	Vacuum ___7.5___ psia Temperature ___250___ °F Design for steam out Yes __x__ No_____
OPERATING CONDITIONS	Pressure ___15___ psig Normal liquid level ___43___ ins.	Temperature ___100___ °F
CORROSION ALLOWANCES	Shell ⅛ in., water boot ¼ in.	
MATERIALS	Shell CS Lining (Note 3) Mist eliminator: Yes	Internals CS Packing None
Type of packing	None	
Type of mist eliminator	Wire mesh min, 4 in. thick	Insulation Yes_____ No__x__ Fireproffing Yes __x__ No_____
SG of liquid(s)	0.682 at 100 °F	Stress relieve Yes_____ No__x__
	Notes and ref. drawings 1. Supply 36" T-T x 24" I/D boot as shown 2. Splash distribution 8" SQ x 4" deep 3. Gunnite line to HLL (1½" thick) shell only. No lining in boot	

This diameter will be specified as 41″ ID, however. This can be met by a standard 42″ schedule "X" pipe and this will reduce the cost of the vessel. The overall dimension for the tower is completed by setting the height of the tower from tan to tan. In the example here, this has been done as follows:

$$\text{Height of trayed section} = (\text{No. of trays} - 1.0) \times \text{tray spacing.} \quad (55)$$

a

Ref	Description	Size, in inch	RTG
A 1	Feed inlet nozzle	6	150 RF
A 2	Reflux inlet nozzle	4	150 RF
A 3	Inlet from reboiler	6	300 RF
B 1	O/Head vapor outlet	8	150 RF
B 2	Outlet to reboiler	4	300 RF
B 3	Bottom product outlet	3	300 RF
L1 L2	Instrument nozzles	¾	300 RF
MW	Manways	24	150 RF

b

Vessel no	C401	
Vessel name	Reformate stabilizer	
Description of material	Unstabilized light hydrocarbons	
Section	Top trays 21–36	Bottom trays 1–20
Total trays in section	16	20
Max ΔP per tray, psi	0.25	0.25
Conditions on tray	Top Tray No 30	Bottom Tray No 1
Vapor		
Temp, °F	167	440
Pressure, psig	205	212
Density, lbs/cuft	1.69	2.0
Rate, lbs/hr	47,700	71,021
ACFS	7.83	9.81
Liquid		
Temp, °F	162	430
Viscosity, Cps	0.3	0.85
Density, lbs/cuft	27.3	38.2
Mole weight	57	100
Rate, lbs/hr	33,273	104,950
Rate, cuft/min	20.34	45.79
Tower diameter, ft	3' 5"	
Number of tray passes,	One	
Type of tray	Valve	
Tray spacing, ins	24	

Fig. 13 (a, b) Attachments 1 and 2. Nozzle schedule and tray data sheet

The fractionation calculation has determined 36 actual trays for this tower. Thus, the trayed height is $(36 - 1) \times 24'' = 70$ ft.

Add another 3 f. to accommodate the feed inlet distributor on tray 20. Then total trayed height is 73 ft.

The bottom of the tower must accommodate the liquid surge requirement. As the tower diameter is relatively small, a swaged section of 4 f. diameter will be considered below the bottom tray for surge. The liquid product goes to storage; therefore, the surge requirement need not be more than 3 min on product.

From the unit's material balance, the bottom product is as follows:

Weight per hour = 117,513 lbs.
Temperature = 440°F.
Density at 440°F = 40 lbs/cuft.

Then hot cuft/min of product = 48.96.

The product goes to storage; therefore, only 2–3 min surge is required. This will be set at 3 min surge to NLL.

$$\text{Total surge to NLL} = 48.96 \times 3 = 146.9 \text{ cuft.}$$
$$\text{Cross - sectional area of surge section} = \pi/4 \times 4^2$$
$$= 12.6 \text{ sqft.}$$
$$\text{Height of NLL above tan} = \frac{146.9}{12.6}$$
$$= 11.7 \text{ ft, make it 12 ft.}$$

Level range will be 24''; then, HLL will be $12 + 1 = 13$ f. above NLL.

Let the reboiler inlet nozzle be 3 f. above HLL = 16 f. above tan.

Allow another 3 f. between reboiler inlet nozzle and bottom tray. This provides adequate space for vapor separation from the high liquid level.

At the top of the column space must be provided between the top tray and the top vapor outlet nozzle to accommodate the reflux return distributor and vapor disengaging from the top tray. Let this be 8 f. from top tray to the tower top tan line. Then,

Total tower height:

$$\text{From bottom tan to bottom tray} = 19.0 \text{ ft.}$$
$$\text{The trayed section} = 73.0 \text{ ft.}$$
$$\text{From top tray to top tan} = 8.0 \text{ ft.}$$
$$\text{Total} = 100.0 \text{ ft tan to tan.}$$

These overall dimensions are now inserted on the vessel diagram as shown in Fig. 11. An attached sheet will be included to give the nozzle description, size, and flange rating referring to those shown in the small circles on the sketch. A schedule of flange ratings for carbon steel is given in Table 10. It is advisable to verify these values with the most recent code information.

Table 10 Schedule of flange ratings for carbon steel

Flange class, #	150#	300#	400#	600#	900#
Service temp, °F	Max operating pressure, psig				
100	275	720	960	1,440	2,160
150	255	710	945	1,420	2,130
200	240	700	930	1,400	2,100
250	225	690	920	1,380	2,070
300	210	680	910	1,365	2,050
350	195	675	900	1,350	2,025
400	180	665	890	1,330	2,000
450	165	650	870	1,305	1,955
500	150	625	835	1,250	1,875
550	140	590	790	1,180	1,775
600	130	555	740	1,110	1,660
650	120	515	690	1,030	1,550
700	100	470	635	940	1,410
750	100	425	575	850	1,275
800	92	365	490	730	1,100
850	82	300	400	600	900
900	70	225	295	445	670
950	55	155	205	310	465
1,000	40	85	115	170	255

The only other dimension that will be shown on the sketch is that of the skirt. Now the vessel is installed on-site supported by a metal skirt fixed to the concrete foundation of the vessel. The height of this skirt is fixed by a few criteria. If the product from the bottom of the tower is to be pumped (as is usual), the skirt height must be such as to accommodate the suction conditions for the pump. The most important of these conditions is the head required to meet the pump's net positive suction head (NPSH). See section "Pumps" of this chapter for details. Usually a skirt height of 15 f. meets most NPSH requirements. The second consideration dictating skirt height may be the head required by a thermosiphon reboiler. If the vessel is new and being designed, the skirt height of 15 f. remains adequate with properly designed piping to and from the reboiler.

Design Conditions

This particular tower operates at 212 psig and 440°F in the bottom and 205 psig and 158°F at the top. It is therefore classed as a pressure vessel and will be fabricated to meet a pressure vessel code. The most common of these codes is the ASME code either Section 1 or Section 8. Most vessels in the chemical industry are fabricated to ASME Section 8. When fabricated, inspected, and approved, it will be stamped to certify that its construction conforms to this pressure vessel code.

Among the data required by the code and for complete vessel engineering and fabrication are the design conditions of temperature and pressure for the vessel.

Table 11 Typical operating versus design pressures for vessels

Maximum:	
Operating pressure, psig	Design pressure, psig
Full or partial vacuum	50
0–5	50
6–35	50
36–100	Operating + 15
101–250	Operating + 25
251–500	Operating + 10 %
501–1,000	Operating + 50
Over 1,000	Operating + 5 %

Table 12 Typical operating versus design temperatures for vessels

Maximum:	
Operating temperature, °F	Design temperature, °F
Ambient −200	250
201–450	Operating + 50
Over 450	Divide into zones add 50 to each operating zone
Vessels	
Up to 225	250
226–600	Operating + 25
Over 600	Operating + 50
Minimum:	Minimum:
Operating temperatures, °F	Design temperatures, °F
15 to Ambient	Operating − 25
14–10	Operating − 20
−10–80	Operating − 10
Below −80	Operating

Design Pressure

The design pressure is based on the maximum operating pressure at which the relief valve will open plus a suitable safety increment. Table 11 provides a guide to this increment.

In cases where vessels relieve to a flare header, it may be necessary to add a little more to the differential between operating and design pressures to accommodate for the flare back pressure.

Design Temperatures

Table 12 may be used as a guide to the max and min design temperatures. Some companies may use the maximum temperature which does not reduce the allowable stress values for design purposes. This avoids having to re-rate a vessel later.

Note: Very often companies will have their own standards for these design criteria. The table given here may be used if there are no company standards.

A vacuum condition can exist in a tower during normal steam out if the tower is accidentally shut in and the steam valve closed. Normally a design vacuum pressure of 7 psia is specified at the steam saturated temperature to cover this contingency. Some companies may also specify full vacuum rating if it does not create any other problems.

Low Temperature
This applies to towers in cryogenic services (such as demethanizers and LNG plants). There may be a situation in a non-cryogenic service where rapid depressurizing causes subzero temperatures to exist. If this is a situation that can exist for several hours and occurs frequently, this condition should be entered. Otherwise make an appropriate remark in "Notes and Special Conditions."

Max Liquid Level
This is the liquid level under operating conditions that will:

Either activate the high liquid level alarm
Or shut down the feed pump.

Whichever system is applicable to protect the plant operation. Usually this is quoted as the HLL and 1–2 % of surge.

SG of Liquid
Quote this as the SG of the liquid on which the surge volume was based. This SG is usually quoted at 60 °F. If another temperature is used, be sure to include that temperature.

Operating Conditions
In most fractionation towers, there will be two distinctly different conditions of temperature and pressure – those for the tower top and those for the bottom of the tower. Both these conditions must be quoted in this case. The same situation may not necessarily arise in an absorption column.

Operating Temperatures and Pressures
Quote the calculated data as they will appear also on the process flow diagram. Show the tower top pressure and temperature first, followed by the bottom set of conditions. If the tower has been sized on data for more than one design case, show the highest numbers calculated for top and bottom. Also make a note in the "Notes and Special Conditions" section of the cases used as a basis.

Other Operating Data
Vacuum conditions in this case only apply if the tower operates normally at subatmospheric pressure. In this case, quote the lowest pressure the tower will be operated on together with the normal operating temperature(s). Note in many vacuum fractionators, there will be a spectrum of these conditions along the tower; these should

be quoted for critical locations in the tower. Such locations would be feed inlet (flash zone), side stream, and pump-around, draw off, tower top.

Low temperatures and the associated pressure apply only to cryogenic plants in this case.

Hydrogen Partial Pressure

This item is important to the metallurgist who will select the grade of metal to be used in fabrication of the vessel. Generally the hydrogen partial pressure that will be quoted will be the one that exists at the tower top under normal operating conditions. For example, the dew point calculation used in the sizing of the tower given in Fig. 11 was based on the following tower top vapor composition.

	Mole fraction
H_2	0.005
C_1	0.021
C_2	0.117
C_3	0.378
iC_4	0.207
nC_4	0.268
iC_5	0.004
Total	1.000

The tower top pressure is 220 psia and the temperature is 158 °F.

$$\text{Hydrogen partial pressure} = \frac{\text{moles}\,H_2}{\text{total moles gas}} \times \text{system pressure} \qquad (56)$$

$$= \frac{0.005}{1.0} \times 220 = 1.1\,\text{psia}.$$

Materials and Corrosion Allowance

The process engineer, working with the materials engineer, will make an initial specification for the types of materials required to meet the process conditions. For example, where carbon steel only is to be used, the process engineer indicates "CS." He is not normally required to state the grade of steel to be used; this is the responsibility of the vessel specialist or the metallurgist. However, if the process engineer has a special knowledge of the material to be used and its specifics, he should note it on the data sheet.

The same applies to the corrosion allowance. Normally 1/8″ is used for this allowance; however, there may be some mild corrosive condition existing which may justify using a higher number.

Description of Internals

This is self-explanatory when it refers to packed columns. In the case of trayed towers, a separate data sheet giving sufficient data for tray rating and sizing is attached to the process data sheet front page. This is shown in the attachments to Fig. 13.

Process Equipment for Petroleum Processing 1485

Other common internals, such as distributors, vortex breakers, and the like, are not normally shown on this data sheet. These are normally standard to a particular design office and will be added to the engineering drawings developed from this process data sheet later.

Insulation and Fireproofing

The insulation requirement for heat conservation is specified by the process engineer as required. An approximate thickness is shown. This will be checked later by the vessel specialist. In the case of fireproofing, the process engineer indicates whether or not it is needed. The process engineer's relief valve sizing based on a fire condition takes into consideration any fireproofing specified.

Notes and Special Conditions

This item is a "catch all" and is used to make note of whatever other information the process engineer may wish to add to the data sheet to ensure the equipment item will meet the process requirements. The question of stress relieving of the vessel is an item which is most important to the proper fabrication of the vessel and to its cost. The process engineer usually has knowledge whether this is needed or not to handle the process material at the conditions specified. He must therefore indicate this in this section of the data sheet. Other entries in this item should be a list of the attachments to the data sheet.

Most of the process data used to define the requirements for a horizontal drum are the same as those applied to a column, and these have already been discussed for Fig. 11. The data included in the example given in Fig. 12 have been calculated earlier in this chapter. In the data sheet however, a "boot" measuring 2 f. ID × 3 f. high has been added to the outlet end of the vessel to accumulate the water phase for better control of its level and to allow the disengaging of the hydrocarbon from the water.

Pumps

Pumps in the petroleum and other process industries are divided into two general classifications which are

- Variable head-capacity
- Positive displacement

The variable head-capacity types include centrifugal and turbine pumps while the positive displacement types cover reciprocating and rotary pumps. Brief descriptions and some examples of each pump type follow.

The centrifugal pump

Centrifugal pumps comprise a very wide class of pumps in which pumping of liquids or generation of pressure is effected by a rotary motion of one or more impellers. The impeller or impellers force the liquid into a rotary motion by impelling action, and the pump casing directs the liquid to the impeller at low pressure and leads it away under a higher pressure. There are no valves in centrifugal-type pumps (except, of course, isolation valves for maintenance and check valves for backflow prevention); flow is uniform and devoid of pulsation. Since this type of pump operates by converting velocity head to static head, a pump impeller operating at a fixed speed will develop the same theoretical head in feet of fluid flowing regardless of the density of the fluid. A wide range of heads can be handled. The maximum head (in ft of fluid) that a centrifugal pump can develop is determined primarily by the pump speed (rpm), impeller diameter, and number of impellers in series. Refinements in impeller design and the impeller blade angle primarily affect the slope and shape of the head-capacity curve and have a minor effect on the developed head. Multistage pumps are available which will develop very high heads. This versatility in handling high-pressure head makes the centrifugal pump the most commonly used type in the process industry.

The turbine pump

A turbine or (regenerative turbine) pump is a type of centrifugal pump that uses an impeller with multiple, small blades running in an annular channel around the outside of a rotating wheel. Pumping is achieved by head impulses created by the rapidly spinning blades. This type of pump can also be used for pressure letdown or reduction and power recovery from high pressure systems. The actual energy that can be recovered in a power recovery service is about half the available horsepower. This type of pump is expensive and maintenance intensive, and is therefore not as widely used as the centrifugal pump.

The rotary pumps

Rotary pumps are positive displacement pumps. Unlike the centrifugal-type pump, these types do not throw the pumping fluid against the casing but push the fluid forward in a positive manner similar to the action of a piston. These pumps however do produce a fairly smooth discharge flow unlike that associated with a reciprocating pump. The types of rotary pumps commonly used in a process plant are:

- *The gear pump.* This pump consists of two or more gears enclosed in a closely fitted casing. The arrangement is such that when the gear teeth are rotated, they are unmeshed on one side of the casing. This allows the fluid to enter the void between gear and casing. The fluid is then carried around to the discharge side by the gear teeth, which then push the fluid into the discharge outlet as the teeth again mesh.
- *Screw pumps.* These have from one to three suitably threaded screwed rotors of various designs in a fixed casing. As the rotors turn, liquid fills the space between the screw threads and is displaced axially as the threads mesh.
- *Lobular pumps.* The lobular pump consists of two or more rotors cut with two, three, or more lobes on each rotor. The rotors are synchronized for positive rotation by external gears. The action of these pumps is similar to that of gear pumps, but the flow is usually more pulsating than that from the gear pumps.

- *Vane pumps.* There are two types of vane pumps: those that have swinging vanes and those that have sliding vanes. The swinging vane type consists of a series of hinged vanes which swing out as the rotor turns. This action traps the pumped fluid and forces it into the pump discharge. The sliding vane pump employs vanes that are held against the casing by the centrifugal force of the pumped fluid as the rotor turns. Liquid trapped between two vanes is carried around the casing from the inlet and forced out of the discharge.

Reciprocating pumps

These are positive displacement pumps and use a piston within a fixed cylinder to pump a constant volume of fluid for each stroke of the piston. The discharge from reciprocating pumps is pulsating. Reciprocating pumps fall into two general categories. These are the simplex type and the duplex type. In the case of the simplex pump, there is only one cylinder which draws in the fluid to be pumped on the back stroke and discharges it on the forward stroke. External valves open and close to enable the pumping action to proceed in the manner described. The duplex pump has a similar pumping action to the simplex pump. In this case however there are two parallel cylinders which operate on alternate stroke to one another. That is, when the first cylinder is on the suction stroke, the second is on the discharge stroke. In addition to duplex pumps, there are also reciprocating pumps with 3 or more cylinders.

Reciprocating pumps may have direct acting drives or may be driven through a crankcase and gearbox. In the case of the direct acting drive the pump piston is connected to a steam drive piston by a common piston rod. The pump piston therefore is actuated by the steam piston directly. Reciprocating pumps driven by electric motors, turbines, etc. are connected to the prime mover through a gearbox and crankcase.

Other positive displacement pumps

There are other positive displacement pumps commonly used in the process industry for special services. Some of these are:

Metering or proportioning pumps. These are small reciprocating plunger-type pumps with an adjustable stroke. These are used to inject fixed amounts of fluids into a larger stream or vessel.

Diaphragm pumps. These pumps are used for handling thick pulps, sludge, acid or alkaline solutions, and fluids containing gritty solid suspensions. They are particularly suited to this kind of service because the working parts are associated with moving a diaphragm back and forth to cause the pumping action. The working parts therefore do not come into contact with this type of fluid which would be harmful to them.

Characteristic curves

Pump action and the performance of a pump are defined in terms of their *characteristic curves*. These curves correlate the capacity of the pump in unit volume per unit time versus discharge or differential pressures. Typical curves are shown in Figs. 14, 15, and 16. Figure 14 is a characteristic curve for a reciprocating simplex pump which is direct driven. Included also is this reciprocating pump on a power drive.

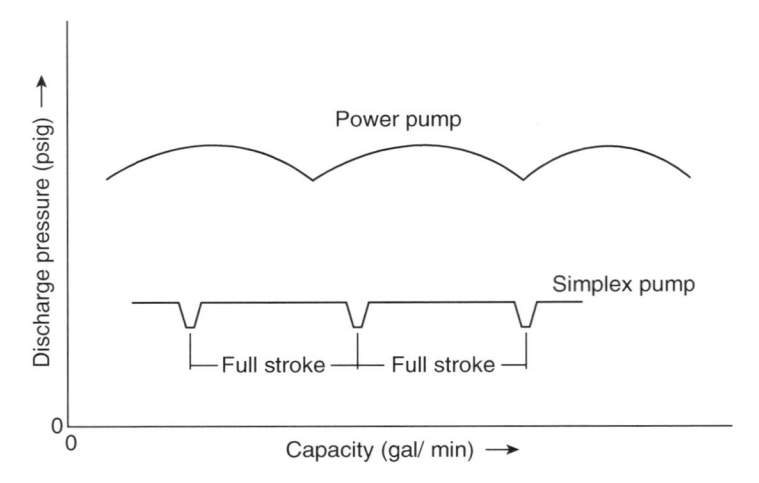

Fig. 14 Characteristic curves for a reciprocating pump

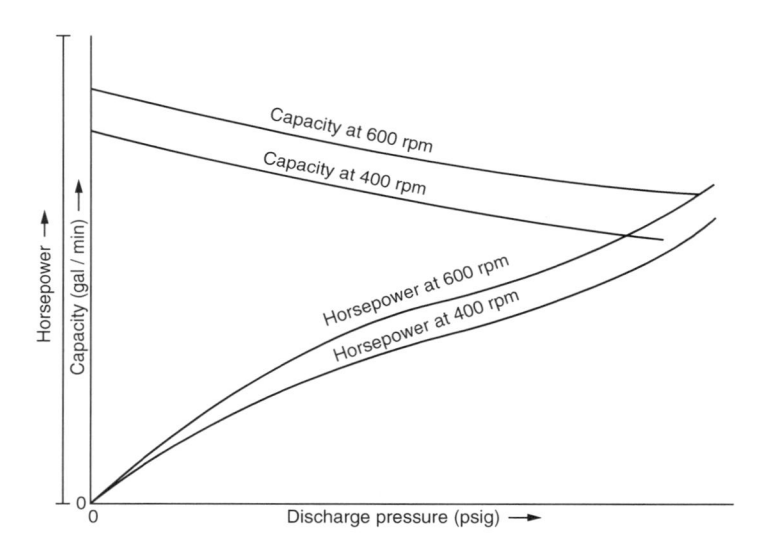

Fig. 15 Characteristic curves for a rotary pump

Figure 15 gives typical curves for a rotary pump. Here the capacity of the pump is plotted against discharge pressure for two levels of pump speed. The curves also show the plot of brake horsepower versus discharge pressure for the two pump speed levels.

Figure 16 is a typical characteristic curve for a centrifugal pump. This curve usually shows four pump relationships in four plots. These are:

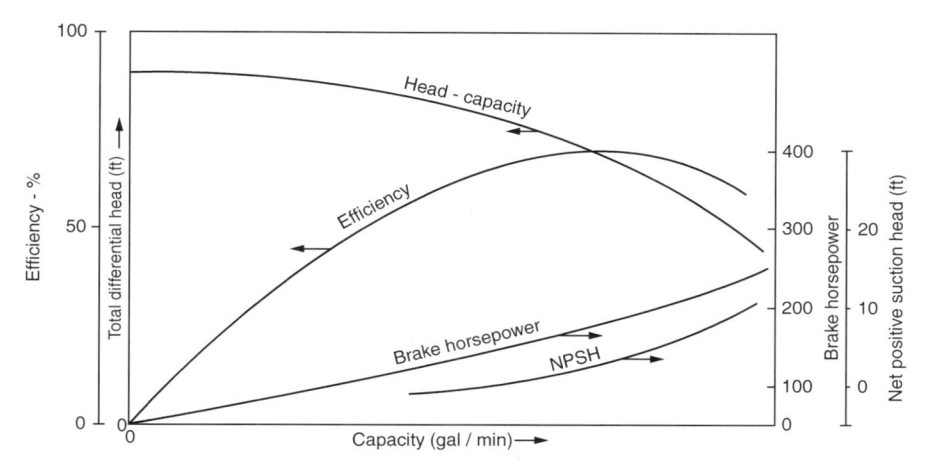

Fig. 16 Characteristic curves for a centrifugal-type pump

- A plot of capacity versus differential head. The differential head is the difference in pressure between the suction and discharge.
- The pump efficiency as a percentage versus capacity.
- The brake horsepower of the pump versus capacity.
- The net positive suction head (NPSH) required by the pump versus capacity. The required NPSH for the pump is a characteristic determined by the manufacturer.

Pump Selection

Most industrial pumping applications favor the use of centrifugal pumps. The prominence of this type of pump stems from its ability to handle a very wide spectrum of fluids at a large range of pumping conditions. It is fitting that in considering pump selection, the first choice has to be the centrifugal pump and all others become a selection by exception. Centrifugal pumps are generally the simplest in construction, lowest initial cost, and simplest to operate and to maintain. This discussion begins with the selection characteristics of the centrifugal-type pump.

The Centrifugal Pump
Before looking at the selection of the centrifugal pump, it is necessary to define the following terminology associated with pumps in general. These are:

- Capacity
- Differential head
- Available NPSH
- Required NPSH

Capacity. This can be defined as the amount of fluid the pump can handle per unit time and at a differential pressure or head. This is usually expressed as gallons per minute at a differential head of so many pounds per square inch or so many feet.

Differential head. This is the difference in pressure between the suction of the pump and the discharge. It is usually expressed as PSI and FEET in specifying a pump. The following formula is the conversion from PSI to feet:

$$\Delta H = \frac{2.31 \times \Delta P}{SG} \tag{57}$$

where:

H = the differential head in feet of fluid being pumped.
ΔP = the differential pressure of the fluid across the pump measured in pounds per square inch.
SG = the specific gravity of the fluid at the pumping temperature.

Available NPSH. The available NPSH is the static head available (in feet or meters) above the vapor pressure of the fluid at the pumping temperature. This is a feature of the design of the system which includes the pump.

Required NPSH. Is the static head above the vapor pressure of the fluid required by the pump design to function properly? The required NPSH must always be less than the available NPSH.

Selection Characteristics

Selection of any pump must depend on its ability to handle a particular fluid effectively and the efficiency of the pump under normal operating conditions. The second of these primary requirements can be determined by the pump's characteristic curves. These have already been described earlier in this part of the chapter, and a further discussion on these now follows.

Capacity range
Normal

Figures 17 and 18 show the normal capacity range for various types of centrifugal pumps in two different speed ranges, 3,550 rpm and 2,950 rpm.

These values correspond to motor full load speeds available with current at 60 and 50 cycles, respectively. Most process applications call for these speed ranges. Lower speeds are for low or medium head and high-capacity requirements and for special abrasive slurries or corrosive liquids. Low-capacity centrifugal pump applications may require special recirculation provisions in the process design to maintain a minimum flow through the pump. Because of practical considerations in impeller construction, the smallest available process-type centrifugal pumps are rated at about 50 gpm.

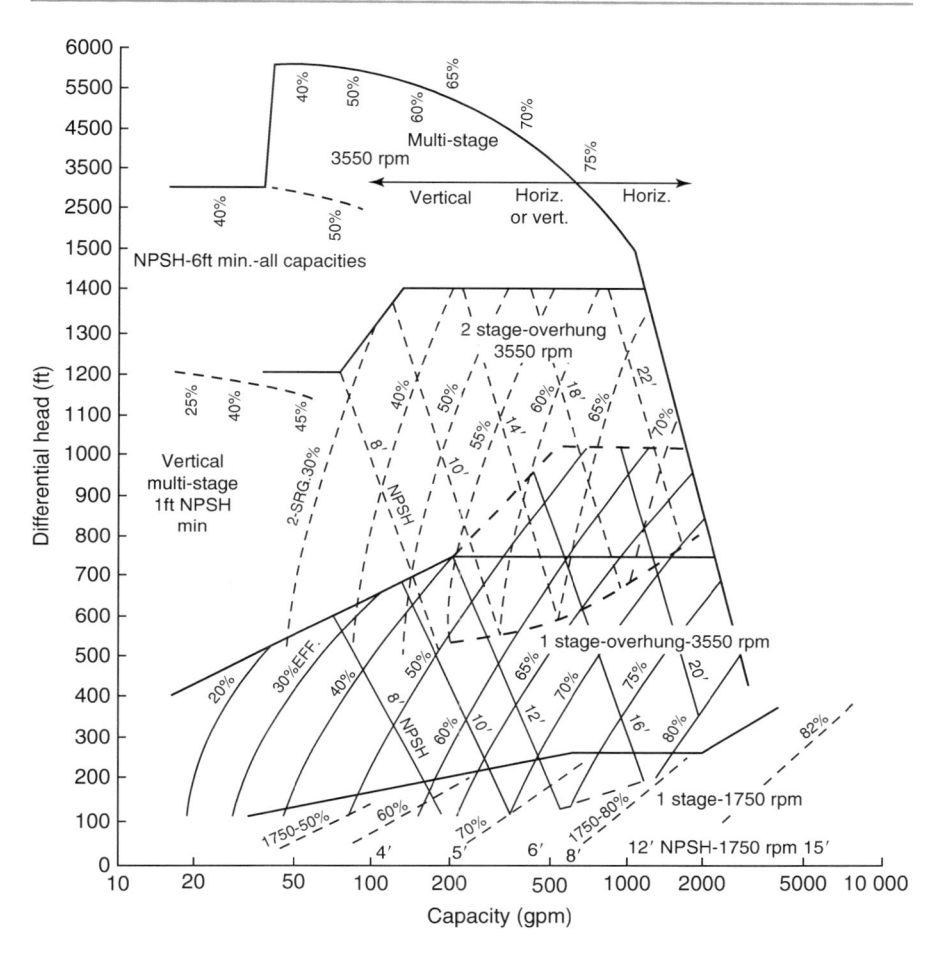

Process Equipment for Petroleum Processing   1491

Fig. 17 Centrifugal pumps at 3,550 rpm

High- and low-capacity ranges

Pumps above the limits shown in Figs. 17 and 18 will normally require large horsepower drivers. Special investigation of efficiency, speed, NPSH requirements, etc., will normally be justified. As an example, heads at or above the limits shown for multistage pumps at standard motor speed may be obtained by speed-increasing gears (motor drive) or turbines to give pump operating speeds above maximum motor speeds (NPSH requirements increase with speed).

In general, centrifugal pumps should not be operated continuously at flows less than approximately 20 % of the normal rating of the pump. The normal rating for the pump is the capacity corresponding to the maximum efficiency point. Table 13 lists minimum desirable flow rates which should be maintained by continuous recirculation, if the required process flow conditions are of lower magnitude:

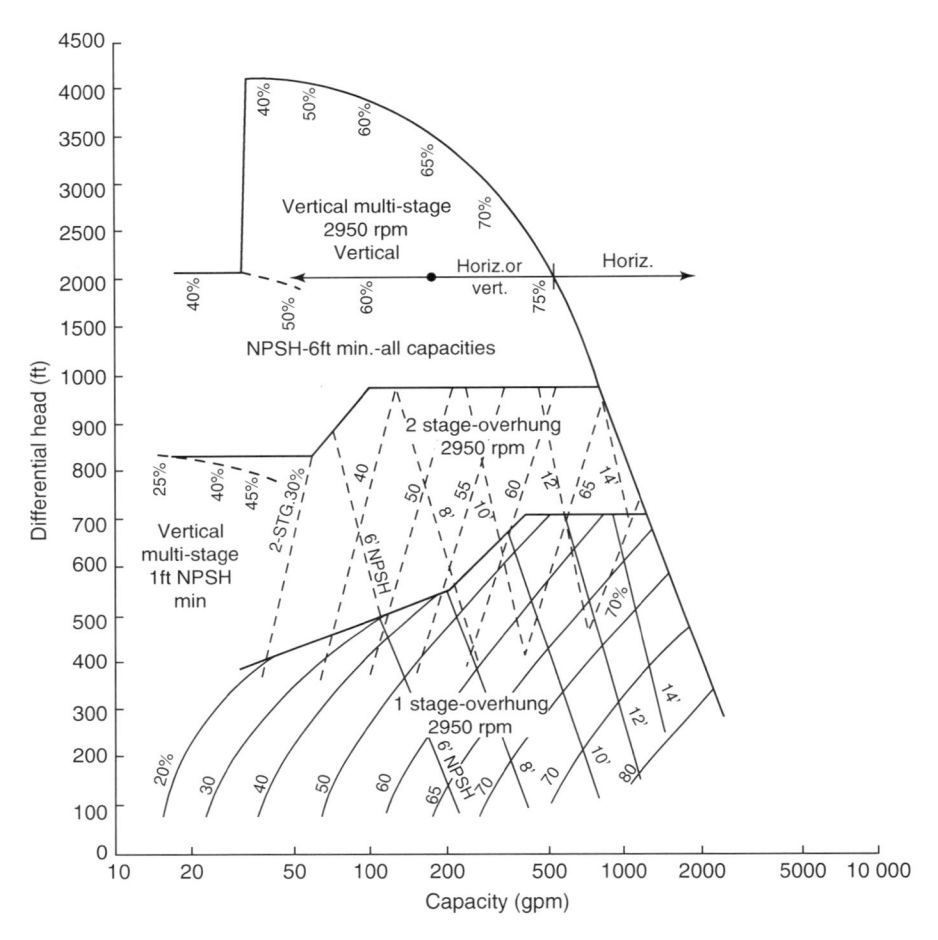

Fig. 18 Centrifugal pumps at 2,950 rpm. High- and low-capacity ranges

Care must be exercised in the design of any recirculation system to insure that the recirculated flow does not increase the temperature of pump suction and cause increased vapor pressure and reduction of available NPSH.

For low-head pumps that can operate at 1,750 or 1,450 rpm, the above-normal and minimum continuous capacities are reduced by 50 %.

Effect of liquid viscosity

When suitably designed, centrifugal pumps can satisfactorily handle liquids containing solids, dirt, grit, and corrosive compounds. Though fluids with viscosities up to 20,000 SSU (440 cSt) can be handled, 3,000 SSU (650 cSt) is usually the practical limitation from an economical operating standpoint.

Effect of suction head

An important requirement is that there be sufficient NPSH at the eye of the first-stage impeller. This is static pressure above the vapor pressure of the

Table 13 Centrifugal pump typical operating limits

Head range feet	Pump type	Minimum continuous capacity rating, gpm	Normal rating of pump, gpm
60 cycle speed (3,550 rpm)			
To 100	1 stg	10	60
100–350	1 stg	15	75–100
350–650	2 stg	30	150
650–1,100	2 stg	40	160
400–1,200	Multistg	15	50
1,200–5,500	Multistg	40	100–120
50 cycle speed (2,950 rpm)			
To 75	1 stg	10	50
75–250	1 stg	15	60–80
250–450	2 stg	25	120
450–775	2 stg	30	130
250–850	Multistg	10	40
850–3,800	Multistg	30	80–100

fluid handled to prevent vaporization at the impeller eye. Flashing of the fluid produces a shock or cavitation effect at the impeller which results in metal loss, noise, lowered capacity and discharge pressure, and rapid damage to the pump. NPSH requirements for various centrifugal pumps will normally vary from 6 to over 20 (in feet of fluid) depending on type, size, and speed. Vertical pumps can be built for practically no NPSH at all at the nozzle. These will have extended barrels in order to provide the required NPSH at the eye of the first-stage impeller.

Efficiency

The efficiency of centrifugal pumps varies from about 20 % for low-capacity (<20 gpm) pumps to a range of 70–80 % for high-capacity (>500 gpm) pumps. Extremely large capacity pumps (several thousand gpm) may have efficiencies up to nearly 90 %.

The rotary pumps

Rotary pumps deliver constant capacity against variable discharge pressure. This is a feature for all positive displacement pumps. Rotary pumps are available for process application over a range of 1–5,000 gpm capacity and a differential pressure up to about 700 psi. Displacement of the pump varies directly as the speed, except in the case where the capacity may be influenced by the viscosity of the fluid. In this case, thick viscous liquids may limit capacity because they cannot flow into the cylinder fast enough to keep it completely filled. Rotary pumps are used mostly in low-capacity service where the efficiency of a centrifugal pump would be very low.

Reciprocating pumps

The liquid discharge from reciprocating pumps is pulsating. The degree of pulsation is higher for simplex pumps than for the duplex type. The pulsation is also higher for direct-driven pumps than for those driven by motor or turbine through a gearbox.

Pulsation is generally not a problem in the case of small low speed pumps of this type. It affects only the associated instrumentation which can be compensated for by local dampeners. However, as the pump speed is increased, the pulsation effect becomes more serious affecting the piping design of the system. Under these conditions of high pulsation, instantaneous piping pressures may often exceed the design pressure of the piping. The piping must then be designed to meet this higher-pressure requirement which invariably results in a higher cost. As an alternative, discharge dampeners are often considered but these too add to the cost of the pump installation.

Reciprocating pumps are used mostly in situations where their low piston speeds will withstand corrosive and abrasive conditions. They are ideal also for pumping at low capacity against high differential head and where it is necessary to maintain a constant flow rate against a gradually increasing discharge pressure.

Evaluating Pump Performance

Many process engineers are involved with the day-to-day operation of process plants. Many of their duties in this respect are concerned with maintaining plant efficiency, locating trouble areas, and solving operational problems. This discussion is directed to these engineers and presents the calculation methods to check pump performance in terms of the pump horsepower and the available NPSH for a pump.

Brake Horsepower Efficiency

Actual running efficiency can be calculated from plant data. This may be compared with a typical expected efficiency to evaluate the pump performance. The steps below are followed to arrive at this efficiency figure.

Step 1. Obtain flow rate from plant readings. Also read discharge pressure and, if available, suction pressure. If this later reading is not available, calculate it from source pressure, height of liquid above pump suction, and frictional loss.

Step 2. Read stream temperature and obtain SG of stream from lab data. Calculate SG at flow conditions.

Step 3. Calculate differential head which is discharge pressure in PSIA − suction pressure in PSIA. Convert to differential head in feet by

$$\frac{\Delta P \times 144}{62.2 \times \mathrm{SG}(\text{at flow cond})}. \tag{58}$$

Step 4. Convert feed rate to pounds per hour. Then calculate hydraulic horsepower from the expression

$$\frac{\mathrm{Head(ft)} \times \mathrm{rate(lbs/h)}}{60\,(\mathrm{min}) \times 33{,}000(\mathrm{ft} - \mathrm{lbs/min})}. \tag{59}$$

Process Equipment for Petroleum Processing

Step 5. From motor data sheet, obtain pump motor running efficiency. From plant data read power usage in kW.

Step 6. Convert pump power to HP by dividing kW by 0.746. Multiply this by pump efficiency expressed as a fraction. This is the brake horsepower.

Step 7. Divide hydraulic horsepower by brake horsepower and multiply by 100 to give efficiency as a percentage.

Step 8. Check against Fig. 17 or 18 to evaluate pump performance. That is, the calculated efficiency within a reasonable agreement with the expected efficiency given by Fig. 17 or 18. You can also compare to the pump curve provided by the vendor, if available. Should there be a large discrepancy, the appropriate mechanical or maintenance engineers should be informed.

Checking Available NPSH

This needs to be done if the pump is showing signs of vibration and losing suction under normal operations.

Step 1. Obtain details of the fluid being pumped (temperature, SG, flow rate, source pressure, and vapor pressure of fluid).

Step 2. Calculate the frictional pressure drop in the suction line.

Step 3. Calculate the suction pressure of the pump by taking source pressure and adding in the static head. For this calculation, take static head as being from the bottom of the vessel (not from the liquid level).

Step 4. From the suction pressure calculated in step 3, take out loss through friction.

Step 5. Calculate NPSH available as being net suction pressure less vapor pressure. This is usually quoted in feet (or meters), so convert using

$$\frac{\text{PSI} \times 144}{62.2 \times \text{SG}} \tag{60}$$

Step 6. Check against manufacturer's data sheet for required NPSH. If the available NPSH is less than the required, the pump will continue to cavitate. Fill tank to a level that the vibration stops and maintain it at that level.

If the problem is really troublesome and maintaining a liquid level as suggested in step 6 above is not practical, contact the pump manufacturer. Very often he is able to make some minor changes to the pump design that will solve the problem.

Example Calculations

Example calculation No. 1 – pump brake HP efficiency:

Flow capacity of pump at flow conditions = 200 gpm.

(From plant data) = 82,632 lbs/h.

$$\text{Suction pressure (plant data)} = 120 \text{ psig.}$$
$$\text{Discharge pressure} = 450 \text{ psig.}$$
$$\Delta \text{pressure} = 330 \text{ psi.}$$
$$\text{SG at flow conditions} = 0.827.$$
$$\text{Differential head ft} = \frac{330 \times 144}{62.2 \times 0.827} = 924 \text{ ft.}$$
$$\text{Hydraulic horsepower} = \frac{82,632 \times 924}{60 \times 33,000}$$
$$= 38.56.$$

From motor rating, motor efficiency is 92 %.
Plant readings show motor power usage = 48.1 kW.

$$\text{Then motor HP input} = \frac{48.1}{0.746} = 64.5 \text{ HP.}$$
$$\text{Motor output} = 64.5 \times 0.92$$
$$= 59.3.$$

This is brake HP.

$$\text{Then pump efficiency} = \frac{38.56 \times 100\%}{59.3} = 65\%.$$

Motor is 50 cycle speed. From Fig. 18 for multistage 2,950 rpm. This efficiency figure is about right.

Example calculation No. 2 – checking available NPSH:

Fluid is gas oil from surge drum at 250 °F and 15 psig. Drum is 12 f. above grade (bottom of drum which is horizontal).
Boiling point at 34 psia is 488 °F.
Vapor pressure of gas oil at 250 °F = 0.8 psia (VP curves).

$$\text{Source pressure} = 29.7 \text{ psia.}$$
$$\text{SG at conditions} = 0.815.$$
$$\text{Head above grade} = 12 \text{ ft (to bottom of drum).}$$
$$\text{Pump center line} = 2 \text{ ft above grade.}$$
$$\text{Liquid head to pump} = 10 \text{ ft} \Rightarrow 3.5 \text{ psi.}$$
$$\text{Friction pressure drop} = 0.4 \text{ psi (calculated from 0.2 psi/100 ft).}$$
$$\text{Less friction} = 0.4 \text{ psi.}$$
$$\text{Less vapor press} = 0.8 \text{ psi.}$$
$$\text{Available NPSH} = 33.2 - (0.4 + 0.8) = 32 \text{ psia}$$
$$= 91 \text{ ft.}$$

Most pumps only require 10 f. or less.

Specifying a Centrifugal Pump

In all engineering companies and in most petroleum refineries, two disciplines are responsible for correctly specifying a pump. These are the mechanical engineer and the process engineer. Most pumps are designed and built in accordance with set and accepted industrial codes, such as the API codes. The mechanical engineer ensures that the mechanical data supplied to the manufacturer for a particular pump meets the requirements of the code and standards to which the pump is to be built. The process engineer develops and specifies precisely the performance required of the pump in meeting the process criteria of the plant. To accomplish this, the mechanical engineer develops a mechanical specification, and the process engineer initiates the pump specification sheet.

Mechanical Specification
This specification is in a narrative form and will contain at least the following topics:

- Scope – introductory paragraph which gives the code to which the pump manufacturer is to conform (such as API 610).
- A list of other standards the pump shall conform to, if these are required.
- Main body of the specification – this covers all additions and any exceptions to the selected code. It provides for the type of drive shaft acceptable if different from code. Items such as impeller size as a percent of maximum allowable by code are given. The need for special bearing arrangements in the case of multistage pumps is detailed in this document.
- Ancillary equipment and piping arrangements – the specification describes in detail the type of cooling medium that shall be used. It provides a guide also to the piping requirements that is required to satisfy the cooling system(s).
- Seal or packing requirements – the mechanical specification details the type of seal or packing that will be installed. It also provides details of the seal arrangement required if this is different from the standard code.
- Pump mounting – some installation guidance is provided by this specification. The method by which the pump is mounted on the base plate is detailed. It also details under what conditions the manufacturer is to provide pedestal cooling facilities.
- Metallurgy – although the process engineer will call out the initial, expected material of construction for the pump (such as carbon steel or cast iron), it is the mechanical engineer who details this. This detail includes the specific grade of the material and in many cases its preoperational treatment.
- Inspection – the mechanical specification will provide details of the inspection that the company will carry out during the manufacture of the pump and before its delivery. This will include dimensional checking during manufacture and some checks on the metallurgy. Prior to shipping, the purchaser may require a running test of the pump and will witness this test. For this purpose, the pump is run in the workshop under specified process conditions.

The mechanical specifications may continue to detail other requirements that the purchaser may wish. Its objective is to ensure that the pump when delivered is mechanically robust, is safe, and is easily maintainable. The mechanical specification must also be aware of the cost implication of the requirements on the pump and to keep them as low as possible.

Process Specification

The data provided by the process engineer must be sufficient to ensure that the pump delivered for the process purpose will meet the duty required of it. These data are furnished to the pump manufacturer in the form of a data sheet similar to the one shown here as Fig. 19. The data sheet collects the essential input from the process engineer, the mechanical engineer, and, later, by the manufacturer to describe fully what is required of the pump and what the manufacturer has supplied. All the data given here will be unique to this pump.

The process input to the pump specification shown on Fig. 19 are those items marked with the "P." Input by other disciplines and the manufacturer are not indicated on the form. The process engineer compiles many of these data from a hydraulic analysis of the piping system (see Appendix 14 and section "Pressure Drop Calculations" of this chapter). A calculation sheet given in Fig. 20 shows the development of this and is described as follows:

Compiling the pump calculation sheet
The pump number, title, and service
    This first section of the calculation sheet is important because it identifies the pump and what it is intended to do. The item number and service description will be unique to this item and will remain as its identification throughout its life. All the data below this section will refer only to this pump and to no other. The item number may contain the suffix "A," "B," "C," etc. This indicates identical pumps in parallel service or as spare or both. This section also shows how many of these pumps are motor driven and how many are turbine (steam) driven. Usually spare pumps in critical service will be turbine driven. The remark column in this section should give any information that will be of benefit to the pump manufacturer or future operators of the pump. For example, if the spare pump is turbine driven, the process engineer may require an automatic start-up of the turbine on a "low flow" of the pumped stream. This should be noted here.

Operating conditions for each pump
    The details of the fluid to be pumped and a summary of the calculations given below are entered here. Starting on the left of this section:
    Liquid: This is a simple definition of the pumped material. In the example given here, this will simply be "vacuum gas oil."
    Pumping temperature (*PT*): This is the temperature of the gas oil at the pump. There are two temperatures called for "normal" and "max." The normal

Fig. 19 A centrifugal pump specification sheet

PUMP CALCULATION.

Item No P 103- A- Unit Crude Vacume Unit   Sheet No 1   Rev 0
Service   HGO Product and BPA   Motor Drive   1
Turbine Drive 1   Remarks Turbine to have Auto Start By PSTT App. J.S

OPERATING	CONDITIONS (Each Pump)			TURBINE CONDITIONS	
Liquid VAL GAS OIL	US GPN &Pt.Min__MOR 13Q Rated 1585			Inlet Stean psig 600	
PTP NOR 545 MAX 740	Dish press Psig 85.5			Temp P 670	
SP GR & PT 0.755	Suct Press Psig Max 50 Rated –0?			Exhaust Psig 50	
Vap Press & Pt.Psig 0.29	Diff Press Psi 86.2			PUMP MATERIALS	
Vis & PT Cp 0.906	Diff Bead FT 264.3			Casing C.S.	
Corrosion\Erosion None	NPSH Available, P 740 Ayd HP 79.7 (1)			Internal Parts C.S.	
ALTERNATES	B.L			SKETCH	
DESTINATION :					
Destination Press	psig	50			
Static Head	psi	5.4			
Line Loss	psi	10.7			
Meter Loss	psi	0.2			
11-E-9/10Δ Ht Exchangers	psi	14.0			
Δ Control Valves	psi	5.20			
TOTAL DISCHARGE PRESS	psig	85.50			
SUCTION.					
Surce Press	psig	–14.4			
Static Head	psi	14.7			
- Ststem Losses	psi	1.0			
TOTAL SUCTION PRESS	psig	–0.7			
NPSH AVAILABLE.		0.29			
Source Press	psia	–1.29			
-(Vap Press ı suct losses) psia sub total Psia\Pt		–1.00			
Elev of liquid - pump CL Pt		45.00			
NPSH Available	Pt	40.40			

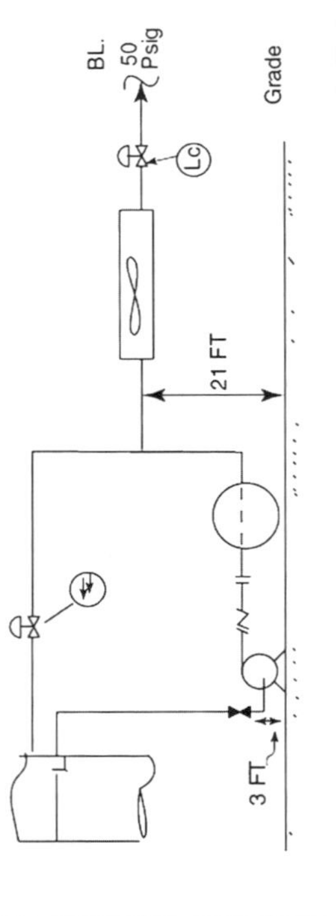

DATED 24.3.92 REV 0 DATED ____ REV ____.
NOTES (1) Based on Rated Flow.

Fig. 20 A centrifugal pump calculation sheet

temperature is that shown on the process flow diagram, while the max temperature is that used for the pump design conditions. It should be the same as the design temperature of the vessel the fluid is pumped from.

Specific gravity at *PT*: This is self-explanatory. Note the item also calls for the SG at 60 °F.

Vapor pressure at *PT*: This is read from the vapor pressure curves given in the Appendix 3 of this chapter. First locate the vapor pressure of the stream at atmospheric pressure (this is the material's normal boiling point). Follow the temperature line down or up to the PT and read off the pressure at that point.

Viscosity at *PT*: This too is self-explanatory. Note this calculation sheet requires the viscosity to be in *centipoise*. This is centistokes × SG.

US gpm at *PT*: This is the pump capacity and three rates are asked for. These are:
- *Minimum rate*: The anticipated lowest rate the pump will operate at for any continuous basis. This rate sets the control valve range.
- *Normal rate*: This is the rate given in the material balance and the basis for the hydraulic analysis.
- *Maximum rate*: This is normally set based on the type of service that the pump will undertake. For example, pumps used only as rundown to storage will have a max rate about 10 % above normal. Those used for reflux to towers will have between 15 % and 20 % above normal.

Discharge pressure, psig: This figure is calculated in the column below. It will also have been determined by the hydraulic analysis of the system (Appendix 14 and section "Pressure Drop Calculations" of this chapter).

Suction pressure, psig: Two pressures are asked for in this item. Rated pressure is that calculated in the column below and in the hydraulic analysis given in item 1. It is based on the "norm" rate. The "max" suction pressure is based on a source pressure at the *design* pressure rating of the vessel the pump is taking suction from.

Differential pressure, psi: This is the discharge pressure minus the rated suction pressure.

Differential head feet: The head is determined from the differential pressure by the equation:

$$\frac{\text{Diff press (psi)} \times 144}{62.2 \times \text{SG @ PT}}. \tag{62}$$

NPSH feet: Calculated in the column below. This is the suction head available greater than the fluid vapor pressure (at the PT) at the pump impeller inlet.

Hydraulic horsepower: This is calculated from the weight per unit time (usually minutes or seconds) of fluid being pumped times the differential head in feet divided by 550 f. lb/s or 33,000 f. lb/min. The differential head is always based on the rated suction pressure and the weight of the rated capacity (gpm) for this calculation.

Corrosion/erosion: The process engineer notes any significant characteristic of the fluid regarding its corrosiveness or abrasiveness here.

Turbine conditions

Although this item is not strictly part of a pump definition, it should be included for completeness in the case of turbine drives. The data required to complete this item is self-explanatory.

Pump material

The process engineer indicates here the expected acceptable material for the pump in handling the fluid. For example, carbon steel, or cast iron, etc.; it is not necessary to specify grade of steel, etc.

The calculation columns

The objective of this section of the calculation sheet is to itemize all the data that are used to provide the figures given in the operating conditions described in section "Operating Conditions for Each Pump" above. The first column lists those items while the other three columns are available for entering the corresponding numbers. These three columns are provided to cater to alternate conditions that may need to be studied. A space is left on the right of the form to sketch the pumping system (it is very advisable to do a sketch).

The first column starts with the destination pressure and continues down with the list of the pressure drops in the system to the pump discharge (see Appendix 14). This section of the column ends with the sum of the pressure drops giving the pump discharge pressure. The items that make up the pump suction pressure are listed next. This starts with the source pressure (usually a vessel) and its static head above the pump. All the pressure drops in the suction side are listed and deducted from the sum of the source pressure and static head to give the pump suction pressure.

The last section in the column itemizes the data that gives the *available* NPSH for the pump. The development of the NPSH is self-explanatory.

Centrifugal Pump Seals

A pump seal is any device around the pump shaft designed to prevent the leakage of liquid out of or air into a pump casing. All industrial pumps have shafts protruding through the casings which require sealing devices. Pump sealing devices are usually either a "packed box" with or without a lantern ring or a mechanical seal. Controlled leakage is a system sometimes used, but fugitive emissions regulations are phasing out this option.

A flushing stream may be introduced into the pump seals for one or more of the following reasons:

- To effect a complete seal
- To provide cooling, washing, or lubrication to the seal
- To keep grit from the seal
- To prevent corrosive liquid from reaching the seal

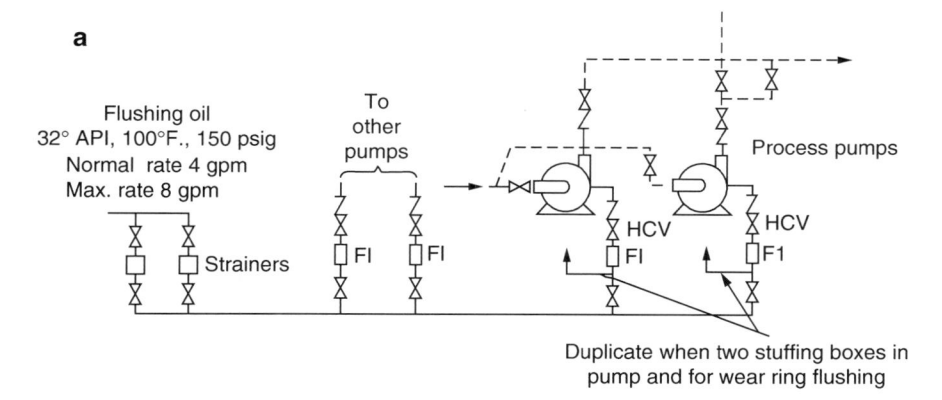

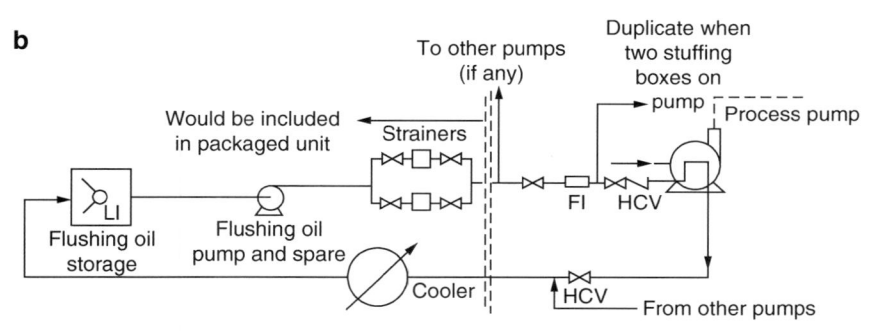

Fig. 21 Typical flushing systems. (**a**) A dead-end system, (**b**) a through – recirculating system

The facilities for accomplishing this are called "flushing system," and there are two types of these in general use.

- A dead-end system
- A through system

In a "dead-end" system, the flushing liquid enters the casing through the stuffing box and combines with the pumped fluid (see Fig. 21a). A "through" system is one in which the flushing liquid is recirculated between a double seal arrangement and does not enter the pump (see Fig. 21b). The liquid source may be external to the pump or, as on most mechanical seals, is a self-flushing system in which the pumped liquid is used as the flushing fluid.

A description of each of the types of sealing devices is presented below and illustrated in Fig. 22.

Packed boxes (without lantern ring) Fig. 22a

This is the simplest type of pump seal. Its principal components are a stuffing box, rings of packing, a throat bushing, and a packing gland. A slight leakage

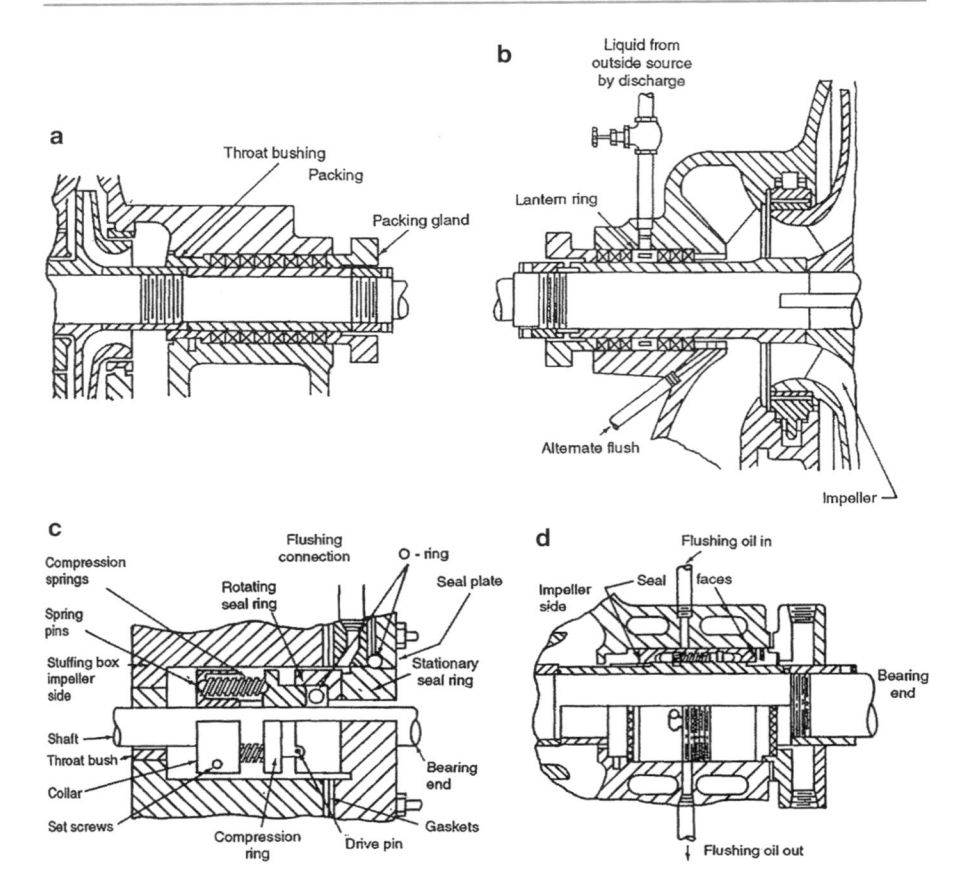

Fig. 22 Pump shaft packing and seals. (**a**) Stuffing box completely filled with packing no lantern ring. (**b**) Externally sealed stuffing box. (**c**) Single mechanical seal. (**d**) Double mechanical seal

through the packing is required at all times to lubricate the packing. A water quench is used at the packing gland if the packing "leakage" is considered flammable or toxic.

Packed box (with lantern ring) Fig. 22b

When a packed box pump seal is used in conjunction with a flushing oil system, a lantern ring is usually provided. This metallic ring provides a flow path for the flushing oil to reach the pump shaft. For very erosive or corrosive services, the lantern ring is often located next to the throat bushing and a liquid is injected into the throat bushing to prevent the pumped fluid from reaching the packing area. For a pump operation with vacuum suction conditions, the lantern ring is installed at the middle of the box and liquid is injected to prevent air entering the system. This type also operates with positive leakage with the same comments as the packed box without lantern ring.

Mechanical seals Fig. 22c, d

Typical basic elements of a single seal are shown in Fig. 22c. Sealing is affected between the precision-lapped faces of the rotating seal ring and stationary seal ring. The stationary seal ring is usually carbon and is mounted in the seal plate by an "O" ring. The two "O" ring packing serves the dual purpose of sealing off any liquid tending to leak behind the seal rings and also to provide flexibility in allowing the seal faces to align themselves exactly so as to compensate for any slight "wobble" of the rotating seal face caused by shaft whip.

The rotating seal ring is usually a stainless steel with a Stellite face. The springs furnish the necessary force to set the "O" ring and hold the seal faces closed under low stuffing box pressures. Any pressure in the box exerts additional force on the rotating sealing ring. The seal is frequently "balanced" so that the face pressure is in correct ratio to the liquid pressure to ensure adequate sealing without excessive loading of the faces. Flushing oil enters the stuffing box through a connection in the seal plate.

A double seal consists of two single seals back-to-back. See Fig. 22d. As a double seal is more expensive and requires a complicated seal-oil system, it is used where single seals are not practical. Regulations regarding fugitive volitile organic vapors (VOCs) are also driving most hydrocarbon applications toward double seals.

Figure 23 summarizes the application of the various sealing systems in the refinery operation.

Pump Drivers and Utilities

Most pumps in the process industry are driven either by electric motors, or by steam, usually in the form of steam turbines. The following discussion deals with the calculation of the driver requirements and driver specifications.

Electric Motor Drivers

Electric motors are by far the most common pump driver in industry. They are more versatile and are cheaper than a comparable size of steam turbine. The electric motors used for pump drivers are normally induction-type motors. They range in size from fractional horsepower to several thousand horsepower. Sizing the required motor for a pump driver takes into consideration the pump brake horsepower, the energy losses occurring in the coupling device between the pump and the motor, and a contingency factor of about 10 %. These are expressed by the equation:

$$\text{Minimum driver BHP} = \frac{\text{maximum pump BHP} \times 1.1}{\text{mechanical efficiency of coupling}}. \quad (63)$$

If the pump is driven through a direct coupling, the efficiency will be 100 %. With gears or fluid coupling, the efficiency will be between 94 % and 97 %.

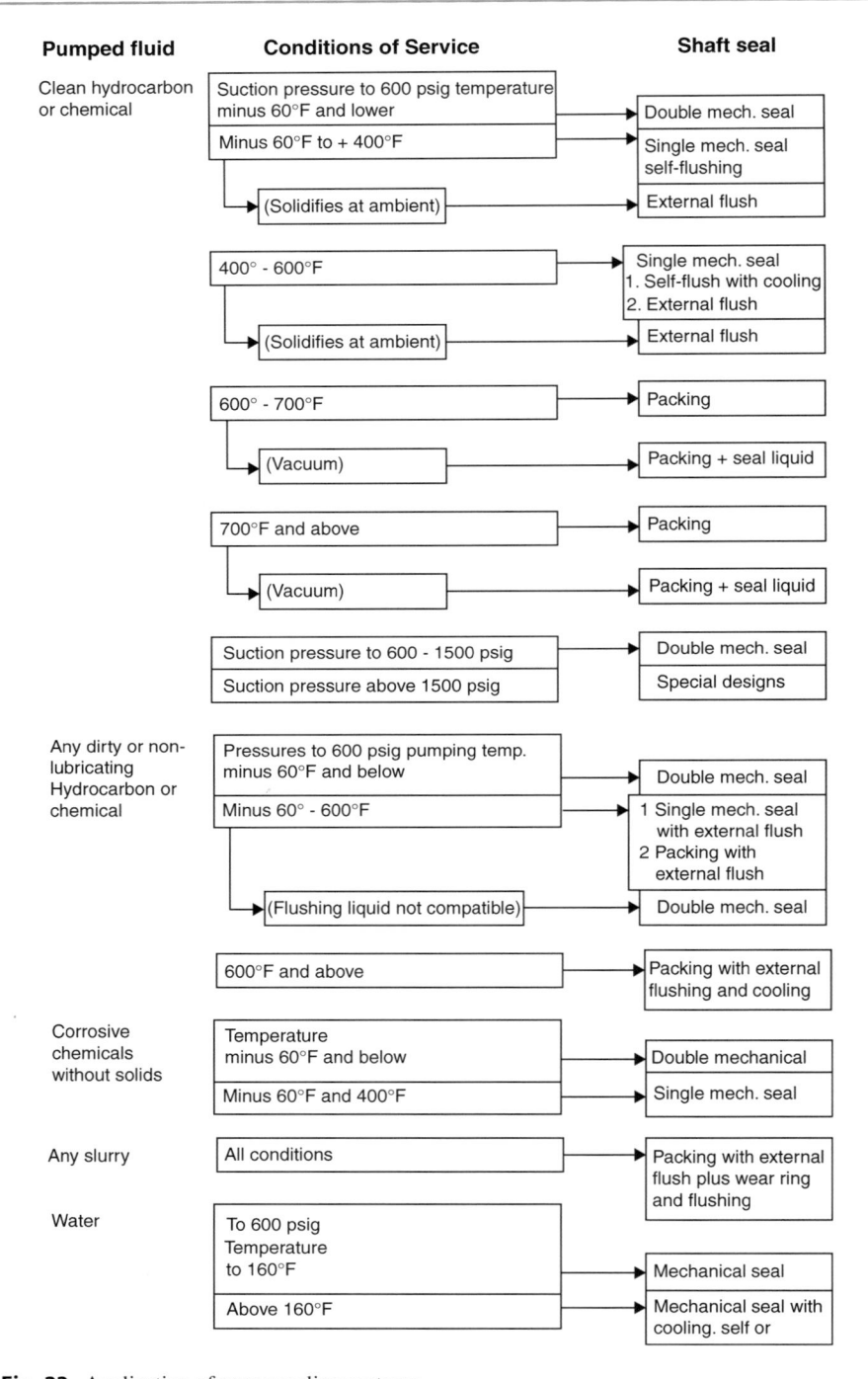

Pumped fluid	Conditions of Service	Shaft seal
Clean hydrocarbon or chemical	Suction pressure to 600 psig temperature minus 60°F and lower	Double mech. seal
	Minus 60°F to + 400°F	Single mech. seal self-flushing
	(Solidifies at ambient)	External flush
	400° - 600°F	Single mech. seal 1. Self-flush with cooling 2. External flush
	(Solidifies at ambient)	External flush
	600° - 700°F	Packing
	(Vacuum)	Packing + seal liquid
	700°F and above	Packing
	(Vacuum)	Packing + seal liquid
	Suction pressure to 600 - 1500 psig	Double mech. seal
	Suction pressure above 1500 psig	Special designs
Any dirty or non-lubricating Hydrocarbon or chemical	Pressures to 600 psig pumping temp. minus 60°F and below	Double mech. seal
	Minus 60° - 600°F	1 Single mech. seal with external flush 2 Packing with external flush
	(Flushing liquid not compatible)	Double mech. seal
	600°F and above	Packing with external flushing and cooling
Corrosive chemicals without solids	Temperature minus 60°F and below	Double mechanical
	Minus 60°F and 400°F	Single mech. seal
Any slurry	All conditions	Packing with external flush plus wear ring and flushing
Water	To 600 psig Temperature to 160°F	Mechanical seal
	Above 160°F	Mechanical seal with cooling. self or

Fig. 23 Application of pump sealing systems

Process Equipment for Petroleum Processing   1507

Specifying Motor Driver Requirements

Process engineers are called upon very often to specify pump driver requirements or to check those already existing. In doing this, two items of data need to be obtained or calculated. These are:

- The actual required horsepower of the pump motor to drive the pump at its specified duty
- What is actually installed in terms of horsepower?

These data are tabulated in terms of power load as follows:

Operating load, kW – power input to the motor a normal operating horsepower.
Connected load, kW – power input to the motor at motor-rated horsepower.

If the pump is spared by another motor-driven pump, then the connected load will be the sum of *both motors*.

$$Operating\ load = \frac{minimum\ required\ driver\ HP \times 0.746}{efficiency\ of\ the\ motor\ at\ its\ operating\ HP}.$$
$$Connected\ load = \frac{rated\ motor\ HP \times 0.746 \times number\ of\ motors}{efficiency\ of\ the\ motors\ at\ 100\%\ full\ load}.$$

(64, 65)

Table 14 gives example motor sizes and standard efficiencies at % of full load. (Higher efficiency and variable-speed motors may also be available.)

Example Calculation

Calculate the operating and connected loads for pump 11-P-3 A and B as specified in Fig. 20 of this chapter.

From the pump calculation sheet, the hydraulic horsepower is calculated:

$$\begin{aligned} HHP &= \frac{lbs/min \times diff\ head\ in\ ft}{33,000} \\ &= \frac{8,665 \times 264.3}{33,000} \\ &= 69.4\ HP. \end{aligned}$$

(66)

From Fig. 20 and assuming 60 cycle pump speed pump, efficiency is 79 %.

$$Then\ brake\ horsepower = \frac{hydraulic\ horsepower}{0.79}$$
$$= 87.8\ HP.$$

(67)

This will be a direct-driven pump thus coupling efficiency is 100 %.

$$Minimum\ motor\ size = BHP \times 1.1 = 96.6\ HP.$$

Table 14 Electric motor size and efficiency

Motor rating BHP	Motor connected load, kW	Motor efficiency at % of full load		
		50	75	100
1	0.98	68	74	76
1.5	1.42	72	76.5	79
2	1.86	73	78	80
3	2.76	77.5	81.5	82
5	4.39	83	82	85
7.5	6.65	81	83.5	84
10	8.78	84	85	85
15	13.0	85	86	86
20	17.05	86.5	87.5	87.5
25	21.0	87.5	88.5	88.5
30	25.1	88	89	89
40	33.5	88	89	89
50	41.7	88	89.5	89.5
75	62.1	89	90	90
100	82.0	84	89	91
125	102.0	85	89.5	91.5
150	123.0	86	89	91
200	161.0	88	91	92.5
250	201.0	90.5	92.5	92.5
300	241.0	90.5	92.5	92.8
350	281.0	90.9	92.6	92.9
400	320.0	91.1	92.8	93.1
450	360.0	91.2	93	93.2

The closest motor size to this requirement is 100 HP (Table 14). This is a little too close so a motor size of 125 HP will be selected.

$$\text{Operating load} = \frac{\text{rated HP} \times 0.746}{\text{efficiency @ \% of full load.}} \tag{68}$$

$$\% \text{ of full load} = \frac{87.8}{125} = 70.2\,\%.$$

Efficiency = 89 % (from Table 14).

$$\text{Operating load} = \frac{87.8 \times 0.746}{0.89}$$
$$= 73.6\,\text{kW.}$$
$$\text{Connected load} = \frac{125 \times 0.746}{0.915}$$
$$= 101.9\,\text{kW say 102 kW.}$$

Note if both regular and spare pumps were motor driven, then the connected load would be $2 \times 102 = 204$ kW.

… Process Equipment for Petroleum Processing

Reacceleration Requirement

To complete the specification for the motor requirement, a degree of process importance of the pump must be established and noted. Voltage drops that can occur in any system may be sufficient to stop the pump. The process engineer must determine how important it is to the process and the safety of the process to be able to restart and reaccelerate the particular pump quickly. The following code of importance has been adopted by some users.

- Reacceleration absolutely necessary – A
- Reacceleration desirable – B
- Reacceleration unnecessary – C

The "A" category involves any pump critical to keeping the process on stream safely and with no possibility of equipment damage. The "B" category applies to those pumps that in operation with the "A" category will maintain the unit "on spec." The "C" category refers to those pumps that can be started manually without any problems.

In the case of the example pump 11-P-3 A and B given here, the service required of the pump is so critical to the operation and orderly shutdown of the process in the case of power failure that the spare pump is turbine driven. Thus, the motor-driven regular pump need only be coded "C" for reacceleration.

Steam Turbine Drivers

Steam turbines are the second most common pump drivers in modern day process industry. Although more expensive than the electric motor, they offer an excellent standby to retain the maximum process "on stream" time. The one big disadvantage with power-driven pumps is the reliability of power availability. Steam turbines therefore offer a good alternative in cases of power failure. Another alternative means of pump drivers is the diesel engine or gas engine, but these require their own fuel storage and are certainly not as reliable as the steam turbine.

Most process plants therefore spare the critical pumps in the process with a turbine-driven unit which may be started automatically on low process flow.

The Principle of the Turbine Driver

Turbines are the most flexible of prime movers in today's industry. Their horsepower output can be varied by the number and size of the steam nozzles used, speeds can be changed readily, and high speeds without gearing are possible. They have a very wide range of horsepower applications. The operation of the steam turbine is analogous to that of a waterwheel where buckets are attached to the wheel which collect the water. The wheel is moved downward by the weight of the water collected and thus causes the rotation of the wheel. In steam turbines, the buckets are replaced by vanes which are impinged by the motive steam to cause the rotating motion. Turbines may consist of one set of vanes keyed to the shaft in the case of a single-stage machine or several sets of vanes in the case of multistage machines. These sets of vanes are called simply "wheels" and the number of stages is referred to as the number of wheels.

In the case of multistage turbines, the steam leaving the first wheel is directed toward a set of stationary vanes attached to the casing. These stationary vanes reverse the steam flow and serve as nozzles directing the steam toward the second wheel attached to the same shaft.

Most turbines used on a regular basis in a process plant are single stage. Multistage machines are more efficient but are also much more expensive. Their use therefore is for drivers requiring horsepower in excess of 300. The power industry is a good example for the use of large multistage turbine drivers. Single or multistage turbines may be operated either condensing or noncondensing. However, pump drivers should not be made *condensing* without a rigorous review to see if other types of drives can be used. The complexity of condensing is hardly worth the small savings in utilities that are made.

Performance of the Steam Turbine

The salient factors in the performance of the steam turbine are:

- Horsepower output
- Speed
- Steam inlet and outlet conditions
- Its mechanical construction (e.g., number of wheels, size of the wheel, etc.)

These factors are interrelated and their effect on the performance of the turbine is reflected by a change in overall efficiency. The *overall efficiency may be defined as the ratio of the energy output to the energy of the steam theoretically available at constant entropy as obtained from a Mollier diagram.* This overall efficiency is the product of mechanical and thermal efficiencies. The losses in turbines are partly due to friction losses of the rotating shaft and partly to thermodynamic losses and turbulence. Figure 24 gives the overall efficiencies plotted against delivered horsepower.

The steam required by a turbine for a given horsepower application is called its "water rate." The actual water rate for a turbine is supplied by the manufacturer from test runs carried out on the actual machine in the workshop. Plant operators and other engineers very often need to be able to estimate these water rates for their work. A typical situation arises when determining the best steam balance for a plant. Such estimates may be obtained from Fig. 25. This and the accompanying notes are self-explanatory.

Cooling Water Requirements for Hot Pumps

Many pumps in process service require water cooling to various parts of the pump. This cooling water is applied to bearings, stuffing boxes, glands, and pedestals.

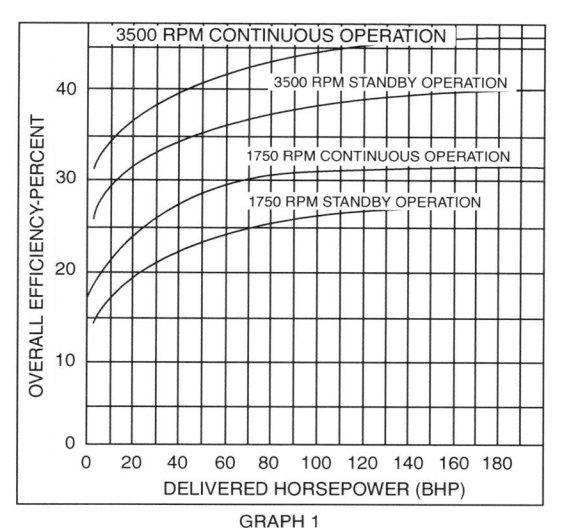

Process Equipment for Petroleum Processing 1511

GRAPH 1
EFFICIENCY OF TURBINES AT 110 PSIG SAT STEAM
INLET WITH 20 PSIG STEAM EXHAUST

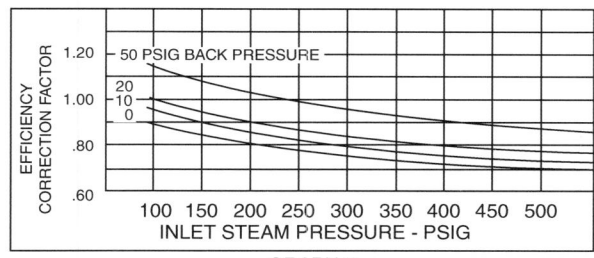

GRAPH 2
STEAM CONDITION CORRECTION FACTOR FOR GRAPH 1

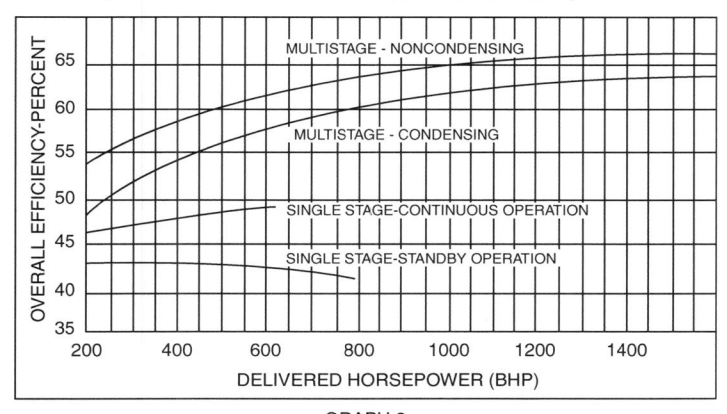

GRAPH 3
EFFICIENCY OF HIGH HORSEPOWER TURBINES

Fig. 24 Steam turbine efficiencies

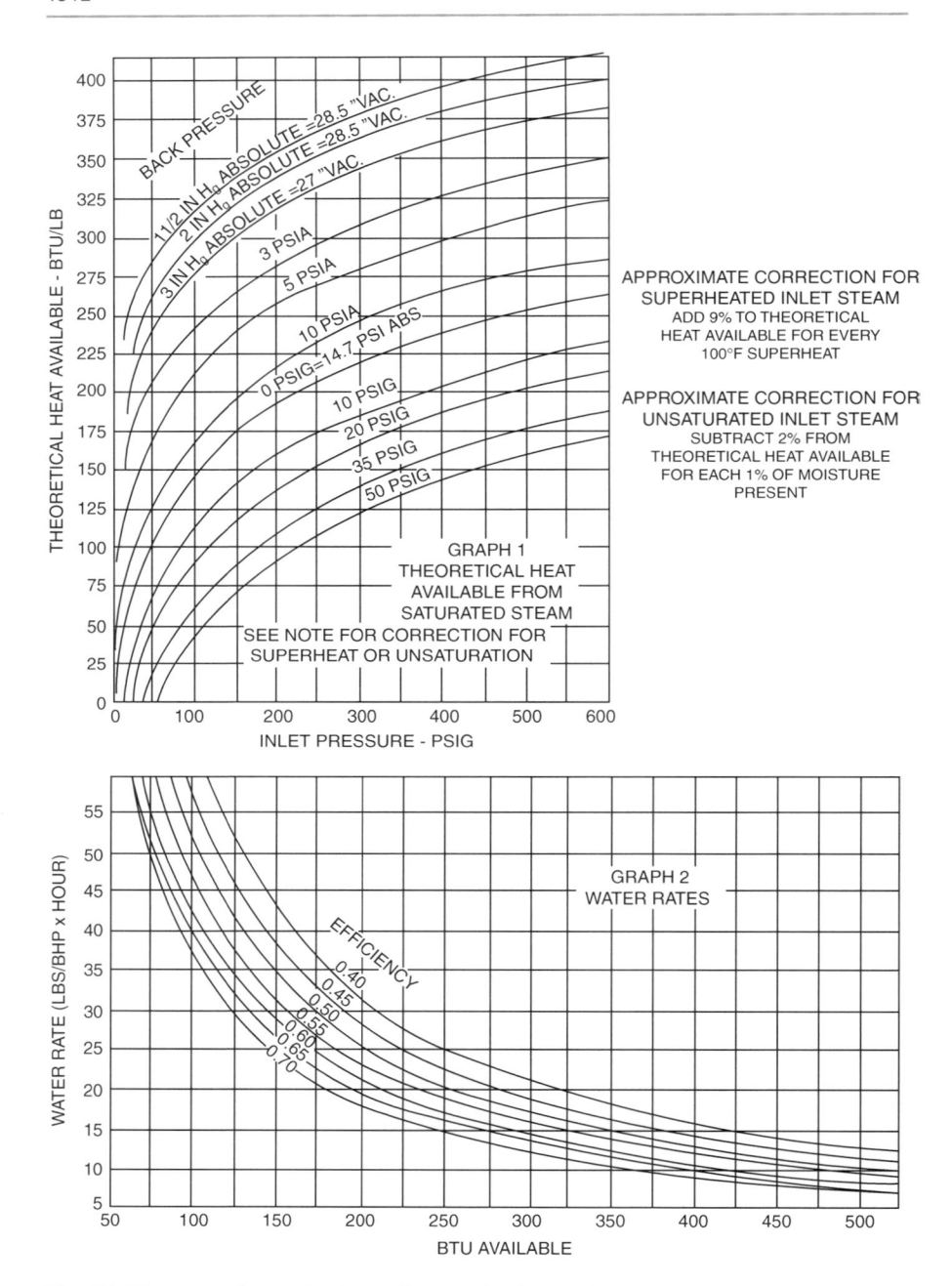

Fig. 25 Water rates for condensing and noncondensing turbines

Process Equipment for Petroleum Processing

Table 15 Typical cooling water requirements for hot pumps

Temperature, pump or steam	Cooling water rate for pump size	
Pumps	To 1,000 gpm	Above 1,000 gpm
Up to 350°F	0 gpm	0 gpm
350–500°F	2 gpm	4 gpm
Above 500°F	3 gpm	6 gpm
Steam turbines		
Up to 450°F	0 gpm	0 gpm
Above 450°F	3 gpm	3 gpm

The application of the cooling water is determined by the manufacturer in accordance with his standard for the service and conditions that the pump must satisfy. Most of the cooling water may be recovered in a closed cooling water system. However, gland cooling water is never recovered but is routed to the wastewater drain. Table 15 lists some typical cooling water requirements for pumps and steam turbines. A vendor will eventually define the requirements for a specific piece of equipment purchased.

Compressors

Types of Compressors and Selection

Compressors are divided into four general types:

- Centrifugal
- Axial
- Reciprocating
- Rotary

The name given to each type is descriptive of the means used to compress the gas. A comparison of the different types of compressors and typical applications is shown in Table 17. A brief description of each of the types follows:

Centrifugal
  This type of compressor consists of an impeller or impellers rotating at high speed within a casing. Flow is continuous and inlet and discharge valves are not required as part of the compression machinery. Block valves are required for isolation during maintenance or emergencies and discharge check valves are normally provided.
  Centrifugal compressors are widely used in the petroleum, gas, and the chemical industries primarily due to the large volumes of gas that frequently have to be handled. Long continuous operating periods without an overhaul make centrifugal compressors desirable for use for petroleum refining and natural gas

applications. Normally they are considered for all services where the gas rates are continuous and above 400 ACFM (actual cubic feet per minute) for a clean gas and 500 ACFM for a dirty gas. These rates are measured at the discharge conditions of the compressor. Dirty gases are considered to be gases similar to those from a catalytic cracker, which may contain some fine particles of solid or liquid material.

The slowly rising head-capacity performance curves make centrifugal compressors easy to control by either suction throttling or variable-speed operation. Spillback control is also common.

The main disadvantage of this type of compressor is that it is very sensitive to gas density, molecular weight, and polytropic compression exponent. A decrease in density or molecular weight results in an increase in the polytropic head requirement of the compressor to develop the required compression ratio.

Axial flow

These compressors consist of bladed wheels that rotate between bladed stators. Gas flow is parallel to the axis of rotation through the compressor. Axial flow compressors become economically more attractive than centrifugal compressors in applications where the gas rates are above 70,000 ACFM at *suction* conditions. The compressors are extremely small relative to capacity and have a slightly higher efficiency than the centrifugal. Axial flow compressors are widely used as air compressors for jet engines and gas turbines.

Reciprocating

Reciprocating compressors are widely used in the petroleum and chemical industries. They consist of pistons moving in cylinders with inlet and exhaust valves. They are cheaper and more efficient than any other type in the fields in which they are used. Their main advantages are that they are insensitive to gas characteristics and they can handle intermittent loads efficiently. They are made in small capacities and are used in applications where the rates are too small for a centrifugal. Reciprocating compressors are used almost exclusively in services where the discharge pressures are above 5,000 psig.

When compared with centrifugal compressors, the reciprocating compressors require frequent shutdowns for maintenance of valves and other wearing parts. For critical services this requires either a spare compressor or a multiple compressor installation to maintain plant throughput. In addition, they are large and heavy relative to their capacity.

Rotary

Recent developments in the rotary compressor field have opened up areas of application in the process industry with the use of the following types of rotary compressors:

1. *High-pressure screw*

   These compressors have been developed into heavy duty-type machines. They consist of two rotating helices in a casing without actual contact. Rotary compressors are lower in cost and have a higher efficiency than centrifugal compressors. They are not sensitive to gas characteristics since they are

Process Equipment for Petroleum Processing   1515

positive displacement machines. Parts are standardized production items so that a spare rotor is not generally required to be stocked for emergency replacement.

This compressor is noisy and sensitive to temperature rise along the screws due to the close clearances involved. They are good for fouling services where the fouling material forms a soft deposit. This decreases the clearances and leakage along the screws and casing. They are not recommended for use in fouling services in which the deposits are hard.

Variation in speed and a discharge bypass to suction are the only types of control that can be used.

2. *Low-pressure screw, lobe, and sliding vane*

   These compressors should be used only for low-pressure, light-duty, noncritical applications. They operate on the same principle as the high-pressure screw type but have different mechanical designs. The same advantages and disadvantages apply as those for rotary high-pressure screw compressors. They are even lower in cost than the high-pressure screw compressors but contain parts having limited life, thus requiring more maintenance.

   Only centrifugal and reciprocating compressors will be discussed further in this book.

Calculating Horsepower of Centrifugal Compressors

Centrifugal compressors are used in process service where high-capacity flows are required. A typical example is the recycle compressor for handling a hydrogen-rich stream in some oil refining and petrochemical processes. Tables 16, 17, and 18 give some idea of the centrifugal compressor capacity ranges and efficiencies.

In general the head or differential pressure levels served by centrifugal compressor are considerably lower than that for reciprocal. Figure 26 illustrates this feature.

Process engineers are often required to establish the capability of a centrifugal compressor in a particular service or to assess the machine's capability to handle a different service. In conducting these studies, it is necessary to determine the machine's horsepower under the study conditions. This item provides a procedure where the gas horsepower (and thereafter the brake horsepower) of a compressor can be calculated.

This procedure is as follows:

Step 1. Establish the duty required from the compressor in terms of:
- Capacity in cft/min at inlet conditions
- Design inlet temp
- Design inlet pressure
- The mole wt of the gas to be handled
- Compression ratio (P_2/P_1) P_2 being the discharge pressure and P_1 the inlet pressure

Table 16 Typical operating efficiencies and heads for centrifugal compressors

Centrifugal compressor flow range			
Nominal flow range (inlet acfm)	Average polytropic efficiency	Average adiabatic efficiency	Speed to develop 10,000 f. head/wheel
500–7,500	0.74	0.70	10,500
7,500–20,000	0.77	0.73	8,200
20,000–33,000	0.77	0.73	6,500
33,000–55,000	0.77	0.73	4,900
55,000–80,000	0.77	0.73	4,300
80,000–115,000	0.77	0.73	3,600
115,000–145,000	0.77	0.73	2,800
145,000–200,000	0.77	0.73	2,500

Step 2. Establish the "K" value for the gas. If this is a pure gas (such as oxygen), the "K" value can be read from data books. Otherwise, the "K" value is the ratio C_P/C_V. See Appendix 5 of this chapter. (*Note*: Do not confuse this "K" factor with equilibrium constants.)

Step 3. Calculate volume of the gas in SCF/min. This is the inlet CFM times inlet pressure times 520 divided by 14.7 psia times inlet temperature in °R, thus

$$\text{SCFM} = \frac{1 \text{ CFM} \times \text{inlet press} \times 520}{14.7 \times \text{inlet temp °R}}. \tag{69}$$

Step 4. Calculate number of moles gas/min by dividing SCF/m by 378. Multiply number of moles by mole wt for lbs/min of gas.

Step 5. Read off the estimated discharge temperature from Fig. 27. Using this and the discharge pressure calculate the volume in cft/min at discharge.

Step 6. Calculate the density of gas at suction and discharge using the weight calculated in step 4 and the CFM for suction and the cft/min calculated in step 5 for discharge. This density will be in lbs/cuft.

Step 7. The average value for Z is taken as Z at suction + Z at discharge divided by 2. Z (compressibility factor) is calculated by the expression

$$Z = \frac{MP}{T\rho_v \times 10.73} \tag{70}$$

where:

M = mole weight.
P = pressure at psia.
T = °R (°F + 460).
ρ_V = density in lbs/cuft.

Step 8. Calculate the adiabatic head in ft lbs/lb using the expression

$$H_{ad} = \frac{Z_{ave} \times R \times T_i}{(K-1)/K} \left[\left(\frac{P_2}{P_1}\right)^{(K-1)/K} - 1 \right] \tag{71}$$

Table 17 Comparison of compressors and typical applications

Type and controllable range	Percent availability	Operating speed, volumetric capacity, compression ratio per stage[f]	Compression efficiency	Advantages	Disadvantages	Usual drivers	Common applications
Centrifugal 70–100 %	99.5–100 %[a]	3,000–15,000 RPM[b]	70–78 %	1. Long continuous operating periods	1. Pressure ratio is sensitive to gas density and molecular weight	Steam turbine	Large refrigeration system
		400–500 ACFM minimum at discharge		2. Low maintenance costs	2. Spare rotor required	Gas turbine	Cat cracker air
		150,000 ACFM max suction volume[d]		3. Small size relative to capacity		Electric motor	Reformer and hydroprocessing unit recycle gas
		80–100,000 ft. polytropic head per casing		4. Ease of capacity control		Waste gas expander	
Axial Flow 80–100 %	99.5–100 %	4,000–12,000 RPM[b]	75–82 %	1. Very high throughputs possible	1. Capacity flexibility limited by steep head-capacity curve and short stable operating range, except when variable pitch stators are used	Steam turbine	Cat cracker air (large)
		70,000 min. ACFM		2. Extremely small size	2. Performance and efficiency are sensitive to fouling	Gas turbine	

(continued)

Table 17 (continued)

Type and controllable range	Percent availability	Operating speed, volumetric capacity, compression ratio per stage[f]	Compression efficiency	Advantages	Disadvantages	Usual drivers	Common applications
		2–4 compression ratio per casing		relative to capacity 3. Higher efficiency than centrifugals 4. Good for parallel operation with other axials or centrifugals	3. Spare rotor and spare stator blading are required	Electric motor	
						Waste gas expander	
Reciprocating; See description for controllable range	98 % clean gas[c]	300–1,000 RPM	75–85 %	1. Handles intermittent loads efficiently	1. Short continuous operating periods require spare or multiple machine installations if service is critical	Synchronous motor	Instrument air
	95 % dirty gas[c]	5 max compression ratio or 330–380 °F max. discharge temperature		2. Lower cost for small capacities	3. Pulsation and vibration require engineered piping arrangement	Coupled or integral electric motor	Refinery air
	95 % clean gas[d]			3. Used for very high discharge pressures (up to 50,000 psig)	4. Availability decreases when non-lubricated machines required to avoid lubricating oil in gas discharge	Coupled or integral engine	Fuel gas

Process Equipment for Petroleum Processing

	93 % dirty gas[d]		4. Higher efficiency than centrifugal in lower capacity ranges		Synthesis gas
			5. Insensitive to gas characteristics		Crude gas
					Small hydroprocessing unit recycle gas
					Hydroprocessing unit makeup gas
					Small refrigeration system
Rotary high pressure screw 55–100 %	99–99.5 %	2,500–10,000 RPM	1. Lower cost than centrifugals	1. Noisy, require inlet and discharge silencers	Electric motor	Refinery air
		1,000–20,000 ACFM at suction	2. Higher efficiency than centrifugals	2. Sensitive to temperature rise due to close tolerances	Steam turbine	Fuel gas
		4–7 compression ratio max. per casing, but not exceeding 100 psi differential pressure	3. Not sensitive to gas characteristics	3. Not recommended for use where fouling produces hard deposits	Gas turbine	Cat cracker air (small)

(continued)

Table 17 (continued)

Type and controllable range	Percent availability	Operating speed, volumetric capacity, compression ratio per stage[f]	Compression efficiency	Advantages	Disadvantages	Usual drivers	Common applications
				4. Parts are standardized production items, so no spare rotor required	4. Speed or bypass control are only types applicable	Waste gas expander	
Low pressure screw, lobe-, and vane-type rotaries, fixed capacity	Not recommended for continuous service	1,500–3,600 RPM	75–80 %	1. Low first cost	1. Limited life	Electric motor	Low pressure, light duty, non-critical services
		100–12,000 ACFM suction		2. Low maintenance cost	2. Speed or bypass control are only types applicable	Steam turbine	
		2 compression ratio per stage with 50 psi max. discharge pressure		3. Parts are standardized production so no spare rotor is required	3. Very noisy		

[a]Clean service machines have the highest reliability
[b]Large machines run at lower speeds
[c]Between turnarounds of 3 days every 8–12,000 h with electric drive. 95–98 % includes 8 h shutdowns every few months for valve maintenance
[d]Between turnarounds of 2 weeks every 8–12,000 h with engine drive. 93–95 % includes 8 h shutdowns every month for maintenance checks on compressor valves and engine driver
[e]Axial flow compressors should be considered at gas rates above 70,000 ACFM at suction
[f]Stages can be compounded in series for higher rates and pressures

Process Equipment for Petroleum Processing 1521

Table 18 Centrifugal compressor categories by head and capacity

	Speed, RPM	Suction, ACFM	Polytropic head, ft lb/lb
Small standard multistage	3,000–3,600	100–1,000	to 8,500
Standard single stage	3,000–3,600	700–60,000	1,000–6,700
Special single stage	3,000–15,000	1,000–60,000	6,700–11,500
Special multistage casing, uncooled	3,000–15,000	1,000–140,000	6,700–100,000
Special multistage, multi-casing, intercooled	3,000–15,000	2,500–140,000	37,000 up

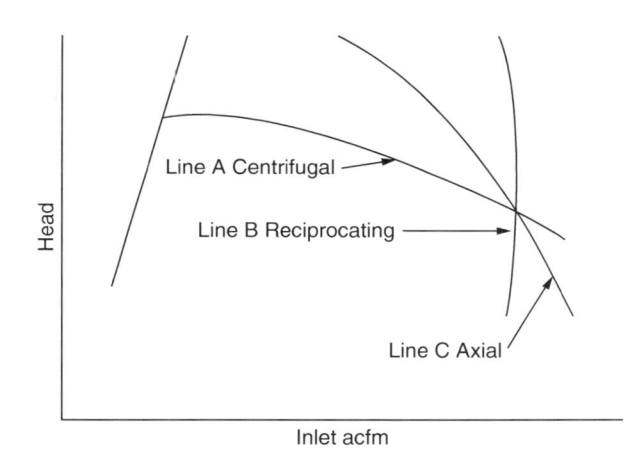

Fig. 26 Head-capacity relationships for difference compressor types

where:
H_{ad} = the adiabatic head in ft lbs/lb.
Z_{ave} = average compressibility factor.
R = gas constant = 1,545/mole wt.
K = adiabatic exponent C_P/C_V.
P_2 = discharge pressure psia.
P_1 = suction pressure psia.
T = inlet temperature °R.

Step 9. The gas HP is obtained using the expression

$$\text{HP} = \frac{W \times H_{ad}}{\eta_{ad} \times 33,000} \tag{72}$$

where:
W = weight in lbs/min of gas.
H_{ad} = adiabatic head in ft lbs/lb.
η_{ad} = adiabatic efficiency (0.7–0.75).

Step 10. Check GHP using Fig. 28.

An example calculation now follows:
Example calculation

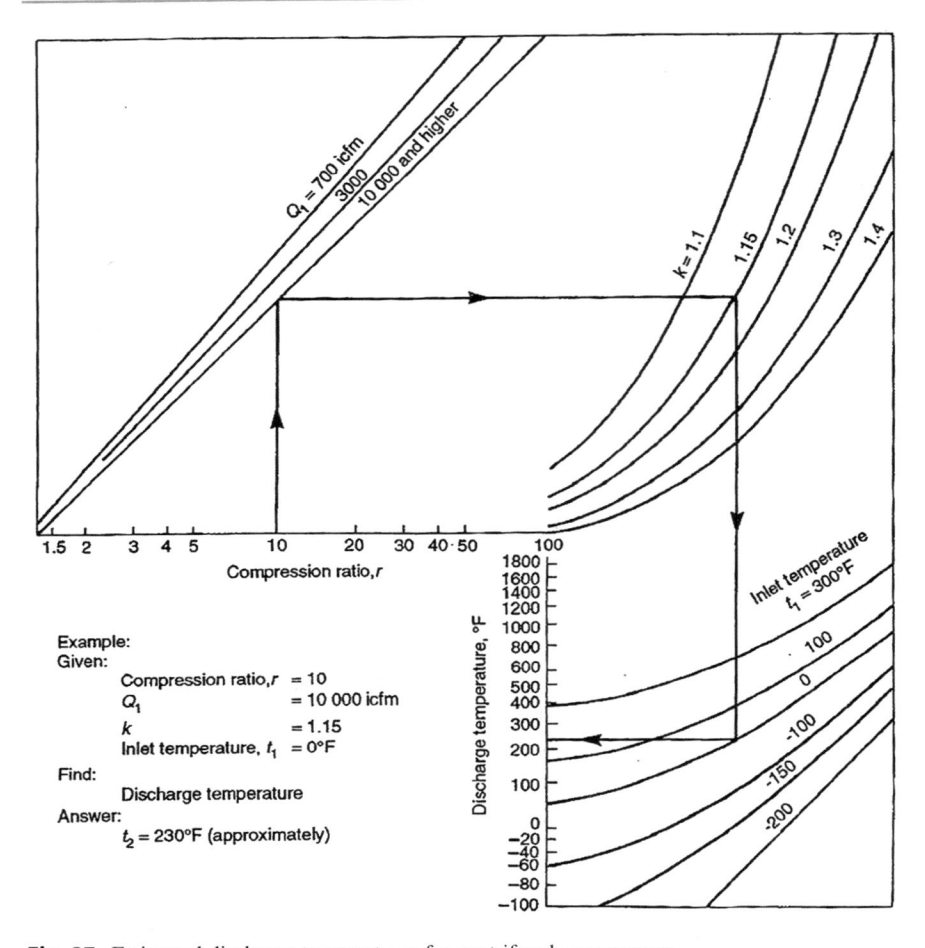

Fig. 27 Estimated discharge temperatures for centrifugal compressors

Determine the Gas HP of a centrifugal compressor assuming isentropic compression:

Compression ratio = 10.0.
Capacity (actual inlet CF/min) = 10,000.
K_{ave} = 1.15.
T_1 °F = 100.
P_1 psia = 100.
Mole wt = 30.
Lbs/min = 5,013.

For isentropic compression

$$H_{ad} = \frac{Z_{ave} \times R \times T_i}{(K-1)/K}\left[\left(\frac{P_2}{P_1}\right)^{(K-1)/K} - 1\right] \tag{73}$$

Process Equipment for Petroleum Processing

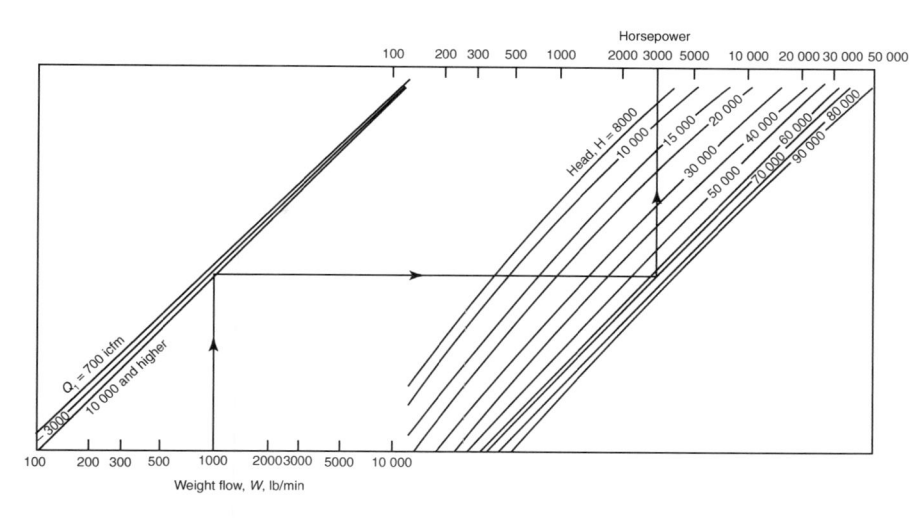

Fig. 28 Determination of centrifugal compressor horsepower

where:
- H_{ad} = adiabatic head in ft lbs/lb.
- Z = compressibility factor (avg).
- R = gas constant = 1,545/MW.
- K = adiabatic exponent $C_P/C_V = 1.15$.
- T = temperature in °R = °F + 460 °F.
- P_2 = discharge pressure psia.
- P_1 = suction pressure psia.

Z at inlet conditions:

$$\rho_V = 0.5 \text{ lbs/cuft}$$

$$Z = \frac{MP}{T\rho_V} \times 10.73 = \frac{30 \times 100}{560 \times 0.5} \times 10.73 = 0.998.$$

Estimated discharge temp (Fig. 27) = 400 °F.

$$Z_{dis} = \frac{30 \times 1,000}{860 \times 3.26} = 0.997.$$

Use 0.998.

$$H_{ad} = \frac{0.998 \times 51.5 \times 560}{\frac{0.15}{1.15}} \left[\left(\frac{1.000}{100}\right)^{0.13} - 1 \right]$$

$$= 220,664 \, (1.349 - 1).$$

$$= 77,004 \text{ ft lbs/lb}$$

$$\text{Gas HP} = \frac{W \times H_{ad}}{\eta_{ad} \times 33,000}.$$

Let η_{ad} be 0.75.

$$\frac{5,013 \times 77,004}{0.75 \times 33,000} = 15,598 \text{ GHP}.$$

This compares well with the estimate based on Fig. 28.

Centrifugal Compressor Surge Control, Performance Curves, and Seals

Centrifugal compressors can be counted on for uninterrupted run lengths of between 18 and 36 months after the initial shakedown run. The 18-month run corresponds to a compressor handling dirty gas, such as furnace gas, and the 36-month run corresponds to a clean gas service, such as refrigerant.

Spare compressors are not usually provided. A spare rotor, however, is required to be stocked as insurance against an extended downtime. Since this rotor is part of the capital cost of the equipment, it is not accounted for as spare parts. Only reliable drivers such as an electric motor, steam, or gas turbine can be used where long continuous run lengths are required. In the case of steam and gas turbines, the drivers will probably dictate the maximum possible run length. The high operating speed of a centrifugal compressor also favors the selection of these types of high-speed drivers. The speed of these drivers can be specified to be the same as those of the compressor. For electric motor drives, a speed-increasing gear is normally required. Centrifugal compressors can be broadly classified with regard to head and capacity as shown in Table 18.

As a guide, the maximum head per impeller is about 10,000 ft. Normally, about eight impellers can be used in a casing.

The minimum allowable volume of gas at the compressor discharge is about 400 ACFM for a clean gas and 500 ACFM for a dirty gas. Dirty gases are considered to be similar to the gas from a steam or catalytic cracking unit.

The discharge temperature is limited to about 250 °F for gases that may polymerize and 400 °F for other gases. Normally intercoolers will be used to keep the discharge temperature within these limits. These temperature limitations do not apply to special centrifugal flue gas recirculators which can be obtained to operate at over 800 °F. There is also a temperature rise limitation of about 350 °F per casing. This is the maximum temperature rise that can be tolerated due to thermal expansion considerations.

Use of cast iron as a casing material is limited to 450 °F maximum. Temperatures of −150 °F to −175 °F can be tolerated in conventional designs. Lower temperatures are not common and will require consulting on individual design features.

Surge

A characteristic peculiar to centrifugal and axial compressors is a minimum capacity at which the compressor operation is stable. This minimum capacity is

Table 19 Typical surge points for centrifugal compressors

Wheels/compression stage	% of normal capacity at surge – maximum
1	55
2	65
3 or greater	70

Table 20 Typical maximum increases in pressure ratio from normal to surge for centrifugal compressors

Wheels/compression stage	Minimum % of rise in pressure ratio from normal to surge flow
1	$3\frac{1}{2}$
2	6–7
2 or greater	$7\frac{1}{2}$

referred to as the surge or pumping point. At surge, the compressor does not meet the pressure of the system into which it is discharging. This causes a cycle of flow reversal as the compressor alternately delivers gas and the system returns it.

The surge point of a compressor is nearly independent of its speed. It depends largely on the number of wheels or impellers in series in each stage of compression. Reasonable reductions in capacity to specify for a compressor are shown below (Table 19).

An automatic recirculation bypass is required on most compressors to maintain the minimum flow rates shown. These are required during start-up or when the normal load falls below the surge point. Cooling is required in the recycle circuit if the discharge gas is returned to the compressor suction.

Performance Curves

The rise of performance curves should be specified for a compressor. This is normally done by specifying the pressure ratio rise to surge required in each stage of compression. A continuously rising curve from normal flow rate to surge flow is required for stable control.

The pressure ratio rise to surge is largely a function of the number of impellers per compression stage. Reasonable pressure ratio rises to specify are shown in Table 20.

Frequently, the performance curves for a compressor have to be plotted to determine if all anticipated process operations will fit the compressor and its specific speed control. Three points on the head-capacity curve are always known. These are the normal, surge, and maximum capacity points. The normal capacity is always considered to be on the 100 % speed curve of the compressor. The surge point and the compression ratio rise to surge have been specified. From this the head produced by the compressor at the surge point can be back calculated using the head-pressure ratio relationship. The maximum capacity point is specified to be at least 115 % capacity at 85 % of normal head.

The head-capacity curve retains its characteristic shape with changes in speed. Curves at other speeds can be obtained from the three known points on the 100 % speed curve by using the following relationships:

1. The polytropic head varies directly as the speed squared.
2. The capacity varies directly as speed.
3. The efficiency remains constant.

Figure 29 shows a typical centrifugal compressor performance curve.

Control

Speed
Speed control is the most efficient type of control from an energy consideration. It requires, however, that a variable-speed driver, such as a steam turbine or gas turbine, or a variable-speed electric motor be used. The compressor is controlled by shifting its performance curve to match the system requirements.

Suction Throttling
- *Adjustable inlet guide vanes.* Adjustable inlet guide vanes are the most efficient method of adjusting the capacity of a constant speed compressor to match the system characteristics. They consist of a venetian blind device that is positioned by a rack and pinion linkage. While the guide vanes do some throttling, their main effect is to change the velocity of the gas to that of the impeller vane by changing the direction of flow. This changes the head produced and in effect changes the characteristic of the machine.
- *Suction throttle.* This control consists of a control valve located in the compressor suction which regulates the suction pressure to the compressor. The control valve results in a greater power loss compared to adjustable inlet guide vane control since it is a pure throttling effect. Suction throttle valves are lower in cost than adjustable inlet guide vanes.

Discharge throttling
This control consists of a control valve located in the compressor discharge. Discharge throttle valves are seldom used since they offer relatively little power reduction at reduced capacity. The effect is simply to "push" the compressor back on its curve.

Seals
Table 21 shows the types of seals that are commonly used in centrifugal compressors. The start-up as well as the operating conditions of the compressor should be considered in selecting a seal. Often the system is evacuated when hydrocarbons are handled prior to its start-up. This requires that the seal be good for vacuum conditions.

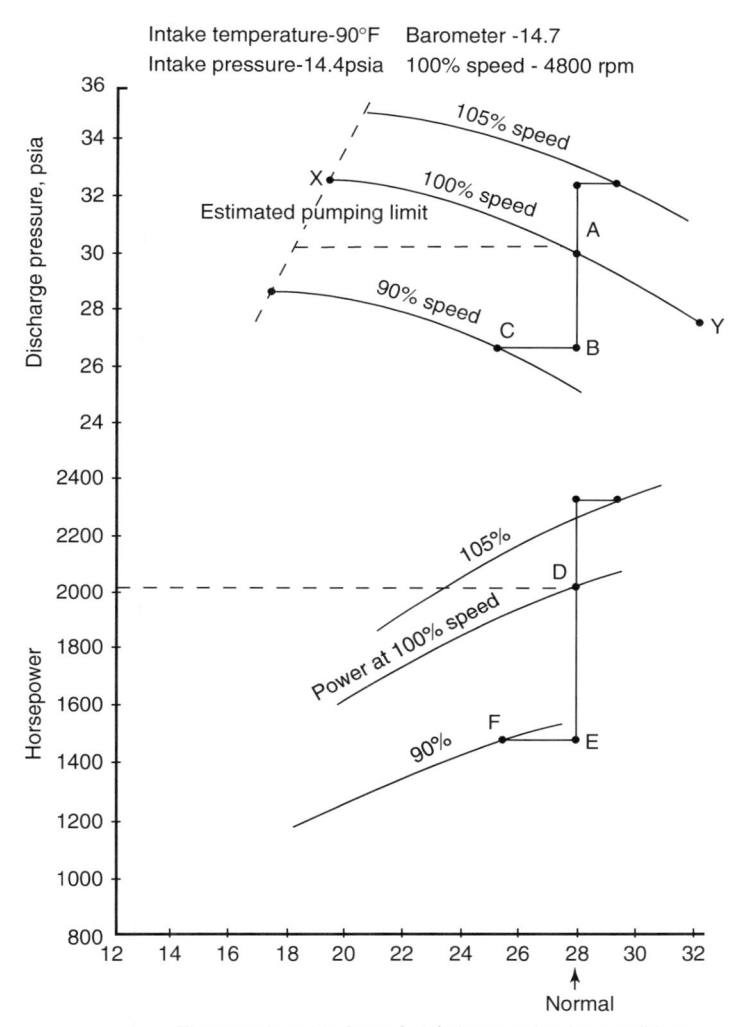

Fig. 29 An example of a centrifugal compressor performance curve

Specifying a Centrifugal Compressor

The process specification must give all the information concerning the gas that is to be handled, its inlet and outlet conditions, the utilities that are available, and the service that is required of the compressor. The process specification sheet for a centrifugal compressor will be similar to that shown in Fig. 34 for a reciprocating compressor. An explanation of this specification now follows covering each of the items in the specification.

Table 21 Centrifugal compressor seals

No	Application	Gas being handled	Inlet pressure, psia	Seal arrangement
1	Air compressor	Atmospheric air	Any	Labyrinth
2	Gas compressor	Noncorrosive	Any	Labyrinth
		Nonhazardous		
		Non-fouling		
		Inexpensive		
3	Gas compressor[a]	Noncorrosive or	10–25	Labyrinth with injection or ejection of fluid being handled
		Corrosive		
		Nonhazardous or		
		Hazardous		
		Non-fouling or fouling		
4	Gas compressor	Noncorrosive	All pressures	Oil seal combined with lube oil system Labyrinth gas seal
		Nonhazardous		
		Non-fouling		
5	Gas compressor	Corrosive	All pressures	Oil seal with seal oil separate from lube oil system
		Nonhazardous or hazardous		
				Labyrinth gas seal

[a]Where some gas loss or air induction is tolerable

Title Block
This requires the item to be identified by item number and its title. The number of units that the specification refers to is also given here. For a centrifugal compressor, this will normally be just one as very seldom is a spare machine required.

Normal and Rated Columns
More often than not, the conditions and quantities required to be handled will vary during the operation of the machine. The two (or more) columns therefore will be completed showing the average normal data in the first column and the most severe conditions and duty required by the compressor in the second. The severe conditions in column two are for a continuous length of operation not instantaneous peaks (or troughs) that may be encountered.

Gas
The composition and gas stream identification must be included as part of the process specification. Usually the composition of the gas is listed on a separate sheet as shown in the example. Note in many catalytic processes that utilize a recycle gas, the composition of the gas will change as the catalyst in the process ages. Thus, it will be necessary to list the gas composition at the start of the run (SOR) and at the end of the run (EOR).

The compressor may also be required to handle an entirely different gas stream at some time or other. This too must be noted. For example, in many petroleum refining processes, a recycle compressor normally handling a light predominately hydrogen gas is also used for handling air or nitrogen during catalyst regeneration, purging, and start-up.

Volume Flow
This is the quantity of gas to be handled stated at 14.7 psia and 60 F.

Weight Flow
This is the weight of gas to be handled in either lbs/min or lbs/h.

Inlet Conditions
Pressure: This is the pressure of the gas at the inlet flange of the compressor in psia.
Temperature: This is the temperature of the gas at the inlet flange of the compressor.
Mole weight: The mole weight of the gas is calculated from the gas composition given as part of the specification.
C_p/C_v: This is the ratio of specific heats of the gas again obtained from the gas mole wt and Appendix 5 of this chapter.
Compressibility factor (z): Use the value at inlet conditions calculated as shown in step 7 of section "Calculating Horsepower of Centrifugal Compressors" in this handbook chapter.
Inlet volume: This is the actual volume of gas at the conditions of temperature and pressure existing at the compressor inlet. Thus,

$$\text{ACFM} = \frac{\text{SCFM} \times 14.7 \times (\text{inlet temp F} + 460)}{(60\ \text{F} + 460) \times \text{inlet press psia}}. \tag{74}$$

Discharge Conditions
Pressure: This is the pressure at the compressor outlet flange and is quoted in either psia or psig.
Temperature: This is estimated using Fig. 27.
C_p/C_v: This will be the same as inlet.
Adiabatic efficiency: This will be as given in Fig. 30.

Approximate Driver Horsepower
This item will include the adiabatic (or gas) horsepower as calculated in the section on calculating the horsepower of centrifugal compressors in this chapter plus the following losses:

Leakage loss – 1 % of adiabatic HP
Seal losses – Allow 35 HP for all HP ranges
Bearing loss – Allow 35 HP for all HP ranges

The remainder of the spec sheet contains all the essential data and requirements that may affect the duty and performance of the compressor. Much of this is self-explanatory; however, there are some items that require comment. These are:

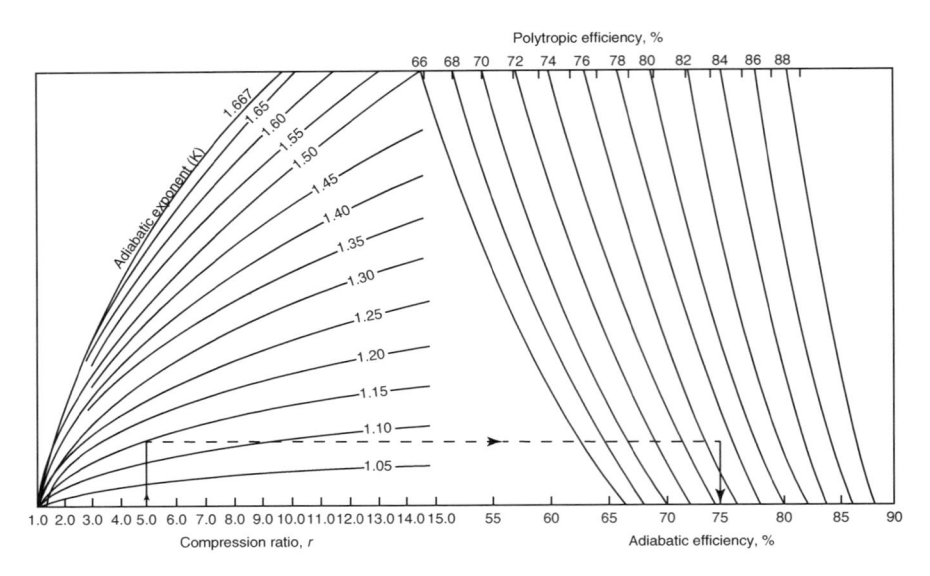

Fig. 30 Adiabatic efficiencies for centrifugal compressors

1. Many compressor installations today are under an open-sided shelter with a small overhead gantry crane assembly for maintenance.
2. Usually the lube and seal oil assemblies have their own pump and control systems. Consequently even if the compressor itself is to be steam driven, there may still be need to give details of utilities for the ancillary equipment.
3. Details of the gas composition is essential for any development of the compressor. This is listed on the last page of the specification together with any notes of importance concerning the machine and its operation.

An example calculation for a specification sheet follows.

Example Calculation

Prepare a process specification sheet for a compressor to handle the hydrogen recycle stream in an o-xylene isomerization plant. Details are as follows:

Fresh feed rate	5,000 bpsd of m-xylene
Recycle gas rate	7,000 Scf of hydrogen per Bbl of fresh feed
Gas composition mole %	

Process Equipment for Petroleum Processing 1531

	Start of run (SOR)	End of run (EOR)
H_2	85.00	68.78
C_1	4.4	9.17
C_2	4.2	8.86
C_3	3.6	7.36
iC_4	0.78	1.62
nC_4	0.99	2.05
C_5s	1.03	2.16

Suction pressure	150 psig
Reactor pressure	500 psig
Suction temperature	100 °F

Step 1. Calculate the mole weight of the gas

	SOR				EOR	
	Mole %	MW	wt factor		Mole %	wt factor
H_2	85.0	2	170		68.78	138
C_1	4.4	16	70		9.17	147
C_2	4.2	30	126		8.86	266
C_3	3.6	44	158		7.36	324
iC_4	0.78	58	45		1.62	94
nC_4	0.99	58	57		2.05	119
C_5s	1.03	72	74		2.16	156
Total	100.00		700		100.00	1,244
MW	7.0				12.44	

Step 2. Calculate volume flow of gas in SCF/min.

$$\text{Total volume of } H_2 \text{ required} = 5,000 \text{ BPSD} \times 7,000 \text{ SCF}$$
$$= 35.00 \text{ MMScf/day}$$
$$= 24,306 \text{ Scf/min.}$$
$$\text{For SOR volume gas flow} = \frac{24,306}{0.85}$$
$$= 28,595 \text{ Scf/min.}$$
$$\text{For EOR volume gas flow} = \frac{24,306}{0.6878}$$
$$= 35,339 \text{ Scf/min.}$$

Step 3. Calculate weight flow in lbs/min.

$$\text{Moles/min of gas} = \frac{\text{Scf/min}}{378}.$$

For SOR moles/min = 75.6.
For EOR moles/min = 93.5.
  lbs/min for SOR = 75.6 × 7.0 = 529 lbs/min.
  lbs/min for EOR = 93.5 × 12.44 = 1,163 lbs/min.

Step 4. Calculate ACFM at inlet conditions.

$$\text{Compressor inlet pressure} = 165 \text{ psia}$$
$$\text{temp} = 100° \text{ F}.$$
$$\text{For SOR, ACFM} = \frac{28,597 \times 14.7 \times 560}{520 \times 165}$$
$$= 2,744 \text{ cft/min}.$$
$$\text{For EOR, ACFM} = \frac{35,339 \times 14.7 \times 560}{520 \times 165}$$
$$= 3,391 \text{ cft/min}.$$

Step 5. Estimate the C_p/C_v ratio.

The molal proportions will be used for this purpose. The ratio for each component will be taken from Appendix 5.

SOR				EOR	
	C_p/C_v	Mole %	C_p/C_v fact	Mole %	C_p/C_v fact
H_2	1.40	85.0	119	68.78	96.29
C_1	1.30	4.4	5.7	9.17	11.92
C_2	1.22	4.2	5.12	8.86	10.81
C_3	1.14	3.6	4.10	7.36	8.39
iC_4	1.11	0.78	0.87	1.62	1.80
nC_4	1.11	0.99	1.10	2.05	2.28
C_5s	1.09	1.03	1.12	2.16	2.35
Total		100.00	131.9	100.00	133.84

Then C_p/C_v for the gas is
SOR = 1.319
EOR = 1.338

Step 6. Calculate compressibility factors.

$$Z = \frac{MW \times P_1}{T \times \rho_v \times 10.73}. \qquad (75)$$

$$\text{For SOR flows } Z = \frac{7.0 \times 165}{560 \times 0.193 \times 10.73}$$
$$= 0.996.$$

$$\rho_v = \frac{\text{wt lbs/min}}{\text{ACFM}}.$$

$$\text{For EOR flows } Z = \frac{12.46 \times 165}{560 \times 0.343 \times 10.73}$$
$$= 0.998.$$

Step 7. Calculate outlet temperature.

Approximate discharge temperature is read from Fig. 27 in this chapter using the following:

$$\text{ACFM for SOR} = 2,744.$$
$$\text{ACFM for EOR} = 3,391.$$
$$\text{Compression ratio} = \frac{515}{165}$$
$$= 3.12.$$
$$\text{Inlet temp F} = 100.$$

Then:

$$\text{Discharge temp for SOR} = 370\,°F.$$
$$\text{Discharge temp for EOR} = 340\,°F.$$

Step 8. Calculate the approximate driver HP.

$$H_{ad} = \frac{Z_{ave} \times R \times T_i}{(K-1)/K} \left[\left(\frac{P_2}{P_1}\right)^{(K-1)/K} - 1 \right] \qquad (76)$$

where:
H_{ad} = adiabatic head in ft lbs/lb.
Z = compressibility factor (avg).
R = gas constant = 1,545/MW.
K = adiabatic exponent C_P/C_V = 1.15.
T = temperature in °R = °F + 460 °F.
P_2 = discharge pressure psia.
P_1 = suction pressure psi.
Then:
H_{ad} for SOR conditions = 161,063.
H_{ad} for EOR conditions = 91,546.

Step 9. The gas HP is obtained using the expression

$$\text{Gas HP} = \frac{W \times H_{ad}}{\eta_{ad} \times 33,000} \tag{77}$$

where:
W = mass flow in lbs/min of gas.
H_{ad} = adiabatic head in ft lbs/lb.
η_{ad} = adiabatic efficiency (0.7–0.75).
Let η_{ad} be 0.73.
For SOR gas, HP = 3,536.
For EOR gas, HP = 4,420.

Step 10. The driver HP is as follows:

	SOR	EOR
Gas HP	3,536	4,420
Leakage losses	35	44
Bearing losses	35	35
Seal losses	35	35
Driver HP	3,641	4,534

Calculating Reciprocating Compressor Horsepower

Reciprocating compressors are used extensively in the process industry. They vary in size from small units used for gas recovery (such as those on a crude distillation overhead system) to fairly large complex machines used for recycle gas streams and for transporting natural gas. Engineers are frequently required therefore to assess the horsepower of these machines and their capability to handle various streams. This item describes a method used to determine horsepower and proceeds with the following steps:

Step 1. Obtain the capacity and the properties of the gas to be handled. Fix the ultimate (discharge) pressure level.

Step 2. From the machine data sheet, ascertain the number of stages.

Step 3. Estimate the brake horsepower from the expression

$$\text{BHP} = 22 \times (\text{compr. ratio/stage}) \times \text{no. of stages}$$
$$\times \text{capacity} \left(\text{in cuft/day} \times 10^6\right) \times F \tag{78}$$

where:

$F = 1.0$ for 1 stage
1.08 for 2 stages
1.10 for 3 stages.

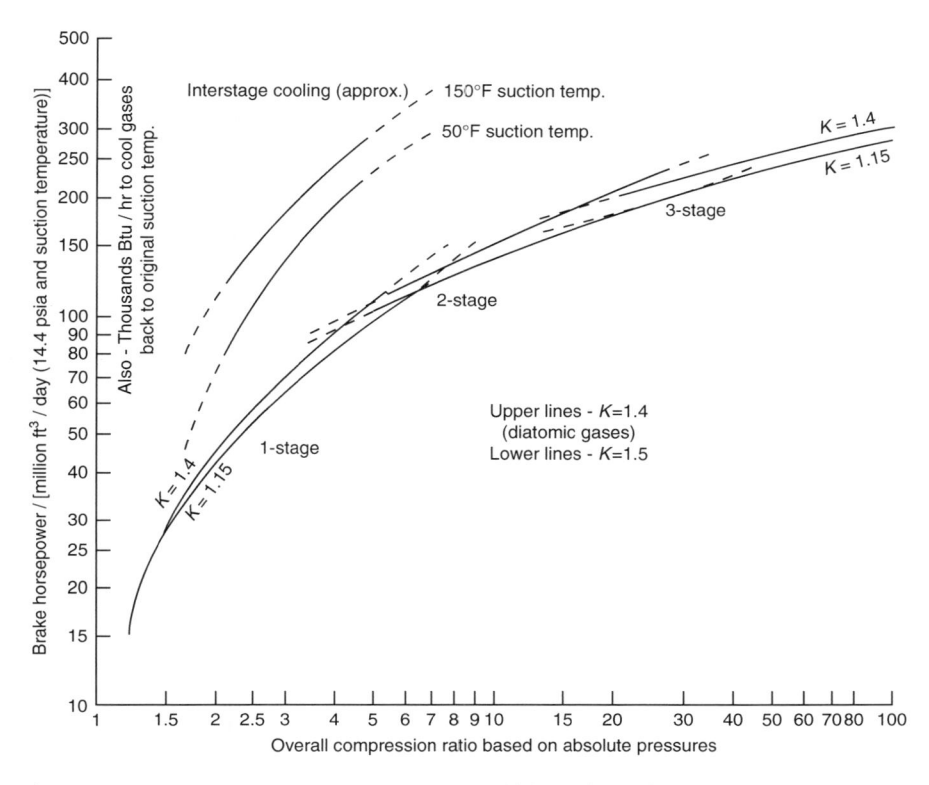

Fig. 31 An estimate of brake horsepower per MMcfd for reciprocating compressors

Ratio/stage = $\sqrt{}$ ratio for two stages and $\sqrt[3]{}$ ratio for three stages.

Step 4. Check the estimate with Fig. 31.

Step 5. Confirm actual suction conditions and compression ratio required (discharge pressure).

Step 6. Calculate compression ratio/stage.

Step 7. Calculate first-stage discharge pressure. This will be suction pressure times compression ratio per stage from step 6.

Step 8. Allow about 3 % for interstage pressure drop then calculate second-stage discharge pressure. Check that overall compression ratio/stage is close to that calculated for step 6.

Step 9. Calculate the "K" value of the gas. "K" value is C_p/C_v of the gas. If the gas is a mixture of components, "K" value may be calculated as the sum of each component mole fraction multiplied by each of their "K" values given in Appendix 5. Alternatively for a good approximation, data in Fig. 32 may be used.

Step 10. Calculate discharge temperature from first stage using Fig. 33. Assume some intercooling (or calculate intercooling from plant data) and fix second-stage discharge temperature also using Fig. 33.

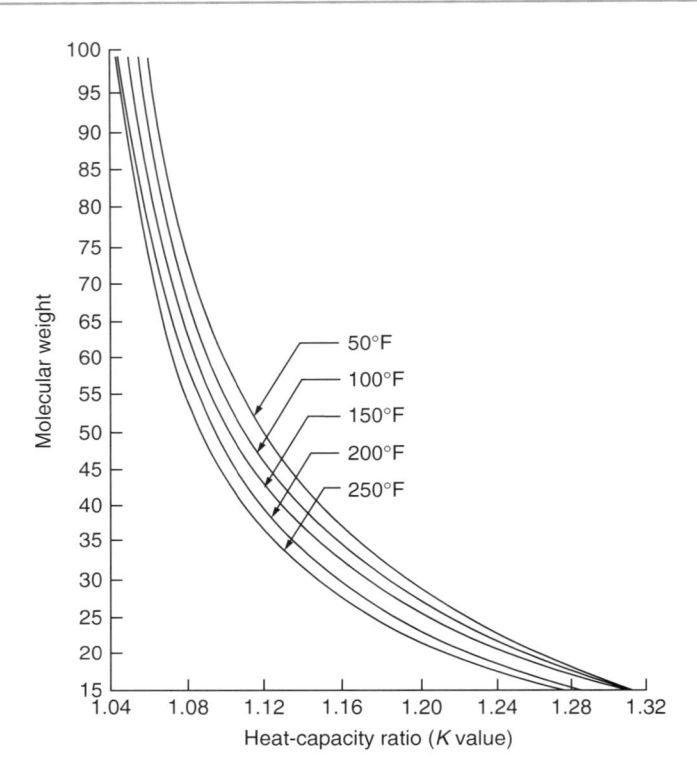

Fig. 32 Approximation of "K" from mole weights

Step 11. Calculate the compressibility factor Z at suction and discharge from the expression

$$\rho_v = \frac{MP}{T \times Z \times 10.73} \tag{79}$$

where:
  ρ_v = gas density in lbs/cuft at condition.
  T = °Rankine (°F + 460 °F).
  Z = compressibility factor.
  M = mole weight.
  P = pressure in psia.
  Use average value at suction and discharge for each stage.

Step 12. Read off BHP/MMcfd at the compression ratio/stage (from step 7) and "K" from step 9 for each stage.

Step 13. Calculate BHP per stage from the expression

$$\text{BHP} = (\text{BHP/MMcfd}) \times \frac{P_L}{14.4} \times \frac{T_S}{T_L} \times Z_{\text{ave}} \times \text{MMscf/D} \tag{80}$$

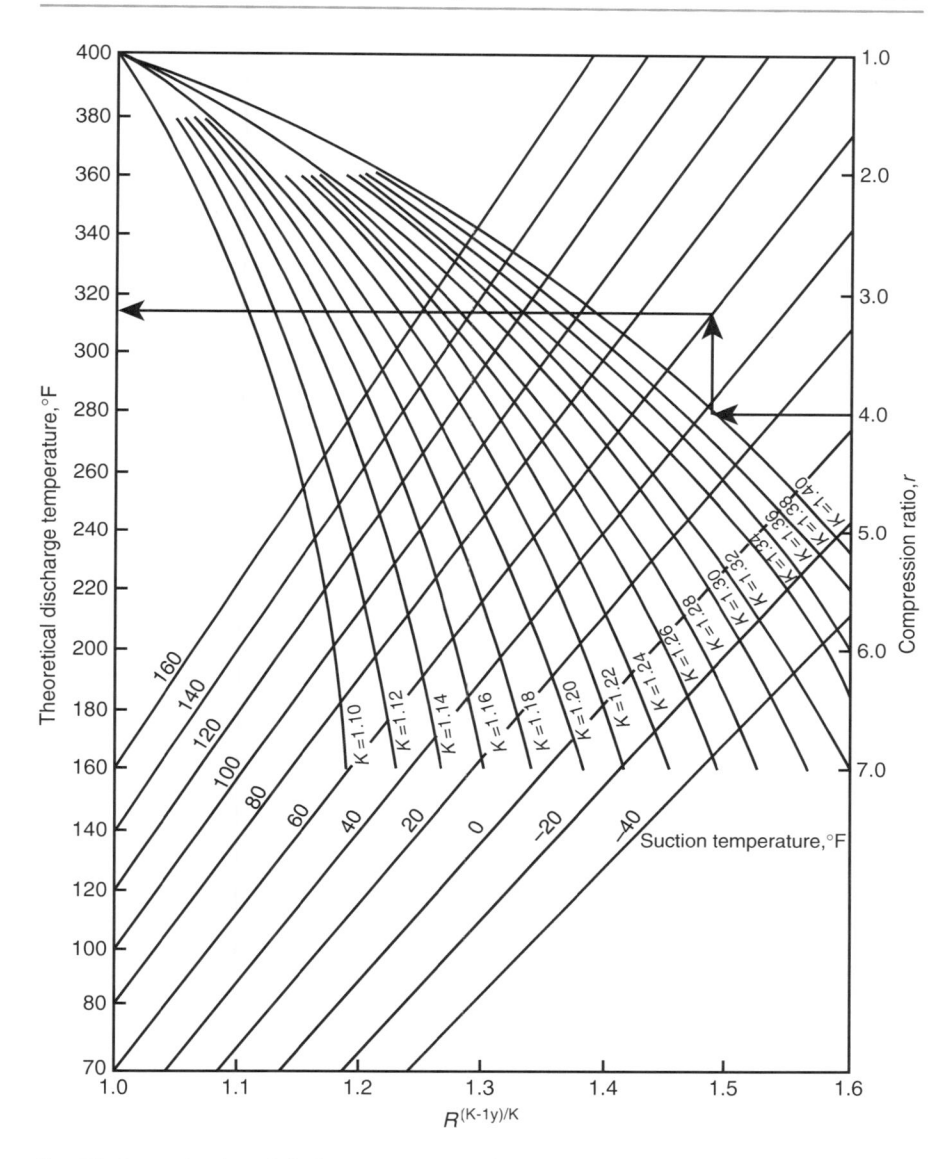

Process Equipment for Petroleum Processing 1537

Fig. 33 Determination of discharge temperature for reciprocating compressors

where:
  BHP/MMcfd = from Fig. 31.
  P_L = pressure base used in psia.
  T_S = intake temperature °R.
  T_L = temperature base used in °R (usually 520 °R).
Step 14. Brake horsepower for the machine is the sum of the BHP calculated for each stage in step 13 above.

Reciprocating Compressor Controls and Intercooling

A reciprocating compressor is a constant displacement-type compressor. It compresses the same volume of gas to the same pressure level without regard to whether the gas is hydrogen or butane. This characteristic makes them desirable for use in services where the gas will have a widely varying composition. In some cases when an extremely low-density gas will be compressed, a reciprocating compressor may be more economical than a centrifugal compressor, even though the flow rate may be very high, due to the large number of stages required for the centrifugal.

Reciprocating compressors are widely used in process services where the flow rates are too small for centrifugal compressors. These units can be obtained with integral or coupled electric motor in sizes from a few HP to 12,000 HP and separate or integral gas drivers varying in sizes from 100 to 5,500 HP.

A range of air-cooled light-duty compressors is available for intermittent service. They range in size from 1/4 to about 100 HP at pressures up to 300 psig and are usually single acting. A primary process use of such equipment is for starting air compressors on gas engine-driven machines. Reciprocating compressors can be designed to handle intermittent loads efficiently. This is done by using cylinder unloaders such as clearance pockets or suction valve lifters. Power losses are low at part load operation with these devices.

Reciprocating parts and pulsating flow present several engineering problems. The foundation and piping system must be constructed to withstand the vibrations produced by the compressor. The pulsating flow produced by the compressor must be dampened by the use of properly engineered suction and discharge bottles. These problems do not arise with the use of other types of compressors.

Reciprocating Compressor Control

Control of the compressor to prevent driver overload can be accomplished with clearance pockets, suction valve lifters, a throttling valve in the suction line, or a control valve in a bypass around the compressor. A hand-operated bypass without cooler is usually furnished inside the block valves for start-up purposes.

1. *Clearance pockets*

   Of the above types of regulation, control by opening fixed clearance pockets gives the smoothest and most efficient control within its range of application. It has the following advantages:
   - Minimizes the intake pulsation as the gas flow is not reversed in the intake lines to the cylinder.
   - Results in lower bearing loads as all inertia loads are cushioned.
   - Results in very efficient part load operation. When the gas compressed into the pockets is expanded, it follows the adiabatic line of compression and results in little power loss.

   Clearance control has the following disadvantages which sometimes completely eliminates it from consideration:

- When low ratios of compression are combined with high suction pressure, clearance pockets of sufficient size to unload the compressor cannot physically be installed in the machine.
- Clearance control is designed for one set of pressure conditions and any variation in either suction or discharge pressure affects the amount of unloading accomplished by a given pocket.
- Condensable corrosive gases sometimes cause corrosion and liquid slugging problems.

2. *Suction valve lifters*

    Suction valve lifters are the other type of internal unloading devices for compressor cylinders and have characteristics that make them applicable when clearance control is not. Suction valve lifters completely unload their end of the cylinder whenever they are opened, regardless of the pressure. They do result in increased bearing loads due to unbalanced inertial forces. Suction line pulsation may increase because the single acting cylinder may excite a different frequency in the gas.

3. *Suction throttle valve*

    A throttle valve in the suction line should be considered only for small reciprocating compressors. For large-size machines, the suction valve cannot give tight enough shutoff to permit unloading the compressor for starting.

4. *Bypass control*

    External bypass control around the compressor is applicable to all sizes of compressors. It results in a loss of power since the full compressor capacity must be compressed to and delivered at the full discharge pressure before being bled back to suction pressure. Care must be taken with this type of control to ensure that the bypassed gas is cooled sufficiently to prevent increasing the discharge temperature. This type of control is preferred for installations up to several hundred horsepower because of its smoothness and lack of complexity. Individual machines can be shut down for large process variations.

5. *Variable-speed reciprocating compressors*

    With a variable-speed driver, cylinder control can usually be eliminated and speed control used to obtain desired process conditions. However, start-up unloading must be furnished, usually consisting of a hand-operated bypass within the machine, or cylinder block valves. On turbine-driven reciprocating compressors, economics usually dictate that the compressor be run at constant speed and that cylinder controls or system bypasses be used to obtain the required control.

Reciprocating Compressor Intercooling

Intercooling for multistage compressors is advisable whenever there is a large adiabatic temperature rise within the cylinder and the cylinder discharge temperature would exceed 350 °F. When intercooling is employed, the inlet temperature to the higher stage should be as close to the cooling water temperatures as practical. On standard commercial air intercoolers, approach temperatures of 15–20 °F are commonly used. Cooling to first-stage inlet temperature is usually economical on process gas compressors.

Intercooling is employed for two basic reasons:

1. For mechanical reasons whereby discharge temperature must be limited to 350 °F for lubrication purposes.
2. An economic reason as intercooling will save from 3 % to 5 % of the required BHP.

In general, on process compressors handling low "n" value gases, intercooling is not employed unless the temperature limitation is exceeded. On high "n" value diatomic gas mixtures, such as air, intercooling is the rule above about a 4 compression ratio and ambient temperature at suction.

In general, cooling water for electric-driven compressors can be any water available, including saltwater. (If the compressor is tied into a plant having gas engine-driven compressors, the electric-driven machine should be tied into the closed system.) The cooling water should be available at a minimum pressure of 25 psig.

For estimating purposes, cooling water temperature rise across the cylinders can be taken as 15 °F and that across the inter- and aftercoolers can be taken as 15 °F. For estimating purposes, cooling water requirements are as follows:

Jacket water cooling	500 Btu/BHP/h
Inter cooler	1,000 Btu/BHP/h
After cooler	1,000 Btu/BHP/h

Use motor rating for the horsepower required.

If the discharge temperature of the gas does not exceed 180 °F, it is common practice to eliminate cooling water on the cylinder and operate with cooling passages which are filled with oil. Any jacketed cylinder must be filled with some fluid to ensure even temperature distribution.

Specifying a Reciprocating Compressor

As in the case of the centrifugal compressor, all data necessary to give a precise requirement for the duty and performance required of a reciprocating compressor must be given in the specification sheet. Many of these data are the same as those given in a specification for a centrifugal compressor (discussed earlier in this chapter). For completeness, all the items in a reciprocating compressor are included below. Fig. 34 illustrates a completed reciprocating compressor specification sheet.

Title block

This requires the item to be identified by item number and its title. The number of units that the specification refers to is also given here. For a centrifugal compressor, this will normally be just one as very seldom is a spare machine required. This may not be so in the case of a reciprocating compressor.

Normal and rated columns

More often than not, the conditions and quantities required to be handled will vary during the operation of the machine. The two columns therefore will be completed showing the average normal data in the first column and the most severe conditions and duty required by the compressor in the second. The severe conditions in column two are for a continuous length of operation not instantaneous peaks (or troughs) that may be encountered. Other "off-design" cases may also be shown.

Gas

a

ITEM No. __C101 A&B__   TITLE __Hydrotreater Recycle Gas Compressor__

Number of units required   __3 (2 + 1 spare)__

	Normal				Rated			
GAS (see attached composition)	75% H_2				65.8% H_2			
VOL. flow scf/min	4337				4937			
WEIGHT lb/min	74.1				136.3			
INLET CONDITIONS (each stage)	Stage 1	Stage 2	Stage 3	Stage 4	Stage 1	Stage 2	Stage 3	Stage 4
Pressure psia	65	198*			65	198*		
Temperature °F	80	100*			80	100*		
Mol weight	6.44				10.44			
C_P/C_V	1.362				1.347			
Compressibility factor	0.994	0.996			0.996	0.997		
Inlet vol acf/min	1019	347			1159	394		
DISCHARGE CONDITIONS (each stage)								
Pressure psia	198	615			198	615		
Temperature °F	270	299			262	292		
C_P/C_V	1.362				1.347			
Compressibility factor	0.991	0.992			0.995	0.995		
APPROX. DRIVER HORSEPOWER	1055				1190			

COMPRESSOR SERVICE REQUIRED: *After intercooling
    Length of uninterrupted service: __8000 hours__

    Type of compressor: Lubricated   __Yes__
        Non Lubricated   _____

    Discharge RV setting (each stage)   __250 and 750 psig__

Fig. 34 (continued)

b

ITEM No. <u>C 101 A&B</u>   TITLE <u>Hydrotreater Recycle Gas Compressor</u>

General Data and Requirements

Enclosure:
- Open to weather _____
- Under shelter _____Yes_____
- In building _____

Corrosiveness and remarks concerning gas _____
_____None_____

Type of driver: Motor: _____2 normal operating_____
                    Stream turbine:
                        Condensing _____
                        Non-condensing _____Spare machine_____

Utilities:
    Power:
        Voltage _____
        Cycle _____
        Phase _____
    Steam:
        Inlet: Pressure (°F)   _____600_____
                  Temperature (°F) _____710_____
        Condensing exhaust
            Pressure (psia)   _____N/A_____
            Temperature (°F)   _____N/A_____
        Non-condensing exhaust
            Pressure (psig)   _____50_____
            Temperature (°F)   _____(by vendor)_____

Cooling water: Pressure _____60_____ psig: Temperature _____40 °F_____
                Allowable temperature rise _____30 °F_____

Materials of construction:
    Cylinder _____CS_____   Type _____By vendor_____
    Piston   _____Ni.Cr_____   Type _____By vendor_____

Type of shaft seal: _____Labyrinth_____

Fig. 34 (continued)

c

ITEM No. ___C 101 A&B___ TITLE ___Hydrotreater Recycle Gas Compressor___

General Data and Requirements (cont.)

Gas Composition

Mol %	Start of run	End of run
H_2	74.9	65.80
C_1	14.17	19.31
C_2	5.85	7.97
C_3	2.43	3.31
iC_4	1.13	1.54
nC_4	1.00	1.36
c_5s	0.52	0.71
Total	100.00	100.00

Remarks and notes:

1. Vendor to provide:

   Intercoolers: ___Yes___

   Aftercooler: ___Yes___

   Dampeners: ___Yes___

   Flushing and sealing oil systems: ___Yes___

2. If motor driven Re acceleration required class ___A___

3. Suction line: Size ___6"___   RTG ___300# RF___

   Discharge line: Size ___4"___   RTG ___600# RF___

Fig. 34 (a–c) An example of a process specification for a reciprocating compressor

The composition and gas stream identification must be included as part of the process specification. Usually the composition of the gas is listed on a separate sheet as shown in the example. Note in many catalytic processes that utilize a recycle gas, the composition of the gas will change as the catalyst in the process ages. Thus, it will be necessary to list the gas composition at the SOR and at the EOR.

The compressor may also be required to handle an entirely different gas stream at some time or other. This too must be noted. For example, in many petroleum refining processes, a recycle or makeup compressor normally handling a light predominantly hydrogen gas is also used for handling air or nitrogen during catalyst regeneration, purging, and start-up.

Volume flow
This is the quantity of gas to be handled stated at 14.7 psia and 60 F.
Weight flow
This is the weight of gas to be handled in either lbs/min or lbs/h.
Inlet conditions
In the case of multistage compressors, the conditions for each stage must be shown. Where interstage cooling is used, the effect must be reflected in the conditions specified.
Pressure: This is the pressure of the gas at the inlet of the compressor stage in psia. Note if intercooling is used, this pressure must include the intercooler pressure drop.
Temperature: This is the temperature of the gas at the inlet of the compressor stage – after the intercooler if applicable.
Mole weight: The mole weight of the gas is calculated from the gas composition given as part of the specification.
C_p/C_v: This is the ratio of specific heats of the gas again obtained from the gas mole wt and Appendix 5.
Compressibility factor (z): Use the value at inlet conditions calculated as shown in step 7 of item on horsepower calculation for reciprocating compressors.
Inlet volume: This is the actual volume of gas at the conditions of temperature and pressure existing at the compressor stage inlet. Thus

$$\text{ACFM} = \frac{\text{SCFM} \times 14.7 \times (\text{inlet temp F} + 460)}{(60\,°\text{F} + 460) \times \text{inlet press psia}}. \tag{81}$$

Discharge conditions
Pressure: This is the pressure at each stage outlet and is quoted in either psia or psig.
Temperature: This is estimated for each stage using Fig. 33.
C_p/C_v: This will be the same as in the inlet.
Approximate driver horsepower
The brake horsepower for the reciprocating compressor is calculated using the method described earlier. This is *brake horsepower* and includes an allowance for mechanical inefficiencies. The approximate minimum driver horsepower is 1.1 × brake horsepower, but the approximate driver HP will be calculated using the inefficiencies for leakage, seals, etc. as for centrifugal compressors.
The reminder of the spec sheet contains all the essential data and requirements that may affect the duty and performance of the compressor. Much of this is self-explanatory; however, there are some items that require comment. These are:
1. Many compressor installations are under an open-sided shelter with a small overhead gantry crane assembly for maintenance.
2. Usually the lube and seal oil assemblies have their own pump and control systems. Consequently even if the compressor itself is to be steam driven, there may still be need to give details of utilities for the ancillary equipment.

3. Details of the gas composition are essential for any development of the compressor. This is listed on the last page of the specification together with any notes of importance concerning the machine and its operation.
4. For environmental purposes, many reciprocating compressors today include sealed and purged distance pieces and crankcases. These are vented to flare or another vapor control system with a purge gas like nitrogen sweeping the volume. The purge requirements must be specified.

An example calculation for a specification sheet follows:

Example calculation

A hydrotreater makeup compressor is required to handle a gas stream such as to provide the unit with 260 SCF per barrel of feed of pure hydrogen. The composition of the gas varies as follows:

Mole %	Start of run	End of run
H_2	74.9	65.80
C_1	14.17	19.31
C_2	5.85	7.97
C_3	2.43	3.31
iC_4	1.13	1.54
nC_4	1.00	1.36
C_5s	0.52	0.71
Total	100.00	100.00

The fresh feed throughput is fixed at 30,000 BPSD (barrels per stream day). It is proposed to use 3 × 60 % machines of which one will be standby and turbine driven. The inlet pressure of the gas is 50 psig at a temperature of 80 °F. The gas is to be delivered at a pressure of 600 psig and 100 °F. Prepare a process specification for reciprocating compressors to meet these requirements.

Step 1. Calculate volume flows.

$$\text{SOR conditions total flow required} = \frac{260}{0.749}$$
$$= 347 \text{ Scf of gas per Bbl of feed}$$
$$= \frac{347 \times 30,000}{24 \times 60} = 7,229 \text{ Scf/min.}$$
$$\text{EOR conditions total flow required} = \frac{260}{0.658}$$
$$= 395 \text{ Scf/Bbl}$$
$$= \frac{395 \times 30,000}{24 \times 60} = 8,229 \text{ Scf/min.}$$

Volume flow per machine (60 % machines):
  SOR = 7,229 × 0.6 = 4,337
  EOR = 8,229 × 0.6 = 4,937

Step 2. Calculate mole wt of gas.

		SOR			EOR	
	MW	Mole %	wt factor		Mole %	wt factor
H_2	2	74.9	149.8		65.8	131.6
C_1	16	14.17	226.7		19.31	309.0
C_2	30	5.85	175.5		7.97	239.1
C_3	44	2.43	106.9		3.31	145.6
iC_4	58	1.13	65.5		1.54	89.3
nC_4	58	1.00	58.0		1.36	78.9
C_5s	72	0.52	37.4		0.71	51.1
Total		100.0	644.3[a]		100.00	1,044.6[b]

[a]SOR gas mole wt = 6.44 (644.3/100)
[b]EOR gas mole wt = 10.44 (1,044.6/100)

Step 3. Weight of gas lbs/min per machine

One mole of any gas occupies about 378 cft at 60 °F and 14.7 psia. Then,

For SOR conditions, moles/min of gas per machine = $\frac{4,337}{378} = 11.5$

and

$$Lbs/min = 11.5 \times 6.44$$
$$= 74.06 \text{ lbs/min.}$$

For EOR conditions, moles/min of gas per machine = $\frac{4,937}{378} = 13.06$

and

$$Lbs/min = 13.06 \times 10.44$$
$$= 136.30 \text{ lbs/min.}$$

Step 4. Inlet conditions.

$$\text{Inlet pressure} = 50 \text{ psig} = 65 \text{ psia.}$$
$$\text{Required outlet pressure} = 600 \text{ psig} = 615 \text{ psia.}$$
$$\text{Overall compression ratio} = \frac{615}{65} = 9.46$$

This will be a two-stage compressor.

Note: At this level of compression in reciprocating compressors, the compression ratio should not exceed 4:1 for any stage.

$$\text{Compression ratio per stage} = \sqrt{9.46} = 3.07.$$
$$\text{Discharge pressure stage 1} = 65 \times 3.07 = 199.6 \text{ psia}$$

Process Equipment for Petroleum Processing

Allowing 2 psi for the pressure drop across the intercooler, the suction pressure of stage 2 is 197.6; call it 198 psia.

Check the compression ratio of stage 2:

$$\text{Required discharge pressure} = 615 \text{ psia}$$

$$\text{Compression ratio} = \frac{615}{198} = 3.1$$

which is close to the originally predicted of 3.07.

Step 5. Calculate ratio C_p/C_v.

	SOR			EOR	
	C_p/C_v	Mole %	Factor	Mole %	Factor
H_2	1.4	74.9	1.049	65.8	0.921
C_1	1.3	14.17	0.184	19.31	0.251
C_2	1.22	5.85	0.071	7.97	0.097
C_3	1.14	2.43	0.028	3.31	0.038
iC_4	1.11	1.13	0.013	1.54	0.017
nC_4	1.11	1.00	0.011	1.36	0.015
C_5s	1.09	0.52	0.006	0.71	0.008
Total		100.0	1.362	100.00	1.347

C_p/C_v SOR gas = 1.362
C_p/C_v EOR gas = 1.347

Step 6. Calculate inlet ACFM per stage
  SOR.
  Inlet volume for first stage:

$$\text{ACFM} = \frac{\text{Scf/min} \times 14.7 \times \text{inlet temp °R}}{(60 + 460) \times \text{inlet press psia}}$$

$$= \frac{4,337 \times 14.7 \times 540}{520 \times 65} \tag{82}$$

$$= 1,019 \text{ cft/min.}$$

Inlet volume for second stage: (intercooled to 100 °F)

$$\text{ACFM} = \frac{4,337 \times 14.7 \times 560}{520 \times 198}$$

$$= 347 \text{ cft/min.}$$

EOR.
  Inlet volume for first stage:

$$\text{ACFM} = \frac{4,937 \times 14.7 \times 540}{520 \times 65} \tag{83}$$

$$= 1,159 \text{ cft/min.}$$

Inlet volume for second stage:

$$\text{ACFM} = \frac{4,937 \times 14.7 \times 560}{520 \times 198}$$
$$= 488.5 \text{ cft/min.}$$

Step 7. Calculate inlet compressibility factor (Z).

$$Z = \frac{\text{MW} \times P_i}{T_i \times \rho \times 10.73} \tag{84}$$

where:

$$\rho = \frac{\text{wt/min}}{\text{ACFM}}. \tag{85}$$

SOR conditions:

$$1\text{st} - \text{stage } Z = \frac{6.44 \times 65}{540 \times 0.0727 \times 10.73}$$
$$= 0.994.$$
$$2\text{nd} - \text{stage } Z = \frac{6.44 \times 198}{560 \times 0.213 \times 10.73}$$
$$= 0.991.$$

EOR conditions:

$$1\text{st} - \text{stage } Z = \frac{10.44 \times 65}{540 \times 0.1176 \times 10.73}$$
$$= 0.996.$$
$$2\text{nd} - \text{stage } Z = \frac{10.44 \times 198}{560 \times 0.345 \times 10.73}$$
$$= 0.997.$$

Step 8. Determine discharge temperature.
 From Fig. 33
 For SOR conditions:
 1st-stage comp ratio = 3.07.
 $C_p/C_v = 1.362$.
 Suction temp = 80 °F.
 Discharge temp reads as 270 °F.
 2nd-stage comp ratio = 3.1.
 $C_p/C_v = 1.362$.
 Suction temp = 100 °F.
 Discharge temp reads as 299 °F.

For EOR conditions:
1st-stage comp ratio = 3.07.
$C_p/C_v = 1.347$.
Suction temp = 80 °F.
Discharge temp reads as 262 °F.
2nd-stage comp ratio = 3.1.
$C_p/C_v = 1.347$.
Suction temp = 100 °F.
Discharge temp reads as 292 °F.

Step 9. Compressibility factors for discharge conditions.
SOR conditions:
First-stage ACFM on discharge (before intercooler)

$$= \frac{4,337 \times 14.7 \times 730}{520 \times 200 \text{ (neglect IC pressure drop)}}$$
$$= 447.5 \text{ cft/min.}$$
$$\rho = \frac{74.1}{447.5} = 0.166 \text{ lb/cft.}$$
$$Z = \frac{6.44 \times 200}{730 \times 0.166 \times 10.73}$$
$$= 0.991.$$

Second-stage ACFM on discharge
$$= \frac{4,337 \times 14.7 \times 759}{520 \times 615}$$
$$= 151.3 \text{ cft/min.}$$
$$\rho = \frac{74.1}{151.3} = 0.490 \text{ lb/cft.}$$
$$Z = \frac{6.44 \times 615}{759 \times 0.49 \times 10.73}$$
$$= 0.992.$$

EOR conditions:
These are calculated in the same way as those above and give the following results:
1st-stage $Z = 0.995$.
2nd-stage $Z = 0.995$.

Step 10. Approximate driver horsepower.
Use the expression:

BHP = 22 × (comp ratio/stage) × No. of stages × capacity × factor F.
Comp ratio/stage = 3.07.
No. of stages = 2
Capacity per machine in MMcft/day at suction temperature (= 10.81 MMscf/d × 0.6).
Factor for 2 − stage machine = 1.08.

For SOR conditions:

$$BHP = 22 \times 3.07 \times 2 \times (0.6 \times 10.81) \times 1.08$$
$$= 946.$$

For EOR conditions:

$$BHP = 22 \times 3.07 \times 2 \times 7.38 \times 1.08$$
$$= 1,077.$$

Use the efficiency factors as given in Fig. 30 for centrifugal compressors. In this case, there will be a gear assembly between compressor and driver. Use the efficiency of this as 97 %. Thus:

	SOR	EOR
BHP	946	1,077
Gear losses (3 % of BHP)	29	33
Leakage	10	10
Seal	35	35
Bearings	35	35
Driver HP	1,055	1,190

Compressor Drivers, Utilities, and Ancillary Equipment

This item covers details on various compressor drivers, the utilities associated with operating the compressors, and their ancillary equipment.

Compressor Drivers
Table 22 gives a listing of the more common types of compressor drivers. It provides some of the data that would influence the choice of the driver. The most common drivers by far in a process plant are the electric motor and the steam turbine. For very large machines as encountered in handling natural gas, the gas turbine or gas engine becomes the more prominent prime mover.

Sizing Drivers
As a basic rule, drivers are sized for the most severe duty required of the compressor plus a factor as an operating contingency. In general the most severe duty is that design case which has the highest suction temperature, the maximum ratio of specific heats, the lowest suction pressure, and the highest required discharge pressure and the gas molecular weight which gives the highest HP. The driver-rated horsepower will then be greater than

Process Equipment for Petroleum Processing 1551

Table 22 Comparison of compressor drivers

Driver	HP range	Available speed, RPM	Efficiency %	Common applications
Synchronous motor	100–20,000	250–3,600	90–97	Reciprocating compressors
Induction motor	1–15,000	3,600	86–94	All types of compressors
Wound rotor induction motor	–	–	–	Normally not used
Steam engine	10–4,000	400–140	60–80	All types of rotary equip
Steam turbine	10–2,000	2,000–15,000	50–76	Centrif, axial, and recip
Combustion gas turbine	3,000–35,000	10,000–3,600	19–24[a]	All types of compressors (except recip)
Gas and oil engines	100–5,000	1,000–300	35–45	Reciprocating compressors

[a]The efficiency given here does not include for waste heat recovery. With WHR the efficiency can be increased to between 28 % and 35 %

$$\text{Driver brake HP} = \frac{\text{max compressor BHP at the most severe duty}}{\text{mechanical efficiency of the power transmission}}. \qquad (86)$$

The mechanical efficiency in this case includes energy losses for bearings, seals, lube oil, etc., in the case of centrifugal compressors and gears in the case of reciprocating compressors.

Electric Motor Drivers

Squirrel cage motors are preferred for this type of duty. These may be drip-proof open type where the location is not a fire or explosion hazard. Where it is required that the units must be explosion or fire proof, these motors must be totally enclosed type. In sizing the motor, efficiencies for squirrel cage motors up to 450 HP given in Table 14 for pumps may be used. Table 23 is used for motors above 500 HP.

The driver-rated brake horsepower is the compressor horsepower times a load factor divided by a service factor. Normally the load factor is 10 % and a service factor for an enclosed squirrel cage motor is between 1.0 and 1.15 for an open type.

Example Calculation

Calculate the operating load and the connected load for the driver of a 4,000 HP centrifugal compressor (includes leakage, seal, and bearing losses). A gear is used and this has a 97 % efficiency. The load factor is 10 % and the motor is open-type squirrel cage with a service factor of 1.15. There will be a normal operating unit and a spare, both motor driven.

Table 23 Motor efficiencies

Motor-rated HP	Motor efficiencies, % full load		
	50	75	100
500	91.4	93.1	93.4
1,000	92.1	93.8	94.1
1,500	92.4	94.1	94.4
2,000	92.7	94.4	94.7
2,500	92.9	94.6	94.9
3,000	93.0	94.7	95.0
3,500	93.0	94.7	95.0
4,000	93.1	94.8	95.1
4,500	93.1	94.8	95.1
5,000	93.2	94.9	95.2

$$\text{Minimum required driver HP} = \frac{4,000 \times 1.1}{0.97}$$
$$= 4,536.$$
$$\text{Driver nameplate rating} = \frac{4,536}{1.15} = 3,944.$$

Call it 4,000 HP.

Connected load for the motor is

$$\text{Motor nameplate rating} \times 1.15 = 4,000 \times 1.15$$
$$= 4,600 \text{ HP (rated HP)}$$
$$= \frac{4,600 \times .746}{0.951 \text{ (at } 100\% \text{ load)}}$$
$$= 3,608 \text{ kW}.$$

There are two units then total connected load $= 3,608 \times 2$
$$= 7,216 \text{ kW}.$$

Operating load for the motor is

$$\frac{4,000}{0.97} = 4,124 \text{ HP}.$$
$$\% \text{ load} = \frac{4,124}{4,600} = 90\% \text{ (use } 75\% \text{ eff)}.$$
$$\text{Operating load} = \frac{4,124 \times 0.746}{0.948}$$
$$= 3,245 \text{ kW}.$$

Steam Turbine Drivers

Next to the motor drivers steam turbines are the most common form of drivers for rotary equipment in general and compressors in particular. The two most

Table 24 Steam turbine efficiencies

Driver BHP	Adiabatic efficiencies %	
	Inlet pressures (psig)	
	900	100
500	48	59
800	53	64
1,000	56	67
1,200	58	68
1,500	60	71
2,000	63	73
2,500	65	74
3,000 and up	67	76

common types of these are turbines that exhaust to a lower pressure, but the exhaust steam is not condensed and those in which the exhaust steam is condensed. Normally the latter is only used in the case of large driver horsepower 5,000 and above. It is far more expensive than the noncondensing type as the exhaust is normally subatmospheric in pressure and the cost of the condenser must be included.

Typical steam turbine approximate efficiencies are listed in Table 24.

These efficiencies are based on the exhaust pressure of the steam being 50 psig for noncondensing type and 2″ Hg Abs for the condensing type.

Determining the rated horsepower of the steam turbine driver follows closely to the method for motor horsepower. First determine the minimum horsepower required of the turbine. Thus:

Step 1. Determine the *minimum* driver horsepower by multiplying the compressor BHP by 1.1.

Step 2. Now the turbine will deliver the normal HP at the normal speed. A contingency in the form of additional speed is added to the driver capability. This will be controlled in practice by a steam governor. This contingency is usually 5 % above-normal speed.

Step 3. Horsepower capability varies as the cube of the speed. Thus, the rated horsepower of the turbine will be

Rated HP = minimum HP $\times (1.05)^3$.

Step 4. The amount of steam that will be used is calculated by the change in enthalpy of the inlet steam to the outlet steam at constant entropy. The change in enthalpy for the two conditions is read from the steam Mollier diagram.

Step 5. The theoretical steam rate is:

$$\frac{2,544}{\text{Inlet enthalpy } - \text{ outlet enthalpy(in Btu/lb)}} \tag{87}$$

This figure divided by the turbine efficiency gives the steam rate in lbs/BHP/h.

Example Calculation
Calculate the turbine horsepower requirements and the theoretical steam rates to drive a 4,000 BHP centrifugal compressor. No gears are included in this case. Steam is available at 650 psig and 760 F. The steam will exhaust into the plant's 125 psig header.

$$\text{Minimum driver horsepower} = 4,000 \times 1.1$$
$$= 4,400 \text{ BHP.}$$
$$\text{Rated turbine HP at } 105\% \text{ speed} = 4,400 \times (1.05)^3$$
$$= 5,094 \text{ HP.}$$

Enthalpy of steam at 650 psig and 760 F = 1,390 (entropy 1.62).
  Enthalpy of steam at 125 psig = 1,225 (entropy 1.62).
  Difference in enthalpy = 165.
  Efficiency of turbine (from Table 24) = 67 %.

$$\text{Theoretical steam rate} = \frac{2,544}{165 \times 0.67}$$
$$= 23 \text{ lbs/BHP/h.}$$

Gas Turbine Drivers
These items of equipment are the most expensive, and because they require a high capital investment, their use can only be justified as compressor drivers where the continual load on the compressor is also very high. These drivers therefore are met mostly in the natural gas industry. They are used extensively in recompressing natural gas after treating for dew point control or desulfurizing.

The thermal efficiencies of gas turbines are low (about 16–20 %), but it is common practice to use the exhaust gases, which are usually at a temperature of above 800 °F, in waste heat recovery. This involves exchanging the waste heat of the exhaust gases with boiler feedwater to generate steam or to preheat a process stream, for example, distillation. Table 25 gives some gas turbine sizes and data. It should be noted that considerable development work is continuing in the field of gas turbines and consequently the data given here may be subject to revision or updating.

To obtain the gas turbine rated horsepower for a specific compressor duty follows closely the same calculation route as the steam turbine. Thus:

Step 1. Obtain the *minimum driver horsepower* by multiplying the compressor brake horsepower by 1.05 (BHP includes seals, leakage, etc.).

Step 2. The rated turbine horsepower is the minimum driver HP divided by the gear efficiency.

Step 3. The horsepower of the turbine selected must equal or slightly exceed the horsepower calculated in step 2. This HP must be corrected for site conditions as shown in step 4.

Table 25 Gas turbine sizes and data

HP rating at 80 °F, 1,000 ft	Fuel consumption LHV lbs/HP-h	Exhaust Flow, lb/s	Temp, °F	RPM
430	1.25	10.3	950	19,250
1,000	0.66	11.1	960	19,500
1,080	0.63	13.7	860	22,300
1,615	0.84	23.6	1,000	13,000
2,500	0.76	43.0	795	9,000
3,800	0.75	53.0	900	8,500
5,500	0.76	77.3	945	5,800
7,000	0.80	101.0	938	5,500
8,000	0.65	102.0	935	5,800
9,000	0.70	130.0	850	5,000
10,000	0.59	123.5	805	6,000
12,000	0.65	160.0	720	4,750
13,500	0.62	187.0	800	4,860
15,000	0.61	188.0	835	4,860
24,000	0.61	258.0	850	3,600

Step 4. The HPs given in Table 25 are at an ambient temperature of 80 °F, and at an elevation of 1,000 ft. Correction for any specific site is given by the following expression:

$$\text{SITE HP} = \text{Quoted HP} \, (1.00 + A \times 10)^{-2}(1.00 - B \times 10)^{-2} \\ \times (1.00 - C \times 10)^{-2} \times \frac{\text{Site atmos psia}}{14.7} \quad (88)$$

where:
A = temp adjustment of % per °F.
B = inlet press loss, % per inches water gauge.
C = discharge press loss, % per inches water gauge.
"A" will be positive for ambient temperatures above 80 °F and negative for ambient temperatures below 80 °F.

Ancillary Equipment

Reciprocating compressor dampening facilities. Dampening facilities are used in conjunction with reciprocating compressors to smooth out the pulsation effect of the compressor action. These facilities are simply in-line bottles sized larger than the gas line which cushion the gas motion. These are essential to minimize expensive piping designs that would be necessary without them. Calculating the size of these bottles is important in the design of the compressor facilities.

The following calculation technique is used to determine the size of new dampers or to evaluate the adequacy of an existing facility. This calculation is described by the following steps:

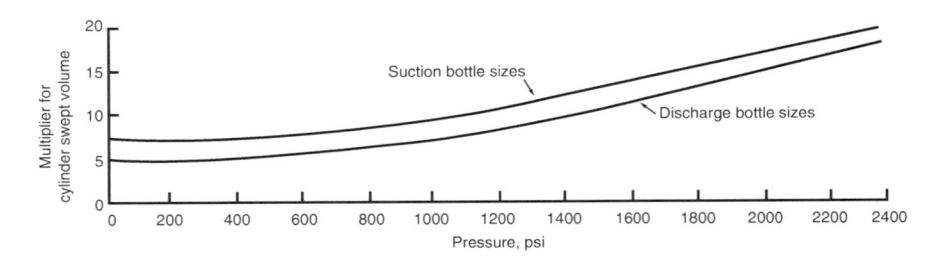

Fig. 35 Dampener bottle sizing

Step 1. From compressor data sheet, obtain cylinder diameter and stroke dimensions.

Step 2. Calculate the swept volume per cylinder using the expression

$$\frac{\pi D^2}{4} \times S \qquad (89)$$

where:
 D = cylinder diameter.
 S = stroke length.

Step 3. Knowing the suction and discharge pressures, the pulsation bottle capacity (both suction and discharge) is obtained from Fig. 35 in terms of a multiple of swept volume.

Step 4. Use the rule of thumb that pulsation bottle diameter equals 1½ times the compressor cylinder diameter. Calculate the suction and discharge bottle length.

Example Calculation

To determine the dimensions of the compressor pulsation bottle of a reciprocating compressor having a 6″ diameter cylinder and a stroke of 15″. The compressor delivers 3.0 MMscf/d gas at a suction pressure of 100 psia and 100 °F and a discharge pressure of 1,200 psia.

The cylinder diameter is 6″ and stroke is 15″.

$$\text{Then swept volume} = \pi/4 \times 6^2 \times 15$$
$$= 424 \text{ cu.in.}$$
$$\text{Capacity of machine} = 3 \text{ MMscf/D.}$$
$$\text{In a MM CF/D} = \frac{3 \times 14.7 \times 560}{520 \times 100}$$
$$= 0.475 \text{ MM ACFD}$$
$$\Rightarrow 330 \text{ ACF/min}$$
$$\Rightarrow 570240 \text{ AC in./min.}$$
$$\text{Machine speed} = \frac{570240}{424} = 1,345 \text{ RPM.}$$

From Fig. 35:

Suction bottle size should be 7 × swept volume (at 100 psia); discharge bottle size should be 7 × swept volume (at 1,200 psia) = 2,968 cu inches or 1.718 cu. ft. As a rule of thumb, diameter of bottle should be 1½ × cylinder diameter = 0.75 f. (9″) length = 2,968/63.6 = 47 in. or 4 ft.

Heat Exchangers

Types and Selection of Heat Exchangers

Heat exchange is the science that deals with the rate of heat transfer between hot and cold bodies. There are three methods of heat transfer, they are:

- Conduction
- Convection
- Radiation

In a heat exchanger, heat is transferred by conduction and convection with conduction usually being the limiting factor. The equipment used in heat exchanger service is designed specifically for the duty required of it. That is, heat exchange equipment cannot be purchased as a stock item for a service but has to be designed for that service.

The types of heat exchange equipment used in the process industry and their selection for use are as follows:

Shell and tube exchanger
This is the type of exchanger most commonly used in a process plant. It consists of a bundle of tubes encased in a shell. It is inexpensive and is easy to clean and maintain. There are several types of shell and tube exchangers and most of these have removable bundles for easier cleaning. The shell and tube exchanger has a wide variety of services for which it is normally used. These include vapor condensation (condensers), process liquid cooling (coolers), exchange of heat between two process streams (heat exchangers), and reboilers (boiling in fractionator service). Most of this chapter will be dedicated to the uses and design specification of the shell and tube exchanger.

Double-pipe exchanger
A double-pipe exchanger consists of a pipe within a pipe. One of the fluid streams flows through the inner pipe while the other flows through the annular space between the pipes. The exchanger can be dismantled very easily and therefore be easily cleaned. The double-pipe exchanger is used for very small process units or where the fluids are extremely fouling. Either true concurrent or countercurrent flows can be obtained, but because the cost per square foot is relatively high, it can only be justified for special applications. Table 26 gives the heat transfer area for various pipe lengths and diameters.

Table 26 Heat transfer areas for various double-pipe exchanger designs

No of tubes	Shell size, in.	Tube size, in.	Surface area, sqft, for indicated length tubes		
			L=10 ft	L=20 ft	L=30 ft
1	2	1	5.8	11.0	16.3
1	3	1.5	10.9	20.9	30.9
1	4	2	13.7	26.1	38.5

Extended surface or fin tubes

This type of exchanger is similar to the double pipe but the inner pipe is grooved or has longitudinal fins on its outside surface. Its most common use is in the service where one of the fluids has a high resistance to heat transfer and the other fluid has a low resistance to heat transfer. It can rarely be justified if the equivalent surface area of a shell and tube exchanger is greater than 200–300 sqft.

Finned air coolers

These are the more common type of air coolers used in the process industry. Air cooling for process streams gained prominence during the early 1950s. In a great many applications and geographic areas, they had considerable economic advantage over the conventional water cooling. Today it is uncommon to see process plants of any reasonable size without air coolers.

Air coolers consist of a fan and one or more heat transfer sections mounted on a frame. In most cases, these sections consist of finned tubes through which the hot fluid passes. The fan located either above or below the tube section induces or forces air around the tubes of the section.

The selection of air coolers over shell and tube is one of cost. Usually air coolers find favor in condensing fractionator overheads to temperatures of about 90–100 °F and process liquid product streams to storage temperatures. Air coolers are widely used in most areas of the world where ambient air temperatures are most times below 90 °F. At atmospheric temperatures above 100 °F, humidifiers are incorporated into the cooler design and operation. The cost under these circumstances is greatly increased and their use is often not justified.

In very cold climates, the air temperature around the tubes is controlled to avoid the skin temperature of the fluid being cooled falling below a freezing criteria or, in the case of a petroleum product, its pour point. This control is achieved by louvers installed to recirculate the air flow and/or by varying the quantity of air flow by changing the fan pitch. Steam coils may also be used.

Box coolers

These are the simplest form of heat exchanger. However, they are generally less efficient and more costly and require a large area of the plant plot. They consist of a single coil or "worm" submerged in a bath of cold water. The fluid flows through the coil to be cooled by the water surrounding it. The box cooler found use in the older petroleum refineries for cooling heavy residuum to storage temperatures. Modern day practice is to use a tempered water system where the heavy oil is cooled on the shell side of a shell and tube exchanger against water at a controlled temperature flowing in the tube side. The water is recycled through an air cooler to

Process Equipment for Petroleum Processing   1559

control its temperature to a level which will not cause the skin temperature of the oil in the shell and tube exchanger to fall below its pour point.

Direct-contact condensers

In this exchanger, the process vapor to be condensed comes into direct contact with the cooling medium (usually water). This contact is made in a packed section of a small tower. The most common use for this type of condenser is in vacuum producing equipment. Here the vapor and motive steam for each ejector stage are condensed in a packed direct-contact condenser. This type has a low pressure drop which is essential for the vacuum producing process.

General Design Considerations

Basic Heat Transfer Equations

The following equations define the basic heat transfer relationships.

These equations are used to determine the overall surface area required for the transfer of heat from a hot source to a cold source.

Overall heat transfer equation

The usual heat transfer mechanisms are conduction, natural convection, forced convection, condensation, and vaporization. When heat is transferred by these means, the overall equation is as follows:

$$Q = UA(\Delta t_m) \tag{90}$$

where:
Q = heat transferred in Btu/h.
U = overall heat transfer coefficient, Btu/h/sqft/°F.
A = heat transfer surface area, sqft.
Δt_m = corrected log mean temperature difference, °F.

Overall heat transfer coefficient U

This coefficient is the summation of all the resistances to the flow of heat in the transfer mechanism. These resistances are the resistance to heat transfer contained in the fluids, the resistance caused by fouling, and the resistance to heat transfer of the tube wall. The resistance to the flow of heat from the liquid outside the tube wall is measured by the film coefficient of that fluid. The resistance of the flow of heat from the fluid inside the tube is similarly the film coefficient of the inside fluid. These film coefficients are products of dimensionless numbers which include:

- The Reynolds number
- The Graetz number
- The Grashof number
- The Nusselt number
- The Peclet number
- The Prandtl number
- The Stanton number

The formats of these numbers and their use are found in all standard textbooks on heat transfer. For example, Kern's *Process Heat Transfer* and McAdams *Heat Transmission*.

These resistances are defined therefore by the following expression:

$$\frac{1}{U_o} = \frac{1}{h_o} + \frac{1}{h_i} \times \frac{A_o}{A_i} + \frac{1}{h_w} + (rf)_o + (rf)_i \times \frac{A_o}{A_i} \qquad (91)$$

where:

U_o = overall heat transfer coefficient based on outside tube surface, in Btu/h/sqft/°F.
h = the film coefficient in Btu/h/sqft/°F.

rf = fouling factors in $\dfrac{1}{\text{Btu/h/sqft/°F}}$.

h_w = heat transfer rate through tube wall in Btu/h/sqft/°F.
A = surface area in sqft.

Subscripts "o" and "i" refer to outside surface and inside surface, respectively.

Flow Arrangements

The two more common flow paths are concurrent and countercurrent. In concurrent (or cocurrent) flow, both the hot fluid and the cold fluid flow in the same direction. This is the least desirable flow arrangement and is only used in those chemical processes where there is a danger of the cooling fluid congealing, subliming, or crystallizing at near-ambient temperatures.

Countercurrent flow is the most desirable arrangement. Here the hot fluid enters at one end of the exchanger and the cold fluid enters at the opposite end. The streams flow in opposite directions to one another. This arrangement allows the two stream exit temperatures to approach one another.

Logarithmic Mean Temperature Difference Δt_m

This may also be designated as LMTD or Δt_m. In either countercurrent or concurrent flow arrangement, the log mean temperature difference used in the overall heat transfer equation is determined by the following expression:

$$\Delta t_m = (\Delta t_1 - \Delta t_2)/(\ln(\Delta t_1/\Delta t_2)). \qquad (92)$$

The Δt's are the temperature differences at each end of the exchanger and Δt_1 is the larger of the two. In true countercurrent flow, the Δt_m calculated can be used directly in the overall heat transfer equation. However, such a situation is not common and true countercurrent flow rarely exists. Therefore, a correction factor needs to be applied to arrive at the correct Δt_m. These are given in Appendix 6 of this chapter.

Table 27 Typical velocities and pressure drops in various exchanger services

	Tube side		Shell side	
	Velocity, ft/s	Press drop, psi	Velocity, ft/s	Press drop, psi
Nonviscous liquids	6–8	10	1.5–2.5	10
Viscous liquids	6	20	3.0 max	15–20
Clean cooling water	6–8	10–15	–	–
Dirty cooling water	3 min	10+	–	–
Suspended solids, in	2–3 min	10	1.5 min	15 liquids[a]
Gases and vapors	$\frac{100}{\sqrt{\text{Gas density}}}$ max	3–5	–	3
Condensing vapors	–	–	–	3–5

[a]Normally erosion by suspended solids in liquids occurs at velocities of above 6 ft/s

Table 28 Allowable tube velocities for steam in exchanger tubes

Pressure	Velocity ft/s
Below atmospheric	225
Atmos. to 100 psig	175
Above 100 psig	150

Fluid Velocities and Pressure Drops

Film coefficients are a function of fluid velocity, density (vapor), and viscosity (liquids). Within limits increasing the velocity of a fluid reduces its resistance to heat transfer (i.e., it increases its heat transfer coefficient). Increasing the fluid velocity however increases its pressure drop. An economic balance needs to be sought between the cost of heat transfer surface and pumping cost. This exercise should be undertaken to find a pay-out balance of 2–4 years. The exercise has been done many times, and the data in Table 27 are considered to be reasonable balances between velocity and pressure drop for some common cases.

For condensing steam, pressure drop is usually not critical but a minimum steam pressure drop is desirable. Allowable steam velocities in tubes are in Table 28.

Choice of Tube Side Versus Shell Side

There are no hard and fast rules governing which fluid flows on which side in a heat exchanger. Much is left to the discretion of the individual engineer and his experience. There are some guidelines and these are as follows:

Tube side flow
Fouling liquids. Tube cleaning is much easier than cleaning the outside of the tubes.
  Also fouling can be reduced by higher tube side velocities.
Corrosive fluids. It is cheaper to replace tubes than shells and shell baffles, so as a general
  rule corrosive fluids are put on the tube side. There are exceptions and a major one is
  for corrosive fluids that become more corrosive at high velocities. An example of this
  is naphthenic acids, which are present in some crude oils and their products.

High pressure. Fluids at high pressure are usually put on the tube side as only the tubes, tube sheet, and channel need to be rated for high pressure in the unit design. This reduces the overall cost of the exchanger. Tube failure scenarios need to be considered in shell-side design, however.

Suspended solids. Fluids containing suspended solids should, whenever possible, be put on the tube side. Shell side flows invariably have "dead spaces" where solids come out of suspension and build up, causing fouling. If not feasible to put the fouling service on the tube side, the dead zones can be minimized by special baffle arrangements, such as spiral baffling.

Cold boxes. These are exchangers used in cryogenic processes where condensing of a vapor on one side of the exchanger is accompanied by boiling of a liquid on the other side. The condensing fluid is preferred on the tube side. Better control of the refrigerant flow is accomplished by the level control across the shell side.

Shell side flow

Available pressure drop. Shell side flows generally require lower pressure drop than tube side. Therefore, if a stream is pressure drop limited, it should be routed to the shell side.

Condensers. Condensing vapors should flow on the shell side wherever possible. The larger free area provided by the shell side space permits minimum pressure drop and higher condensate loading through better film heat transfer coefficients.

Large flow rates. In cases where both streams are of a similar nature with similar properties, the stream with the largest flow rate should be sent to the shell side where the difference in flow rates is significant. The shell side provides more flexibility in design by baffle arrangements to give the best heat transfer design criteria.

Boiling service. The boiling liquid, as in the case of reboilers, waste heat recovery units, and the like, should be on the shell side of the exchanger. This allows space for the proper disengaging of the vapor phase and provides a means of controlling the system by level control of the liquid phase.

Types of shell and tube exchangers

Figure 36 gives some of the more common arrangements in shell and tube exchanger design. The arrangements shown here are all one shell pass and one or two tube passes. Equipment with more than two tube passes (up to five) is also fairly common particularly in petroleum refining. Shell arrangements are however left at one if at all possible. Where multi-pass shell side is required, companies prefer to use complete exchangers in series or in parallel or both rather than making two or more shell passes using horizontal baffling in one exchanger.

Estimating Shell and Tube Surface Area and Pressure Drop

There are many excellent computer programs available that calculate exchanger surface area and pressure drops from simple input. The actual calculation when

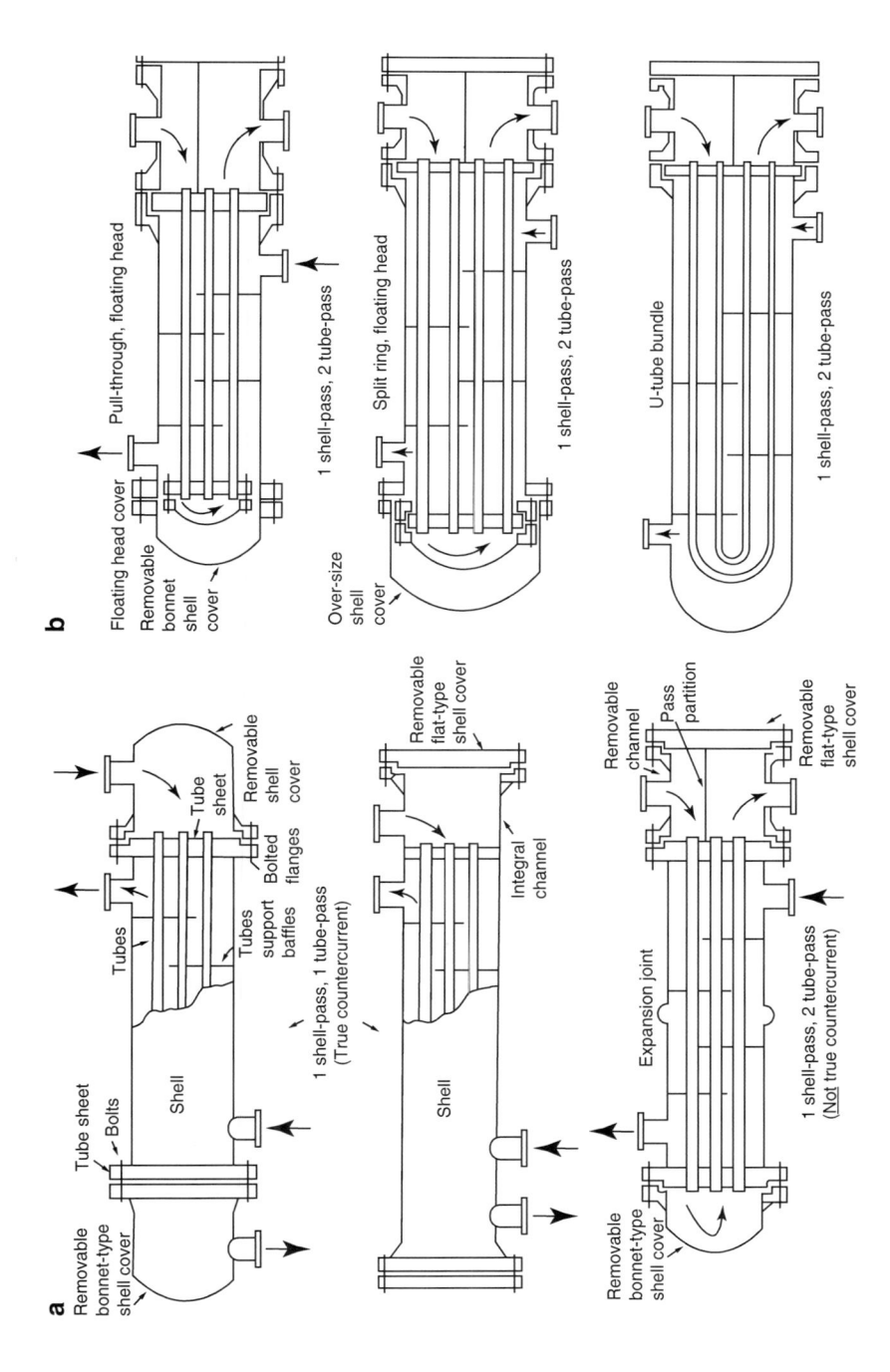

Fig. 36 Some common types of shell and tube exchangers, (**a**) fixed tube sheet, (**b**) removable bundle

done manually is tedious and long. However, to understand a little of the importance of the input required by these computer programs, it is educational to at least view a typical manual calculation. The one given here is for a shell and tube cooler with no change of phase for either tube side or shell side fluids.

The calculation follows these steps:

Step 1. Establish the following data by heat balances or from observed plant readings:
- The inlet and outlet temperatures on the shell side and on the tube side.
- The flow of tube side fluid and that for the shell side. It may be necessary to calculate one or the other from a heat balance over the exchanger.
- Calculate the duty of the exchanger in heat units per unit time (usually hours).
- Establish the stream properties for tube side and shell side fluids. The properties required are: SG, viscosity, specific heats, and thermal conductivity.

Step 2. Calculate the log mean temperature difference (Δt_m). Assume a flow pattern (i.e., either cocurrent or countercurrent). Most flows will be a form of countercurrent. Then show the temperature flow as follows:

Shell In	$\longrightarrow$	Shell Out
Tube out	$\longleftarrow$	Tube In
Temp difference: Δt_1		Δt_2

The log mean temperature difference is then calculated using the expression

$$\Delta t_m = \frac{\Delta t_1 - \Delta t_2}{\log_e \frac{\Delta t_1}{\Delta t_2}} \tag{93}$$

This temperature needs to be corrected for the flow pattern, and this is done using the correction factors given in Appendix 6 of this chapter. The use of these is given in the Appendix figures.

Step 3. Calculate the approximate surface area.

From Appendix 7 of this chapter, select a suitable overall heat transfer coefficient U in Btu's/h sqft. F. Use the expression to calculate "A", the area.

$$Q = UA\Delta t_m \tag{94}$$

where:
Q = Heat transferred in Btu/h (the exchanger duty).
U = overall heat transfer coefficient.
A = exchanger surface area in sqft.
Δt_m = log mean temperature difference (corrected for flow pattern) in °F.

From the surface area calculated, select the tube size and pitch. Usually 3/4 in. on a triangular pitch for clean service and 1" on a square pitch for dirty or fouling service. A single standard shell will hold about 4,100 sqft of surface per pass.

Many companies prefer to avoid multi-pass shells and instead prefer sets of shells in series if this becomes necessary. The "norm" therefore is single pass shells each containing up to 4,100 sqft of surface.

Step 4. Calculate the tube side flow and the number of passes. If it cannot be read from plant data, estimate the tube side flow in cuft/h by heat balance. Select the tube gauge and length. The tube data are given in Appendix 8 of this chapter and standard lengths of tubes are 16 and 20 ft. Calculate the number of selected tubes per pass from the expression

$$N_p = \frac{F_t \times 144}{3,600 \times A_t \times V_t} \qquad (95)$$

where:
N_p = number of tubes per pass.
F_t = tube side flow in cuft/h.
A_t = cross-sectional area of one tube.
V_t = linear velocity in tube in ft/s.

See earlier item on "Fluid Velocities and Pressure Drop" for recommended fluid velocities. The number of tube passes is arrived at by dividing the total surface area required by the total (external) surface area of the number of tubes per pass calculated above.

Step 5. Calculate tube side film coefficient corrected to outside diameter (h_{io}). The tube side film coefficient may be calculated for water by the expression

$$h_{io} = \frac{300 \times (V_t \times \text{tube ID ins})^{0.8}}{\text{tube OD ins}} \qquad (96)$$

where:
h_{io} = inside film coefficient based on outside tube diameter in Btu/h sqft °F.
V_t = linear velocity of water tube side in ft/s.

For fluids other than water flowing tube side, use the expression

$$h_{io} = \frac{K}{D_o}(C\mu/K)^{1/3}(\mu/\mu_w) \cdot \phi(DG^{0.14}/\mu) \qquad (97)$$

where:
h_{io} = inside film coefficient based on outside diameter in Btu/h sqft °F.
K = thermal conductivity of the fluid in Btu/h/sqft/(°F per ft). See Maxwell *Data Book on Hydrocarbons* or Perry *Chemical Engineers Handbook.*
D = inside tube diameter in inches.
D_o = outside tube diameter in inches.
C = specific heat in Btu/lb/°F.
G = mass velocity in lbs/s sqft.
μ = absolute viscosity cPs at average fluid temp.
μ_w = absolute viscosity cPs at average tube wall temp.
$\phi(DG/\mu)$ = from Fig. 37.

Step 6. Calculate shell side dimensions. First determine the shell side average film temperature as follows:

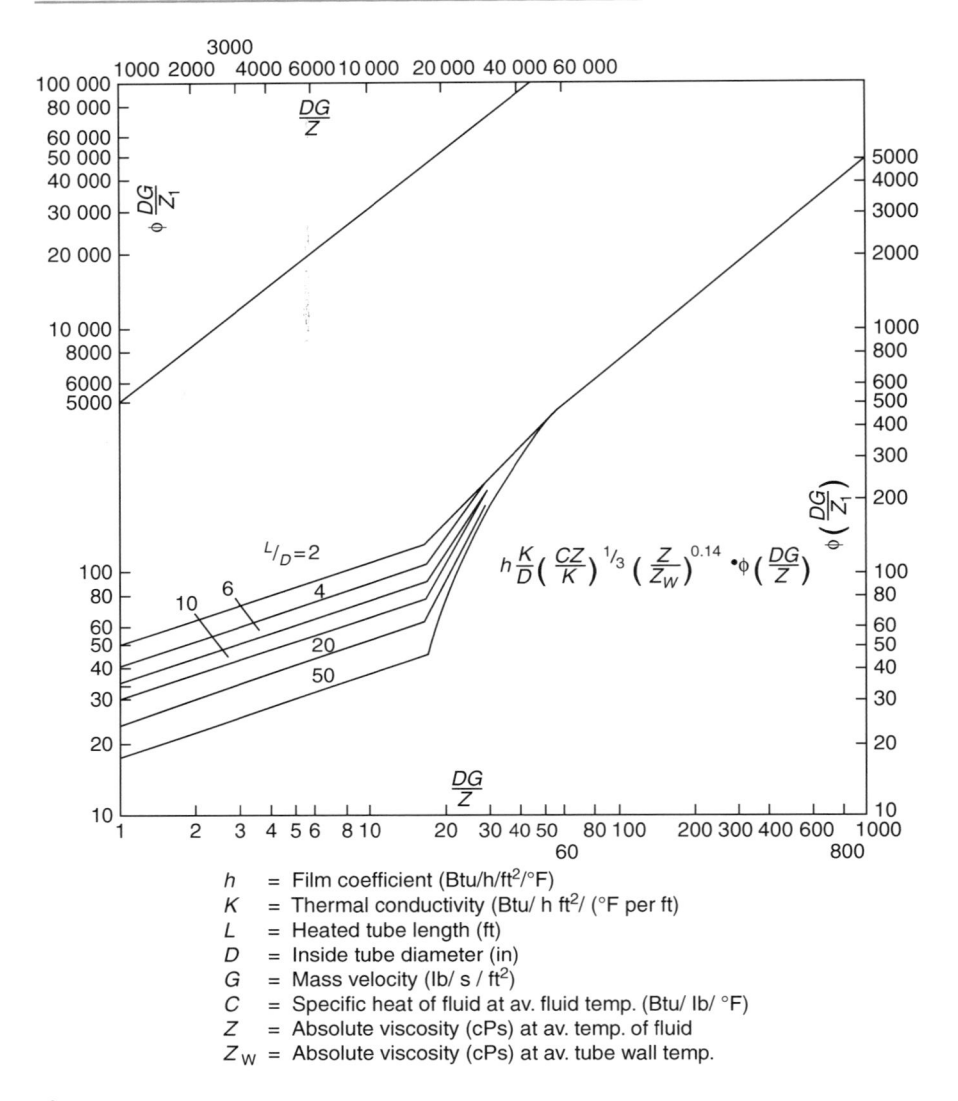

Fig. 37 Heat transfer inside tubes

h = Film coefficient (Btu/h/ft^2/°F)
K = Thermal conductivity (Btu/ h ft^2/ (°F per ft)
L = Heated tube length (ft)
D = Inside tube diameter (in)
G = Mass velocity (lb/ s / ft^2)
C = Specific heat of fluid at av. fluid temp. (Btu/ lb/ °F)
Z = Absolute viscosity (cPs) at av. temp. of fluid
Z_W = Absolute viscosity (cPs) at av. tube wall temp.

$$\text{Inlet ave} = \frac{T_1 + T_2}{2} \quad \text{Outlet ave} = \frac{T_3 + T_4}{2} \tag{98}$$

where:

T_1 = shell fluid inlet temperature.
T_2 = tube outlet temperature.
T_3 = shell outlet temperature.
T_4 = tube inlet temperature.

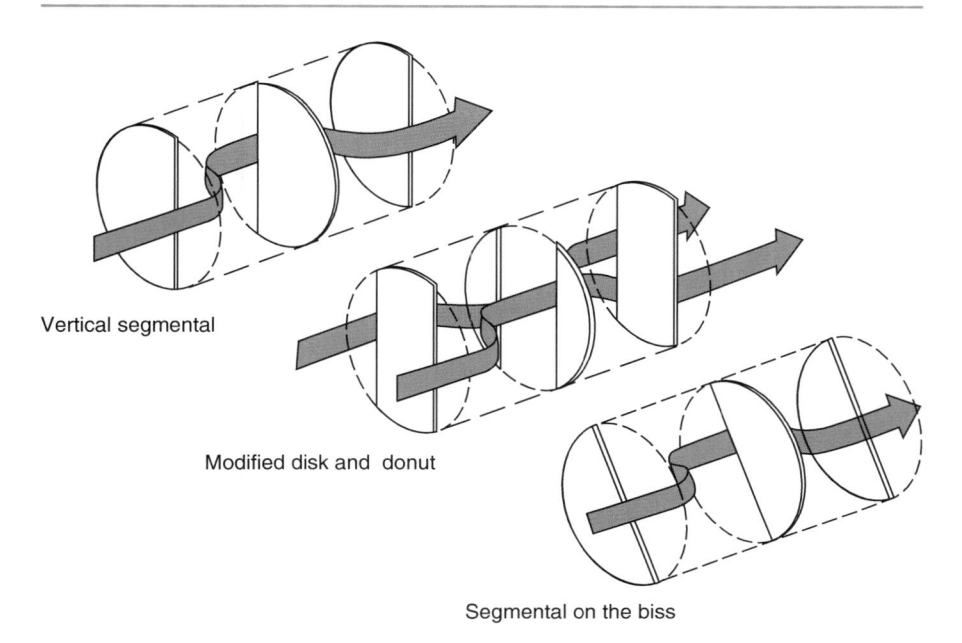

Process Equipment for Petroleum Processing 1567

Vertical segmental

Modified disk and donut

Segmental on the biss

Fig. 38 Types of baffles

Average shell side film temperature:

$$\frac{\text{Inlet ave} + \text{outlet ave}}{2} \tag{99}$$

Use this temperature to determine density and viscosity used in the shell side film coefficient calculations.

The shell diameter. Next calculate the diameter of the tube bundle and the shell diameter. For this use one of the following equations to calculate the number of tubes across the center line of the bundle:
1. For square-pitch tube arrangement:
   $T_{cl} = 1.19 \text{ (number of tubes)}^{0.5}$.
2. For triangular pitch tube arrangement:
   $T_{cl} = 1.10 \text{ (number of tubes)}^{0.5}$.

Note these equations use the *total* number of tubes, namely, those calculated in step 4 times number of tube passes.

Set number of baffles and their pitch. The types of baffles usually used are shown in Fig. 38. Disc- and donut-type baffles are only used where pressure drop available is very small and there is a pressure drop problem. Baffles on the bias are used in sq-pitch tube arrangement and baffles perpendicular to the tubes are usual for triangular tube arrangements.

Other types of baffling are applied in severely fouling services. These types include "rod-baffle" and spiral baffle patterns.

The minimum baffle pitch should not be less than 16 % of the shell diameter. Pitch in this case is the space between two adjacent baffles. Normally 20 % of shell ID is used for the baffle pitch. The number of baffles is calculated from the expression

$$N_B = \frac{10 \times \text{tube length}}{\text{baffle pitch\%} \times \text{diameter of shell}} \qquad (100)$$

Free area of flow between baffles. The space available for flow on the shell side is calculated as

$$W = D_i - (d_o \times T_{cl}) \qquad (101)$$

where:
 W = space available for flow in sq inches.
 D_i = shell inside diameter in inches.
 d_o = tube outside diameter in inches.
 T_{cl} = number of tubes across centerline.

The free area of flow between baffles is now calculated as follows:

$$A_f = W \times (B_p - 0.187) \qquad (102)$$

where:
 A_f = free flow area between baffles in sq inches.
 B_p = baffle pitch in inches.

Step 7. Calculate the shell side film coefficient h_o. The following expression is used to determine the outside film coefficient:

$$h_o = \frac{K}{d_o}(C\mu/K)^{1/3} \cdot \phi(d_o G_m/\mu_f) \cdot \frac{4P_b}{D} \qquad (103)$$

where:
 h_o = outside film coefficient in Btu/h sqft °F.
 G_m = maximum mass velocity in lbs/s sqft.
 d_o = outside tube diameter in inches.
 K = thermal conductivity of fluid in Btu/h/sqft/(°F/ft) .
 C = specific heat of fluid in Btu/lb/F.
 μ_f = viscosity at mean film temperature in cPs.
 P_b = baffle pitch in inches.
 D = shell internal diameter in inches.
 $\phi(d_o G_m/\mu_f)$ is a function of the Reynolds number read from Fig. 39. The Reynolds number in this case is

Process Equipment for Petroleum Processing

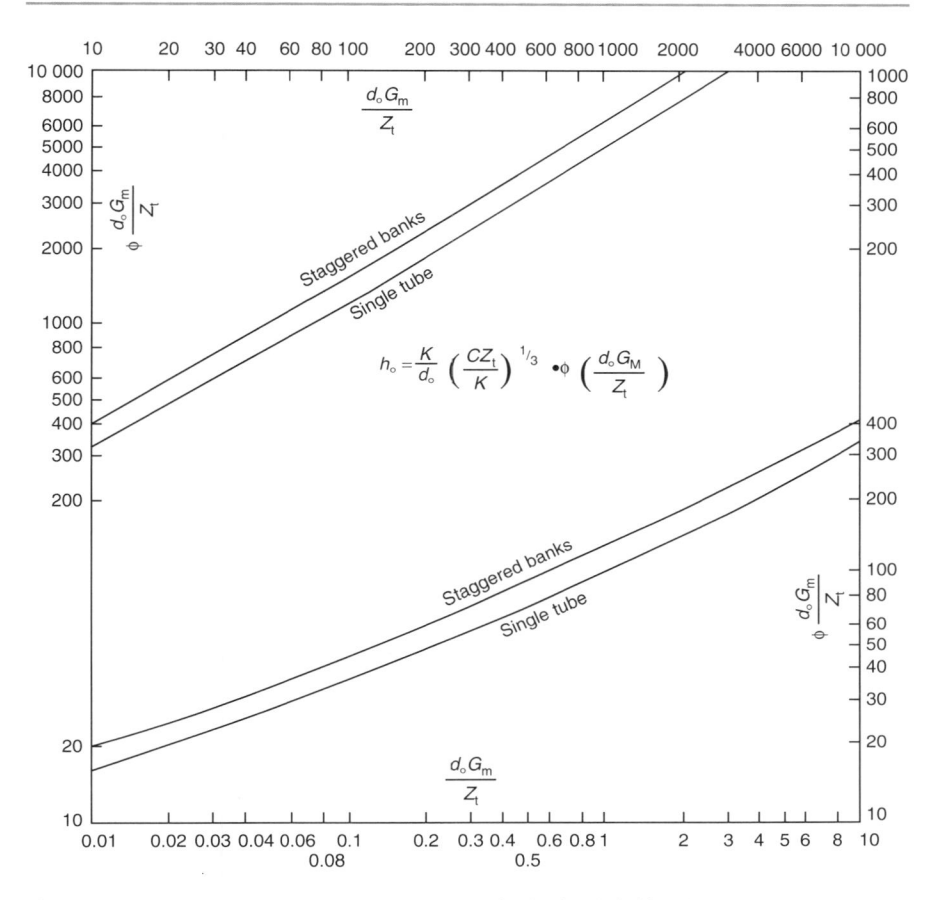

Fig. 39 Heat transfer to fluids outside tubes (See Fig. 37 for definitions.)

The chart shows:

$$h_o = \frac{K}{d_o} \left(\frac{CZ_t}{K}\right)^{1/3} \cdot \phi \left(\frac{d_o G_M}{Z_t}\right)$$

with axes plotting $\dfrac{d_o G_m}{\phi Z_t \cdot N}$ versus $\dfrac{d_o G_m}{Z_t}$ for Staggered banks and Single tube curves.

$$\text{Re} = \frac{d_o G_m}{\mu_f} \tag{104}$$

where: G_m = lbs/s sqft.

This film coefficient is corrected for the type of baffle and tube arrangement by multiplying it by one of the following factors:

For square pitch vertical to tube rows	0.50
Square pitch on the bias	0.55
Triangular tube pitch	0.70

Step 8. Calculate the overall heat transfer coefficient U_o. The film coefficients calculated in steps 5 and 7 are now used in the expression

Table 29 Thermal conductivities of tube metals

Material	K, Btu/h sqft F/ft
Admiralty brass	64
Aluminum brass	58
Aluminum	117
Brass	57
Carbon steel	26
Copper	223
Cupronickel	41
Lead	20
Monel	15
Nickel	36
Red brass	92
Type 316 alloy steel	9
Type 304 alloy steel	9
Zinc	65

$$\frac{1}{U_o} = \frac{1}{h_{io}} + r_{io} + \frac{1}{h_o} + r_o + r_w \qquad (105)$$

where:
U_o = overall heat transfer coefficient in Btu/h sqft °F.
r_{io} and r_o = tube side and shell side fouling factors, respectively, in h sqft °F/Btu. For clean tubes, this is 0.001 as a sum of both factors. For fouling services, 0.005 or higher may apply.
r_w = tube wall resistance to heat transfer in h sqft °F/Btu, which is expressed as

$$r_w = \frac{t_w \cdot d_o}{12 \times K \times (d_o - 2t_w)} \qquad (106)$$

where:
t_w = tube wall thickness inches.
d_o = outside tube diameter inches.
K = thermal conductivity Btu/h sqft °F/ft. See Table 29.
Compare the calculated value of the overall heat transfer coefficient with the assumed one in step 3.
If there is agreement within ±10 %, then the calculated one will be used for revising the calculation for surface area and the other dimensions. If there is no agreement, repeat the calculation using a new value for the assumed U until they agree within tolerance.
Step 9. Calculate tube side pressure drop. Using the adjusted dimensional values from the calculated U_o, calculate the tube side pressure drop using one of the following equations:
For water only:

$$\Delta P_t = 0.02 F_t \times N_p \times \left(V^2 + \left(0.158 L V^{1.73}/d_i\right)^{1.27}\right). \qquad (107)$$

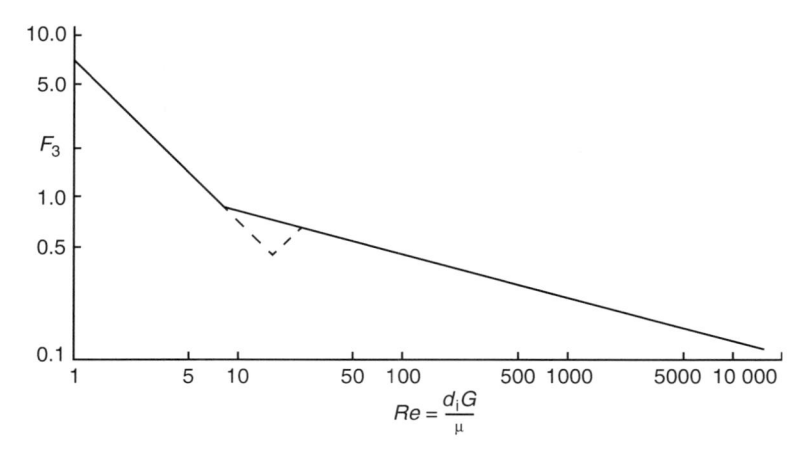

Fig. 40 Pressure drop factor F_3 for flow inside tubes

Table 30 Values of F_t for various tube sizes and materials

Tube OD, inches	Tube metal	F_t
0.75	Steel	1.50
1.00	Steel	1.40
1.50	Steel	1.20
0.75	Ad brass	1.20
1.00	Copper	1.15

For fluids other than water, use

$$\Delta P_t = F_t \times N_p \times (\Delta P_{tf} + \Delta P_{tr}) \qquad (108)$$

where:

$$\Delta P_{tf} = F_3 \cdot \frac{L}{d_i} \cdot (\rho_m \times V^2/9{,}270) \cdot (\mu_w)^{0.14}. \qquad (109)$$

$$\Delta P_{tf} = 3 \times (\rho_m \times V^2/9{,}270). \qquad (110)$$

F_3 = factor based on Reynolds number; see Fig. 40
ρ_m = density in lbs/cuft at mean fluid temperature.
μ_w = viscosity of fluid at tube wall temperature in cPs (use mean film temperature).
V = linear velocity in ft/s.
F_t = pressure drop fouling factor (dimensionless) from Table 30.

The pressure drop figure calculated by these equations is for one unit. Where there are more shells in series, multiply the figures by the number of shells.

Step 10. Calculate the shell side pressure drop. Using the revised dimensions calculated in step 8, the total shell side pressure drop is calculated using the following equation:

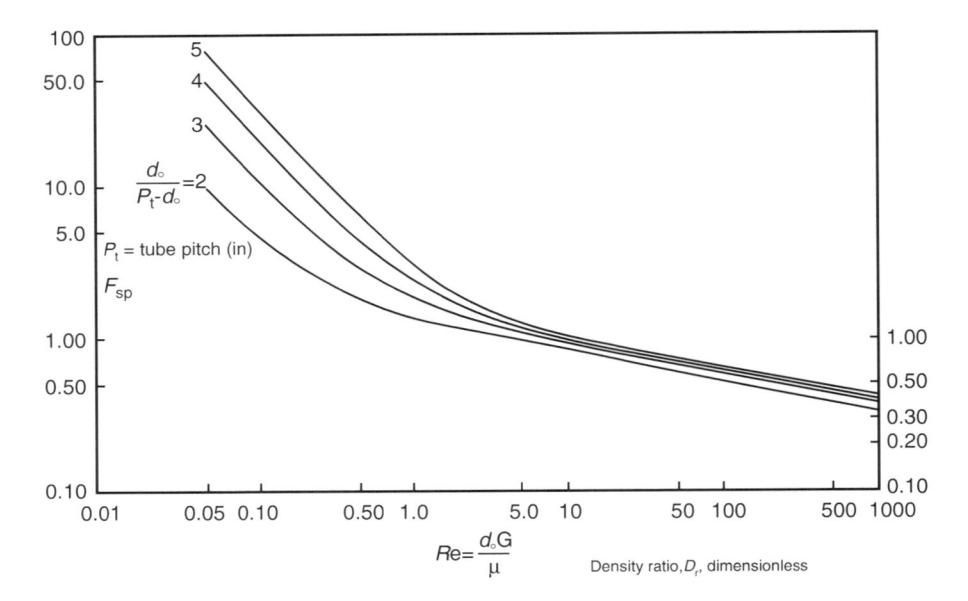

Fig. 41 Pressure drop factor F_{SP} for flows across banks of tubes

$$\Delta P_s = F_s \times (\Delta P_{sr} + \Delta P_{sf}) \qquad (111)$$

where:

$\Delta P_{sf} = B_2 F_{sp} N_{tc} N_b (m \times V^2 / 9{,}270).$

ΔP_{sr} = pressure drop due to turns given by

$$(N_b + 1) \cdot (3.5 - 2P_b/D) \cdot \frac{(m \times V^2)}{9{,}270}. \qquad (112)$$

B_2 = factor as follows:

Baffle position	Tube layout	B_2
Vertical	Square	0.30
Bias at 45°	Square	0.40
Vertical	Triangular	0.50

F_{sp} = factor based on Reynolds number. See Fig. 41.
N_{tc} = number of tubes on center line.
N_b = number of shell baffles.
P_b = space between baffles, inches.
D = shell ID in inches.

The pressure drop calculated here is for one shell. If there are more shells in series, then multiply these pressure drops by the number of shells.

Air Coolers and Condensers

Air cooling of process streams or condensing of process vapors is more widely used in the process industry than cooling or condensing by exchange with cooling water. Cooling water is normally reserved for smaller "trim" coolers or when lower temperatures are required. The use of individual air coolers for process streams using modern design techniques has economized in plant area required. It has also made obsolete those large cooling towers and ponds associated with product cooling. This item of the chapter describes air coolers in general and outlines a method to estimate surface area, motor horsepower, and plant area required by the air coolers.

As in the case of shell and tube exchangers, there are many excellent computer programs that can be used for the design of air coolers. The method given here for such calculation may be used in the absence of a computer program or for a good estimate of a unit. The method also emphasizes the importance of the data supplied to manufacturers for the correct specification of the units.

General Description of Air Coolers/Condensers

Figure 42 shows the two common types of air coolers used in the process industry. Both units consist of a bank of tubes through which the fluid to be cooled or condensed flows. Air is passed around the tubes either by a fan located below the tubes forcing air through the tube bank or a fan located above the tube bank drawing air through the tube bank. The first arrangement is called "forced draft" and the second "induced draft."

Air in both cases is motivated by a fan or fans driven by an electric motor or a steam turbine or in some cases a gas turbine. The fan and prime driver are normally connected by a "V" belt or by a shaft and gearbox. Electric motor drives are by far the most common prime drivers for air coolers.

The units may be installed on a structure at grade or, as is most often the case, on a structure above an elevated pipe rack. Most air coolers in condensing service are elevated above pipe racks to allow free flow of condensate into a receiving drum.

Thermal Rating

Thermal rating of an air cooler is similar in some respects to that of a shell and tube exchanger described in the previous item. The basic heat transfer equation

$$Q = U \Delta T A \qquad (113)$$

is used to determine the surface area required. The calculation for U is different in that it requires the calculation for the air-side film coefficient. This film coefficient is usually based on an extended surface area which is formed by adding fins to the bare surface of the tubes. Thermal rating, surface area, fan dimensions and horsepower are calculated by the following steps:

Step 1. Calculate the heat duty and the tube side material characteristics.
Step 2. Calculate the log mean temperature for the exchanger. Using the following equation, determine the temperature rise for the air flowing over the tubes:

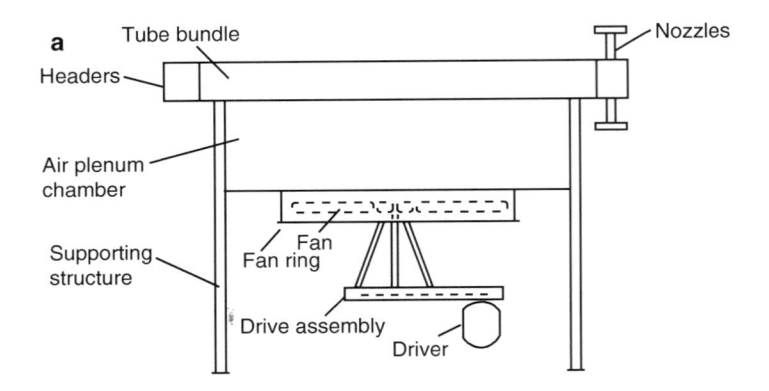

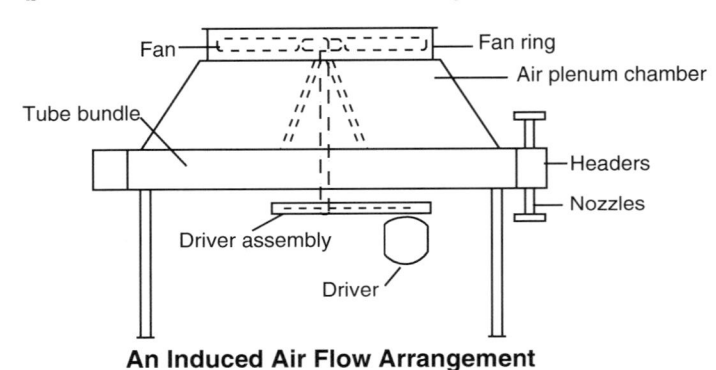

Fig. 42 Air coolers. (**a**) Forced draft. (**b**) Induced draft

$$\Delta t_m = ((U_e + 1)/10)) \cdot ((\Delta t_m/2) - t_1)) \tag{114}$$

where:

Δt_m = air temperature rise °F.
U_e = overall heat transfer coefficient assumed (From Table 31).
Δt_m = mean tube side temperature °F
t_1 = inlet air temperature °F.

Calculate the log mean temperature difference (LMTD) as in step 2 of the previous item on shell and tube exchanger design.

Step 3. Determine an approximate extended surface area using the expression:

$$A_E = \frac{Q}{U_E \cdot \Delta t_m} \tag{115}$$

where:

A_E = extended surface area in sqft.
Q = exchanger duty in Btu/h.
U_E = overall heat transfer coefficient based on extended surface from Table 31.
Δt_m = log mean temperature difference corrected for number of passes, in °F.

Process Equipment for Petroleum Processing

Table 31 Some common overall transfer coefficients for air cooling

	1/2" by 9 Fin ht. by fin/in.		5/8" by 10 Fin ht. by fin/in.	
Service	U_e	U_o	U_e	U_o
Process water	95	6.5	110	5.2
Hydrocarbon liquids				
Visc at avg temp cps				
0.2	85	5.9	100	4.7
1.0	65	4.5	75	3.5
2.5	45	3.1	55	2.6
6.0	20	1.4	25	1.2
10.0	10	0.7	13	0.6
Hydrocarbon gases				
at pressures psig				
50	30	2.1	35	1.6
100	35	2.4	40	1.9
300	45	3.1	55	2.6
500	55	3.8	65	3.0
1,000	75	5.2	90	4.2
Hydrocarbon condensers				
Cooling range 0 °F	85	5.9	100	4.7
10 °F	80	5.5	95	4.4
60 °F	65	4.5	75	3.5
100 + °F	60	4.1	70	4.2
Refrigerants				
Ammonia	110	7.6	130	6.1
Freon	65	4.5	75	3.5

U_e is transfer coefficient for finned surface
U_o is transfer coefficient for bare tubes

Step 4. Calculate the number of tubes from the expression

$$N_t = \frac{A_E}{A_f \times L} \tag{116}$$

where:
 N_t = total number of tubes.
 A_E = extended surface area in sqft.
 A_f = extended area per ft of fin tube read from Table 32.
 L = length of tube (30 f. is standard).

Step 5. Fix the number of passes (usually 3 or 4) and calculate the mass flow of tube side fluid using the expression

$$G = \frac{\text{lbs/h of tube side fluid} \times N_p \times 144}{N_t \times A_t \times 3{,}600} \tag{117}$$

Table 32 Fin tube to bare tube relationships based on 1″ O/D tubes

Fin Ht. by fins/in.	1/2″ by 9		5/8″ by 10	
Area/ft fin tube	3.8		5.58	
Ratio of areas, fin/bare tube	14.5		21.4	
Tube pitch in.	2Δ	2¼Δ	2¼Δ	2½Δ
Bundle area sqft/ft[a]				
3 rows	68.4	60.6	89.1	80.4
4 rows	91.2	80.8	118.8	107.2
5 rows	114.0	101.0	148.5	134.0
6 rows	136.8	121.2	178.2	160.8

[a]Bundle area is the external area of the bundle face area in sqft/ft

where:
 G = Mass velocity in lbs/s sqft.
 N_p = Number of tube passes.
 A_t = inside cross-sectional area of tube in sq inches

Step 6. Calculate the Reynolds number for tube side using the expression

$$\mathrm{Re} = \frac{d_i \cdot G}{\mu} \qquad (118)$$

where:
 R_e = Reynolds number (dimensionless).
 d_i = tube ID in inches.
 μ = tube side fluid viscosity at average temperature in cPs.

Step 7. Calculate the inside film coefficient from the expression

$$h_{io} = \frac{K}{D}(C\mu/K)^{1/3} \cdot (\mu/\mu_w)^{0.14} \cdot \phi(DG/Z) \qquad (119)$$

where:
 h_i = inside film coefficient in Btu/h·sqft °F.
 K = thermal conductivity of the fluid in Btu/h/sqft/(° F per ft).
 See Maxwell *Data Book on Hydrocarbons*.
 D = inside tube diameter in inches.
 C = specific heat in Btu/lb/°F.
 G = mass velocity in lbs/s sqft.
 μ = absolute viscosity cPs at average fluid temp.
 μ_w = absolute viscosity cPs at average tube wall temp.
 $\phi(DG/\mu)$ = from Fig. 37

Step 8. Calculate the mass velocity of air and the film coefficient on the air side, thus

$$\text{Weight of air} = \frac{Q}{C_{\text{Air}} \times \Delta t_{\text{Air}}} \qquad (120)$$

where:
 Q = exchanger duty in Btu/h.
 C_{Air} = specific heat of air (use 0.24).
 Δt_{Air} = temperature rise of the air °F.

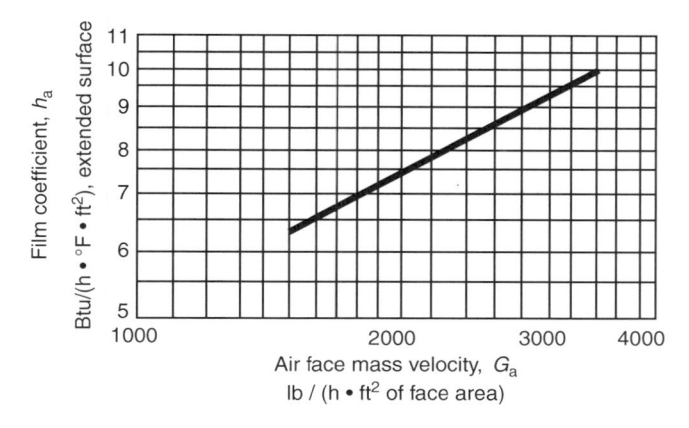

Process Equipment for Petroleum Processing 1577

Fig. 43 Air film coefficients

Face area of tubes A_f is calculated as follows:
Set the OD of the tubes (usually 1″), length, fin size (usually 5/8″ at 10 fins to the inch or 1/2″ at 9 fins to the inches), Pitch (see Table 32), and number of tube rows (start with 3 or 4). Then face area is

$$A_f = \frac{\text{Total extended surface area } A_E}{\text{External area per ft of bundle (from Table 32)}}. \quad (121)$$

Mass velocity of air is calculated from the expression

$$G_a = \frac{\text{lbs per hour of air flow}}{\text{face area } A_f}. \quad (122)$$

The film coefficient for the air side is read from Fig. 43.
Step 9. Calculate the overall heat transfer coefficient as follows:
Area ratio of bare tube outside to fins outside is read from Table 32. Then, the factor to convert all heat flow resistance to outside tube diameter basis is

$$F_t = \frac{A_r \times \text{tube OD}}{A_t} \quad (123)$$

where:
A_r = area ratio.
A_t = inside tube cross-sectional area sq inches.
then:

$$\frac{1}{U_o} = \frac{1 + (r_t \times F_t) + r_w + 1}{h_i h_o} \quad (124)$$

where:
r_t = inside fouling factor.
r_w = tube metal resistance (normally ignored).

If the calculated U is within 10 % of the assumed, there will be no need to recalculate with a new assumed value for the U. The dimensions and data are adjusted however using the calculated value for U.

Step 10. Calculate the required fan area and the fan diameter as follows:

$$\text{Fan area} = \frac{0.4 \times \text{Face Area } A_f}{\text{Assumed Number of Fans}}. \tag{125}$$

Begin by assuming 2 fans and continue with multiples of 2 until a reasonable fan diameter (about 10–12 ft) is obtained. On very large units, fans can be maximized at 16 ft.

$$\text{Fan diameter} = \sqrt{(\text{Fan area} \times 4/\pi)}. \tag{126}$$

Step 11. Calculate air-side pressure drop and actual air flow in cuft/min.

$$\text{Average air temperature} = \frac{t_1 + t_2}{2}. \tag{127}$$

From Fig. 44, $D_r =$ relative density factor for air at elevations of site.
From Fig. 45, $\Delta P_a =$ pressure drop of air in inches of H_2O.

$$\Delta P_a \text{ corrected} = \frac{\Delta P_a \times \text{No of rows}}{D_r}. \tag{128}$$

Density of air at corrected ΔP_a:

$$\frac{29}{(378 \times 14.7 \times T_2)/(T_1 \times (\text{Corr } \Delta P_a + 14.7))} \tag{129}$$

where T_1 and T_2 are absolute temperatures.

ACFM of air therefore is

$$\frac{\text{lbs/h of air}}{\text{Density} \times 60}. \tag{130}$$

ΔP of air at the fan is obtained by the expression

$$\Delta P_m = \left[\frac{\text{ACFM}}{(4,000(\pi d)/4)}\right]^2 \tag{131}$$

in inches of water gauge.

Step 11. Calculate the fan horsepower as follows:

$$\text{Hydraulic HP} = \frac{\text{ACFM} \times \text{density of air} \times \text{diff head in ft}}{33,000}$$

$$\text{Differential head} = \frac{\text{total } \Delta P \text{ at fan in in. } H_2O \times 5.193}{\text{Density}} \tag{132, 133, 134}$$

$$\text{BHP} = \frac{\text{hydraulic HP}}{\eta_r}$$

where η_r is the fan efficiency (usually 70 %).

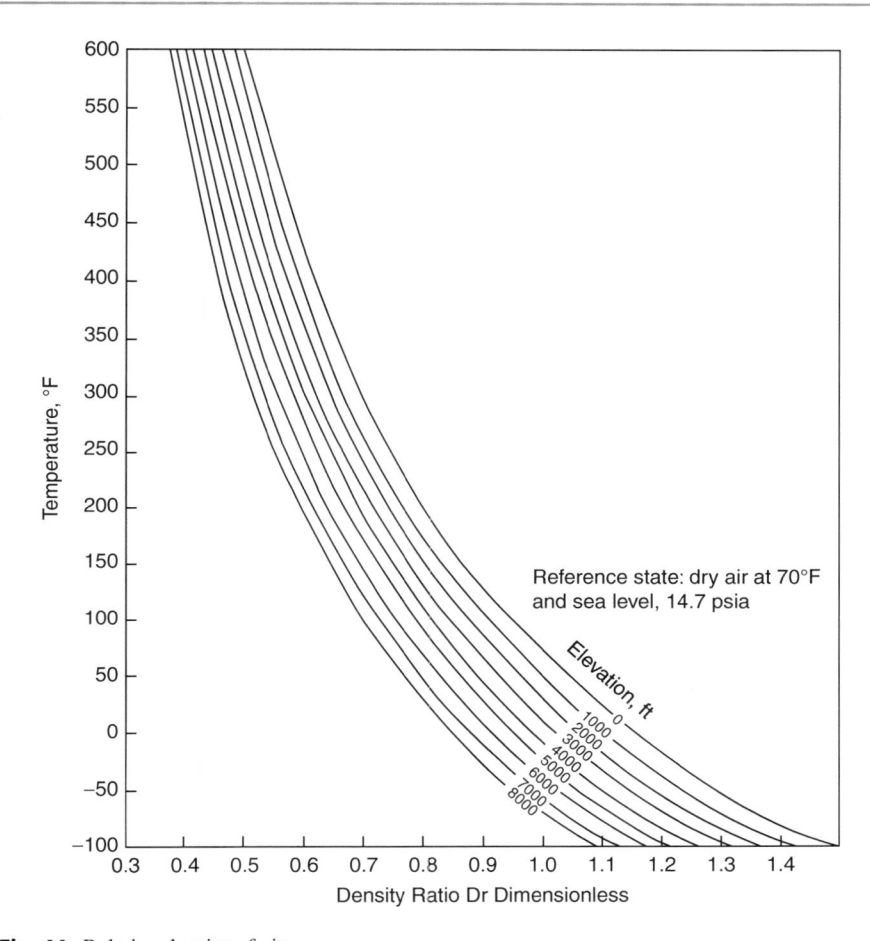

Fig. 44 Relative density of air

Condensers

In petroleum refining and most other chemical process plants, vapors are condensed either on the shell side of a shell and tube exchanger, the tube side of an air cooler, or by direct contact with the coolant in a packed tower. By far the most common of these operations are the first two listed. In the case of the shell and tube condenser, the condensation may be produced by cooling the vapor by heat exchange with a cold process stream or by water. Air cooling has overtaken the shell and tube condenser in the case of water as coolant in popularity as described in the previous item.

In the design or performance analysis of condensers, the procedure for determining thermal rating and surface area is more complex than that for a single-phase

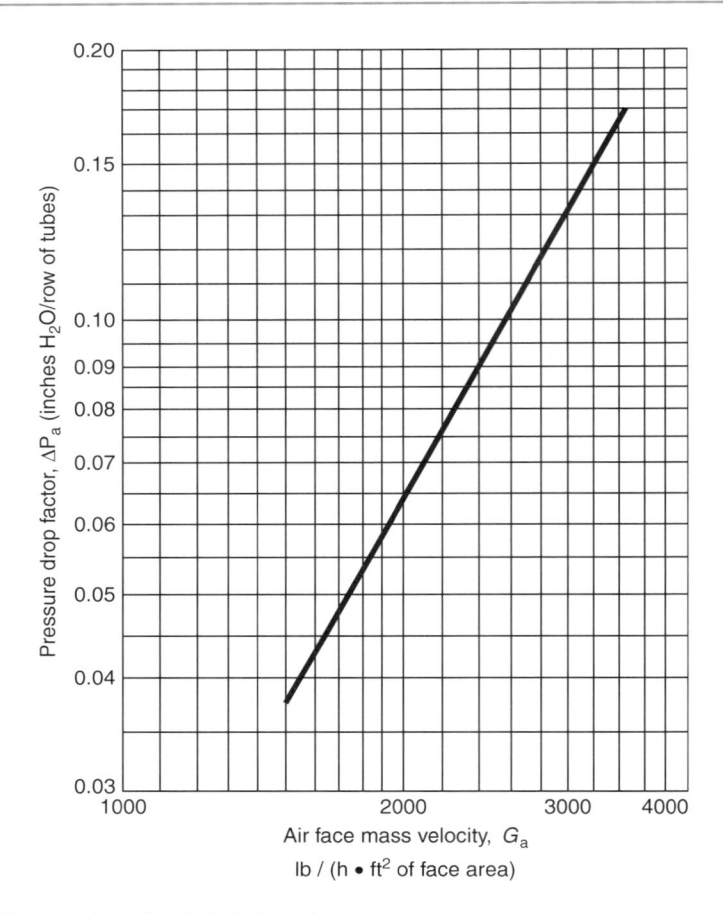

Fig. 45 Pressure drop air side in inches of water

cooling and heating. In condensers there are three mechanisms to be considered for the rating procedure. These are:

- The resistance to heat transfer of the condensing film
- The resistance to heat transfer of the vapor cooling
- The resistance to heat transfer of the condensate film cooling

Each of these mechanisms is treated separately and along preselected sections of the exchanger. The procedure for determining the last two of the mechanisms follows that described earlier for single-phase heat transfer. The following expression is used to calculate the film coefficient for the condensing vapor:

$$h_c = \frac{8.33 \times 10^3}{(M_c/L_c \cdot N_s)^{0.33}} \times k_f \times \left[\frac{SG_c^2}{\mu_f}\right]^{0.33} \tag{135}$$

where:

h_c = condensing film coefficient.
M_c = mass condensed in lbs/h
L_c = tube length for condensation.

$$= \frac{A_{zone}}{A} \times (L - 0.5). \tag{136}$$

$N_s = 2.08\, N_t^{0.495}$ for triangular pitch.
k_f = thermal conductivity of condensate at film temperature.
SG_c = specific gravity of condensate.
μ_f = viscosity of condensate at film temperature in cPs.

Again there are many excellent computer programs that calculate condenser thermal ratings, and these of course save the tedium of the manual calculation. However, no matter which method of calculation is selected, there is required one major additional piece of data over that required for single-phase heat exchange. That item is the enthalpy curve for the vapor.

Enthalpy curves are given as the heat content per lb or per hour contained in the mixed-phase condensing fluid plotted against temperature. An example of such a curve is given in Fig. 46. These enthalpy curves are developed from the vapor/liquid or flash calculations described in the handbook chapter "▶ Atmospheric and Vacuum Crude Distillation Units in Petroleum Refineries."

Briefly the calculation for the curve commences with determining the dew point of the vapor and the bubble point of the condensate. Three or more temperatures are selected between the dew and bubble points and the V/L calculation of the fluid at these temperatures carried out. Enthalpy for the vapor phase and the liquid phase are added for each composition of the phases at the selected temperatures. These together with the enthalpy at dew point and bubble point are then plotted.

As in the case of the shell and tube exchanger and the air cooler, a manual calculation for condensers is described here. Again this is done to provide some understanding of the data required to size such a unit and its significance in the calculation procedure. Computer-aided designs should however be used for these calculations whenever possible.

The following calculation steps describe a method for calculating the film coefficient of a vapor condensing on the shell side of an S and T exchanger. The complete rating calculation will not be given here as much of the remaining calculation is simply repetitive.

Step 1. Calculate the dew point of the vapor stream at its source pressure. Estimate the pressure drop across the system. Usually 3–5 psi will account for piping and the exchanger pressure drop. Calculate the bubble point of the condensate at the terminal pressure. Select three or more temperatures between dew point and bubble point and calculate the vapor/liquid quantities at these conditions of temperature and pressure.

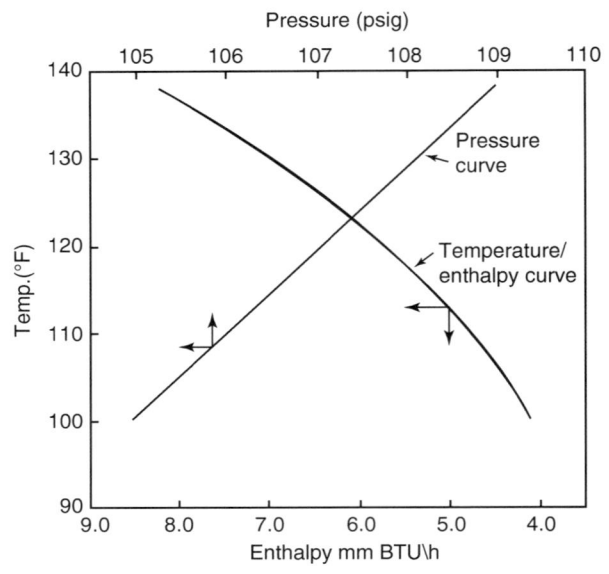

Fig. 46 Enthalpy curve for a debutanizer overhead condenser

- Pressure curve
- Temperature/enthalpy curve

X-axis (top): Pressure (psig) — 105, 106, 107, 108, 109, 110
X-axis (bottom): Enthalpy mm BTU\h — 9.0, 8.0, 7.0, 6.0, 5.0, 4.0
Y-axis: Temp.(°F) — 90, 100, 110, 120, 130, 140

Step 2. Calculate the enthalpy of the vapor and liquid at these temperatures. Plot the total enthalpies against temperature to construct the enthalpy curve. Establish the properties of the vapor phase and liquid phase for each temperature interval. The properties mostly required are SG, viscosity, mole wt, thermal conductivity, and specific heat.

Step 3. In the case of a water cooler, calculate the duty of the exchanger and the quantity of water in lbs/h. Commence the heat transfer calculation by assuming an overall heat transfer coefficient (use the data given in Appendix 7), calculating the corrected LMTD and the surface area.

Step 4. Using the surface area calculated in step 3, define the exchanger geometry in terms of number of tube passes, number of tubes on the center line, shell diameter, baffle arrangement, and the shell free flow area. Calculate also the water flow in feet per sec.

Step 5. Divide the exchanger into three or four zones by selecting the zone temperatures on the enthalpy curve. Calculate the average weight of vapor and the average weight of condensate in each zone. Using these averages, calculate the average heat transferred for:

Cooling of the vapor Q_v
Cooling of the condensate Q_L
Condensing of the vapor which will be:
Total heat in the zone (from the enthalpy curve) less the sum of Q_v and Q_L

Step 6. Calculate the film coefficient for the tube side fluid. See previous item "Estimating Shell and Tube Surface Area and Pressure Drop."

Step 7. Starting with zone 1 and knowing the outlet temperature of the coolant fluid, the total heat duty of the zone, and the shell side temperatures, calculate the coolant inlet temperature. Using this calculate the LMTD for the zone and, assuming a zone overall heat transfer coefficient U, calculate a surface area for the zone. Using this and the total exchanger area estimated in step 4, establish L_c in feet.

Step 8. Calculate the condensing film coefficient from the equation given earlier. This will be an uncorrected value for h_c. This will be corrected to account for turbulence by the expression

$$h_{c(\text{corr})} = h_c \times (G_v/5) \quad (137)$$

where:
G_v = average vapor mass velocity in lbs/h·sqft.

Step 9. Calculate the value of G_v using the free flow area allocated to the vapor γ_v. The following expressions are used for this:

$$\gamma_v = 1 - \gamma_L.$$
$$\frac{1}{\gamma_L} = 1 + \frac{\text{ave mass vapor}}{\text{ave mass liquid}} \times (\mu_v/\mu_L)^{0.111} \times (\rho_L/\rho_v)^{0.555}. \quad (138, 139, 140)$$
$$G_v = \frac{\text{ave mass vapor}}{25 \times \text{free flow area} \times \gamma_v}.$$

Step 10. Calculate the film coefficient h_v for the vapor cooling mechanism. This will be the procedure used for a single-phase cooling given in a previous item. This is corrected to account for resistance of the condensate film by the expression

$$\frac{1}{h_{v\,\text{corr}}} = -\frac{1}{h_c} + \frac{1}{1.25 h_v}. \quad (141)$$

Step 11. Calculate the film coefficient for the condensate cooling mechanism. Again this is the procedure described in the item for single-phase cooling on the shell side. This is corrected for drip cooling that occurs over a tube bank.

$$\text{Drip cooling } h_{dc} = 1.5 \times h_c$$
$$\text{and } h_L \text{ corrected} = \frac{2 \times h_{dc} \times h_L}{h_{dc} + h_L}. \quad (142, 143)$$

Step 12. Calculate the total zone film coefficient h_o using the following expression:

$$h_o = -\frac{Q_{\text{zone}}}{\frac{Q_c}{h_c} + \frac{Q_v}{h_v} + \frac{Q_L}{h_L}} \quad (144)$$

where:
Q_c, Q_v, and Q_L are the enthalpies for condensing, vapor cooling, and condensate cooling, respectively.

Step 13. Calculate the overall heat transfer coefficient neglecting the shell side coefficient from step 12. Thus,

$$\frac{1}{U_x} = r_o + r_w + r_{io} + R_{io} \tag{145}$$

where:
r_o, r_w, and r_{io} are fouling factors for shell fluid, wall, and tube side fluid, respectively. R_{io} is the tube side film coefficient calculated in step 6.

Step 14. Calculate the overall heat transfer coefficient U_{zone} for the zone using the expression

$$U_{zone} = \frac{h_o \times U_x}{h_o + U_x}. \tag{146}$$

Check the calculated U against that assumed for the zone. Repeat the calculation if necessary to make a match.

Step 15. Calculate the zone area using the acceptable calculated U. Repeat steps 7 through 14 for the other zones. The total surface area is the sum of those for each zone.

Reboilers

Reboilers are used in fractionation to provide a heat source to the system and to generate a stripping vapor stream to the tower. Reboilers are operated either by the natural circulation of a fluid or by forced circulation of the fluid to be reboiled. This chapter deals only with natural circulation reboilers.

There are three common types of reboilers and these are:

- The kettle-type reboiler
- The once-through thermosiphon reboiler
- The recirculating thermosiphon reboiler

The Kettle Reboiler

This type of reboiler (Fig. 47) is extremely versatile. It can handle a very wide range of vaporization loads (e.g., when used as LPG vaporizer for fuel gas purposes, it vaporizes 100 % of the feed). The equipment consists of a large shell into which is fitted a tube bundle through which the heating medium flows. The liquid to be reboiled enters the bottom of the shell at the end adjacent to the tube inlet/outlet chamber. The liquid is boiled and partially vaporized by flowing across the tube bundle. The diameter of the shell is sized such that there is sufficient space above the tube bundle and the top of the shell to allow some disengaging of the liquid and vapor. A baffle weir is installed at the end of the tube bundle furthest from the inlet. This baffle weir establishes a liquid level over the tube bundle in the shell. The boiling liquid flows over this weir to the shell outlet nozzle, while the vapor generated is allowed to exit from the top of the shell through one or two nozzles.

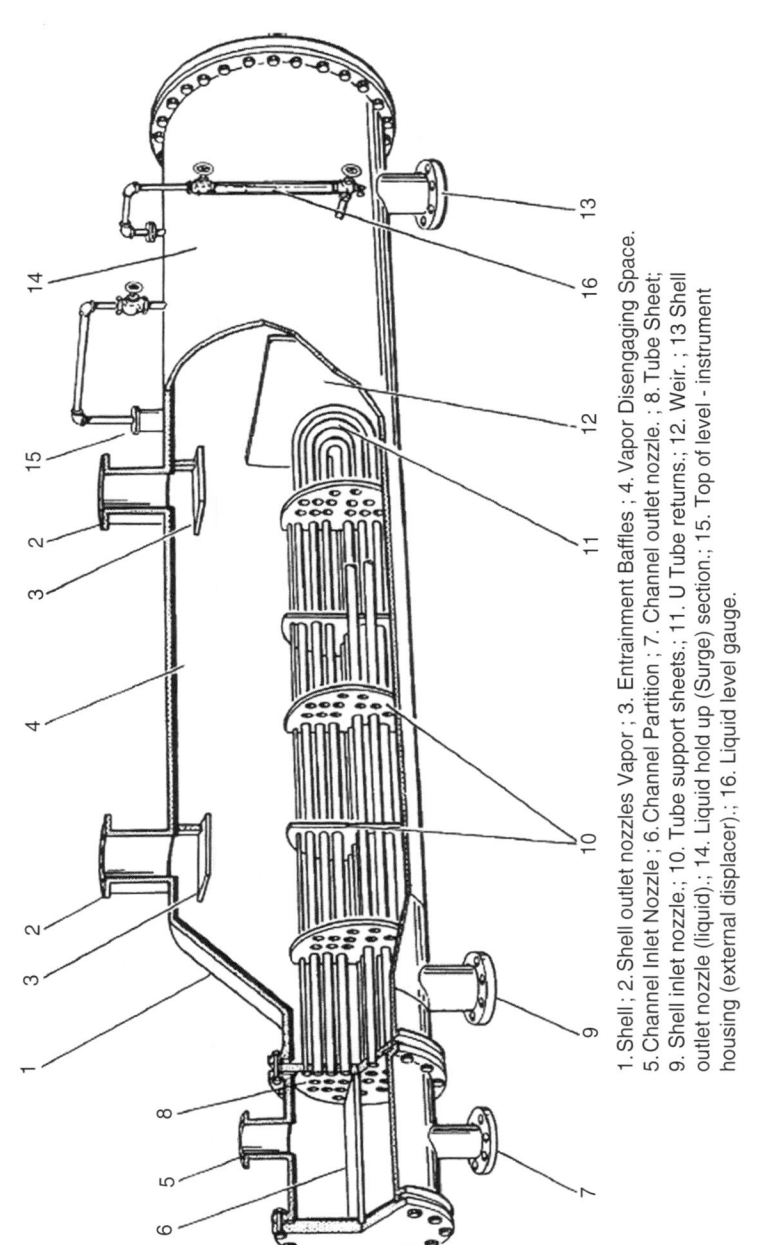

Fig. 47 The components of a kettle reboiler

1. Shell; 2. Shell outlet nozzles Vapor ; 3. Entrainment Baffles ; 4. Vapor Disengaging Space. 5. Channel Inlet Nozzle ; 6. Channel Partition ; 7. Channel outlet nozzle. ; 8. Tube Sheet; 9. Shell inlet nozzle,; 10. Tube support sheets,; 11. U Tube returns,; 12. Weir. ; 13 Shell outlet nozzle (liquid);.; 14. Liquid hold up (Surge) section.; 15. Top of level - instrument housing (external displacer).; 16. Liquid level gauge.

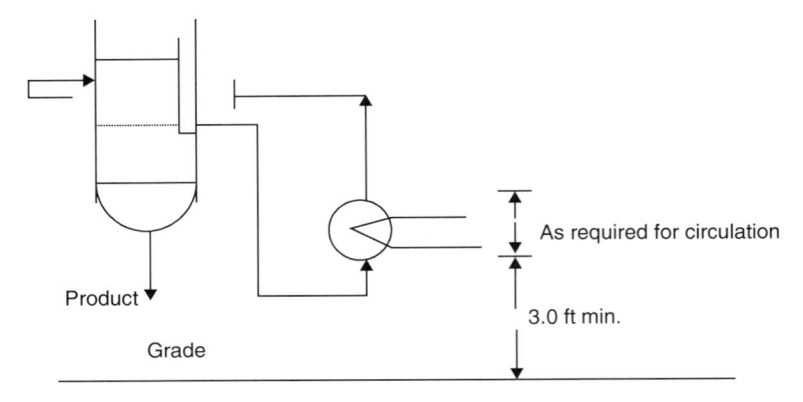

Fig. 48 Once-through thermosiphon reboiler arrangement

The space downstream of the weir is sized for liquid holdup to satisfy the surge requirements for the product. Thus, it is not necessary to provide space in the bottom of the tower for product surge. If the heating medium is non-fouling, it is permissible to use U tubes for the tube bundle. Otherwise, the tube bundle is often of the floating head type. New, smaller hydroblast nozzles and cleaning methods for U-tubes, however, allow use of these even in fouling services. The kettle-type reboiler should always be the first type to be considered if there are no elevation constraints to pumping the bottoms product away.

Once-Through Thermosiphon Reboiler

This type of reboiler and its location relative to the tower are illustrated Fig. 48.

This type of reboiler should be considered when a relatively high amount of surge is required for the bottom product and when it is necessary to provide head for the product pump (NPSH requirement).

This type of reboiler takes the liquid from the bottom tray of the fractionator as feed. This stream enters the shell side of a vertical single tube pass shell and tube exchanger by gravity head to the bottom of the shell. The heating medium flows tube side to partially vaporize the liquid feed. A siphoning effect is caused by the difference in density between the reboiler feed and the vapor/liquid effluent. This allows the reboiler effluent to exit from the top of the shell side and reenter the tower where the vapor disengages from the liquid phase. The liquid is the bottom product of the fractionator and is discharged from the bottom of the tower.

Both the kettle type and the once-through thermosiphon type constitute a theoretical tray as regards fractionation. Unlike the kettle reboiler, the once-through thermosiphon is limited to a vaporization of not more than 60 % of the feed. The low holdup of the feed from a tray results in severe surging through the reboiler at high vaporization rates.

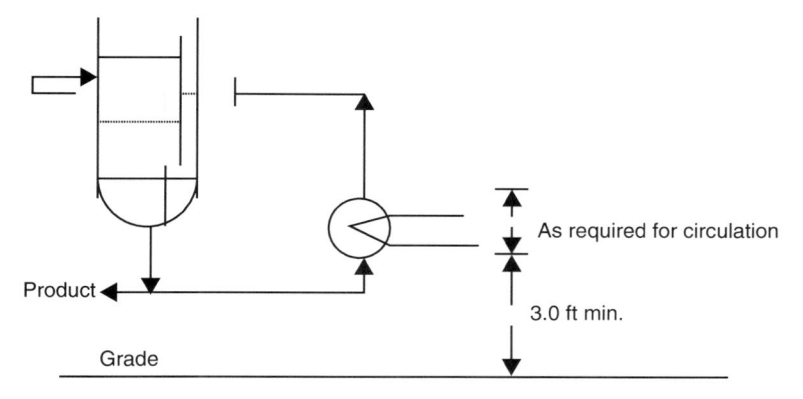

Fig. 49 Recirculating thermosiphon reboiler arrangement

Recirculating Thermosiphon Reboiler

When vaporization rates higher than 60 % of reboiler feed are required and a kettle reboiler is unsuitable, a recirculating thermosiphon-type reboiler should be considered. A sketch showing this type of reboiler is given in Fig. 49.

This reboiler is similar to the once-through thermosiphon in that it operates by flowing a liquid feed through the shell side of the vertical reboiler by the siphon mechanism. In the case of the recirculating reboiler; however, the feed to the reboiler is a stream of the bottom product from the fractionator. This is vaporized as described earlier and the liquid/vapor effluent returned to the tower. The vaporization by this reboiler can exceed 60 % without danger of surging. However, vaporization in this type of reboiler should not exceed 80 %. Its action is directed solely to providing heat to the tower and, because it recycles the same composition stream to the tower bottom, it cannot be considered as a theoretical fractionating tray (although some amount of fractionation does occur in this system).

Note in the description of both the thermosiphon-type reboilers, the heating fluid is shown as flowing tube side. There may be cases where this stream will be routed shell side and the reboil fluid directed tube side. Some guidance to this selection is provided by the following preference for tube side fluid:

1. Corrosive or fouling fluids
2. The less viscous of the two fluids
3. The fluid under the higher pressure
4. Condensing steam

Reboiler Sizing

As in the case of most heat exchangers, the sizing calculation is quite rigorous and complex. Normally process engineers rarely need to compute this in detail. There will be need however to estimate the size of these items for cost purposes or for plot layout studies. This sizing is greatly simplified by applying heat flux quantities to the predetermined reboiler duty. Heat flux is the value of heat transferred per unit time per sqft of surface. The following list gives a range of heat fluxes that have been used in design and observed in operating units.

	Design (Btu/h·sqft)	Observed
Kettle type	12,000	15,000–20,000
Once through	15,000	17,500+
Recirculating	15,000	up to 20,000
Forced circulation	20,000	–

The duty of the reboiler is obtained by the overall heat balance over the tower. This is accomplished by equating the total heat out of a fractionating tower to the heat supplied, making the reboiler duty the unknown in the heat supplied statement. Now the heat out of the fractionator is the total heat in the products leaving plus the condenser duty. The heat supplied to the tower is the heat brought in with the feed and the heat supplied by the reboiler.

Example Calculation

The feed to a fractionator is 87,960 lbs/h of mixed hydrocarbons. It enters the tower as a vapor and liquid stream and has a total enthalpy of 15.134 MMBtu/h. The overhead products are a distillate and a vapor stream at 95°F. The vapor is 1,590 lb/h with an enthalpy of 320 Btu/lb. The distillate is 8,028 lbs/h with an enthalpy of 170 Btu/h. The bottom product from the tower is 78,342 lbs/h and leaves as a liquid at its boiling point at 440 °F. Its enthalpy is 370 Btu/lb. The overhead condenser duty is 4.278 MMBtu/lb. Calculate the reboiler duty.

Calculation

Calculate the reboiler duty from the overall tower heat balance as in the following:

	V/L	°API	°F	lb/h	Enthalpy Btu/lb	MMBtu/h
In						
Feed	V/L	–	300	87,960	–	15.134
Reboiler duty						x
Total in				87,960		$15.134 + x$
Out						
Bottom prod	L	–	440	78,342	370	28.986
Overhead dist	L	–	95	8,028	170	1.364
Overhead vapor	V	–	95	1,590	320	0.508
Total out				87,960		35,136

Heat in = heat out
Then:

$$15.134 + x = 35.136$$

$$\text{Reboiler duty } x = 20.002 \text{ MMBtu/h}.$$

Using a heat flux of 15,000 Btu/h sqft, the surface area for the reboiler becomes

$$\frac{20.002 \times 10^6}{15,000} = 1,333.5 \text{ sqft}.$$

Estimating the Liquid and Vapor Flow from the Reboiler

It is necessary to know the vapor and liquid flow leaving the reboiler and entering the tower for the following reasons:

- To establish that there is sufficient vapor rising in the tower to strip the bottom product effectively
- To establish the vapor loading to the bottom tray for calculating the tray loading
- To be able to calculate the driving force for flow through the exchanger in the case of thermosiphon reboilers

The calculation of this flow is again based on a heat balance. In this case, it is the heat balance across the reboiler itself. With the duty of the reboiler now established by the overall tower heat balance, as described above, the balance over the reboiler can proceed as follows:

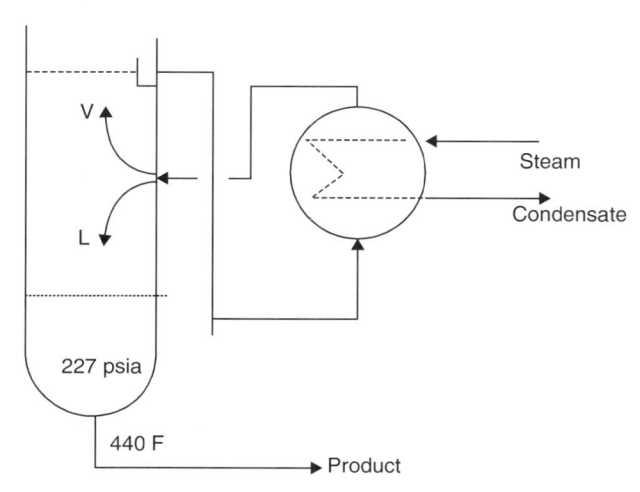

	V/L	°API	°F	lb/h	Enthalpy Btu/lb	MMBtu/h
In						
Liquid from tray 1	L	–	430	78,342 + V	369	28.908 + 369 V

(continued)

	V/L	°API	°F	lb/h	Enthalpy Btu/lb	MMBtu/h
Reboiler duty						20.002
Total in				78,342 + V		48.910 + 369 V
Out						
Bottom prod	L	–	440	78,342	370	28.986
Vapor to tray 1	V	–	440	V	458	458 V
Total out				78,342 + V		28.986 + 458 V

The temperature of the bottom tray (430 °F) is estimated from a straight-line temperature profile of the tower. As a rule of thumb, for a 30–40 tray tower, the bottom tray will be about 10 °F lower than the bottom temperature.

$$\text{Again heat in} = \text{heat out.}$$
$$\text{Then}: 48.910 + 369\ V = 28.986 + 458\ V.$$
$$V = 223,865\ \text{lbs/h}.$$

Now the mole weight of the vapor is determined from the bubble point calculation of the bottom product used to determine the tower bottom temperature (see the introductory chapter of this handbook). In the case of the calculation, example given above the bubble point calculation for the bottom product was as follows:

$$\text{Pressure at bottom of tower} = 220 + (30 \times 0.25)$$
$$= 227\ \text{psia.}$$

	X_w	1st trial = 400°F		2nd trial = 435 °F		MW	Weight factor	lbs/gal	vol factor
		K	Y = XK	K	Y = XK				
nC_4	0.017	2.7	0.046	3.1	0.053	58	3.1	4.86	0.64
iC_5	0.047	1.9	0.089	2.2	0.103	72	7.4	5.20	1.42
									(°API 53.1)
nC_5	0.055	1.7	0.094	1.9	0.105	72	7.6	5.25	1.45
C_6	0.345	0.96	0.331	1.3	0.449	81	36.4	6.83	5.33
C_7	0.322	0.44	0.142	0.59	0.190	102	19.4	6.84	2.84
C_8	0.214	0.17	0.036	0.25	0.054	128	6.9	6.94	0.99
Total	1.000		0.738		0.954	84.7	80.8	6.38	12.67

Actual temp = 440 °F

The mole weight of the vapor is that calculated for the "y" column in the above table, which is 84.7.

Calculating the Pressure Head Driving Force Through a Thermosiphon Reboiler

The big advantage of thermosiphon reboilers is that there are no working parts, such as pumps, that can fail. However, a major cause of failure in a thermosiphon reboiler is loss of driving force to move the fluid over or through the tube bundle. This problem mostly occurs during commissioning when the reboiler has been incorrectly positioned relative to the tower nozzles or during start-up where debris left after maintenance blocks one or other of the nozzles. In both these cases, the problem is really the loss of pressure head that drives the fluid to be reboiled through the exchanger. The calculation to determine the theoretical driving force is based on the density of the incoming liquid, the head of that liquid to the inlet nozzle, the density of the outflowing liquid/vapor fluid, and its head. An example of a pressure driving force calculation based on a once-through thermosiphon (as shown in the diagram below) is as follows. The flow data are based on the heat balance given earlier in this item.

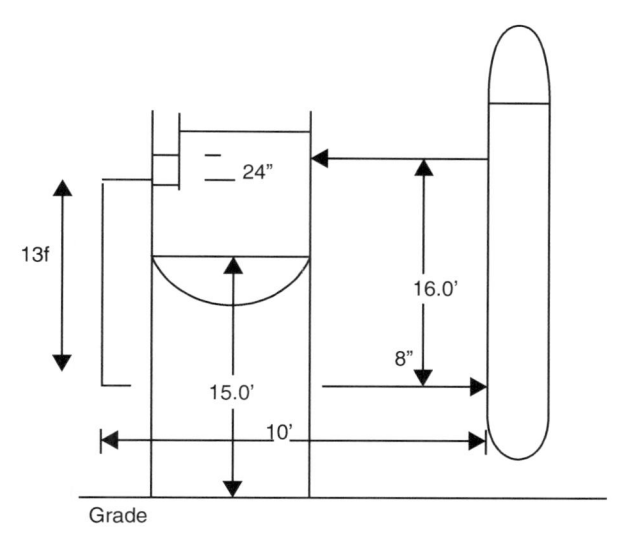

The density of the liquid to the reboiler is 38.2 lbs/cuft at 430 °F and the total flow is 302,207 lbs/h (223,865 lbs/h V + 78,342 lbs/h L).

$$\text{Hot cuft/h} = 7,911.$$
$$\text{Hot gpm} = 966.$$

The transfer line from the tower to the bottom nozzle of the reboiler is an 8″ schedule 40 seamless steel pipe. The head between the bottom of the tower draw-off pot and the reboiler nozzle is 13 ft. The equivalent horizontal line length including fittings is 15 ft.

From the friction loss tables in Appendix 9 of this chapter, head loss due to friction = 66.4 ft/100 f. of line (viscosity is taken as 1.1 cSt).

Total equivalent line length to the reboiler is $13 + 15$ ft $= 28$ ft.

$$\text{Head loss due to friction} = \frac{28 \times 66.4}{1,000} = 1.85 \text{ ft.}$$

Head of liquid in draw off pot $= 2$ ft $(24'')$.
Head of liquid to the reboiler inlet nozzle $= 13 + 2$ ft. $= 15$ ft.
Pressure head at the reboiler inlet $= 15 - 1.85$ ft $= 13.15$ ft

$$= \frac{13.15 \times 38.2}{144}$$
$$= 3.49 \text{ psi.}$$

The density of the vapor/liquid stream leaving the reboiler is calculated as follows:

$$\frac{\text{Total mass of fluid}}{\text{cuft liquid} + \text{cuft vapor}}. \tag{147}$$

Lbs/cft of liquid (this is bottom product) $= 39.4$ at $440\,°\text{F}$.
Mole wt of vapor (see bubble point calculation above) $= 84.74$.

$$\text{Cuft/h of liquid} = \frac{78,342 \text{ lbs/h}}{39.9} = 1,963.5.$$

$$\text{Cuft/h of vapor} \Rightarrow \frac{223,865}{84.5} \text{ lbs/h} = 2,643 \text{ moles/h}$$

$$= \frac{2,643 \times 378 \times 14.7 \times (460 + 440\,°\text{F})}{227 \times 520}$$

$$= 111,974.6 \text{ cuft/h.}$$

$$\text{Density of fluid from reboiler} = \frac{302,207}{1,963.5 + 111,974.6}$$
$$= 2.65 \text{ lbs/cuft.}$$

In this case, the fluid to be reboiled flows on the shell side of the exchanger. The manufacturer's certified shell side pressure drop based on all vapor flow is 1.5 psi. The mixed-phase pressure drop is calculated using Fig. 50, thus

$$m_v = \text{average mass of vapor} = \frac{223,865}{2} = 111,933 \text{ lbs/h.}$$

$$m_l = \text{average mass of liquid} = \frac{302,207 + 78,342}{2} = 190,275 \text{ lbs/h.}$$

$$\rho_v = \text{average density of vapor} = \frac{2.0}{2} = 1.0 \text{ lbs/cuft.}$$

$$\rho_l = \text{average density of liquid} = \frac{38.8 + 39.9}{2} = 39.35 \text{ lbs/cuft.}$$

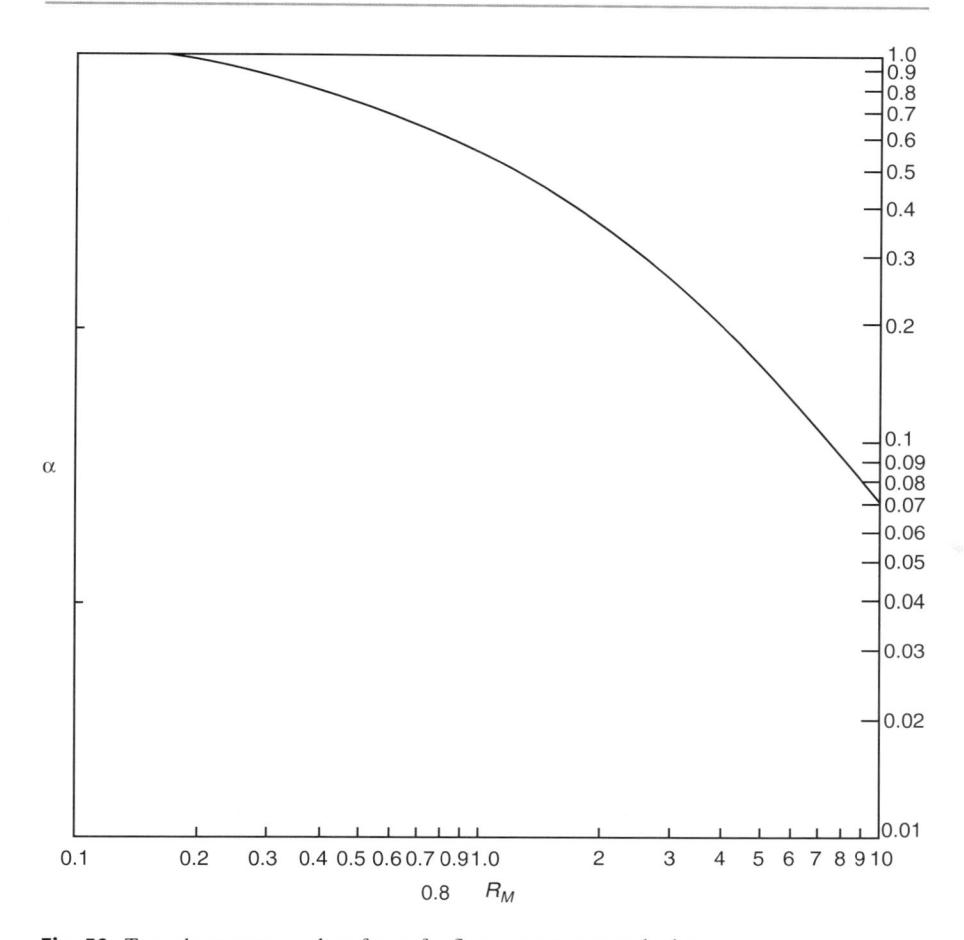

Fig. 50 Two-phase pressure drop factor for flow across staggered tubes

Referring to Fig. 50,

$$R_m = \cfrac{1}{\cfrac{m_v}{m_l} + \cfrac{\rho_v}{\rho_l}}$$

$$= \cfrac{1}{\cfrac{111,933}{190,275} + \cfrac{1.0}{39.32}} \tag{148}$$

$$= 1.63.$$

From Fig. 50,

$$\alpha = 0.42$$

$$\Delta P_{\text{mixed phase}} = \alpha \times \Delta P_{\text{gas}} \quad (149)$$
$$= 0.42 \times 1.5 \text{ psi}$$
$$= 0.63 \text{ psi}.$$

To calculate the total head of liquid from exchanger inlet to outlet nozzle.
Tube length is standard 16 ft.
Assume bottom 20 % of length is all liquid phase at a density of 38.8 lbs/cuft.

$$\text{Then head is} \frac{16 \times 0.2 \times 38.8}{144} = 0.862 \text{ psi.}$$

Remaining head is mixed phase at a density of 2.65 lbs/cuft.

$$\text{Then head is} \frac{16 \times 0.8 \times 2.65}{144} = 0.24 \text{ psi.}$$

Neglecting the small pressure drop due to friction in the 4 f. long return line to tower,

$$\text{Friction loss through exchanger} = 0.630 \text{ psi.}$$
$$\text{Lower section head} = 0.862 \text{ psi.}$$
$$\text{Upper section head} = 0.240 \text{ psi.}$$
$$\text{Total} = 1.732 \text{ psi.}$$

Then driving force = pressure head available − pressure head required
$$= 3.49 \text{ psi} - 1.732 \text{psi}$$
$$= 1.758 \text{ psi which is satisfactory for good flow.}$$

Note the height of the tower above grade is usually fixed by pump suction requirements in the first place. It may be adjusted upward if necessary to accommodate a head to a reboiler. However, this necessity is quite rare. The transfer line to the reboiler should have its horizontal section at least 3.0 f. above grade to allow for maintenance.

Fired Heaters

Types of Fired Heaters

This chapter provides some features and detail of fired heaters.
Most chemical plants and all petroleum refineries contain fired heaters as a means of providing heat energy into a system. Because the equipment utilizes an outside source of fuel, it is usually supported and enhanced by a heat exchange system to minimize the quantity of fuel required.

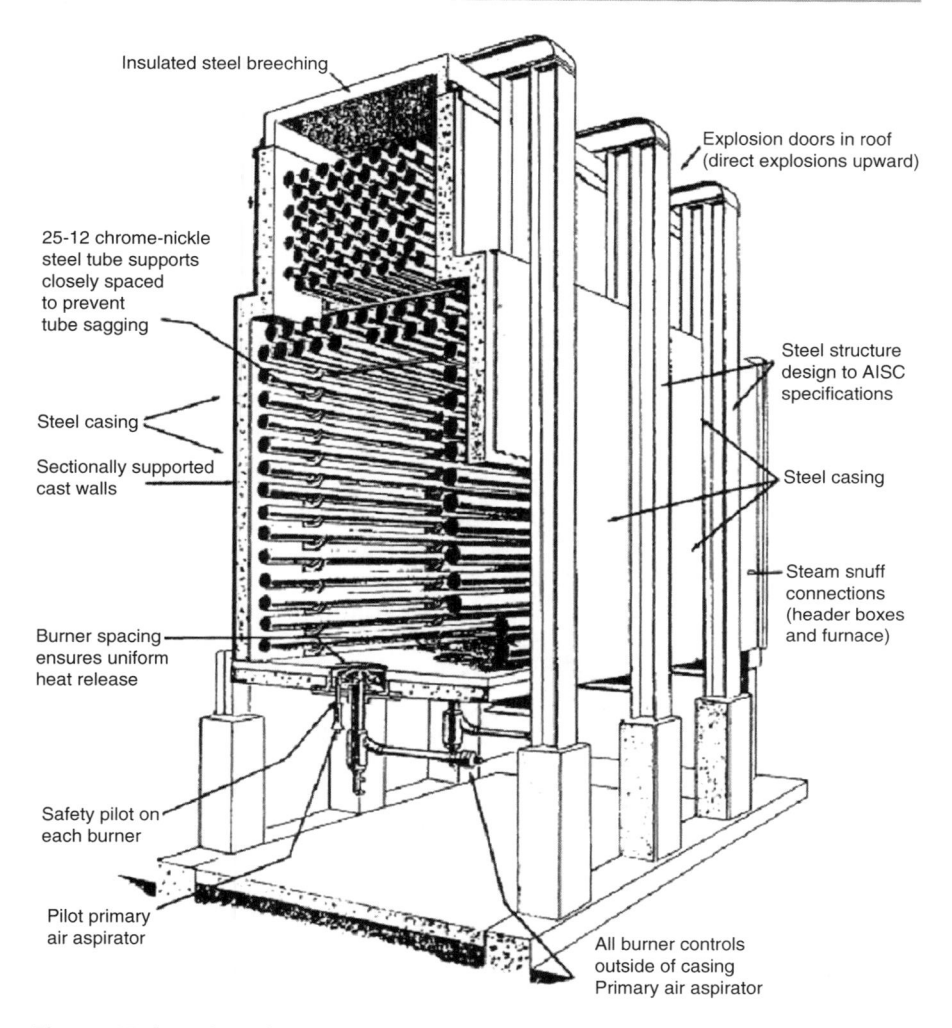

Process Equipment for Petroleum Processing   1595

Fig. 51 Horizontal-type heater

Generally fired heaters fall into two major categories:

- Horizontal type
- Vertical type

The horizontal-type heater usually means a box-type heater with the tubes running horizontally along the walls. Vertical type is normally a cylindrical heater containing vertical tubes. Figures 51 and 52 show examples of these two types of heaters.

These figures also give some nomenclature used in describing these items of equipment. Other terms used in connection with fired heaters are as follows:

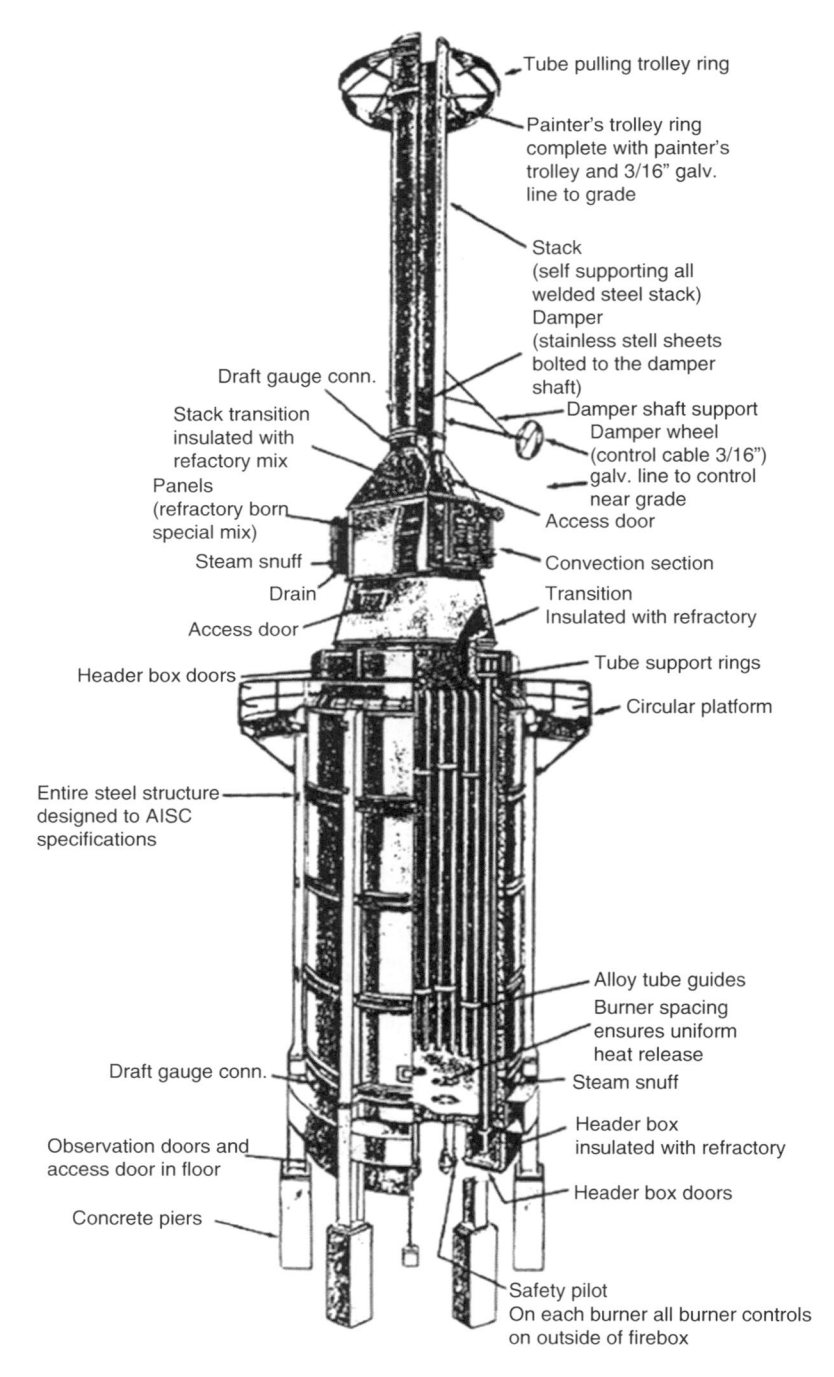

Fig. 52 Vertical-type heater

Process Equipment for Petroleum Processing

Headers and return bends are the fittings used to connect individual tubes.
Terminals are the inlet and outlet connections.
Crossovers are the piping used to connect the radiant with the convection section, usually external to the heater.
Manifold is the external piping used to connect the heater passes to the process piping and may be furnished with the heater.
Setting is any and all parts that form:
- Coil supports
- Enclosure (housing)

Casing is the steel shell which encloses the heater.
Bridge wall or partition wall is the refractory wall inside the heater that divides the radiant section into separately fired zones.
Shield tubes or shock tubes are the first two or three rows of tubes in the convection section. They protect or shield the convection tubes from direct radiant heat and normally must have the same metallurgy as the radiant tubes with no fins.
Air plenum is the chamber enclosing burners under the heater and having louvers to control the air flow.

Cylindrical heaters require less plot space and are usually less expensive. They also have better radiant symmetry than the horizontal type.

Horizontal box types are preferred for crude oil heaters, although vertical cylindrical heaters have been used in this service. Vacuum unit heaters should have horizontal tubes to eliminate the static head pressure at the bottom of vertical tubes and to reduce the possibility of two-phase slugging in the large exit tubes.

Occasionally, several different services ("coils") may be placed in a single heater with a cost saving. This is possible if the services are closely tied to each other in the process. Catalytic reforming preheater and reheaters in one casing are examples. Reactor heater and stripper reboiler in one casing is another example. This arrangement is made possible by using a refractory partition wall to separate the radiant coils. The separate radiant coils may be controlled separately over a wide range of conditions by means of their own controls and burners. If a convection section is used, it is usually common to the several services. If maintenance on one coil is required, the entire heater must be shut down. Also, the range of controllability is less than with separate heaters.

Each of these types may be shop fabricated if size permits. Shop fabrication reduces costs. However, shop fabrication should not be forced to the extent of getting an improperly proportioned heater.

Codes and Standards

Fired heaters have a "live" source of energy. That is, they use a flammable material in order to impart heat energy to a process stream. Because of this, the design, construction, and operation of process fired heaters and boilers are strictly

Table 33 Typical heater tube sizes and center-to-center dimensions from API standard 630

Tube OD, in.		Header C-C, in.	
Primary	Secondary	Group A	Group B
2.375	–	4.00	4.75
2.875	–	5.00	5.25
3.50	–	6.00	–
4.00	–	7.00	6.50
4.50	–	8.00	7.25
–	5.00	9.00	7.75
5.563	–	10.00	8.50
–	6.00	11.00	9.00
6.625	–	12.00	10.00
–	7.625	14.00	12.00
8.625	–	16.00	14.00

controlled by legislative and other codes and standards. This item outlines some of the more important of these codes and standards which need to be recognized by engineers dealing with fired heaters in any way.

Codes and standards directly applicable to fired heaters are listed below. In addition, there are many codes and standards covering such factors as materials, welding, refractories, structural steel, etc., which apply to fired heaters. This discussion reflects the standards at the time of this writing. For detailed design, current versions of the applicable standards should be used.

API RP-530
   Calculation of heater tube thickness in refineries
   This recommended practice sets forth procedures for calculating the wall thickness of heater tubes for service at elevated temperatures in petroleum refineries.

API Standard 630
   Tube and heater dimensions for fired heaters for refinery service
   This standard establishes certain standard dimensions for heater tubes and for cast and wrought headers. It also provides guidelines on maximum heat release per heater volume.
   Tube sizes and header center-to-center dimensions covered by the standard are in Table 33 as of this writing.
   Groove dimensions and tolerances for rolled headers are also given. Much of this standard is also used in chemical and petrochemical plants.

API RP-2002 Fire protection in natural gasoline plants
   This practice contains a brief statement about the use of snuffing steam. This system provides the piping of a steam source to the heater firebox which in an emergency can introduce steam into the box to quench any uncontrolled fire. This system may be automatically controlled or activated manually.

API guide for inspection of refinery equipment, chapter IX, fired heaters, and stacks
This reference gives a general description of fired heaters and describes how to inspect them, what damage to look for, and how to report the results of the inspection.

ASME boiler code and boiler codes of the USA
These are applicable in process plants if steam is generated and superheated or boiler feedwater preheated in the convection section. Special materials are required according to ASME Section I. The external piping and pressure relieving devices must also be in accordance with ASME Section I.

Contractors' and company standards
These will be covered in the "Narrative Specification" for the particular job and/or heater. The narrative specification is written by the heat transfer engineer specialists in the Contractors' Mechanical Equipment Group. These specifications detail all of the pertinent aspects required in the manufacture of the equipment. They will encompass all of the requirements that are applicable.

Thermal Rating

Refinery process engineers are seldom if ever required to thermal rate a fired heater or indeed check the thermal rating. This is a procedure that falls in the realm of specialist mechanical engineers with extensive experience in heater design and fabrication. Process engineers are however required to specify the equipment so that it can be designed and installed to meet the requirements of the process heat balance. To do this effectively, it is desirable to know something about the mechanism of heater thermal rating.

A fired heater is essentially a heat exchanger in which most of the heat is transferred by radiation instead of by convection and conduction. Rating involves a heat balance between the heat-releasing and heat-absorbing streams and a rate relationship.

Fuel is burned in a combustion chamber to produce a "flame burst." The theoretical flame burst temperature may vary from 4,000 °F when burning refinery gases with 20 % excess air preheated to 460 °F down to 2,300 °F when burning residual fuel oils with 100 % excess air at 60 °F. Heat is transferred from the flame burst to the gases in the firebox by radiation and mixing of the products of combustion. Heat is then transferred from the firebox gases to the tubes mainly by radiation.

The common practice is to assume a single temperature for the firebox gases for the purpose of radiation calculations. This temperature may be the same as the exit gas temperature from the firebox to the convection section (bridge wall temperature), or it may be different due to the shape of the heater and to the effect of convection heat transfer in the radiant section. Experience with the particular type of heater is required in order to select the effective firebox temperature accurately.

While this chapter does not detail the rating procedure or give an example calculation, the following steps summarize the rating procedure:

Step 1. Calculate net heat release and fuel quantity burned from the specified heat absorption duty and an assumed or specified efficiency.

Step 2. Select excess air percentage and determine flue gas rates.

Step 3. Calculate duty in the radiant section by assuming 70 % of total duty is radiant. This is a typical figure and will be checked later in the calculations. For very high process temperatures, such as in steam-methane reforming heaters, the radiant duty may be as low as 45 % of the total.

Step 4. Calculate the average process fluid temperature in the radiant section and add 100 °F to get the tube wall temperature. The figure of 100 °F is usually a good first guess and can be checked later by using the calculated inside film coefficient and metal resistance. This temperature is sensitive to tube flow rates and fouling.

Step 5. Calculate the radiant surface area using the average allowable flux. Convection surface is usually about equal to the radiant surface.

Step 6. Select a tube size and pass arrangement that will give the required total surface and meet specified pressure drop limitations.

Step 7. Select a center-to-center spacing for the tubes from the API 630 standard or from dimensions of standard fittings from other standards, and calculate firebox dimensions. Long furnaces minimize the number of return bends and thus reduce cost. Shorter and wider fireboxes usually give more uniform heat distribution and lessen the probability of flame impingement on the tubes. For vertical cylindrical heaters, the ratio of radiant tube length to tube circle diameter should not exceed 2.7.

Step 8. The remainder of the calculation involves determining the firebox exit temperature from assumption (3) above, applying an experience factor for the type of furnace to obtain the average firebox temperature, and then checking if this temperature will transfer the required radiant heat.

Step 9. The average heat flux (proportional to radiant surface) and the percent of total duty of the radiant section (which affects average tube wall temperature) are varied until a balance is obtained. The convection section surface and arrangement can now be calculated.

Step 10. The heater is normally designed to allow adequate draft at the burners with at least 125 % of design heat release and an additional 10 % excess air at the maximum and minimum ambient air temperatures.

Heat Flux

Although a process engineer is not normally required to thermally rate a process heater, he is often required to estimate the heater size, for preliminary cost estimates or plot layout and the like. This can be accomplished quite simply by the use of "heat flux." While this figure is quoted as Btu/h sqft, it is not however an overall heat transfer coefficient; it lacks the driving force ΔT for this. Heat flux is the rate of heat transmission through the tubes into the process fluid.

Process Equipment for Petroleum Processing   1601

The maximum film temperature and tube metal temperature are a function of heat flux and the inside film heat transfer coefficient.

The heat flux varies around the circumference of the tube, being a maximum on the side facing the fire. The value depends upon the sum of the heat received directly from the firebox radiation and the heat reradiated from the refractory.

Single-fired process heaters are usually specified for a maximum average heat flux of 10,000–12,000 Btu/h sqft. The maximum point heat flux is about 1.8 times greater.

Double-fired heaters are usually specified for about 13,500–18,000 Btu/h sqft *average* heat flux with the maximum point flux being about 1.2 times greater.

The following are typical flux values for heaters in hydrocarbon service:

Horizontal, fired on one side	8,000–12,000
Vertical, fired on one side from bottom	9,000–12,000
Vertical, single row, fired on both sides	13,000–18,000

Heater Efficiency

The efficiency of a fired heater is the ratio of the heat absorbed by the process fluid to the heat released by combustion of the fuel expressed as a percentage.

Heat release may be based on the lower heating value (LHV) of the fuel or higher heating value (HHV). Process heaters are usually based on LHV and boilers on HHV. The HHV efficiency is lower than the LHV efficiency by the ratio of the two heating values.

Heat is wasted from a fired heater in two ways:

- Hot stack gas
- Radiation and convection from the setting

The major loss is by the heat contained in the stack gas. The temperature of the stack gas is determined by the temperature of the incoming process fluid unless an air preheater is used. The closest economical approach to process fluid is about 100 °F. If the major process stream is very hot at the inlet, it may be possible to find a colder process stream to pass through the convection section to improve efficiency, provided plant control and flexibility are adequately provided for. A common method of improving efficiency is to generate and/or superheat steam and preheat boiler feedwater.

The lowest stack temperature that can be used is determined by the dew point of the stack gases. See the section on stack emissions.

Figures 53 and 54 may be used to estimate flue gas heat loss.

The loss to flue gas is expressed as a percentage of the total heat of combustion available from the fuel. These figures also show the effect of excess air on efficiency. Typically excess air for efficiency guarantees is 20 % when firing fuel gas and 30 % when firing oil.

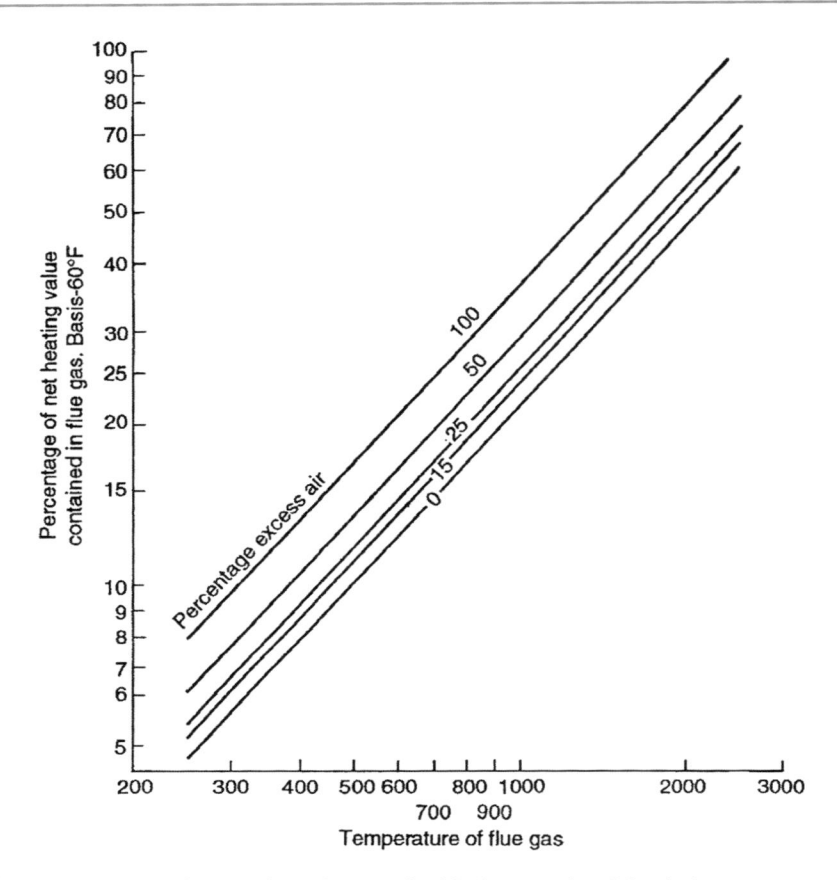

Fig. 53 Percentage of net heating value contained in flue gas when firing fuel gas

Heat loss from the setting, called radiation loss, is about 1½–2 % of the heat release. The range of efficiencies is approximately as follows:

Very high	90 %+. Large boilers and process heaters with air preheaters
High	85 %. Large heaters, low process inlet temps and/or air preheaters
Usual	70–80 %
Low	60 % and less. All radiant

Engineers are often required to check the efficiencies of the process heaters on operating units assigned to them. This can be done using the heats of combustion given in Appendices 10 and 11 to this chapter. The steps used to carry out these calculations are as follows:

Step 1. Obtain details of the heater from the manufacturer's data sheet or drawings. The data required are:

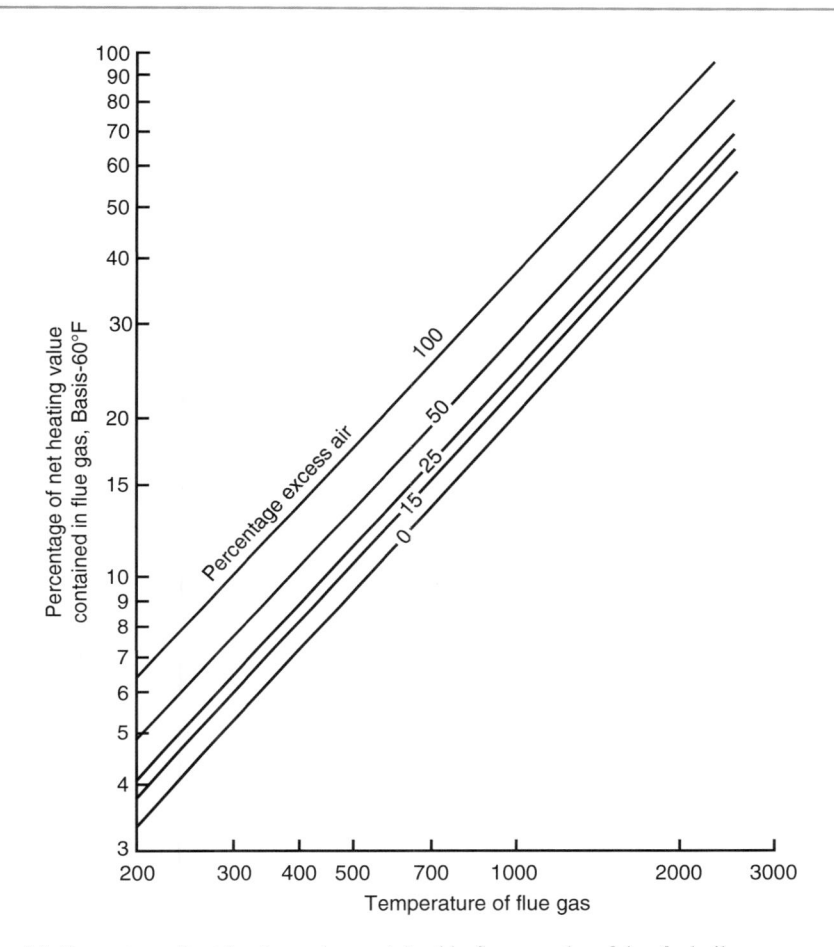

Fig. 54 Percentage of net heating value contained in flue gas when firing fuel oil

- Tube area
- Layout (is it vertical and how are the burners located?)

Step 2. From plant data, obtain coil inlet and outlet temperatures, pressures, and flows. Calculate the outlet flash (i.e., vapor or liquid or a mixture of vapor and liquid) condition, then calculate its enthalpy. Do the same for the inlet flow. Usually this will be a single phase either liquid or vapor.

Step 3. Again from plant data, obtain the quantity of fuel fired and its properties (API gravity in particular).

Step 4. The difference between the enthalpies calculated in step 2 is the enthalpy absorbed by the feed in Btu per hour.

Step 5. Divide this absorbed enthalpy by the tube area to give the heat flux in Btu/h/ft^2. Heat fluxes are generally as follows:

Horizontal, fired one side	8,000–12,000 Btu/h/ft^2
Vertical, fired from bottom on one side	9,000–12,000 Btu/h/ft^2
Vertical, single row, fired on both sides	13,000–18,000 Btu/h/ft^2

If the heat flux falls outside this range, there could be excessive fouling. Check the pressure drop – if this is far above manufacturer's calculated value, then fouling is certainly present. Fouling can be confirmed by an infrared scan of an operating heater.

Step 6. Check the thermal efficiency of the heater by giving the fuel fired a heating value. These are available in Appendices 10 and 11. Use the LHV (lower heating valve) in Btu/lb and multiply it by the lbs/h of the fuel.

Step 7. Divide the heat absorbed by the heat released calculated in step 4 to give the thermal efficiency. For most heaters, this should be between 70 % and 80 %. If it falls below this range, note should be taken of burner operation and the amount of excess air being used.

Burners

The purpose of a burner is to mix fuel and air to ensure complete combustion. There are about 12 basic burner designs. These are based on the factors in Table 34.

Various combinations of the above types are available.

Gas Burners

The two most common types of gas burners are the "premix" and the "raw gas" burners.

Premix burners are preferred because they have better "linearity," i.e., excess air remains more nearly constant at turndown. With this type, most of the air is drawn in through an adjustable "air register" and mixes with the fuel in the furnace firebox. This is called secondary air. A small part of the air is drawn in through the "primary air register" and mixed with the fuel in a tube before it flows into the furnace firebox. A turndown of 10:1 can be achieved with 25 psig hydrocarbon fuels. A more normal turndown is 3:1.

Oil Burners

An oil burner "gun" consists of an inner tube through which the oil flows and an outer tube for the atomizing agent, usually steam. The oil sprays through an orifice into a mixing chamber. Steam also flows through orifices into the mixing chamber. An oil-steam emulsion is formed in the mixing chamber and then flows through orifices in the burner tip and then out into the furnace firebox. The tip, mixing chamber and inner and outer tubes can be disassembled for cleaning.

Oil pressure is normally about 140–150 psig at the burner but can be lower or higher. Lower pressure requires larger burner tips; the pressure of the available atomizing steam may determine the oil pressure.

Table 34 Types of burner design parameters and burners

Categories	Options
Firing direction	Vertical upward
	Vertical downward
	Horizontal
Capacity	High
	Low
Fuel type	Gas
	Oil
	Combination
Flame shape	Normal
	Slant
	Thin, fan-shaped
	Flat
	Adaptable pattern
Hydrogen content	High
	Normal or none
Excess air	Normal
	Low
Type of atomization (oil)	Steam
	Mechanical
	Air-assisted mechanical
Boiler types	Various styles
Low NOx or ultra-low NOx (LN or ULN)	
High intensity	

Atomizing steam should be at least 100 psig at the burner valve and at least 20–30 psi above the oil pressure. Atomizing steam consumption will be about 0.15–0.25 lbs steam/lb oil, but the steam lines should be sized for 0.5.

Combination Burners

This type of burner will burn either gas or oil. It is better if they are not operated to burn both fuels at the same time because the chemistry of gas combustion is different from that of oil combustion. Gases burn by progressive oxidation and oils by cracking. If gas and oil are burned simultaneously in the same burner, the flame volume will be twice that of either fuel alone.

Pilots

Pilots are usually required on oil-fired heaters. Pilots are fired with gas. Natural gas is preferred, but they can be fired on very clean fuel gas.

Pilots are not required when heaters are gas fired only, but most gas-fired heaters today have them. Minimum flow bypasses around the fuel gas control valves are also used to prevent the automatic controls from extinguishing burner flames. Pilots normally are self-contained, with their own aspirated air intake, separate from the main burners.

Excess Air and Burner Operation

The excess air normally used in process-fired heaters is about 15–25 % for gas burners and about 30 % for oil burners. Excess oxygen then runs from 2-5 % for gas burners and about 6% for oil burners. These excess air rates permit a wide variation in heater firing rates which can be effectively controlled by automatic controls without fear of "starving" the heater of combustion air. There has been considerable work lately to reduce this excess air considerably mostly to minimize air pollution. This practice has not been used in process heaters to date. It has however been adopted in the operation of large power station-type heaters with some success.

Normally companies specify that burners be sized to permit operation at up to 125 % of design heat release with a turndown ratio of 3:1. This gives a minimum controllable rate of 40 % of design without having to shut down burners.

Burner Control

Burner controls become very important from safety and operation considerations. Most systems include an instrumentation system with interlocks that prohibit:

- Continuing firing when the process flow in the heater coil fails
- The flow of fuel into the firebox on flame failure

Under normal operating conditions, the amount of fuel that is burnt is controlled by flow controllers operated on the coil outlet temperature. With combination burners, the failure of one type of fuel automatically introduces the second type. Such a switch over can also be effected manually. This aspect is usually activated on pressure control of the respective fuel system, that is, on low pressure being sensed on the fuel being fired automatically switches to the second fuel.

Most companies operate their own specific controls for the heater firing system. Refer also to the discussion of burner management systems in the topic "▶ Utilities in Petroleum Processing" of this handbook.

Heater Noise

All heaters are noisy and this noise is the result of several mechanisms. Among these is the operation of the burners. Gas burners at critical flow of fuel emit a noise. This can be minimized by designing for low pressure drop in the system. Intake of primary and secondary air is another source of noise. Forced draft burners are generally quieter than natural draft if the air ducting is properly sized and insulated. The design of the fan can also reduce noise in this mechanism. Low-tip-speed fan favors low noise levels. Simple, cost-effective techniques are available to mitigate heater noise, but they cannot eliminate all noise.

Refractories, Stacks, and Stack Emissions

Refractories are used on the inside walls of the heater firebox, floor, and through the convection side of the heater. The purpose of this refractory lining is to conserve the heat

Process Equipment for Petroleum Processing 1607

by limiting its loss to atmosphere by convection. It is also necessary for personal safety of those working on or about the heater who may accidentally touch the heater walls.

Good insulation has the following qualities:

- It has good high temperature strength.
- It is resistant to abrasion, spalling, chemical reaction, and slagging.
- It has good insulating properties.

Among the most common refractories that meet some if not all of the above criteria are silica refractories, high alumina, and fire clay brick. These have high resistance to spalling and to thermal shock. Their insulating qualities are also good.

Silica refractories tend to form slag with metal oxide dust and ashes. Compounds of sodium and potassium attack most refractories while refractories containing magnesium react with acids and acid gases. Carbon monoxide and other reducing chemicals that may be present in the firebox reduce the life of refractories, particularly fire clay brick and silica.

Dense refractories with low porosity are the strongest but have the poorest insulation qualities. Castable refractories containing a mixture of cement and refractory aggregate are cheap and relatively easy to install. They are not very rugged however. Normally in process heaters, conditions are such that the use of a light insulating refractory will satisfy all that is required from a refractory lining.

Today, many furnaces use refractory fiber in modular blocks and blankets for insulation. This material is lighter, meaning the furnace wall supports much less weight, and it has much better insulating properties than more dense materials. The blanket refractory can be installed in the shop before the furnace is shipped, with limited need for finish work in the field. These refractories do not require curing, which is a big time savings in start-ups. Blanket refractory can be eroded by high flue gas velocities, so it is common to "rigidize" the blanket and block modules in high-velocity areas. The rigidizing is accomplished simply by spraying the refractory with a proprietary solution which, when dried and fired, bonds and stiffens the refractory fibers together so that they are less affected by velocities.

The ASTM, standard part 13, gives more detail on refractories. This standard provides the classification of refractories and describes their characteristics and composition. It also offers a procedure for calculating the heat loss through the insulation and thus its thickness.

Preparing Refractories for Operation

All new hard refractories need to be "cured" after installation and before use. Refractories contain moisture, some due to the installation procedure and some in the form of water of crystallization. Curing the refractory means removing this moisture by applying a slow heating mechanism. New heater manufacturers will usually provide details of the curing procedure they recommend. The procedure must not damage any part of the furnace or process system. The following procedure may however be used as a guide to refractory curing when manufacturers' procedures are not available:

Step 1. Raise the flue gas temperature at the arch 50 °F/h to a temperature of 400 °F and hold for 8 h at that temperature. Maintain a cooling flow on inside the tubes with minimal pressure – see step 6 below.

Step 2. Then increase the temperature again at 50 °F/h from 400 to 1,000 °F and hold for another 8 h.

Step 3. If necessary continue heating at 100 °F/h to the operating temperature if higher than 1,000 °F. Hold at the operating temperature for a further 8 h.

Step 4. Cool at 100 °F/h to about 500 °F and hold ready for operation.

Step 5. On start-up the heater can be heated up to its operating temperature at a rate of 100 °F/h.

Step 6. During the curing of the refractory, it will normally be necessary to pass some fluid through the heater coils to protect them from overheating. Steam or air may be circulated through the coils for this purpose. In certain cases such as in the catalytic reforming of petroleum stock, it may be necessary to circulate the nitrogen or the hydrogen that was used to purge and pressure test the unit. Air and steam in this process would not be desirable.

Following refractory repairs, it may be necessary to cure the repair material. Normally, if the repairs were minor, you can heat up at 50 °F/h and adequately cure the refractory. If the refractory work was extensive, a regular cure needs to be built into the start-up schedule.

Stacks

Stacks are used to create an updraft of air from the firebox of a heater. The purpose of this is to cause a small negative pressure in the firebox and thus enable the introduction of air from the atmosphere. This negative pressure also allows for the removal of the products of combustion from the firebox. The stack therefore must have sufficient height to achieve these objectives and overcome the frictional pressure drop in the firebox and the stack itself.

The height required for a stack to achieve good draft can be estimated from the following equation:

$$D = 0.187H(\rho_a - \rho_g) \tag{150}$$

where:

D = draft in inches of water.
H = stack height in feet.
ρ_a = density of atmospheric air in lbs/cuft.
ρ_g = density of stack gases in lbs/cuft at stack conditions.

For stack gas temperature, use 100 °F lower than gases leaving the convection section. Stack gases have a molecular weight close to that of nitrogen. For this calculation, use 28 as the mole weight. Burner draft requirements range from 0.2 to 0.5 in. of water. Use 0.3 in. of water as a good design value.

Stacks must also be designed to handle and disperse stack emissions. This usually results in having to build stack heights greater than that required for obtaining draft. There are available specific computer programs relating plume height of the stack gases above the stack outlet to the probable ground level fallout of the impurities in the gas. These programs also produce a map of the relative concentrations of these impurities at ground level. Such data are usually available from government authority offices in most countries. The predicted ground concentration and map calculated are checked against local legal requirements. The results usually form part of the government approval to build and/or operate a facility.

The stack diameter is based on an acceptable velocity of stack gases in the stack. This is generally taken as 30 ft/s. Some allowance must be made for fractional losses; these are:

1.5 velocity heads for inlet and outlet losses
1.5 velocity head for damper
1.0 velocity head for each 50 f. of stack height

Stack Emissions

The obnoxious compounds in stack gas emissions arise from:

- Impurities in the fuel
- Chemical reactions resulting from the fuel combustion with air

The three major impurities in oil or gas fuels which produce undesirable emissions are:

- Sulfur
- Metals
- Nitrogen

Sulfur

All gas and oil fuels contain sulfur at some level of concentration. When these fuels are burned, the sulfur reacts with air to form SO_2 and SO_3. These compounds are objectionable because they cause:

- Air pollution in the form of smog.
- They contribute to the corrosion of heater tubes and stack.
- SO_3 lowers the dew point of the flue gases resulting in an objectionable visible plume at the stack exit.

Figure 55 shows the effect of sulfur in the feed on the flue gas dew point. This dew point also varies with the amount of excess air and the relative partial pressure of the combustion gases.

There is no precise way of determining the amount of SO_3 that is formed in the flue gas. Nor can it be determined with any degree of accuracy where the SO_3 is

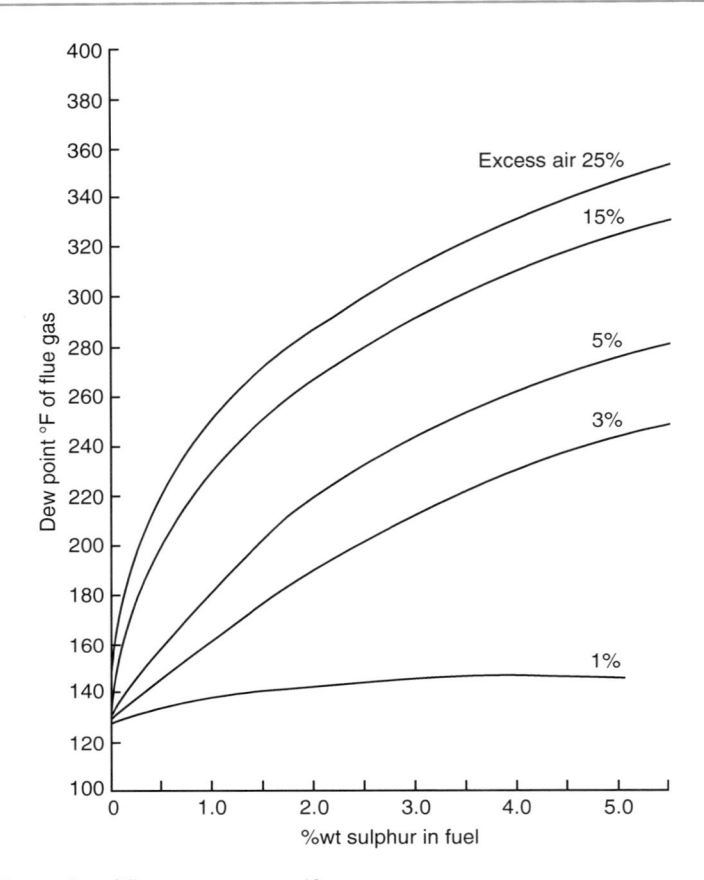

Fig. 55 Dew point of flue gases versus sulfur content

formed in the system. The total amount of the sulfur oxides is determined simply from the sulfur content of the feed. Because of this uncertainty, it is important to restrict the minimum allowable convection side metal temperatures to 350 °F when firing fuels containing sulfur. Also the minimum temperatures to the stack should be 320 °F when firing fuel gas and 400 °F when firing fuel oil. In the case of metal stacks, the use of a noncorrosive lining must be used in the colder section of the stack if the flue gas temperature falls below those stated above.

Metals

The most objectionable metal impurities in fuel oil are sodium and vanadium. These can cause severe corrosion of tubes and refractory lining. By high vanadium content is meant concentrations between 200 and 400 ppm, above 400 ppm is considered very high and should not be used as a fuel. Vanadium in the presence of sodium and oxygen readily forms a corrosive compound:

Process Equipment for Petroleum Processing 1611

$$Na_2O \cdot 6V_2O_5.$$

This compound attacks iron or steel to form ferric oxides according to the equation:

$$Na_2O \cdot 6V_2O_5 + Fe \rightarrow Na_2OV_2O_4 \cdot 5V_2O_5 + FeO. \qquad (152)$$

This vanadium product is oxidized back to the original corrosive compound according to the reaction:

$$Na_2OV_2O_4 \cdot 5V_2O_5 + O_x \rightarrow Na_2O \cdot 6V_2O_5. \qquad (153)$$

These reactions occur at temperatures between 1,070 and 1,220 °F with the vanadium compound being continually regenerated to its corrosive state.

Sodium and sulfur also combine to form undesirable corrosive compounds without vanadium being present. This reaction forms sodium iron trisulfate $Na_3Fe(SO_4)_3$ at temperatures of around 1,160 °F. The critical temperature for both vanadium and sodium corrosion is around 1,100 °F. Vanadium is not a major problem at this temperature but becomes so at temperatures above this level.

Nitrogen

Some nitrogen oxides will be formed in combustion gases even if there are no nitrogen compounds in the fuel. The presence of the nitrogen compounds in the fuel significantly increases the nitrogen oxide content of the flue gas. There are six oxides of nitrogen present in flue gases but only two are present in any appreciable amount. These are:

NO nitrogen oxide
NO₂ nitrogen dioxide

Nitrogen oxide is a poisonous gas which readily oxidizes to nitrogen dioxide on entering the atmosphere. Nitrogen dioxide is a yellowish gas which readily combines with the moisture in the air to form nitric acid. In the presence of sunlight and oxygen, nitrogen dioxide also contributes to the formation of other air pollutants. The production of NO_x can be reduced by limiting the amount of excess air, by washing the flue gases with aqueous ammonia, or by catalytically reducing NO_x in the presence of ammonia.

The environmental chapter of this handbook further discusses emissions and emissions control.

Specifying a Fired Heater

Some basic data concerning a fired heater must be made known before the equipment can be designed for fabrication or even costed. These data are provided by a specification sheet or sheets. In the case of a fired heater, as in the case of compressors or exchangers, this specification can run into several sheets or forms. Such sheets will define the unit in terms of:

- Process requirement
- Mechanical detail
- Civil engineering requirement
- Operational requirement
- Environmental requirement

This discussion will deal only with the "process requirement" and the duty of the heater. A typical process originated specification sheet is shown here as Fig. 56. The data provided on this sheet are the minimum that will be necessary for a heater manufacturer or a heater specialist to begin to size the heater or price out the item. Usually these data would also be supported with a vaporization curve of the feed if there is a change of phase taking place in the heater coils. The process engineer developing this specification would also provide details of services required in addition to the main duty of the heater. For example, if it is intended that the convection side of the heater is to be used for steam generation, preheating, or steam superheating, the system envisaged must be properly described by additional diagrams and data.

The example specification sheet in Fig. 56 is completed for a crude oil vacuum distillation unit heater with a steam superheater coil located in the convection side of the heater. The following paragraphs describe the content on a line-by-line basis.

1. Flash curve of the reduced crude feed versus % volume distilled is attached for atmos pressure and at 35 mmHg Abs would be attached.
2. Gravity and mole wt curves for the reduced crude versus mid volume % would be attached.
3. Soot blowers are to be considered in this package. Steam is available at 600 psig and 750 °F.
4. Studded tubes to be considered for the convection side.

Line 1. *Design duty* refers to the total duty required of the unit. In the case of the example given here, it includes the duty required to heat and partially vaporize the oil in the radiant section and the duty required to superheat the steam. The data sheet will be split into two sections to reflect each of these duties.

Service describes the main purpose for which the heater will be used. In this case, it will be used to heat and vaporize hydrocarbons.

Line 2. *No. of heaters.* This is self-explanatory. In this case, there is only one heater required. Should there have been more than one identical unit, this would be reflected here.

Unit. This is the title given to this unit of equipment as it appears on an equipment list. It will correspond to the item number also given in the equipment list. In the case of the example, it is "vac unit preheater."

Line 3. *Item No.* This is the reference number given to the item in the equipment list. This reference number and unit title identifies the equipment on all drawings where it appears and all documents used in its purchase, costing, maintenance, etc.

Process Equipment for Petroleum Processing 1613

Type. The type of heater (if decided on) is given here. In the case of the example heater, a cabin (horizontal tubes) type has been selected for vacuum services considerations.

Line 4. This is the first line of the specification sheet proper. It commences with the service of the heater or section of the heater. In the case of this example, only two columns have been provided. These are designated for the "Radiant" section and "Convection" section. On most preprinted specification forms, there would be at least four columns.

Line 5. *Heat absorption.* This line divides the duty of the heater into that required from the radiant coils and that required from the convection side. In this example, the oil is routed through the radiant section only while saturated steam from a waste heat boiler is superheated in the convection coils. Both are measured in million Btu/h.

a   FIRED HEATER.

Design Duty 40.886 MMBtu/hr Btu\hr Service Hydrocarbons

No Heaters ___1___   Unit Vac Unit Preheater

Item No ___H 201___   Type Horizontal

DESIGN DATA:-	Radiant sec	Conv sec
Service	Red Crude	Steam s.heat
Heat Absorption MMBtu\hr	38.501	2.385
Fluid	Hydrocarbon	Sat Steam
Flow Rate lbs/hr	197085	30040
Allowable Pressure drop PSI	300	30
Allowable average Flux Btu/hr. sqft	15000	—
Maximum Inside Film Temp °F	800	—
Fouling Factor °F. sqft.hr/Btu	0.004	0.001
Residence Time sec	N/A	N/A
Inlet Conditions		
Temperature °F	554	368
Pressure PSIG	270	155
Liquid Flow lbs/hr	197085	nil
Vapor Flow lbs/hr	nil	30040
Liquid Density lbs/cuft	48.4	—
Vapor Density lbs/cuft	_____	Steam
Visc Liq\Vap cPs	2.31 /__	__ /__
Specific Heats liq\vap Btu\lb	0.65 /__	/
Thermal Cond liq\vap Btu\hr.sqft.°F\Ft	0.0671 /__	/

Fig. 56 (continued)

Partial vaporization of the oil occurs in the radiant coil. Therefore, a flash curve or a phase diagram of the oil must accompany this specification sheet. In the example, the duty to the oil is calculated from data developed in the material balance and heat balance of the process. Thus:

Temperature of the feed into the heater is 554 °F and that of the coil outlet is 750 °F. From the flash curve at the outlet pressure of 35 mmHg abs (0.68 psia) and the material balance, the weight per h of vapor is calculated to be 127,444 lbs/h and the liquid portion is 69,641 lbs/h. The heat absorbed in the radiant section is

b

DESIGN DATA(Cont)	Rad Sec	Conv Sec
Outlet Conditions.		
Temperature °F	750	500
Pressure (Psia) PSIG	(0.68)	125
Liquid Flow lbs\hr	69641	nil
Vapour Flow lbs\hr	127444	30040
Liquid Density lbs\cuft	47.3	
Vapour Density lbs\cuft	0.023	0.246
Visc of Liq\Vap cPs	1.4/0.002	_/_
Specific Heat Liq\Vap Btu\lb	0.64/0.069	_/_
Thermal Cond Liq\Vap Btu\hr.sqft.°F\Ft	0.062/0.023	/
FUEL DATA		
Type (Gas or Oil)	OIL	GAS
LHV Gas Btu\cuft Oil Btu\lb	17560	2320
HHV Gas Btu\cuft Oil Btu\lb	18580	2520
Pressure at Burner PSIG	82	25
Temp at Burner °F	176	68
Mol Wt of Gas		44
Visc of Oil @ Burner cPs	23.3	
Atomising Steam Temp °F Press PSIG	500 125	
Composition of Gas Mol%		
H2		0.2
C1		12.0
C2		28.2
C3		31.3

Fig. 56 (continued)

c

FUEL DATA (Cont)		
Composition of Gas (Cont)		
C4's		
C5's		
C6+		
Properties of Fuel oil		
°API	15.2	
Visc @ 100°F cSt	175	
Visc @ 210°F cSt	12.5	
Flash Pt °F	200	
Vanadium ppm	12	
Sodium ppm	32	
Sulphur %wt	2.4	
Ash %wt	<1.0	

Fig. 56 (a–c) Typical specification sheet for a fired heater

$$\text{Heat in with feed} = 197,085 \text{ lbs/h} \times 268 \text{ Btu/lb (all liquid)}$$
$$= 52.819 \text{ mm Btu/h.}$$

Heat out in feed:

$$\text{Liquid portion} = 69,641 \text{ lbs/h} \times 378 \text{ Btu/lb}$$
$$= 26.324 \text{ mm Btu/h.}$$
$$\text{Vapor portion} = 127,444 \text{ lbs/h} \times 510 \text{ Btu/llb}$$
$$= 64.996 \text{ MMBtu/h.}$$
$$\text{Total heat out} = 91.320 \text{ MMBtu/h.}$$
$$\text{Duty of radiant sect} = 91.320 - 52.819$$
$$= 38.501 \text{ MMBtu/h.}$$

For the convection section, the duty is calculated as follows:
Weight of saturated steam at 155 psig is 30,040 lbs/h.
Temperature of 155 psig steam = 368 °F.
From steam tables steam enthalpy = 1,195.9 Btu/lb.
Temperature of steam out = 500 °F.
Pressure of steam out = 125 psig.

From steam tables steam enthalpy = 1,275.3 Btu/lb.
Duty of convection side = 30,040 (1,275.3–1,195.9)
= 2.385 MMBtu/h.

Line 6. *Fluid.* This refers to the material flowing in the coil. In the case of this example, it will be hydrocarbons in the radiant coil and steam in the convection coil.

Line 7. *Flow rate.* This is the total flow rate in lbs per hour entering the respective section of the heater. Thus, for the radiant side, the figure will be 197,085 lbs/h, and for the convection side it will be 30,040 lbs/h.

Line 8. *Allowable pressure drop.* The process engineer enters the required pressure drop calculated from the hydraulic analysis of the system (see Appendix 14 and section "Pressure Drop Calculations)." This pressure drop is measured from the heater side of the inlet manifold downstream of the balancing control valves and the coil outlet downstream of the outlet manifold.

Line 9. *Allowable average flux.* This is usually a standard set by the company for its various heaters. In the example here, this value would be between 13,500 and 18,000 Btu/h sqft (a horizontal heater fired on both sides). It is specified as 15,000 Btu/h sqft for this example and refers only to the radiant section.

Line 10. *Maximum inside film temperature.* It is important to notify the heater manufacturer of any temperature constraint that is required by the process. In the case of this example, temperatures of the oil above 800 °F may lead to the oil cracking. Such a situation could adversely affect the performance of the downstream fractionation equipment, and therefore high temperatures in excess of 800 °F must be avoided. There is no constraint on the convection coil.

Line 11. *Fouling factor.* The fouling factors used in heat exchanger rating can be used here also. In this example, the radiant side would have a fouling factor of about 0.004 °F sqft h/Btu for the oil and 0.001 for the steam.

Line 12. *Residence time.* This becomes important when a chemical reaction of any kind takes place in the heater tubes. In the case of this example, this item does not apply. If the example were a thermal cracking heater or a visbreaker, the appropriate kinetic equations and calculations would be attached to the specification sheet to support this item.

Line 13–30. These are self-explanatory. The only comment here is that the data are quoted at the inlet or outlet conditions of temperature and pressure.

Line 31. *LHV.* The section of the specification sheet that follows deals with the characteristics of the fuel that will be used in the heater. This section is divided into oil and gas which are the usual fuels used in modern day processes. This item requires the "lower heating value" of the fuel. This can be read off charts such as Appendices 10 and 11 of this chapter.

Line 32. *HHV.* This is the other heating value data required by the heater designer. This "higher heating value" data can also be read off charts such as Appendices 10 and 11.

Line 33. *Pressure at burner.* This normally refers to the oil fuel and is measured at the heater fuel oil manifold.

Process Equipment for Petroleum Processing   1617

Line 34. *Temperature at the burner.* This item too is self-explanatory and these measurements are also taken at the respective manifolds.

Line 35. *Mol wt of gas.* This refers to the gas stream normally expected to be used. Obviously in practice, this will vary with the process operation from day to day.

Line 36. *Viscosity of the oil at the burner.* This viscosity is quoted at the burner temperature and may be arrived at from the two viscosity figures given later in the specification sheet. This is important for the best design or selection of the burner itself.

Line 37. *Atomizing steam.* This item calls for the temperature and pressure of the steam that will be used for atomizing the oil fuel. This is also required for the best design or selection of the oil burner.

Line 38. *Composition of gas.* This section requires the composition of the gas fuel in terms of mol percent. This is the normal expected fuel gas that will be used in the heater. If there is likely to be a wide variation in the quality of the fuel gas that will be used, this should be noted here as a range of two or even three compositions. This situation is particularly common in petroleum refining.

Line 46. *Properties of fuel oil.* This final section of the specification sheet requires details of the fuel oil that will be used. These details are:
Gravity of the oil at 60 °F (in °API)
Viscosity of the oil at 100 °F and at 210 °F
Flash point in °F

From the two viscosities, the "Refutus" graph can be used to determine the viscosity at any other temperature. This graph can be found in most data books that carry viscosity data. Note the viscosity requested in this section is in centistokes (kinematic viscosity). This is because most suppliers quote in centistokes. To convert to centipoise, multiply by the specific gravity (g/cm^3).

The flash point required is that determined by the Pensky-Marten method and is a measure of the oil's flammability.

This completes the explanation of the main body of the specification sheet. It represents the minimum data required to commence the sizing of the item. The last part of the specification sheet is given to "Remarks." In this section, the engineer should provide all the other data that may influence the design of the heater, e.g., Fig. 57.

Pressure Drop Calculations

Incompressible Flow Pressure Drop in Piping

The technical literature provides numerous methods for calculating pressure drop for incompressible flow. We will look here at a couple of relatively simple methods for estimating pressure drop. More sophisticated methods, sometimes proprietary, are normally used for plant design, but the approaches described here give good

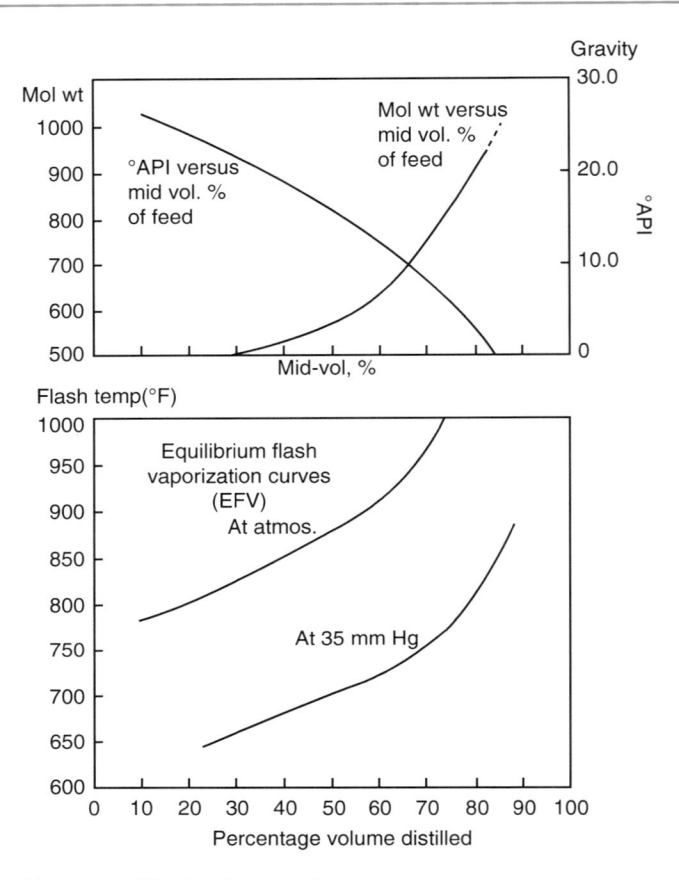

Fig. 57 Fired heater specification sheet attachment

order of magnitude estimates. For now, we will consider only horizontal piping pressure drops. In sections "Fittings and Piping Elements" and "Overall Pressure Drop, Including Elevation and Velocity Changes" we will include the effects of piping elements (fittings, valves, etc.) and elevation changes, respectively.

Appendix 14 illustrates the calculation of pressure drop through a piping system. This calculation is frequently required when specifying or analyzing process systems.

For our sign convention here, we will use a positive pressure drop to indicate a reduction in pressure through a system $(\Delta P = P_{inlet} - P_{outlet})$. A negative pressure drop then indicates the pressure increases through a system, which can occur due to liquid head, for instance). If you use a method for calculating pressure drop other than those described here, be sure you understand the sign convention that method uses. In some texts, the pressure drop is defined as outlet minus inlet.

Appendix 9 Charts

The charts in Appendix 9 of this chapter provide a good, quick method for calculating pressure drop for viscous fluids. This would apply to most hydrocarbons. These are based on the Darcy equation.

To use the charts, you would calculate the volumetric flow rate at actual conditions in US gpm (or barrels per hour, BPH). You also need the kinematic viscosity of the fluid in cSt or SSU and the nominal pipe size in inches. The frictional pressure drop is then read off the chart where the flow rate meets the viscosity for the appropriate pipe size. This dP is for Schedule 40, steel pipe, which is the most widely used. The pressure drop on the charts is in feet of liquid per 1,000 f. of pipe. To convert to psi, multiply the chart value by the fluid density in lb/cft and divide the result by 144 in.^{2}/ft^2. This would give psi/1000 ft.

The dP can be adjusted to other pipe schedules by multiplying the dP from the table by the ratio of d_2^5/d_1^5, where:

d_2 = internal diameter of the actual pipe schedule, inches.
d_1 = diameter of the Schedule 40 pipe used in the table, inches.

General Calculation Method

Another simple approach to estimating incompressible flow dP was adapted from a book by Carl Branan (1976).

In this approach, we again apply the Darcy equation in the form

$$\Delta P_{100} = (W/370)^2 \left[1/(\rho * d^5)\right] (f/0.0055) \tag{154}$$

where:

ΔP_{100} = pressure drop per 100 f. of pipe, psi/100 ft.
W = mass flow rate in lb/h.
ρ = fluid density in lb/cft.
d = piping ID in inches.
f = friction factor (assume 0.0055 unless you want to adjust).

The correlation assumes commercial steel pipe, Reynolds number of 10^5 or higher (turbulent flow), and a C factor of 110. This equation can also be used for compressible flows if the pressure drop is less than 10 % of the total pressure.

If you want to apply a friction factor correction:

- Calculate the Reynolds number (we will use Re here) for the flow:

$$Re = \rho V D_H / \mu = V D_H / \nu = Q D_H / \nu A \tag{155}$$

where:

$$\text{Re} = \text{Reynolds number (dimensionless)}.$$
$$\rho = \text{fluid density in lb/cft}.$$
$$V = \text{velocity in ft/s}.$$
$$D_H = \text{hydraulic diameter (ID) in ft}.$$
$$\mu = \text{dynamic viscosity in lb/ft/s}.$$
$$\nu = \mu/\rho = \text{kinematic viscosity in ft}^2/\text{s}.$$
$$A = \text{pipe internal cross} - \text{sectional area, ft}^2.$$
$$Q = \text{volumetric flow rate in cft/s}.$$

(156)

If you are working with metric numbers, the units for the above parameters are listed below with conversion factors. The Re comes out the same as long as you use consistent units.

$$\rho = \text{fluid density in kg/m}^3 \, (1 \text{ kg/m}^3 = 0.0624 \text{ lb/cft}).$$
$$V = \text{velocity in m/s} (1 \text{ m/s} = 3.281 \text{ ft/s}).$$
$$D_H = \text{hydraulic diameter(ID) in m} (1 \text{ m} = 3.281 \text{ ft} = 39.37 \text{ in}).$$
$$\mu = \text{dynamic viscosity in kg/m/s} (1 \text{ kg/m/s} = 0.672 \text{ lb/ft/s}).$$
$$v = \text{kinematic viscosity in m}^2/\text{s} (1 \text{ m}^2/\text{s} = 10.765 \text{ ft}^2/\text{s}).$$
$$A = \text{pipe internal cross} - \text{sectional area, m}^2 \, (1 \text{ m}^2 = 10.765 \text{ ft}^2).$$
$$Q = \text{volumetric flow rate in m}^3/\text{s} (1 \text{ m}^3/\text{s} = 35.31 \text{ cft/s}).$$

- The friction factor, f, can then be estimated using the Reynolds number and the appropriate equation below:

$$\text{Laminar flow}(Re < 2100) : f = 16/Re. \quad (157)$$
$$\text{Commercial pipe} : f = 0.054/Re^{0.2}. \quad (158)$$
$$\text{Smooth tubes} : f = 0.046/Re^{0.2}. \quad (159)$$
$$\text{Rough pipe} : f = 0.013. \quad (160)$$

To apply the calculation method, follow these steps:

1. Calculate the pipe ID and area in ft and ft^2, respectively, for the size of interest.
2. Calculate the flow rate in lb/h.
3. Calculate the fluid density in lb/cft.
4. Estimate the dynamic viscosity in lb/ft/s.
5. Calculate the fluid velocity in the pipe in ft/s at flowing pressure and temperature.
6. Calculate the Reynolds number for the specific case.
7. Estimate the friction factor using the appropriate equation.
8. Plug the variables into the pressure drop equation to get the psi/100 f. frictional pressure loss.

Process Equipment for Petroleum Processing   1621

Example Using the Two Incompressible Fluid Approaches
Let us consider the following:

- Flow rate at P and T = 900 gpm of liquid.
- The flowing density is 62 lb/cft.
- The pipe is new, nominal 10 in., Sch. 40, commercial pipe.
- The kinematic viscosity is 2.1 cSt.
- Estimate the pressure drop per 100 f. for these conditions.

Using Appendix 9 Charts
From the charts in Appendix 9, we find the chart for 10 in. pipe. Read down the left side to 900 gpm and then read across to 2.1 cSt viscosity. The chart value is 4.49 f. of liquid per 1,000 f. of pipe.
Converting to psi/100 ft:

$$4.49 \text{ ft}/1000 \text{ ft} \times 62 \text{ lb/c ft} \times 1/144 \text{ ft}^2/\text{in}^2 \times 100 \text{ ft}/1000 \text{ ft} = 0.193 \text{ psi}/100 \text{ ft.}$$

Using the General Calculation Approach
1. Pipe ID and area in ft and ft^2
   For 10″ Sch. 40 pipe: ID = 10.02″ (0.835 ft) and flow area = 0.5475 ft^2.
2. Flow rate in lb/h.

   $$900 \text{ gpm} \times 0.1337 \text{ cft/gal} \times 60 \text{ min/h} \times 62 \text{ lb/cft} = 447,628 \text{ lb/h.}$$

3. Fluid density is given as 62 lb/cft.
4. Estimate the dynamic viscosity in lb/ft/s.

   $$2.1 \text{c St}/(62/62.4 \text{ SG}) = 2.087 \text{ cP.}$$
   $$2.087 \text{ cP} \times (6.72 \times 10^{-4}) = 0.00140 \text{ lb/ft/s.}$$

5. Fluid velocity at flowing pressure and temperature.

   $$447,628 \text{ lb/h} \times 1\text{h}/3600\text{s} \times 1\text{cft}/62 \text{ lb} \times 1/0.5474 \text{ ft}^2$$
   $$= 3.663 \text{ ft/s.}$$

6. Reynolds number.

   $$Re = \rho V D/\mu = 62 \text{ lb/cft} \times 3.663 \text{ ft/s} \times 0.835 \text{ ft} \div 0.00140 \text{ lb/ft/s}$$
   $$= 1.35 \times 10^5.$$

7. Friction factor.
   We will use the f equation for commercial pipe, turbulent flow.

   $$f = 0.054/Re^{0.2} = 0.054/(1.35 \times 105)^{0.2} = 0.00508.$$

8. Pressure drop calculation.

   $$\Delta P_{100} = (447,628/370)^2 \left[1/\left(62 * (10.02)^5\right)\right] (0.00508/0.0055\ )$$
   $$= 0.216 \text{ psi}/100 \text{ ft.}$$

The pressure drops determined by the two methods compare fairly well:

Using the Appendix 9 charts: 0.193 psi/100 ft
Using the calculation approach: 0.216 psi/100 ft

The difference is about 10 %, which is within the error for these approaches to pressure drop. If it was found that the pressure drop or the velocity was too high or too low, a different pipe diameter would be selected and the calculations repeated until an acceptable combination of conditions was determined.

Compressible Flow Pressure Drop in Piping

For the compressible flow case, we will make the assumption that the flow is essentially isothermal, i.e., the flowing pressure drop does not produce a significant change in the fluid temperature. For most cases, this is valid.

As in the case of incompressible flow, we will consider only horizontal piping pressure drops. In sections "Fittings and Piping Elements" and "Overall Pressure Drop, Including Elevation and Velocity Changes," we will adjust for the effects of piping elements (fittings, valves, etc.) and elevation changes, respectively.

We will again examine two approaches to this calculation: the chart in Appendix 13 in this chapter and a generalized calculation approach.

Appendix 13 Pressure Drop Estimating

The chart in Appendix 13 provides an estimate using the "Weymouth formula." To use the chart, determine the volumetric flow rate in Mscfd (1,000 actual cubic feet per day) and the pipe diameter. Then, starting with the Y-axis, you move across the chart to the appropriate line for the pipe size chosen and read vertically downward to obtain the value of $P_1^2 - P_2^2$. P_1 is the upstream pressure in psia and P_2 is the downstream absolute pressure. The pressure drop is determined from the X-parameter.

The chart is based on a specific gravity of 0.9 (air = 1.0), 90 °F, and 14.65 psia base pressure. The results may need to be adjusted to other conditions, but the chart provides a good first estimate of dP.

General Calculation Method

For a general calculation approach, we will again follow an approach described by Branan in his book (1976).

We will apply the general equation for compressible flow:

$$\Delta P = 2P_1/(P_1 + P_2)\{0.323[(fL/d) + \ln(P_1/P_2)/24]SG_1U_1^2\} \quad (161)$$

where:

P_1 = initial pressure, psia (absolute).
P_2 = final pressure, psia.
SG_1 = specific gravity of vapor vs water = $0.0015 \, M \, P_1/T$.
d = pipe ID, in.
U_1 = upstream velocity, ft/s. (162)
f = friction factor(assume 0.005 for estimates).
L = length of pipe, ft.
M = molecular weight of gas.
T = upstream temperature, °R.

Example Using the Two Compressible Fluid Approaches

Let's consider the following case:

- 10.5 MMscfd of vapor with a molecular weight of 26.0 (scfd are at 60 °F, 1 Atm).
- Piping is 8 in., Sch. 40 nominal commercial pipe.
- Initial temperature is 125 °F.
- Initial pressure is 260 psig.
- Estimate the pressure drop per 100 f. for these conditions.

We will do the estimate using both the Appendix 13 chart and the calculation approaches.

Using Appendix 13 Chart

$$Mcfd(chart) = Mcfd(base) \times P/14.65(SG/0.9 \times T/550 \times Z/1)^{0.5}. \quad (163)$$

Now:

Our $Mcfd$ in this case is 10,500 scfd.
P is 14.7 for our scf basis.
SG of the gas is 26 MW/29 = 0.9.
T for the scf basis is 520 °R.
Assume $Z = 1.0$.

So the Mcfd to use on the chart is

$$Mcfd(chart) = 10,500 \times 14.7/14.65 \times (0.9/0.9 \times 520/550 \times 1.0/1)^{0.5}$$
$$= 10,244.$$

Picking 10,200 for the Y-axis of the chart and moving across to the line for 8″, Sch. 40 pipe, the value of $P_1^2 - P_2^2$ is ~1,000.

We know P_1 is 274.7, so solving for P_2 gives 272.9 psia. Pressure drop is the difference or 1.8 psi/1,000 f. or 0.18 psi/100 ft.

Using the General Calculation Method
1. $P_1 = 260 + 14.7 = 274.7$ psia.
2. $P_2 = P_1$ – initially assumes negligible pressure drop change. We will test this assumption later.
3. Absolute temperature $= 125\,°F + 460 = 585\,°R = T_1$.
4. Use 100 f. of pipe as a basis $= L$.
5. Gas-specific gravity versus water.

$$SG_1 = 0.0015(26\text{ mw})(274.7\text{ psia})/585\,°R = 0.0183.$$

6. Pipe diameter $= 7.981$ in. and flow area $= 0.3474$ sqft.
7. Assume a friction factor of 0.005 for estimating work.
8. Calculate the velocity:

$$\begin{aligned}U_1 &= 10.5 \times 10^6 \times (14.7\text{ psia}/274.7\text{ psia}) \times (585\,°R/520\,°R) \\ &\quad \times 1\text{ day}/24\text{ h} \times 1\text{ h}/3600\text{ s} \div 0.3474\text{ ft}^2 \\ &= 21.1\text{ ft/s}.\end{aligned}$$

9. Calculate the pressure drop using the equation:

$$\begin{aligned}\Delta P &= 2P_1/(P_1+P_2)\{0.323[(fL/d) + \ln(P_1/P_2)/24]SG_1 U_1^2\} \\ &= 1(274.7\text{ psia})/(274.7\text{ psia} + 274.7\text{ psia}) \\ &\quad \times \{0.323[((0.005 \times 100\text{ ft})/7.981) + \ln(274.7\text{ psia}/274.7\text{ psia})/24] \\ &\quad \times (0.0183\text{ SG})(21.1)^2\} = 1\{0.323[0.0626 + 0/24](0.0183)(445.2)\} \\ &= 0.17\text{ psi}/100\text{ ft}.\end{aligned}$$

This is a small enough pressure drop that there is no need for iteration. If the pressure drop was significant, this dP would have been used for the difference between P_1 and P_2 in a second iteration.

Comparing Results of Compressible Flow
The answers we got using the two methods of calculation varied by only 0.01 psi per 100 f. or about 6 %. This is about as accurate as you can expect from these types of estimates. For detailed design work, more sophisticated methods are normally applied to increase accuracy.

You can also break a piping system into shorter segments and perform the dP calculation for each segment to improve accuracy.

Process Equipment for Petroleum Processing   1625

Fittings and Piping Elements

We have so far dealt with pressure drops in straight, horizontal piping, but seldom is piping all horizontal or all straight. There are fittings, elbows, and valves. We will look at the simplest approach to allowing for these elements next, starting with pipe fittings, valves, and the like. In the section after this, we will review the impacts of elevation and velocity changes.

The most common and simplest approach to account for fittings in piping uses equivalent lengths defined for each type of element. In this approach, for instance, the equivalent length of an 8 in standard elbow in a line is about 20 ft. The nomograph of Appendix 12 to this chapter provides a fast, easy method of determining the equivalent lengths for many types of fittings and pipe configurations.

An alternative to Appendix 12 is Table 35 translated from the GPSA Engineering Data Book after Branan (1976; GPSA 1987).

To determine the total pressure drop due to valves and fittings:

1. Count the numbers of each type of valve and fitting.
2. Look up the equivalent length of each type of fitting in Appendix 12 or in Table 35.
3. Multiply the number of each type of element by the equivalent length of that element.
4. Add the total equivalent lengths of all the elements. The answer, in the units used here, would be in equivalent feet of pipe.
5. Multiply the equivalent length by the calculated ΔP_{100} for straight pipe that has been calculated using the other described methods for single- or two-phase flows to get the total dP due to the piping elements.
6. The total dP for the pipe elements would be added to the dP for the total straight pipe to get the overall system pressure drop from frictional losses.

We will review an example of this that also includes elevation and velocity changes a little later in the next section.

Overall Pressure Drop, Including Elevation and Velocity Changes

Appendix 14 provides an example of a system pressure drop calculation. Here we will review some background and another approach.

Piping is seldom all at the same elevation and piping diameters can be changed along the way. If we define pressure drop as the difference in pressure between two points, then the overall pressure drop for a system would be defined as

$$\Delta P_{overall} = \Delta P_{friction} + \rho \Delta Z + \rho \left(\Delta U^2 / 2g_c \right) \qquad (164)$$

Table 35 Equivalent lengths of valves and fittings in feet (GPSA 1987)

Nominal pipe size, in.	Gate valve or ball check valve	Angle valve	Swing check valve	Plug cock valve	Gate or ball valve	45° elbow Welded	45° elbow Threaded	Short rad 90° elbow Welded	Short rad 90° elbow Threaded	Long rad 90° elbow Welded	Long rad 90° elbow Threaded	Hard tee Welded	Hard tee Threaded	Soft tee Welded	Soft tee Threaded	90° Mitered bend 2 miters	90° Mitered bend 3 miters	90° Mitered bend 4 miters	Enlargement Sudden d/D=1/4	Enlargement Sudden d/D=1/2	Enlargement Sudden d/D=3/4	Enlargement Std reducer d/D=1/2	Enlargement Std reducer d/D=3/4	Contraction Sudden d/D=1/4	Contraction Sudden d/D=1/2	Contraction Sudden d/D=3/4	Contraction Std reducer d/D=1/2	Contraction Std reducer d/D=3/4
1½	55	26	13	7	1	1	2	3	5	2	3	8	9	2	3				5	3	1	4	1	3	2	1	1	
2	70	33	17	14	2	2	3	4	5	3	4	10	11	3	4				7	4	1	5	1	3	3	1	1	
2½	80	40	20	11	2	2		5		3		12		3					8	5	2	6	2	4	3	2	2	
3	100	50	25	17	2	2		6		4		14		4					10	6	2	8	2	5	4	2	2	
4	130	65	32	30	3	3		7		5		19		5					12	8	3	10	3	6	5	3	3	
6	200	100	48	70	4	4		11		8		28		8					18	12	4	14	4	9	7	4	4	1
8	260	125	64	120	6	6		15		9		37		9					25	16	5	19	5	12	9	5	5	2
10	330	160	80	170	7	7		18		12		47		12					31	20	7	24	7	15	12	6	6	2

12	400	190	95	170	9	9	22	14	55	14	28	21	20	37	24	8	28	8	18	14	7	7	2
14	450	210	105	80	10	10	26	16	62	16	32	24	22	42	26	9			20	16	8		
16	500	240	120	145	11	11	29	18	72	18	38	27	24	47	30	10			24	18	9		
18	550	280	140	160	12	12	33	20	82	20	42	30	28	53	35	11			26	20	10		
20	650	300	155	210	14	14	36	23	90	23	46	33	32	60	38	13			30	23	11		
22	688	335	170	225	15	15	40	25	100	25	52	36	34	65	42	14			32	25	12		
24	750	370	185	254	16	16	44	27	110	27	56	39	36	70	46	15			35	27	13		
30				312	21	21	55	40	140	40	70	51	44										
36						25	66	47	170	47	84	60	52										
42						30	77	55	200	55	98	69	64										
48						35	88	65	220	65	112	81	72										
54						40	99	70	250	70	126	90	80										
60						45	110	80	260	80	190	99	92										

where:

$\Delta P_{overall}$ = overall pressure drop, lb/ft^2 (Note this is per square foot).
$\Delta P_{friction}$ = frictional pressure loss, lb/ft^2 (calculated for pipe and pipe elements using single- or two-phase methods).
ρ = flowing fluid density, lb/cft.
ΔZ = net elevation change, ft.
ΔU = net change in flowing fluid velocity, ft/s.
g_c = gravitational constant, 32.17 f. lb$_m$/(lb$_f$ s^2).

The application of the above equation, including all the various elements, is best illustrated by an example.

Suppose we have the following case:

- 250 gpm water flowing at 60 °F
- 10 in., Sch. 40 pipe
- From the piping layout, we determine:
  - The net elevation change is 25 f. upward.
  - There are three standard elbows in the line.
  - There are two fully open gate valves.
  - The line ends in an empty tank (sudden enlargement).
  - There are 265 f. of straight line.
- What is the total system pressure drop for this configuration?

1. Determine the water density = approximately 62.4 lb/cft.
2. Determine flow rate in lb/s and cft/s:

$$250 \text{ gpm} \times 8.34 \text{ lb/cft} \times 1 \min/60 \text{ s} = 34.75 \text{ lb/s}.$$
$$34.75 \text{ lb/s}/62.4 \text{ lb/cft} = 0.557 \text{ cft/s}.$$

3. Velocity in line:

$$0.557 \text{ cft/s}/0.5475 \text{ ft}^2 (\text{flow area of } 10'' \text{ pipe}) = 1.02 \text{ ft/s or fps}.$$

4. In this case, we are going from 1.02 fps velocity to zero in the tank, so the $\Delta U = 1.02$ fps and the velocity term of the pressure drop equation is

$$62.4 \text{ lb/ft3} \times (1.02 \text{ ft/s})^2 / (2 \times 32.17 \text{ ft lb}/(\text{lb s}^2)) = 1.01 \text{ lb/ft}^2.$$

This is less than 0.1 psi pressure drop; so, it is negligible in this case.

5. The elevation change is +25 ft, so the elevation change term is

$$25 \text{ ft} \times 62.4 \text{ lb/cft} = 1560 \text{ lb/sqft} (10.8 \text{ psi}).$$

6. Using the techniques described earlier for single-phase, incompressible flow, the ΔP_{100} frictional loss for straight pipe is determined to be 0.019 psi or 2.75 lb/ft² per 100 f. of pipe. The calculation is not detailed here.
7. Calculating the equivalent line length for frictional loss, we have:

Straight pipe	265 ft
3 std. elbows (26 f. each, Appendix 12)	78 ft
2 open gate valves (6 f. each, Appendix 12)	12 ft
1 sudden enlargement (15 ft, Table 35)	15 ft
Total equivalent line length (sum)	370 ft

8. Frictional pressure loss overall is

$$2.75 \text{ lb/ft}^2/100 \text{ ft} \times 370 \text{ equiv. ft}/100 = 10.2 \text{ lb/ft}^2 = \Delta P_{friction}.$$

9. Overall system pressure drop is then

$$\Delta P_{overall} = \Delta P_{friction} + \rho \Delta Z + \rho(\Delta U^2/2g_c)$$
$$= 10.2 \text{ lb/ft}^2 + 1560 \text{ lb/ft}^2 + 0 \text{ lb/ft}^2$$
$$= 1570.2 \text{ lb/ft}^2 \text{ or } 10.9 \text{ psi for the system}.$$

In this case, the elevation change dominated the overall pressure drop across the system. This would not necessarily be the case if the flow rate were significantly higher or the piping system was more complex and long.

Also, note that if the final destination tank was below the start of the line, the pressure drop would be negative or the pressure would actually be higher at the tank than at the start of the line (which would require introduction of some sort of flow control).

If the tank were not empty, the tank level would have to be added to the elevation change for the calculation.

Two-Phase (Liquid + Vapor) Pressure Drop and Flow Regimes in Piping

So far, we have just dealt with single-phase flows, but most petroleum refining processes involve two-phase, mixed liquid/gas flows. Here we will discuss the general approach to estimating pressure drop for two-phase horizontal flows using the method developed by Ovid Baker and described in an excellent article by Robert Kern (1969). The approach here is a simplification described by A.K. Coker (1990). There are other approaches also available with some simplifications to enable easy computing. We will focus here on Baker's approach, however.

We begin by defining the Baker parameters B_x and B_y:

$$B_x = 531(W_L/W_G)\left[\left((\rho_L\rho_G)^{0.5}\right)/\left(\rho_L^{0.667}\right)\right]\left[\left(\mu_L^{0.333}\right)/\sigma_L\right] \quad (165)$$

$$B_y = 2.16(W_G/A)\left[1/\left((\rho_L\rho_G)^{0.5}\right)\right] \quad (166)$$

where:

B_x = Baker chart X-parameter.
B_y = Baker chart Y-parameter.
W_L, W_G = flow rate of liquid and gas phases, respectively, lb/h.
ρ_L, ρ_G = densities of liquid and gas phases, respectively, at op. conditions lb/cft.
μ_L = viscosity of liquid phase in cP at operating conditions.
σ_L = surface tension of liquid at operating conditions in dynes/cm.
A = pipe flow area, ft^2.

A log-log plot of the Baker parameters defines the type of flow occurring in the piping or the flow regime as shown in Fig. 58. Within each flow regime, a different approach to the pressure drop applies.

The types of two-phase flow that are generally identified are:

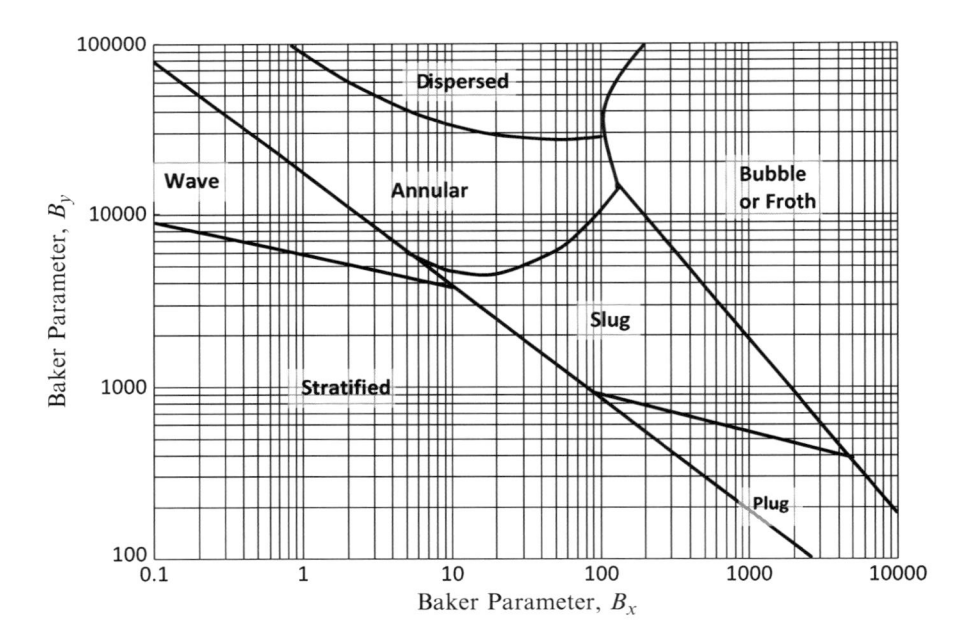

Fig. 58 Two-phase flow regime map for horizontal piping

- Dispersed – In this regime, nearly all the liquid is entrained as a spray in a bulk gas phase. This regime may also be referred to as mist flow. The gas and liquid move at the same velocity. This type of flow works well in upward flow as well as horizontal flow.
- Bubble or froth – Bubbles of gas are dispersed in a primarily liquid phase with the gas moving about the same velocity as the liquid. This regime works for upward or horizontal flows.
- Annular – There is a liquid film flowing along the wall of the pipe with gas or dispersed flow at a higher velocity in the center core of the pipe. If properly sized, operation in this regime is fine for upward or horizontal flows.
- Stratified – This is normally a horizontal flow phenomenon when the gas and liquid separate long horizontal pipes. The gas moves at a higher velocity than the liquid, with the liquid flowing along the bottom of the pipe. The interface between the liquid and gas is relatively smooth. This is a regime that is a concern for vertical flow because it can cause slugging or pipe hammer.
- Wave – This is similar to stratified flow, but the gas velocity is higher, producing a wavy interface between gas and liquid layers. Again, this regime can give rise to problems when flow is vertical.
- Slug – This is a more extreme form of wave flow where the waves are periodically picked up and "slugged" through the pipe. This regime is undesirable as it causes excessive pipe movement and hammer that is potentially damaging.
- Plug – Alternate plugs of liquid and gas move through the pipe. Again, this is an undesirable regime generally, that can result in damage.

Now, to pull this information into a pressure drop value for two-phase flow, we will define some additional parameters:

- The basic two-phase flow equation is

$$\Delta P_{100,2\phi} = \Delta P_{100,G} \times \Phi^2 \tag{167}$$

where:

$\Delta P_{100,2\phi}$ = two-phase pressure drop, convenient units.
$\Delta P_{100,G}$ = pressure drop for the gas phase only, consistent units.
Φ = two-phase flow modulus that depends on the flow regime (see below).
- Lockhart-Martinelli modulus, X

$$X = (\Delta P_L/\Delta P_G)^{0.5} \tag{168}$$

where:

ΔP_L, ΔP_G = pressure drop for each phase in the pipe separately, convenient units (calculate this using the single-phase approach discussed previously).

- Two-phase flow modulus, Φ, depends on flow regime. The relevant empirical equations are:
  - Bubble or froth:

$$\Phi = (14.2X^{0.75})/(WL/A)^{0.1}. \quad (169)$$

  - Plug:

$$\Phi = (27.315X^{0.855})/(WL/A)^{0.17}. \quad (170)$$

  - Stratified:

$$\Phi = (15400X)/(WL/A)^{0.8}. \quad (171)$$

  - Slug:

$$\Phi = (1190X^{0.815})/(WL/A)^{0.5}. \quad (172)$$

  - Annular:

$$\Phi = (4.8 - 0.3125d)X^{(0.343-0.021d)}. \quad (173)$$

  For pipe diameters greater than 10'', use 10'' as the basis.
  - Dispersed or spray flow:

$$\Phi = \exp\left[1.4659 + 0.49138 \ln X + 0.04887(\ln X)^2 - 0.000349(\ln X)^3\right] \quad (174)$$

where:
  W = total L + G mass flow in lb/h at operating conditions.
  L = equivalent piping length, ft (use 100 for pressure drop per 100 ft).
  A = pipe flow area, ft^2.
  X = Lockhart-Martinelli modulus as defined above.
- For wavy flow, a slightly different approach is used:

$$H_x = (W_L/W_G)(\mu_L/\mu_G) \quad (175)$$

$$\ln(f_H) = 0.211 \ln(H_x) - 3.993 \quad (176)$$

$$\Delta P_{100, 2\phi} = \left(0.000336 f_H (W_G)^2\right)/(d^5 \rho_G) \quad (177)$$

where:
  W_L, W_G = as defined above, lb/h.
  μ_L, μ_G = viscosities of the liquid and gas phases, cP at op. conditions.
  d = pipe ID in inches.
  ρ_G = gas density, lb/cft.

Process Equipment for Petroleum Processing

So for the general two-phase pressure drop calculation, the following approach is defined:

1. Calculate the Baker parameters, B_x and B_y.
2. Determine the flow regime.
3. Calculate the pressure drop per 100 f. for each phase as though that was the only phase flowing in the pipe ($\Delta P_{L,100}$, $\Delta P_{G,100}$). You can use any method previously described or an alternate, preferred method.
4. Calculate the Lockhart-Martinelli modulus, X.
5. Calculate the appropriate flow modulus value, Φ (or follow the wavy flow approach).
6. For flow regimes other than wave flow, use the calculated flow modulus in the equation below to get the final two-phase pressure drop per 100 ft.

$$\Delta P_{100,2\phi} = \Delta P_{100,G} \times \Phi^2. \tag{167}$$

7. The remaining calculations to get total system pressure drop follow the same pattern as for single phases using equivalent lengths of pipe, elevation changes, and velocity changes. For elevation changes, you will have to estimate the average flowing density. This may be a little more complex for flows where there is significant phase separation.

We will illustrate the two-phase pressure drop calculation with an example.
Two-phase pressure drop example
Let's say we are given the following from a simulation:

- Two-phase flow in an 8 in., Sch. 40 pipe.
- Liquid flow rate = W_L = 59,033 lb/h.
- Vapor flow rate = W_G = 9,336 lb/h.
- Liquid density at flowing conditions = ρ_L = 31.2 lb/cft.
- Vapor density at flowing conditions = ρ_G = 1.85 lb/cft.
- Liquid surface tension at flowing conditions = σ_L = 5.07 dynes/cm.
- Liquid viscosity at flowing conditions = μ_L = 0.11 cP.
- Initial pressure = P_I = 338 psig.
- Temperature = T = 250 °F.
- Molecular weight of the gas is 40 = M.

Now we follow the solution step-by-step:

1. Calculate the Baker parameters:
  (a) $B_x = 531(W_L/W_G)\left[\left((\rho_L\rho_G)^{0.5}\right)/\left(\rho_L^{0.667}\right)\right]\left[(\mu_L^{0.333})/\sigma_L\right]$
  $= 531(59,033/9,336)\left[(31.2*1.85)^{0.5}/31.2^{0.667}\right]\left[0.11^{0.333}/5.07\right]$
  $= 531(6.323)[0.766][0.0945].$
  $B_x = 243.$

(b)
$$B_y = 2.16(W_G/A)\left[1/\left((\rho_L\rho_G)^{0.5}\right)\right]$$
$$= 2.16(9,336/0.3474 \text{ ft}^2)\left[1/(31.2 * 1.85)^{0.5}\right]$$
$$= 2.16(26,874)[0.1316].$$
$$B_y = 7,639.$$

2. Looking at the Baker chart (Fig. 58), this puts us on the boundary between slug and bubble/froth flow. Let us say the regime is bubble/froth. (The regimes are not perfectly defined anyway.)
3. Calculate the single-phase pressure drops for both the liquid and gas using one of the previous methods. This calculation is not detailed here, but the resulting pressure drops are

$$\Delta P_{L,100} = 0.025 \text{ psi}/100 \text{ ft for liquid phase.}$$
$$\Delta P_{G,100} = 0.0108 \text{ psi}/100 \text{ ft for gas phase.}$$

4. Calculate the Lockhart-Martinelli flow modulus:

$$X = (0.025/0.011)^{0.5} = 1.51.$$

5. Calculate the two-phase flow modulus using the formula for froth/bubble flow.

$$\Phi = \left(14.2X^{0.75}\right)/(WL/A)^{0.1}$$
$$= \left(14.2 * 1.51^{0.75}\right)/[((59,033 + 9,336) * 100)/0.3474]^{0.1}$$
$$= 19.343/5.363.$$
$$\Phi = 3.607.$$

6. Calculate the two-phase pressure drop.

$$\Delta P_{100,2\phi} = \Delta P_{100,G} \times \Phi^2$$
$$= 0.0108 \times (3.607)^2$$
$$= 0.141 \text{ psi}/100 \text{ ft.}$$

7. No additional data were given, but from here the solution follows the previous discussion for determining overall system pressure drop.

The calculations for two-phase pressure drops are only approximate order of magnitude. For design, more detailed, proprietary methods are normally recommended.

General Comments on Pressure Drops in Beds of Solids

The next few sections of this handbook provide techniques to estimate pressure drop through beds of solids, like reactors or sorbent beds (adsorbents, absorbents). There are a few types of cases to consider:

- Single-phase, cocurrent upflow or downflow
- Two-phase, cocurrent upflow or downflow
- Two-phase counterflow
- Moving solids bed flow

For our purposes here, we will focus on the concurrent flow cases, which are the most common. Discussion of countercurrent pressure drop is beyond the scope of this book. We will only briefly touch on flows where the solids bed moves with some references. The techniques here are also not intended to apply to distillation or absorber tower packings, which are discussed in section "Vessels and Towers" of this chapter.

The approaches described here are for approximate dP only (+/−20 to 30 %). If used for design, it would be advisable to perform some experimental work to fine-tune the parameters required in the calculation. Many licensors have more accurate, proprietary correlations for design. Still, if you have an operating unit with known pressure drops at different rates, this approach could be fine-tuned to reflect the actual dP and can be used for predicting system dP response.

Single-phase flow results are combined together to determine two-phase flow results using various empirical correlations, similar to the way single-phase piping flow pressure drops are combined into two-phase dPs for piping.

Fixed bed dP calculations are largely empirical in nature. The two most critical factors are the void fraction of the bed and the equivalent particle diameter. Unfortunately, these are the two most difficult parameters to determine.

Flow direction plays a role in these techniques. As with piping, if the flow is downward through a bed, the mass of the flowing materials reduces the pressure drop through the bed. Conversely, if the flow is upward, the mass of the fluids increases the pressure drop by fighting gravity.

We will begin with single-phase flow through beds of solids.

Single-Phase Pressure Drop in Beds

The primary equation used for single-phase pressure drop in beds is the Ergun equation. For our purposes here, we will use the form:

$$(\Delta P/L)_{friction} = [(\alpha(1-\varepsilon)/Re_B + \beta][G^2/(g_c D_p \rho(\varepsilon^3/(1-\varepsilon)))] \quad (178)$$

where:

$(\Delta P/L)_{friction}$ = frictional pressure loss in lb/ft² per foot of bed.

α = Ergun constant that depends on size, shape, and packing of bed.

β = Ergun constant that depends on size, shape, and packing of bed.

ε = void fraction of the bulk bed, dimensionless.

Re_B = bed Reynolds number = $D_p G/\mu = D_p V \rho/\mu$.

G = mass flux through the bed = lbs/s ÷ empty vessel flow area.

g_c = gravitational constant = $32.17 \text{ ft} * \text{lb}/(\text{lb} * \text{s}^2)$.

D_p = equivalent particle diameter in feet = $6/S_v$.

ρ = fluid density, lb/cft at operating conditions.

S_v = particle average surface area/volume, ft²/cft

(for shaped particles, use smoothed/projected particle dimensions).

V = superficial velocity of fluid in empty vessel, ft/s.

μ = fluid viscosity in lb/(ft*s), $(= \text{cP} \times 6.72 \times 10^{-4})$.

(179)

Some typical values of α and β are in Table 36.

To obtain the overall pressure drop in a bed, the fluid density must be factored in, so the final overall pressure drop through the bed itself would be

$$(\Delta P/L)_{bed} = (\Delta P/L)_{friction} - / + \rho \qquad (180)$$

where:

$(\Delta P/L)_{bed}$ = overall bed pressure drop, lb/ft².
$(\Delta P/L)_{friction}$ = frictional pressure drop from above.
ρ = fluid density at flowing conditions, lb/cft (be careful of the sign).
  The density is subtracted if flow is downward (negative sign).
  The density is added if flow is upward (positive sign).

Table 37 illustrates some typical void fractions for various solids shapes that are randomly packed in a bed. These values should be decreased by 10–15 % if the bed has been dense loaded.

Use of these equations is best illustrated by an example:

Example single-phase bed pressure drop calculation

Consider a downflow reactor in vapor phase. The following parameters are given:

Process Equipment for Petroleum Processing

Table 36 Typical values of Ergun α and β for various shapes (for design work, these should be determined experimentally for the specific solid being used)

Size/shape	α	β
1/16" extrudate	350	2.83
1/8" extrudate	350	1.75
3/8" sphere	118	1.00
3/8" Raschig ring	266	2.33

Table 37 Typical values of void fraction for various shapes (For design work, these should be determined experimentally for the specific solid being used)

Shape	ε
Cylinder	0.33
Extrudate	0.35
Sphere, ball	0.15
Rings, shapes	~0.42
High void toppings	~0.5

- Catalyst was sock loaded.
- The main catalyst bed is 1/8" diameter cylindrical extrudate.
- The average L/D (length-to-diameter ratio) for the catalyst is 4.0.
- Reactor internal diameter is 96 in.
- Vapor flow rate is 150,000 lb/h.
- Vapor density is 2.10 lb/cft.
- Vapor viscosity is 0.02 cP.
- Bed depth is 35 ft.
- What is the dP through the reactor bed, assuming negligible density change?

We follow this sequence for the calculation:

1. Assemble the terms required for the Ergun equation:
   (a) For 1/8" extrudate, sock loaded, we pick the following from Tables 36 and 37:

$$\alpha = 350.$$
$$\beta = 1/75.$$
$$\varepsilon = 0.35.$$

   (b) Calculate the equivalent particle diameter, D_p:
   - Cross-sectional area of particle

$$(0.125 \text{ in.}/12 \text{ in.}/\text{ft})^2 \times \pi/4 = 8.522 \times 10^{-5} \text{ ft}^2.$$

   - Circumference of particle (assume round)

$$(0.125 \text{ in.}/12 \text{ in.}/\text{ft}) \times \pi = 3.272 \times 10^{-2} \text{ ft}.$$

- Superficial surface area of particle

  Ends : $2 \times 8.52 \times 10^{-5}$ ft^2 = 1.704×10^{-4} ft^2.
  Sides : 3.272×10^{-2} ft $\times$ ((0.125 in $\times$ 4.0)/12 in/ft) = 1.363×10^{-3} ft^2.
  Total area : $1.704 \times 10^{-4} + 1.363 \times 10^{-3} = 1.533 \times 10^{-3}$ ft^2.

- Average volume of a particle

  8.52×10^{-5} ft^2(end) $\times$ ((0.125 in $\times$ 4.0)/12 in/ft) = 3.551×10^{-6} cft.

- Surface to volume ratio

  $S_v = 1.533 \times 10^{-3} / 3.551 \times 10^{-6} = 4.317 \times 10^2$ ft^{-1}.

- $D_p = 6/S_v = 6/4.317 \times 10^2$ ft^{-1} = 1.39×10^{-2} ft(~ 0.167 in.).

(c) Calculate the mass flux for the reactor

Reactor diameter of 96 in. $\rightarrow$ Flow area = 50.67 ft^2
Flow rate = 150,000 lb/h or 41.67 lb/s.
Mass flux, $G = (41.67 \text{ lb/s})/(50.27 \text{ ft}^2) = 0.8289$ lb/ft^2/s.

(d) Convert fluid viscosity to the desired units

$$0.02 \text{ cP} \times 6.72 \times 10^{-4} = 1.34 \times 10^{-5} \text{lb/ft/s} = \mu.$$

(e) Superficial velocity, V

$$V = (150{,}000 \text{ lb/h})/(2.10 \text{ lb/cft})/(50.27 \text{ ft}^2)/3600 = 0.395 \text{ ft/s}.$$

2. Calculate the Reynolds number for the bed

$$Re_B = D_p G/\mu = (1.39 \times 10^{-2} \text{ft})(0.8289 \text{ lb/ft}^2/\text{s})/(1.34 \times 10^{-5} \text{lb/ft/s})$$
$$= 859.8.$$

3. Calculate the frictional pressure drop using the Ergun equation

$$(\Delta P/L)_{friction} = [(\alpha(1-\varepsilon)/Re_B + \beta] [G^2/(g_c D_p \rho(\varepsilon^3/(1-\varepsilon)))]$$
$$= \left[(350(1-0.35)/859.8 + 1.75](0.8289)^2/((32.17)(1.39 \times 10^{-2})(2.10) \right.$$
$$\left. \times (0.35^3/(1-0.35)) \right] = 22.35 \text{ lb/ft}^2/\text{ft } (0.155 \text{ psi/ft depth}).$$

4. Calculate total bed pressure drop

$$(\Delta P/L)_{bed} = (\Delta P/L)_{friction} - \rho$$
$$= 22.35 \text{ lb/ft}^2/\text{ft} - 2.10 \text{ lb/cft/ft}$$
$$= 20.25 \text{ lb/ft}^2/\text{ft}(0.141 \text{ psi/ft of bed}).$$
$$\text{Total dP} = 20.25 \text{ lb/ft}^2/\text{ft} \times 35 \text{ ft depth} = 709 \text{ lb/ft}^2 (4.9 \text{ psi}).$$

This excludes any fouling allowances or internals.

Two-Phase Pressure Drop in Beds

For two-phase pressure drop in solid beds, we will adopt the technique developed by Larkins, White, and Jeffery in 1961. There are other methods, but this approach is relatively simple to apply and it uses the Ergun equations already discussed.

The method developed by Larkins et al. follows these steps:

1. Calculate the $(\Delta P/L)_{friction}$ for each phase (liquid and gas) using the Ergun equation or another approach as though that phase was the only phase flowing in the reactor. This is similar to the approach in two-phase piping pressure drop.
2. Calculate the X modulus, similar to the Lockhart-Martinelli modulus in pipe flows:

$$X = \left[(\Delta P/L)_{fric, L}/(\Delta P/L)_{fric, G}\right]^{0.5}. \tag{181}$$

3. Calculate the frictional two-phase pressure drop in lb/ft²/ft of bed using the modulus and the following empirical correlation:

$$(\Delta P/L)_{fric, 2\phi} = \left[(\Delta P/L)_{fric, L} + (\Delta P/L)_{fric, G}\right] \left[10^{(0.416/(\text{Sq}(\log X)+0.666)]}. \tag{182}$$

4. Calculate the liquid saturation factor, S

$$S = 10^{(-0.774+0.525\log X - 0.109 \text{Sq}(\log X))}. \tag{183}$$

5. Calculate the flowing mixture density at operating conditions, ρ_{mix}

$$\rho_{mix} = \rho_L S + \rho_G (1 - S). \tag{184}$$

6. Adjust the pressure drop for density of the mixture

$$(\Delta P/L)_{total, 2\phi} = (\Delta P/L)_{fric, 2\phi} - / + \rho_{mix} \text{ lb/ft}^2/\text{ft of depth}. \tag{185}$$

As before:
- The sign for the density is negative for downflow.
- The sign of the density is positive for upflow.

So let's consider an example.
Example two-phase bed pressure drop calculation
We will simplify this calculation somewhat by assuming we have already calculated the single-phase frictional losses through the bed for gas and liquid individually. Our input data are then:

$$(\Delta P/L)_{fric,L} = 2.00 \text{ lb/ft}^2/\text{ft}(0.014 \text{ psi/ft}).$$
$$(\Delta P/L)_{fric,G} = 20.1 \text{ lb/ft}^2/\text{ft}(0.14 \text{ psi/ft}).$$
$$\rho_L = 40 \text{ lb/cft at flowing conditions}.$$
$$\rho_G = 2.0 \text{ lb/cft at flowing conditions}.$$

So following the step-by-step approach outlined above:

1. Calculate the $(\Delta P/L)_{fric}$ for each phase:
   These were given as 2.00 and 20.1 lb/ft²/ft depth for liquid and gas phases, respectively.
2. Calculate the X modulus

$$X = \left[(\Delta P/L)_{fric,L}/(\Delta P/L)_{fric,G}\right]^{0.5}$$
$$= (2.00/20.1)^{0.5} = 0.315.$$

3. Calculate the frictional two-phase pressure drop

$$(\Delta P/L)_{fric,2\phi} = \left[(\Delta P/L)_{fric,L} + (\Delta P/L)_{fric,G}\right]\left[10^{(0.416/(\text{Sq}(\log X)+0.666)}\right]$$
$$= [(2.00) + (20.1)]\left[10^{(0.416/(\text{Sq}(\log(0.315))+0.666)}\right]$$
$$= (22.01)(10^{0.453}) = 62.5 \text{ lb/ft}^2/\text{ft depth}(0.43 \text{ psi/ft}).$$

4. Calculate the liquid saturation factor, S

$$S = 10^{(-0.774+0.525\log X - 0.109\text{Sq}(\log X))}$$
$$= 10^{(-0.774+0.525\log(0.315)-0.109\text{Sq}(\log(0.315)))}$$
$$= 10^{-1.065} = 0.0861.$$

5. Calculate the flowing mixture density at operating conditions

$$\rho_{mix} = \rho_L S + \rho_G(1-S)$$
$$= 40 \text{ lb/cft}(0.0861) + 2.00 \text{ lb/cft}(1-0.0861)$$
$$= 5.3 \text{ lb/cft}.$$

6. Adjust the pressure drop for density of the mixture

$$(\Delta P/L)_{total, 2\phi} = (\Delta P/L)_{fric, 2\phi} - /+\rho_{mix}$$
$$= 62.5 - 5.3 = 57.2 \text{ lb/ft}^2/\text{ft depth}(0.40 \text{ psi/ft}).$$

This is a feasible order of magnitude. If the dP per foot was, say, 10 psi or 0.001 psi per foot, these extremes would not be credible for a two-phase system. Practical pressure drops are normally in the 0.1–2.0 psi/ft range. If you get an answer far outside the normal range, check your work and, if it is correct, consider a different vessel/reactor diameter or a different particle size or shape.

The dP per foot is multiplied by the total feet of bed depth to get the final overall dP for the bed.

Estimating Overall Pressure Drop in Beds

In practice, the total pressure drop through a reactor or sorbent bed includes more than just the bed pressure drop itself. For a commercial fixed bed, total dP includes drop generated by the hardware or internals of the reactor. For a typical downflow reactor (see Fig. 59), for instance, the internals include:

- Inlet diffuser or distributor – you can normally assume about 1 psi for this element at full rate and the dP will vary with rates and phase properties.
- Distribution tray (for two phases) – you can assume another 1 psi for this tray at design conditions; again, the dP will vary depending on conditions.
- Quench zones (if required), including bed support grid, mixing system, and redistribution tray – assume 2–4 psi per zone at design rate; again, the dP will vary depending on conditions.
- Outlet collector (elephant stool) – assume 1 psi at design rates.

In addition to the internals, the overall bed dP will be affected by fouling from coke, corrosion products, dust, etc.. i.e., accumulation of solids in the interstitial spaces, thus reducing the void fraction. In design, a typical fouling allowance ranges from 20 % to 100 % of the frictional pressure drop in a bed. Alternately, you can assume a certain loss of void fraction (say 20 %) from start of run to end of run and recalculate the frictional dP for the bed to get the fouled bed dP.

To illustrate how the bed design pressure drop is estimated, consider the following case. Say:

- You have a two bed reactor with two-phase flow.
- The overall bed dPs, clean, are:
  - Bed 1–20 psi
  - Bed 2–35 psi

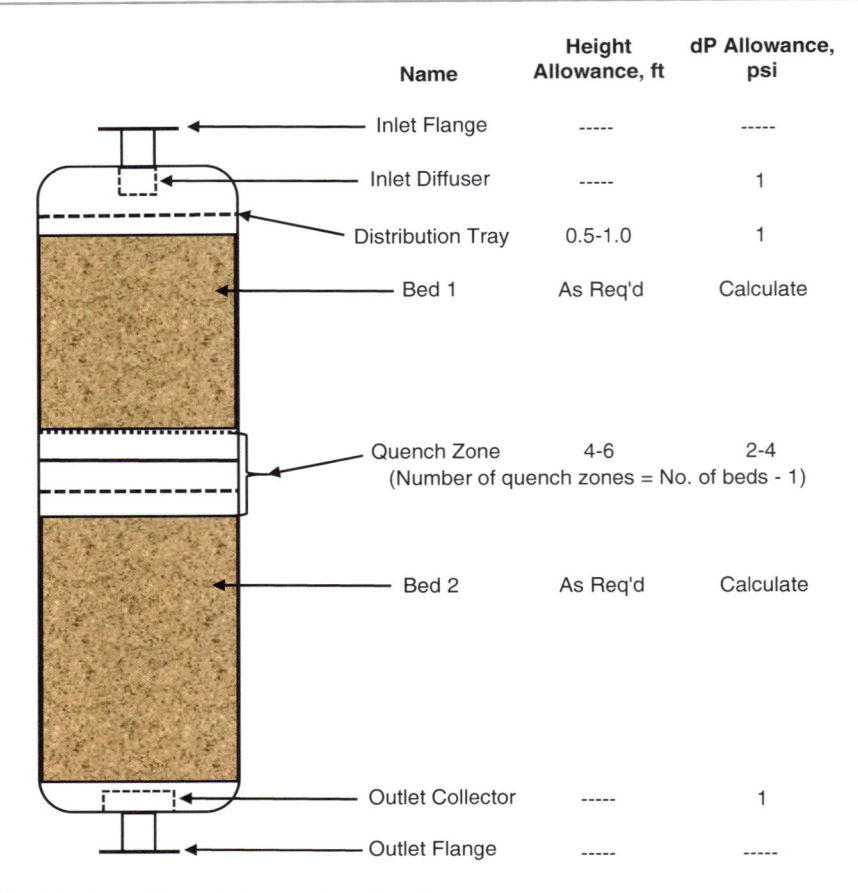

Name	Height Allowance, ft	dP Allowance, psi
Inlet Flange	-----	-----
Inlet Diffuser	-----	1
Distribution Tray	0.5-1.0	1
Bed 1	As Req'd	Calculate
Quench Zone (Number of quench zones = No. of beds - 1)	4-6	2-4
Bed 2	As Req'd	Calculate
Outlet Collector	-----	1
Outlet Flange	-----	-----

Fig. 59 Typical internals in a two-bed, downflow reactor

These would have been calculated using average vapor and liquid flows, densities, and viscosities in each bed and would include the effects of fluid head.
- The reactor inlet pressure is 627 psig.
- What is the pressure profile in this reactor on an end-of-run design basis?

Solution:

1. Assume a fouling allowance of 50 % dP increase in bed 1 and 20 % increase in bed 2.

2. The reactor will have:
   - One inlet diffuser (assume 1 psi)
   - One top distribution tray (assume 1 psi)
   - One quench zone (assume 2 psi)
   - One outlet collector (assume 1 psi)
3. Set up a table for the pressure profile:

	dP, psi	Psig after element
Inlet to reactor	–	**627 (inlet flange)**
Inlet diffuser	1	626
Top distribution tray	1	625
Bed 1		
Clean bed dP	20	
Fouling (allow 50 %)	10	
Total bed 1	30	595
Quench zone	2	593
Bed 2		
Clean bed dP	35	
Fouling (allow 20 %)	7	
Total bed 2	42	551
Outlet collector	1	550
Overall reactor	**77**	**550 (outlet flange)**

Moving Solids Bed Pressure Drops

Moving solids beds are typically found in fluid catalytic crackers, ebullated and bubbling bed reactors, and certain other solids handling systems in a refinery or petroleum processing facility. The details of pressure drop estimates in these services are beyond the scope of this book. Many of the best methods for dealing with the moving bed dPs are proprietary and are used for design by technology licensors.

One good available reference on these calculations is the book: Hydrodynamics of Gas-Solids Fluidization by Cheremisinoff and Cheremisinoff (1984).

Appendices

Appendix 1: Chord Height, Area, and Length for Distillation Trays and Circular Cross-Sections

See Table 38.

Table 38 Chord height, area, and length for distillation trays and circular cross-sections

Weir length and downcomer area

R*	L*	A*	R*	L*	A*	R*	L*	A*	R*	L*	A*	R*	L*	A*
0.070	0.511	0.306	0.120	0.650	0.680	0.170	0.751	0.113	0.220	0.828	0.163	0.280	$	0.230	0.390	0.977	0.361
1	0.514	0.0315	1	0.652	0.0688	1	0.753	0.114	1	0.829	0.164	5	0.903	0.236	6	0.979	0.367
2	0.517	0.0321	2	0.654	0.0697	2	0.755	0.115	2	0.831	0.165	0.290	0.908	0.241	0.400	0.980	0.374
3	0.521	0.0328	3	0.657	0.0705	3	0.756	0.116	3	0.832	0.166	5	0.913	0.247	5	0.982	0.380
4	0.524	0.0335	4	0.659	0.0714	4	0.758	0.117	4	0.834	0.157	0.300	0.917	0.252	0.410	0.984	0.386
5	0.527	0.0342	5	0.661	0.722	5	0.760	0.117	5	0.835	0.169	5	0.921	0.258	6	0.986	0.392
6	0.530	0.0348	6	0.663	0.0731	6	0.762	0.118	6	0.836	0.170	0.310	0.925	0.264	0.420	0.967	0.398
7	0.533	0.0355	7	0.665	0.0739	7	0.763	0.119	7	0.838	0.171	5	0.930	0.2780	5	0.969	0.405
8	0.536	0.0362	8	0.668	0.0748	8	0.765	0.120	8	0.839	0.172	0.320	0.933	0.276	0.430	0.991	0.412
9	0.539	0.0368	9	0.670	0.0756	9	0.766	0.122	9	0.841	0.173	5	0.937	0.282	5	0.993	0.418
0.080	0.542	0.0375	0.130	0.672	0.0765	0.180	0.768	0.122	0.230	0.842	0.174	0.330	0.941	0.288	0.440	0.994	0.424
1	0.545	0.0382	1	0.674	0.0774	1	0.770	0.123	1	0.843	0.175	5	0.945	0.294	5	995	0.430
2	0.548	0.0389	2	0.677	0.0782	2	0.772	0.124	2	0.845	0.176	0.340	0.948	0.300	0.450	0.996	0.437
3	0.552	0.0396	3	0.679	0.0791	3	0.773	0.125	3	0.846	0.177	5	0.951	0.306	5		
4	0.555	0.0403	4	0.682	0.0799	4	0.775	0.126	4	0.848	0.178	0.350	0.955	0.312	0.460	0.997	0.450
5	0.558	0.0410	5	0.684	0.0808	5	0.777	0.127	5	0.849	0.179	5	0.958	0.318	0.470	0.998	0.462
6	0.551	0.0418	6	0.686	0.0817	6	0.778	0.128	6	0.850	0.180	0.360	0.961	0.324	0.480	0.998	0.475
7	0.554	0.0425	7	0.688	0.0825	7	0.780	0.129	7	0.851	0.181	5	0.964	0.330			
8	0.567	0.0432	8	0.691	0.0834	8	0.781	0.130	8	0.853	0.182	0.370	0.967	0.337	0.490	0.999	0.488
9	0.570	0.0439	9	0.693	0.0842	9	0.783	0.131	9	0.854	0.183	5	0.969	0.343	0.500	1.0	0.50
0.090	0.573	0.0446	0.140	0.695	0.0851	0.190	0.784	0.132	0.240	0.855	0.184	0.360	0.961	0.348			
1	0.576	0.0454	1	0.697	0.0860	1	0.785	0.133	5	0.860	0.190	5	0.977	0.354			
2	0.578	0.0461	2	0.699	0.0869	2	0.787	0.134									
3	0.581	0.0469	3	0.700	0.0878	3	0.789	0.135	0.250	0.866	0.196	0.370					
4	0.583	0.0476	4	0.702	0.0887	4	0.790	0.136	5	0.872	0.202	5					
5	0.586	0.0484	5	0.704	0.0896	5	0.782	0.137	0.260	0.878	0.207	0.380					
6	0.589	0.0491	6	0.706	0.0905	6	0.794	0.138	5	0.583	0.213	5					

R			L							
7	0.592	0.0499	7	0.708	0.0914	7	0.795	0.139		
8	0.594	0.0506	8	0.710	0.0923	8	0.797	0.140	0.270	0.888	0.218
9	0.597	0.0514	9	0.712	0.0932	9	0.798	0.141	5	0.893	224
0.100	0.600	0.0521	0.150	0.714	0.0941	0.200	0.800	0.142			
1	0.603	0.0529	1	0.716	0.0950	1	802	0.143			
2	0.605	0.0537	2	0.718	0.0959	2	0.803	0.144			
3	0.608	0.0645	3	0.720	0.0969	3	0.805	0.145			
4	0.610	0.0553	4	0.722	0.0978	4	0.806	0.146			
5	0.613	0.0551	5	0.724	0.0987	5	0.808	0.148			
6	0.615	0.0568	6	0.726	0.0996	6	0.809	0.149			
7	0.618	0.0576	7	0.728	0.1005	7	0.810	0.150			
8	0.620	0.0584	8	0.729	0.1015	8	0.812	0.151			
9	0.623	0.0592	9	0.731	0.102	9	0.813	0.152			
0.110	0.625	0.0600	0.160	0.733	0.103	0.210	0.814	0.153			
1	0.628	0.0608	1	0.735	0.104	1	0.816	0.154			
2	0.630	0.0616	2	0.737	0.105	2	0.817	0.155			
3	0.633	0.0624	3	0.738	0.106	3	0.819	0.156			
4	0.635	0.0632	4	0.740	0.107	4	0.820	0.157			
5	0.638	0.0640	5	0.742	0.108	5	0.822	0.158			
6	0.640	0.0648	6	0.744	0.109	6	0.823	0.159			
7	0.043	0.0656	7	0.746	0.110	7	0.824	0.160			
8	0.645	0.0654	8	0.747	0.111	8	0.826	0.161			
9	0.648	0.0672	9	0.749	0.112	9	0.627	0.162			

$R = \dfrac{*\text{Downcomer Rise}}{\text{Diameter}} = \dfrac{r}{\text{Dia.}}$ $L = \dfrac{*\text{Weir Length}}{\text{Diameter}} = \dfrac{l_o}{\text{Dia.}}$ $A = \dfrac{*\text{Downcomer Area}}{\text{Tower Area}} = \dfrac{A_o}{A_s}$

*This table relates the downcomer area, the weir length, and the height of the circular segment formed by the weir

Appendix 2: Valve Tray Design Details

Valve tray design principles			
Design feature	Suggested value	Alternate values	Comment
1. Valve size and layout			
(a) Value diameter	–		Valve diameter is fined by die vendor
(b) Percent hole area. A_u/A_n	12	8–15	Open area should be set by the designer. In general, the lower and open area, the higher the efficiency and flexibility, and the lower the capacity (due to increased pressure drop). At values of open area toward the upper end of the range (say 15 %), the flexibility and efficiency are approaching sieve tray values. At the lower end of the range, capacity and downcomer filling becomes limited
(c) Valve pitch/diam. ratio	–		Value pitch is normally triangular. However, this variable is usually fixed by the vendor
(d) Valve distribution	–		On trays with flow path length $\geq 5'$ and for liquid rates $>5,000$ GPH/ft (diameter) on trays with flow path length $<5'$, provide 10 % more valves on the inlet half of the tray than on the outlet half
(e) Bubble area. A_u	–		Bubble area should be maximized
(f) Plate efficiency	–		Valve tray efficiency will be about equal to sieve tray efficiency provided there is not a blowing or flooding limitation
(g) Valve blanking	–		This should not generally be necessary unless tower is being sized for future service at much higher rates. Blanking strips can then be used. Blank within buddle area, not around periphery to maintain best efficiency

(*continued*)

Process Equipment for Petroleum Processing

Valve tray design principles

Design feature	Suggested value	Alternate values	Comment
2. Tray spacing, inches	–	12–36	Generally economic to use min. Values given on p. III-E-2 which are set by maintenance requirements. Other considerations are downcomer filling and flexibility. Use of variable spacings to accommodate loading changes from section to section should be considered
3. Number of liquid passes	11	to 2 M	Multi-passing improves liquid handling capacity at the expense of vapor capacity for a given diameter column and tray spacing. Cost is apparently no greater – at least, for tower diameters <8 ft
4. Downcomers and weirs			
(a) Allowable downcomer inlet velocity. ft/s of clear liq.		0.3–0.4	Lower value recommended for absorbers or other systems of known high frothiness
(b) Type downcomer	Chord	Chord. arc	Min. chord length should be 65 % of tray diameter for good liquid distribution. Sloped downcomers can be used for high liquid rates – with maximum outlet velocity =0.6 ft/s. Arc downcomers may be used alternatively to give more bubble area (and higher capacity) but are somewhat more expensive. Min. width should be 6 in. for latter
(c) Inboard downcomer width (inlet and outlet)		Min. 8 in.	Use of a 14–16″ "jump baffle" suspended lengthwise in the center of the inboard downcomer and extending the length of the downcomer is suggested to prevent possible bridging over by froth entering the downcomer form opposite sides. Elevation of the base of jump baffle should be level with outlet weirs. Internal accessway must be provided to allow passage from one side to another during inspection

(*continued*)

Valve tray design principles

Design feature	Suggested value	Alternate values	Comment
(d) Outlet weir height	2″	1–4″	Weir height can be varied with liquid rate to give a total liquid head on the tray (h_g) in the range of 2.5–4″ whenever possible. Lower values suggested for vacuum towers, higher ones for long residence time applications
(e) Clearance under downcomer, in.	1.5″	1″ min	Set clearance to give head loss of approximately 1 in. Higher values can be used if necessary to assure sealing of downcomer
(f) Downcorner seal (inlet or outlet weir height minus downcomer clearance)	Use outlet weir to give min.1/2″seal in plate liquid	Inlet weir or recessed inlet box	In most cases, plate liquid level can be made high enough to seal the downcomer through use of outlet weir only. Inlet weirs add to downcomer buildup; in some cases, they may be desirable for 2-pass trays to ensure equal liquid distribution. Recessed inlets are more expensive but may be necessary in cases where an operating seal would require an excessively high outlet weir
(g) Downcomer filling, % of tray spacing		40–50	Use the lower value for high-pressure towers, absorbers, vacuum towers, known foaming systems, and also for tray spacings of 18″ or lower

Appendix 3: Pressure-Temperature Curves for Hydrocarbon Equilibria

See Figs. 60 and 61.

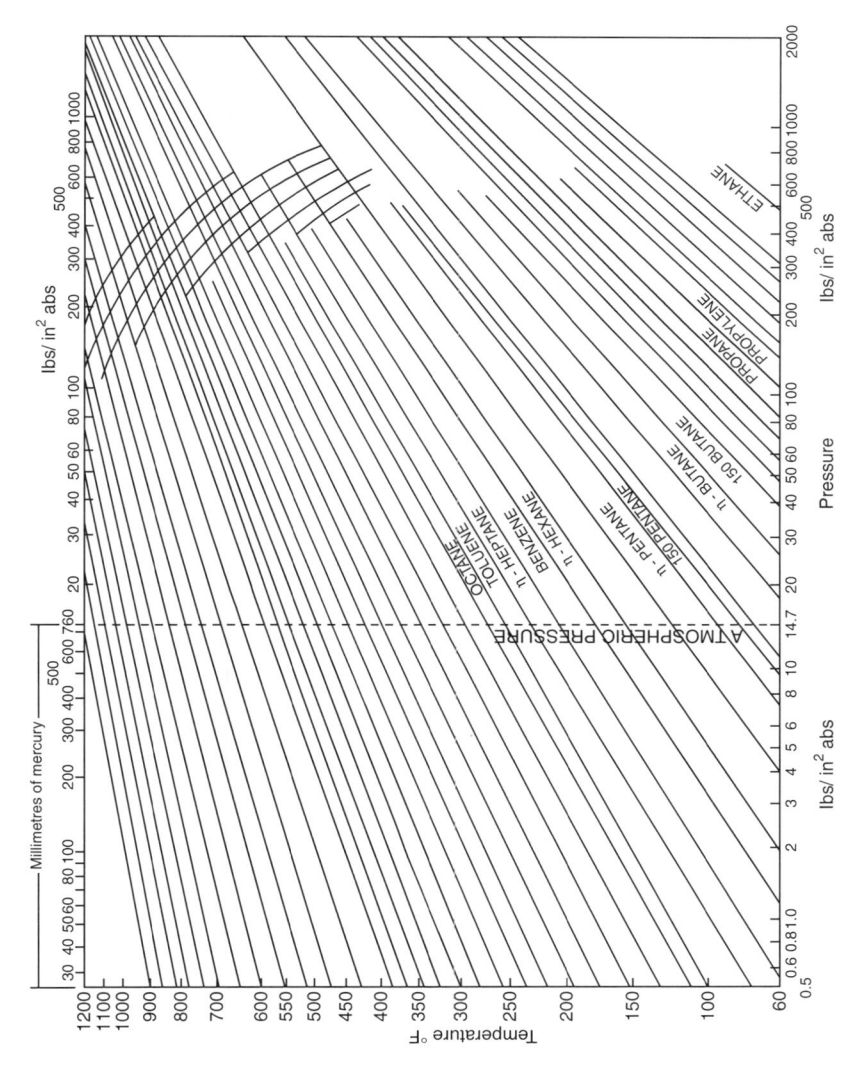

Fig. 60 Pressure-temperature curves for hydrocarbon equilibria (chart 1 of 2)

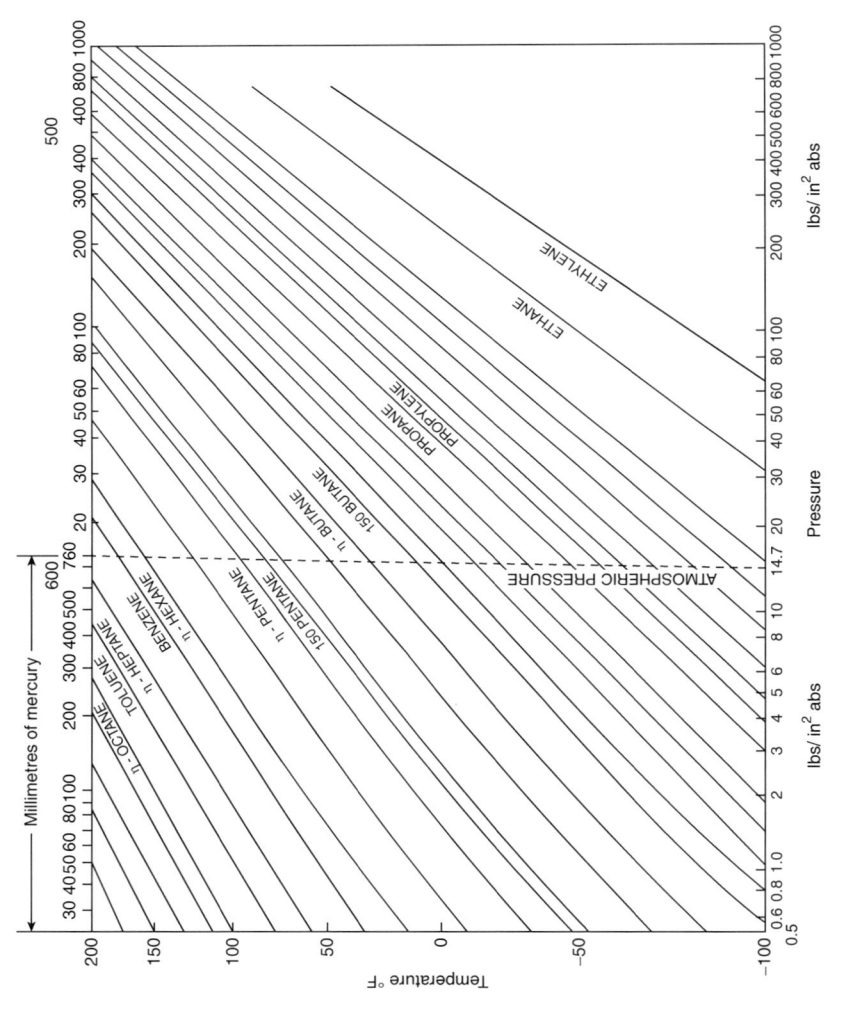

Fig. 61 Pressure–temperature curves for hydrocarbon equilibria (chart 2 or 2)

Appendix 4: ASTM Gaps and Overlaps

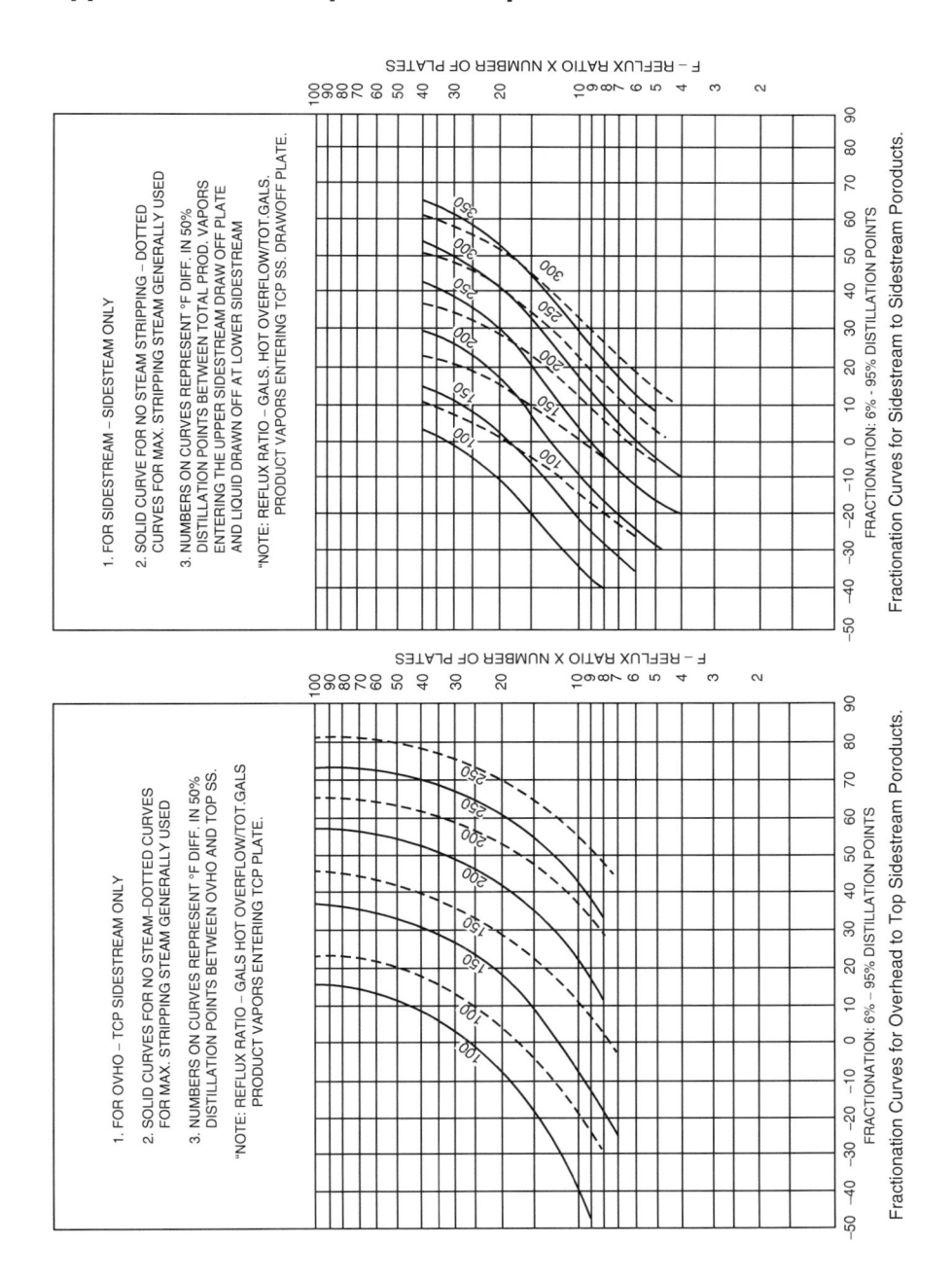

Fractionation Curves for Sidestream to Sidestream Poroducts.

Fractionation Curves for Overhead to Top Sidestream Poroducts.

Appendix 5: Values of Coefficient C for Various Materials

Values of coefficient $C = 520\sqrt{k\left(\frac{2}{k+1}\right)^{\frac{k+1}{k-1}}}$; $k = \frac{C_p}{C_v}$

k	C	k	C	k	C	k	C	k	C	k	C
0.41	219.28	0.71	276.09	1.01	316.56[a]	1.31	347.91	1.61	373.32	1.91	394.56
0.42	221.59	0.72	277.64	1.02	317.74	1.32	348.84	1.62	374.09	1.92	395.21
0.43	223.86	0.73	279.18	1.03	318.90	1.33	349.77	1.63	374.85	1.93	395.86
0.44	226.10	0.74	280.70	1.04	320.05	1.34	350.68	1.64	375.61	1.94	396.50
0.45	228.30	0.75	282.20	1.05	321.19	1.35	351.60	1.65	376.37	1.95	397.14
0.46	230.47	0.76	283.69	1.06	322.32	1.36	352.50	1.66	377.12	1.96	397.78
0.47	232.61	0.77	285.16	1.07	323.44	1.37	353.40	1.67	377.86	1.97	398.41
0.48	234.71	0.78	286.62	1.08	324.55	1.38	354.29	1.68	378.61	1.98	399.05
0.49	236.78	0.79	288.07	1.09	325.65	1.39	355.18	1.69	379.34	1.99	399.67
0.50	238.83	0.80	289.49	1.10	326.75	1.40	356.06	1.70	380.08	2.00	400.30
0.51	240.84	0.81	290.91	1.11	327.83	1.41	356.94	1.71	380.80	2.01	400.92
0.52	242.82	0.82	292.31	1.12	328.91	1.42	357.81	1.72	381.53	2.02	401.53
0.53	244.78	0.83	293.70	1.13	329.98	1.43	358.67	1.73	382.25	2.03	402.15
0.54	246.72	0.84	295.07	1.14	331.04	1.44	359.53	1.74	382.97	2.04	402.76
0.55	248.62	0.85	296.43	1.15	332.09	1.45	360.38	1.75	383.68	2.05	403.37
0.56	250.50	0.86	297.78	1.16	333.14	1.46	361.23	1.76	384.39	2.06	403.97
0.57	252.36	0.87	299.11	1.17	334.17	1.47	362.07	1.77	385.09	2.07	404.58
0.58	254.19	0.88	300.43	1.18	335.20	1.48	362.91	1.78	385.79	2.08	405.18
0.59	256.00	0.89	301.74	1.19	336.22	1.49	363.74	1.79	386.49	2.09	405.77
0.60	257.79	0.90	303.04	1.20	337.24	1.50	364.56	1.80	387.18	2.10	406.37
0.61	259.55	0.91	304.33	1.21	338.24	1.51	365.39	1.81	387.87	2.11	406.96
0.62	261.29	0.92	305.60	1.22	339.24	1.52	366.20	1.82	388.56	2.12	407.55
0.63	263.01	0.93	306.86	1.23	340.23	1.53	367.01	1.83	389.24	2.13	408.13
0.64	264.72	0.94	308.11	1.24	341.22	1.54	367.82	1.84	389.92	2.14	408.71
0.65	266.40	0.95	309.35	1.25	342.19	1.55	368.62	1.85	390.59	2.15	409.29
0.66	268.06	0.96	310.58	1.26	343.16	1.56	369.41	1.86	391.26	2.16	409.87
0.67	269.70	0.97	311.80	1.27	344.13	1.57	370.21	1.87	391.93	2.17	410.44
0.68	271.33	0.98	313.01	1.28	345.08	1.58	370.99	1.88	392.59	2.18	411.01
0.69	272.93	0.99	314.19a	1.29	346.03	1.59	371.77	1.89	393.25	2.19	411.58
0.70	274.52	1.00	315.38a	1.30	346.98	1.60	372.55	1.90	393.91	2.20	412.15

Values of C for gases

	Mol wt	k = C_p/C_v	C	C/356		Mol wt	k = C_p/C_v	C	C/356
Acetylene	26	1.28	345	0.969	Hydrochloric acid	36.5	1.40	356	1.000
Air	29	1.40	356	1.000	Hydrogen	2	1.40	356	1.000
Ammonia	17	1.33	351	0.986	Hydrogen sulfide	34	1.32	348	0.978
Argon	40	1.66	377	1.059	Isobutane	58	1.11	328	0.921

(*continued*)

Compound	MW	k	C	a	Compound	MW	k	C	a
Benzene	78	1.10	327	0.919	Methane	16	1.30	346	0.972
Carbon disulfide	76	1.21	338	0.949	Methyl alcohol	32	1.20	337	0.947
Carbon dioxide	44	1.28	345	0.969	Methyl chloride	50.5	1.20	337	0.947
Carbon monoxide	28	1.40	356	1.000	N-butane	58	1.11	328	0.921
Chlorine	71	1.36	352	0.989	Natural gas	19	1.27	345	0.969
Cyclohexane	84	1.08	324	0.910	Nitrogen	28	1.40	356	1.000
Ethane	30	1.22	339	0.952	Oxygen	32	1.40	356	1.000
Ethylene	28	1.20	337	0.947	Pentane	72	1.09	325	0.913
Helium	4	1.66	377	1.059	Propane	44	1.14	331	0.930
Hexane	86	1.08	324	0.910	Sulfur dioxide	64	1.26	342.	0.961

[a]Interpolated values, since C becomes indeterminate as k approaches 1.00 (Reproduced by permission of Gas Processors Suppliers Association)

Appendix 6: LMTD Correction Factors

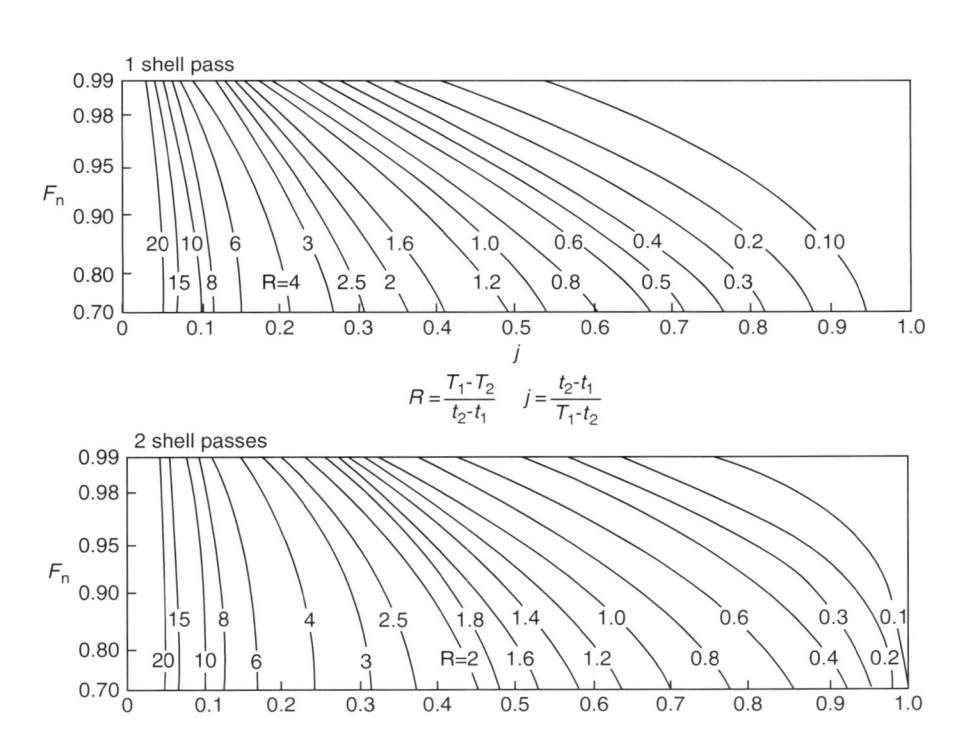

$$R = \frac{T_1 - T_2}{t_2 - t_1} \qquad j = \frac{t_2 - t_1}{T_1 - t_2}$$

Appendix 7: Common Overall Heat Transfer Coefficients – U_O

Fluid being cooled	Fluid being heated	$\dfrac{U_0}{(\text{BTU/h.ft}^2 \cdot {}^\circ\text{F})}$
Exchangers		
C_4s and lighter	Water	75–110
C_4s and lighter	LPG	75
Naphtha	Naphtha	75
Naphtha	Water	80–100
BPT 450 (kero)	Heavy oil (crude)	70–75
Gas oils	Crude	40–50
Gas oils	Water	40–70
Light fuel oil	Crude	20–30
Waxy distillates	Heavy oils	30–40
Slurries	Waxy distillate	40
MEA or DEA	Water	140
MEA or DEA	MEA or DEA	120–130
Water	Water	180–200
Air	Water	20–30
Lt HC vapor	H_2-rich stream	35–40
Lt HC vapor	Naphtha	38
Condensers		
Full-range naphtha	Water	70–80
Amine stripper O/heads	Water	100
C_4s and lighter	Water	90
Reformer effluent (Lt HC)	Water	65

Appendix 8: Standard Heat Exchanger Tube Data

d_0 = OD of tubing (in.)	BWG gauge	l = thickness (ft)	d_i = ID of tubing (in.)	Internal area (in.^{2})	External surface per foot length (ft^2)
$\frac{3}{4}$	10	0.0112	0.482	0.1822	0.1963
$\frac{3}{4}$	12	0.00908	0.532	0.223	0.1963
$\frac{3}{4}$	14	0.00691	0.584	0.268	0.1963
$\frac{3}{4}$	16	0.00542	0.620	0.302	0.1963
$\frac{3}{4}$	18	0.00408	0.652	0.334	0.1963
1	8	0.0137	0.670	0.355	0.2618
1	10	0.0112	0.732	0.421	0.2618
1	12	0.00908	0.782	0.479	0.2618
1	14	0.00691	0.834	0.546	0.2618
1	16	0.00542	0.870	0.594	0.2618
1	18	0.00408	0.902	0.639	0.2618
$1\frac{1}{2}$	10	0.0112	1.232	1.192	0.3927
$1\frac{1}{2}$	12	0.00908	1.282	1.291	0.3927
$1\frac{1}{2}$	14	0.00691	1.334	1.397	0.3927
$1\frac{1}{2}$	16	0.00542	1.37	1.474	0.3927

Appendix 9: Line Friction Loss for Viscous Fluids

See Tables 39, 40, 41, 42, 43, 44, 45, 46, and 47.

Table 39 Line friction loss

Friction Loss for Viscous Liquids
(Based on Darcy's Formula)
Loss in Feet of Liquid per 1000 Feet of Pipe
1 Inch (1.049" inside dia) Sch 40 New Steel Pipe

Flow		Kinematic viscosity—centistokes									
US gal per min	Bbl per hr (42 gal)	0.6	1.1	2.1	2.7	4.3	7.4	10.3	13.1	15.7	20.6
					Approx SSU viscosity						
			31.5	33	35	40	50	60	70	80	100
.5	.71	.29	.28	.55	.70	1.12	1.93	2.68	3.41	4.08	5.35
1	1.4	.96	1.13	1.09	1.41	2.24	3.86	5.36	6.82	8.16	10.7
2	2.9	3.23	3.72	4.41	4.80	4.48	7.72	10.7	13.6	16.3	21.4
3	4.3	6.84	7.63	9.04	9.48	10.8	11.6	16.1	20.5	24.5	32.1
4	5.7	11.4	12.2	14.9	15.9	17.6	15.4	21.5	27.3	32.6	42.8
5	7.1	17.2	19.2	22.1	23.4	26.3	19.3	26.8	34.1	40.8	53.5
6	8.6	24.2	26.8	30.5	32.1	36.8	41.3	32.2	40.9	49.0	64.2
7	10.0	32.3	35.4	40.8	42.6	47.4	53.8	37.5	47.7	57.2	74.9
8	11.4	41.6	45.5	51.1	54.3	59.9	68.2	75.3	54.5	65.2	85.6
9	12.9	51.8	56.2	63.5	66.3	73.4	84.9	91.7	61.4	73.4	96.3
10	14.3	62.7	68.1	76.2	80.1	88.2	101	111	115	81.6	107
12	17.1	89.3	95.3	106	111	122	142	151	162	167	129
14	20.0	120	129	140	147	160	185	198	212	221	150
16	22.8	155	164	181	188	205	234	257	268	279	295
18	25.7	194	202	224	233	254	286	305	326	341	365
20	28.6	237	250	272	281	308	342	372	394	405	437
25	35.7	368	383	410	429	464	501	551	583	599	651
30	42.9	523	545	582	600	640	712	759	803	842	904
35	50.0	708	735	780	795	852	933				

Flow		Kinematic viscosity—centistokes									
US gal per min	Bbl per hr (42 gal)	26.4	32.0	43.2	65.0	108.4	162.3	216.5	325	435	650
					Approx SSU viscosity						
		125	150	200	300	500	750	1000	1500	2000	3000
.1	.14	1.37	1.66	2.25	3.38	5.65	8.45	11.3	16.9	22.6	33.8
.3	.43	4.12	4.98	6.75	10.2	17.0	25.3	33.8	50.7	67.8	102
.5	.71	6.86	8.32	11.3	16.9	28.3	42.3	56.4	85	113	169
1	1.4	13.7	16.6	22.5	33.8	56.5	84.5	113	169	226	338
2	2.9	27.5	33.2	45.0	67.6	113	169	226	338	452	676
3	4.3	41.2	49.8	67.5	102	170	253	338	507	678	
4	5.7	55.0	66.5	90.0	136	226	338	452	677	904	
5	7.1	68.7	83.2	113	169	283	423	564	845		
6	8.6	82.4	99.7	135	203	339	507	677			
7	10	96.2	117	158	237	395	591	790			
8	11.4	110	133	180	271	452	676	903			
9	12.9	124	150	203	303	508	760				
10	14.3	137	167	225	338	565	845				
12	17.1	165	200	270	406	678					
14	20.0	192	233	315	474	792					
16	22.8	220	266	360	541	904					
18	25.7	248	299	405	609						
20	28.6	470	332	460	677						

Friction Loss for Viscous Liquids (Continued)
(Based on Darcy's Formula)

Loss in Feet of Liquid per 1000 Feet of Pipe
1½ Inch (1.610" inside dia) Sch 40 New Steel Pipe

Flow		Kinematic viscosity—centistokes								
		0.6	1.1	2.1	2.7	4.3	7.4	10.3	13.1	15.7	20.6
US gal per min	Bbl per hr (42 gal)		Approx SSU viscosity								
			31.5	33	35	40	50	60	70	80	100
1	1.4	.13	.10	.20	.25	.41	.69	.97	1.23	1.47	1.93
2	2.9	.42	.49	.39	.51	.83	1.39	1.93	2.46	2.94	3.86
3	4.3	.86	.98	1.17	1.25	1.24	2.08	2.89	3.68	4.41	5.79
4	5.7	1.43	1.63	1.92	2.07	1.65	2.78	3.86	4.91	5.88	7.72
5	7.1	2.11	2.42	2.83	3.05	2.06	3.47	4.82	6.14	7.35	9.65
6	8.6	2.90	3.36	3.89	4.18	4.69	4.17	5.79	7.37	8.82	11.6
8	11.4	4.97	5.80	6.44	6.87	7.77	9.02	7.72	9.83	11.8	15.5
10	14.3	7.51	8.34	9.58	10.1	11.6	13.2	9.65	12.3	14.7	19.3
12	17.1	10.4	11.6	13.4	14.0	15.6	17.9	19.8	14.7	17.6	23.2
15	21.4	16.0	17.4	19.6	20.7	23.2	26.5	29.1	31.1	21.0	29.0
20	28.6	27.2	29.5	32.9	34.6	38.2	43.9	47.6	51.1	53.8	38.6
25	35.7	41.4	44.8	49.5	51.8	57.5	64.6	70.1	75.2	78.9	84.7
30	42.9	58.8	63.0	69.1	72.0	79.0	89.3	97.1	103	109	116.5
40	57.1	102	107	117	122	132	150	160	170	178	191
50	71.4	157	164	178	183	198	222	237	251	263	281
60	85.7	224	233	249	259	279	306	330	347	362	388
70	100	300	312	333	343	369	402	436	457	477	508
80	114	389	403	427	440	470	516	551	580	602	643
90	129	498	508	536	550	585	634	681	715	746	792
100	143	601	624	656	670	714	774	820	863	898	949

Flow		Kinematic viscosity—centistokes								
		26.4	32.0	43.2	65.0	108.4	162.3	216.5	325	435	650
US gal per min	Bbl per hr (42 gal)		Approx SSU viscosity								
		125	150	200	300	500	750	1000	1500	2000	3000
1	1.4	2.47	3.00	4.14	6.09	10.2	15.2	20.3	30.4	40.8	69.0
2	2.9	4.95	6.00	8.28	12.2	20.3	30.4	40.6	60.8	81.5	122
3	4.3	7.42	9.00	12.4	18.3	30.4	45.6	60.9	91.3	122	183
4	5.7	9.90	12.0	16.6	24.4	40.6	60.8	81.2	122	163	244
5	7.1	12.4	15.0	20.7	30.4	50.7	76.0	102	152	204	304
6	8.6	14.9	18.0	24.8	36.5	60.8	91.2	122	183	244	365
8	11.4	19.8	24.0	33.1	48.7	81.2	122	163	243	326	487
10	14.3	24.7	30.0	41.4	60.9	102	152	203	304	406	609
12	17.1	29.7	36.0	49.7	73.2	122	182	244	365	490	732
15	21.4	37.1	45.0	62.2	91.4	152	228	304	457	612	914
20	28.6	49.5	60.0	82.8	122	203	304	406	608	815	
25	35.7	61.9	75.0	103	152	254	380	507	760		
30	42.9	124	90.0	124	183	304	456	609	913		
40	57.1	204	216	186	244	406	608	812			
50	71.4	302	317	342	304	507	760				

Table 40 Line friction loss

Friction Loss for Viscous Liquids *(Continued)*
(Based on Darcy's Formula)

Loss in Feet of Liquid per 1000 Feet of Pipe
2 Inch (2.067" inside dia) Sch 40 New Steel Pipe

Flow		Kinematic viscosity—centistokes								
		.6	1.1	2.1	2.7	4.3	7.4	10.3	13.1	15.7	20.6
US gal per min	Bbl per hr (42 gal)				Approx SSU viscosity						
		31.5	33	35	40	50	60	70	80	100
1	1.4	.04	.04	.07	.09	.15	.26	.36	.45	.54	.71
2	2.9	.13	.15	.15	.19	.30	.51	.71	.90	1.08	1.42
4	5.7	.43	.50	.58	.63	.59	1.02	1.42	1.81	2.17	2.84
6	8.6	.87	1.00	1.20	1.27	1.46	1.53	2.13	2.71	3.25	4.26
8	11.4	1.47	1.68	1.97	2.13	2.38	2.04	2.84	3.61	4.33	5.68
10	14.3	2.20	2.55	2.90	3.09	3.52	2.56	3.58	4.52	5.42	7.11
12	17.1	3.06	3.46	3.97	4.23	4.78	5.57	4.27	5.43	6.51	8.53
14	20.0	4.07	4.51	5.22	5.51	6.26	7.28	4.98	6.33	7.59	9.96
16	22.8	5.17	5.79	6.65	7.01	7.92	9.16	9.95	7.23	8.67	11.4
18	25.7	6.44	7.16	8.18	8.63	9.87	11.2	12.6	13.1	9.76	12.8
20	28.6	7.82	8.64	9.77	10.4	11.6	13.5	14.8	15.5	10.8	14.2
25	35.7	11.9	13.0	14.7	15.4	17.2	19.9	21.6	22.8	24.1	17.8
30	42.9	17.0	18.2	20.4	21.5	23.8	27.2	29.9	31.2	33.0	35.6
35	50.0	22.3	24.1	27.0	28.2	31.2	35.6	38.6	40.4	43.4	47.3
40	57.1	28.8	31.0	34.2	36.0	39.6	44.6	48.8	52.9	54.1	60.7
50	71.4	44.1	47.2	52.0	54.0	59.2	66.4	72.0	76.9	78.4	86.2
60	85.7	62.7	66.5	72.2	74.2	82.3	91.8	98.6	105	111	119
70	100	84.1	83.5	95.8	99.4	108	120	130	137	145	156
80	114	109	114	123	127	138	154	166	174	182	195
90	129	137	143	154	158	171	192	204	215	225	239
100	143	167	176	188	193	208	230	244	260	269	289
110	157	202	211	225	231	246	275	290	307	319	335
120	171	238	249	265	273	290	321	341	358	375	396
130	186	277	290	307	316	335	372	392	411	432	459
140	200	320	347	352	364	383	424	449	472	491	521
150	214	366	382	403	415	437	479	510	529	553	586
160	228	414	431	457	469	494	536	572	595	621	659
170	243	467	485	513	522	553	601	639	665	694	729
180	257	524	543	572	583	619	665	714	743	767	804
190	271	584	602	634	649	688	733	792	825	846	884
200	286	643	666	699	716	756	808	851	901	927	977
210	300	709	731	768	786	826	880	935	975		
220	314	778	798	838	858	902	958				
230	328	851	873	912	934	982					
240	343	922	945	988							

Friction Loss for Viscous Liquids *(Continued)*
(Based on Darcy's Formula)

Loss in Feet of Liquid per 1000 Feet of Pipe
2 Inch (2.067" inside dia) Sch 40 New Steel Pipe

Flow		Kinematic viscosity—centistokes								
		26.4	32.0	43.2	65.0	108.4	162.3	216.5	325	435	650
US gal per min	Bbl per hr (42 gal)	Approx SSU viscosity									
		125	150	200	300	500	750	1000	1500	2000	3000
1	1.4	.91	1.10	1.49	2.24	3.74	5.60	7.48	11.2	15.0	22.4
2	2.9	1.82	2.21	2.98	4.48	7.49	11.2	15.0	22.4	30.0	44.9
3	4.3	2.73	3.31	4.47	6.73	11.2	16.8	22.4	33.6	45.0	67.4
4	5.7	3.64	4.42	5.96	8.98	15.4	22.4	29.9	44.8	60.0	89.9
5	7.1	4.56	5.52	7.45	11.2	18.7	28.0	37.4	56.0	75.0	112
6	8.6	5.47	6.63	8.95	13.5	22.5	33.6	44.8	67.2	90.0	135
7	10.0	6.38	7.73	10.4	15.7	26.2	39.2	52.3	78.4	105	157
8	11.4	7.29	8.84	11.9	18.0	30.0	34.8	59.8	89.6	120	180
9	12.9	8.20	9.94	13.4	20.2	33.7	50.4	67.3	101	135	202
10	14.3	9.11	11.0	14.9	22.4	37.4	56.0	74.8	112	150	224
12	17.1	10.9	13.3	18.9	26.9	44.9	67.3	89.7	135	180	269
14	20.0	12.7	15.5	20.9	31.4	52.4	78.4	105	157	210	314
16	22.8	14.6	17.7	23.9	35.9	59.9	89.6	120	179	240	359
18	25.7	16.4	19.9	26.8	40.3	67.4	101	135	202	270	404
20	28.6	18.2	22.1	29.8	44.9	74.9	112	150	224	300	449
25	35.7	22.8	27.6	37.3	56.1	93.6	140	187	280	375	562
30	42.9	27.3	33.1	44.7	67.3	112	168	224	336	450	674
35	50.0	31.9	38.7	52.2	78.5	131	196	262	392	525	786
40	57.1	63.0	44.2	59.6	89.8	150	224	299	448	600	899
45	64.3	70.8	80.2	67.1	101	168	252	336	503	675	
50	71.4	92.8	97.1	74.5	112	187	280	374	560	750	
60	85.7	127	134	146	135	225	336	448	672	900	
70	100	162	176	189	157	262	392	523	784		
80	114	208	219	238	180	300	448	598	896		
90	129	257	270	293	327	337	504	673			
100	143	309	322	352	388	374	560	748			
110	157	364	379	412	465	412	617	823			
120	171	423	445	463	537	449	673	898			
130	186	487	510	549	618	487	728				
140	200	548	580	627	703	524	784				
150	214	622	655	705	792	909	840				
160	228	697	737	792	887		896				
170	243	775	808	882			952				
180	257	858	871	962							
190	271	947	986								

Table 41 Line friction loss

Friction Loss for Viscous Liquids *(Continued)*
(Based on Dàrcy's Formula)
Loss in Feet of Liquid per 1000 Feet of Pipe
2½ Inch (2.469" inside dia) Sch 40 New Steel Pipe

Flow		Kinematic viscosity—centistokes									
US gal per min	Bbl per hr (42 gal)	.6	1.1	2.1	2.7	4.3	7.4	10.3	13.1	15.7	20.6
					Approx SSU viscosity						
			31.5	33	35	40	50	60	70	80	100
10	14.3	.92	1.05	1.23	1.31	1.48	1.26	1.75	2.22	2.65	3.50
12	17.1	1.28	1.47	1.70	1.80	2.04	1.51	2.10	2.67	3.19	4.19
14	20.0	1.68	1.92	2.23	2.37	2.65	3.09	2.45	3.11	3.73	4.89
16	22.8	2.15	2.43	2.81	2.99	3.34	3.86	2.80	3.56	4.26	5.59
18	25.7	2.68	2.99	3.46	3.68	4.21	4.77	5.26	4.00	4.80	6.28
20	28.6	3.23	3.61	4.27	4.42	4.94	5.73	6.26	4.44	5.33	6.98
25	35.7	4.88	5.39	6.17	6.55	7.31	8.42	9.20	9.79	6.66	8.73
30	42.9	6.87	7.57	8.55	9.07	10.1	11.5	12.6	13.5	14.1	10.5
35	50.0	9.18	9.97	11.3	11.8	13.3	15.1	16.4	17.7	18.5	20.0
40	57.1	11.8	12.8	14.3	15.0	16.8	19.0	20.7	22.1	23.2	25.0
45	64.3	14.8	15.9	17.7	18.7	20.8	23.5	25.6	27.1	28.4	30.6
50	71.4	18.1	19.3	21.4	22.4	24.9	28.3	30.6	32.5	34.1	36.8
60	85.7	25.6	27.2	29.9	31.1	34.2	39.1	42.1	44.7	46.7	50.4
70	100	34.2	36.2	39.6	41.4	45.2	51.4	55.4	58.5	61.5	66.0
80	114	44.1	46.7	50.6	53.0	57.3	64.5	69.7	74.2	77.5	82.7
90	129	55.2	58.8	63.2	65.4	70.9	80.0	86.6	91.2	95.3	103
100	143	67.2	72.3	76.7	79.4	85.8	96.3	104	109	115	123
110	157	80.9	86.2	92.4	94.8	103	113	121	130	135	146
120	171	95.7	102	108	112	121	133	143	151	158	169
130	186	112	118	125	130	139	155	166	176	181	194
140	200	129	136	144	150	160	176	188	198	209	221
150	214	147	155	165	170	181	198	212	223	234	250
160	228	167	175	187	191	203	224	238	253	262	281
170	243	188	196	210	214	226	248	267	283	290	312
180	257	210	219	234	239	253	277	297	311	322	345
190	271	233	243	260	265	280	306	328	339	356	378
200	286	258	269	286	292	308	334	357	373	391	417
220	314	310	322	343	351	369	400	427	448	461	486
240	343	367	381	404	416	436	469	494	522	539	573
260	371	429	445	470	482	505	543	575	599	621	660
280	400	497	513	540	556	580	630	657	686	710	758
300	429	568	586	617	632	659	705	748	775	803	849
320	457	643	663	695	716	747	799	837	875	903	952
340	486	725	745	776	800	839	894	933	980		
360	514	809	835	866	892	936	994				

Friction Loss for Viscous Liquids *(Continued)*
(Based on Darcy's Formula)
Loss in Feet of Liquid per 1000 Feet of Pipe
2½ Inch (2.469" inside dia) Sch 40 New Steel Pipe

Flow		Kinematic viscosity—centistokes								
		26.4	32.0	43.2	65.0	108.4	162.3	216.5	325	435	650
US gal per min	Bbl per hr (42 gal)	\multicolumn{10}{c}{Approx SSU viscosity}
		125	150	200	300	500	750	1000	1500	2000	3000
1	1.4	.45	.54	.73	1.10	1.84	2.75	3.67	5.52	7.38	11.0
2	2.9	.90	1.09	1.47	2.20	3.68	5.50	7.35	11.0	14.8	22.0
4	5.7	1.79	2.17	2.93	4.41	7.36	11.0	14.7	22.1	29.5	44.1
6	8.6	2.69	3.26	4.40	6.62	11.0	16.5	22.0	33.1	44.3	66.2
8	11.4	3.58	4.34	5.87	8.82	14.7	22.0	29.4	44.1	59.1	88.2
10	14.3	4.48	5.43	7.33	11.0	18.4	27.5	36.7	55.2	73.8	110
12	17.1	5.38	6.51	8.80	13.2	22.1	33.0	44.1	66.2	88.6	132
14	20.0	6.27	7.60	10.3	15.4	25.7	38.5	51.4	77.2	103	154
16	22.8	7.16	8.68	11.7	17.6	29.4	44.0	58.8	88.2	118	176
18	25.7	8.06	9.77	13.2	19.8	33.1	49.5	66.1	99.3	133	198
20	28.6	8.96	10.9	14.7	22.0	36.8	55.0	73.4	110	148	220
25	35.7	11.2	13.6	18.3	27.6	46.0	68.8	91.8	138	185	276
30	42.9	13.4	16.3	22.0	33.1	55.2	82.5	110	165	222	331
35	50.0	15.7	19.0	25.6	38.6	64.4	96.3	129	193	258	386
40	57.1	17.9	21.7	29.3	44.2	73.6	110	147	221	295	441
45	64.3	33.0	24.4	33.0	49.6	82.8	124	165	248	332	496
50	71.4	39.2	27.2	36.6	55.2	92.0	138	184	276	369	551
60	85.7	54.0	56.5	44.0	66.2	110	165	220	331	443	662
70	100	70.0	73.5	51.3	77.2	129	193	257	386	517	772
80	114	87.7	93.4	101	88.3	147	220	294	441	591	882
90	129	110	115	125	99.3	166	248	330	497	665	993
100	143	130	137	148	110	184	275	367	552	738	
110	157	154	164	176	197	202	303	403	607	812	
120	171	180	188	205	226	221	330	441	662	886	
130	186	206	216	232	263	239	358	477	717	960	
140	200	234	247	267	299	257	385	514	772		
150	214	265	279	305	333	276	413	551	827		
160	228	296	312	338	374	294	440	588	882		
170	243	328	345	373	415	312	468	624	937		
180	257	364	384	412	461	530	495	661	993		
190	271	403	420	454	514	587	523	699			
200	286	438	457	493	550	628	550	734			
220	314	522	540	586	658	752	605	808			
240	343	612	633	682	760	866	660	881			
260	371	711	732	782	867		715	955			

Table 42 Line friction loss

Friction Loss for Viscous Liquids *(Continued)*
(Based on Darcy's Formula)
Loss in Feet of Liquid per 1000 Feet of Pipe
3 Inch (3.068" inside dia) Sch 40 New Steel Pipe

Flow		Kinematic viscosity—centistokes									
		.6	1.1	2.1	2.7	4.3	7.4	10.3	13.1	15.7	20.6
US gal per min	Bbl per hr (42 gal)	\multicolumn{10}{c}{Approx SSU viscosity}									
		31.5	33	35	40	50	60	70	80	100
8	11.4	.22	.25	.29	.32	.24	.42	.59	.74	.89	1.18
10	14.3	.32	.37	.43	.47	.54	.53	.73	.93	1.11	1.47
15	21.4	.70	.76	.89	.94	1.07	.79	1.10	1.40	1.67	2.20
20	28.6	1.12	1.27	1.47	1.57	1.78	2.07	1.46	1.86	2.23	2.93
25	35.7	1.69	1.93	2.23	2.31	2.61	3.01	3.29	2.33	2.79	3.66
30	42.9	2.36	2.64	2.99	3.22	3.60	4.12	4.50	4.83	3.35	4.40
35	50.0	3.13	3.48	3.97	4.21	4.66	5.41	5.89	6.35	6.61	5.13
40	57.1	4.03	4.42	5.02	5.29	5.90	6.80	7.46	7.93	8.37	5.87
50	71.4	6.10	6.70	7.50	7.93	8.76	10.1	10.9	11.7	12.3	13.2
60	85.7	8.57	9.32	10.4	11.0	12.0	13.7	15.0	16.0	16.8	18.0
70	100	11.5	12.4	13.8	14.5	15.9	18.0	19.6	20.9	21.9	23.6
80	114	14.7	15.9	17.5	18.4	20.3	22.9	24.6	26.4	27.7	29.8
90	129	18.4	19.9	21.8	22.8	25.0	28.0	30.4	32.4	33.8	36.3
100	143	22.4	24.2	26.3	27.5	30.2	33.7	36.4	39.0	40.8	43.6
120	171	31.8	34.1	36.9	38.6	41.9	46.8	50.5	53.4	56.4	60.0
140	200	42.4	45.6	49.4	50.9	55.4	65.5	66.0	70.0	73.2	78.6
160	228	54.8	58.0	63.3	65.4	70.4	79.1	83.8	87.9	92.3	98.2
180	257	69.0	72.7	78.7	81.6	87.2	97.2	104	109	114	122
200	286	84.7	88.9	95.7	99.4	106	117	125	131	137	146
225	322	107	112	120	124	132	145	155	164	169	180
250	357	131	137	147	151	160	175	188	195	204	218
275	393	158	164	175	180	191	208	226	233	243	258
300	429	187	193	204	212	225	244	260	273	281	298
325	464	218	225	238	247	261	283	300	316	325	345
350	500	253	260	275	283	300	324	344	361	373	396
375	536	288	298	314	322	341	367	388	407	424	448
400	571	328	339	354	363	385	414	436	458	478	498
425	607	368	381	397	407	432	463	488	511	529	550
450	643	410	427	443	455	480	515	543	568	587	619
475	679	457	473	493	504	532	571	599	625	646	681
500	714	504	524	544	555	587	627	658	684	707	750
525	750	555	574	597	609	644	688	720	748	770	821
550	786	606	627	651	665	703	748	783	814	838	890
575	822	663	685	708	723	761	814	852	886	912	962
600	857	721	742	767	783	820	882	919	960	989	

Friction Loss for Viscous Liquids *(Continued)*
(Based on Darcy's Formula)

Loss in Feet of Liquid per 1000 Feet of Pipe
3 Inch (3.068" inside dia) Sch 40 New Steel Pipe

Flow		Kinematic viscosity—centistokes								
		26.4	32.0	43.2	65.0	108.4	162.3	216.5	325	435	650
US gal per min	Bbl per hr (42 gal)	Approx SSU viscosity								
		125	150	200	300	500	750	1000	1500	2000	3000
4	5.7	.75	.91	1.23	1.85	3.08	4.62	6.16	9.25	12.4	18.5
6	8.6	1.13	1.37	1.84	2.77	4.62	6.92	9.24	13.9	18.5	27.7
8	11.4	1.50	1.82	2.45	3.70	6.16	9.23	12.3	18.5	24.7	36.9
10	14.3	1.88	2.28	3.06	4.62	7.70	11.5	15.4	23.1	30.9	46.2
12	17.1	2.25	2.73	3.68	5.55	9.24	13.8	18.5	27.7	37.1	55.5
14	20.0	2.63	3.18	4.29	6.47	10.8	16.2	21.5	32.3	43.3	64.7
16	22.8	3.00	3.64	4.90	7.39	12.3	18.5	24.6	37.0	49.5	73.9
18	25.7	3.38	4.09	5.52	8.31	13.9	20.8	27.7	41.6	55.6	83.2
20	28.6	3.76	4.55	6.13	9.24	15.4	23.1	30.8	46.2	61.8	92.4
25	35.7	4.69	5.69	7.67	11.5	19.3	28.8	38.5	57.7	77.3	115
30	42.9	5.63	6.83	9.20	13.9	23.1	34.6	46.2	69.3	92.7	139
35	50.0	6.57	7.97	10.7	16.2	27.0	40.3	53.8	80.9	108	162
40	51.1	7.61	9.10	12.3	18.5	30.8	46.2	61.6	92.5	124	185
50	71.4	9.39	11.4	15.3	23.1	38.5	57.7	77.0	115	154	231
60	85.7	19.2	13.7	18.4	27.7	46.2	69.2	92.4	139	185	277
70	100	25.3	26.8	21.5	32.3	53.9	80.8	108	162	216	323
80	114	31.6	33.6	24.7	37.0	61.6	92.3	123	185	247	369
90	129	38.9	40.9	44.6	41.6	69.3	104	139	208	278	416
100	143	46.1	49.5	53.0	46.2	77.0	115	154	231	309	462
120	171	64.0	67.8	72.9	55.5	92.4	138	185	277	371	555
140	200	83.9	89.1	94.9	108	108	162	215	323	433	647
160	228	106	111	120	135	123	185	246	370	495	739
180	257	131	137	148	164	139	208	277	416	556	832
200	286	157	163	179	198	154	231	308	462	618	924
225	322	191	204	223	242	279	260	346	520	696	
250	357	229	242	261	291	332	288	385	577	773	
275	393	271	285	311	343	396	317	423	635	850	
300	429	316	331	361	398	456	346	462	693	927	
325	464	364	381	416	458	527	375	500	751		
350	500	415	435	467	523	593	672	538	809		
375	536	469	493	528	586	672	746	577	867		
400	571	526	550	592	656	751	843	616	925		
425	607	587	612	656	728	834	937	654	982		
450	643	652	675	728	802	928					
475	679	718	744	801	689						

Process Equipment for Petroleum Processing

Table 43 Line friction loss

Friction Loss for Viscous Liquids *(Continued)*
(Based on Darcy's Formula)

Loss in Feet of Liquid per 1000 Feet of Pipe
3½ Inch (3.548" inside dia) Sch 40 New Steel Pipe

Flow		*	Kinematic viscosity—centistokes								
		.6	1.1	2.1	2.7	4.3	7.4	10.3	13.1	15.7	20.6
US gal per min	Bbl per hr (42 gal)					Approx SSU viscosity					
			31.5	33	35	40	50	60	70	80	100
20	28.6	.56	.83	.72	.78	.88	1.03	.82	1.04	1.25	1.64
25	35.7	.82	.93	1.08	1.14	1.29	1.52	1.67	1.30	1.56	2.05
30	42.9	1.15	1.29	1.50	1.57	1.78	2.09	2.30	1.58	1.87	2.46
35	50.0	1.53	1.70	1.94	2.08	2.31	2.68	2.97	3.18	2.18	2.87
40	57.1	1.95	2.17	2.49	2.68	2.91	3.35	3.70	4.00	4.21	3.28
45	64.3	2.46	2.68	3.03	3.22	3.59	4.11	4.57	4.91	5.19	3.69
50	71.4	2.95	3.25	3.68	3.90	4.32	4.98	5.42	5.87	6.20	6.64
60	85.7	4.17	4.54	5.12	5.38	6.00	6.82	7.41	7.97	8.41	9.21
70	100	5.78	6.02	6.78	7.11	7.84	8.87	9.76	10.4	10.9	11.9
80	114	7.19	7.72	8.57	8.79	9.81	11.2	12.3	13.0	13.7	14.9
90	129	8.92	9.59	10.6	11.2	12.5	13.9	15.0	16.1	16.8	18.3
100	143	10.8	11.7	12.9	13.6	14.8	16.8	18.0	19.3	20.2	21.7
120	171	15.4	16.4	18.1	18.7	20.5	22.9	25.0	26.3	27.7	29.7
140	200	20.5	22.0	23.9	24.9	27.3	30.5	33.1	34.6	36.0	39.1
160	228	26.4	28.3	30.9	31.9	35.0	38.4	41.8	43.9	45.6	49.0
180	257	33.0	35.0	38.0	39.6	42.5	47.7	50.6	54.0	56.3	60.1
200	286	40.3	43.0	46.3	48.0	51.4	57.1	61.8	65.5	68.0	72.2
225	322	50.7	53.2	57.7	59.7	64.7	71.2	75.9	80.1	84.3	88.8
250	357	62.6	65.0	70.9	72.7	78.9	86.5	92.7	96.9	101	107
275	393	75.4	77.9	84.6	87.1	93.3	102	110	115	120	127
300	429	89.2	92.2	99.6	103	109	119	128	135	140	148
325	464	104	108	116	120	127	138	147	154	161	171
350	500	121	124	133	138	146	159	169	178	183	194
375	536	138	142	152	157	167	181	191	200	207	220
400	571	156	161	171	178	188	203	213	225	233	248
425	607	176	181	192	200	211	225	238	252	259	279
450	643	196	203	213	221	234	250	265	280	287	309
475	679	219	225	235	244	257	276	292	310	319	336
500	714	241	249	259	268	282	304	321	340	350	368
550	786	290	300	311	323	340	365	385	407	415	440
600	857	343	355	367	377	399	428	452	466	480	510
650	929	400	414	428	440	461	496	522	540	557	587
700	1000	464	480	494	505	532	572	597	621	637	675
750	1070	532	548	567	576	604	651	682	704	725	769
800	1140	606	624	641	652	684	730	765	794	815	861

Friction Loss for Viscous Liquids *(Continued)*
(Based on Darcy's Formula)
Loss in Feet of Liquid per 1000 Feet of Pipe
3½ Inch (3.548" Inside dia) Sch 40 New Steel Pipe

Flow		Kinematic viscosity—centistokes								
		26.4	32.0	43.2	65.0	108.4	162.3	216.5	325	435	650
U S gal per min	Bbl per hr (42 gal)	\multicolumn{9}{c}{Approx SSU viscosity}
		125	150	200	300	500	750	1000	1500	2000	3000
10	14.3	1.05	1.27	1.72	2.58	4.32	6.46	8.62	12.9	16.9	25.8
15	21.4	1.57	1.91	2.58	3.88	6.47	9.68	12.9	19.4	25.4	38.8
20	28.6	2.10	2.54	3.44	5.17	8.63	12.9	17.3	25.8	33.8	51.7
25	35.7	2.62	3.18	4.29	6.47	10.8	16.1	21.6	32.3	42.3	64.7
30	42.9	3.15	3.82	5.15	7.78	13.0	19.4	25.9	38.8	50.7	77.6
35	50.0	3.67	4.45	6.01	9.05	15.1	22.6	30.2	45.3	59.2	90.6
40	57.1	4.20	5.09	6.87	10.3	17.3	25.8	34.5	51.7	67.6	103
45	64.3	4.72	5.73	7.73	11.6	19.4	29.0	38.8	58.2	76.1	116
50	71.4	5.25	6.36	8.59	12.9	21.6	32.3	43.1	64.7	84.5	129
60	85.7	6.30	7.64	10.3	15.5	25.9	38.8	51.8	77.6	101	155
70	100	12.8	8.91	12.0	18.1	30.2	45.2	60.4	90.6	118	181
80	114	16.1	16.9	13.7	20.7	34.5	51.6	69.0	103	135	207
90	129	19.6	20.8	15.5	23.3	38.8	58.1	77.6	116	152	233
100	143	23.6	24.8	17.2	25.9	43.2	64.6	86.2	129	169	258
120	171	31.9	34.0	37.1	31.0	51.8	77.5	104	155	203	310
140	200	41.5	43.8	48.2	36.2	60.4	90.4	121	181	238	362
160	228	52.3	55.3	60.5	67.7	69.1	103	138	207	270	413
180	257	64.6	67.3	73.9	83.4	77.7	116	155	233	304	466
200	286	77.3	81.5	87.9	99.9	86.3	129	173	258	338	517
225	322	94.4	99.8	108	122	97.1	145	194	291	380	520
250	357	114	120	129	146	108	161	216	323	422	647
275	393	134	141	153	172	199	177	237	356	465	711
300	429	157	164	178	198	233	194	259	388	507	776
325	464	181	188	205	227	266	210	280	420	549	840
350	500	206	215	233	258	301	228	302	452	592	906
375	536	234	245	262	291	339	242	323	485	634	970
400	571	261	274	291	326	379	423	345	517	676	
425	607	291	303	325	363	419	473	367	550	718	
450	643	322	337	359	402	462	525	388	582	761	
475	679	353	368	395	440	505	572	410	614	803	
500	714	389	403	433	482	551	625	431	647	845	
550	786	459	481	514	566	648	733	795	712	930	
600	857	538	566	600	658	749	851	923	776		
650	929	620	648	690	755	865	978		841		
700	1000	708	735	789	863	987			906		

Table 44 Line friction loss

Friction Loss for Viscous Liquids *(Continued)*
(Based on Darcy's Formula)
Loss In Feet of Liquid per 1000 Feet of Pipe
6 Inch (6.065" inside dia) Sch 40 New Steel Pipe

Flow		Kinematic viscosity—centistokes									
		.6	1.1	2.1	2.7	4.3	7.4	10.3	13.1	15.7	20.6
US gal per min	Bbl per hr (42 gal)				Approx SSU viscosity						
		31.5	33	35	40	50	60	70	80	100	
75	107	.45	.49	.58	.61	.68	.80	.86	.93	.98	1.72
100	143	.77	.85	.96	1.01	1.14	1.30	1.42	1.52	1.62	1.74
125	178	1.14	1.27	1.43	1.51	1.68	1.95	2.10	2.23	2.35	2.57
150	214	1.61	1.78	2.01	2.09	2.32	2.66	2.86	3.08	3.20	3.46
175	250	2.13	2.37	2.63	2.79	3.04	3.52	3.74	3.97	4.24	4.51
200	286	2.75	3.00	3.34	3.55	3.85	4.41	4.76	5.02	5.31	5.89
225	322	3.42	3.74	4.17	4.38	4.78	5.39	5.89	6.16	6.45	7.05
250	357	4.15	4.55	5.07	5.21	5.76	6.94	7.11	7.47	7.77	8.41
275	393	4.99	5.42	6.02	6.28	6.88	7.60	8.35	8.66	9.18	9.81
300	429	5.87	6.38	7.06	7.37	8.09	8.94	9.69	10.3	10.8	11.4
350	500	7.90	8.45	9.38	9.80	10.5	11.8	12.6	13.5	14.2	15.0
400	571	10.2	10.9	11.8	12.5	13.4	15.0	16.0	17.0	18.0	19.1
450	643	12.8	13.6	14.8	15.5	16.7	18.5	19.7	20.1	21.8	23.7
500	714	15.6	16.6	18.0	18.7	20.4	22.6	24.0	25.1	26.4	28.5
550	786	18.8	19.8	21.5	22.3	24.3	26.8	28.5	29.6	29.9	33.4
600	857	22.1	23.3	25.1	26.2	28.4	31.1	33.2	35.0	36.2	38.8
650	929	25.8	27.2	29.2	30.4	32.8	36.1	38.5	40.4	41.8	44.7
700	1000	29.7	31.2	33.5	34.9	37.5	41.1	44.2	46.3	47.9	51.3
750	1070	33.9	35.6	38.2	39.7	42.5	46.7	49.9	51.8	54.0	57.2
800	1140	38.3	40.5	43.2	44.4	47.8	52.7	56.1	58.5	60.9	64.3
900	1285	48.5	50.7	54.4	55.8	59.3	65.4	69.1	72.8	74.6	79.9
1000	1430	59.5	62.2	66.4	67.5	72.4	79.3	83.4	87.8	91.0	95.9
1100	1570	71.6	74.8	79.4	80.8	86.7	94.5	99.6	104	109	114
1200	1715	84.6	87.9	93.4	85.6	102	111	117	121	126	133
1400	2000	115	118	126	128	135	146	155	161	167	177
1600	2285	150	153	162	164	173	187	199	207	213	224
1800	2570	188	193	203	206	216	232	246	256	264	276
2000	2860	231	237	247	253	264	284	296	311	320	334
2200	3140	277	286	297	303	316	338	354	371	382	398
2400	3430	330	341	352	356	374	395	417	430	448	470
2600	3710	387	395	408	418	433	461	485	500	520	543
2800	4000	449	458	470	482	497	526	553	574	595	621
3000	4285	515	526	536	550	567	597	628	655	666	706
3250	4640	605	613	629	641	665	701	729	757	777	817
3500	5000	697	711	729	739	771	808	841	869	897	938

Friction Loss for Viscous Liquids *(Continued)*
(Based on Darcy's Formula)
Loss in Feet of Liquid per 1000 Feet of Pipe
6 Inch (6.065" inside dia) Sch 40 New Steel Pipe

Flow		Kinematic viscosity—centistokes								
		26.4	32.0	43.2	65.0	108.4	162.3	216.5	325	435	650
US gal per min	Bbl per hr (42 gal)	Approx SSU viscosity								
		125	150	200	300	500	750	1000	1500	2000	3000
50	71.4	.62	.74	1.00	1.51	2.52	3.78	5.04	7.57	10.1	15.1
75	107	.92	1.12	1.51	2.27	3.78	5.66	7.56	11.4	15.2	22.7
100	143	1.23	1.49	2.01	3.03	5.05	7.55	10.1	15.1	20.3	30.2
125	178	2.75	1.86	2.51	3.79	6.31	9.45	12.6	18.9	25.3	37.8
150	214	3.75	3.96	3.01	4.54	7.58	11.3	15.1	22.7	30.4	45.4
175	250	4.90	5.17	5.62	5.30	8.84	13.2	17.6	26.5	35.5	53.0
200	286	6.10	6.51	7.07	6.06	10.1	15.1	20.2	30.3	40.5	60.6
225	322	7.43	7.93	8.68	6.82	11.4	17.0	22.7	34.1	45.6	68.1
250	357	8.91	9.43	10.4	7.57	12.6	18.9	25.2	37.8	50.7	75.7
275	393	10.6	11.1	12.2	13.7	13.9	20.8	27.7	41.7	55.8	83.2
300	429	12.3	12.9	14.2	15.9	15.1	22.6	30.2	45.4	60.9	90.8
350	500	15.9	17.1	18.3	20.8	17.7	26.4	35.3	53.0	71.0	106
400	571	20.1	21.3	23.1	26.2	20.2	30.2	40.3	60.6	81.1	121
450	643	24.7	26.0	28.6	31.9	36.9	34.0	45.3	68.2	91.3	136
500	714	30.0	31.3	34.1	28.0	44.2	37.8	50.4	75.7	101	151
550	786	35.6	36.9	40.2	44.6	52.1	41.6	55.4	83.3	112	166
600	857	41.5	43.1	46.4	51.7	59.1	45.3	60.5	90.9	122	182
650	929	47.7	50.0	53.4	59.6	69.4	49.1	65.5	98.5	132	197
700	1000	54.1	57.0	60.8	68.6	78.8	88.3	70.6	106	142	212
750	1070	60.8	64.4	68.5	76.8	88.5	99.4	75.6	114	152	227
800	1140	68.0	72.1	76.9	85.7	97.8	111	80.6	121	162	242
900	1285	83.9	88.5	95.2	105	120	136	148	136	183	272
1000	1430	101	106	115	126	144	163	177	151	203	302
1100	1570	120	125	136	148	171	192	208	167	223	333
1200	1715	140	146	158	173	200	220	242	182	243	363
1400	2000	184	193	206	230	258	287	316	353	284	424
1600	2285	234	244	260	288	323	363	393	445	324	484
1800	2570	292	299	322	350	399	452	480	543	591	545
2000	2860	350	364	387	425	481	535	576	652	707	605
2200	3140	417	435	459	510	573	628	683	771	833	666
2400	3403	487	507	535	585	668	730	799	885	968	726
2600	3710	564	587	620	677	769	841	913			787
2800	4000	645	669	714	773	874	954				
3000	4285	734	751	805	867	993					
3200	4570	827	850	909	982						

Table 45 Line friction loss

Friction Loss for Viscous Liquids *(Continued)*
(Based on Darcy's Formula)
Loss in Feet of Liquid per 1000 Feet of Pipe
8 Inch (7.981" inside dia) Sch 40 New Steel Pipe

Flow		Kinematic viscosity—centistokes									
		.6	1.1	2.1	2.7	4.3	7.4	10.3	13.1	15.7	20.6
US gal per min	Bbl per hr (42 gal)	Approx SSU viscosity									
		31.5	33	35	40	50	60	70	80	100	
150	214	.42	.47	.53	.56	.63	.72	.79	.84	.88	.96
200	286	.71	.78	.89	.94	1.05	1.15	1.30	1.38	1.45	1.57
250	357	1.07	1.18	1.33	1.40	1.56	1.77	1.92	2.04	2.14	2.30
300	429	1.50	1.65	1.85	1.94	2.15	2.43	2.65	2.80	2.93	3.15
350	500	2.01	2.19	2.45	2.57	2.81	3.20	3.46	3.69	3.85	4.13
400	571	2.58	2.78	3.12	3.26	3.58	4.04	4.37	4.64	4.83	5.21
450	643	3.21	3.48	3.85	4.05	4.42	5.08	5.39	5.70	5.96	6.38
500	714	3.94	4.23	4.69	4.90	5.33	5.98	6.49	6.82	7.18	7.91
600	857	5.54	5.95	6.61	6.82	7.44	8.27	8.89	9.48	9.83	10.6
700	1000	7.44	7.96	8.71	9.04	9.84	10.9	11.9	12.4	13.0	13.9
800	1140	9.66	10.2	11.1	11.7	12.6	13.9	14.8	15.8	16.4	17.4
900	1285	12.1	12.8	13.8	14.4	15.5	17.1	18.4	19.4	20.2	21.5
1000	1430	14.8	15.5	16.8	17.4	18.7	20.8	22.2	23.3	24.4	26.0
1200	1715	21.0	22.0	23.7	24.5	26.4	28.9	30.7	32.4	33.5	35.9
1400	2000	28.3	29.6	31.8	32.6	35.6	38.2	40.7	42.6	44.3	47.5
1600	2285	36.7	38.4	40.6	42.1	44.5	48.7	51.9	54.1	56.1	59.3
1800	2570	46.1	48.0	50.8	52.3	55.4	60.4	64.5	67.0	69.4	73.5
2000	2860	56.5	58.8	61.9	63.8	67.3	73.4	77.7	81.1	83.8	80.8
2200	3140	67.9	70.2	74.4	76.3	80.5	87.5	92.1	96.8	99.6	105
2400	3430	80.8	83.0	88.0	90.2	94.7	103	108	114	117	123
2600	3710	94.2	97.5	103	105	110	119	125	131	135	142
2800	4000	109	112	118	121	127	136	144	149	155	163
3000	4285	125	129	135	138	145	155	164	170	176	184
3200	4570	142	146	153	156	162	174	184	191	197	208
3400	4860	160	164	172	174	182	196	206	213	220	232
3600	5140	178	183	192	196	204	217	228	237	244	258
3800	5425	199	204	212	217	226	240	251	262	269	285
4000	5715	220	225	234	238	249	265	277	289	295	311
4500	6425	276	284	294	300	311	331	345	358	368	385
5000	7145	341	348	350	365	380	404	418	433	447	466
5500	7855	410	419	433	439	457	480	500	518	532	555
6000	8570	488	498	512	519	540	567	592	609	623	654
6500	9280	573	581	601	609	630	662	686	707	723	755
7000	10000	664	673	692	702	725	763	791	810	829	867
7500	10700	760	773	789	806	827	865	897	925	946	984

Friction Loss for Viscous Liquids *(Continued)*
(Based on Darcy's Formula)
Loss in Feet of Liquid per 1000 Feet of Pipe
8 Inch (7.981" inside dia) Sch 40 New Steel Pipe

Flow		Kinematic viscosity—centistokes								
		26.4	32.0	43.2	65.0	108.4	162.3	216.5	325	435	650
US gal per min	Bbl per hr (42 gal)	Approx SSU viscosity									
		125	150	200	300	500	750	1000	1500	2000	3000
50	71.4	.21	.25	.34	.50	.84	1.26	1.68	2.52	3.30	5.06
100	143	.41	.50	.67	1.01	1.68	2.52	3.36	5.04	6.70	10.1
150	214	1.03	.75	1.01	1.51	2.52	3.78	5.04	7.56	10.1	15.1
200	286	1.67	1.78	1.34	2.02	3.37	5.04	6.72	10.1	13.5	20.2
250	357	2.46	2.60	2.85	2.52	4.21	6.30	8.40	12.6	16.9	25.3
300	429	3.37	3.56	3.89	3.03	5.05	7.56	10.1	15.1	20.3	30.3
350	500	4.39	4.63	5.04	5.69	5.89	8.82	11.8	17.7	23.6	35.3
400	571	5.54	5.83	6.35	7.18	6.73	10.1	13.5	20.2	27.0	40.4
450	643	6.79	7.16	7.75	8.76	7.58	11.3	15.1	22.7	30.4	45.4
500	714	8.17	8.58	9.32	10.4	8.42	12.6	16.8	25.2	33.8	50.5
550	786	9.65	10.1	11.0	12.3	9.26	13.9	18.5	27.8	37.2	55.5
600	857	11.2	11.8	12.8	14.3	16.6	15.1	20.2	30.3	40.5	60.6
700	1000	14.7	15.5	16.7	18.6	21.7	17.6	23.5	35.3	47.3	70.6
800	1140	18.6	19.4	21.0	23.4	27.1	20.1	26.9	40.3	54.0	80.7
900	1285	22.9	24.0	25.9	28.7	33.1	37.4	30.2	45.4	60.5	90.8
1000	1430	27.4	28.8	31.0	34.4	39.7	44.9	33.9	50.4	67.6	101
1200	1715	37.8	39.4	42.8	47.3	54.3	60.9	66.4	60.5	81.0	121
1400	2000	49.7	52.0	56.1	62.0	70.8	78.3	88.7	70.6	94.5	141
1600	2285	63.2	65.7	70.6	78.0	89.3	99.4	109	80.7	108	161
1800	2570	77.9	81.3	86.9	96.2	110	122	132	150	122	182
2000	2860	93.4	98.0	105	116	132	147	159	180	135	202
2200	3140	111	116	124	137	155	173	186	211	149	222
2400	3430	130	135	145	159	181	201	217	244	266	242
2600	3710	149	155	168	183	208	231	250	279	306	262
2800	4000	170	178	191	210	236	262	283	317	347	282
3000	4285	193	201	215	236	267	296	319	357	389	303
3200	4570	217	225	240	265	299	332	357	396	433	323
3400	4860	242	251	268	294	333	369	397	441	480	343
3600	5140	268	284	296	326	369	407	438	488	529	598
3800	5425	296	307	328	359	404	447	482	536	582	656
4000	5715	325	337	358	394	441	488	526	586	635	718
4500	6425	405	417	442	485	543	601	646	718	777	876
5000	7145	488	505	536	582	656	726	776	860	932	
5500	7855	578	605	634	689	773	855	913			
6000	8570	678	708	744	810	910	993				

Table 46 Line friction loss

Friction Loss for Viscous Liquids *(Continued)*
(Based on Darcy's Formula)
Loss in Feet of Liquid per 1000 Feet of Pipe
10 Inch (10.02" inside dia) Sch 40 New Steel Pipe

Flow		Kinematic viscosity—centistokes								
		.6	1.1	2.1	2.7	4.3	7.4	10.3	13.1	15.7	20.6
US gal per min	Bbl per hr (42 gal)					Approx SSU viscosity					
		31.5	33	35	40	50	60	70	80	100
400	571	.83	.92	1.03	1.09	1.19	1.35	1.46	1.55	1.63	1.75
500	714	1.27	1.38	1.53	1.60	1.78	2.00	2.16	2.30	2.40	2.59
600	857	1.78	1.91	2.14	2.24	2.47	2.77	2.98	3.15	3.31	3.54
700	1000	2.39	2.55	2.84	2.97	3.26	3.62	3.93	4.14	4.32	4.67
800	1140	3.06	3.29	3.63	3.79	4.12	4.63	4.99	5.25	5.46	5.86
900	1285	3.84	4.12	4.49	4.72	5.09	5.72	6.14	6.46	6.74	7.19
1000	1430	4.68	4.99	5.42	5.70	6.13	6.90	7.36	7.83	8.10	8.63
1100	1570	5.63	5.97	6.49	6.82	7.34	8.20	8.76	9.25	9.62	10.3
1200	1715	6.61	7.05	7.63	7.85	8.65	9.58	10.3	10.8	11.3	11.9
1300	1855	7.71	8.18	8.85	9.16	9.95	11.0	11.8	12.4	13.0	13.7
1400	2000	8.88	9.42	10.2	10.6	11.4	12.6	13.5	14.2	14.7	15.7
1500	2140	10.1	10.8	11.7	12.0	12.9	14.3	15.3	16.1	16.6	17.8
1600	2285	11.5	12.2	13.2	13.6	14.6	16.0	17.2	18.1	18.7	20.0
1800	2570	14.3	15.1	16.2	16.7	17.9	19.7	20.9	22.1	22.9	24.3
2000	2860	17.8	18.6	19.8	20.6	21.8	24.0	25.5	27.0	28.0	29.5
2200	3140	21.3	22.2	23.7	24.6	26.1	28.6	30.3	32.1	31.8	35.0
2400	3430	25.2	26.3	28.0	28.9	30.7	33.4	35.5	37.3	38.9	41.0
2600	3710	29.6	30.6	32.5	33.5	35.6	38.7	41.0	42.9	44.8	47.3
2800	4000	34.1	35.3	37.4	38.4	40.8	44.5	47.1	49.0	51.0	54.1
3000	4286	39.1	40.2	42.7	43.5	46.6	50.7	53.2	55.7	57.7	61.3
3500	5000	52.5	54.4	57.4	58.9	62.3	66.4	70.6	73.6	76.2	80.8
4000	5715	68.0	70.5	73.9	75.9	79.9	85.8	90.2	94.2	97.1	102
4500	6430	86.1	88.6	92.3	94.8	99.2	107	112	117	120	127
5000	7145	106	109	113	116	122	130	136	142	146	153
5500	7855	128	131	136	139	145	156	162	169	173	182
6000	8570	152	154	161	164	172	183	191	197	204	213
6500	9280	177	180	187	191	201	212	221	228	236	246
7000	10000	205	208	217	220	231	243	255	263	269	282
7500	10700	236	239	248	251	262	277	291	298	303	321
8000	11400	266	272	282	286	296	314	329	337	345	360
8500	12100	301	307	318	321	334	352	367	378	387	403
9000	12900	337	341	354	359	372	392	407	422	429	447
10000	14300	416	422	434	441	453	478	492	511	524	542
11000	15700	503	511	522	533	544	574	593	611	626	649
12000	17150	599	603	617	630	643	679	701	719	737	763

Friction Loss for Viscous Liquids *(Continued)*
(Based on Darcy's Formula)

Loss in Feet of Liquid per 1000 Feet of Pipe
10 Inch (10.02" inside dia) Sch 40 New Steel Pipe

Flow		Kinematic viscosity—centistokes								
		26.4	32.0	43.2	65.0	108.4	162.3	216.5	325	435	650
U S gal per min	Bbl per hr (42 gal)	Approx SSU viscosity									
		125	150	200	300	500	750	1000	1500	2000	3000
150	214	.25	.30	.40	.61	1.02	1.52	2.03	3.04	4.06	6.09
200	286	.58	.40	.54	.81	1.35	2.03	2.71	4.06	5.43	8.12
300	429	1.15	1.22	1.33	1.22	2.03	3.04	4.06	6.09	8.15	12.2
400	571	1.83	1.98	2.17	1.62	2.71	4.06	5.41	8.12	10.9	16.2
500	714	2.75	2.91	3.18	3.60	3.39	5.07	6.77	10.1	13.6	20.3
600	857	3.78	3.97	4.34	4.89	4.06	6.08	8.12	12.2	16.3	24.4
700	1000	4.94	5.19	5.66	6.37	4.74	7.10	9.47	14.2	19.0	28.4
800	1140	6.21	6.55	7.10	7.97	9.31	8.12	10.8	16.2	21.7	32.5
900	1285	7.66	8.04	8.71	9.76	11.4	9.13	12.2	18.3	24.5	36.5
1000	1430	9.21	9.61	10.5	11.7	13.6	10.1	13.5	20.3	27.2	40.6
1100	1570	10.9	11.4	12.3	13.8	16.0	18.0	14.9	22.3	29.9	44.8
1200	1715	12.6	13.4	14.4	16.0	18.6	21.0	16.2	24.4	32.6	48.7
1300	1855	14.5	15.4	16.4	18.4	21.2	24.0	17.6	26.4	35.3	52.8
1400	2000	16.6	17.5	18.7	20.9	24.1	27.2	18.9	28.4	38.0	56.8
1500	2140	18.7	19.6	21.2	23.6	27.2	30.6	33.3	30.4	40.8	60.9
1600	2285	21.0	21.9	23.7	26.3	30.4	34.1	37.2	32.4	43.5	64.9
1800	2570	26.0	27.3	29.2	32.2	37.1	41.7	45.5	36.5	48.9	73.1
2000	2860	31.3	32.7	35.0	38.6	44.4	49.6	54.4	40.6	54.3	81.2
2200	3140	37.0	38.6	41.3	45.8	52.3	58.8	64.0	71.9	59.8	89.3
2400	3430	43.1	45.1	48.3	53.4	60.9	68.3	74.2	83.8	85.2	97.4
2600	3710	49.8	52.1	55.4	61.4	70.0	78.3	85.0	96.2	70.7	105
2800	4000	56.8	59.2	63.5	70.1	79.7	88.9	96.1	109	76.1	114
3000	4285	64.3	66.6	71.8	78.8	89.8	99.8	108	122	133	122
3500	5000	84.9	88.3	94.4	103	117	131	142	159	180	142
4000	5715	108	112	119	131	149	165	178	199	217	162
4500	6430	133	139	147	162	182	202	218	244	266	300
5000	7145	160	168	179	195	218	241	262	293	318	360
5500	7855	191	199	212	231	258	286	309	345	373	423
6000	8570	223	232	247	268	302	334	359	399	435	489
6500	9280	258	267	286	310	348	384	411	460	499	560
7000	10000	296	305	326	355	396	438	466	522	566	637
7500	10700	335	347	369	402	447	492	529	589	638	766
8000	11400	377	389	414	452	505	550	594	659	710	797
9000	12900	469	482	512	557	624	679	729	809	869	976
10000	14300	567	582	619	686	743	817	872	964		

Table 47 Line friction loss

Friction Loss for Viscous Liquids *(Continued)*
(Based on Darcy's Formula)
Loss in Feet of Liquid per 1000 Feet of Pipe
12 Inch (11.938" inside dia) Sch 40 New Steel Pipe

Flow		Kinematic viscosity—centistokes								
US gal per min	Bbl per hr (42 gal)	,0.6	1.13	2.1	2.7	4.3	7.4	10.3	13.1	15.7	20.6
						Approx SSU viscosity				
			31.5	33	35	40	50	60	70	80	100
300	429	.21	.24	.27	.28	.31	.36	.38	.41	.43	.47
400	571	.36	.40	.45	.47	.51	.59	.64	.67	.70	.77
500	714	.54	.60	.67	.70	.75	.87	.95	.99	1.03	1.12
600	857	.76	.83	.93	.98	1.04	1.19	1.30	1.39	1.41	1.53
700	1000	.98	1.11	1.23	1.29	1.37	1.56	1.70	1.82	1.84	2.00
800	1140	1.26	1.42	1.57	1.64	1.74	1.98	2.15	2.30	2.36	2.51
900	1285	1.57	1.76	1.94	1.96	2.15	2.44	2.65	2.82	2.94	3.08
1000	1430	1.92	2.07	2.36	2.38	2.61	2.94	3.19	3.40	3.57	3.70
1200	1715	2.73	2.91	3.18	3.32	3.62	4.07	4.41	4.68	4.91	5.08
1400	2000	3.67	3.90	4.24	4.41	4.80	5.37	5.79	6.14	6.43	6.65
1600	2285	4.75	5.02	5.43	5.64	6.12	6.83	7.35	7.78	8.14	8.51
1800	2570	5.96	6.29	6.77	7.02	7.59	8.44	9.07	9.59	10.0	10.6
2000	2860	7.32	7.69	8.25	8.54	9.21	10.2	11.0	11.6	12.1	12.9
2500	3570	11.3	11.8	12.6	13.0	13.9	15.3	16.4	17.3	18.0	19.2
3000	4285	16.1	16.8	17.7	18.3	19.5	21.4	22.8	23.9	24.9	26.5
3500	5000	21.8	22.6	23.8	24.4	26.0	28.3	30.1	31.6	32.9	34.9
4000	5715	28.3	29.2	30.7	31.5	33.3	36.2	38.4	40.3	41.8	44.3
4500	6430	35.7	36.8	38.5	39.4	41.6	45.0	47.7	49.9	51.7	54.8
5000	7145	44.0	45.2	47.1	48.2	50.7	54.7	57.8	60.4	62.6	66.2
5500	7855	53.1	54.4	56.6	57.8	60.7	65.3	68.9	71.9	74.4	78.7
6000	8570	63.0	64.5	66.9	68.3	71.6	76.8	80.9	84.3	87.2	92.1
6500	9260	73.8	75.4	78.1	79.6	83.3	89.2	93.8	97.7	101	106
7000	10000	85.4	87.2	90.1	91.8	95.9	102	108	112	116	122
7500	10700	97.9	99.8	103	105	109	117	122	127	131	138
8000	11400	111	113	117	119	124	132	138	143	148	155
9000	12850	141	143	147	149	155	164	172	178	183	192
10000	14300	173	176	180	183	190	200	209	217	223	233
11000	15700	209	212	217	220	228	240	250	269	266	278
12000	17150	249	252	258	261	269	283	294	304	312	326
13000	18550	291	295	301	305	314	330	342	353	363	373
14000	20000	338	342	348	353	363	380	394	406	416	434
15000	21400	387	392	399	403	414	433	449	462	473	493
16000	22850	440	445	453	457	469	490	507	522	534	556
18000	25700	557	561	571	577	590	614	634	651	666	692
20000	28600	687	692	703	709	725	752	775	795	812	842

Friction Loss for Viscous Liquids *(Continued)*
(Based on Darcy's Formula)

Loss in Feet of Liquid per 1000 Feet of Pipe
12 Inch (11.938" inside dia) Sch 40 New Steel Pipe

Flow		Kinematic viscosity—centistokes									
US gal per min	Bbl per hr (42 gal)	26.4	32.0	43.2	65.0	108.4	162.3	216.5	325	435	650
		\multicolumn{10}{c}{Approx SSU viscosity}									
		125	150	200	300	500	750	1000	1500	2000	3000
100	143	.08	.10	.13	.20	.34	.51	.68	1.00	1.35	2.00
200	286	.16	.19	.27	.40	.67	1.00	1.37	2.05	2.74	4.01
300	429	.49	.53	.41	.62	1.01	1.51	2.00	3.08	4.11	6.16
400	571	.81	.86	.94	.82	1.34	2.02	2.68	3.98	5.46	8.21
500	714	1.22	1.25	1.37	1.02	1.71	2.50	3.37	5.01	6.84	10.3
600	857	1.66	1.71	1.87	2.12	1.97	3.02	4.05	6.03	7.99	12.3
700	1000	2.15	2.30	2.43	2.75	2.37	3.66	4.88	7.05	9.36	13.9
800	1140	2.70	2.88	3.05	3.45	2.83	4.04	5.36	8.09	10.7	15.9
900	1285	3.31	3.52	3.74	4.21	4.93	4.44	6.05	9.30	12.1	18.0
1000	1430	3.97	4.22	4.48	5.04	5.89	5.13	6.84	10.0	13.5	20.0
1200	1715	5.43	5.77	6.33	6.88	8.01	5.91	7.89	12.1	16.2	24.1
1400	2000	7.10	7.53	8.24	8.96	10.4	11.8	9.47	14.6	19.9	28.2
1600	2285	8.96	9.48	10.4	11.3	13.1	14.8	10.5	16.2	21.7	32.4
1800	2570	11.0	11.6	12.7	13.8	16.0	18.0	19.7	17.7	24.2	36.5
2000	2860	13.2	14.0	15.2	16.6	19.1	21.5	23.6	20.5	27.4	40.1
2500	3570	19.6	20.6	22.4	25.3	28.0	31.5	34.3	26.7	34.2	51.3
3000	4285	27.2	28.4	30.8	34.6	38.4	43.0	46.8	53.1	41.1	61.6
3500	5000	36.2	37.3	40.3	45.2	50.2	56.0	60.9	68.8	47.9	71.9
4000	5715	43.4	47.3	51.0	57.0	63.2	70.5	76.5	86.3	94.5	82.1
4500	6430	57.6	58.3	62.8	69.9	77.6	86.4	93.6	105	115	92.4
5000	7145	69.8	70.3	75.6	84.1	93.3	104	112	126	138	103
5500	7855	82.8	83.6	89.5	99.4	110	122	132	148	162	183
6000	8570	96.8	98.2	104	116	128	142	154	172	188	212
6500	9280	112	114	120	133	148	164	176	197	215	243
7000	10000	128	131	137	152	174	186	201	224	244	275
7500	10700	145	148	155	172	196	210	226	253	274	310
8000	11400	163	167	174	192	220	235	253	282	306	345
9000	12850	202	208	215	237	270	289	311	346	375	422
10000	14300	204	253	260	286	325	347	373	415	450	505
11000	15700	290	301	309	338	384	411	441	490	530	594
12000	17150	341	354	361	395	448	479	514	569	616	689
13000	18550	394	409	417	456	516	551	591	655	707	790
14000	20000	452	469	477	521	588	628	673	745	804	898
15000	21400	513	532	541	589	664	710	760	840	906	
16000	22850	578	598	608	662	745	796	851	940		

Appendix 10: Heats of Combustion for Fuel Oils

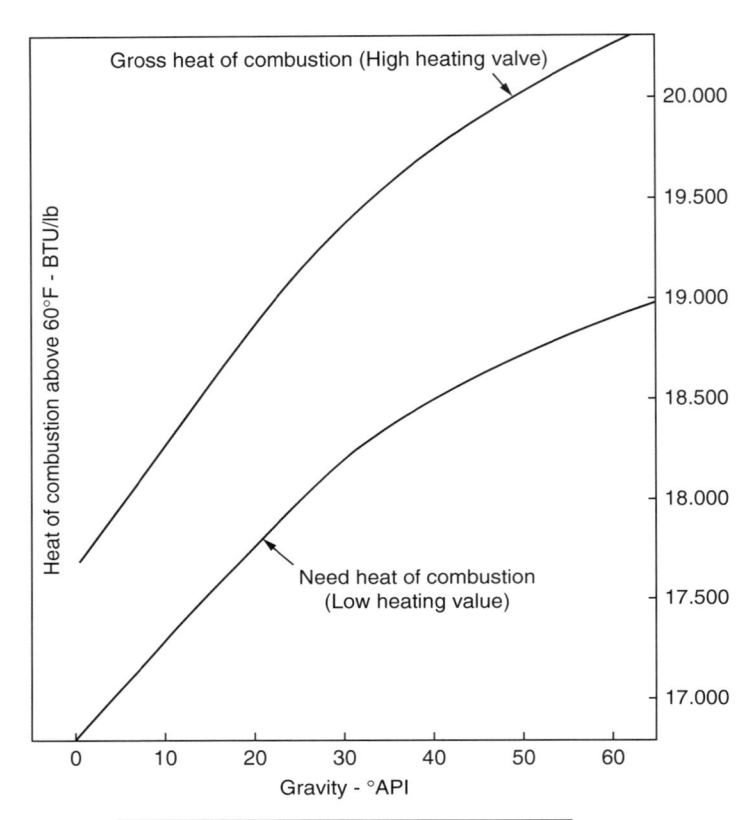

Graph: Heat of combustion above 60°F - BTU/lb vs. Gravity - °API, showing Gross heat of combustion (High heating valve) and Need heat of combustion (Low heating value) curves, ranging from 17.000 to 20.000 BTU/lb over 0 to 60 °API.

°API	%S	%inerts	Total impurity
Impurities in average fuels			
Residual fuel oil and crudes			
0	2.95	1.15	4.10
5	2.35	1.00	3.35
10	1.80	0.95	2.75
15	1.35	0.65	2.20
20	1.00	0.75	1.75
Crude oils			
25	0.70	0.70	1.40
30	0.40	0.65	1.10
35	0.30	0.60	0.90

Appendix 11: Heats of Combustion for Fuel Gases

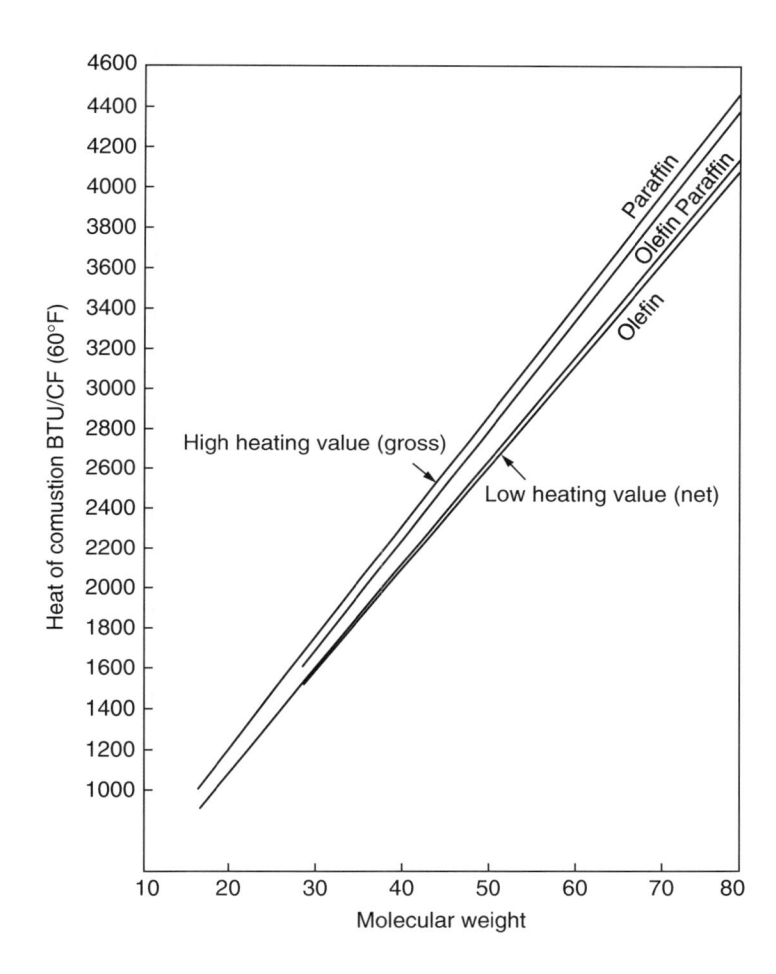

Appendix 12: Resistances of Valves and Fittings

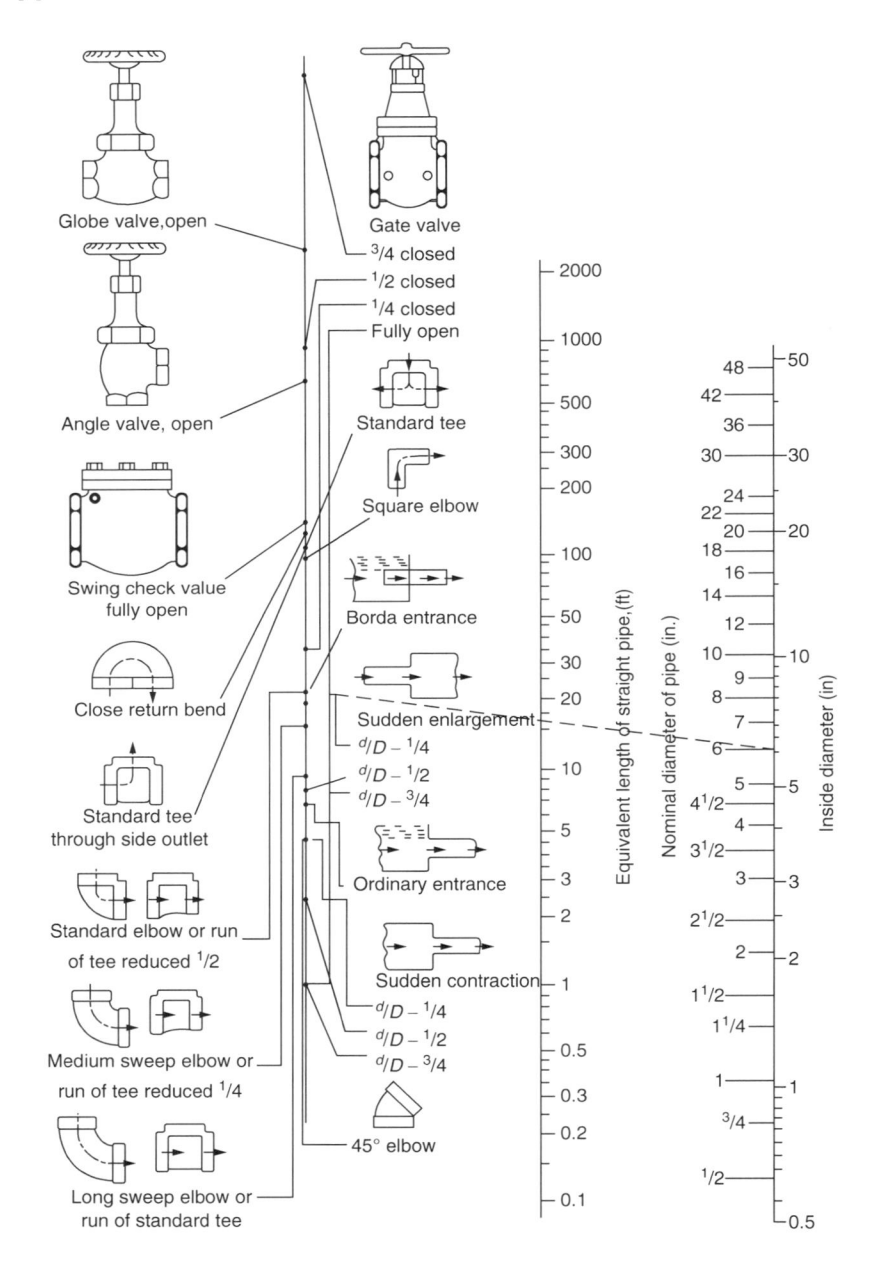

Appendix 13: Flow Pressure Drop for gas Streams

See Figs. 62 and 63.

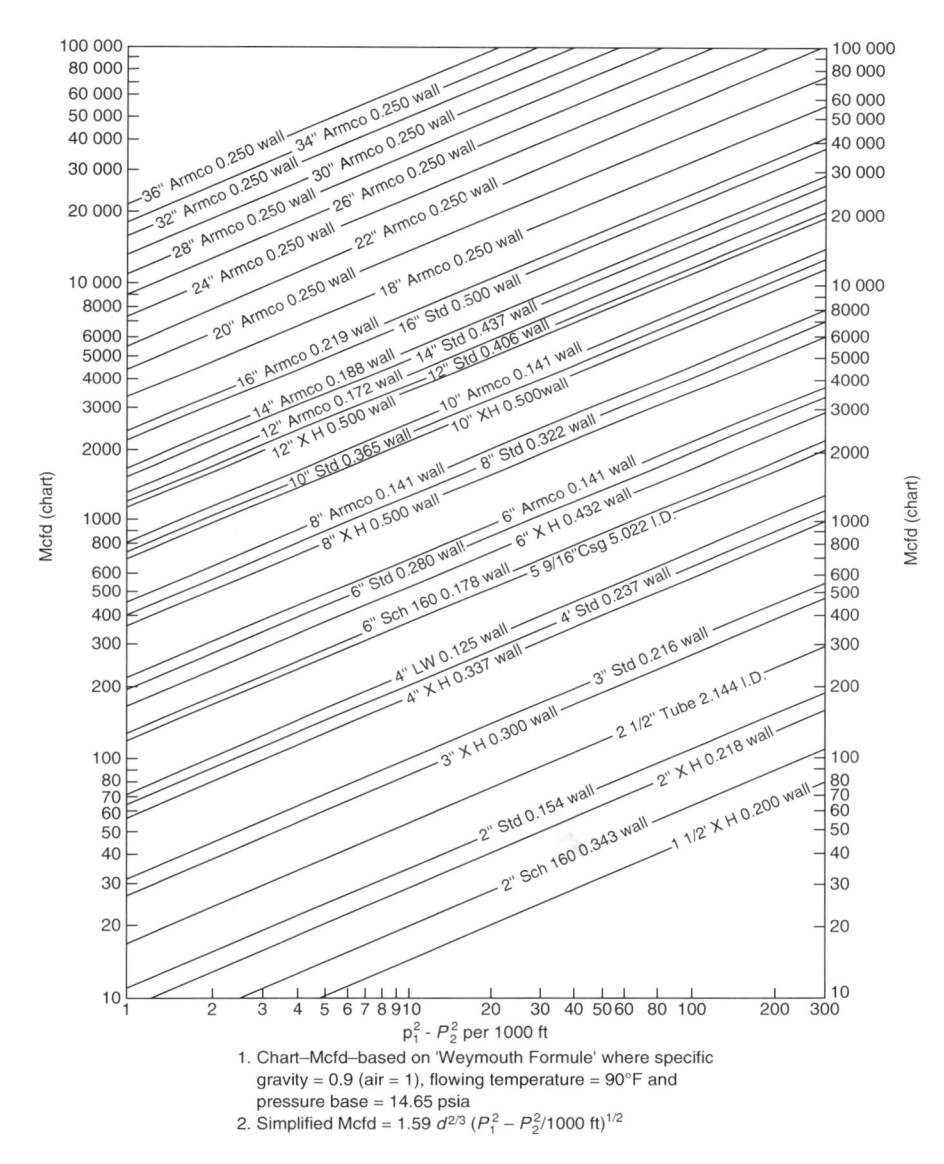

1. Chart–Mcfd–based on 'Weymouth Formule' where specific gravity = 0.9 (air = 1), flowing temperature = 90°F and pressure base = 14.65 psia
2. Simplified Mcfd = 1.59 $d^{2/3}$ $(P_1^2 - P_2^2/1000 \text{ ft})^{1/2}$

Fig. 62 Flow pressure drop for gas streams

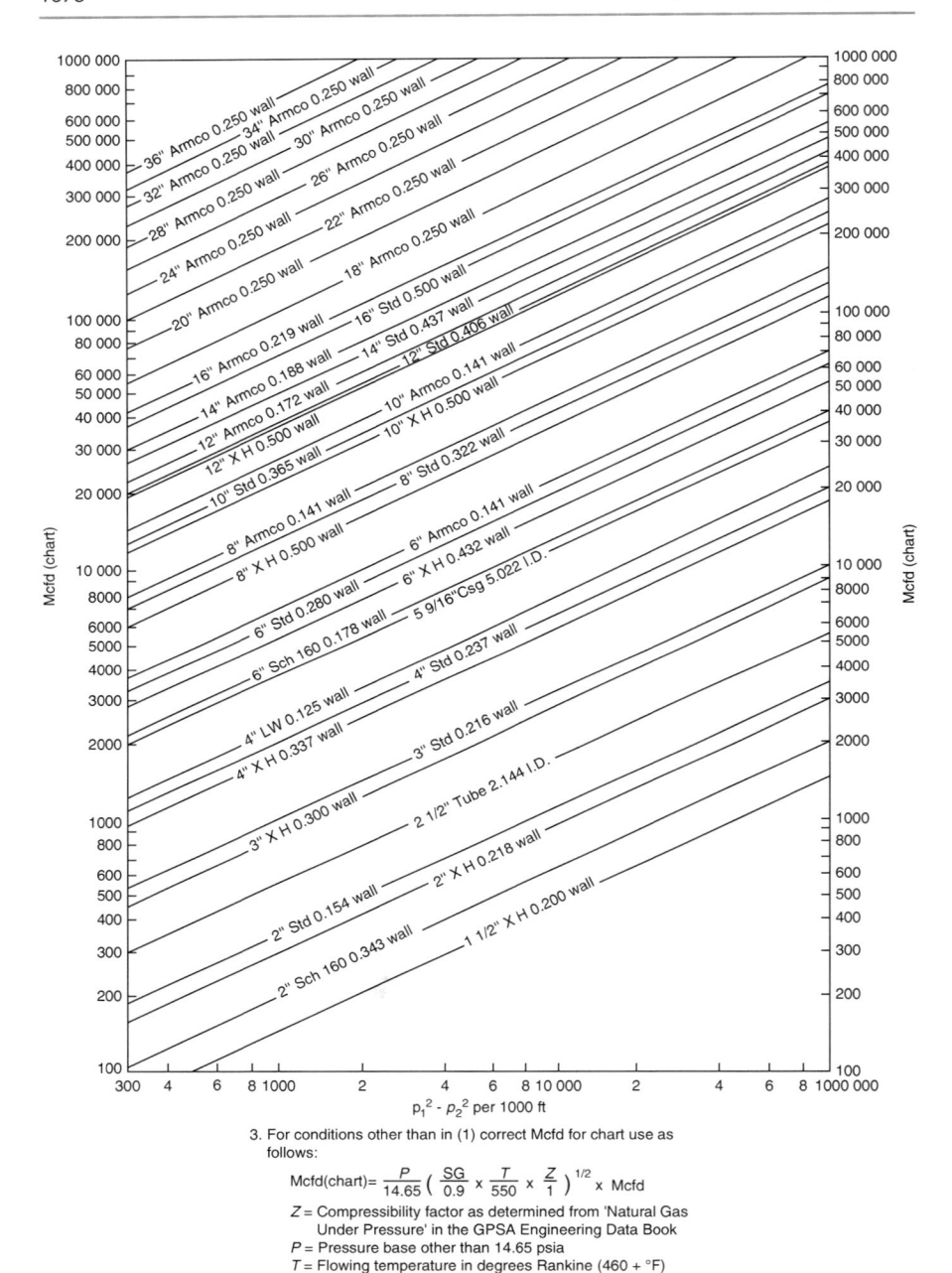

3. For conditions other than in (1) correct Mcfd for chart use as follows:

$$\text{Mcfd(chart)} = \frac{P}{14.65} \left(\frac{SG}{0.9} \times \frac{T}{550} \times \frac{Z}{1} \right)^{1/2} \times \text{Mcfd}$$

Z = Compressibility factor as determined from 'Natural Gas Under Pressure' in the GPSA Engineering Data Book
P = Pressure base other than 14.65 psia
T = Flowing temperature in degrees Rankine (460 + °F)

Fig. 63 Flow pressure drop for gas streams

Process Equipment for Petroleum Processing 1679

Appendix 14: Example Hydraulic Analysis of a Process System

The following calculation is an example of a typical system that process engineers' encounter in a design of a plant or in checking out an operating plant's process flow. The attached diagram, Fig. 64, is used as the basis for this example.

Example Calculation Basis
Total flow to P-103 A and B = 519,904 lbs/h.
API gravity = 20.7 = SG of 0.930 at 60 °F.
Stream temperature = 545 °F.
SG at 545°F = 0.755 = 6.287 lbs/gal.
Gallons per minute at temperature = 519,904/(6.287 × 60) = 1,378 gpm.
Viscosity of the oil at 545°F = 1.2 cSt.
Suction line to P-103 A and B is 10″; Sch. 40.

From Appendix 9 of this chapter:
Friction loss in feet/1,000 f. of pipe is 9.3 f. (equivalent to 0.30 psi/100 ft).

Suction Line Equivalent Length
The following information would normally be obtained from a piping general arrangement drawing (a piping GA). For this calculation, this information is fictional:

- Number of standard elbows in the line = 8.
- Number of gate valves (all open) in the line = 3.
- Total straight length of 10″ line = 85 ft.

From Appendix 12 of this chapter:

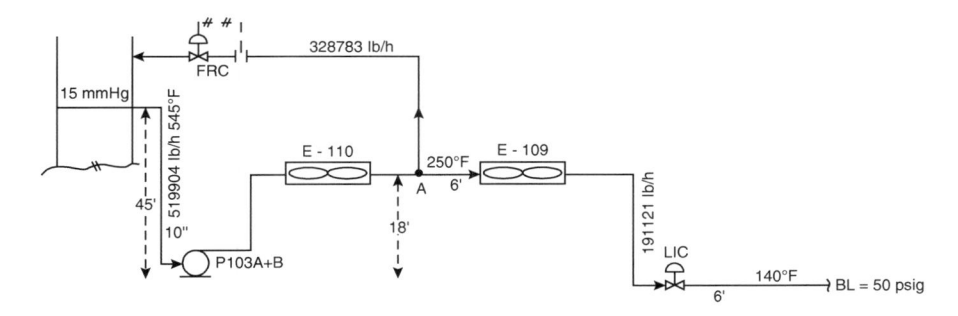

Fig. 64 Example system pressure drop calculation flow sheet

The equivalent length for 10" elbows is 22 ft per elbow
$= 22 \times 8 = 176$ ft.

The equivalent length for open 10" gate valves is 5 ft per valve.
$= 5 \times 3 = 15$ ft.

Total equivalent line length : $85 + 176 + 15 = 276$ ft.

Head loss to pump suction due to friction
$= (9.3 \times 276)/1000 = 2.57$ ft.

In terms of pounds per square inch this is
$= (2.57 \times 62.2 \times 0.755)/144 = 0.83$ psi.

Pump Suction Pressure

Source pressure at vacuum tower draw off $= 15$ mmHg $= 0.29$ psia.
Static head $= 45$ ft. $= (45 \times 62.2 \times 0.755)/144 = 14.7$ psi.
Line loss $= 0.83$ psi.
Total pressure at pump suction flange $= 0.29 + 14.7 - 0.83 = 14.2$ psia
$= -0.5$ psig.

Calculating the Pressures at the Pump Discharge
Destination pressure at battery limits $= 50$ psig.
Temperature of the oil at battery limits $= 140\,°F$.
Viscosity of the oil at $140\,°F = 20$ cSt.
SG of the oil at $140\,°F = 0.900$; lbs/gal $= 7.495$.

From a material balance or from plant data: Flow of oil $= 191{,}121$ lbs/h.

Flow rate $= 191{,}121/(7.495 \times 60) = 425$ gpm.

Line Pressure Drop from E-109 to Battery Limits
Equivalent length of line:

Straight line $= 126$ ft.
Number of elbows $= 18$ equiv. length $= 18 \times 16 = 288$ ft.
Number of gate valves $= 6$ equiv. length $= 6 \times 3.5 = 21$ ft.
Number of Tee's $= 1$ equiv. length $= 1 \times 30 = 30$ ft.
Total equivalent length $= 465$ ft.

Line to battery limit (BL) is a 6" Schedule 40. Then, from Appendix 9, loss due to friction is 21.4 ft/1,000 f. which is equivalent to 0.83 psi/100 ft.

Then line friction loss is

$$(21.4 \times 245)/1{,}000 = 9.95 \text{ ft.}$$

or

$$(9.95 \times 62.2 \times 0.900)/144 = 3.86 \text{ psi.}$$

Control Valve Pressure Drop

There is a battery limit level control valve (LICV) between E-109 and the BL. It is required to calculate the pressure drop for this valve at design flow. The valve pressure drop will be estimated as 20 % of the circuit frictional pressure drop plus 10 % of the static head of the receiving vessel.

The oil discharges into a surge drum of another downstream unit. This drum is pressurized by a blanket of inert gas. The net static head to this drum is 15 f. above valve outlet flange. This is equivalent to 6 psi.

The total line pressure drop for the whole circuit is estimated at three times that calculated above. (This will be checked and may be revised when the analysis of the whole circuit is complete.) This pressure drop therefore is $3 \times 3.86 = 11.58$ psi for line losses. In addition to this line loss, there are also two air coolers which have pressure drops as follows:

E-109 = 6 psi (from data sheets).
E-110 = 8 psi.

Then total system pressure drop is estimated as $11.58 + 14 = 25.58$ psi.
And control valve pressure drop is

$$(0.1 \times 6.0) + (0.2 \times 25.58) = 5.2 \text{ psi.}$$

Calculate Pressure at Point "A" Which Is the Reflux Stream Take Off

Equivalent length of line between point "A" and inlet to E-109 is as follows:

Straight line = 121 ft.
4 elbows = $4 \times 16 = 64$ ft.
2 valves = $2 \times 3.5 = 7$ ft.
Total 6″ sched 40 line equiv. = 192 ft.

Temperature of stream at this point is 250 °F.
Viscosity at 250 °F = 3.5 cSt.
SG at 250 °F is 0.860 which is 7.16 lbs/gal.
Rate of flow into E-109 = $191{,}121/(7.16 \times 60) = 445$ gpm.
Friction loss in 6″ pipe = 16.1 ft/1,000 ft.

Total line loss:

$$(192 \times 16.1)/1000 = 3.09 \text{ ft or } 1.15 \text{ psi.}$$

Pressure at point "A" therefore is the sum of:

Destination pressure at BL = 50 psig.
Pressure drop for E-109 = 6 psi.
Loss in line from E-109 to battery limits. = 3.86 psi.
Loss in line from point "A" to E-109 = 1.15 psi.
Control valve pressure drop = 5.2 psi.
Flow meter (not shown) say = 0.2 psi.
Total pressure at point "A" = 66.41 psig.

Calculating the Pressure at the Pump Discharge Flange

Total flow from the pump is 519,904 lbs/h.
   Oil temperature at outlet of E-110 is 250 °F.
   Gpm of flow from E-110 is

$$519,904/(7.16 \times 60) = 1,210 \text{ gpm.}$$

Line size at this point is 8″ Sch. 40.
   Head loss in this line is 31.1 ft/1,000 ft.
   Equivalent length of line:

      Straight line = 82 ft.
      1 Tee = 30 ft.
      Total equivalent length = 112 ft.
      Total head loss in line = $(112 \times 31.1)/1,000 = 3.5$ ft or 1.3 psi.

Pressure at E-110 inlet will be:

Pressure at point "A" = 66.41 psig.
Head loss in line = 1.3 psi.
Pressure drop across 11-E-10 = 8 psi
=75.7 psig.

This pressure is 18 f. above grade as these air coolers are located above the pipe rack. Allowing 1.5 f. from grade to pump center line, the static head at pump discharge flange is 16.5 ft.
   Equivalent length of line from pump to E-110 is:

      Straight length = 155 ft.
      12 elbows = $12 \times 20.5 = 246$ ft.
      3 gate valves = $3 \times 4.8 = 14.4$ ft.
      1 nonreturn valve = $1 \times 51 = 51$ ft.
      Total equivalent length = 466.4 ft.

Process Equipment for Petroleum Processing   1683

Head loss in 8″ Sch. 40 pipe at a flow rate of 1,378 gpm:

Pump temperature is 545 °F.
Viscosity at pump temperature is 1.2 cSt.
SG at pump temperature is 0.755.
Head loss = 28.8 ft/1,000 ft.
Total line head loss = (28.8 × 466.4)/1,000 = 13.4 f. or 4.38 psi.

Then the pump discharge pressure is the sum of:

Pressure at E-110 inlet = 75.7 psig.
Line pressure drop = 4.38 psi.
Static head (16.5 ft) = 5.38 psi.
Total discharge pressure = 85.46 psig.

Calculating the Pressures in the Reflux Line from Point "A"

At point "A," the pressure has been calculated as 66.41 psig.
Flow of the reflux stream (from the material balance) is:

519,904 − 191,121 = 328,783 lbs/h.
Temperature of the stream is 250 °F.
Viscosity at 250 °F = 3.5 Cs.
SG at 250 °F = 0.860 and lbs/gal is 7.16
Rate of flow = 328,783/(7.16 × 60) = 765 gpm.

Line to tower from point "A" is a 6″ Sch. 40 and head loss in this line is found to be 40.2 ft/1,000 ft.
Equivalent line lengths:
To the flow controller inlet flange:

> Straight line = 16 ft.
> 2 elbows = 2 × 16 = 32 ft.
> 1 valve = 1 × 3.5 = 7 ft.
> Total equivalent length = 81.5 ft.

From the flow controller to tower:

> Straight line = 71ft.
> 6 elbows = 6 × 16 = 96 ft.
> 1 valve = 1 × 3.5 = 3.5 ft.
> Total equivalent length = 171 ft.

Total line length from point "A" to tower:

$$81.5 + 171 = 252.5 \text{ ft.}$$

Total head loss in line due to friction:

$$(252.5 \times 40.2)/1,000 = 10.15 \text{ ft. or } 3.77 \text{ psi.}$$

The pressure required to deliver 765 gpm of reflux to the tower excluding the pressure drop across the control valve at this rate is the sum of the following:

Destination pressure = 0.29 psia.
Static head = 14.11 psi.
Distributor (tower internals) = 2.0 psi (from data sheet).
Flow meter pressure = 0.5 psi (from data sheet).
Head loss in line = 3.77 psi.
Total required = 20.67 psia or 5.97 psig.

Then valve pressure drop at design flow of 765 gpm is:

$$\begin{aligned}\text{Pressure at point "A"} &- \text{required pressure} \\ &= 66.41 - 5.97 \text{ psig} \\ &= 60.44 \text{ psi.}\end{aligned}$$

Note: Line pressure drops are given in Appendix 9 of this chapter.

References

C. Branan, *The Process Engineer's Pocket Handbook* (Gulf Publishing Company, Houston, 1976) (This book contains a great deal of very useful, practical information on several subjects in addition to pressure drop)

N.P. Cheremisinoff, P.N. Cheremisinoff, *The Hydrodynamics of Gas-Solids Fluidization* (Gulf Publishing Company, Houston, 1984)

A.K. Coker (H&G Engineering, 1990), Article: Understand two-phase flow in process piping. Chem. Eng. Prog. November, pp. 60–65 (1990)

Gas Processors Suppliers Association (GPSA, 1987), Engineering Data Book (2 volumes), GPSA, Tulsa, OK, USA (This is another outstanding reference for many petroleum processing subjects.)

R. Kern (Hoffmann-La Roche Inc., 1969), Article: How to size process piping for two-phase flow. Hydrocarb. Process. October, pp. 105–116 (1969)

Dictionary of Abbreviations, Acronyms, Expressions, and Terms Used in Petroleum Processing and Refining

David S. J. Jones, Peter R. Pujadó, and Steven A. Treese

Abstract
This dictionary provides definitions and background information on hundreds of abbreviations, acronyms, expressions, and terms used in the oil industry. Special emphasis is placed on refining and petroleum processing. References are provided for more detailed discussions of most of these terms elsewhere in this handbook.

Keywords
Petroleum • Refining • Dictionary

Miscellaneous

#2 fuel oil Similar boiling range as diesel. Can be low sulfur or ULS.
#6 fuel oil Heavy residual, high-sulfur fuel oil. AKA bunker fuel oil
3:2:1 crack Generic per barrel crack based on prices of 2 barrels gasoline + 1 barrel diesel minus 3 barrels of WTI crude. This is an indicator of gross refining margin.

David S. J. Jones: deceased.
Peter R. Pujadó has retired.
Steven A. Treese has retired from Phillips 66.

D.S.J. Jones
Calgary, AB, Canada
e-mail: dai.jones.06@shaw.ca

P.R. Pujadó
UOP LLC, A Honeywell Company, Kildeer, IL, USA
e-mail: prpujado@gmail.com

S.A. Treese (✉)
Puget Sound Investments LLC, Katy, TX, USA
e-mail: streese256@aol.com

A

AA (H, HH, L, LL) Analyzer alarm high, high–high, low, or low–low – e.g., AAHH is a high–high analyzer alarm

AACE Association for the Advancement of Cost Engineering

Abel flash points This is a test procedure for determining the flash point of light distillate such as kerosene. The flash points determined by this method will be in the range of 85–120 °C. The apparatus consists of a water bath with a heating source into which is suspended a cup containing the material to be tested. The water bath is heated and retained at a fixed temperature. The temperature of the sample being tested in the cup is measured by a thermometer. The lid of the cup contains a shutter and a small gas burner from which a flame of determined length exits. The rate of temperature rise of the test material is noted, and at predetermined temperatures, the shutter is opened and the burner flame exposed to the space above the test liquid in the cup. The temperature at which a flame is observed crossing the surface of the oil sample when the burner is dipped into the cup is the flash point. The ASTM name and number for this test is D56-01 Standard Test Method for Flash Point by Tag Closed Tester.

Abel tester Apparatus used to determine the Abel flash point of an oil. See "Abel Flash Points".

ABMA American Boiler Manufacturers Association

ABS (1) Ammonium bisulfide; (2) American Bureau of Shipping; (3) acrylonitrile–butadiene–styrene polymer; (4) absolute (as in absolute pressure)

Absolute pressure Pressure above total vacuum. Normally gauge pressure in psig plus 14.696 psi (result expressed as psia) or as appropriate for the measurement system being used: bar absolute (bara), kg/cm^2 abs, etc.

Absorbent Material capable of removing specific component(s) from a stream by chemical means (absorbing the material) or taking the component(s) into solution

Absorber A type of liquid/vapor or solid/vapor contactor designed to remove or absorb one or more components from the vapor phase into the liquid or solid phase. Liquid/vapor contactors are used to remove H_2S from fuel gas or to remove heavier hydrocarbons from a vapor using a sponge oil, for example. In a solid/vapor absorber example, canisters of activated carbon may be used to remove volatile organic compounds from vapors vented to the atmosphere.

Absorption units Absorption units usually consist of a trayed or packed tower in which a gas stream is contacted with a lean solvent in a countercurrent flow. Usually the gas stream enters below the bottom tray or packed bed and rises up the tower, meeting the solvent liquid stream which enters the tower above the top tray or packed bed flowing down the tower. Undesirable material in the gas stream is selectively absorbed into the liquid stream. This liquid stream then enters a stripping column (usually a Steam Stripper), where the absorbed material is stripped off and leaves as an overhead product. The stripped solvent stream is then rerouted to the absorber tower to complete the cycle. These processes are used in petroleum refining to remove heavy hydrocarbons from a light gas stream. More commonly, this type of unit is used in gas treating for the

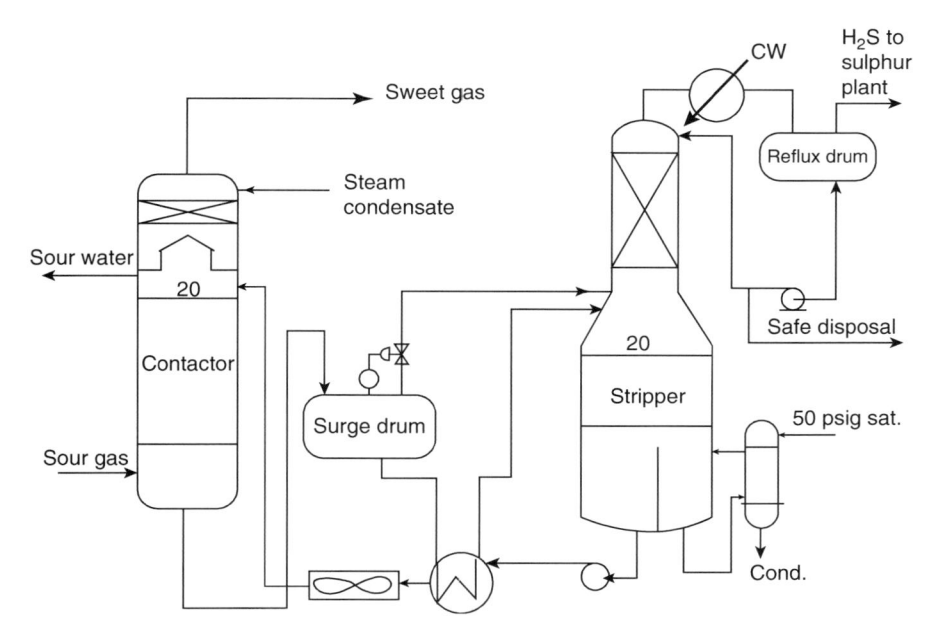

Dictionary of Abbreviations, Acronyms, Expressions, and Terms Used in... 1687

Fig. A.1 A schematic of a typical absorption unit

removal of H_2S from a gas stream (see the topic in chapter "▶ Refinery Gas Treating Processes" of this handbook). Figure A.1 is an example of a typical absorption (gas treating) unit.

ABT Average bed temperature

AC (1) Alternating current; (2) analyzer controller

Accumulation In a relief valve, accumulation is the pressure increase over the maximum allowable working pressure of the vessel during discharge through the pressure relief device expressed as a percent of that pressure, or in psi or bar.

Accumulator A type of surge vessel, normally in distillation or hydraulics. Refer to the equipment chapter.

Acid gas A vapor containing a compound that can form an acidic aqueous solution when dissolved in water. In refining, typical acid gases include vapors containing H_2S or CO_2. The feed gas for a sulfur plant is an "acid gas" because it contains high H_2S concentrations, for instance.

Acid number An ASTM test which determines the organic acidity of a refinery stream. Normally determined by titration and expressed as mg KOH/g oil (or sometimes per 100 g oil - check context).

Acid suit PPE suit intended to prevent acid contact with personnel.

ACFD Actual cubic feet per day

ACFM Actual cubic feet per minute at flowing temperature and pressure

ACGIH American Conference of Governmental Industrial Hygienists

ACS American Chemical Society

Activated sludge unit (ASU) Secondary wastewater treatment unit using aerobic organisms to reduce BOD. See the environmental chapter.

Actuator The device which moves a final control element. Normally a pneumatic diaphragm or a pneumatic or hydraulic piston. See the discussion in the controls chapter.

Adiabatic No external heat input or removal

Adsorbent Material capable of removing specific component(s) from a stream by adsorbing the component(s) onto the surface of the material. For example, the materials used in a PSA are adsorbents. No chemical reaction occurs with adsorbents as compared to absorbents.

Ad valorem tax A tax based on the quantity of an item sold or retained as property. This tax may be levied on inventory in some jurisdictions, for instance. The tax is intended to cover property tax, municipal service costs, etc.

Advanced process control (APC) A control system overlaying the DCS or basic process control system that performs more complex control functions for a process unit. Refer to the chapter on controls.

Aerobic Requiring oxygen – generally refers to organisms.

AFPM American Fuel and Petrochemical Manufacturers – formerly NPRA

AGA American Gas Association

AGO Atmospheric gas oil. Typically boils between about 500 and 850 °F. Sometimes called heavy diesel or heavy D

AHS Albian Heavy Synthetic crude – from Canada

AI (1) Artificial intelligence; (2) analyzer indicator; (3) analog input

AIChE American Institute of Chemical Engineers

AIChE-DIERS AIChE Design Institute for Emergency Relief Systems

AIHA American Industrial Hygiene Association

Air condensers and coolers An air cooler or condenser transfers heat from a process stream by passing air over a large, flat tube bundle, similar to an automobile radiator. The tubes may have fins wrapped around them to enhance heat transfer. The use of air cooling saves cooling water. These units are normally elevated above the pipe racks in a refinery. The air is usually blown over the tubes by large fans. Figure A.2 illustrates a typical air cooler or condenser layout. Refer to the chapter "▶ Process Equipment for Petroleum Processing" of this handbook for a detailed discussion under heat exchangers.

Air/fuel ratio The ratio of air weight to fuel weight consumed in an internal combustion engine or a furnace. This ratio is often expressed as a volume or molar ratio for gaseous fuels.

Air-line respirator A type of fresh-air-supply respirator with the air provided through a tube to a full-face mask. The air is generally provided through a regulator from banks of compressed breathing air bottles.

Air systems All chemical and petroleum plants require a supply of compressed air to operate the plant and for plant maintenance. There are usually two separate systems: (1) plant air and (2) instrument air. Plant air is used for general utility services, such as air tools. Instrument air is a clean, dry motive air supply for

Fig. A.2 A forced airflow arrangement

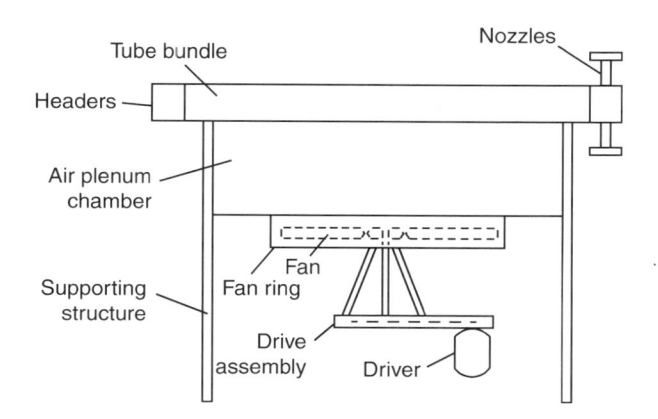

operation of control valves and pneumatic instruments. Refer to the chapter titled "▶ Utilities in Petroleum Processing" of this handbook for a detailed discussion of these critical systems.

AISC American Institute of Steel Construction

AKA Also known as

ALARP As low as reasonably practical

Alcohol An organic chemical compound composed of carbon, hydrogen, and oxygen. Alcohols vary in chain length and are composed of a hydrocarbon plus a hydroxyl group, $CH_3\text{-}(CH_2)_n\text{-}OH$ (e.g., methanol, ethanol, and tertiary butyl alcohol).

Alcohol blend A fuel with alcohol blended into the hydrocarbon base stock. Examples include reformulated gasoline, where ethanol may be blended with conventional hydrocarbon stocks to provide the regulatory required oxygen content.

Algal oil Renewable distillate-range oil derived from processing algae. See the discussion of unconventional stocks and renewables.

Aliphatic Refers to a nonaromatic hydrocarbon. Aliphatic compounds may be saturated (like paraffins and naphthenes) or unsaturated (olefins).

Alkylate The hydrocarbon product from an alkylation unit. Normally used as a gasoline blendstock.

Alkylation Alkylation is a process which reacts a small paraffin molecule (e.g., isobutane) with a small olefin molecule (e.g., 1-butene) to make larger molecules. The normal product of alkylation (alkylate) is a high-octane gasoline blending component with a relatively low vapor pressure (RVP) and almost no sulfur. The reaction is catalyzed by strong liquid and solid acids. The most common catalysts are liquid anhydrous hydrofluoric acid and concentrated sulfuric acid. Figure A.3 illustrates a typical alkylation unit employing anhydrous hydrofluoric acid catalyst in the UOP configuration. Alkylation is provided on a licensed basis by a handful of licensors. Refer to the topic "▶ Alkylation in Petroleum Processing" of this handbook for a detailed discussion of this technology.

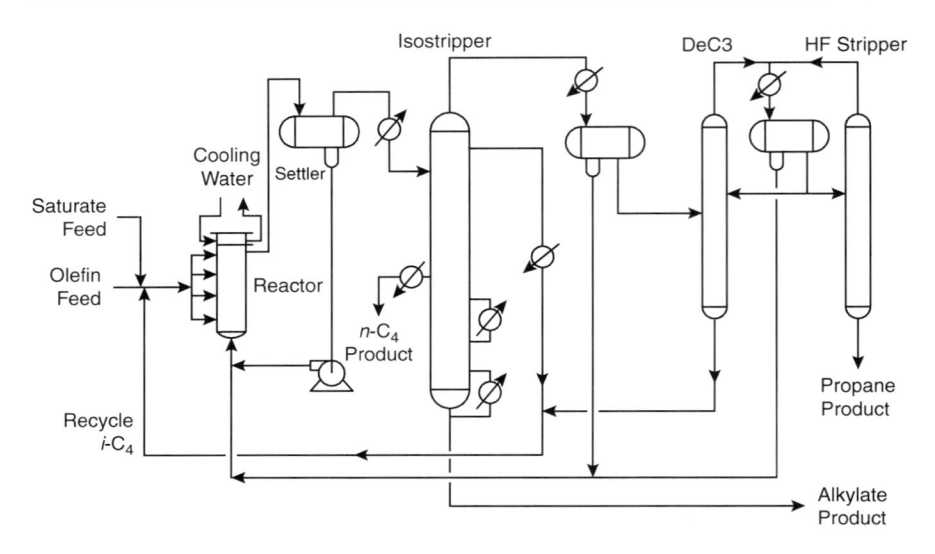

Fig. A.3 AHF alkylation process

Alkylation unit HF An HF alkylation unit uses anhydrous hydrofluoric acid to catalyze the alkylation reaction, which builds smaller paraffins and olefins into larger molecules. The product alkylate is normally a high-octane gasoline blending component with low vapor pressure and sulfur content. Figure A.3 is a simplified process flow diagram for the UOP propylene–butane HF alkylation process. Refer to the topic "▶ Alkylation in Petroleum Processing" of this handbook for a detailed discussion of this technology.

Alkylation unit H_2SO_4 A sulfuric acid alkylation unit uses concentrated sulfuric acid to catalyze the alkylation reaction, which builds smaller paraffins and olefins into larger molecules. The product alkylate is normally a high-octane gasoline blending component with low vapor pressure and sulfur content. There are two primary processes licensed for H_2SO_4 alkylation: (1) the cascade process licensed by ExxonMobil and M.W. Kellogg and (2) the Stratco effluent refrigerated process. Figure A.4 is a simplified process flow diagram for the cascade alkylation process. Refer to the topic "▶ Alkylation in Petroleum Processing" of this handbook for a detailed discussion of this technology.

American National Standards Institute Private, nonprofit organization that develops and administers voluntary consensus standards for products, services, processes, systems, and personnel in the United States. Abbreviated, ANSI

American Petroleum Institute US trade association for the oil and natural gas industry. Abbreviated, API. Develops and administers consensus standards for refining, among other activities. Advocates for the petroleum industry

American Society for Testing and Materials International standards organization. Develops and publishes voluntary, consensus technical standards for a wide

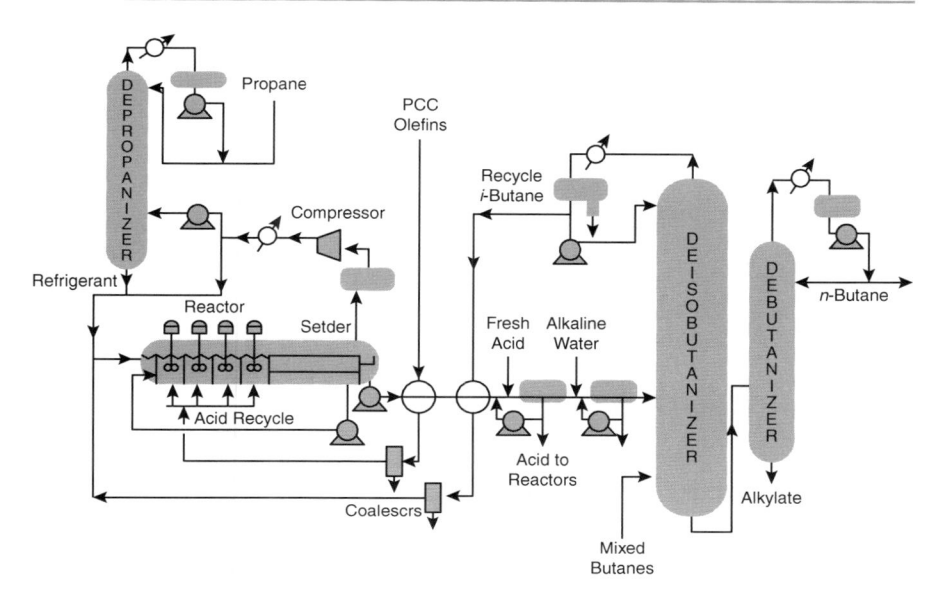

Fig. A.4 H_2SO_4 alkylation process

range of materials, products, systems, and services. Abbreviated, ASTM. Many fuel standards are set based on specific ASTM procedures.

Amine An amine is an organic nitrogen compound based on substitution of one or more organic groups around the nitrogen in ammonia. $NR^1R^2R^3$ where R^1, R^2, and R^3 are either hydrogen or an organic group like methane or ethanol. An example is diethanolamine, $(HOC_2H_4)_2NH$.

Amine solvents Several amines are used, in aqueous solutions, as solvents in the removal of hydrogen sulfide or carbon dioxide from refinery gas streams. There are many amine compounds used for this purpose. The more common of these are monoethanolamine (MEA), diethanolamine (DEA), diglycolamine (DGA), methyl diethanolamine (MDEA), and diisopropanol amine (DIPA). Amine mixtures may also be used. The handbook chapter titled "▶ Refinery Gas Treating Processes" provides a detailed discussion of amine applications in refineries. The topic in the handbook on "▶ Hazardous Materials in Petroleum Processing" discusses handling of these chemicals.

Amine units These are process units which use chemically "basic" aqueous amine solutions to remove "acid gases," such as H_2S and CO_2, from refinery streams. The typical amine unit has an absorber tower coupled with a reboiled regenerator tower, with the amine solution circulating between the two towers. The handbook chapter titled "▶ Refinery Gas Treating Processes" details the processes and applications of amine units in refineries. The topic in the handbook on "▶ Hazardous Materials in Petroleum Processing" discusses handling of these chemicals.

Fig. A.5 A typical amine absorber for MEA or DEA unit

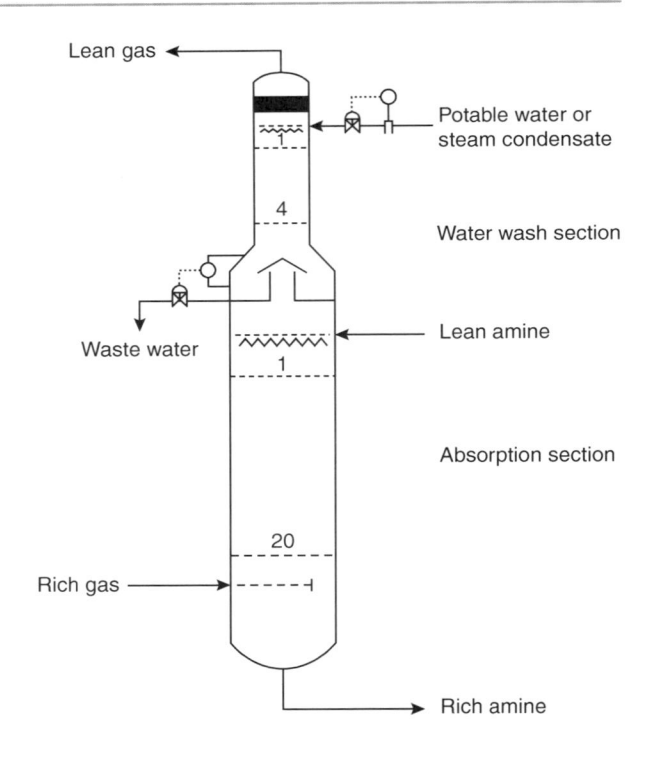

Amine absorber An amine absorber is a tower in which a "lean" amine solution is contacted countercurrently with a stream containing a contaminated gas to scrub an "acid" gas contaminant from the stream. The "rich" solution produced is sent to a regenerator where the "acid" gas is stripped from the amine for further processing. The regenerated "lean" amine is recycled to the absorber. A typical amine absorber tower is illustrated in Fig. A.5.

Ammonia NH_3
Anaerobic Not requiring oxygen – generally refers to specific organisms
Analyzer A device for determining a specific chemical or physical property of a material. The analysis is then used for controlling the process or for ensuring that product specifications are met. Analyzers are found both offline (in the lab) and online. Online analyzers are often connected directly to the process control system.
Ancillary equipment Attachments and support systems required to make a main piece of equipment function properly, such as lubrication system, frame cooling, etc. Refer to the equipment chapter for some examples.
Anhydrous Refers to a chemical that is not in a water solution or a material that contains no water.

Anhydrous ammonia Ammonia is used in the refinery to neutralize vapor containing HCl at the overhead section of the atmospheric crude distillation tower. The ammonia is injected into the vapor spaces above the top four (or five) trays of the tower and into the overhead vapor line. The ammonia may be in the anhydrous form and introduced directly from cylinders or may be in the form of an aqueous solution. The aqueous form is injected from a storage bullet by means of metering pumps. Anhydrous ammonia is also used for NOx control in SCR and SNCR systems on flue gases.

Anhydrous hydrofluoric acid Anhydrous hydrofluoric acid (AHF) is a colorless, mobile liquid that boils at 67 °F at atmospheric pressure. It is used in the HF alkylation process. Refer to the topic "▶ Alkylation in Petroleum Processing" of this handbook for a discussion of its application in a refinery. The acid is hygroscopic, and its vapor combines with the moisture of air to form "fumes." It attacks glass, concrete, and some metals – especially cast iron and alloys that contain silica (e.g., Bessemer steels). The acid also attacks such organic materials such as leather, natural rubber, and wood. The handbook chapter titled "▶ Hazardous Materials in Petroleum Processing" provides a detailed discussion on handling this chemical.

Aniline point Minimum temperature at which equal volumes of aniline ($C_6H_5NH_2$) and the oil are miscible. The oil and aniline are heated until no more solids are present and then cooled until the first indication of cloudiness occurs. This is recorded as the aniline point. A low aniline point is indicative of a stock high in aromatics. A high aniline point indicates a paraffinic stock.

Annular flow A flow regime where a liquid film flows along the wall of the pipe with gas or dispersed flow at a higher velocity in the center core of the pipe. Refer to the chapter on equipment, section on two-phase pressure drop.

ANS Alaskan North Slope crude oil

ANSI American National Standards Institute

Anti-backflow Preventing the flow of fluids backwards through a process system. Backflow may result in overpressure or other undesired consequences. Refer to the discussions in the chapters on utilities and safety.

Anti-cross-contamination Preventing fluids from mixing together and contaminating each other. A major concern with utility systems. Refer to the chapter on utilities.

Anti-surge or antisurge In centrifugal compressors, this is a control system or technique that automatically keeps the compressor away from the unstable surge point, which can damage the machine.

APC Advanced process control

API (1) American Petroleum Institute; (2) application programming interface

API codes The American Petroleum Institute provides the industry with a set of standards which defines the design and measurement parameters that will be used in the petroleum industry. These codes cover such items as vessel design, oily water separators, boiler design, safety items, etc., and a number of laboratory test procedures for feed and petroleum products.

API gravity This item is used in the compilation of most crude assays (see chapter "▶ Introduction to Crude Oil and Petroleum Processing" in this handbook). Although not a laboratory test, as such it is derived from the standard test to determine the specific gravity of a liquid. The correlation between specific gravity and degrees API is as follows:

Sp. Gr. or $SG = 141.5/(131.5 + °API)$

The specific gravity and the API are at 60 °F. Note API is always quoted in degrees.

API separator A device for gravity separation of oil and sludge from facility effluent water. Refer to the chapter on offsites.

Aqueous Water based. Normally refers to water solutions

Aqueous or aqua ammonia A water solution of ammonia, available in various concentrations

AR Analyzer recorder

ARAMCO Arabian American Oil Company

Arbitrage In economics, taking advantage of a price difference between two or more markets

Arbor coil A furnace design where the furnace tubes form an arch over the top of the firebox – like a garden arbor

ARC Analyzer recorder and controller

Arch The roof of a fired heater above the firebox in an upflow firebox

Aromatics Aromatics are present throughout the entire boiling range of crude oil above the boiling point of benzene, the compound with the lowest boiling point in the homologue. These compounds consist of one or more closed, conjugated carbon rings with one or more alkyl groups attached. The lighter aromatics such as benzene, toluene, and the xylenes are removed as products in the petroleum chemical plants (see the chapter on "▶ Non-energy Refineries in Petroleum Processing" in this handbook). In the energy petroleum refinery, these lighter aromatics are included in the finished gasoline products to enhance the octane rating of the products, but their maximum content in gasoline is usually limited (especially benzene). Catalytic reforming is aimed at converting the lower-octane compounds (predominately naphthenes) into the high-octane light aromatics. The heavier aromatic compounds are undesirable compounds in many products, such as kerosene, jet fuel, and many lube oils. In these cases, the aromatic compounds are either converted (dearomatizing hydrotreater for kerosenes) or removed by solvent extraction as in the case of lube oils. Figure A.6 illustrates a typical chemical plant configuration for aromatics production.

ARU Amine regeneration unit

As Arsenic

Ash or ash content (petroleum) Petroleum ash content is the noncombustible residue of a lubricating or fuel oil determined in accordance with ASTM D582 – also D874 (sulfated ash).

ASM Abnormal situation management – responding to an emergency

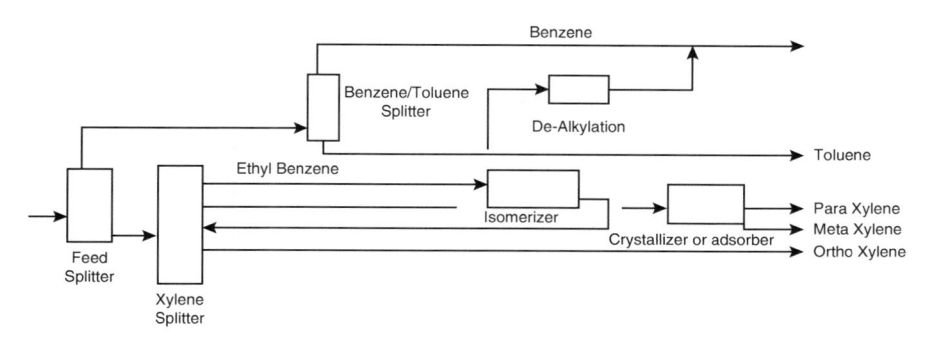

Fig. A.6 A typical aromatic plant configuration

ASME (American Society of Mechanical Engineers) The American Society of Mechanical Engineers has been a world leader in codes, standards, accreditation, and certification for over a century. These programs have now been extended to include the registration (certification) of quality systems in conformance with the standards set by the International Organization of Standardization (ISO). In the petroleum industry, this organization sets the quality requirements for vessel fabrication, piping, and in particular the boiler code among many other standard definitions.

ASO Acid-soluble oil(s) in the alkylation process

Asphalt Asphalt is a group of products produced from the vacuum distillation residue of crude oil. The products that make up this group have distinct properties that must be met by the treating of the vacuum residue. There are two major categories of asphalt products: (1) paving and liquid asphalt and (2) roofing asphalt. Refer to the topic "▶ Upgrading the Bottom of the Barrel" in this handbook for a detailed discussion of asphalt types and production.

Asphaltene Insoluble, semisolid or solid particles which are combustible and are highly aromatic (AKA "polyvinyl chicken wire"). Asphaltenes contain a high carbon/hydrogen ratio and entrap water, fuel ashes, and other impurities. They are primarily characterized by being insoluble in n-heptane and soluble in toluene.

Aspiration Normally refers to inhalation of a substance that is not a gas, such as aspiration of droplets of a chemical. May result in impaired lung function

Assay The crude oil assay is a compilation of laboratory and pilot plant data that defines the properties of a specific crude oil. At a minimum, the assay should contain a distillation curve for the crude and a specific gravity curve. Most assays however contain data on pour point (flowing criteria), sulfur content, viscosity, and many other properties. The assay is usually prepared by the company selling the crude oil. It is used extensively by refiners in their plant operation, development of product schedules, and examination of future processing ventures. Engineering companies use the assay data in preparing the process designs for petroleum plants they are bidding on or, having been awarded the project, they are now building. Refer to the topic "▶ Introduction to Crude Oil and Petroleum Processing" in this handbook for a detailed discussion of crude assays. The Appendix of this handbook titled "▶ Selection of Crude Oil Assays for Petroleum Refining" provides examples of assays for several common crudes. The crude assay should not be confused with the Fischer assay used for characterizing the potential yields from destructive distillation of oil shale and similar solids. Fischer assay is discussed in the topic "▶ Unconventional Crudes and Feedstocks in Petroleum Processing."

ASSE American Society of Safety Engineers

AST Aboveground storage tank

ASTM (American Society for Testing and Materials) ASTM is an organization that standardizes test methods for many types of materials. The ASTM tests and test methods provided by this body define and establish the quality of petroleum products and provide data on petroleum intermediate streams. This latter provision is used as a basis for refinery planning, operation, and engineering work associated with the refinery. Further details are given in the chapter on "▶ Quality Control of Products in Petroleum Refining" in this handbook. Among the more important tests of interest to us in petroleum processing are:

- D56 – Tag closed flash point (the Abel flash)
- D86 – Standard Test for Distillation of Petroleum Products
- D93 – The Pensky–Martens flash-point, closed-cup test
- D97 – Cloud and pour points
- D129 – Sulfur content (Bomb Method)
- D189 – Conradson carbon content
- D323 – Reid vapor pressure
- D445 – Kinematic viscosity
- D613 – Cetane number
- D908 – Octane Number, Research
- D1160 – Gas oil distillation (subatmospheric)
- D1298 – Specific gravity (by hydrometer)
- D1837 – Weathering test for LPGs
- D2163 – Component analysis of LPGs (by gas chromatography)
- D2887 – Simulated distillation

ASTM distillation Refers normally to a D86 distillation, but may refer to any ASTM method of distillation testing. Be sure you know which distillation test is being used.

ASTM ON Motor octane number of a gasoline determined using the standard ASTM test apparatus. AKA MON and F-1

ASU (1) Air separation unit, separates nitrogen and oxygen out of the air; (2) activated sludge unit, for microbio treatment of wastewater

AT Analyzer transmitter

ATB Atmospheric tower bottoms – see atmospheric resid

Atmospheric crude distillation unit Often referred to simply as the "crude unit," the atmospheric crude distillation unit provides the initial separation of crude oil

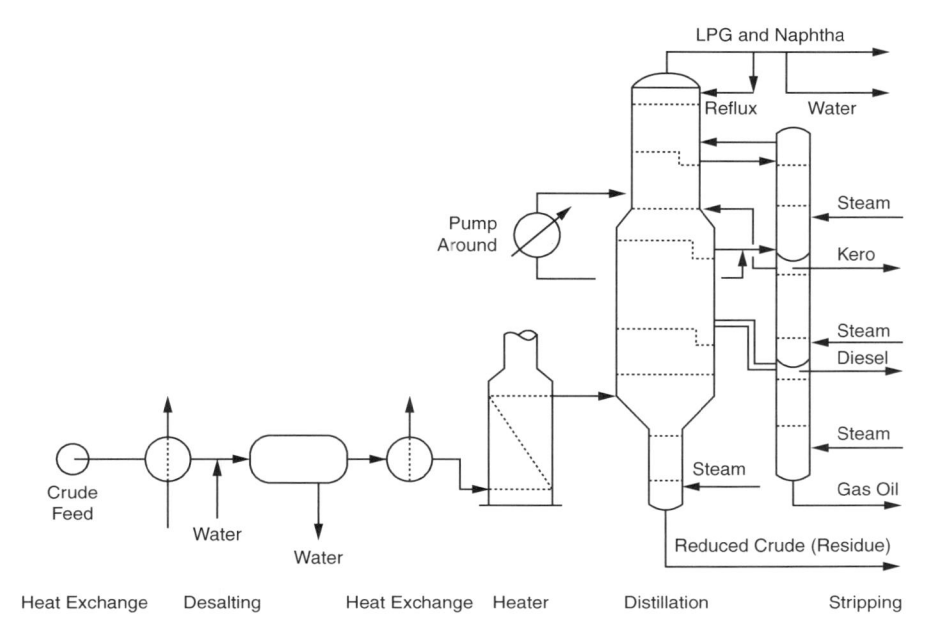

Fig. A.7 A typical atmospheric crude distillation unit

into distinct boiling ranges to be further refined into finished products. It is called "atmospheric" because it operates at a slightly positive gauge pressure, as opposed to the vacuum crude distillation unit that operates at low, negative gauge pressure. In practice, the atmospheric unit will operate at pressures between about 20 and 50 psig. The chapter titled "▶ Atmospheric and Vacuum Crude Distillation Units in Petroleum Refineries" discusses these units in detail. Figure A.7 illustrates a typical atmospheric crude unit configuration.

Atmospheric discharge In reference to relief systems, release of vapors and gases from pressure relief and depressurizing devices to the atmosphere

ATU Atmospheric tower unit. See "Atmospheric Crude Distillation Unit."

Autoignition and autoignition temperature Autoignition refers to spontaneous ignition of a compound or stock when exposed to air at a specific temperature without the presence of an ignition source. The autoignition temperature is the temperature at which autoignition will occur.

Auxiliary See "Ancillary Equipment." Often this refers to the utilities required to operate a major piece of process equipment.

Aviation gasoline A complex mixture of relatively volatile hydrocarbons with or without small quantities of additives, blended to form a fuel suitable for use in aviation reciprocating engines meeting ASTM or military specifications. The octane requirement for aviation gasoline is very high compared to road gasoline.

Aviation gasoline blending components Naphthas used for blending into finished aviation gasoline. May include straight-run gasoline, alkylate, reformate, benzene, toluene, and xylene

Axial (flow) compressor These compressors consist of bladed wheels that rotate between bladed stators. Gas flow is parallel to the axis of rotation through the compressor. A jet engine air compressor is an axial compressor example. Refer to the equipment chapter discussion of compressors.

AXL Arab extra-light crude oil

B

Back pressure In relief devices, back pressure is the pressure existing at the outlet of the pressure relief device due to pressure in the discharge system at the time the device must be relieved.

BACT Best available control technology – for the environment. Refer to the environmental chapter.

BACT box Operating limits on a heater to stay within emissions and permit limits

Baffle A dividing wall to separate two compartments or regions in equipment. For instance, in a separator, a baffle may be used at one end to keep a collected oil phase separated from an oil/water interface or two-phase zone in the separator. Baffles are also used to direct flow into desired patterns, such as in a heat exchanger. Refer to the chapter on equipment.

Baker correlations Calculation method for two-phase liquid/vapor flow in piping systems. Refer to the equipment chapter discussion of two-phase pressure drop.

Balanced safety relief valves A balanced safety relief valve incorporates means for minimizing the effect of back pressure on the performance characteristics-opening pressure, closing pressure, lift, and relieving capacity.

Barge A large, flat-bottomed boat used to move oil or other materials. A barge is not normally powered but must be pushed or pulled using a tug or towboat.

Barrels In the petroleum industry, the barrel is a standard form of measuring liquid volume. A barrel of oil is defined as 42 US gallons (1 US gallon equals 231 in^3). It is still used extensively in most countries but is being replaced particularly in European countries by the metric measures of cubic meters or liters or by weight measures like kg or tonnes (1,000 kg).

Barrels per calendar day See "BPCD and BPSD."

Barrels per stream day See "BPCD and BPSD."

Base lube oils This refers to a lube product that meets all lube specifications and is suitable for blending to meet performance specifications.

Basic process control system (BPCS) Primary control system that takes input from the process variables, compares the input to the setpoints, and outputs the appropriate control moves to the control elements in the field. See the chapter on controls.

BAT Best available technology. Similar to BACT for Europe. Refer to the environmental chapter.

Batch blending Adding components to a blend tank and then mixing the tank contents to achieve the target product properties. Refer to the chapter on offsites.

Battery limit (station) The interface between a process system and the rest of the facility. Often all the piping connecting a unit to external locations passes through a localized battery limit station where it can be isolated when necessary.

Bayonet heater A coil that is inserted into the side of a tank or tower to supply heat. The bayonet can be used for keeping the contents warm or for reboiling. Also called a stab-in heater. See the chapter on offsites.

BB Butane–butylene

Bbl or BBL or bbl Abbreviation for barrel (42 US gallons)

BCF Billion cubic feet

BCFD Billion cubic feet per day

Benfield process A licensed process employing a regenerable hot potassium carbonate solution to remove carbon dioxide from an acid gas stream. This process is used in steam-methane-reforming hydrogen plants and partial oxidation/gasification plants to clean up product gases. Similar to Catacarb process. The process is licensed from UOP.

BenSat Benzene saturation process

Benzene The simplest aromatic compound with a six-member carbon ring: C_6H_6. Benzene is found in many processes and petroleum products within a refinery and related facilities. In some cases, the product may be benzene that is deliberately recovered. In other streams, such as gasoline or reformate, the benzene is present as part of the hydrocarbon mixture.

BEP Best efficiency point (compressor or pump)

Berl saddle A type of mass transfer packing shaped somewhat like a saddle. Refer to the chapter on equipment, discussion of packed towers.

BFD Block flow diagram

BFOE Barrel of fuel oil equivalent – about 6.05 MMBtu, although the number may vary. The US IRS considers it 5.8 MMBtu. Other sources use 6.4 MMBtu.

BFW Boiler feed water

BHP Brake horsepower

Biochemical oxygen demand (BOD) Amount of oxygen required for biological organisms to digest the impurities in an effluent water stream. See the environmental chapter.

Biocide A chemical used to prevent microbiological growth. The most common biocides are chlorine and bleach. Uses in refineries include cooling water, potable water, sour water storage tanks, and certain product storage tanks (e.g., kerosene).

Biocrude Renewable oil made by destructive distillation of biomass – such as wood or corn stover pyrolysis

Biodiesel Applies to long-chain methyl, ethyl, and propyl esters, like FAME. These are made by transesterification of lipids with methanol. Refer to the discussions on renewables and unconventional stocks.

Bioethanol Ethanol made from a renewable source, usually corn or sugar cane.

Biofuel Any fuel derived from recently fixed carbon, usually plant or animal sources; this is in contrast to petroleum or other oils where carbon was fixed in the distant past.

Biooil See "Biocrude."

Bitumen Bitumen is the term often given to a sticky, black, highly viscous liquid or semisolid form of petroleum (essentially tar). It is also applied to untreated asphalt from the vacuum distillation of crude and the extract from the deasphalting unit. This is before the stream has been treated, with cut backs or by air blowing to make the various asphalt product grades. This term is also used for the extra heavy oils produced from tar or oil sands and other, similar deposits.

Blanket or blanket gas Providing a compatible vapor above a storage tank or vessel, normally to prevent contact with air or development of an explosive mixture. See offsites chapter.

Bleach Sodium hypochlorite solution

BLEVE Boiling liquid expanding vapor explosion

Blend component A stock that is to be used for blending a finished product.

Blank A solid metal disk inserted into a pipe (usually between flanges) to isolate equipment or a system and prevent flow. The terms blank and blind are often used interchangeably.

Blind or blind flange Like a blank, this is a solid metal disk inserted between flanges to isolate equipment or a system and prevent flow. The terms blank and blind are often used interchangeably.

BLM Bureau of Land Management

Blocked Preventing flow

Blocked discharge Applies to pumps or compressors when flow is prevented because a discharge valve is still closed or there is some other flow obstruction (like a blind) in the discharge piping. This is a controlling relief scenario in many cases.

Blocked-in Isolated by valve(s) so that flow is prevented.

Blowdown (or blow-down) (1) Purging of part of a process stream to manage buildup of impurities in the stream, e.g., blowdown from a cooling tower or steam generation system. Refer to the discussions in the chapter on utilities. (2) In some plants, this refers to the flare or waste gas system where relief valves and process vents can be safely routed. Refer to the discussion in the chapter on offsites. (3) In relief valves, blow-down is the difference between the set pressure and the resealing pressure of a pressure relief valve, expressed as a percent of the set pressure, or in psi or bar.

BLS Bureau of Labor Statistics

BMS Burner management system

BOD Biochemical oxygen demand – in wastewater. The BOD is a measure of the oxygen depletion due to wastes which are biologically oxidizable. Refer to the environmental chapter section on water.

BOE Barrel of oil equivalent – about 6.05 MMBtu HHV. See also "BFOE."

BOG Boil-off gas

Boiler feed water (BFW) Water that has been treated to make it suitable for use in a boiler to make steam. Refer to the chapter on utilities.

Boiling point analyzer This analyzer can be used to determine the 5 % or 95 % points of products for fractionation control purposes, for instance. Refer to the chapter on controls.

Boiling points and boiling range It is not feasible or necessary to separate the components of crude oil into individual chemical compounds. However, groups of these component mixtures are grouped together and identified by the boiling point at atmospheric pressure of the lightest component in the group and the boiling point of the heaviest component in the group. The group itself is called a cut, and the range of temperatures that identify it is called the boiling range or cut range. These can be related to the crude TBP curve to determine its yield on that particular crude (see the topic "▶ Introduction to Crude Oil and Petroleum Processing" in this handbook).

BOM Bill of materials

BOQ Bill of quantities

Bottoms or botts Refers to the material produced from the bottom of a distillation column. In an atmospheric crude distillation unit, this would be the atmospheric residuum. In a propane–butane splitter, the butane product would be the bottoms or botts.

Bourdon tube A mechanical pressure measuring device used in most local pressure gauges. Refer to the chapter on controls.

Box cooler A type of exchanger with tubes immersed in a water bath. Refer to the equipment chapter discussion of heat exchangers.

BPC (1) Basic process control – normally the DCS system or individual loop controllers that are active all the time as opposed to systems that act only in hazardous situations. (2) Blend property control

BPCD and BPSD BPCD is the measure of throughput or stream flow based on an operation over 1 year of 365 days. BPSD is the rated throughput of a plant or the rate of a stream over the total operating days in the year. BPCD is barrels per calendar day, and BPSD is barrels per stream day. BPSD is defined as BPCD divided by the service factor as a fraction. The service factor is the fraction of time over a calendar year that a unit is operating (see "Service Factor").

BPCS Basic process control system

BPD Barrels per day

BPSD Barrels per stream day. See "BPCD and BPSD."

BRC (1) Basic regulatory control – the normal control system for a unit; (2) blend ratio control

Breakeven The point at which product value = cost of production

Bridgewall In a fired heater – see Arch and the discussion in the equipment chapter section on fired heaters.

Bright stocks (lube oils) These are processed from the raffinate of the vacuum residue deasphalting unit. (See the chapter on non-energy refineries in this handbook.)

Bromine number The bromine number of petroleum distillates is determined by electrometric titration in accordance with ASTM laboratory test D1159. It is a measure of olefins in the sample according to the equation:

% Olefins = (Bromine number × molecular wt. of olefins)/160

BS&W Bottom sediment and water – settles out of oil during storage. Standard test by centrifuge

BTEX Benzene, toluene, ethylbenzene, and xylene

Btu or BTU British thermal unit – how we measure heat in refineries in the United States

BTX Benzene, toluene, and xylenes

Bubble cap A mass transfer/contacting device consisting of an inverted, slotted cup (or cap) fitted over a riser with an annular gap. Vapor is forced to move through a path that contacts it with a standing liquid level around the cap. Refer to the chapter on equipment, distillation design discussion.

Bubble flow (AKA froth flow) Flow regime in which bubbles of gas are dispersed in a primarily liquid phase with the gas moving about the same velocity as the liquid. Refer to the chapter on equipment, two-phase pressure drop discussion.

Bubble point Bubble point is the temperature and pressure at which a hydrocarbon begins to boil. Refer to the topic "▶ Introduction to Crude Oil and Petroleum Processing" in this handbook for an example of how this definition is used in setting a reflux drum pressure.

Bulk terminal Facility used primarily for the storage and/or marketing of petroleum products with stores of oil in bulk tankage

Bullet or storage bullet A horizontal, cylindrical vessel generally used for storage of high-vapor-pressure materials such as propane or ammonia

Bunker fuel Heavy residual fuel oil used in ships. See "#6 Fuel Oil."

Burner The purpose of a burner is to mix fuel and air to ensure complete combustion in a fired heater. Refer to the handbook topic "▶ Process Equipment for Petroleum Processing" for a detailed discussion of burner designs and critical factors. Additional discussion on burner design factors affecting emissions are included in the topic "▶ Environmental Control and Engineering in Petroleum Processing" in this handbook.

Burner management system A safety system to avoid conditions that can result in an explosion or other problems in a fired heater or boiler

Burner tip The tube through which fuel is introduced into a burner – normally, it has several small holes in it of specific sizes and orientations

Bernoulli principle The observation that a stream's velocity and pressure are inversely related. High flow velocity results in reduced pressure in the high-velocity streamline.

Burst pressure Burst pressure is the value of inlet static pressure (or the differential pressure, considering backpressure) at which a rupture disk device functions.

BWON Benzene waste operations NESHAP

Bz Shorthand for benzene

C

C/H ratio Carbon/hydrogen molar or weight ratio - be careful of the units used
Ca Calcium
CAA US Federal Clean Air Act. Refer to the environmental chapter.
CAAA US Federal Clean Air Act amendment
CAD Computer-aided design
CADD Computer-aided drafting design
CAE Computer-aided engineering
CAF Vapor capacity factor in a distillation tower. Refer to the equipment chapter, section on distillation.
California Air Resources Board (CARB) State agency in California responsible for regulating motor fuels in that state. Generally, CARB sets the fuel specification trends others follow somewhat later.
Calorie The amount of heat required to raise the temperature of 1 g of water by 1 °C, at or near maximum density.
Calorific value Amount of heat produced by the complete combustion of a unit weight of fuel. Usually expressed in calories per gram or BTUs per pound; the latter being numerically 1.8 times the former
Capacity-factored estimate Capital facility cost estimate method that uses estimated unit capacities and a scaling factor applied to known unit costs to estimate cost for a new facility. This method requires the least information to apply but provides the least accurate estimate. It is normally used very early in screening studies for capital projects.
Car seal A numbered plastic or metal seal that provides a method for tracking proper positioning of a valve or other equipment. The car seal must be broken to operate the sealed device. Normally these are administrative controls.
CARB diesel Term which refers to the diesel standard mandated for sale by the California Air Resources Board. It includes tough standards for sulfur and for very low aromatics.
CARBOB California version of RBOB – meets additional specs
Carbon monoxide CO
Carbon residue Carbon residue is a measure of the coke-forming tendencies of oil. It is determined by destructive distillation in the absence of air of the sample to a coke residue. The coke residue is expressed as weight percent of the original sample. There are two standard ASTM tests, Conradson carbon residue (CCR) and Ramsbottom carbon residue (RCR).
Carbonyl sulfide (COS) Sulfur compound found in cracked products (like FCC or coker) and other sour gases with some oxygen exposure. COS must sometimes be removed from these gases when they are used as fuel.
Carcinogen or carcinogenic Substance (or property of a substance) that can cause cancer
CAS Chemical abstract services

Cascade control A control system with two or more controllers. The "master controller(s)" provides a setpoint to the "slave controller."
Casing The outer shell of a fired heater
Cat cracker See "FCC" or fluid catalytic cracking.
Catacarb A licensed process employing a regenerable hot potassium carbonate solution to remove carbon dioxide from an acid gas stream. This process is used in steam-methane-reforming hydrogen plants and partial oxidation/gasification plants to clean up product gases. Similar to Benfield process. Catacarb is licensed by Eickmeyer & Associates.
Catalyst A substance that speeds up a reaction and directs the course of the reaction, but is not consumed by the reaction.
Catalyst fines Hard, abrasive crystalline particles of alumina, silica, and/or alumina–silica that can be carried over from the fluidic catalytic cracking process of residual fuel stocks. Particle size can range from submicron to greater than 60 microns in size. These particles become more common in the higher-viscosity marine bunker fuels from FCC products.
Catalytic hydrocracking See "Hydrocracking."
Catalytic hydrotreating See "Hydrotreating."
Catalytic reforming unit A refining process that rearranges naphtha-range hydrocarbons to increase octane. Hydrogen is a by-product. The reformate product contains no sulfur and has an increased aromatic content. It is applied to naphtha boiling range stocks to produce gasoline blendstock. There is detailed discussion of this process in the chapter "▶ Catalytic Reforming in Petroleum Processing" of this handbook. Figure C.1 illustrates a typical semi-regenerative catalytic reformer.
Caustic or caustic soda Sodium hydroxide (NaOH) in solution or solid form
CBO Carbon black oil
CBOB Conventional gasoline blendstock intended for blending with oxygenates downstream of the refinery where it was produced. CBOB must become conventional gasoline after blending with oxygenates. Motor gasoline blending components that require blending other than with oxygenates to become finished conventional gasoline are reported as all other motor gasoline blending components. Excludes reformulated blendstock for oxygenate blending (RBOB)
CBS Cost breakdown structure
CBT Computer-based training
CCB Central control building
CCPS Center for Chemical Process Safety – a group under the American Institute of Chemical Engineers that defines process safety practices – sets de facto standards for process safety.
CCR (1) Continuous catalytic reformer; (2) central control room; (3) Conradson carbon residue
CCTV Closed circuit television
CD Consent decree
CDCIR Community Documentation Centre on Industrial Risk

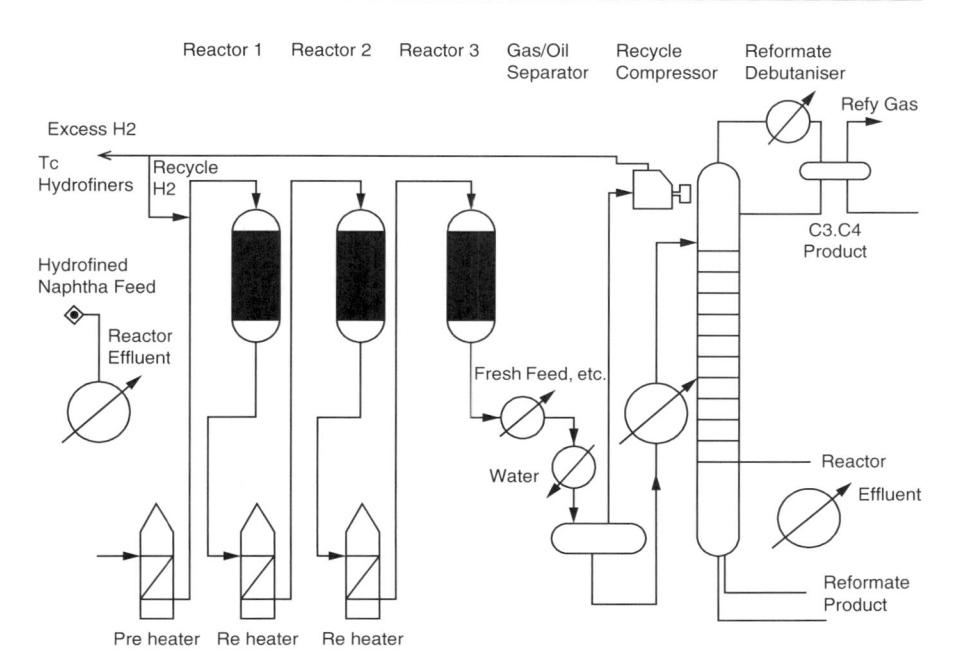

Fig. C.1 A typical catalytic reforming unit

CE Civil engineer or engineering

CELD Cause–effect logic diagram

Celsius Europeans use this term instead of centigrade (see below) to honor physicist Anders Celsius who developed a temperature scale that uses the freezing and boiling points of water as references.

CEM Continuous emission monitor

CEMS Continuous emission monitoring system

Centigrade Temperature scale based on 0° for the temperature at which water freezes and 100° for the temperature at which water boils. See "Celsius" above. This term is accepted and used in North American chemical textbooks; so which term you use may depend on your location, but both are abbreviated with a degree symbol and capital C (°C).

Centipoise 0.01 poise or centistokes times specific gravity at the test temperature. Abbreviated cP or cPs

Centistoke 0.01 stoke (see "Stoke"). Abbreviated cS or cSt

Centrifugal compressors Compressors which use centrifugal force or action to increase the pressure of a gas. This type of compressor consists of an impeller or impellers rotating at high speed within a casing. Centrifugal compressors are widely used in the petroleum, gas, and chemical industries primarily due to the large volumes of gas that frequently have to be handled. Long continuous

Fig. C.2 Typical performance curves for a centrifugal compressor

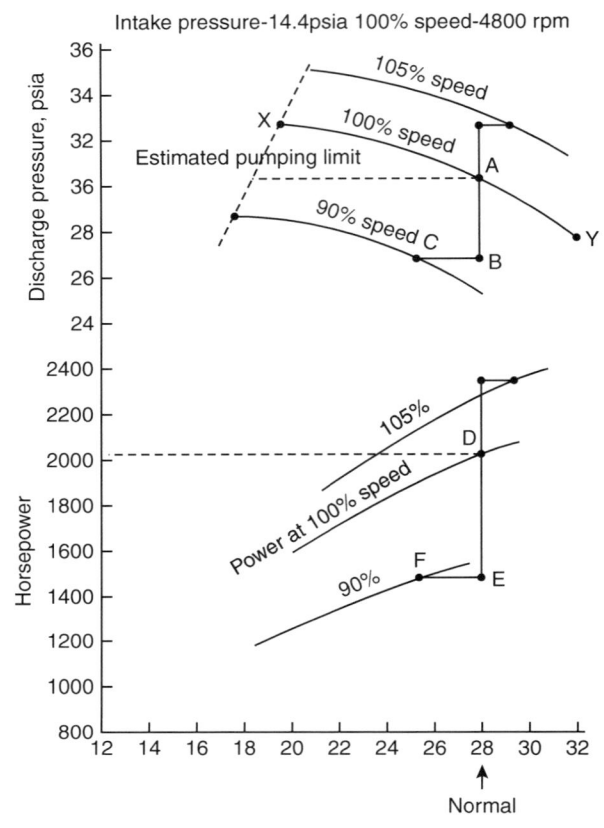

Intake pressure-14.4psia 100% speed-4800 rpm

Thousands cubic feet of air/minute at intake conditions

operating periods without an overhaul make centrifugal compressors desirable for use in petroleum refining and natural gas applications. Refer to the handbook topic "▶ Process Equipment for Petroleum Processing" for a detailed discussion of centrifugal compressors. A typical centrifugal compressor performance curve is illustrated in Fig. C.2.

Centrifugal pump A very wide class of pumps in which pumping of liquids or generation of pressure is effected by a rotary motion of one or several impellers. The impeller or impellers force the liquid into a rotary motion by impelling action, and the pump casing directs the liquid to the impeller at low pressure and leads it away under a higher pressure. Refer to the chapter on equipment, discussion of pumps.

Centrifuge A machine using centrifugal force produced by high-speed rotation for separating materials of different densities. Applied to diesel engine fuels and lubricating oils to remove moisture and other extraneous materials, among other uses.

CEO Chief executive officer
CEP Chemical Engineering Progress magazine (from AIChE)
CERCLA US Comprehensive Environmental Response, Compensation, and Liability Act
Cetane A measure of the engine combustion characteristics of a fuel in a diesel engine. Higher cetane is desired. Can be measured in an engine (cetane number) or calculated (cetane index)
Cetane index An empirical measure of ignition quality. Defined as the percentage by volume of cetane in a mixture of cetane and methyl naphthalene which has the same ignition quality when used in an engine as a fuel under test.
Cetane number Normally refers to cetane values measured by a CFR diesel testing engine. See "Cetane."
CFC Chlorofluorocarbons
CFL Compact florescent light
CFPP Cold filter plugging point in diesel fuels. This is a measure of a fuel's cold weather plugging potential, which can cause engine failure from lack of fuel.
CFR US Code of Federal Regulations
CFR diesel testing unit A standard engine employed in making cetane number tests of diesel engine fuels
CFS Cubic feet per second
CH_4 Methane
Channel In a shell-and-tube exchanger, the channel bolts onto the tube side of the tube sheet and provides the route into and/or out of the tubes.
Characteristic curve Graph or curve that relates pump or compressor flow rate to head, discharge pressure, or differential pressure at various speeds or other parameters. Refer to the equipment chapter, sections on pumps and compressors.
Charge Feedstock to a unit
Charge capacity The input (feed) capacity of the refinery processing facilities
ChE Chemical engineer or engineering
Check valve or check A valve that allows fluid flow in only one direction and stops or "checks" any flow in the reverse direction
Chemical sewer Sewer system designed to accept chemical wastes from a facility and convey them to treatment or reuse
Chemiluminescence analyzer This type of analyzer works for compounds like NO, NO_2, NO_x, and O_2. The analyzer makes use of the luminescence when the compounds are reacted with ozone.
Chiksan joint Mechanical swivel joint that seals piping segments, normally used inside a storage tank. FMC brand name
CHOPS Cold heavy oil production with sand. See the chapter on unconventional crudes.
CHPS Cold high-pressure separator
CHPV Cold high-pressure vapor
Chronic Long-term or lingering
CII Construction Industry Institute

CIPS Cold intermediate pressure separator

Cl Chloride

Claus process The most common process for sulfur recovery to reduce air emissions. Uses the reaction between H_2S and SO_2 to produce a pure sulfur product. Refer to the environmental chapter section on air pollution.

Clearance pocket In a reciprocating compressor, a dead-ended, fixed volume that can be opened or closed by a valve to change the equivalent displacement and thus capacity of the compressor. Refer to the chapter on equipment, section on reciprocating compressors.

Cloud point(s) The cloud point of a transparent or semitransparent oil sample is the temperature at which a cloud or mist forms in the sample. The method is ASTM D2500 and is carried out in the laboratory where a sample of the oil is reduced in temperature by submerging the sample in its container first in ice then in iced salt and finally in a solid CO_2 bath. The container is a cylindrical glass vessel about 19 mm in diameter by 100 mm deep. The vessel is filled to a mark about 80 mm deep and a low-range thermometer inserted so the bulb is about 10–12 mm below the sample surface. Temperature readings are taken at every stage. Readings are more frequently taken in the last cooling stage, and that temperature at which the sample becomes misty is taken as its cloud point.

Clarifier A machine used for a liquid-sludge separation in which the particles with a higher specific gravity are separated from the lower specific gravity of the liquid. A clarifier bowl has one outlet for the light phase oil; the heavier phase particles are retained on the bowl wall. Similar to a centrifuge.

CLPS Cold low-pressure separator

CMA Chemical Manufacturers Association

CMS Continuous monitoring system

CO (1) Carbon monoxide; (2) change order

Co Cobalt

Coagulation Process to cause suspended impurities in effluent or wastewater to form larger particles and drop out. Refer to the chapter on offsites.

Coal A readily combustible black or brownish-black rock whose composition, including inherent moisture, consists of more than 50 % by weight and more than 70 % by volume of carbonaceous material. It is formed from plant remains that have been compacted, hardened, chemically altered, and metamorphosed by heat and pressure over geologic time.

COBRA Consolidated Omnibus Budget Reconciliation Act

COD (1) Chemical oxygen demand – in wastewater. The COD is a measure of the oxygen depletion due to organic and inorganic wastes which are chemically oxidizable. (2) Cutoff date

COE (1) Center of excellence; (2) common operating environment

Co-generation (or cogeneration or cogen) Generating both power and steam (normally) in a utility system. Refer to the chapter on utilities.

COGS Cost of goods sold

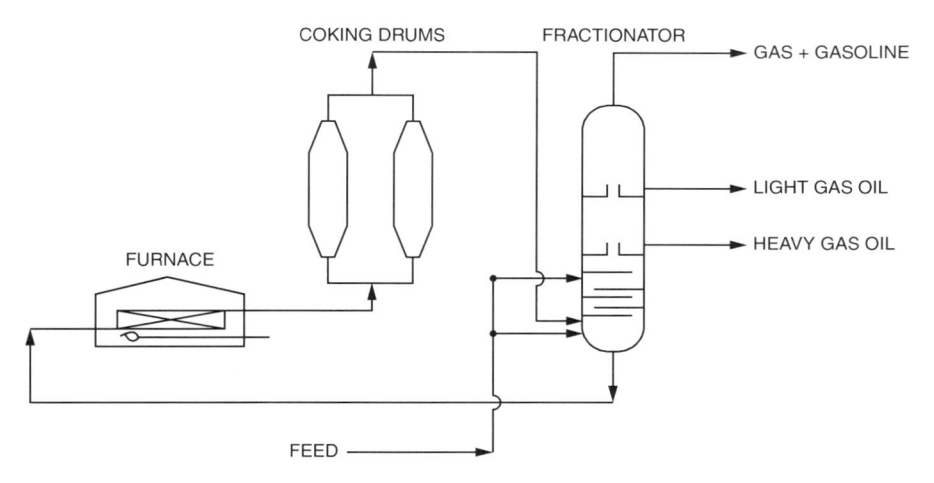

Fig. C.3 A delayed coker

Coke Coke is formed in the processes to convert the residuum fuels to the more desirable distillate products of naphtha and lighter through to the middle distillates. By far, the largest production of coke is the sponge coke from the delayed coking process. Uncalcined sponge coke has a heating value of about 14,000 Btu/lb and is used primarily as a fuel. High-sulfur sponge coke is popular for use in cement plants since the sulfur reacts to form sulfates. Sponge coke is calcined to produce a coke grade suitable for anodes in the aluminum industry.

Coker Resid thermal cracking process. See "Coking Process."

Coking process Coke is formed in the processes to convert the residuum fuels to the more desirable distillate products of naphtha and lighter through to the middle distillates. Coking is essentially destructive distillation of heavy, residual oil (usually vacuum resid) at high temperature and low pressure. The process produces gases, oils, and a residual solid "petroleum coke." There are two coking processes: delayed (Fig. C.3) and fluid coking (Fig. C.4). Delayed coking is the simplest and most common. Refer to the chapter on "▶ Upgrading the Bottom of the Barrel" in this handbook for additional details.

Coking or cracking propensity Tendency of an oil to form coke or thermally crack when heated, as in a furnace

Cold flash separator In many high-temperature and high-pressure hydrocracker or hydrotreater units, the reactor effluent is reduced in pressure and temperature in several stages. The last of these stages is the cold flash separation stage. Figure C.5 shows a common cold flash separator arrangement in a hydrotreater.

Cold properties (or cold flow properties) Refer generically to hydrocarbon properties at low temperatures. These are important in middle distillate fuels (jet, diesel) and fuel oils to permit the materials to be handled at low or winter

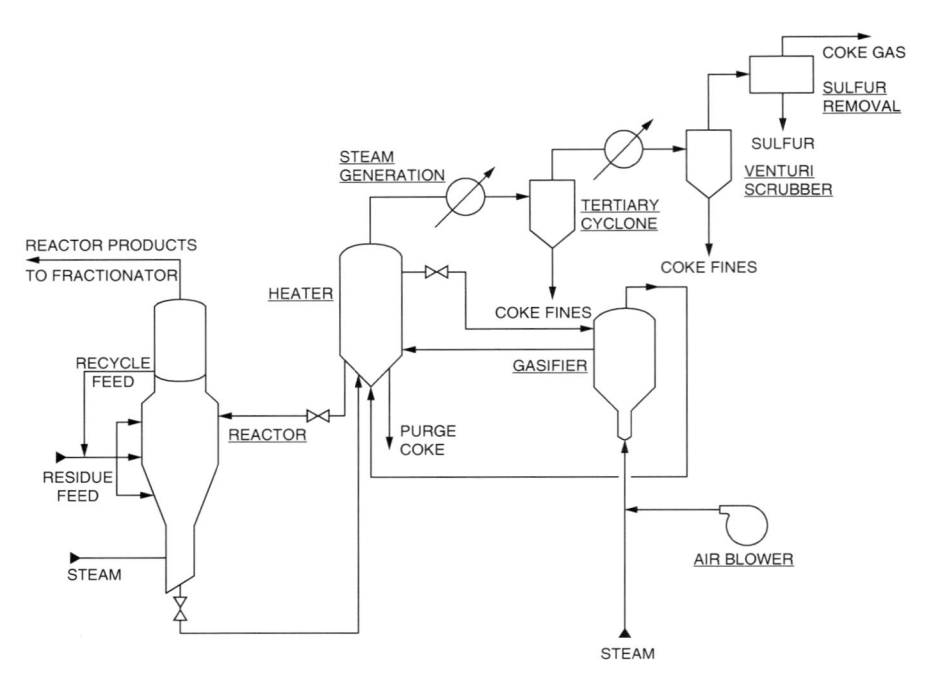

Fig. C.4 A flexi (fluid) coking process

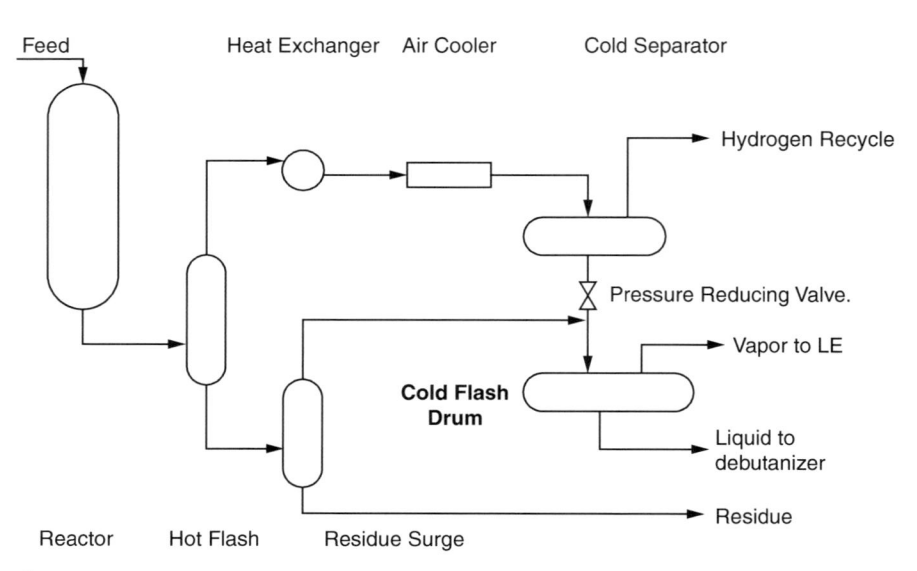

Fig. C.5 Typical residue hydrocracker with cold flash drum

temperatures and to avoid engine failure due to lack of fuel flow. Typically, cold properties include pour point, cloud point, freeze point, viscosity, and the cold filter plugging point.

Column Generally refers to a distillation, absorption, or stripping tower/vessel because of the tall cylindrical shape of this equipment (they look like columns). Refer to the equipment chapter, section on tower and distillation design.

Commingling Term which generally applies to the mixing of two petroleum products with similar specifications. Most branded gasoline firms require that their product not be commingled to preserve the integrity of the brand. May also refer to mixing two streams within a unit to achieve specific properties

Commissioning In projects, commissioning describes the activities necessary to prepare a process unit for operation. This includes line cleaning, refractory curing, catalyst/sorbent loading, etc.

Common carrier A pipeline or transport company which has government authority to move product for hire, operating like a public utility with standard rates for various shipments.

Complex refinery Also known as a "full conversion" refinery; this is a type of refinery configuration with process capability for conversion of nearly all crude oils to high-value products. Complex refineries may also make petrochemicals in addition to fuels and other petroleum products. These refineries include crude distillation, hydroprocessing, reforming, cracking, resid conversion, octane/cetane improvement, and other processes.

Component balances Component balances are derived from the TBP of the material requiring the balance. These component balances are derived by splitting the TBP into mid-boiling point pseudo components (see the topic chapter "▶ Introduction to Crude Oil and Petroleum Processing" in this handbook for definition of pseudo components and mid-boiling points). The purpose of component balances is to calculate more accurately fractions of the feed material properties in terms of specific gravity, sulfur content, mole weight, cloud and pour points, and the like. Figure C.6 illustrates a hypothetical TBP curve of a middle distillate fraction. This fraction has been broken up into six pseudo components, with mid-boiling points of 410–591 °F. By referencing the crude feed assay from which this fraction originates, the SG of each component can be read off as °API. If it is a well-produced assay, the component sulfur and cloud and pour points can also be read off for each pseudo component. The volume of each pseudo component forming the fraction is shown as the "x" of the TBP curve.

Condensate (1) Material condensed from vapor anywhere in a process. (2) A naturally occurring gaseous hydrocarbon that liquefies when cooled to surface temperature. Condensate is usually considered to be a part of crude oil production. (3) Frequently refers to steam that has been condensed to liquid water.

Condensers (e.g., shell and tube) In the chemical process, plant vapors are condensed either on the shell side of a shell and tube exchanger, the tube side of an air cooler, or by direct contact with the coolant in a packed tower. By far

Fig. C.6 A typical component breakdown

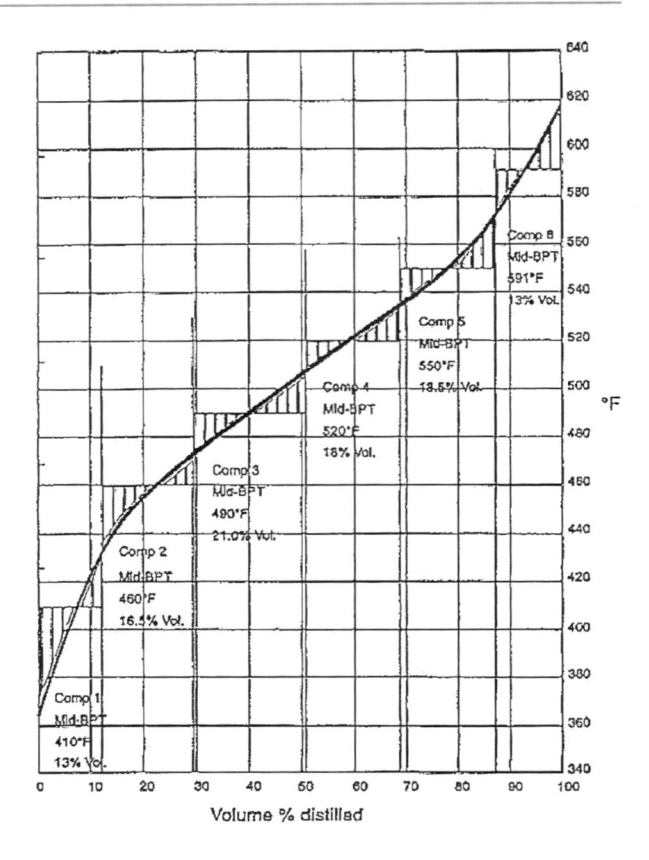

the most common of these operations are the first two listed. In the case of the shell and tube condenser, the condensation may be produced by cooling the vapor by heat exchange with a cold process stream or by water. Refer to the handbook topic chapter "▶ Process Equipment for Petroleum Processing" for a detailed discussion of heat exchangers and exchanger design.

Conductivity A measure of the ability of a material to conduct electricity. Measurement units are usually mhos/cm. Conductivity prevents the accumulation of static electric charge during handling, which can lead to explosions with hydrocarbon stocks. Middle distillates, like jet fuel and diesel, often require conductivity additives for handling. Crudes with very high conductivity, like shale oil, are often hard to desalt because the conductivity interferes with the development of electric charges on the suspended water droplets.

Cone-roof tank A storage tank having a fixed, conical-shaped roof. Refer to the chapter on offsites.

Configuration In a refinery, the arrangement of refinery process units to make the desired products from crudes. Configuration establishes the capabilities of the refinery and greatly determines potential profitability.

Confined space A space large enough to enter with limited access and not intended for occupancy. Such spaces also include spaces with configurations that could trap personnel, such as sloping bins. Refer to the chapter on safety.

Conradson carbon (ASTM D189) A measure of the coke generated from an oil when heated at low pressure in the absence of air. This is a rough indication of potential coker yield and the tendency to form coke (carbon deposits) on a catalyst when hydrotreated or cracked. See the chapter on "▶ Quality Control of Products in Petroleum Refining" in this handbook.

Contactor A generic term normally applied to a mass transfer tower, for intimately mixing and separating two streams to transfer components from one stream to another in an equilibrium stage process. An H_2S absorber for fuel gas is an example of a vapor/liquid contactor. There are liquid–liquid contactors as well, such as those used for removing H_2S with amine solutions from liquefied petroleum gases (propane, butane, etc.).

Control element A device that performs a control function, such as a control valve or variable speed controller.

Control loop The basic device arrangement from process sensing element in the field to the process controller to the final field control element (like a control valve)

Control system architecture Refers to essentially a map of how the process control system is assembled. Architecture would include what devices are tied into the process control system or network.

Control valves Control valves are specially designed valves with actuators which enable automatic, remote control of the valves. They are used throughout the process and oil refining industries to control operating parameters, such as flow, temperature, pressure, level, and composition. Refer to the handbook chapter on "▶ Process Controls in Petroleum Processing" for a detailed discussion of these critical control elements in petroleum processing. Figure C.7 illustrates the key parts of a typical control valve.

Control valve characteristic The relationship between a control valve position (% open) and the amount of flow the control valve can pass (% of maximum). Refer to the chapter on controls.

Control valve response The time required for a control valve to respond to a control system output signal.

Convection section The portion of a fired heater through which the hot flue gases travel on their way to the stack and exchange heat with other fluids

Convection tube or coil Tubes in the convection section of a fired heater that are heated by exchange with hot flue gases

Conventional crude Crude produced from the types of sources that have been traditionally used where a producer can just drill, pump, and separate the produced fluids

Conventional safety relief valve A conventional safety relief valve is a closed bonnet pressure relief valve that has the bonnet vented to the discharge side of the valve. Refer to the safety chapter.

Fig. C.7 A doubled seated control valve

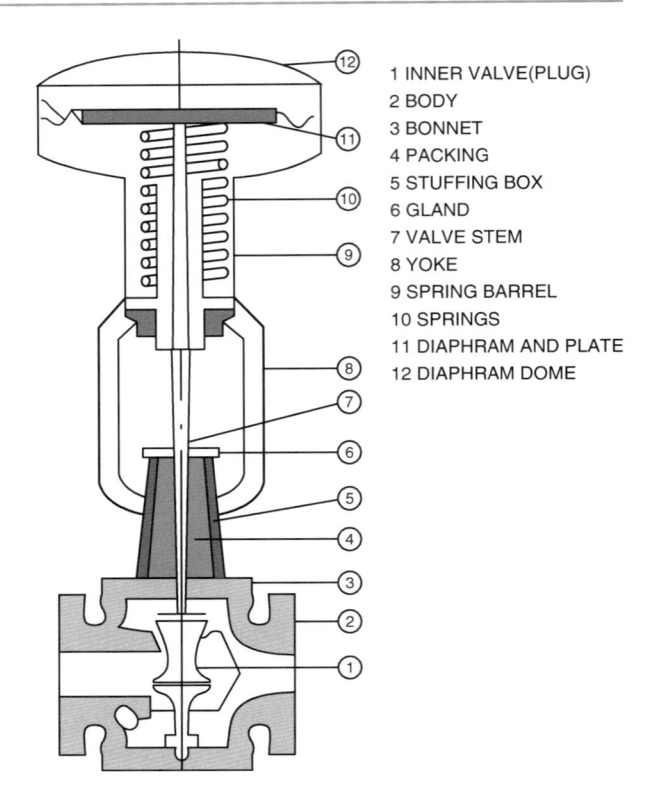

1 INNER VALVE(PLUG)
2 BODY
3 BONNET
4 PACKING
5 STUFFING BOX
6 GLAND
7 VALVE STEM
8 YOKE
9 SPRING BARREL
10 SPRINGS
11 DIAPHRAM AND PLATE
12 DIAPHRAM DOME

Coprocessing Processing two or more dissimilar materials together in the same process units. For instance, running conventional diesel and vegetable oil in the same hydrotreater at the same time to produce a green diesel stock is coprocessing.

Cooling water A circulating or once-through water system used for process or equipment cooling. Refer to the discussion in the chapter on utilities.

Cooling tower Equipment that generates cool/cold water for use in process cooling services through evaporation of part of the water. Refer to the chapter on utilities.

Coproduct A by-product or additional product from a process

Correlation coefficient (1) A statistical measure of how well a correlation fits a specific data set. A correlation coefficient of 1 would indicate a perfect 1-to-1 relationship; (2) more specifically, a statistical factor measuring how well any two markets (i.e., a cash market and a futures market) move in unison.

Corrosion Detrimental change in the size or characteristics of material under conditions of exposure or use. It usually results from chemical action either regularly and slowly, as in rusting (oxidation), or rapidly, as in metal pickling.

Corrosivity Hazardous waste characteristic of being corrosive to materials or skin – generally refers to waste caustic or acidic materials. Refer to the environmental chapter section on hazardous wastes.

Corrugated plate interceptor A device for enhancing gravity separation of oil from facility effluent water using multiple, parallel plates. Refer to the chapter on offsites.

Cost estimating Estimating the cost of facilities or services prior to purchase or construction. There are several types of cost estimates used. The scope and accuracy of the different types of estimates depend on the available detail on which to base the estimate. Details about cost estimating as it applies to process plants, including oil refineries, are discussed in the chapter titled "▶ Petroleum Processing Projects" in this handbook.

Cost-plus A pricing mechanism, commonly used by contracting and transportation firms. Creates a buying price for a good or service by starting with actual cost to supply the good or service and adding a markup

COT (1) Coil outlet temperature from a fired heater; (2) critical operating task

Counterflow cooling tower A cooling tower where air flows upward through a downward shower of water to cool the water. Refer to the chapter on utilities.

Coupon A method for monitoring corrosion by subjecting a piece of test metal to the same conditions as the process and testing the metal for properties.

CPE Continuing professional education

CPI (1) Continuous process (or performance) improvement; (2) chemical process industries; (3) cost performance index; Corrugated plate interceptor for oil/water separation

CPM (1) Critical path method; (2) certified purchasing manager

CPQRA Chemical process quantitative risk analysis

CPR Cardiopulmonary resuscitation

CPU Central processing unit

CPVC Chlorinated polyvinyl chloride

CQI Continuous quality improvement

Cr Chrome

Crack Difference between product value and cost in $/bbl. Gross measure of margin. Tracked generically. (1) Light oil crack/3:2:1 = 2 RU gasoline bbls +1 ULSD bbl – 3 WTI bbls prices converted to per bbl crude; (2) gas crack =1 bbl gasoline – 1 bbl WTI prices; (3) heat crack = ULSD price – WTI crude price per bbl; (4) various other cracks are defined.

Crack spread See "Crack"

Cracked Refers to a petroleum product produced by a secondary refining process such as thermal cracking or visbreaking processes which yield very-low-quality residue and gasoil and lighter stocks

Cracking refinery A refinery configuration which includes gasoil conversion to high-value products in addition to crude distillation, hydrotreating, and reforming. Cracking refineries do not substantially convert any residuum instead produce resid as fuel oil or tar/asphalt.

Cradle A device to support a tube in a fired heater.

CRC Chemical Rubber Company - original publishers of the Handbook of Chemistry and Physics

Crossflow cooling tower A cooling tower where air flows horizontally through a packed section with water flowing downward through the packing. Refer to the chapter on utilities.

Crossover Piping used to connect the radiant with the convection section of a fired heater; usually external to the heater. Refer to the chapter on equipment, section 6.

CRU Catalytic reforming unit

Crude distillation unit Initial separation unit in a refinery where the crude oil is fractionated into components by boiling ranges for further processing

Crude oil or crude Crude oil is a mixture of hydrocarbon compounds produced from underground geological formations by various production methods. These compounds range in boiling points and molecular weights from methane as the lightest compound to those whose molecular weight will be in excess of 500. In addition to the hydrocarbons, there are impurities like sulfur, nitrogen, and metals. Refining and petroleum processing take the raw crude and convert it to useful products. The topic chapter "▶ Introduction to Crude Oil and Petroleum Processing" in this handbook describes crude in some detail. The balance of this book focuses on the processes to convert the wide variety of crudes to products.

Crude unit See "Crude Distillation Unit" and "Atmospheric Crude Distillation Unit."

Cryogenic (AKA cryo) Processing fluids at very low temperatures where things that are normally gases can be liquefied

CS Carbon steel

CSA Canadian Standards Association

CSC (1) Car sealed closed; (2) commercial subcommittee; (3) customer support center

CSCC Caustic stress corrosion cracking

CSO Car sealed open

cSt or CST or CS Centistokes at 50 °C (or other temperatures as defined

CTE Coefficient of thermal expansion

CTL Coal to liquids

CTU Crude topping unit

Cu Copper

CUI Corrosion under insulation

Cumulative cash flow and present worth There are several methods of assessing profitability based on discounted cash flow (DCF), but the most reliable yard stick is a return on investment method using the present worth (or net present value) concept. This concept equates the present value of a future cash flow as a product of the present interest value factor and the future cash flow. Based on

Fig. C.8 Typical cut points on a TBP curve

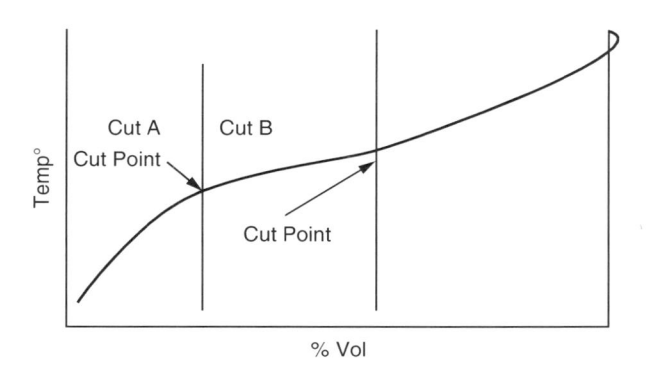

this concept, the return on investment is that interest value or discount factor which forces the cumulative present worth value to zero over the economic life of the project. Refer to the chapter titled "▶ Petroleum Processing Projects" in this handbook for a detailed discussion of this evaluation.

Curing Preparation of refractory for operation by careful dryout. Refer to the chapter on equipment, section on fired heaters.

Cut Material boiling in a specific temperature range. See also "Cut Point."

Cut point(s) A cut point is defined as that temperature on the whole crude (or any other) TBP curve that represents the limits (upper and lower) of a fraction to be produced. Consider the curve shown in Fig. C.8 of a typical crude oil TBP curve. A fraction with an upper cut point of 100 °F produces a yield of 20 vol% of the whole crude as that fraction. The next adjacent fraction has a lower cut point of 100 °F and an upper one of 200 °F; this represents a yield of 50−20 % = 30 vol% on crude.

Cutter stock or flux stock A petroleum stock which is used to reduce the viscosity of a heavier residual stock by dilution.

CUWI Corrosion under wet insulation

CV (1) A measure of valve size/capacity [C_v], valve flow coefficient (see the chapter on controls and the appendix to that chapter); (2) control valve abbreviation; (3) control volume in fluid calculations; (4) controlled variable in control systems

CVU Crude vacuum unit

CWA US Federal Clean Water Act

CWMS Crinkled wire mesh screen – for liquid demisting in vapor phase or water coalescing in liquid phase. Refer to the chapter on equipment, section on vessels and drums.

CWO Chain wheel operated

CWR Cooling water return

CWS Cooling water supply

Cycloparaffin A paraffin molecule with a ring structure. A naphthene

D

DBA or dBA A-weighted decibels – sound pressure level measurement. Refer to the section on noise control in the environmental chapter.

DT or dT (1) Differential temperature (rise or drop, also ΔT); (2) design temperature

D1160 Distillation curve (vol% distilled vs temperature) using a specific ASTM method. This method is for hydrocarbons boiling up to about 1,150 °F and employs vacuum.

D2887 See "SimDist"

D86 Distillation curve (vol% distilled vs temperature) using a specific ASTM method. This method is for hydrocarbons boiling up to about 700–750 °F.

DAF Dissolved air flotation – for wastewater treatment

Damper A large, loose-fitting butterfly valve or set of louvers that control draft in a fired heater firebox by creating a small back pressure – usually at the base of the stack or ahead of an induced draft fan.

Darcy equation Equation used to estimate pressure drop in piping due to single-phase flow. Refer to the equipment chapter, section on pressure drop.

Dashboard Collection of key variables for display

Data historian The computer system and files used to collect and access historical operating and lab data in a refinery

DC (1) Direct current; (2) delivery capacity; (3) document control

DCAT Drug, chemical, and allied trades

DCS Distributed control system – computer-based control system for a process unit

DCSC Distributed control system controller

DCU Delayed coking unit – resid conversion process. See "Coking Process."

DDC Direct digital control

DEA See Diethanolamine

Dead time (or dead band) The amount of time before a process starts changing after a disturbance or control action in the system

Deaerator A type of process equipment used to eliminate or strip air from boiler feed water. Refer to the chapter on utilities.

Dearomatization process The dearomatization process is a hydrotreating process that converts aromatic compounds to naphthenes and paraffins. Its most common use is to raise the smoke point of kerosene, particularly if the product is to be blended to make aviation turbine gasoline. Initially, the increase of kerosene smoke point was effected by the removal of aromatic compounds by extraction using liquid SO_2 (Edeleanu process). A typical kerosene dearomatizing unit is shown in Fig. D.1. Pressure at the separator is 350–450 psig, and kerosene recovery is 85–90 % of feed. The feed must be nearly sulfur-free. Catalyst options include nickel and precious metals.

Deasphalting and deasphalting process Deasphalting is an extraction process used in petroleum refining to remove the asphaltene portions of residue to prepare a suitable feedstock for catalytic conversion units. This eliminates the

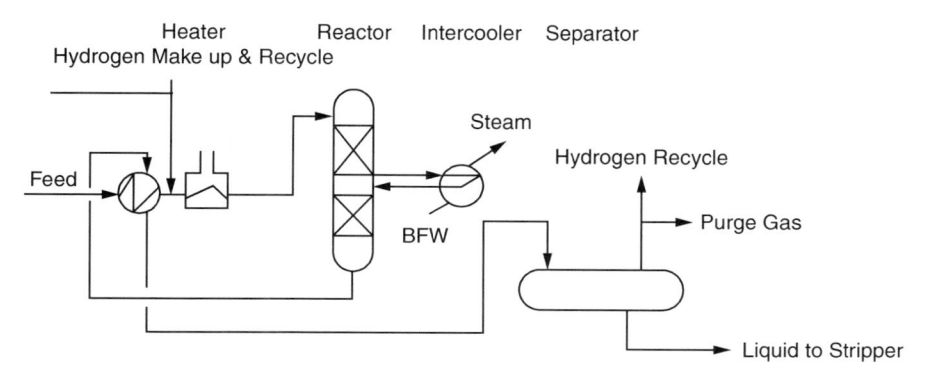

Fig. D.1 Typical hydro-dearomatizer process

heavy metals and the high Conradson carbon content of the residue feed that affect the catalytic units. In lube oil production, the light liquid phase resulting from the extraction of the asphalt makes excellent lube base oil. Details of the deasphalting process are in the chapter "▶ Non-energy Refineries in Petroleum Processing" of this handbook.

Debottlenecking Increasing the capacity of a process unit by removing or modifying parts of the system that limit feed or production rates

Debutanizer (or de-butanizer) Debutanizers are used in refineries to remove butanes and lighter compounds from product streams. The most notable of the feed streams is the overhead distillate from the atmospheric crude distillation tower. Other uses of this process are in the removal of these light ends from hydrotreaters (kero and heavier products) and the reactor effluent liquids from catalytic reforming and hydrocracking. The overhead from the fluid catalytic cracker main tower is also debutanized, but the butane compounds from this process will also contain the olefinic compounds from the cracking process. A typical process configuration for a debutanizer is shown in Fig. D.2. Refer to the topic "▶ Distillation of the "Light Ends" from Crude Oil in Petroleum Processing" in this handbook for details on debutanizers and debutanizer design.

Decarboxylation Eliminating oxygen from an organic molecule via partial hydrotreating with CO/CO_2 by-products

Decarburization Damage to steel caused by removal of carbon from grain boundaries

Decoking Removal of accumulated coke deposits, normally from heater tubes, by burning the coke off using air and a moderator, such as steam. May also be done mechanically without burning

Deethanizer (or de-ethanizer) The purpose of the deethanizer is to remove ethane from the product stream of LPG. Normally, it will be the last tower in a light ends distillation configuration. Refer to the topic "▶ Distillation of the "Light Ends" from Crude Oil in Petroleum Processing" in this handbook for a discussion of where the deethanizer fits into light ends fractionation.

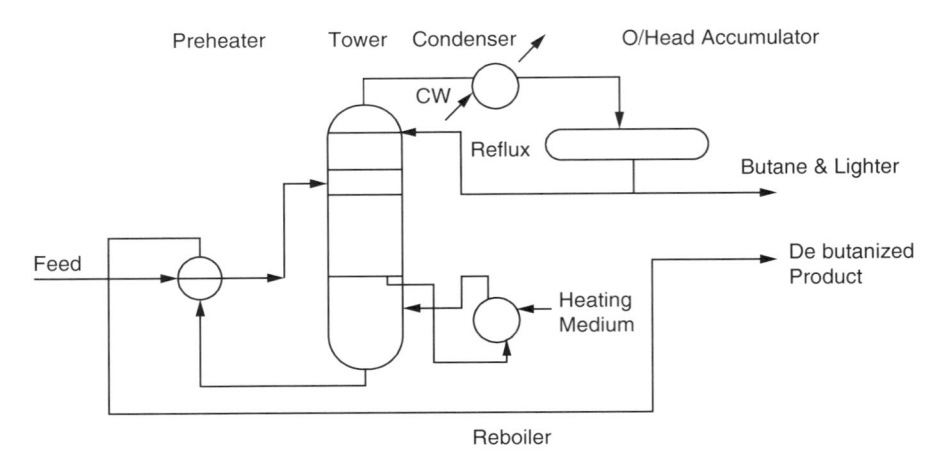

Fig. D.2 A typical debutanizer configuration

A deethanizer tower operates at an overhead accumulator pressure below 450 psia at 100 °F. This ensures that the operating pressure at the bottom of the tower, which is propane LPG, will be below its critical pressure. In the design example given in the topic "▶ Distillation of the "Light Ends" from Crude Oil in Petroleum Processing" in this handbook, the pressure in the accumulator was calculated to be 350 psia at 100 °F, and this was the dew point of the overhead vapor product. Deethanizers operating without overhead refrigeration facilities have partial condensers, and in this case, only sufficient overheads from the tower are condensed to meet the reflux required in the tower. Thus, the accumulator becomes a theoretical tray itself and there will be no liquid distillate product as such.

Definitive estimate A capital cost estimate-based quoted equipment prices and completion of most engineering work. The accuracy is around +/−5–10 %.

Degasifier Similar to a deaerator, a degasifier is used to strip out CO_2 or other gases from an aqueous stream. Commonly found in hydrogen plants. Refer to the chapters on utilities and hydrogen production (SMR section).

Degree days Temperature times number of days at temperature used for tracking aging in catalysts or materials

Degrees API or °API A measure of hydrocarbon density. 10 °API = 1.000 specific gravity or 349.776 lb/bbl. Normally measured at 60 °F. Specific gravity = 141.5/(131.5 +°API).

Delayed coking See "Coking Process."

Deluge system A permanently installed firefighting system which sprays copious amounts of water on specific equipment to help manage a fire. Deluge systems are sometimes found around pump seals, for instance. A sprinkler system in a building is another example of a deluge system. They can be automatically or manually actuated. Refer to the chapter on firefighting.

Demethanizer Distillation tower for stripping methane from a stream. Normally cryogenic.

Demethylation Thermal decomposition of a hydrocarbon to methane.

Demulsibility The resistance of an oil to emulsification or the ability of an oil to separate from any water with which it is mixed. The better the demulsibility rating, the more quickly the oil separates from water.

Demurrage Charge for or cost of holding a bulk carrier (ship, truck, or iso-container) for an extended period of time beyond a normally expected period

Denitrogenation Removing nitrogen from an oil. See also "Hydrodenitrogenation."

Density Mass of a material per unit volume at specified conditions. Be very careful of the actual conditions (pressure and temperature) used for the basis.

Department of Energy The US government, cabinet-level agency concerned with energy development, policies, and safety. Includes responsibility for hydrocarbon and alternative energy sources plus nuclear energy

Department of Transportation The US government, cabinet-level agency concerned with transportation policies and safety. Promulgates regulations for fuels, as well as movement of fuels

Depropanizer In the refinery configuration of light ends distillation, this unit is usually located between the debutanizer and the deethanizer. The process flow is similar to the debutanizer, with total overhead product and reflux being condensed. The tower operates at a reflux drum pressure of between 200 and 250 psia at 100 °F. The tower contains 35–40 actual trays, and the bottom product will be butane LPG. The overhead distillate will be fed to the deethanizer whose bottom product will be propane LPG.

Derivative control A control mode where the controller output is determined by the rate of change of the deviation from the setpoint. Acts like a dampening term. Normally used as a fine-tuning parameter in a proportional control mode

Desalter The desalter extracts salts from a hydrocarbon stream by mixing the stream with a small amount of fresh water (e.g., 10 % by volume) forming a water-in-oil emulsion. The resulting emulsion is subjected to an electric field wherein the water is coalesced as an underflow from the upper flow of a relatively water-free, continuous hydrocarbon phase. The desalted hydrocarbon stream is produced at relatively low cost and has a very small residual salt content.

Design basis The operating conditions, yields, etc., upon which a process unit is designed

Design pressure Pressure used in the design of equipment to determine the minimum permissible thickness or physical characteristics of the different parts of the equipment. Works with the design temperature

Design temperature Temperature used in the design of equipment to determine the minimum permissible thickness or physical characteristics of the different parts of the equipment. Sets the stress limits for the materials of construction. Works with the design pressure

Desulfurization Removing sulfur from an oil. See also "Hydrodesulfurization."

Desuperheater A device for reducing the temperature of superheated steam by injecting a controlled amount of boiler feed water into the stream

Detonation An explosion that generates an expanding, supersonic shock wave from the point of initiation. An unconfined vapor cloud explosion is an example. A controlled detonation also occurs in internal combustion engines when the fuel–air mixture in the cylinders is ignited by a spark or heat of compression.

Dew point(s) Is the temperature and pressure condition at which a hydrocarbon vapor begins to condense. In a calculation, the sum of the mole fraction composition of the vapor divided by the equilibrium constant of each compound must equal the sum of the sum of the mole fraction of the liquid phase at the dew point condition of temperature and pressure. Table D.1 is a dew point calculation example based on a tower top condition for an atmospheric crude distillation unit. The dew point calculation here is carried out at 8.3 psia, which is the partial pressure of the hydrocarbons in the overhead vapor.

Dewaxing Removal of wax from a hydrocarbon stock. May refer to solvent processes or specific types of hydroprocessing. Used extensively in lube oil production

DGA See diglycolamine

DHA Detailed hydrocarbon analysis

Diaphragm A relatively thin metal or elastomer disk that is moved by differential pressure between the two sides of the membrane. These are used extensively in control valve actuators and certain types of pump. One side of the membrane normally has an actuating fluid whose pressure and/or frequency is varied as needed. The other side of the membrane may be connected mechanically to the device to be moved or may have a fluid to be pumped.

Diaphragm pump A type of pump used for moving thick pulps, sludge, acid or alkaline solutions, and fluids containing gritty solid suspensions. The pump can use a diaphragm material compatible with the pumpage.

DIB De-isobutanizer

Diesel index Product of the API gravity and the aniline point (in degrees Fahrenheit) of a diesel fuel divided by 100; an indication of the ignition quality of the fuel

Diesel oil or diesel Diesel oil (sometimes called automotive gas oil) is used as fuel for heavy internal combustion engines such as heavy trucks (lorries) and rail locomotives, as well as automobiles. Its main components are light gas oils from the atmospheric crude distillation unit and hydrotreated light gas oils. In some cases, light FCC cycle oil may be included. This product has a nominal ASTM boiling range of around 480–610 °F but may boil over a 350–700 °F range. Hydrocarbon cuts in this boiling range are hydrotreated to remove sulfur (and the process does change pour and cloud points to some extent) and blended with kerosene and some heavier middle distillate stocks to meet the diesel oil specification. Sulfur in diesel ranges from less than 10 ppm for diesel used in vehicles on the road in the United States and Europe up to 500 ppm for some

Table D.1 Example dew point calculation

COMP	Mole frac Y	1st trial @ 220		2nd trial @ 225 °F		MOL wt	Weight factor	SG	vol factor	Liquid prop
		K	X = Y/K	K	X = Y/K					
C_2	0.008	–	NEG		NEG					
C_3	0.054	84.3	0.001	93.9	0.001	44	0.044	0.508	0.009	
iC_4	0.021	38.6	0.001	39.8	0.001	58	0.053	0.563	0.010	Mol wt = 130.7
nC_4	0.084	29.52	0.003	30.1	0.003	58	0.174	0.584	0.030	
C_5	0.143	12.53	0.011	12.65	0.011	72	0.792	0.629	0.126	SG = 0.766
C_6	0.155	4.70	0.033	4.94	0.031	85	2.635	0.675	0.390	
C_7	0.175	2.17	0.081	2.19	0.080	100	8.00	0.721	1.110	°API = 53
Mid-BP 260	0.124	1.00	0.124	1.16	0.107	114	12.193	0.743	1.642	
Mid-BP 300	0.124	0.506	0.245	0.518	0.239	126	30.114	0.765	3.936	K = 12
Mid-ESP 340	0.075	0.229	0.328	0.253	0.293	136	39.848	0.776	5.135	
Mid-BP 382	0.037	0.108	0.343	0.126	0.294	152	44.688	0.788	5.671	
Totals	1.000		1.170		1.06	130.7	138.551	0.767	18.059	

$$K = \frac{\text{Vapor press at selected temperature}}{\text{Total system pressure}}$$

2nd trial = 0.108 × 1.170 (K for mid – BP 362 component)

New K = 0.126 then VP = 8.3 psia × 0.126 = 1.05 psia = 225°F

The 2nd trial is close enough to $\sum x_i = 1.00$

Notes:
a. In estimating for the 2nd trial and final temperature, the "K" of the highest X component is multiplied by the total value of X function. Then vapor pressure curves are used to give the component temperature corresponding to this new vapor pressure.
b. The molar composition of the final "X" is the composition of the liquid in equilibrium with the product vapor. The tower top conditions are 229 °F at 15 psig (this is the total pressure and includes the steam effect).

off-road uses and higher for marine diesel. Directionally, sulfur levels in all grades of diesel are being driven lower in most major markets. At the same time, cetane requirements for diesels are being pushed higher. For more information, see the chapters "▶ Petroleum Products and a Refinery Configuration" and "▶ Quality Control of Products in Petroleum Refining" of this handbook and consult the current diesel specifications for the product grade of interest.

Diethanolamine (DEA, $(HOC_2H_4)_2NH$) One of the common, regenerable amine solvents used for acid gas removal (H_2S or CO_2) from vapor streams

Differential pressure (dP) The actual pressure difference between two points. This pressure is not referenced to atmospheric pressure.

Diglycolamine (DGA, $HOCH_2CH_2OCH_2C_2H_4NH_2$) A regenerable amine acid gas solvent normally used to remove H_2S or CO_2 from gases. Developed by Fluor Corporation

DIH Deisohexanizer. A DIH fractionation column is commonly installed in light naphtha isomerization process units on the stabilized reactor product to separate a side-cut stream rich in lower octane C_6 compounds for recycle to the reactor from an overhead product stream containing C_5 compounds and the higher octane isohexane isomers. A small drag stream of heavy compounds is taken from the bottom.

Diisopropanol amine (DIPA, $(HOC_3H_6)_2NH$) One of the regenerable amine solvents used for acid gas removal (H_2S or CO_2) from vapor streams

Dilbit A blend of some diluent (normally naphtha) and bitumen. Many Canadian tar/oil sands "crudes" are dilbits.

DIN Deutsches Institut für Normung e.V. (The Standards Institute of Germany)

DIP Deisopentanizer. A DIP fractionation column separates isopentane from heavier compounds and is usually fed a light naphtha stream containing C_5 and C_6 compounds.

DIPA See diisopropanol amine.

Direct contact condenser A condenser where the cooling medium is mixed directly into the process stream being cooled. Sometimes used in small towers. Refer to the equipment chapter section on heat exchangers.

Discounted cash flow (DCF) A method for analyzing potential projects by examining the expected cash flow for the project over several years and "discounting" future cash value based on an assumed inflation rate or time value of money back to a present value for the investment. The DCF rate of return for a project is the interest rate that forces the cumulative present net worth of the project to zero over its economic life. See the chapter on economics for additional detail.

Dispersed flow In this flow regime, nearly all the liquid is entrained as a spray in a bulk gas phase. This regime may also be referred to as mist flow. Refer to the equipment chapter section on two-phase pressure drop.

Distillate Kerosene or diesel boiling range materials. Generically, something distilled out of another material, usually crude oil

Distillate hydrocracking Hydrocracking is a versatile catalytic refining process that upgrades petroleum feedstocks by adding hydrogen, removing impurities,

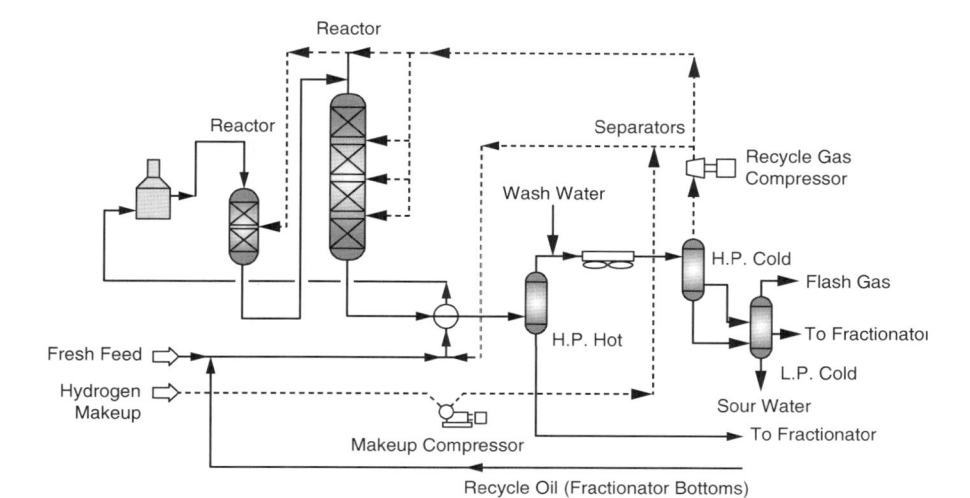

Fig. D.3 A distillate hydrocracker reactor section

and cracking to a desired boiling range. Hydrocracking requires the conversion of a variety of types of molecules and is characterized by products that are of significantly lower molecular weight than the feed. The process is carried out at elevated temperatures (550–825 °F) and hydrogen partial pressures from 1,000 to 2,500 psig using a variety of catalysts suitable for the feedstock and desired selectivity. Hydrocracking feeds can range from heavy vacuum gas oils and coker gas oils to atmospheric gas oils. Products usually range from heavy diesel to light naphtha. Figure D.3 illustrates one hydrocracker flow sheet. The chapter on "▶ Hydrocracking in Petroleum Processing" in this handbook reviews the hydrocracking process in detail.

Distillation Also known as fractionation, there are numerous applications of distillation in petroleum processing. After heat transfer, distillation is the most common unit operation in a refinery. Considered generically, distillation is simply a process that separates materials by boiling range. Examples of distillation applications are discussed in detail in multiple sections of this handbook. For perspective here, the following are offered:

Types of distillation curves:

ASTM distillation curve

While the TBP curve is not produced on a routine basis, the ASTM distillation curves are. Rarely, however, is an ASTM curve conducted on the whole crude. This type of distillation curve is used on a routine basis for plant and product quality control. This test is carried out on crude oil fractions using a simple apparatus designed to boil the test liquid and to condense the vapors as they are produced. Vapor temperatures are noted as the distillation proceeds and are plotted against the distillate recovered. Because only one equilibrium stage is used and no reflux is returned, the separation of components is poor and mixtures are distilled. Thus, the initial boiling point for ASTM is higher than the corresponding TBP point, and the final boiling point of the ASTM is lower than that for the TBP curve. There is a correlation between the ASTM and the TBP curve, and this is dealt with in the topic "▶ Introduction to Crude Oil and Petroleum Processing" in this handbook.

True boiling point curve (TBP)
This is a plot of the boiling points of almost pure components, contained in the crude oil or fractions of the crude oil. In earlier times, this curve was produced in the laboratory using complex batch distillation apparatus of a hundred or more equilibrium stages and a very high reflux ratio. Nowadays, this curve is produced by mass spectrometry or gas chromatography techniques much more quickly and accurately than by batch distillation.

Equilibrium flash vaporization curve (EFV)
The EFV curve of an oil is determined in a laboratory using an apparatus that confines liquid and vapor together until the required degree of vaporization is achieved. The percentage vaporized is plotted against temperature for several runs to produce the EFV curve. Separation is poorer for this type of distillation than for an ASTM or TBP. Therefore, the initial boiling point will be higher for the EFV than for the ASTM. The final boiling point of the EFV will be lower than that for the ASTM. This test is rarely done, but the EFV curve is calculated either from a TBP curve or an ASTM curve. These methods are given in the introductory and atmospheric and vacuum crude distillation topics of this handbook.

Types of distillation processes:
The separation of products by distillation falls into three general categories:
- Total vapor condensation such as the atmospheric crude distillation
- Vacuum distillation processes, such as the vacuum distillation of atmospheric residue
- Light ends distillation

Total vapor condensation processes
The best example of this type of distillation unit in modern refining is the atmospheric crude distillation unit. In these types of units, the feed is heated and vaporized to a temperature above the total product cut point. This mixed vapor/liquid is produced by an external heater (and heat exchanger system) and the mixed stream flashed in the lower section of a tower. The vapor rises in the tower and is condensed by a cooled reflux stream at various stages up the tower according to the various distillate boiling points. Full details of this type of distillation are given in the chapter titled "▶ Atmospheric and Vacuum Crude Distillation Units in Petroleum Refineries" of this handbook. Similar distillation systems are also used in the primary separation of fluid catalytic cracking unit effluent and the reactor effluent from a hydrocracker.

Vacuum distillation processes
These processes are designed to operate as total vapor condensation similar to the atmospheric units. However, the feed to these units is usually heavy residual oils, which if heated to vaporize the product distillate required at atmospheric pressure (or near-atmospheric pressure), would cause the feed to crack and coke. The system is therefore set at low vacuum pressure so that the vaporizing temperature is well below the feed's cracking properties. The distillates are produced in the same way as that for the atmospheric units, that is, by selective cooled reflux streams. Full details of a crude oil vacuum process are also given in the chapter titled "▶ Atmospheric and Vacuum Crude Distillation Units in Petroleum Refineries" of this handbook. Vacuum distillation is also used for the distillation of light residues from hydrocrackers to produce light vacuum distillates and heavy residuum for further thermal cracking or lube oil production.

Light ends distillation
The most common of this type of distillation is the crude unit light ends process. The feed to this unit is the overhead distillate from the atmospheric crude unit. This feed may also include overhead distillate from other processes (such as the catalytic reformer). The products from the light ends process are light and heavy naphtha, butane LPG, propane LPG, and refinery gas. These processes are discussed more fully in the topic "▶ Distillation of the "Light Ends" from Crude Oil in Petroleum Processing" in this handbook.

Distributed control system (DCS) This is a computer-based control system used for basic process control. The basic control loops are in the background. The DCS displays the process and control-loop information for the operator. It also records the information and communicates with any advanced control systems. The operator and APC change the settings in the DCS, which then sends the moves on to the process loops. DCS is the most common type of control system today in refineries.

Distributor (1) A device for even distribution of fluids in process equipment, such as at the top of a reactor or inside a distillation column. (2) An organization which disburses products to retail or wholesale locations for sale

dll Dynamic-link library

DMDS Dimethyl disulfide

DME Dimethyl ether

DMS (1) Document management system; (2) dimethyl sulfide

DMSO Dimethyl sulfoxide

DNAPL Dense nonaqueous phase liquid

Doctor test A qualitative method for detecting undesirable sulfur compounds in petroleum distillates using a lead/caustic solution to determine whether an oil is "sour" or "sweet" (positive or negative, respectively)

DOE (1) US Federal Department of Energy; (2) design of experiment – technique for optimizing a unit; (3) direct operating efficiency

DOT US Federal Department of Transportation

Double-pipe exchanger A double-pipe exchanger consists of a pipe within a pipe. One of the fluid streams flows through the inner pipe, while the other flows through the annular space between the pipes. There are also multiple tube versions of these exchangers. Refer to the equipment chapter section on heat exchangers.

Downcomer In a distillation tower or stage contactor, this is a free-flow path for liquid from one tray to the tray below. Normally, the downcomer is a curtained-off area next to the column wall where liquid can fall from one tray to the next without resistance from vapor flowing upward. The downcomer is separated from the active tray areas by overflow weirs. Refer to the equipment chapter section on towers and columns.

Downstream (1) In the petroleum industry, refers to refining and marketing activities and may also include transportation or anything else not associated with exploration and production; (2) refers to processes or activities that occur subsequent to a specific process in refining – e.g., the blender is downstream of hydrotreating, cat cracking, and cat reforming

DP Design pressure

DP or dP or ΔP Differential pressure or pressure drop

dP cell A diaphragm device for measuring small differences in pressure with the higher pressure on one side of the diaphragm and the lower pressure on the other side. The difference may be directly indicated or may be transmitted.

Draft Movement of air or flue gases based on density differences, normally caused by temperature differences, but may be mechanically forced or induced

Drawoff In a distillation tower, refers to pulling a stream from the side of the tower (not the overhead or bottom). Also referred to as a sidestream draw. Within a tower, there is normally an internal sump from which the stream can be drawn.

Drums Drums are horizontal or vertical vessels used to separate gas and liquid phases. Generally drums do not contain complex internals like fractionating trays or packing as in the case of towers. They are used however for removing material from a bulk material stream and often use simple baffle plates or wire mesh to maximize efficiency in achieving this. Drums are used in a process principally for (1) removing liquid droplets from a gas stream (knockout pot) or separating vapor and liquid streams, (2) separating a light from a heavy liquid stream (separators), (3) surge drums to provide suitable liquid holdup time within a process, and (4) reduction of pulsation in the case of reciprocating compressors. A detailed discussion of the design of drums is included in the handbook topic "▶ Process Equipment for Petroleum Processing" with discussion of surge times in drums in the handbook section "▶ Process Controls in Petroleum Processing."

DSO Disulfide oil from caustic treating of oils (e.g., or Merichem processes)

Dumbbell distillation Refers to a stock, normally a blend, with high proportions of light material and heavy material, but little material in the boiling range between the two extremes. An example of such a stock is a dilbit, which consists of a bitumen (tar) that is mixed with a naphtha to facilitate transportation.

E

E&C Engineering and construction
E&I Electrical and instrumentation
E&P Exploration and production
EAM Ergonomic accident model
EBIT Earnings before interest and taxes
EBITDA Earnings before interest, taxes, depreciation, and amortization
Economic evaluation Economic evaluation is used for most aspects of the refinery planning and its operation. The methods used in these evaluations may differ from company to company, but the end product must reflect the profitability of the present and often the future profitability of a proposed venture or operation. This measure is reflected in terms of the return on investment of the item. This is described and discussed in the chapter titled "▶ Petroleum Refinery Planning and Economics" in this handbook.
Economic life Number of years a project is expected to yield the projected profit and pay for its installation
Edmister correlations A series of correlations which relates an ASTM distillation test to the TBP of a petroleum cut is presented in W. Edmister's publication titled: "*Applied Thermodynamics.*" The correlations relating to the ASTM and TBP curves from this book are used in this handbook and are given in the topic "▶ Introduction to Crude Oil and Petroleum Processing". Similar correlations have also been prepared by Edmister for the relationship between TBP and the equilibrium flash vapor curve (EFV).
EDS Emergency depressuring system – for dumping a high-pressure unit quickly in an emergency to prevent damage
Eductor A vacuum-producing device; also referred to as an ejector. It uses a high-velocity motive stream, such as steam or compressed air, to create a vacuum applying the Bernoulli effect.
EE Electrical engineer or engineering
EEBSS Emergency escape breathing support system
EEP Enabling event probability – in LOPA
Effluent Refers to materials leaving a process system or facility
Effluent water Effluent water is the wastewater that comes from the petroleum processing units. The effluent water may be contaminated with oil and grease, solids, metals, sulfur compounds, nitrogen compounds, and other undesirable materials. Effluent water must generally be processed to remove the contaminants before it can be discharged into a waterway or an external, finishing treatment plant. Treatment of effluent water is discussed in more detail in the topic "▶ Environmental Control and Engineering in Petroleum Processing" in this handbook. The refinery sources and contaminants that are normally of concern are listed in Table E.1.
EFT Electronic funds transfer
EFV Equilibrium flash vaporization

Table E.1 Effluent water sources and contaminants

Process	Wastewater	Air
Atmospheric and vacuum distillation	Sour water (NH$_3$ and H$_2$S)	Furnace
	Desalter water spent caustic	Flue gases – SO$_2$
	Process area wastewater (pump glands, area drains, etc.)	
Thermal cracking	Sour water	Furnace
Delayed coking	Decking water (oil)	Flue gases
	Process area wastewater	
Fluid cat cracking	Sour water (NH$_3$, H$_2$S, phenols)	Furnace
Unsat gas plant	Spent caustic	Flue gases SO$_2$, CO, particulates
	Process area wastewater	
Hydrocracker	Sour water (inc Phenols)	Furnace
Hydrogen plant	Process area wastewater	Flue gases SO$_2$
Sat gas plant alkylation	Spent caustic	Nil
	Process area wastewater	
Naphtha hydrotreater	Sour water	Furnace
Cat reformer	Process area wastewater	Flue gases
Sulfur plant	Nil	Incinerator flue gas – SO$_2$ hydrocarbons – flare
Tankage area	Tank dike area drains, noncontaminated rain runoff	Tank vents, hydrocarbons

EHSRMA Extremely Hazardous Substances Risk Management Act
EI Energy Institute of Petroleum (UK)
EIA Environmental impact assessment
EII Energy Intensity Index as defined by Solomon Associates
EIS (1) Environmental impact study or statement; (2) executive information system
EIT Engineer in training
Ejector A vacuum-producing device; also referred to as an educator or steam-jet ejector. It uses a high-velocity motive stream, normally steam, to create a vacuum applying the Bernoulli effect.
Electrolytic process A process using electrical energy to cause a chemical reaction. An example is electrolysis of water to generate hydrogen. Note that desalting would *not* be considered electrolytic.
Electrostatic precipitator or separator A pollution control or solids removal system which uses an electrically induced charge on suspended particles or droplets to remove the suspended material from a stream. Examples would include removal of catalyst fines from FCC flue gas and removal of acid droplets from alkylation plant hydrocarbon streams. Desalters also apply a form of electrostatic separation.
Element The sensor for a process variable or condition
ELSA Emergency life-saving apparatus

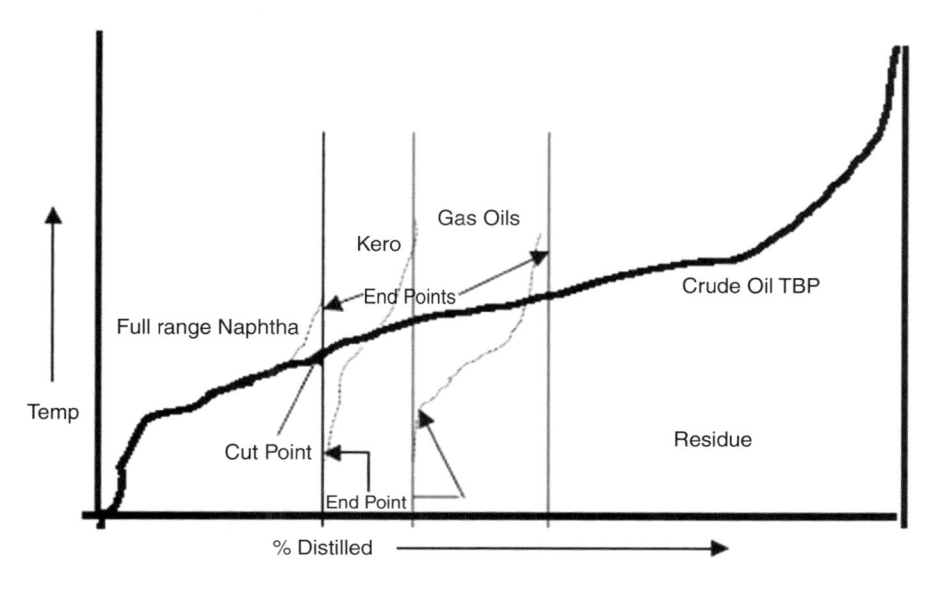

Fig. E.1 Definition of end points

ELV In the EU, emissions limit value for air pollutants. Refer to the environmental chapter.

Embrittlement Loss of metal toughness or ductility due to low temperatures or process exposure

EMI Electromagnetic interference

Emission(s) Generally refers to gases (including pollutants) sent into the atmosphere. Refer to the environmental chapter.

EMS (1) Emergency medical services; (2) environmental management system

EMT Emergency medical technician

Emulsion A liquid mixture of two or more substances not normally dissolved in one another, one liquid held in suspension in the other. Water-in-oil emulsions have water as the internal phase and oil as the external, while oil-in-water have oil as the internal phase and water as the external.

End point(s) Whereas a cut point is an ideal temperature on a TBP curve to define the yield of a fraction, the end points define the shape of the fraction distillation curve when produced commercially. In actual processes, the initial boiling point of a fraction will be much lower than its front end cut point. The final fraction boiling point will be higher than the corresponding cut point. This is demonstrated by Fig. E.1.

There is a correlation between the TBP cut point and the ASTM end point. This is described in the topic "▶ Introduction to Crude Oil and Petroleum Processing" in this handbook. A relationship also exists for the 90 % ASTM point and the TBP 90 % cut point. With these two ASTM data and using an ASTM probability graph (see the topic "▶ Introduction to Crude Oil and Petroleum Processing" in this handbook), a full ASTM curve with its end points can be drawn. This, converted to a TBP curve, is used to define the cut's properties.

Endothermic reaction Reaction that absorbs heat.

Energy Information Administration (EIA) The US federal agency responsible for collecting, analyzing, and providing information about energy. The goals of the agency are to promote sound policies, efficient markets, and public understanding of energy use and impacts.

Energy Intensity Index (EII) From Solomon Associates. Generally expressed as the Actual Btus of energy used per barrel of feedstock in a specific process divided by the standard Btu/bbl for that process developed by Solomon based on their worldwide survey of refineries. The ratio is multiplied by 100 for the final EII number.

Engineering flow diagrams Diagrams that are used extensively by all disciplines of engineers to convey ideas and data. These include: process flow diagram (PFD), mechanical flow diagram (MFD, sometimes called the piping and instrumentation diagram (P&ID)), auxiliary flow diagram (AFD), and utilities diagram. Refer to the chapter titled "▶ Petroleum Processing Projects" in this handbook for further discussion. Examples of sections of these engineering drawings are given in the appendix titled "▶ Examples of Working Flow Sheets in Petroleum Refining" of this handbook as illustration.

Engler viscosity A viscosity obtained by dividing the outflow time in seconds for 200 ml of the material being tested by the time in seconds for 200 ml of water at 68 °F (20 °C) to flow out of an Engler viscosimeter

Enterprise control An overall control system that factors in economics, market, and other factors to optimize and coordinate several refinery units simultaneously

Entrainment Refers to physical carryover of solid particles or liquid droplets in a vapor or liquid stream, respectively, due to velocity

Environmental Protection Agency (EPA) US government agency that promulgates and enforces environmental regulations at a federal level. See the chapter on environment.

EOL (1) Environmental operating limit; (2) end of life (equipment)

EOP Emergency operating procedure

EOR End of run

EP End point

EPA US Federal Environmental Protection Agency. Refer to the environmental chapter of this handbook.

EPBC Ethane, propane, butane, and condensate

EPC (1) In projects – engineer, procure, and construct; (2) event-driven process chain

EPCI Engineering, procurement, construction, and installation

EPCM Engineering, procurement, and construction management

EPCRA US Emergency Planning and Community Right-to-Know Act

EPF Engineering, procurement, and fabrication
EPR Ethylene–propylene chloride
E-PRTR European Pollutant Release and Transfer Register
EQE Earthquake engineering
EQSD EU Environmental Quality Standards Directive
Equilibrium flash vaporization (EFV) When a mixture of compounds vaporizes or condenses, there is a unique relationship between the composition of the mixture in the liquid phase and that in the corresponding phase at any condition of temperature and pressure. This relationship is termed the equilibrium flash vaporization for the mixture or simply the "flash." The topic is detailed under the topic "▶ Introduction to Crude Oil and Petroleum Processing" in this handbook. Refer to Appendix 1 of that chapter for an example calculation procedure.
Equipment-factored estimate A method for estimating capital cost for new facilities that uses known equipment costs scaled to new applications. This method requires rough equipment sizing and completion of significant process engineering. Accuracy is $\sim +/-20\ \%$.
ER Emergency response. Refer to the chapter on "▶ Safety Systems for Petroleum Processing" in this handbook.
ER probe Electrical resistance corrosion probe – basically a corroding wire where resistance changes are monitored to indicate corrosion.
Ergun equation Primary equation to calculate single-phase pressure drop through beds of discrete solid materials. Refer to the equipment chapter section on pressure drop.
Error This is usually defined in process control as the difference between the setpoint and the process variable or measurement.
ERP (1) Emergency response plan or program – refer also to the chapter on "▶ Safety Systems for Petroleum Processing" in this handbook; (2) enterprise resource planning
ERT Emergency response team – refer also to the chapter on "▶ Safety Systems for Petroleum Processing" in this handbook.
ES Expert system
ESD (1) Emergency shutdown system – may apply to a piece of equipment or a unit to prevent damage; (2) emergency shutdown
ESP (1) Electrostatic precipitator; (2) emergency shutdown procedure
ETA (1) Estimated time of arrival; (2) event tree analysis
ETBE Ethyl tertiary-butyl ether
ETD In the EU, energy taxation directive. Refer to the environmental chapter topic on air emissions.
Ethane (C_2H_6) A normally gaseous straight-chain hydrocarbon. It is a colorless paraffinic gas that boils at a temperature of $-127.48\ °F$. It is extracted from natural gas and refinery gas streams.
Ethanol Ethyl alcohol, which is most often derived from corn or other renewable organic sources. Ethanol is blended with gasoline to produce a cleaner burning

fuel and is an accepted oxygenate component for the oxygenated seasons mandated by the EPA.

Ether A generic term applied to a group of organic chemical compounds composed of carbon, hydrogen, and oxygen, characterized by an oxygen atom attached to two carbon atoms (e.g., methyl tertiary-butyl ether).

Etherification Conversion of a compound into an ether by reaction. An example of etherification is the reaction of methanol and butene to make methyl tertiary-butyl ether (MTBE).

Ethyl tertiary-butyl ether [$(CH_3)_3COC_2H_5$] An oxygenate blendstock formed by the catalytic etherification of isobutylene with ethanol

Ethylene (C_2H_4) An olefinic hydrocarbon recovered from refinery processes or petrochemical processes. Ethylene is used as a petrochemical feedstock for numerous chemical applications and the production of consumer goods.

EU European Union

EuReDatA European Reliability Data Association

EV Earned value

EVP Executive vice president

EWP Environmental work practice – requirements to control pollution

Ex situ Refers to action or processes that are performed on materials that have been removed from their original location. In common use, this term refers to oil shale retorting processes that process rock which has been removed from the shale formation to a surface retort.

Excess O_2 or air In a fired heater, this is the surplus air supplied to the firebox (or leaking in) that is un-combusted. An indication of heater combustion efficiency. Low excess O_2 is more efficient than high O_2 because the furnace has to heat up less air.

Exothermic reaction Reaction that releases heat

Exploration Organization focused on finding crude oil and gas

Exports Normally refer to materials shipped out from a country commercially, but may also refer to materials sent out of an individual unit or refinery, e.g., a hydrogen plant may export surplus steam to a refinery

Extra-heavy oil (EHO) Term used for naturally occurring hydrocarbon deposits where the oil is present as a tar-like substance that must be heated or dissolved with a flux in order to be produced. Tar sands would be considered an EHO.

Extrudate In refineries, this generally refers to a particle produced by extrusion of a mud or clay-like material through a die. The extruded material is then chopped or broken into shorter particles. Used extensively for catalyst and absorbent production

Eyewash A deluge system designed for flushing the eyes, usually located near a potential chemical exposure area, where personnel can flush their eyes with copious amounts of clean water to eliminate or control damage from chemical exposure. Refer to the safety chapter.

F

F1 Research octane number
F2 Motor octane number
FA (H, HH, L, LL) Flow alarm high, high–high, low, or low–low, e.g., FAHH is a high–high flow alarm
FAA US Federal Aviation Authority
Fahrenheit Temperature scale based on 32 °F for the temperature at which water freezes and 212 °F for the temperature at which water boils (180 °F difference). Conversion to Fahrenheit from Celsius (centigrade) temperature scale is by the following formula: $F = 1.8C + 32$, where C is the temperature in Celsius degrees.
Fail-safe Failure of a device toward the safest position. Refer to the chapter on controls.
Failure position For a control device, this is the position the device moves to if the control signal or motive means is compromised. For instance, control valves will normally fail closed, open, or in last position. The failure position is selected based on what position is most safe for the process. Refer to the chapter on controls.
Falldown Unplanned shutdown caused by a failure
FAQ Frequently asked question
FAS Federal accounting standards
FASB Financial and Accounting Standards Board
FAME Fatty acid methyl ester
FAT Factory acceptance test
Fatigue Loss of metal strength due to repetitive strain cycles
Fatty acid methyl ester (FAME) Renewable diesel fuel derived from fats by conversion to an ester in the diesel boiling range. Refer to the chapters on unconventional stocks and renewables.
FBP Final boiling point or end point
FC (1) Flow controller; (2) fail closed valve designation
FCC Fluid catalytic cracker – heavy oil conversion process
FCE Final control element – normally a control valve but can be other devices
FCPA US federal Foreign Corrupt Practices Act
FD (1) Forced draft – in a firebox, a fan that blows air into the firebox; (2) filter drum; (3) fire detection
FE (1) Flow element – like an orifice plate; (2) fundamentals of engineering exam in professional engineer licensing; (3) facilities engineer
Fe Iron
Feed-forward control A control scheme where changes in an upstream variable initiate changes in downstream controls in anticipation of upcoming deviations. For example, if the feed rate to a unit changes, the appropriate controllers can also be changed downstream so they are set correctly for the new feed rate.

Feedstock Any of the raw or semifinished materials which move to the various units of a refinery or petrochemical plant. Crude is a feedstock, but the term is mainly used to describe raw materials after the distillation process which in turn go on to more sophisticated units at the refinery. VGO, cat feed, naphtha, condensate, and straight-run residual fuel are commonly referred to as feedstocks.

FEL Front-end loading in project execution. See the chapter on projects.

Fenceline Refers to the very outer property limits of a processing facility – generally where there is a fence that encompasses the whole facility

FEP Fluorinated ethylene–propylene

FERC US Federal Energy Regulatory Commission

FeS Iron sulfide, frequent corrosion product, may be pyrophoric

FFC Feed-forward controller

FGT Fuel gas treating

FHR Flame and heat resistant

FIFO First in–first out (inventory control)

Filming amine Any of several amines that prevent corrosion by coating the surface of metal equipment – e.g., distillation tower overhead corrosion inhibitors

Filtration A mechanical process for removing solids from a liquid or vapor stream by passing the stream through a fine media to collect the solids. Filtration and ultrafiltration can be used to fractional micron sizes.

Final boiling point, FBP The highest temperature indicated on the thermometer inserted in the flask during a standard laboratory distillation. This is generally the temperature at which no more vapor can be driven over into the condensing apparatus. Usually the end point

Final control element The device which actually controls the desired operating condition in a control loop. Refer to the chapter on controls.

Finned tube A fired heater or air cooler tube that has fins protruding from the surface to enhance heat transfer

Fines Refers to small particles and dust

Fire brick Refractory used in furnaces and other high-temperature equipment that is in the form of a brick. Refer to the chapter on equipment, section on fired heaters.

Fire extinguisher A relatively small, normally hand-operated device for putting out smaller fires quickly at the incipient stage. There are multiple types of fire extinguishers for different services. Refer to the chapter on firefighting.

Fire foam A chemical mixed with water and air to create foam for firefighting. Refer to the chapter on firefighting.

Fire main The main pressurized firefighting water supply piping system in a facility. Branches, laterals, or submains/sublaterals come off the fire main to the various firefighting equipment. Refer to the firefighting chapter.

Fire point The lowest temperature at which an oil vaporizes rapidly enough to burn for at least 5 s after ignition, under standard conditions

Fire relief Relief scenario or relief valve intended to prevent overpressure of a vessel or other equipment when exposed to an external fire. Refer to the safety chapter.

Firebox The chamber in a fired heater where fuel and air are burned to release useful heat. The firebox contains the coils where fluids are heated.

Fired heater Device for providing process heat by burning a fuel in a firebox and transferring the heat generated to a process fluid contained in tubes within the heater. Generally fired heaters fall into two major categories: horizontal and vertical. There is a detailed discussion of fired heaters or furnaces and their design in the handbook topic "▶ Process Equipment for Petroleum Processing." Emissions and emission controls for fired heaters are covered in the topic "▶ Environmental Control and Engineering in Petroleum Processing" of this handbook.

Fireproofing Application of fire-resistant materials, such as insulating concrete, to structures or vessel skirts to help prevent collapse in the event of fire.

Firewater (or firefighting water) Water used for extinguishing or cooling service in a fire. Normally supplied by a secure, dedicated system. Refer to the chapters on utilities and firefighting.

Fischer assay A more-or-less standard test to determine the potential products from retorting or destructive distillation of a solid material. The test involves heating a known amount of solid to high temperature (800 °F plus) at atmospheric pressure and collecting all the products. Yields are determined by material balance. This test is used for shale oil yield comparisons. See the chapter on unconventional crudes.

Fischer–Tropsch process Process for conversion of a synthesis gas (CO + H2) to hydrocarbon products. Refer to the chapters on unconventional crudes, lube oil, and coal or gas to petrochemicals.

Fixed cost A cost that does not depend significantly on rate. For example, the cost of a turnaround does not depend very much on how many barrels were run.

Fixed equipment Vessels, reactors, separators, exchangers, etc. Process equipment that does not move to do its job – hopefully.

FL Fail last (or locked) position – control valve

Flammability Refers to the ease of combustion or ignition of a material

Flammability limits The range of concentrations of a material in air that can support ignition of the material, possibly explosively. The flammability limits are usually expressed as upper and lower explosive or flammability limits as volume % in air at atmospheric pressure.

Flame impingement Flames in a firebox burning directly on a tube's surface – often leads to tube failure

Flame retardant (or fire retardant) Material that does not continue to burn when an ignition source is removed (self-extinguishing). Usually refers to types of clothing, like Nomex, used in refineries to help protect workers in the event of fires or flash fires

Flameout Loss of flame in a fired heater

Flare System for collecting and safely disposing of combustible vapors by burning them. Refer to the offsites chapter.

Flare tip The device at the end of the flare where fluids to be burned are vented. The tip contains multiple ignition sources and usually steam to promote mixing with air. Many flare tips also contain molecular seals to prevent intrusion into the flare system by air. Refer to the offsites chapter.

Flash point(s) The temperature to which a product must be heated under prescribed conditions to release sufficient vapor to form a mixture with air that can be readily ignited. Flash point is used generally as an indication of the fire and explosion potential of a product. For a rough estimate of flash point, the equation below is useful. Refer to the handbook chapters on "▶ Introduction to Crude Oil and Petroleum Processing," "▶ Petroleum Products and a Refinery Configuration," and "▶ Quality Control of Products in Petroleum Refining" for more details on flash point.

$$\text{Flash Point } °F =\sim 0.77(\text{ASTM } 5 \% - 150°F).$$

Flash zone A flash zone is associated with the distillation of crude oil, both atmospheric and vacuum, and the main fractionating towers of the fluid catalytic cracking unit, visbreaker unit, or thermal cracking unit. The flash zone is the area in these distillation towers where the distillate vapors are allowed to separate from the unvaporized liquid. The transfer line from the heater enters the flash zone. The vapors rise up through the tower to be condensed by cold reflux streams coming down. Steam and some hot vapors enter the flash zone from the bottom product stripper section located below the flash zone. The chapter titled "▶ Atmospheric and Vacuum Crude Distillation Units in Petroleum Refineries" of this handbook deals with a calculation procedure to establish these flash zone conditions for atmospheric and vacuum distillation crude distillation towers. The handbook topic "▶ Process Equipment for Petroleum Processing" provides further information.

Flashing A condition where liquid is vaporizing to reach a vapor/liquid equilibrium

Flexicoking Proprietary fluid coking process. A thermal cracking process which converts heavy hydrocarbons such as crude oil, tar sands bitumen, and distillation residues into light hydrocarbons. Feedstocks can be any pumpable hydrocarbons including those containing high concentrations of sulfur and metals.

Flight scraper In an API separator, flight scrapers are used to sweep or scrape the sludge into the sludge pit and along the top of the separator level to move the accumulated oil toward the oil skim pipe.

Floating-roof tank (or floater) A storage tank in which the surface of the fluid is covered by a roughly flat roof that floats on top of the commodity in the tank to prevent vapor losses and contamination. Refer to the chapter on offsites.

Flocculation Process for removal of suspended solids or other undesirable materials from effluent water using chemicals to cause the particles of material to

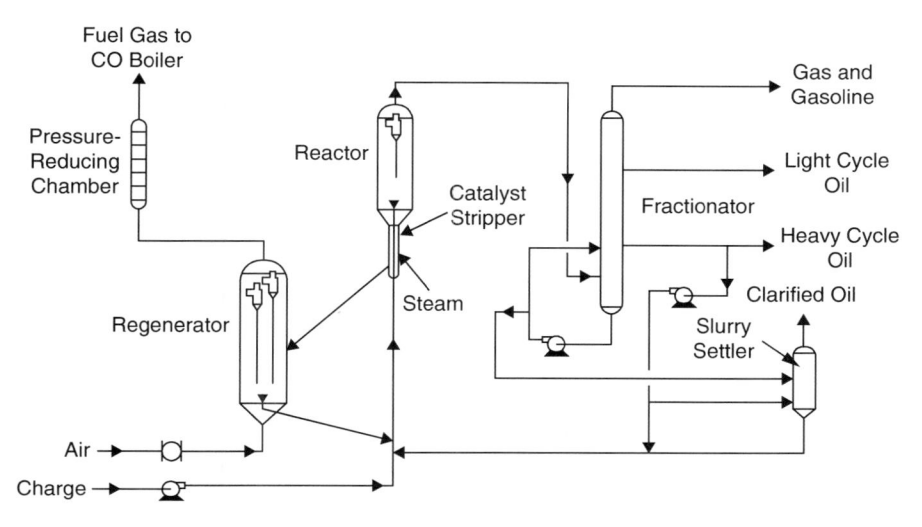

Fig. F.1 A typical "side-by-side" fluid catalytic cracking unit

stick together until they become too heavy to stay suspended. Refer to the chapter on offsites.

Flooding In a distillation tower or absorber, buildup of liquid in the tower producing high-pressure drop and loss of efficiency. Flooding may also mechanically damage the tower internals. Flooding is normally caused by excessive liquid or vapor rates.

Flow meter A device to measure fluid or solid flow rates. Refer to the discussion in the chapter on process controls.

Flow regime Category for the type of flow occurring inside a pipe or other piece of flowing equipment. Refer to the equipment chapter section on two-phase flow.

Flue gas tunnel(s) In a fired heater where the burners are on the roof of the firebox and flow is downward (downfired), these are the refractory tunnels that collect the flue gases evenly and convey them out of the firebox.

Fluid catalytic cracking unit (FCCU or FCC) The fluid catalytic cracking unit (FCCU) is a process for converting middle and heavy distillates to high-octane gasoline and olefin-rich light gases. It is one of the most important conversion processes in a refining. There are several licensed and proprietary versions of this technology and the equipment involved. Generally, the process contacts hot fluidized solid catalyst of very small size with preheated heavy oil. The oil is thermally and catalytically cracked in a very short contact time. The product vapors are separated from the catalyst. Products are fractionated from the vapors, and the catalyst, coated with coke formed in the process, is regenerated with air by burning off the coke. The regeneration process reheats the catalyst for its next trip through the reactor. Figure F.1 shows a simplified flow sheet for an FCC. The process is detailed in the chapter "▶ Fluid Catalytic Cracking (FCC) in Petroleum Refining" of this handbook.

Fluid coking A thermal cracking process utilizing the fluidized-solids technique to remove carbon (coke) for continuous conversion of heavy, low-grade oils into lighter products. Flexicoking is an example.

Flushing Using a liquid, such as water or a solvent, to wash out equipment or a system

Flushing oil A low pour point oil used to wash out or "flush" a heavy oil process system or equipment. See the chapter on utilities.

Flux A rate of exposure divided by the area exposed. For instance, a flow rate of 20 lb/h over a 3 square-foot area results in a mass flux of 20 $lb/h \div 3$ $ft^2 = 6.7$ $lb/ft^2 - h$. Flux can be applied to rate of heat transfer, rate of mass or volume movement, and so on.

FM Factory Mutual, Inc.

FMEA Failure modes and effects analysis

FMECA Failure modes, effects, and criticality analysis

FO (1) Fail open – control valve; (2) framework order (purchase order type)

Foam generator In firefighting, a device to mix air into a flowing foam chemical stream to create a foam stream to be directed onto a fire. Foam generators may be permanently installed (as on tanks) or portable with a fire truck. Refer to the chapter on firefighting.

Foaming In a tower or separator, foaming is the generation of a stable layer of vapor trapped in liquid film bubbles on top of liquid which can interfere with equipment operation or result in poor performance. Foam can be created by a number of factors, including presence of fine solids or surfactant contaminants in the streams.

FOB Free on board. Terms of a transaction where the seller agrees to make the product available within an agreed-upon time period at a given location. Any subsequent costs are the responsibility of the buyer.

FOE or FOEB Fuel oil equivalent or fuel oil equivalent barrels. A way of accounting for light hydrocarbons by converting their heating value to the equivalent barrels of fuel oil. Normal basis is about 6.05 MMBtu/bbl equivalent, but the basis sometimes varies by user. Used primarily in planning and optimization

Force majeure The legal cancellation of a delivery obligation due to the occurrence of natural acts beyond the direct control of the seller (i.e., operating problems with tankers or refineries or weather disruptions)

Forced draft In a firebox, a fan that blows air into the firebox

Foul (1) See "Fouling." (2) Refers to streams that contain H_2S and other "foul"-smelling compounds

Fouling Deposition of undesirable materials on the surface of equipment or interstitial spaces of packed beds that inhibits operations

FP (1) Flash point; (2) fire protection; (3) freeze point

FPS Feet per second

FQD In the EU, fuel quality directive. Refer to the environmental chapter section on air emissions.

FR Flow recorder

Fracking Crude oil and gas production process used extensively for tight shale formations where a mixture of sand, water, and chemicals is pumped into a well at high pressures to hydraulically fracture the rock formation and prop the fractures apart with sand. On reversing the well flow, oil and gas can go back to the well and be produced.

Fraction A separate identifiable part of crude oil; the product of a refining operation. The different cuts of petroleum products that come off a distillation column contingent on their volatility or boiling range

Fractionation This is a unit operation in chemical engineering which separates components from mixtures in which they are contained by boiling points (AKA distillation). In petroleum refining, this process is a major means of separating precise groups of petroleum components from the crude oil feed and other intermediate refining processes. Separation by fractionation is accomplished by heat and mass transfer on successive stages represented by carefully designed trays. These trays are designed to enhance the heat and mass transfer by good mixing of hot vapors rising through the trays with colder liquid entering the tray. The mixing tends to achieve a phase equilibrium between the liquid and vapor traffic. Refer to the handbook topics "▶ Atmospheric and Vacuum Crude Distillation Units in Petroleum Refineries," ▶ Distillation of the "Light Ends" from Crude Oil in Petroleum Processing," and "▶ Process Equipment for Petroleum Processing" for details about some of the applications and design calculations for fractionation units.

FRC (1) Fire (or flame)-resistant (or retardant) clothing (e.g., Nomex); (2) flow recorder and controller

Fresh feed input Represents input of material (crude oil, unfinished oils, natural gas liquids, other hydrocarbons and oxygenates, or finished products) to processing units at a refinery that is being processed (input) into a particular unit for the first time. Examples:
- Unfinished oils coming out of a crude oil distillation unit which are input into a catalytic cracking unit are considered fresh feed to the catalytic cracking unit.
- Unfinished oils coming out of a catalytic cracking unit being looped back into the same catalytic cracking unit to be reprocessed are not considered fresh feed.

Front-end loading (FEL) An approach to project execution that uses more detailed engineering during the early phases of the project evaluation to provide a more accurate project cost and better economic evaluation. This approach is discussed in the chapter on projects.

Froth flow (AKA bubble flow) Flow regime in which bubbles of gas are dispersed in a primarily liquid phase with the gas moving about the same velocity as the liquid. Refer to the chapter on equipment, two-phase pressure drop section.

FSA (1) Flexible spending account; (2) food services administration; (3) formal safety assessment

FSSL Fail-safe solid-state logic

FT (1) Flow transmitter; (2) flash trend

FTA Fault-tree analysis

FTE (1) Full-time equivalent person in hours; (2) in controls = fault-tolerant Ethernet

FTIR Fourier transform infrared analysis method
FTZ Foreign trade zone
Fuel coke Grade of petroleum coke used in fired heaters in place of coal
Fuel ethanol An anhydrous denatured aliphatic alcohol intended for gasoline blending as described in oxygenates definition
Fuel gas Any gas used as fuel in a facility
Fuel oil The heavy distillates from the oil refining process; used as fuel for power stations and marine boilers. Heavy fuel oils produced from the nonvolatile residue from the fractional distillation process. Heavy oils that are "leftovers" from various refining processes. Heavy black oils used in marine boilers and in heating plants
Fuel oil equivalent (FOE) A measure of heat content of a fuel by comparison with a standard barrel of fuel oil, nominally containing 6.05 MMBtu of energy. The basis can vary by user, however.
Fuel oil equivalent barrels (FOEB) See fuel oil equivalent.
Fuel tip See "Burner Tip."
Fuel rich A firebox condition where there is more fuel than combustion air – very dangerous – potential for explosion
Fuel solvent deasphalting A refining process for removing asphalt compounds from petroleum fractions, such as reduced crude oil. The recovered stream from this process is used to produce fuel products.
Fugitive Generally refers to air emissions of pollutants, such as "fugitive" dust generated by material-handling activities
Fungible Term which refers to the likeness or at least "interchangeability of a petroleum product." Material shipped on a pipeline must be "fungible," i.e., meets a common set of specifications acceptable to various shippers, and the same holds true for future contracts. The less fungible the product, the less likely it is to succeed in the futures arena and the more problem it is likely to create in the distribution process. Various elements of the Clean Air Act have made several petroleum products less fungible.
Full-conversion refinery Also known as a "complex" refinery; this is the type of refinery configuration with process capability for conversion of nearly all of a crude oil to high-value products. These refineries may also make petrochemicals in addition to fuels and other petroleum products. Such refineries include crude distillation, hydroprocessing, reforming, cracking, resid conversion, octane/cetane improvement, and other processes.
Fungible crude See "Fungible."
Furfural or furfuraldehyde Furfural is a heterocyclic aldehyde, with a ring structure containing four carbons and an oxygen atom. The chemical formula is OC_4H_3CHO. Used for solvent refining of lube oils. Refer to the chapters on nonfuel refineries and hazardous materials.
Furnace oil Canadian term used to describe high-sulfur No. 2 oil. So, furnace oil in Canada is the equivalent of our high-sulfur, off-road, fuel or home heating oil.
FV Full vacuum
FZGO Flash zone gas oil or "fuzzy gas oil"

G

GAAP Generally accepted accounting principles

Gallon Measurement of volume in the oil industry (42 gal = 1 barrel)

Galvanic corrosion Corrosion of metal caused by electrochemical interaction of incompatible materials

Gap control A process control mode where the controller takes action if the process variable moves outside a specified rage, but takes no action as long as the variable is within the allowable range.

Gaps and overlaps In distillation, gaps and overlaps normally refer to the differences between the temperatures at 95 vol% recovery in a lighter cut and 5 vol% recovery of the adjacent heavier cut based on their ASTM distillation. Figure G.1 illustrates this concept.

The ASTM gap shown in Fig. G.1 would be typical in the separation between naphtha and kero in the atmospheric distillation unit. The 5 % temperature of the kero (cut 2) is higher than the naphtha (cut 1) 95 % temperature. This is a good separation because there are few kero components in the heavy end of the naphtha. The overlap illustration is typical of the separation between the heavier products of the atmospheric distillation of crude oil such as between light gas oil cut and the heavy gas oil cut. A gap then is when the numeric difference between the 5 % of the heavier cut and the 95 % of the lighter cut is positive. An overlap is when this difference is negative.

Gas chromatograph An analyzer type that uses differential diffusion rates of compounds through a tiny packed column to determine a material's composition. Refer to the chapter on controls.

Gas – refinery fuel The gas fractions in petroleum refining may be taken as that fraction of crude or produced in a process that boils below propane. This vapor fraction is usually used as a fuel to the refinery's fired heaters. The fuel gas is normally treated to remove sulfur compounds to control SOx emissions and any

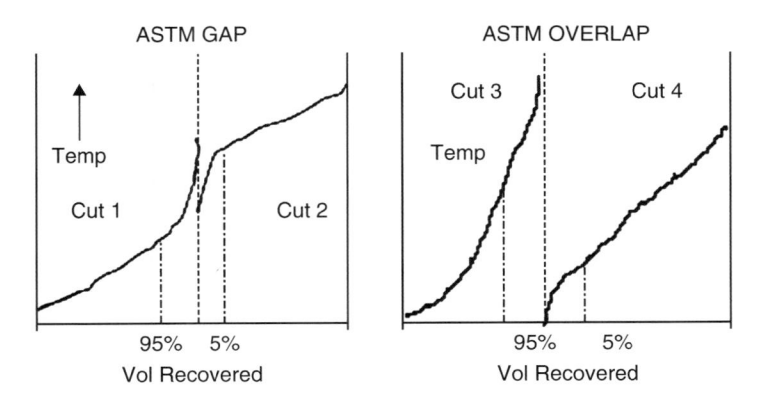

Fig. G.1 Defining gaps and overlaps in fractionation

liquids are knocked out ahead of the heaters. Gas treating is discussed in the chapter "▶ Refinery Gas Treating Processes," and fuel gas utility handling is in the handbook chapter "▶ Utilities in Petroleum Processing."

Gas oil (gasoil) A medium-heavy hydrocarbon boiling typically between about 500 and 1,200 °F. Gasoil is normally cracked to lighter products or used as fuel. There are multiple grades: AGO (atmospheric gas oil), VGO (vacuum gas oil), SRVGO (straight-run VGO), LGO (light gas oil), etc. Usually, there will be two gas oil sidestreams, a light gas oil sidestream and, below this take-off, a heavy gas oil sidestream. Both these sidestreams are normally stripped using steam or reboiling to meet their respective flash point specifications (usually 150 °F minimum). The lighter sidestream cut of about 480–610 °F on crude is the principal precursor for the automotive diesel-grade finished product; this sidestream is desulfurized to meet the diesel sulfur specification in a hydrotreater (see the chapter "▶ Hydrotreating in Petroleum Processing" in this handbook). The lower gas oil stream can be a guard stream to correct the diesel distillation end point. This heavy gas oil may also be hydrodesulfurized and routed to either the fuel oil pool (e.g., as a precursor for marine diesel), a finished heating oil product from the gas oil pool, or cracked and lighter products. See also the topic "▶ Petroleum Products and a Refinery Configuration" in this handbook.

Gas to liquids (GTL) A process that converts hydrocarbon gas to liquid hydrocarbon products. See the discussion of "xTL" in the chapters on unconventional crudes and feedstocks and chemicals from natural gas and coal.

Gas treating processes Refinery gas treating usually refers to the process used to remove the so-called acid gases (hydrogen sulfide and carbon dioxide) from the refinery gas streams. These acid gas removal processes used in the refinery are required either to purify a gas stream for further use in a process or for environmental reasons associated with the use of the gas for fuel. Clean air regulations now being practiced in most industrial countries require the removal of these acid gases to very low concentrations. The treating generally involves scrubbing the gas with a chemically "basic" solution to remove the acidic gases. The solution is then regenerated to release the acid gases for further processing, and the lean solution is recycled. See the chapter "▶ Refinery Gas Treating Processes" and the definition for amine absorption processes.

Gas blanketed Using an inert or relatively inert gas to prevent air contact with a petroleum stock during storage or in surge drums. Common blanket gases include natural gas and nitrogen. In ships, frequently engine exhaust is used for blanketing; although it still contains some oxygen, it will not easily support combustion.

Gasification Generic term for conversion of heavy liquids or solids to gaseous products. Gasification includes partial oxidation of solids or tars to form synthesis gas. Examples of gasification include the initial steps of the coal-to-liquids or resid-to-liquids processes.

Gasohol A blend of finished motor gasoline containing alcohol (generally ethanol but sometimes methanol) at a concentration of 10 % or less by volume. Data on gasohol that has at least 2.7 % oxygen, by weight, and is intended for sale inside carbon monoxide nonattainment areas are included in data on oxygenated gasoline. See oxygenates.

Gasoline(s) Fuel generally boiling in the 50–400 °F range used for automotive transportation. Several subgrades exist. They are readily and widely used by the general public. The current trend in gasoline specifications is toward lower sulfur (<10 ppm), lower vapor pressure (RVP <7), and fewer aromatics and air toxics. These enable cleaner engines.

Two major gasoline products are produced in most petroleum refineries. Their specifications and standard quality are discussed in multiple chapters of this handbook. The two finished grades shipped from final blending facilities are usually (1) a regular grade with an octane number of 87 and (2) a premium grade with an octane number of 93. Because of oxygenate blending requirements, the blended gasoline actually leaving a refinery will have a lower octane value because the final grade, with the oxygenate addition, is made at a product terminal. These octane levels may differ slightly from country to country, but these are the key quality for North America, with octane numbers defined as (RON + MON)/2.

Gasoline blending components Naphthas which will be used for blending or compounding into finished aviation or motor gasoline (e.g., straight-run gasoline, alkylate, reformate, benzene, toluene, and xylenes). Excludes oxygenates (alcohols, ethers), butane, and natural gasoline

Gauge (or gage) pressure Pressure referenced to atmospheric pressure. Atmospheric pressure is taken as 0 psig.

GC (1) Gas chromatograph or chromatography – analytical method; (2) Gulf Coast

GC/MS Combined analytical technique using both GC and MS to characterize a hydrocarbon stock

GFCI Ground fault circuit interrupter

GHG Greenhouse gas – refers to CO_2, methane, NO_x, etc., in the atmosphere that can affect global warming

GHV Gross heating value. Essentially the same as higher heating value (HHV)

Gilliland correlations An empirical correlation between the number of stages in a distillation or contacting column at finite reflux ratio and the minimum number of stages required at the minimum reflux ratio. Refer to the handbook chapter on light ends.

GOHDS Gas oil hydrodesulfurization unit

GOPA Gas oil pump around in a fractionator

GOR Gas-to-oil ratio

GPD or gpd Gallons per day

GPM or gpm Gallons per minute

GPSA Gas Processors Suppliers Association. Produce a very good technical data book

Green diesel Renewable diesel stock produced by hydrotreating animal fats or vegetable oils. The category includes stand-alone processing as well as coprocessing. Refer to the discussions on unconventional stocks and renewables.

Green gasoline Renewable gasoline stock produced by coprocessing vegetable oils and animal fats with gas oils in an FCC unit. See the chapters on renewables and unconventional stocks.

Green olefins Renewable petrochemical feedstock produced from an FCC unit by coprocessing vegetable oils and animal fats with gas oils. See the chapters on renewables and unconventional stocks.

Grid (1) Grids are used as low-pressure drop packing in certain fractionation towers. They came into prominence with the development of the crude oil "dry vacuum" units. See the chapter titled "▶ Atmospheric and Vacuum Crude Distillation Units in Petroleum Refineries" of this handbook and the relevant section of the chapter on "▶ Process Equipment for Petroleum Processing" for details of this type of packing and its use. (2) The electrical anodes in a desalter, electrostatic precipitator, or other electrostatic separation processes

Gross Apparent overall difference. For example, "Gross Margin."

GRP Gas recovery plant

GTL Gas-to-liquids abbreviation

Gulf Coast spot market Large-volume transactions (from 25,000 barrels to full tankers of petroleum products) bought or sold for a stipulated delivery in the near future. Although this market might entail several pipeline or waterborne transaction points in the Texas and Louisiana area, unless specified otherwise, it reflects the delivery of the product the same month at a Pasadena, Texas, origin on Colonial Pipeline.

GUI Graphical user interface

Gum A heavy polymer that forms in a hydrocarbon stock from polymerization of olefinic components in the stock. It may form in storage, during transit, or when a stock is heated in a refining process. Gums are a significant source of fouling resulting in loss of heat transfer and increased pressure drop/energy consumption in refining units.

GWD EU Ground Water Directive

H

H&MB Heat and material balance

H_2 or H2 (1) Hydrogen; (2) how to

H_2S Hydrogen sulfide

HAZ waste or Haz Waste Hazardous waste

Hazardous waste A waste that is either defined as a hazardous waste by regulation or possesses certain hazardous characteristics. Refer to the environmental chapter.

HAZCOM Hazard communication

HAZID Hazard identification

HAZMAT Hazardous materials

HazOp or HAZOP Hazard and operability review – one of the process hazards analysis methods most commonly used
HAZSUB Hazardous substance
HAZWOPER Hazardous Waste Operations and Emergency Response
HC Hydrocarbon – generic
HCGO Heavy coker gas oil
HCK Sometimes used as abbreviation for hydrocracker
HCO Heavy cycle oil – from an FCC
HCR Human cognitive reliability model
HCU Hydrocracking unit
HDC Hydrocracker
HDF Hydrofinisher
HDN Hydrodenitrification or hydrodenitrogenation
HDPE High-density polyethylene
HDS (1) Hydrodesulfurization; (2) hydrodesulfurization unit – a hydrotreater
HDT Hydrotreater
HDW Hydrodewaxer
Header A piping system used for collection/distribution of a process fluid from/to multiple points. For example, a steam header is used to distribute steam to various steam users within an area.
Heat and material balance A material balance with all the heat inputs, outputs, chemical releases, and other thermodynamic factors included
Heat of combustion, gross Total heat evolved during complete combustion of unit weight of a substance with water vapor condensed, usually expressed in Btu per pound
Heat of combustion, net Gross heat of combustion minus the latent heat of condensation of any water produced
Heat pump A system for moving heat from a lower temperature to a higher temperature using a working fluid
Heater(s) Heaters are used extensively in petroleum refining to provide heat energy to the process plants utilizing an independent energy source, namely, fuel oil or fuel gas. A heater or fired heater provides process heat by burning a fuel in a firebox and transferring the heat generated to a process fluid contained in tubes within the heater. Generally, fired heaters fall into two major categories: horizontal and vertical. There is a detailed discussion of fired heaters or furnaces and their design in the handbook topic "▶ Process Equipment for Petroleum Processing." Emissions and emission controls for fired heaters are covered in the topic "▶ Environmental Control and Engineering in Petroleum Processing" of this handbook.
Heater burners (or burners) The purpose of a heater burner is to mix fuel and oxidizer (normally air) to ensure complete combustion in a fired heater. Refer to the handbook topic "▶ Process Equipment for Petroleum Processing" for a detailed discussion of burner designs and critical factors. Additional discussion

on burner design factors affecting emissions is included in the topic "▶ Environmental Control and Engineering in Petroleum Processing" of this handbook.

Heater coil The tubes in a fired heater through which the process fluid flows as it is being heated

Heating oil A distillate used for home or commercial heating. Widely used as a synonym for No. 2 fuel oil or diesel

Heavy crude Crude with a high specific gravity, i.e., high density (low API gravity, e.g., <15 °API); due to the presence of a high proportion of heavy hydrocarbon fractions and metallic content

Heater efficiency The efficiency of a fired heater is the ratio of the heat absorbed by the process fluid to the heat released by combustion of the fuel expressed as a percentage. Heat release may be based on the LHV (lower heating value) of the fuel or HHV (higher heating value). Process heaters are usually based on LHV and boilers on HHV. The HHV efficiency is lower than the LHV efficiency by the ratio of the two heating values. Refer to the handbook topic "▶ Process Equipment for Petroleum Processing."

Heat exchangers Devices for transferring heat from a hot fluid to a cooler fluid. Heat exchange is the science that deals with the rate of heat transfer between hot and cold bodies, in this case, different flowing streams. There are three modes of heat transfer typically encountered in petroleum processes: conduction, convection, and radiation. Refer to the handbook topic "▶ Process Equipment for Petroleum Processing" for a detailed treatment of heat exchange and heat exchangers.

Heavy oil cracking (HOC) This is a process similar to a fluid catalytic cracker (FCC) that is designed to crack residual oils that were beyond the heat balance capability of FCCs. The primary difference in the process compared to an FCC is the introduction of steam generation coils into the regenerator vessel to enable control of the regeneration temperature. Please refer to the chapter on "▶ Upgrading the Bottom of the Barrel" in this handbook for more details on this process.

HEP (1) Height equivalent to a theoretical tray for packing in a mass transfer or fractionation tower. Refer to the equipment chapter, section on distillation and packed towers. (2) Human error probability

HETP Height equivalent to a theoretical tray (or "plate") for packing in a mass transfer or fractionation tower. Refer to the equipment chapter section on distillation and packed towers.

HF Hydrofluoric Acid

Hg Mercury

HGO Heavy gas oil – normally refers to vacuum gas oil material (650 °F and heavier) but may also refer to heavy diesel (normally 500–650 °F). Refiner (and sometimes refinery) specific.

HHC Highly hazardous chemical

HHPS Hot high-pressure separator

HHPV Hot high-pressure vapor

HHV Higher heating value – assumes all combustion products are cooled back to 25 °C
HID Hot inlet temperature, design
High-TAN crude Crude with total acid number (TAN) >1.0
HIPPS (1) High-integrity pressure protection system; (2) high-integrity pipeline protection system
HIPS Hot-intermediate-pressure separator
HIT Hot inlet temperature, normal
HK Hydrocracker or hydrocracking
HKO Hydrocracked oil
HLL High liquid level
HLSD High-level shutdown
HMI Human–machine interface – the way we interact with a control system
HOD (1) Heating oil distillate; (2) hot outlet temperature, design
Homogenizer A mechanical device which is used to create a stable, uniform dispersion of an insoluble phase (asphaltenes) within a liquid phase (fuel oil)
Horton sphere A type of spheroidal pressure storage tank designed for products having vapor pressures higher than atmospheric pressure but relatively limited vapor pressure – such as pentanes.
Hose reel or hose station Firefighting equipment consisting of a coiled reel of hose with a fire nozzle. In the event of a fire, the hose can be unrolled and generally operated by one person. Refer to the chapter on firefighting.
HOT (1) Hot outlet temperature, normal; (2) head of terms – agreement in principle for a business arrangement
Hot and cold flash separators Hot and cold flash separators are used in high-pressure, high-temperature hydrocracking, gasoil hydrotreating, and similar applications where heavy oils are present along with lighter oils. Purposes of the hot/cold separations include (1) avoiding heavy oils setting up at the low temperatures in the cold separator, (2) recovering high purity hydrogen for recycle in as economically as possible, (3) increasing energy efficiency by reducing heat rejected to air, and (4) reducing the size of the effluent train exchangers and cold separator. The hot separator is normally located in the process effluent at a temperature of ~400–600 °F. Liquid from the hot flash is dry and can go directly to fractionation or another unit. The hot flash vapor is cooled further and sent to the cold-high-pressure separator where the recycle hydrogen gas and lighter oils are drawn and sour effluent wash water is removed. See the chapter on "▶ Upgrading the Bottom of the Barrel" in this handbook for an example of a hot/cold separator system in a resid hydrotreater. Some units also have warm separators at ~300–400 °F interposed between the hot and cold separators, but these are not common.
Hot lime process Method for purifying water using lime to remove hardness components. Used for boiler feed water. Refer to the chapter on utilities.
Hot work Work in a facility that produces sparks or sufficient heat to ignite flammable materials

HP (1) Horsepower; (2) high pressure; (2) human performance
HRA Human reliability analysis
HRF Hazard reduction factor
HRSG Heat recovery steam generator – waste heat boiler
HSE (1) Health, safety, and environmental; (2) Health and Safety Executive (UK)
HSM Horizontal support members
HSR (1) Heavy straight run; (2) high-sulfur recarburizer (coke)
HSS Heat stable salt
HSSR Heat stable salt removal
HSWA Hazardous and solid waste amendments to the US RCRA
HT Hydrotreater or hydrotreating
HTHA High-temperature hydrogen attack
HTML HyperText Markup Language
HTRI Heat Transfer Research, Inc.
HTSD High-temperature simulated distillation
HTTP Hypertext Transfer Protocol
Huff and puff Production method for heavy oils that involves heating a formation with superheated steam or other means followed by a soak period. The heated hydrocarbon in the formation is then produced as crude oil. See the discussion of bitumens and extra heavy oils in the chapter on unconventional crudes and feedstocks.
HVAC Heating, ventilation, and air conditioning
HVGO Heavy vacuum gas oil
HX Heat exchanger
Hydrant Firefighting water connection point for hoses and for direct use on a fire or for pumper truck suction. Hydrants are usually located at defined intervals along the mains and in the loops. See the chapter on firefighting.
Hydraulic analysis of process systems Hydraulic analysis is used to calculate a pressure profile of a process system. Such a calculation is used to size pipelines and to determine the pumping requirements for the system, as well as setting operating and design pressures for equipment within the system. Refer to Appendix 14 of the chapter entitled "▶ Process Equipment for Petroleum Processing" for an example of how a hydraulic analysis can be performed. That chapter contains a significant discussion of pressure drop calculations for various scenarios.
Hydraulic fracturing An oil field production process which creates fractures or cracks in the rock formation to allow oil and gas to flow out of the formation. See "Fracking" and the discussion of shale crudes in the chapter on unconventional crudes.
Hydraulic horsepower Power required for pumping liquid to the required pressure at the specified flow rate. Neglects efficiency and other losses. Only the amount of power that must be put into the fluid. Refer to the equipment chapter section on pumps.
Hydroclone An in-line device for separating solids from a liquid using centrifugal force and density differences. This is a liquid version of a gas cyclone.

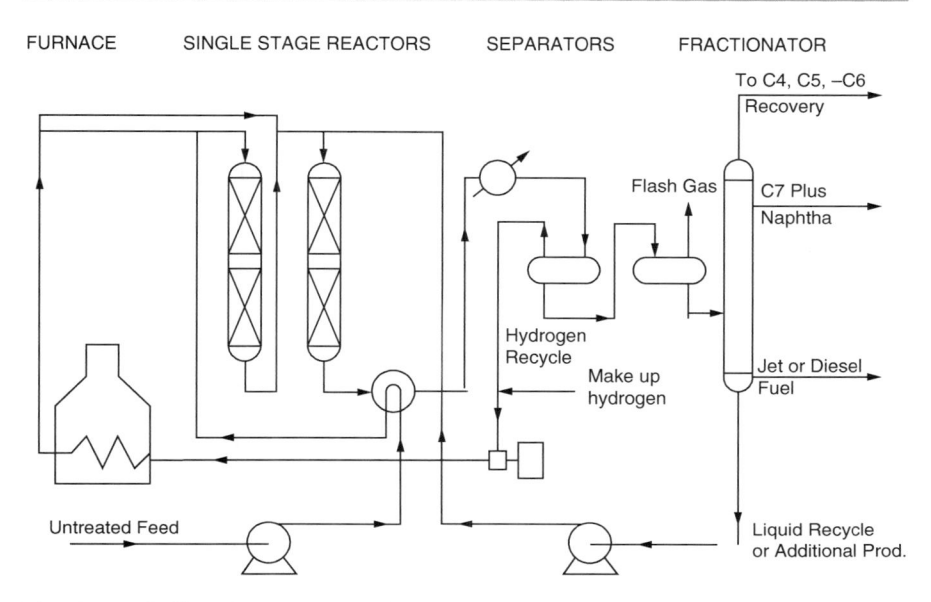

Fig. H.1 A distillate hydrocracker configuration

An example application would be removal of fine particulates from a flushing oil stream going to a centrifugal pump seal.

Hydrocracker Process using high-pressure hydrogen at high temperatures with acidic catalyst to reduce the size of various hydrocarbon feed molecules to smaller, more valuable products. Products typically include diesel, kerosene, and naphtha with nearly zero sulfur and nitrogen. The unconverted oil can be reduced to a very small stream in a hydrocracker. See the chapter on "▶ Hydrocracking in Petroleum Processing" in this handbook for additional details. Also, see Fig. H.1.

Hydrocracking (distillate) See "Hydrocracker" above. Details of this process are given in the chapter on "▶ Hydrocracking in Petroleum Processing" in this handbook. Figure H.1 shows the reaction section for a typical hydrocracking unit for distillates. Figure H.2 is a typical block flow diagram for recovery of distillate hydrocracker products.

Hydrodenitrogenation Removing nitrogen from an oil by hydrotreating it to ammonia

Hydrodesulfurization Removing sulfur from an oil by hydrotreating it to hydrogen sulfide

Hydrofluoric acid HF. Refer to "Anhydrous hydrofluoric acid."

Hydrogen Hydrogen gas is used in several petroleum processes. The primary users in a refinery are the hydrotreater and hydrocracker processes. It also finds smaller uses in reformers for startup and in some sulfur plants. Hydrogen is produced by catalytic naphtha reformers as a by-product of the reforming reactions and by steam-methane-reforming hydrogen plants. Many refineries

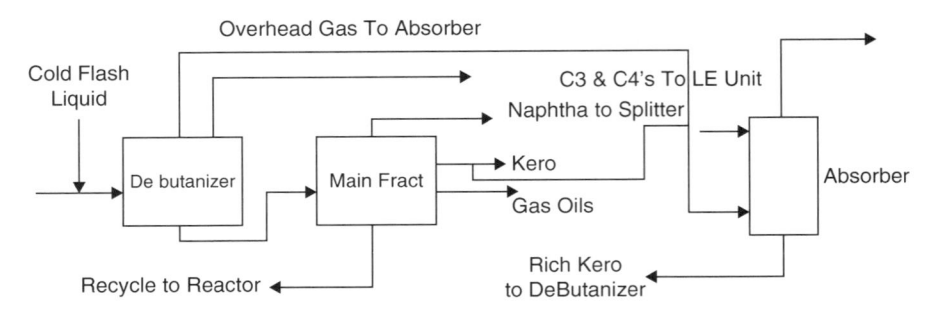

Fig. H.2 Block flow diagram of a typical hydrocracker recovery side

buy hydrogen externally, often delivered by pipeline from a producer. Unused hydrogen finds its way into the fuel gas system. Many facilities recover and recycle hydrogen from the fuel gases. Refer to the chapter "▶ Hydrogen Production and Management for Petroleum Processing" in this handbook for a more detailed discussion of hydrogen in a refinery.

Hydrogenation Adding hydrogen to an unsaturated organic molecule

Hydrogen sulfide (H_2S) Hydrogen sulfide is a colorless, extremely poisonous gas that has a very disagreeable odor, much like that of rotten eggs, in low concentrations. In high concentrations that are IDLH, it deadens the sense of smell. In a refinery, it is usually formed during the desulfurizing processes used to sweeten product streams as well as other processes. In practice, H_2S can be assumed to be present throughout a refinery in most processing streams at varying concentrations.

Hydrolysis Breaking down a substance by addition of water

Hydrometer An instrument (float) for determining the gravity of a liquid

Hydroprocessing The generic term for refining processes employing hydrogen in reactions. They operate at elevated temperatures and pressures using a selective catalyst. The term includes hydrotreating, hydrocracking, isomerization, reforming, hydrodewaxing, and other processes that use recycle or once-through hydrogen-containing streams.

Hydroskimming refinery A simple refinery configuration in which straight-run atmospheric crude distillation products are processed to meet product specifications. Atmospheric resid is normally sold as fuel oil. Processing may include reforming, hydrotreating, isomerization, and other product property improvement processes in addition to crude distillation. There is no cracking or resid conversion in hydroskimming facilities.

Hydrotreater Process using hydrogen at high temperatures and elevated pressures with catalyst to remove sulfur, nitrogen, and other contaminants from hydrocarbon stocks. See "Hydrotreating."

Hydrotreating The hydrotreating family of processes removes sulfur, nitrogen, oxygen, and other impurities from petroleum stocks to yield clean, finished

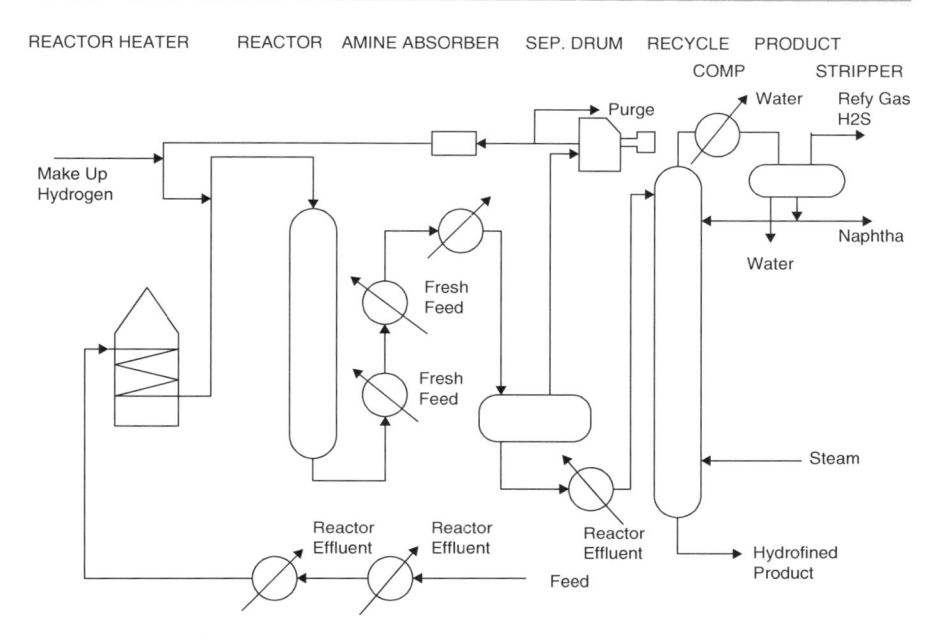

Fig. H.3 A simple gas oil hydrodesulfurizer

blendstocks or feeds for other processes. Hydrotreating can also improve distillate product yield properties. The flavors of hydrotreating are based on the boiling range of the feedstock and the product objectives. Hydrotreating applications include naphtha, kerosene, diesel, gas oil, and residuum. Refer to the chapter "▶ Hydrotreating in Petroleum Processing" in this handbook for a discussion of hydrotreating for most of these stocks. Figure H.3 illustrates a simple gasoil hydrotreater flow.

I

I/O Input/output for a control system
I/P Current to pneumatic transducer or converter
IA Instrument air
IAP (1) Incident action plan; (2) integrated asset planning
IBP Initial boiling point. In a standard laboratory distillation, the temperature on the distillation thermometer at the moment the first drop of distillate falls from the condenser
i-component The "i" as a prefix to a chemical component indicates that the compound is an isomer. There are several isomers in petroleum refining, and the most common relate to the light components of the structure. Notably these are:

- Isobutane: i C_4 or i-butane
- Isopentane: i C_5 or i-pentane
- Isohexane: iC_6 or i-hexane

and so on through the homologues. When isomers exist and are quoted together with the normal compound, this normal compound will be identified by a prefix n.

IC Initiating cause

ICC (1) International Chamber of Commerce; (2) US Interstate Commerce Commission (eliminated by legislation in 1995 and replaced with the DOT)

ICFM Inlet cubic feet per minute

IChE Institute of Chemical Engineers (UK)

ICL Independent control layer

ICS Incident command system. See the chapter on firefighting in this handbook.

ID (1) Inside diameter; (2) induced draft (in a firebox – a fan that sucks out the flue gases)

IDLH Immediately dangerous to life and health

IDW Isodewaxer

IEA International Energy Agency (Paris)

IEC International Electrotechnical Commission (Switzerland)

IED Industrial Emissions Directive – in the EU. Refer to the environmental chapter.

IEEE Institute of Electrical and Electronics Engineers

IFP (1) Institute Francais do Petrole (France); (2) invitation for prequalification or proposal

IG Index of gravity

IGCC Integrated gasification combined cycle

Ignitability Hazardous waste characteristic of being pyrophoric or self-heating on exposure to air. Refer to the environmental chapter section on hazardous wastes.

IHS (1) Industrial health and safety; (2) a company providing database services in the IHS areas; (3) industrial hygiene standard; (4) information handling services

Immediately Dangerous to Life and Health (IDLH) Concentration or exposure to a substance which can result in immediate or imminent death. For example, exposure to 1,200 ppm of CO is likely to result in death unless the person receives medical attention immediately.

IMO International Maritime Organization. Refer to the environmental chapter section on air emissions.

Impeller speeds (pumps) This term refers to the revolutions per minute (rpm) at which a centrifugal pump impeller spins. Higher-speed centrifugal pumps normally operate at impeller speeds of ~3,550 rpm (60 cycles/s power) or ~2,950 rpm (50 cycles/s power). Lower-speed pumps operate at ~1,750 or ~1,450 rpm on 60 and 50 cycles/s power, respectively. Refer to the handbook topic "▶ Process Equipment for Petroleum Processing" for a more detailed discussion.

Impingement Refers to direct contact or impact of one material on another, for example, flame impingement on heater tubes when the flame "licks" the tube.

In another example, fluid entering the shell side of a heat exchanger may impinge on the tubes.

Imports Receipts of crude oil and petroleum products into a country or other region. May also refer to imports into a refinery or process units

IMT Incident Management Team (for emergency responses)

In. W.C. Inches of water column (pressure measure)

Incident command system (ICS) A standardized organization and responsibilities structure for responding to emergencies that represents best practices. It is a key feature of the US National Incident Management System (NIMS). This system can be used for fires, releases, medical emergencies, or any other type of situation where fast response is required. Refer to the chapter on firefighting for a detailed discussion.

Incompatible Refers to materials or stocks that have adverse results when mixed together. For instance, mixing a very aromatic crude and a very paraffinic crude together will usually result in precipitation of asphaltene compounds from the aromatic crude or paraffin wax from the paraffinic crude. May also refer to tolerance of construction materials in contact with stocks. For instance, a high-TAN crude is not compatible with carbon steel at elevated temperatures.

Indices Indices are used extensively in petroleum refining technology to correlate one set of data with another. The handbook discussions under the topics "▶ Introduction to Crude Oil and Petroleum Processing" and "▶ Atmospheric and Vacuum Crude Distillation Units in Petroleum Refineries" provide indices that relate the properties of various components to temperature, viscosities, flash point blending, and the like. A number of these indices are used in the blending of petroleum fractions to give the properties of the blended product. For example, the components listed in Table I.1 are to be blended in the proportions given and the viscosity of the blended product determined.

Induced draft In a firebox, a fan that sucks out the flue gases

Inert Incapable of reacting

Table I.1 An example of viscosity blending using viscosity blending indices

Component	Vol %	Mid BPt °F	Viscosity cSt 100 °F	Blending index	Viscosity factor
	(A)			(B)	(A × B)
1	13.0	410	1.49	63.5	825.5
2	16.5	460	2.0	58.0	957
3	21.0	489	2.4	55.0	1,155
4	18.0	520	2.9	52.5	945
5	18.5	550	3.7	49.0	906.5
6	13.0	592	4.8	46.0	598
Total		100.0			5,387.0

Overall viscosity index $= \frac{5,387}{100} = 53.87$

An index of 53.87 corresponds to 2.65 cSt (actual plant test data was 2.7 cST). Indices are covered in the introductory chapter of this handbook

Inert entry Personnel entry into equipment that is full of inert gas (normally nitrogen) used to prevent undesirable reactions. Inert entry requires an air supply for those entering and strict, extensive procedures to protect the entrant. Refer to the chapter on safety.

Inferred properties In control systems, physical and chemical properties in a stream can be estimated fairly accurately over a limited range using the plant operating conditions. This approach is used extensively in advanced control, especially around fractionation systems. Development of the inferred property correlations requires an initial set of controlled conditions and analyses. See the chapter on controls.

Infrared analyzer An analyzer (online or offline) which determines composition of a sample by measuring absorption or transmittance of infrared light through the sample. Refer to the chapter on controls.

Initial boiling points Initial boiling points or IBPs refer to the temperature at which a petroleum cut begins to boil. Usually, this temperature is taken as that at atmospheric pressure. These are determined in the refinery's laboratory from the ASTM distillation carried out as a routine test. Details of these tests are given in the chapter on "▶ Quality Control of Products in Petroleum Refining" in this handbook.

Inlet diffuser A device at the inlet to a reactor or other equipment to dissipate the incoming fluid momentum and provide a rough distribution of the incoming fluid over the equipment cross section

Inlet guide vane In a compressor, a set of vanes that control the velocity of incoming gas to affect the compressor equivalent capacity. Refer to the equipment chapter section on centrifugal compressors.

In-line blending System which produces a blended product by continuously mixing controlled proportions of the blend components in a common line going to a finished blend tank. Refer to the chapter on offsites.

Innage Space occupied in a product container

INPNA Analytical method that determines iso and normal paraffins, naphthenes, and aromatics in a hydrocarbon stock

INPO Institute of Nuclear Power Operations

Input (1) Total of all streams entering a refinery or process unit. (2) The measured parameter or parameters provided to a controller. These may be direct readings, 3–15 psi pneumatic signals, 4–20 ma electrical signals, or digital.

In situ Refers to processes that are performed with raw materials still in place, for example, processes that retort oil shale while it is still in place in the ground

Instrument air Compressed, dried air supplied by a secure system that is used for motive force in pneumatic instruments. Refer to the chapter on utilities.

Instrumentation This term refers generically to the control systems used in managing the petroleum refining processes. These systems are used to sense and maintain the correct, stable operating conditions (such as flow, temperature, pressure, level, and composition) in process equipment and piping.

These are described and discussed in detail in the chapter "▶ Process Controls in Petroleum Processing."

Insulation Material to prevent heat loss

Intalox saddle A mass transfer packing shaped somewhat like a saddle with additional structure to increase surface area and contacting. Refer to the chapter on equipment, section on packed towers.

Integral control A control mode where the controller output is determined by the deviation from the setpoint and how long that deviation has existed. Acts like an amplification term. Normally used as a fine-tuning parameter in a proportional control mode

Integrated oil company A company involved in all aspects of the petroleum business from exploration to retail sales of refined petroleum products

Intercooling Cooling between stages of a reciprocating compressor to improve compression efficiency. Refer to the chapter on equipment, section on reciprocating compressors. More generically, refers to cooling between two stages of any stage-wise process system

Interface A mixture of petroleum products occurring when batches of different products are shipped consecutively through a pipeline

Intermediate Partially refined product between process units. Intermediates may also be traded between refineries to manage short-term needs or bought and sold

Intermediate law A mathematical relationship that describes the rate of movement of a particle or droplet through another continuous phase in the transition flow regime between laminar and turbulent conditions. Refer to the environmental chapter section on water and the process equipment chapter discussion of separators and drums.

Internals (vessels) This is a generic term used to designate the parts of process equipment that are inside vessels, towers, drums, and other equipment. These internals are primarily aimed at (1) enhancing heat and mass transfer (e.g., fractionating towers), (2) maintaining proper conditions for settling (e.g., condensate drums), (3) promoting good distribution of fluids (vessel or reactor inlet distributors), and (4) preventing vortexing of fluids leaving vessels to pump suction. Vessel internals are discussed in several chapters of this handbook, but a detailed discussion is in the handbook topic "▶ Process Equipment for Petroleum Processing."

International Energy Agency (IEA) An agency in Paris, France, which tracks energy statistics and information on an international level

International Petroleum Exchange Now "ICE Futures"; formerly one of the world's largest energy futures and options exchanges (London). Brent crude pricing

Interstate Commerce Commission (ICC) Former US federal authority regulating interstate commerce that was eliminated by legislation in 1995 and replaced with the Department of Transportation (DOT).

Investment analysis See earlier term: "Economic analysis." Refer to the chapter titled "▶ Petroleum Processing Projects" in this handbook.

IOD Immediate oxygen demand. Oxygen consumption by reducing chemicals, such as sulfides and nitrates where oxidation occurs rapidly. Refer to the environmental chapter section on water.

Ion exchange A process using specifically designed polymer resins which can react with ions in aqueous solution. The resins are preloaded with a desirable ion, like H^+ or OH^- using a strong acid or base, respectively. When a solution is passed through beds of the resin, the more desirable ions in the resin displace the original undesirable ions (ion exchange), improving the solution purity. Refer to the chapter on utilities.

Ionization The process of adding electrons to, or removing electrons from, atoms or molecules, thereby creating ions. High temperatures, electrical discharges, and nuclear radiation can cause ionization.

IP (1) Intellectual Property – company confidential information; (2) British Institute of Petroleum

IPE International Petroleum Exchange, now "ICE Futures" (London)

IPL Independent protection layer – part of LOPA method

IR (1) Infrared; (2) integrated review in projects; (3) individual risk

IRR Internal rate of return

IS (1) Information systems; (2) inherently safer or safe

ISA Instrument Society of America

ISO (1) International Organization for Standardization (Switzerland); (2) short for isometric drawing – a 3-D representation of equipment or piping

Isobutane (i-C_4H_{10}) A normally gaseous branch-chain hydrocarbon. It is a colorless paraffinic gas that boils at a temperature of 10.9 °F. It is extracted from natural gas or refinery gas streams. AKA 2-methyl propane

Isobutylene (i-C_4H_8) An olefinic hydrocarbon recovered from refinery processes or petrochemical processes. AKA 2-methyl propene

Iso-container A special tank used for bulk shipment of chemicals. An iso-container can be connected to a truck bed or be moved on its own. Iso-containers can withstand moderate pressures and are used for shipping many hazardous materials.

Isohexane (C_6H_{14}) A saturated branch-chain hydrocarbon. It is a colorless liquid that boils at a temperature of 156.2 °F.

Isomerization The rearrangement of straight-chain hydrocarbon molecules to form branched-chain products

ISS Instrumented safety system

Isomerization The isomerization process forms branched hydrocarbons from paraffins or less-branched hydrocarbons. The branching increases the octane value of the product. Isomerization can also be used to improve cold-flow properties of distillates. This is a type of hydroprocessing operated at mildly elevated temperature and pressure using a selective catalyst. A simplified isomerization flow diagram is shown in Fig. I.1. Refer to the chapter titled "▶ Isomerization in Petroleum Processing" in this handbook for a detailed discussion of the process.

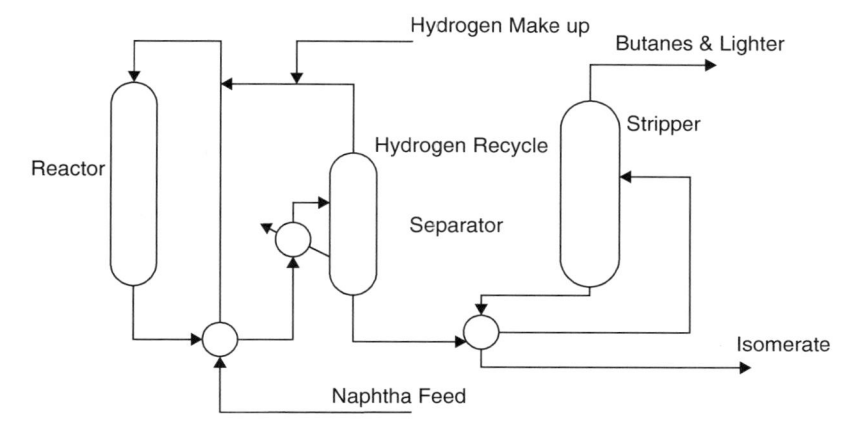

Fig. I.1 A process diagram of a typical isomerization unit

Table J.1 Comparison of some jet fuels

Specification	ASTM D1655	Def Std 91–91
Aromatics vol% max	25	25
Distillation °C		
10 % recovered max	205	205
50 % recovered max	Report	Report
90 % recovered max	Report	Report
End point	300	300
Flash point °C min	38	38
Density at 15 °C kg/m_3	775–840	775–840
Freezing point °C max	−40	−47

IT Information technology – the computer group
ITS (1) Incident tracking system; (2) installation of template
IUPAC International Union of Pure and Applied Chemistry

J

Jet See "Jet Fuels" and "Kerosene."
Jet fuel Jet fuel refers to products meeting the specifications for civilian or military jet aircraft fuel. A simplified comparison of these two types of fuel is in Table J.1. Jet fuel is discussed in detail in the chapter on "▶ Quality Control of Products in Petroleum Refining" in this handbook.
Jet mixer A device, normally inside a storage tank, which promotes mixing of two fluids using a high-velocity stream. Refer to the chapter on offsites.
Jetties Tankers and barges are loaded and unloaded at jetties or docks. In almost all circumstances, these facilities for handling petroleum products are separate

from those used for general cargo. Very often, tankers, particularly modern "super" tankers are loaded and unloaded by submarine pipelines at deepwater anchorage. Only the "onshore" jetty facility is discussed in this handbook. See the chapter on "▶ Off-Site Facilities for Petroleum Processing."

JIT Just in time

Jobber Someone who purchases refined products at the wholesale level and then transfers or resells the product at the retail level. The retail level sale/transfer can occur at facilities owned by the jobber, independent dealers, or commercial accounts.

Joint venture A facility that is operated by two or more companies in partnership. The companies share in the JV profits.

Joule–Thomson effect The observed phenomenon of heating or cooling of a compressed gas if it is allowed to expand adiabatically. Most gases cool on pressure reduction. Hydrogen gas actually heats up when its pressure is reduced ("reverse Joule–Thomson effect"). Hydrogen can even ignite in air when pressure is reduced sufficiently.

JRC Joint Research Centre

JSA Job safety analysis – pre-walk of a job to ensure it can be done safely

JV Abbreviation for joint venture

K

K Watson characterization factor. A factor introduced by Watson and Nelson in 1933 to "quantify" the paraffinicity of petroleum hydrocarbon fractions. It is defined as

$$K = \frac{(T_b^{1/3})}{S}$$

where T_b is the mid-boiling point in degrees Rankine and S is the specific gravity at 60 °F. It is also sometimes referred to as the UOP K (please refer to "UOP K" for details).

KB (1) Thousand barrels; (2) kilobyte

KBPD Thousand barrels per day

Kerosene Fuel boiling in the 350–550 °F range normally used for jet aircraft fuel blending but also used for heating and lamps in some areas. The term kerosene generally refers to the blendstocks within the boiling range required for jet and related uses. Table K.1 gives the general spec for some of the kerosene products. Refer to the chapter on "▶ Quality Control of Products in Petroleum Refining" in this handbook for additional discussion of the product quality tests.

Kerosene-type jet fuel A kerosene-based product having a maximum distillation temperature of 400 °F at the 10 % recovery point and a final maximum boiling point of 572 °F and meeting ASTM specification D1655 and military specifications MIL-T-5624P and MIL-T-83133D (grades JP-5 and JP-8). It is used for

Table K.1 Specifications for some kerosene finished products

Parameters	Reg kero	ATG	TVO	Test
Flash point °F	100	<66	100	D-56
Aromatics vol%	–	20	–	D-1319
Temperature at 20 % max		293°	–	D-86
Temperature at 50 % max		374°	–	D-86
Temperature at 90 % max		473°	540	D-86
Final boiling point	572 °F	572°	–	D-86
Sulfur max wt%	0.04	0.04	0.3	D-1266
Smoke point min	–	25 mm	25 mm	D-1322
Freeze point °C	–	−47	–	D-2386

commercial and military turbojet and turboprop aircraft engines:
- Commercial – kerosene-type jet fuel intended for use in commercial aircraft
- Military – kerosene-type jet fuel intended for use in military aircraft

Kettle A type of heat exchanger used for tower reboilers and steam generation. Refer to the chapters on equipment and utilities.

KEV (1) Key energy variable – for a specific unit or process; (2) thousand electron volts

Kinematic viscosity The ratio of the absolute viscosity of a liquid to its specific gravity at the temperature at which the viscosity is measured. Expressed in stokes or centistokes. Example: viscosity, kinematic, cSt at 100 °F. The Darcy–Weisbach and Colebrook relationships are based on using a Reynolds number which varies inversely with the *kinematic* viscosity. This kinematic viscosity is defined as the dynamic (or absolute) viscosity divided by its density, where dynamic or absolute viscosity is force x (time/length squared) and the unit for this is the poise. The unit most frequently used for the kinematic viscosity is the metric unit: the stoke. Both viscosities are usually quoted in the hundredth unit. Thus, absolute viscosity would be the centipoise, while the kinematic viscosity would be centistokes.

Knockout or knock-out Separating entrained solids or liquid droplets from a gas stream, normally ahead of a compressor or analyzer

Knockout drum or pot Vessel with the primary purpose of preventing undesired or potentially harmful materials from carrying over to downstream equipment. Refer to the chapter on equipment, section on drum design.

KO Knockout, normally in reference to liquid removal from a gas stream in a "knockout drum"

KOH Potassium hydroxide

KPI Key performance indicator

KPM Key performance measure

KS Knowledge sharing

KVA Thousand volt-amps – a way of expressing electrical load or power. Approximately kw

KW or kw Kilowatts – power

L

LA (H, HH, L, LL) Level alarm high, high–high, low, or low–low, e.g., LAHH is a high–high-level alarm

LAB Liquid alive bacteria – a proprietary odor suppressant for sulfur-containing chemicals

LAER Lowest achievable emissions rate

LAH Level alarm high

LAL Level alarm low

Laminar Smooth, nonturbulent flow regime

LAN Local area network

Langelier Saturation Index (LSI) Measure of water fouling potential equal to the difference between the actual pH and the pH at saturation of calcium carbonate. Refer to the chapter on utilities.

Larkins, White, and Jeffery method Technique for estimating two-phase liquid/vapor pressure drop through a bed of discrete solids. Refer to the description in the equipment chapter section on two-phase pressure drop.

Latent heat Heat required to change the state of a unit weight of a substance from solid to liquid or from liquid to vapor without change of temperature

Lateral In sewer systems, refers to a branch. Refer to the chapter on utilities.

Layering This occurs in tanks when a high-density fuel is mixed with a low-density fuel. AKA stratification

Layers of protection analysis (LOPA) Technique for evaluating process hazards protection that considers the number of protective layers of instruments or systems and the relative reliability of each compared to the hazard presented.

LC (1) Level controller; (2) local content

LC50 Lethal concentration of a chemical in air that results in 50 % mortality

LCC Life cycle cost

LCCA Lifecycle cost analysis

LCGO Light coker gas oil

LCL (1) Less-than car load; (2) lower control limit in statistical controls

LCO Light cycle oil (from an FCC). Typically boils between 350 and 700 °F

LD50 Lethal amount of a chemical ingested that results in 50 % mortality

LDAR Leak detection and repair – environmental

Leaded gasolines Until the restrictions on lead compounds imposed by the "Clean Air" Act of the 1960s, tetraethyl lead was used extensively as a gasoline additive to improve octane number.

Tetraethyl lead is a liquid with a gravity of 1.66 and a formula $Pb(C_2H_5)_4$. It is extremely toxic. The restriction of "no lead" in gasolines promoted further development of the catalytic reformer process to obtain higher conversion. It also influenced the use of the alkylation process and the development of isomerization and the oxygenated compounds as octane enhancers in gasoline. Coupled with this, motor manufacturers improved their respective auto engine design to operate efficiently on lower octane number fuel. Tetraethyl lead has all but disappeared from the petroleum industry. See the chapter titled "▶ Petroleum Products and a Refinery Configuration" in this handbook. Details of octane enhancing processes are given in the handbook chapters on alkylation, olefin condensation, and isomerization.

Legionella bacteria Bacteria that grow in moist environments, like cooling towers, and produce potentially life-threatening respiratory illness. Refer to the utilities chapter.

LEL Lower explosive limit

Letdown Reducing pressure of a fluid – sometimes recovering work as the pressure is reduced

Level control range This is the distance between the high liquid level (HLL) and the low liquid level (LLL) in the vessel. When using a level controller, the signal to the control valve at HLL will be to fully open the valve. At LLL, the signal will be to fully close the valve.

LFL Lower flammable limit

LHSV Liquid hourly space velocity

LHV Lower heating value – assumes all combustion products are vapor. Close to actual heat available from a fuel

LIFO Last in–first out in inventory management

Lift In relief valves, lift is the actual travel of the disk away from closed position when the valve is relieving.

Lifting Refers to tankers and barges taking on cargoes of oil or refined product at the terminal or transshipment point

Light crude Crude oil with a low specific gravity and high API gravity due to the presence of a high proportion of light hydrocarbon fractions and low content of metallic compounds.

Light ends Light hydrocarbons and other gases. Normally refers to methane, ethane, propane, and butane produced as by-products of refining. May also include other associated gases, like hydrogen, H_2S, and ammonia

Light ends unit The light ends unit in a refinery produces the light and heavy naphtha cuts, the butane LPG, and the propane LPG products, respectively. The straight-run light ends units take the atmospheric overhead distillate, the overhead distillate from the catalytic reformer stabilizer, the overhead distillate from a thermal cracker fractionator (if there is one in the configuration), and light ends from any other units. Often, a facility has more than one light ends unit, with each unit handling different feedstocks and possibly producing different products. A typical process flow schematic showing a typical process flow for a light ends unit is given in Fig. L.1. Refer to the topic "▶ Distillation of the "Light Ends" from Crude Oil in Petroleum Processing" in this handbook for a thorough discussion of light ends processing.

Light gas oil Normally refers to a straight-run diesel boiling range material (∼300–700 °F). In some facilities, it may also refer to similar boiling range materials in other process units.

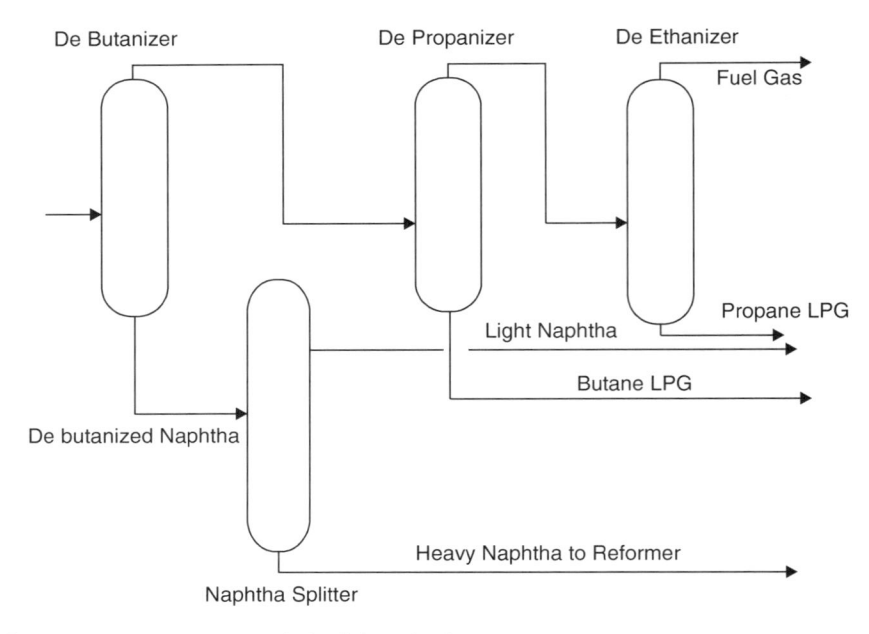

Fig. L.1 Process flow schematic for light ends plant

LIMS Laboratory information system – where a lab collects and disperses lab analyses. Many software options are available.

Linear program(ming) This is a computerized technique that came into prominence during the late 1960s and early 1970s. It is used extensively now by most refiners to (1) optimize new process configurations, (2) plan the refinery operation, and (3) select crude oil feed slate and product slate. The technique uses equations (linear) that represent the properties of the crude feed and the resulting products. These equations also describe the blending characteristics of the components making up the finished product slate. Included also are the cost parameters such as the price of crude feed, the refinery fence selling price of products, operating cost, and any other relevant cost centers (such as licensing fees, interest on loans, etc.). These equations form a mathematical model, and a suitable programmed computer is used to solve these equations simultaneously to meet the objective function subject to the constraints of the analysis. Further details describing linear programs are in the handbook topic "▶ Petroleum Refinery Planning and Economics."

Liquid ring Rotating type of compressor with vanes sealed at the outer ends by a fluid or liquid ring

Liquefied petroleum gas(es) (LPG) A group of hydrocarbon-based gases derived from crude oil refining or natural gas fractionation that are often liquefied, through pressurization, for ease of transport. They include: ethane, propane, normal butane, and isobutane and natural gasoline. Uses of these fuels include: home heating, industrial, automotive fuel, petrochemical feedstocks, and drying purposes in farming. Refineries produce two types of LPG (liquefied petroleum gas) products: butane LPG and propane LPG. LPGs are used widely in industry and domestically as portable fuel source.

Liquefied refinery gas(es) (LRG) Liquefied petroleum gases fractionated from refinery or still gases. Through compression and/or refrigeration, they are retained in the liquid state. The reported categories are ethane/ethylene, propane/propylene, normal butane/butylenes, and isobutane/isobutylene. Excludes still gas

Liter A measure of capacity in the metric system equal to 61,022 cubic inches, 0.908 US quarts dry and 1.0567 US quarts wet

LLL Low liquid level

LME Liquid metal embrittlement

LN2 or LN_2 Liquid nitrogen

LNAPL Light nonaqueous-phase liquid

LNG Liquefied natural gas

Loader or lifter In a reciprocating compressor, a device to open or close the compressor cylinder valves to control compressor capacity. Refer to the equipment chapter section on reciprocating compressors.

LOC (1) Loss of containment – can describe a small leak to a major explosion; (2) letter of commitment

Lockhart–Martinelli modulus Dimensionless number used to relate vapor and liquid rates in two-phase pressure drop calculations in piping. Refer to the equipment chapter section on two-phase pressure drop.

Logarithmic mean temperature difference (Δt_m or ΔT_{LM}) For heat transfer in exchangers, this is defined as

$$\Delta t_m = (\Delta t_1 - \Delta t_2)/\ln(\Delta t_1/\Delta t_2)$$

The Δts are the temperature differences at each end of the exchanger and Δt_1 is the larger of the two.

LOHC Lube oil hydrocracker

LOI (1) Loss on ignition; (2) letter of indemnity; (3) letter of intent

Long ton An avoirdupois weight measure equaling 2,240 lb

LOPA Layer(s) of protection analysis in process safety

LOS Light oil spread – see "Crack." Same as 3:2:1 crack

LOTO Lockout–tagout – part of making equipment safe to work on

LOX Liquid oxygen

LP (1) Linear program. Used for planning refinery operations. (2) Lost production; (3) liquid product

LPG Liquefied petroleum gas(es)

LR Level recorder

LRC Level recorder and controller

LRG Liquefied refinery gas(es)

LSD Low-sulfur diesel – usually refers to diesel with less than 500 wppm sulfur

LSG Low-sulfur gasoline – usually refers to gasoline with less than 30 wppm sulfur

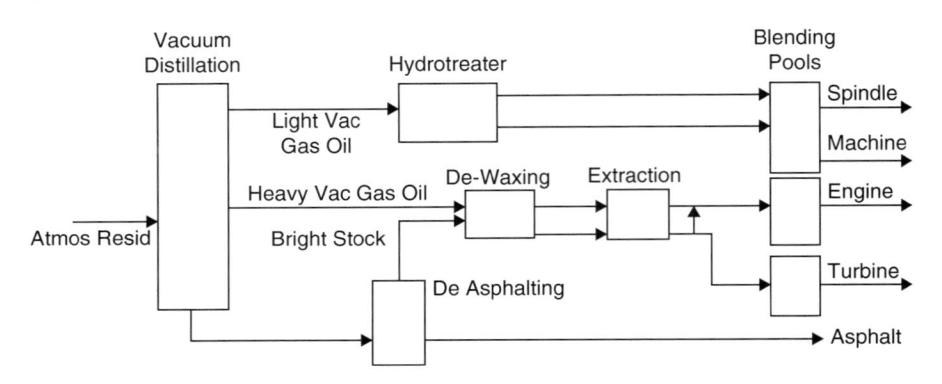

Fig. L.2 A schematic of one lube oil processing configuration

LSR (1) Light straight-run naphtha; (2) low-sulfur recarburizer (coke)
LT (1) Level transmitter; (2) long ton – 2,240 lb
LTL Less-than truck load
LTPD Long tons per day
Lube oils Lube oils are specialized, refined products intended for machinery lubrication. They include automotive oils, greases, and other lubricants derived from petroleum or synthetic products. Not all refineries produce lubricating oils. When they are produced, about three to ten grades of basic lube oil components are made and then blended to meet the many grades of light lube oils, engine lubes, and heavy turbine lube oils. A simple block flow diagram for lube oil production using the solvent approach is shown in Fig. L.2. More recently, high-quality lube oil stocks are being produced by heavy vacuum gas oil hydrocracking followed by hydrodewaxing and hydrofinishing at pressures over 2,000 psig. Some plants are also producing lube stocks via gas-to-liquids processes. The chapter "▶ Non-energy Refineries in Petroleum Processing" in this handbook discusses lube oils in more detail.
Lubricants Substances used to reduce friction between bearing surfaces or as process materials either incorporated into other materials used as processing aids in the manufacture of other products or used as carriers of other materials. Petroleum lubricants may be produced either from distillates or residues. Lubricants include all grades of lubricating oils from spindle oil to cylinder oil and those used in greases.
Lubricity The ability of a material to maintain a lubricating film. There are ASTM tests for lubricity. Hydrocarbon products, such as jet fuel, have lubricity specifications and often require additives.
LVGO Light vacuum gas oil

M

M (1) Maintenance; (2) thousand; (3) million – watch out for context
M&EB Material and energy balance

M&S Materials and supplies or services

MACT Maximum achievable control technology – refer to the environmental chapter.

MAE Major accident event

Manifold Term used to describe a system for connecting multiple sources or destinations. Manifolds are used for sending products to different destinations, for collecting feedstocks entering a unit, for connecting heater passes to the process piping, and so on.

MAOP Maximum allowable operating pressure

Margin Difference between two numbers. (1) Gross margin = gross revenues minus raw material costs – does not include operating or capital costs. (2) Variable margin = gross margin minus variable costs. (3) Net margin = variable margin minus fixed costs.

Marine diesel oil Marine diesel oil is a middle distillate fuel oil which can contain traces, often 10 % or more, residual fuel oil from transportation contamination and/or heavy fuel oil blending. The MDO does not require heated storage.

Marketing Organization that promotes sales of products to the public, the government, or other businesses

MARPOL International convention on the prevention of pollution from ships. Refer to the environmental chapter on air emissions.

Mass spectrometry Mass spectrometry is concerned with the separation of matter according to atomic and molecular mass. It is most often used in the analysis of organic compounds of molecular mass up to as high as 200,000 Daltons and until recent years, was largely restricted to relatively volatile compounds. Continuous development and improvement of instrumentation and techniques have made mass spectrometry one of the most versatile, sensitive, and widely used analytical method available today. One of its major uses in the petroleum refining industry is for the production of distillation curves such as TBP and EFV. This technique does away with the cumbersome distillation apparatus previously used for this purpose. It is also by far the more accurate method.

MAST Maximum allowable skin temperature in a furnace tube

Material balance The weight balance of all inputs to and outputs from a process unit or refinery. Material balances form the basis for plant design and are essential in refinery operation to account for gains and/or losses in the refinery's daily production. They are also essential in process plant audits and troubleshooting. The material balance in all cases is complete and correct when the quantities into the process equal the total quantities out when expressed in mass (weight) units. Volumes and moles in and out may differ because of chemical changes and/or thermal and pressure effects.

Material safety data sheet (MSDS) An information sheet for each substance, including internal refinery streams, that is handled which provides workers and emergency personnel with procedures for handling that substance in a safe manner. These sheets include information such as physical data, fire hazards and control measures, toxicity, health effects, first aid, reactivity, storage, disposal, protective equipment, and spill-handling procedures. AKA safety data sheet (SDS) or product safety data sheet (PSDS)

MAWP Maximum allowable working pressure

MBO (1) Management by objectives – technique of setting goals and tracking against goals; (2) multiblend optimizer

MC (1) Mechanical completion; (2) management committee

MCP (1) Methylcyclopentane; (2) major capital project

MCR Maximum continuous rating

MCSF (1) Thousand standard cubic feet; (2) minimum continuous stable flow; for a pump, the minimum flow limit

MDEA Methyl diethanolamine

MDMT Minimum (or maximum) design metal temperature – watch out for context

MDO Marine diesel oil

MEA Monoethanolamine

Mechanical flow diagram This type of flow diagram is sometimes referred to as a P&ID or the Engineering Flow Diagram. Details of this type of drawing are given under the item flow sheets.

Mechanical integrity Physical condition of equipment to contain materials and perform its required function

MEK Methyl ethyl ketone

Membrane unit A type of process equipment that uses a membrane barrier that allows diffusion of specific gas compounds through the membrane while excluding others. Membrane units are often used to purify hydrogen streams, separating the streams into high purity hydrogen (that diffuses through the membrane) and waste gas containing the impurities. Membrane separations are controlled by diffusion rates and limited by equilibrium so products are generally not chemically pure. Membranes are also applied to liquids in the form of reverse osmosis, dialysis, or electrodialysis to remove impurities.

Merichem Licensor of proprietary hydrocarbon treating and sweetening processes, such as Merifining. Refer to the environmental chapter section on air emissions.

Merox A proprietary UOP process for caustic treating of petroleum stocks to remove mercaptans. Refer to the environmental chapter section on air emissions.

Metals in crude oil Metallic organic compounds have a deleterious effect on some products and are also usually poisonous to catalysts in some processes. The most common metals met with are sodium, nickel, and vanadium. Sometimes other metals are encountered, such as mercury or selenium, in the associated gas streams or arsenic in some crudes. Many of these metallic compounds are found in the asphalt portion of the crude oil and are usually deeply imbedded in the asphaltene molecules. In the production of fuel oil, the metal content of the fuel makes the product problematic to the steel production companies who use fuel oil in their processes. Secondly, in the upgrading of the "bottom of the barrel" using a catalytic process (such as hydrocracking or fluid catalytic cracking), these metallic compounds deteriorate the catalyst life and performance of the processes. Much has been done to improve the catalysts to withstand metals. Today, the metals may be removed by either (1) the coking processes or (2) deasphalting. These processes are described in the handbook topics "▶ Upgrading the Bottom of the Barrel" and "▶ Non-energy Refineries in Petroleum Processing."

Metering pump A positive displacement pump which can be adjusted to deliver a precise amount of flow

Methanation Highly exothermic reaction of carbon oxides with hydrogen to form methane

Methanol (CH_3OH) A light, volatile alcohol intended for gasoline blending as described in oxygenate definition.

Methyl diethanolamine (MDEA, $(HOC_2H_4)_2NCH_3$) One of the common, regenerable amine solvents used for acid gas removal (H_2S or CO_2) from vapor streams

Methyl ethyl ketone (MEK) $CH_3C(O)CH_2CH_3$ colorless liquid with a sweet/sharp, fragrant, acetone-like odor. It is extremely flammable in both the liquid and vapor phase. Used in oil refining for the removal of wax from lube oil stock

Methyl tertiary-butyl ether (MTBE) An ether used at one time in the blending of reformulated gasolines, affecting vapor pressure and octane level. Unlike ethanol, MTBE is fungible and will not separate out during shipment. There is no US market for MTBE due to experience with groundwater contamination; so any production is exported.

Metric ton A weight measure equal to 1,000 kg, 2,204.62 lb, and 0.9842 long tons

Metric(s) (1) Measurement system; (2) a performance measure

MFD (1) Mechanical flow diagram; (2) multifunctional device (a printer/copier/scanner/fax)

MFG Manufacturing

mg/l Milligrams per liter (= ~ppm (parts per million) in water) – expresses a measure of the concentration by weight of a substance per unit volume

MI Mechanical integrity – keeping fluids in the equipment and pipes

MIC Microbiologically influenced corrosion

Micron A unit of length. One millionth of a meter or one thousandth of a millimeter. One micron equals 0.00004 of an inch.

Mid-barrel A term sometimes used to refer to middle distillates, the "middle" of a crude barrel.

Mid-boiling point and mid-volume point components These have been defined and described earlier under "Component balances" and in the topic "▶ Introduction to Crude Oil and Petroleum Processing" in this handbook.

Middle distillate Term applied to hydrocarbons in the so-called middle range of refinery distillation. Examples: heating oil, diesel fuels, and kerosene

Midstream The portion of the value chain concerned with the marketing and transportation of hydrocarbons.
MII Maximum intended inventory
MIS Management information system
MM (1) Materials management; (2) materials manager
mm Millimeter
MMBOE Million barrels of oil equivalent
MMBOPD Million barrels of oil per day
MMBtu Million British thermal units
MMI Man–machine interface. See "HMI."
MMS (1) Maintenance management system; (2) US Minerals Management Service
MMscf/d Million standard cubic feet per day
MMxxx Used in refineries to indicate 1,000,000. For example, 1 MMbbls = one million barrels. Latin derivation
MOC Management of change in Process Safety Management requirements
Molecular sieve (or mol sieve) Material with very small holes of precise and uniform size (Angstrom sizes) which is small enough to block large molecules but allows small molecules to enter. Mol sieves uses in a refinery include drying instrument air, pressure swing adsorption, and paraffin separations.
Molecule The smallest division of a compound that still retains or exhibits all the properties of the substance
Mollier diagram A chart that relates the enthalpy and entropy of steam to its conditions (pressure/temperature)
MON Motor octane number
Monitor (1) In process operations, refers to continuously or regularly reviewing operating data to enable the plant to make appropriate changes or plans for operating/maintaining the facility based on the observed data. (2) A piece of firefighting equipment capable of spraying or streaming water continuously onto a fire or adjacent equipment. A monitor does not require continuous presence of someone to operate it. Refer to the chapter on firefighting.
Monoethanolamine (MEA, $HOC_2H_4NH_2$) One of the common, regenerable amine solvents used for acid gas removal (H_2S or CO_2) from vapor streams
MOR Middle of run
Motor gasoline (AKA mogas) A complex mixture of relatively volatile hydrocarbons, with or without small quantities of additives, which have been blended to form a fuel suitable for use in spark-ignition engines
Motor gasoline blending Mechanical mixing of motor gasoline blending components, and oxygenates when required, to produce finished motor gasoline. Finished motor gasoline may be further mixed with other motor gasoline blending components or oxygenates, resulting in increased volumes of finished motor gasoline and/or changes in the formulation of finished motor gasoline (e.g., conventional motor gasoline mixed with fuel ethanol to produce oxygenated motor gasoline).

Motor gasoline blending components Naphthas (e.g., straight-run gasoline, alkylate, reformate, benzene, toluene, xylenes) used for blending or compounding into finished motor gasoline. These components include reformulated gasoline blendstock for oxygenate blending (RBOB) but exclude oxygenates (alcohols, ethers), butanes, and pentanes. Note: Oxygenates are reported as individual components and are included in the total for other hydrocarbons and oxygenates.

Motor octane number (MON, ASTM ON, F-1) A measure of resistance to self-ignition (knocking) of a gasoline under laboratory conditions which correlates with road performance during highway driving conditions. The percentage by volume of isooctane in a mixture of isooctane and n-heptane that knocks with the same intensity as the fuel being tested. A standardized test engine operating under standardized conditions (900 rpm) is used. This test approximates cruising conditions of an automobile, ASTM D2723.

MPC Model predictive control – a type of advanced control
MPT Minimum pressurization temperature
mpy Mils per year corrosion rate
MRU Methanol recovery unit
MS (1) Mass spectrometer – analytical method; (2) marketing services
MSA (1) Master service agreement; (2) Mine Safety Appliances
MSDS Material safety data sheet
MSFD EU marine strategy framework directive (water pollution)
MSP Monosodium phosphate
MSW Mixed sweet crude – from Canada
MT (1) Thousand tons; (2) metric ton (2,205 lb) in environmental reporting – be careful of context; (3) magnetic testing; (4) management team
MTBE (1) Methyl tertiary-butyl ether (octane booster); (2) mean time between events – electrical
MTBF Mean time between failures
MTDF Mean time to detect failures
MTO (1) Material take-off; (2) Methanol to olefins
MTP Methanol to propylene
MTPA Million tonnes per annum
MTTF Mean time to failure
MTTR Mean time to repair
Multipass tray In distillation, a tray with multiple liquid flow paths and more than one downcomer and overflow weir. Used in large towers. Refer to the chapter on equipment, section on tower design.
MV (1) Millivolts; (2) motor valve – a control valve
MVA Million volt-amps – a way of expressing connected electrical load. Approximately MW
MW Megawatts – power
Mxxx Used in refineries to indicate 1,000. For example, 1 Mbbls = one thousand barrels. Latin derivation

N

Na Sodium
NAAQS National ambient air quality standards
NAC Naphthenic acid corrosion
NACE National Association of Corrosion Engineers
NAICS North American Industry Classification System
Naphtha A volatile, colorless product of petroleum distillation. Used primarily as paint solvent, cleaning fluid, and blendstock in gasoline production to produce motor gasoline by blending with straight-run gasoline. Typically boils at 50 to ~400 °F. There are usually two straight-run naphtha cuts produced from crude: (1) light naphtha (sometimes called light gasoline) and (2) heavy naphtha.
Naphtha-type jet fuel A fuel in the heavy naphtha boiling range having an average gravity of 52.8° API, 20–90 % distillation temperatures of 290–470 °F, and meeting military specification MIL-T-5,624 L (Grade JP-4). It is used primarily for military turbojet and turboprop aircraft engines because it has a lower freeze point than other aviation fuels and meets engine requirements at high altitudes and speeds.
Naphthalene ($C_{10}H_{10}$) An aromatic compound with two adjacent aromatic rings. The simplest polynuclear aromatic compound
Naphthene A cyclic, non-aromatic hydrocarbon. One of the three basic hydrocarbon classifications found naturally in crude oil. Naphthenes are widely used as petrochemical feedstock. Examples are cyclopentane, methyl, ethyl, and propylcyclopentane.
National Pollutant Discharge Elimination System (NPDES) NPDES is the regulatory authority which issues permits to control all discharges of pollutants from point sources into US waterways. NPDES permits regulate discharges into navigable waters from all point sources of pollution, including industries, municipal wastewater treatment plants, sanitary landfills, large agricultural feed lots, and return irrigation flows.
Natural draft In a furnace, flue gas and airflow induced by natural convective circulation
Natural gas A naturally occurring raw material often produced in conjunction with crude oil that is processed through a variety of facilities to yield NGLs. It is a commercially acceptable product for industrial and residential consumption and is shipped via pipeline.
Natural gas liquids Those hydrocarbons in natural gas that are separated from the gas as liquids through the process of absorption, condensation, adsorption, or other methods in gas processing or cycling plants. Generally, such liquids consist of propane and heavier hydrocarbons and are commonly referred to as lease condensate, natural gasoline, and liquefied petroleum gases. Natural gas liquids include natural gas plant liquids (primarily ethane, propane, butane, and isobutane; see "Natural Gas Plant Liquids") and lease condensate

(primarily pentanes produced from natural gas at lease separators and field facilities; see "Lease Condensate").

Natural gas plant liquids Those hydrocarbons in natural gas that are separated as liquids at natural gas processing plants, fractionating and cycling plants, and, in some instances, field facilities. Lease condensate is excluded. Products obtained include ethane, liquefied petroleum gases (propane, butanes, propane–butane mixtures, ethane–propane mixtures), isopentane, and other small quantities of finished products, such as motor gasoline, special naphthas, jet fuel, kerosene, and distillate fuel oil.

Natural gasoline and isopentane A mixture of hydrocarbons, mostly pentanes and heavier, extracted from natural gas, that meets vapor pressure, end point, and other specifications for natural gasoline set by the Gas Processors Association. Includes isopentane

NAV (1) Net asset value; (2) Norton antivirus

NC Normally closed

NCEES National Council of Examiners for Engineering and Surveying – the organization that administers the professional engineering exams

NCM (1) Net cash margin; (2) normal cubic meters at 1 Atm, 15 °C

NDE Nondestructive examination

NDT Nondestructive testing

Near-infrared analyzer An analyzer (online or offline) which determines composition of a sample by measuring absorption or transmittance of near infrared light through the sample. Refer to the chapter on controls.

NEC US National Electrical Code

Needle coke Premium grade of petroleum coke used in metallurgy

NEMA National Electrical Manufacturers Association

NESHAP US National Emissions Standards for Hazardous Air Pollutants

Net investment In a project, this includes the capital cost of the plant, which is subject to depreciation, and the associated costs that are not subject to depreciation. It would also include the capitalized construction loan interest.

Net profit margin A measure of profitability based on the ratio of net income to total operating revenues

Netback The price a refiner receives for the sale of petroleum products after deducting the transportation or affiliated costs in shipping the product from its point of origin (i.e., pipeline tariffs, waterborne freight, storage fees, line loss, cost of capital, etc.)

Neutralization Refers to reacting a material with a reagent to bring the pH of the material closer to 7.0 (or "neutral"). This process is used in water effluent treatment to avoid sending water that is too acidic or caustic to a waterway to public treatment works.

Neutralization number The number that expresses the weight in milligrams of an alkali (normally KOH) needed to neutralize the acidic material in 1 g of oil. The neutralization number of an oil is an indication of its acidity.

Neutralizing amine Any of several amines that prevent corrosion by neutralizing acids in a stream – e.g., steam condensate system corrosion control

New York Mercantile Exchange (NYMEX) US commodities exchange where a number of commodities, including WTI crude, heating oil, and unleaded gasoline, are traded on a future basis

Newton's law (particle settling) One of several of Newton's laws is a mathematical relationship that describes the rate of movement of a particle or droplet through another continuous phase in more turbulent conditions. Refer to the environmental chapter section on water and the equipment chapter section on drum design.

NFPA National Fire Protection Association, Inc.

NGL Natural gas liquids – typically propane and butanes recovered with natural gas

NGO Nongovernmental organization

NH_3 Ammonia

NH_3N Ammonia nitrogen. Nitrogen present as ammonia.

NH_4Cl Ammonium chloride – corrosive salt

NH_4HCO_3 Ammonium bicarbonate – corrosive salt

NH_4HS Ammonium bisulfide – corrosive salt

NHT Naphtha hydrotreater

NHV Net heating value

Ni Nickel

Nickel carbonyl ($Ni(CO)_4$) Nickel carbonyl is one of the most toxic substances which may be encountered in a petroleum processing facility anywhere CO is present along with nickel in a reduced state. Refer to the chapter on hazardous materials.

NIOSH US National Institute for Occupational Safety and Health

NIR Near infrared

NLL Normal liquid level

NMR Nuclear magnetic resonance – analytical method

NNF Normally no flow

NO Normally open

Non-energy product refineries This term refers to refineries which primarily produce petroleum products other than fuels. The two major types of non-energy refineries are (1) the lube oil refinery and (2) the petrochemical refinery. These types of facilities are detailed in the chapter "▶ Non-energy Refineries in Petroleum Processing" in this handbook.

NORM Naturally occurring radioactive material

NOx Nitrogen oxides – generically

NP Normal paraffin

NPDES US National Pollutant Discharge Elimination System

NPDES permit US National Pollutant Discharge Elimination System permit – see "National Pollutant Discharge Elimination System."

NPRA National Petroleum Refiners Association – now AFPM

NPSH Net positive suction head – for a pump. Refer to the equipment chapter section on pumps.
NPSHA Net positive suction head available. Refer to the equipment chapter section on pumps.
NPSHR Net positive suction head required. Refer to the equipment chapter section on pumps.
NPV Net present value
NRC Nuclear Regulatory Commission
NSC National Safety Council
NSPS New Source Performance Standards – US Federal. Refer to the environmental chapter.
NTP Normal temperature and pressure in metric units. Normally at either 0 °C or 15 °C and 1 Atm. Check context.
NTSC National Transportation Safety Board
NYMEX New York Mercantile Exchange
NYSE New York Stock Exchange

O

O&GJ Oil and Gas Journal
O_2 Oxygen
OCAW Oil, Chemical, and Atomic Workers Union. Now combined into the United Steelworkers' Union. Represented refinery union workers in most US refineries for many years
Octane number and octane Octane numbers are a measure of a gasoline's resistance to knock or early detonation in a cylinder of a gasoline engine. The higher this resistance is, the higher will be the efficiency of the fuel to produce work. A relationship exists between the antiknock characteristic of the gasoline (octane number) and the compression ratio of the engine in which it is to be used. The higher the octane rating of the fuel, the higher the compression ratio of engine in which it can be used. By definition, an octane number is that percentage of isooctane in a blend of isooctane and normal heptane that exactly matches the knock behavior of the gasoline. Thus, a 90-octane gasoline matches the knock characteristic of a blend containing 90 % isooctane and 10 % n-heptane. The knock characteristics are determined in the laboratory using a standard single-cylinder test engine equipped with a supersensitive knock meter. Online knock engines have been used during blending for many years. The octane number can also be calculated from near-infrared analysis, which enables improved, continuous online measurement of octane during gasoline blending. The pump octane posted in the United States is (RON + MON)/2. In Europe, often only RON is given.
OE (1) Operations excellence; (2) open-ended; (3) organizational error
OECD Organization for Economic Cooperation and Development

OEL Open-ended lines – environmental

OEM Original equipment manufacturer – usually refers to mechanical or electrical parts that originally were installed or meet the specifications of parts originally installed on equipment

Offsite systems or offsites These are facilities outside the main refining process units that support operation of those units. Among the major facilities found in refinery offsite systems are storage (tanks), product blending, road and rail loading, jetty facilities, waste disposal, and effluent water treating. Refer to the topic "▶ Off-Site Facilities for Petroleum Processing" in this handbook for a more complete description and discussion of these facilities.

OGJ Oil and Gas Journal. Abbreviation is frequently used in reference citations.

OH Over head or overhead

Oil Crude petroleum and other hydrocarbons produced at the wellhead in liquid form

Oil sand crude Heavy hydrocarbons recovered from tar sands or oil sands by thermal methods.

Oily water sewer The sewer system that collects water within a refinery or process facility that may contain oil. The collected water is further processed to remove the oil. Refer to the chapters on utilities and offsites.

OIW Oil in water

Olefins Class of unsaturated paraffin hydrocarbons recovered from petroleum. Typical examples include: butene (butylene), ethene (ethylene), and propene (propylene). They constitute a fourth group of hydrocarbons, in addition to paraffins, naphthenes, and aromatics. They are not naturally present in any great quantity in most crude oils, but are often produced in significant quantities during production processes for heavy oils or refining of the crude oil to products. They are generated in those processes that subject the oil to high temperature for a relatively long period of time. Under these conditions, the saturated hydrocarbon molecules break down permanently, losing one or more of the four atoms attached to the quadrivalent carbon. The resulting hydrocarbon molecule is unstable and readily combines within itself (forming double-bond links) or with similar molecules to form polymers. An example of such an unsaturated compound is ethylene: $H_2C = CH_2$. Note the double bond in this compound linking the two carbon atoms that are the defining characteristic of olefins.

On/off-discrete control See "Gap Control."

OOS Out of service

OP (1) Operating pressure; (2) output – from a controller

OPA 90 US Oil Pollution Act of 1990

Opacity In stacks, a measure of the amount of particulate matter or smoke coming out of a stack

OpCom Britannia Operating Committee

OPEX Operating expense

ORI Octane requirement increase

Orifice meter The most common flow metering device that takes advantage of the Bernoulli effect by introducing a restriction in a pipe to create a *vena contracta*.

The pressure difference between the *vena contracta* and the bulk fluid is a function of the flow rate.

Orsat An analysis method for fired heater flue gases used in determining heater performance. Largely supplanted by analyzers. Refer to the environmental chapter section on air emissions.

OSA Suncor Synthetic A (OSA) crude oil – pre-refined from Canadian bitumen

OSARP Oil Spill Assessment Response Program

OSBL Outside battery limits

OSH Suncor sour synthetic crude – from Canada

OSHA Occupational Safety and Health Administration (US Department of Labor)

OT Operating temperature

OTJ On the job – applied to training

Outage (1) Space left in a product container to allow for expansion during the temperature changes it may undergo during shipment and application. (2) Measurement of space that is *not* occupied in a drum or vessel

Outlet collector The device at the bottom of a catalytic reactor bed where the reactants are collected and exit the reactor. The outlet collector is designed to retain the catalyst in the reactor.

Output (1) The total of all streams from a refinery or process unit. (2) In control systems, the signal coming from a controller to the field control device. These may be 3–15 psi pneumatic signals, 4–20 ma electrical signals, digital, or some other form.

Overflash Overflash is a term normally associated with the design of crude atmospheric or vacuum towers. Its objective is to provide additional heat (over and above that set by the product vaporization) required by the process to generate the internal reflux. It also influences the flash zone conditions of temperature and partial pressure of the hydrocarbon vapor feed. Usually, it is fixed between 3 and 5 vol% on crude. This atmospheric flash temperature is adjusted to the temperature at the previously calculated partial pressure existing in the flash zone. A further description and the purpose of overflash is given in the chapter titled "▶ Atmospheric and Vacuum Crude Distillation Units in Petroleum Refineries" of this handbook.

Overhead (1) In processing, refers to the top product from a distillation column or absorber (abbreviated OH); (2) in economics, refers to the operating costs that are not directly used in making the products, such as administrative or building costs

Overlap A measure of separation quality expressed as the amount of one distillation product that is still present in another after processing. Generally, overlaps are expressed as the difference between two distillation reference points – such as the 95 % point of the lighter product and the 5 % point of the next heavier product. An overlap exists when, say, the 5 % point of the heavier product is less than the 95 % point of the lighter product. See also "Gaps and Overlaps."

Overplus Design allowance above minimum required. Essentially the design margin for equipment and processes

Overpressure Overpressure is the pressure increase over the set pressure of the primary relieving device; it would be termed accumulation when the relieving device is set at the maximum allowable working pressure of the vessel.

Overspeed Exceeding the allowable rpm of a piece of rotating equipment – such as a steam turbine without load on it

OVHD Overhead

Oxidation Combining elemental compounds with oxygen to form a new compound. A part of the metabolic reactions

Oxidation pond A surface impoundment for treatment of wastewater by aerating the water and allowing microbiological digestion of the contaminants. These ponds may also be used to prevent sanitary sewage from becoming septic and smelling while it is held before final treatment.

Oxidizing agent Any substance, such as oxygen or chlorine, which can accept electrons. For example, when oxygen or chlorine is added to wastewater, organic substances are oxidized. These oxidized organic substances are more stable and less likely to give off odors or to contain disease bacteria.

Oxygen deficient Refers to an environment with insufficient oxygen concentration to support life. Refer to the safety chapter.

Oxygen enrichment The use of pure oxygen or air with oxygen added to enhance an oxidation process. Oxygen enrichment is used as a debottlenecking option in sulfur plants, FCCs, and other process units. Refer to the chapter on utilities for discussion of supply.

Ozonation The application of ozone to water, wastewater, or air, generally for the purposes of disinfection or odor control.

Ozone O_3. A pollutant in the atmosphere when generated by photochemical reaction with NO_x. May also be used for microbiological control and in certain reactions

P

P Phosphorous

P&ID Process (or piping) and instrumentation diagram or flow diagram – also called MFD or mechanical flow diagram

P/I Pneumatic to current transducer or converter

PA (H, HH, L, LL) Pressure alarm high, high–high, low, or low–low – e.g., PAHH is a high–high-pressure alarm.

Packed tower A distillation or fractionation column containing specially shaped mass transfer packing materials instead of trays. Packing often enables much greater mass and heat transfer rates in less height than trays and offers lower pressure drop in many cases. Packing is also used for small-diameter towers (less than 3 ft) or very-large-diameter towers (like vacuum fractionators) where pressure drop is critical. The packing in the tower may be stacked in beds on a random basis or in a defined structured basis. For practical reasons, and to avoid crushing the packing at the bottom, the packing is usually installed in beds of 15–20 f. in height. Refer to the handbook topic "▶ Process Equipment for Petroleum Processing" for discussion of packed towers.

Packing (1) The shaped contacting elements (e.g., rings, saddles, etc.) that are placed in a contactor tower to provide surface area for mass transfer; (2) the physical orientation or arrangement of mass transfer packings or catalysts within a tower or reactor; (3) A type of seal around a rotating or reciprocating shaft in a compressor or pump. Refer to the equipment chapter sections on compressors and pumps; (4) a type of seal around a valve shaft to contain process fluids – refer to the chapter on process controls, section on control valves and the environmental chapter section on fugitive emissions controls.

Padding Providing a compatible vapor above a storage tank or vessel, normally to prevent contact with air or development of an explosive mixture. Also known as blanketing. See offsites chapter.

PADD US Federal Petroleum Administration for Defense District – defined regions of the United States

PAH Polycyclic aromatic hydrocarbons (occasionally polyaluminum hydroxide)

Pall ring A type of mass transfer packing for towers. Refer to the chapter on equipment, section on packed towers.

Pantograph An articulated set of beams that are able to extend and contract with the movement of whatever they are attached to. Often used for sample lines and roof drains in tanks

Paraffin A saturated hydrocarbon compound in which all carbon atoms in the molecule are connected by single bonds

Partial oxidation (POX) A process for gasification of hydrocarbons by burning the hydrocarbon sub-stoichiometrically with air or oxygen to produce a gas rich in CO and hydrogen, along with some CO_2. The resulting synthesis gas can be cleaned up and used to make hydrogen, petrochemicals, or paraffinic liquid fuels.

Partial pressure The portion of total pressure exerted by a specific component or compound. Normally equal to the total absolute pressure of the system times the mole fraction of the component. For instance, a gas at 250 psig containing 75 % hydrogen has a hydrogen partial pressure of 75 %/100 % $\times (250 \text{ psig} + 14.7 \text{ psi}) = 198.5$ psi.

Particulate Free suspended solid or a discrete particle

Partition wall In a fired heater, refractory wall inside a heater that divides the radiant section into separately fired zones. Refer to the equipment chapter, section on fired heaters.

PB Propane + butane

PBC Propane–butane casing head

PC (1) Pressure controller; (2) personal computer – watch out for context; (3) paired comparisons

PCB See "Polychlorinated Biphenyls."

PCS Process control system
PD (1) Positive displacement – like a piston; (2) project development
PDA Personal digital assistant
PDC Power distribution center. AKA the switch room
PDH Professional development hours – in Professional Engineer licensing
PE (1) Professional engineer; (2) production efficiency
PEL Permissible exposure limit
Pensky–Martens (PMCC) A closed-cup test for flash points of oil
Pentanes plus A mixture of hydrocarbons, mostly pentanes and heavier, extracted from natural gas. Includes isopentane, natural gasoline, and plant condensate
PERC (1) Perchloroethylene; (2) power emergency release coupling
Peristaltic pump A type of positive displacement pump
Permissible exposure limit (PEL) Time-weighted average maximum concentration of a chemical which is considered to be safe (negligible health hazard) for workers over an 8 h period. Regulatory limits set by US OSHA
Pet coke See "Petroleum Coke."
Petrochemical An intermediate chemical derived from petroleum, hydrocarbon liquids, or natural gas, such as ethylene, propylene, benzene, toluene, and xylene
Petrochemical feedstocks Chemical feedstocks derived from petroleum principally for the manufacture of chemicals, synthetic rubber, and a variety of plastics
Petrochemical refineries See the previous item on "Non-Energy Product Refineries."
Petroleum A generic name for hydrocarbons, including crude oil, natural gas liquids, natural gas, and their products
Petroleum Administration Defense District (PADD) Five geographic areas into which the United States was divided by the Petroleum Administration for Defense for purposes of administration during federal price controls or oil allocation. They are:
- PADD1: Connecticut, Delaware, District of Columbia, Florida, Georgia, Maine, Maryland, Massachusetts, New Hampshire, New Jersey, New York, North Carolina, Pennsylvania, Rhode Island, South Carolina, Vermont, Virginia, and West Virginia
- PADD2: Illinois, Indiana, Iowa, Kansas, Kentucky, Michigan, Minnesota, Missouri, Nebraska, North Dakota, Ohio, Oklahoma, South Dakota, Tennessee, and Wisconsin
- PADD3: Alabama, Arkansas, Louisiana, Mississippi, New Mexico, and Texas
- PADD4: Colorado, Idaho, Montana, Utah, and Wyoming
- PADD5: Alaska, Arizona, California, Hawaii, Nevada, Oregon, and Washington

Petroleum coke A solid residue, high in carbon content and low in hydrogen, which is the final product of thermal decomposition in the condensation process in cracking. This product is reported as marketable coke. The conversion is 5 barrels (of 42 US gallons each) per short ton. Coke from petroleum has a heating value of about 6.024 million Btu per barrel.

Petroleum products Petroleum products are obtained from the processing of crude oil (including lease condensate), natural gas, and other hydrocarbon compounds. Petroleum products include unfinished oils, liquefied petroleum gases, pentanes plus, aviation gasoline, motor gasoline, naphtha-type jet fuel, kerosene-type jet fuel, kerosene, distillate fuel oil, residual fuel oil, petrochemical feedstocks, special naphthas, lubricants, waxes, petroleum coke, asphalt, road oil, still gas, and miscellaneous products.

PFA Perfluoroalkoxy

PFD (1) Process flow diagram; (2) probability of failure on demand

pH pH is an expression of the intensity of the basic or acidic condition of a liquid. Mathematically, pH is the logarithm (base 10) of the reciprocal of the hydrogen ion concentration. The pH may range from 0 to 14, where 0 is most acidic, 14 most basic, and 7 is neutral. Natural waters usually have a pH between 6.5 and 8.5.

PHA Process Hazards Analysis. Refer to the chapter on safety.

Phenol An organic compound that is an alcohol derivative of benzene

Photovoltaic A device for generating electrical energy from light – like a solar cell

PHR Process Hazards Review (MOC related)

PI (1) Pressure indicator – a pressure gauge; (2) process input; (3) performance indicator; (4) process or plant information – system that collects and retrieves process operating data for analysis (may include a licensed system); (5) profitability index; (6) project implementation; (7) proportional–integral (controller)

PIB Product information bulletin. General information on a product

PID (1) Proportional–integral–derivative (controller). See the chapter on controls. (2) See also "P&ID."

Pilot A small, self-contained burner in a fired heater that burns continuously to ensure ignition of any fuel that enters the firebox. Also used on flare tips

Pilot-operated pressure relief valve A pilot-operated pressure-relief valve is one that has the major flow device combined with and controlled by a self-actuated auxiliary pressure relief valve. This type of valve does not use an external source of energy.

PIMS Linear program. Used for planning refinery operations. PIMS = Process Industry Modeling System.

Pinch technique A calculation and analysis method used for optimizing heat recovery and other similar problems

PIP Process Industry Practices

Pipe fitting Refers to elbows, tees, and other items used to connect straight pipe together into a system

Pipe hanger Device for supporting pipe by hanging it off a structure above the pipe. These may be fixed or may be spring loaded or counterweighted to account for pipe movement.

Pipe support Device for supporting pipe from below or alongside the pipe. Pipe supports may be fixed or may be spring loaded or counterweighted to account for pipe movement.

Pipeline(s) A piping network that allows crude oil, refined products, and gas liquids to move across the country, usually from either refineries to terminals or from coastal (import) locations to terminals and refineries further inland

Piping element General term for anything that goes into a piping system, including elbows, tees, flanges, valves, reducers/enlargers, etc.

PIS Protective instrument systems

Pitch (1) Same as vacuum resid – extremely heavy oil. (2) Attack angle of blades in a fan. (3) Layout angle of the tubes or shell-side baffles in a heat exchanger. Refer to the equipment chapter, section on heat exchangers.

PL Protection layer

Planning refinery operations This term refers to the activities required to define and coordinate the products, crudes, unit rates, unit feeds, and economics to ensure optimum refinery operation and profitability within the constraints of the refinery. Refer to the chapter titled "▶ Petroleum Refinery Planning and Economics" in this handbook for a detailed discussion of these activities.

Plant air Also called utility air. Compressed air used for general plant services, such as pneumatic tools. Separated from the instrument air system for security purposes. Refer to the discussion in the utilities chapter.

Plant commissioning The sequence of events starting with a newly completed process facility and ending with an operating plant. Commissioning includes pre-energizing activities, energizing the plant, conditioning equipment, calibrating instruments, setting relief valves, final checking and closure of vessels and towers, loading catalysts, preparation for "startup," startup, lining out, and performance and guarantee test runs. Commissioning is discussed in detail in the chapter titled "▶ Petroleum Processing Projects" in this handbook.

Plant water Also called utility water. This is water available in a facility for general use. May not necessarily be drinkable or potable. See the discussion in the chapter on utilities.

Play In the oil and gas industry, this refers to a general area in which significant quantities of hydrocarbon oil or gas can be found. Development of a "play" results in commercial production, hopefully.

PLC Programmable logic controller

Plenum A chamber enclosing equipment used to conduct flow into the equipment. Normally refers to a preheated air chamber surrounding the burners in a fired heater with louvers to control the airflow. Refer to the equipment chapter, section on fired heaters.

Plug flow Two-phase piping flow regime where alternate plugs of liquid and gas move through the pipe. This is an undesirable regime that can result in damage. Refer to the equipment chapter, section on two-phase pressure drop.

PM (1) Particulate matter – normally refers to dust and the like in the air. When followed by a number, the number refers to the maximum size particles considered (e.g., PM10 means particles less than 10 μm in size). (2) Preventative maintenance. (3) Performance management. (4) Project manager

PMCC Pensky–Martens closed cup – a flash point test

PMI Positive materials identification

PMS (1) Piping material specifications; (2) power management system; (3) platform maintenance system

PNA Polynuclear aromatics – big, multi-ring aromatic compounds – AKA polyvinyl chicken wire or "red death" in hydroprocessing

Pneumatic Operated by air or other gas pressure

Point In finance, 1/100th of a cent ($0.0001)

Poison In refining, in addition to health hazards, this term refers to compounds, elements, or characteristics that reduce catalyst activity over a period of time. Normally, the term applies to permanent activity reductions.

Pollution The impairment (reduction) of air, water, or other environmental condition by agriculture, domestic, or industrial wastes (including thermal and radioactive wastes) to such a degree as to hinder any beneficial use of the resource or render it offensive to the senses of sight, taste, or smell or when sufficient amounts of waste creates or poses a potential threat to human health or the environment

Polyaromatic hydrocarbon A hydrocarbon having an aromatic structure that includes two or more conjugated or conjoined rings. These tend to condense into larger polyaromatic structures when heated to high temperatures.

Polychlorinated biphenyls (PCBs) Difficult to remediate chemical used in old-style transformers. Concentrated PCBs used to be referred to as "1268". Includes polychlorobiphenyls

Polymer A chemical formed by the union of many monomers (a molecule of low molecular weight). See also "Gums."

Polymerization Process of combining two or more simple molecules of similar type, called monomers, to form a single molecule having the same elements in the same proportion as in the original molecules, but having increased molecular weight. The product of the combination is a polymer.

PON Posted octane number; (RON + MON)/2

PONA Analysis method for paraffins, olefins, naphthenes, and aromatics

Porosity (1) Ability of a bed of solids or a rock formation to permit movement of oil and gas through the solids. High porosity means fluids can easily move through the solids. (2) A measure of the % of the internal volume of a solid particle that is open space. Specifically refers to catalyst particles

PORV Pilot-operated relief valve

Potable water Water that is fit for human consumption. Refer to the chapter on utilities.

POTW Publically owned treatment works. For example, the city or county sewer treatment plant

Pour point The "pour point" of an oil is the temperature at which the oil ceases to flow. It is usually a test applied to middle distillates, lube oils, and fuel oils. The test itself is simple and requires the sample oil to be carefully treated before the test and to reduce its temperature in a controlled and orderly way. Unlike most other petroleum properties, pour points of two or more components cannot be blended directly to give a pour point of the blended stock. Blending indices are used with the volumetric composition of the blend components for this purpose. Details of the blending for "pour point" are given in the topic "▶ Introduction to Crude Oil and Petroleum Processing" in this handbook, while details of the test itself are given in the chapter on "▶ Quality Control of Products in Petroleum Refining" in this handbook.

Power recovery A system for recovering power from high-pressure fluids as they are reduced to lower pressure. Power may be recovered through direct coupling of a letdown device to a pump or compressor or by power generation. Power recovery devices include letdown turbines and turboexpanders.

POX See "Partial Oxidation."

PP (1) Propane/propylene; (2) pour point

ppb Parts per billion – normally by weight, but may be by volume in gases

PPD Pollution prevention design. Reducing plant pollution during the design phase by appropriate process modifications – i.e., designing lower pollution into the process

PPE Personal protective equipment – hard hat, safety glasses, steel-toed boots, etc., to protect an individual from hazards

PPG Polypropylene glycol

ppm Parts per million – normally by weight, but may be by volume in gases

PR Pressure recorder

Practical (or Puckorius) Scale Index (PSI) A measurement of fouling potential of water. This index attempts to account for buffering. Refer to the chapter on utilities.

PRC Pressure recorder and controller

Preheat and preheat exchanger train In most processes, heat recovery by heat exchange is of great importance. In the design of major processes, such as the crude oil distillation unit, cracker recovery units, and the like, the optimization of this heat recovery concept is of paramount importance. One method for heat recovery design is described in the chapter titled "▶ Atmospheric and Vacuum Crude Distillation Units in Petroleum Refineries" of this handbook. Briefly, the method consists of examining several configurations of a heat transfer train, applying "pinch" analysis a cost data to the equipment and to determine the terminal feed temperature of each configuration. This end temperature relates to the heater duty required and therefore to the fuel required by the heater. An economic balance may then be made to select the optimum heat exchanger configuration. Developing the various configurations uses the total enthalpy of the feed (in this case, the crude oil feed to the atmospheric distillation unit) and the total enthalpies of the exchanged streams.

Preprocessing See "Pre-refining."

Pre-refining Many unconventional feedstocks require removal of impurities before they can be processed in a normal refinery. This cleanup is referred to as pre-refining or preprocessing.

Pressure regulator Device for controlling a set pressure. A regulator is normally a local controller that is self-actuating.

Pressure relief valve Pressure relief valve (PRV) is a generic term applied to relief valves (RV), safety valves (SV), pressure safety valves (PSV), or safety relief valves (SRV).

Pressure swing adsorption (PSA) A cyclic process where gas containing impurities is contacted with an adsorbent at high pressures. Some compounds in the gas will "stick" to the adsorbent, while others will not be adsorbed significantly and will pass through. When the adsorbent has collected a target amount of material, the adsorbent bed is taken out of service and regenerated by reducing the pressure, which allows adsorbed components to be desorbed and removed. The sorbent is then repressurized and put back in service. PSAs are used extensively in purifying hydrogen streams, producing essentially 100 % hydrogen and a tail gas containing all the impurities. PSAs normally employ multiple beds connected by valves and headers to mimic a continuous process.

Pre-startup safety review (PSSR) A team review to evaluate a new or modified facility before it is operated. The PSSR confirms the changed plant is ready to operate safely.

Presulfiding See "Sulfiding."

Pretreatment or pretreating Refers to steps taken to prepare a stock for a primary refining process. For instance, hydrotreating naphtha to remove sulfur and nitrogen ahead of a reformer is a type of pretreating.

Primary storage Petroleum storage tanks at refineries, pipelines, and oil company terminals. Product inventory changes at these facilities are what constitute API and EIA demand computations.

Prime driver See "Prime Mover"

Prime mover A drive motor or turbine for a compressor, pump, or other piece of equipment

Process Refers generically to a method for physical or chemical conversion applied to a feedstock to produce a desired product

Process control Management of process operating conditions to meet product objectives safely and efficiently within equipment limitations

Process guarantees Among the items of major concern to the operating refinery staff and the engineering contractor in the design, procurement, and construction of a grassroots process or a revamped process are the final process guarantees that are developed and accepted. The process guarantees may begin to be developed as soon as a firm process has been established and manufacturers' guarantees obtained for the performance of the various manufactured items of equipment. The process performance is usually tied to a guarantee of process efficiency, such as utility consumption in the plant while operating on the design throughput and conditions. These items are described and discussed in to the chapter titled "▶ Petroleum Processing Projects" in this handbook with examples of both the performance guarantee and its associated utility guarantee in Appendices 1 and 2 of that chapter.

Process configurations Process configurations are represented in the form of block flow diagrams. They are prepared usually as the first step in deciding the

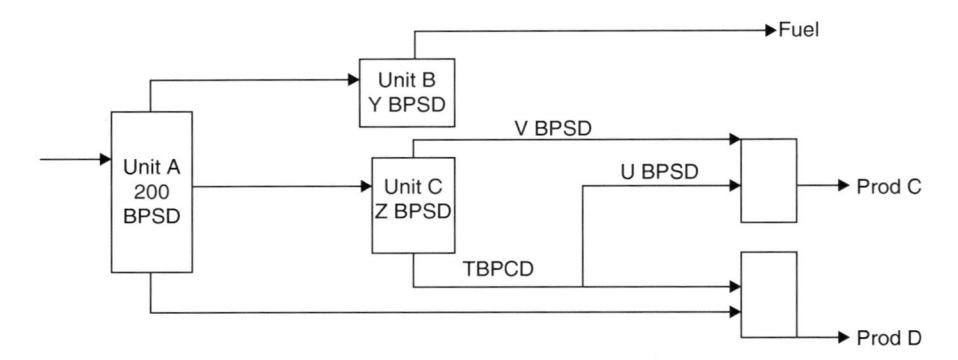

Fig. P.1 A typical block flow diagram

type of process units that will make up a desired complex. The units are shown in sequence to each other by blocks which will be labeled with their throughput size (in the case of a petroleum refinery, in barrels per stream day or in cubic meters or metric tonnes per stream day). The diagram is further developed showing the product and feed lines from and to the unit blocks. The measure of flow is shown on each line. Several block flow diagrams of differing configurations but meeting the end product objectives will be developed prior to the decision on which process route meets all the company's objectives. An example of a simplified block flow diagram is Fig. P.1. Refer to the chapter titled "▶ Petroleum Processing Projects" in this handbook for discussion of process configuration development. Refer to the topic "▶ Petroleum Products and a Refinery Configuration" in this handbook for discussion of the common general refinery types and configurations.

Processing gain or loss The volumetric amount by which total output is greater than or less than input for a given period of time. This difference is due to the processing of crude oil into products which, in total, have a lower or high specific gravity than the crude oil processed, respectively.

Product blending Finished products are blended using two or more components which are rundown stream products from the refinery processes. The blending in most modern refineries is done "in-line"; that is to say that a metered amount of each component is mixed with metered amounts of the other blend components in a pipe that finally enters the respective finished product storage tank. The flow of each of the components is controlled "online" by analyzers which are programmed to fine-tune the component flow control rates to meet the specified product blend recipes. Although these analyzers are quite accurate, the final contents of the finished tanks are always checked by laboratory tests before dispatch out of the refinery. The topic "▶ Petroleum Products and a Refinery Configuration" in this handbook discusses the product properties. That chapter also illustrates the blending recipes in a typical process configuration. Blender design is discussed further in the topic "▶ Off-Site Facilities for Petroleum Processing" in this handbook.

Product information bulletin A summary of product properties for commercial or safety purposes

Product properties The physical and chemical properties of a process product. Normally these are defined by the product specifications and must be measured during operations to ensure the product stays on specification. They may also be estimated or predicted from available information about the stock or operation.

Product property prediction Product properties are predicted from the composition and properties of real components and of pseudo components that make up the product streams. Refer to the handbook topics "▶ Introduction to Crude Oil and Petroleum Processing," "▶ Petroleum Products and a Refinery Configuration," and "▶ Atmospheric and Vacuum Crude Distillation Units in Petroleum Refineries" for more detailed discussions of property predictions.

Product safety data sheet See "Material Safety Data Sheet."

Production In an oil company, the organization that drills wells, removes the oil and gas from the ground, and initially separates the raw production (oil and gas) for transportation to further processing

Production capacity The maximum amount of product that can be made from processing facilities

Programmable logic controller (PLC) An electronic control system that uses programmed logic to carry out repetitive, sequential process tasks. Outputs from the PLC are used to control field elements. It may interface with other control systems but is intended to perform its function stand-alone.

Progressing cavity Using rotation to isolate a volume of fluid and move it from one pressure to another – examples: Moyno pump or liquid-ring compressor

Project duty specification Among the first activities to initiate a refinery project is the development of the "project duty specification," also known as the project basis or the project premises within different companies. This document describes in detail the plant or complex of plants the company wishes to build. Among the major items of the project specification are process specification, general design criteria, any preliminary flow sheets (duly labeled "Preliminary"), utilities specification, basis for economic evaluations, materials of construction, equipment standards and codes, instrument standards required, and company preferences. These items are described more fully in the chapter titled "▶ Petroleum Processing Projects" in this handbook.

Propane (C_3H_8) A normally gaseous straight-chain hydrocarbon. It is a colorless paraffinic gas that boils at a temperature of -43.67 °F. It is extracted from natural gas or refinery gas streams. It includes all products designated in ASTM Specification D1835 and Gas Processors Association Specifications for commercial propane and HD-5 propane.

Propant In oil production, especially hydraulic fracturing, a propant is part of the mixture pumped down a well which will hold open the fissures created and help maintain permeability around the well so oil and gas can get to the well. The most common propant is sand.

Proportional band This determines the response time of the controller. Normally a proportional band is adjustable between 5 % and 150 %. The wider the proportional band, the less sensitive is the control. If a slower response time is required, a wider proportional band is used.

Proportional control A control mode where the controller output is determined by the deviation from setpoint times a constant. See "Proportional Band."

Proportioning pump A positive displacement pump that can be adjusted to deliver a precise flow rate

Propylene (C_3H_6) AKA propene. An olefinic hydrocarbon recovered from refinery processes or petrochemical processes. Nonfuel use propylene includes chemical-grade propylene, polymer-grade propylene, and trace amounts of propane. Nonfuel use propylene also includes the propylene component of propane/propylene mixes where the propylene will be separated from the mix in a propane/propylene splitting process.

PRV Pressure relief valve

PSA Pressure swing absorption

PSD Prevention of significant deterioration – environmental – US Federal requirements. Refer to the environmental chapter.

PSE Pressure safety element – a rupture disk. Refer to the chapter on safety.

Pseudo components See "Product Property Prediction." Essentially, pseudo components are mathematically derived, narrow boiling range cuts of a petroleum stock that are treated as a single compound for calculations using average properties estimated for the specific cut boiling range.

PSI or psi (1)lb/square inch pressure; (2) process safety information; (3) Process Safety Institute

Psia or PSIA Pounds per square inch, absolute (gauge pressure +14.7 psi)

Psig or PSIG Pounds per square inch, gauge (read from a gauge referencing atmospheric pressure – a normal pressure gauge reading)

PSM Process Safety Management. Refer to the chapter on safety.

PSSR Pre-startup safety review – part of the MOC process

PSV Pressure safety valve – a relief valve

PT (1) Pressure transmitter; (2) physical testing; (3) project team; (4) primary tower; (5) process technology; (6) dye penetrant testing

PTASCC Polythionic acid stress corrosion cracking

PTB Pounds per thousand barrels – usually refers to salt in crude

PTOH Primary tower overhead

Pulsation dampener or bottle (PD) A device for reducing the pressure variation around reciprocating pumps or compressors. In some cases, a PD may take the form of a wide spot in the line or an actual vessel. For instruments, PD may be achieved by a flow restriction. Refer to the chapter on process equipment, section on reciprocating compressors.

Pumpage Material being pumped

Pumparound This is the term given to any reflux stream which is created inside a distillation tower by taking off a hot liquid stream, cooling it, and returning the

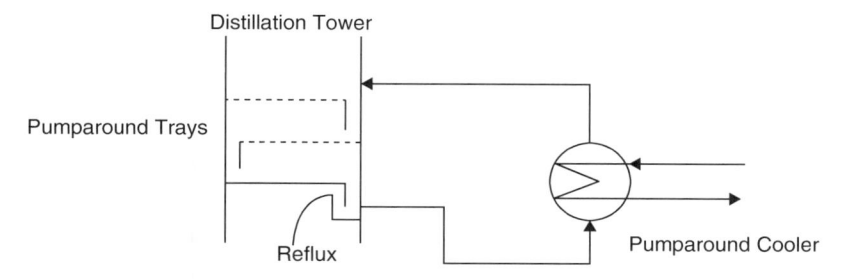

Fig. P.2 A typical pumparound arrangement

stream back into the tower two or three trays above the draw-off tray. Figure P.2 illustrates a typical pumparound system.

Pumps Mechanical device for moving liquids throughout a processing facility. Two types of pumps are typically used in a petroleum processing facility: (1) variable head capacity and (2) positive displacement. Pumps are discussed in detail in the handbook topic "▶ Process Equipment for Petroleum Processing."

Purifier A machine used for a liquid–liquid separation in which the two intermixed liquids which are insoluble in each other have different specific gravities. Solids with specific gravities higher than those of the liquids can be separated off at the same time. A purifier bowl has two outlets, one for the light-phase liquid and one for the heavy-phase liquid. A centrifugal separator

PV (1) Process value; (2) process variable

PVC Polyvinyl chloride

PVDC Polyvinylidene chloride

PVDF Polyvinylidene fluoride

PVF Pipes, valves, and fittings

PWHT Post-weld heat treatment – stress relief

Pyrolysis Breaking down a substance by high temperature

Pyrolysis oil Any oil made by pyrolysis

Q

QA Quality assurance

QC Quality control

QRA Quantitative risk assessment

Quench zone In a multi-bed, exothermic, catalytic reactor, this is a zone or reactor section where reactants are collected, mixed with cooler material (lowering the temperature), and redistributed to the next catalyst bed. Refer to the chapters on hydrotreating, hydrocracking, and equipment (bed pressure drop calculations).

R

R, R', R", etc. Shorthand for any hydrocarbon group

R&D Research and development

Radiant tube or coil In a fired heater, a tube exposed directly to flames

Raffinate The residual product left after a refining process. For instance, the product from solvent extraction, with the material that is removed called the "extract." Also, more generally used in reference to any low-octane product left over after any secondary refining process

RAGAGEP Recognized and generally accepted good engineering practice

RAM Reliability, availability, and maintainability

Ramsbottom coke A carbon residue test originated by Dr. J.R. Ramsbottom in England

Rangeability The rangeability of a control valve is the ratio of the flow coefficient at the maximum flow rate to the flow coefficient at the minimum controllable flow rate.

RAQ In China, Regional Air Quality. Refers to air emissions regulations. Refer to the environmental chapter section on air emissions.

Raschig ring A hollow cylindrical-shaped mass transfer packing or catalyst base. Refer to the chapter on equipment.

RBI Risk-based inspection

RBOB Reformulated blendstock for oxygenate blending – gasoline without the ethanol

RCA Root cause analysis

RCFA Root cause failure analysis

RCM Reliability-centered maintenance – focuses on improving equipment and plant reliability

RCOOH Generic for an organic acid (carboxylic acid)

RCR Ramsbottom carbon residue

RCRA US Federal Resource Conservation and Recovery Act – especially important for disposal of hazardous wastes

RD Rupture disk – pressure relief device. See chapter on safety.

REAC Reactor effluent air cooler – in a hydroprocessing unit

Reacceleration Automatically restarting a piece of rotating equipment after a temporary power loss. Normally refers to electrical equipment that can restart in the event of a power dip without negative consequences or which must be restarted immediately to prevent worse consequences, such as cooling. Reacceleration is sometimes an alternative to providing a spare operated from a different commodity – for instance, a critical service may have one steam turbine-driven pump and one electric motor-driven pump for the service or there may be two electric motors with reacceleration. Refer to the discussion in the equipment chapter.

REACH European Union Registration, Evaluation, Authorization, and Restrictions of Chemicals. Refer to the environmental chapter section on water.

Reactivity Hazardous waste characteristic of being capable of generating a hazardous condition or product when reacted with other materials or air. Refer to the environmental chapter section on hazardous wastes.

Reagent A pure chemical substance that is used to make new products or is used in chemical tests to measure, detect, or examine other substances

Reboiler(s) Reboilers are one of two heat energy input systems to a fractionation unit. The other source is the heat delivered by the feed or feeds to the unit. Reboilers are usually associated with the light ends distillation units and the product stabilizers and splitters on catalytic or thermal cracking units. In most cases, the reboiler is of a shell and tube or a kettle type. In some cases, a fired heater may be used as a reboiler. This reboiler may be fed either by the liquid phase from the bottom tray of the tower or by vaporizing a portion of the bottom product. The first method uses thermosyphon (alt. thermosiphon) as the driving force for flow through the heat exchanger. In the second case, the reboiler feed may be pumped or flow into the kettle section of the exchanger. In both cases, the flow from the reboiler is returned to the tower below the bottom fractionating tray. Exchanger and kettle reboiler arrangements are depicted in Figs. R.1 and R.2. Details on reboiler design are in appropriate sections of this handbook, with additional information in the handbook topic "▶ Process Equipment for Petroleum Processing."

Reciprocating Employing a back-and-forth action – like a locomotive drive cylinder

Reciprocating compressor(s) Reciprocating compressors are positive displacement compressors that are widely used in the petroleum and chemical industries where high discharge pressures are required. They consist of pistons moving in cylinders with inlet and exhaust valves. They are cheaper and more efficient than any other type of compressor where they are applied. Their main advantages are that they are insensitive to gas characteristics and they can handle intermittent loads efficiently. They are made in small capacities and are used in applications where the rates are too small for a centrifugal compressor. They are used almost exclusively in services where the discharge pressures are above 5,000 PSIG. However, reciprocating compressors require more frequent shutdowns for maintenance of valves and other wearing parts than centrifugal compressors. For critical services, this requires either a spare compressor or a multiple compressor installation to maintain plant throughput. The handbook topic "▶ Process Equipment for Petroleum Processing" describes and discusses reciprocating compressors in detail.

Reciprocating pump These are positive displacement pumps that use a piston within a fixed cylinder to pump a constant volume of fluid for each stroke of the piston. In some types, the piston stroke or the cylinder volume may also be mechanically varied to control flow (e.g., metering or proportioning pumps).

Reclaiming (1) In amine absorber systems, reclaiming is the process for removal of contaminants from the circulating amine stream to "reclaim" the amine

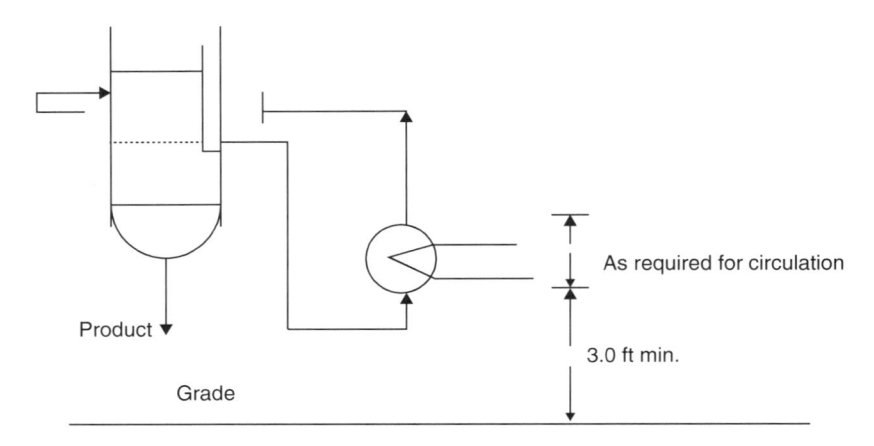

Fig. R.1 A once-through thermosyphon reboiler

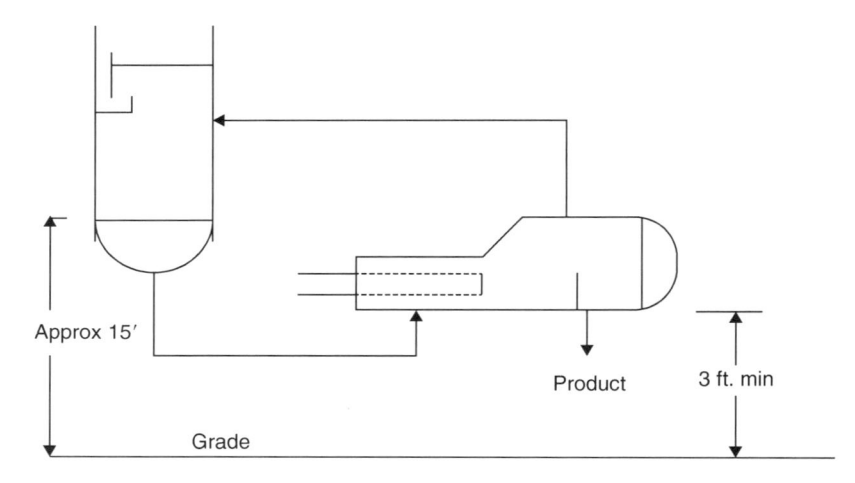

Fig. R.2 A kettle-type reboiler

component (see the chapter on gas treating). (2) In waste management, reclaiming refers to processing of a waste to recover useful metals or other components for reuse and recycle.

Recommended exposure limit (REL) An exposure limit that has been recommended by the US NIOSH to OSHA for adoption as a permissible exposure limit (PEL). Often taken as a limit before official adoption by OSHA. Normally a weighted average concentration over 10 h

Recovered oil See "Slops."

Recycle The reuse of any stream within (internally) a facility: e.g., reusing hydrogen-rich gas in a hydrotreater using a recycle gas compressor

RED In the EU, Renewable Energy Directive. Refer to the environmental chapter section on air emissions.

Reduced crude oil (reduced crude) Crude oil that has undergone at least one distillation process to separate some of the lighter hydrocarbons. Reducing crude lowers its API gravity but increases the handling safety by raising the flash point.

Reducing agent Any substance, such as the base metal (iron) or the sulfide ion that will readily donate (give up) electrons. The opposite of an oxidizing agent

Redwood viscosity The number of seconds required for 50 ml of an oil to flow out of a standard Redwood viscosimeter at a definite temperature; British viscosity standard

Refinery A plant used to separate the various components present in crude oil and convert them into usable products or feedstock for other processes

Refinery-grade butane (C_4H_{10}) A refinery-produced stream that is composed predominantly of normal butane and/or isobutane and may also contain propane and/or natural gasoline. These streams may also contain significant levels of olefins and/or fluorides contamination

Refinery input – crude or total In reference to crude, refers to the total crude oil input to the crude oil distillation units and other refinery processing units (such as direct feed to cokers). Total input refers to all the raw materials and intermediate materials processed at refineries to produce finished petroleum products. These include crude oil, products of natural gas processing plants, unfinished oils or intermediates, other hydrocarbons and oxygenates, motor gasoline and aviation gasoline blending components, and finished petroleum products.

Refinery production Petroleum products produced at a refinery or blending plant. Published production of these products equals refinery production minus refinery input. Negative production will occur when the amount of a product produced during the month is less than the amount of that same product that is reprocessed (input) or reclassified to become another product during the same month. Refinery production of unfinished oils and motor and aviation gasoline blending components appear on a net basis under refinery input.

Refinery types Most of this handbook is focused on the energy refinery, that is, a refinery that converts the crude oil feed to energy products such as gasoline, diesel, turbine fuel, fuel oil, and the like. The topic "▶ Petroleum Products and a Refinery Configuration" in this handbook discusses configurations within this type of refinery. There are two major refinery complexes, however, that convert the crude oil into non-energy products. These are the lube oil refinery and the petrochemical refinery. Very often, these complexes are located adjacent to the energy refinery and they are often integrated into one major refinery complex. The chapter "▶ Non-energy Refineries in Petroleum Processing" in this handbook describes and discusses these two alternate refinery complexes. They are also summarized in the topic "▶ Introduction to Crude Oil and Petroleum Processing" in this handbook.

Refinery yield Refinery yield (expressed as a percentage) represents the percent of finished product produced from input of crude oil and net input of unfinished oils. It is calculated by dividing the sum of crude oil and net unfinished input into the individual net production of finished products. Before calculating the yield for finished motor gasoline, the input of natural gas liquids, other hydrocarbons, and oxygenates and net input of motor gasoline blending components must be subtracted from the net production of finished motor gasoline. Before calculating the yield for finished aviation gasoline, input of aviation gasoline blending components must be subtracted from the net production of finished aviation gasoline.

Refining Converting crude oil into useful products like gasoline, jet fuel, diesel, fuel oil, and olefins

Reflux Returning a cooled liquid stream to a distillation or stripper tower to improve separation and control heat balance in the tower.

Reflux drum A drum that collects fluids to be refluxed to a tower and provides surge volume.

Reflux ratio Normally, volume of reflux per volume of overhead product from a distillation column; although several other definitions are also used. Watch out for the context.

Reformate A naphtha which has been upgraded in octane by catalytic or thermal reforming

Reforming An oil refining process in which naphthas are changed chemically to increase their octane level. Paraffins convert to isoparaffins and naphthenes, and naphthenes change to aromatics. The catalyst used is usually platinum, though sometimes palladium.

Reformulated gasoline The requirements in the United States of the Clean Air Act of 1990 (CAA), and additions to it since, have changed refining requirements to meet this product's need quite significantly. Prior to this date, much of the finished gasoline product recipe consisted of normal light naphtha, reformate, usually some cracked naphtha, and possibly alkylate, with some butane added to meet volatility. The CAA and its subsequent additions forced a reduction of both reformate and cracked stock and replacement of them with oxygenates, such as ethanol, MTBE, and TAME to meet octane number. Oxygenates were used originally simply as additives to improve octane number. However, because of their oxygen content, their addition was also required to reduce the carbon monoxide and hydrocarbons in the emitted gases from combustion. There are a number of oxygenates used in gasoline manufacture; some of the more common are given in Table R.1.

Table R.1 Oxygenates commonly used in gasoline

Name	Formula	RON	RVP, psig	Oxygen, wt%	[a]Water solubility, %
Methyl tertiary-butyl ether (MTBE)	$(CH_3)_3COCH_3$	110–112	8	18	4.3
Ethyl tertiary-butyl ether (ETBE)	$(CH_3)_3COC_2H_5$	110–112	4	16	1.2
Tertiary-amyl methyl ether (TAME)	$(CH_3)_2(C_2H_5)COCH_3$	103–105	4	16	1.2
Ethanol	C_2H_5OH	112–115	18	35	100

[a]wt% soluble in water

The EPA established limits for the use of each oxygenates in gasoline blends. The use of MTBE in the United States was essentially discontinued by the end of 2002 due to concern over groundwater contamination. MTBE is still widely used outside the United States, although the trend in Western Europe is to use ETBE instead. More details on the manufacture of gasolines are given in the topic "▶ Petroleum Products and a Refinery Configuration" in this handbook. More recent regulations are reducing sulfur levels in gasoline in the United States below 10 ppm. This adds more processing requirements.

Refractory (1) Insulating material used in fired heaters (and some other places) to keep heat in and prevent damage to the outer casing. Refer to the equipment chapter, section on fired heaters. (2) Refers to a contaminant that is difficult to remove. For instance, certain nitrogen compounds are much more difficult to remove from oil than others and are termed "refractory nitrogen compounds."

Regeneration (1) In absorber or adsorber systems, regeneration is the process of removing impurities from the sorbent so the sorbent can be reused. (2) In catalysts, regeneration is the process of burning off accumulated coke and impurities from the catalyst to return it to a usable state.

Register A large, loose-fitting butterfly valve or set of louvers that control air entering a burner in a fired heater. These are a primary adjustment for excess O_2 in the firebox.

Regulatory control In control systems, refers to the basic process control system

Reid vapor pressure (RVP) The vapor pressure at 100 °F of a product determined in a volume of air four times the liquid volume. Reid vapor pressure is an indication of the ease of starting and vapor-lock tendency of a motor gasoline as well as explosion and evaporation hazards. This test is the standard test for low-boiling-point distillates. It is used for naphtha, gasoline, light-cracked distillates, and aviation gasoline. For the heavier distillates with vapor pressures expected to be below 26 psig at 100 °F, the apparatus and procedures are different. Only the Reid vapor pressure for those distillates with vapor pressures above 26 psig at 100 °F is described in the chapter on "▶ Quality Control of Products in Petroleum Refining" in this handbook.

REL Recommended exposure limit

Reliability Ensuring equipment is available, functional, and safe to operate when needed

Relief valves These are devices designed to avoid exceeding allowable equipment operating pressures by controlled venting of excess vapors if a pressure vessel's MAWP is exceeded. Full details and discussion on relief systems, which include the relief valves, used in the petroleum industry are given in the chapter on "▶ Safety Systems for Petroleum Processing" in this handbook.

Relieving conditions Relieving conditions pertain to pressure relief device inlet pressure and temperature at a specific overpressure. The relieving pressure is equal to the valve set pressure (or rupture disk burst pressure) plus the overpressure.

Renewable As normally used, a material that can be replaced by natural sources – like renewable diesel from plant oils or animal fats or fuel ethanol produced by fermentation of sugars

Re-refining Refers to reprocessing of hydrocarbon stocks. Normally used in reference to waste lube oils which can be cleaned up and reused through specific processing methods

Research octane number (RON, CFRR, F-1) A measure of antiknock quality of a gasoline stock or blend. Technically, this is the percentage by volume of isooctane in a blend of isooctane and n-heptane that knocks with the same intensity as the fuel being tested. A standardized test engine operating under standardized conditions (600 rpm) is used. Results are comparable to those obtained in an automobile engine operated at low speed or under city driving conditions: ASTM D2722. Alternative near-infrared analysis methods can also be used to estimate the RON.

Reset The rate at which a control system makes changes to a process. A high reset rate will make more frequent changes to the system. In a PID or PI controller, this may result in windup of controller output which causes it to overshoot the setpoint. A low reset rate may never reach its target.

Resid The heaviest portion of crude oil boiling above 650 °F. Atmospheric resid typically boils above 650 °F. Vacuum resid typically boils above 1,050 °F.

Residual fuel oil Heavy fuel oils produced from the nonvolatile residue from the fractional distillation process. Heavy oils that are "leftovers" from various refining processes. Heavy black oils used in marine boilers and in heating plants

Residue conversion There are several processes used to reduce or eliminate the residues from crude oil refining. The most common processes included here are thermal cracking (e.g., coking, visbreaking, Flexicoking), "deep oil" fluid catalytic cracking (e.g., heavy oil cracking or HOC), and residuum hydrocracking and desulfurization (e.g., H-Oil, LC-fining). The processes are discussed in the chapter on "▶ Upgrading the Bottom of the Barrel" in this handbook.

Residue In petroleum refining, the term "residue" refers to the unvaporized portion of the heated crude oil entering either the atmospheric crude oil distillation tower or vacuum tower that leaves these towers as their bottom product. The stream from the atmospheric column is often referred to as the "long" residue, while that from the vacuum unit is often called the "short" residue or bitumen. Both residues are black in color; the atmospheric residue has a specific gravity usually between 0.93 and 0.96, while the vacuum residue will be 0.99 and higher.

Residuum See "Resid" and "Residue."

Resilient seat In a valve, a resilient seat is one in which the plug closes against a rubbery or elastomeric surface.

Respirator Breathing device to prevent personnel exposure to hazardous atmospheres. Includes supplied air and cartridge-type systems. Refer to the safety chapter.

Retort A process for high-temperature thermal decomposition of materials, like oil shale, to produce hydrocarbon oil and gas. Similar to destructive distillation

Return bend A U-shaped pipe fitting or tube to connect two adjacent tubes in a fired heater. Refer to the equipment chapter, section on fired heaters.

Revenues Money taken in

Reverse osmosis A membrane process which purifies a water stream by passing the water through a membrane module at high pressure. The water molecules preferentially pass through the membrane to a lower pressure, leaving the impurities at higher pressure

RFG Reformulated gasoline

RFI (1) Request for information; (2) radio-frequency interface

RFID Radio-frequency identification

RFP Request for proposal

RFQ Request for quotation

RG Recycle gas – in a hydroprocessing unit

RGG Reducing gas generator

RH Relative humidity

RIE Remote instrument enclosure – in the field

RO Restriction orifice

Road and rail loading facilities These are the facilities used to transfer crude, intermediates, or finished products to and from bulk carriers such as tank trucks and rail cars. Refer to the relevant discussion in "▶ Off-Site Facilities for Petroleum Processing" in this handbook.

ROCE Return on capital employed

ROI Return on investment

ROL Reliability operating limit – limit that, if exceeded for an extended time, will result in failure including loss of containment

RON See "Research Octane Number."

ROR Rate of return

Rotary compressor A class of compressors including screws, lobes, and sliding vanes. Refer to the equipment chapter, section on compressors.

Rotary pump A class of positive displacement pumps. Unlike the centrifugal-type pump, these types do not throw the pumping fluid against the casing but push the fluid forward in a positive manner similar to the action of a piston. These pumps however do produce a fairly smooth discharge flow unlike that associated with a reciprocating pump. See the equipment chapter, section on pumps.

Rotating equipment Pumps, compressors, etc. – anything that performs its process function by rotational action, except control valves

Routine maintenance Maintenance done on a daily or schedule basis

RPM or rpm Revolutions per minute – rotational speed

RS Generic for an organic sulfide

RSC US Regional Supply Corridors

RSH Generic for mercaptan or thiol

RSOP Refining standard of practice

RSR' Generic for a conjugated sulfide

RSSR' Generic for a disulfide

RTD Resistance temperature detector – a type of temperature measuring element. See the chapter on controls.

RTR Residual risk and technology review. Refer to the chapter on environmental engineering.

RUL gasoline Regular unleaded gasoline (no ethanol)

Runaway reaction Loss of positive reaction control of an exothermic reaction, resulting in rapidly increasing process temperatures and accompanying increases in reaction rate. Refer to the chapter on safety.

Rupture disk (RD or PSE) A relief device that consists of a thin metal plate or disk designed to burst or fail when a specific pressure differential is imposed on the disk. Once the disk has failed, it must be replaced. It will not stop flowing after the overpressure condition is relieved.

RV Relief valve

RVP See "Reid Vapor Pressure."

Ryznar stability index (RSI) A measure of water fouling potential. Refer to the chapter on utilities.

S

S Sulfur

S/D Shutdown

S/U Startup

S_8 Sulfur – liquid form

Safe park A condition in which a process unit can be held indefinitely without damage to the equipment or catalyst

Safety data sheet See "Material Safety Data Sheet"

Safety integrated system (SIS) AKA safety integrity system, safety instrumented system, or safety interlock system. This is a system that takes secure process information parallel to other control systems and applies specific limits to the measured variables. If one of the variables goes outside the predefined limits, the system first alarms at low deviation and then takes a predefined set of emergency actions at high or sustained deviation. The system is similar to a PLC but is normally triple redundant to ensure high integrity.

Safety shower A deluge system, usually located near a potential chemical exposure area, where personnel can be washed down with copious amounts of clean water to eliminate or control damage from chemical exposure. Refer to the safety chapter.

Safety systems These are the systems that prevent process and other safety incidents in a refinery. Safety systems include fire protection facilities and equipment, emergency shutdown systems, emergency depressuring systems, fired heater safety interlocks, and many other systems. These systems are designed, operated, maintained, and managed to prevent fire, explosion, loss of containment, and other hazardous situations. The chapters on "▶ Safety Systems for Petroleum Processing" and "▶ Fire Prevention and Firefighting in Petroleum Processing" in this handbook discuss the systems in detail. The handbook topic "▶ Hazardous Materials in Petroleum Processing" further highlights the properties and safe handling of several hazardous materials commonly found in refineries.

SAGD Steam-assisted gravity drain. A production process for bitumens and extra-heavy oils where high-pressure, superheated steam is injected into a horizontal well. The hot fluids reduce the viscosity of the oil in the formation and push it toward a lower depth horizontal well from which it is produced. See the discussion on this process in the chapter on unconventional crudes.

Sanitary sewer Sewer system that accepts domestic waste (toilets, showers, etc.) and conveys it to treatment

SAP (1) Sulfuric acid plant; (2) common oil industry enterprise software accounting system offered by SAP AG in Germany

SARA US Superfund Amendments and Reauthorization Act

Saybolt Furol viscosity A viscosity test similar in nature to the Saybolt Universal viscosity test but one more appropriate for testing high-viscosity oils. Certain transmission and gear oils and heavy fuel oils are rated by this method. The results obtained are approximately one tenth the viscosity which would be shown by the Saybolt Universal method.

Saybolt Universal viscosity A measure of kinematic viscosity. The methods determines the time required for 60 cm^3 of oil to flow through a standard, calibrated tube at a controlled temperature. This method is used for oils with flowing time up to 5,600 s (low to medium viscosity). The results are reported as Saybolt Universal Seconds (SUS or SSU). For higher-viscosity oils, the Saybolt Furol viscosity method is used. See "Saybolt Furol Viscosity."

Saybolt viscosity See "Saybolt Universal Viscosity."

SC/RC Sour crude/resid conversion

SCADA Supervisory control and date acquisition

Scavenger In refining, this refers to a chemical used to remove the final traces of an undesirable compound from a material. Examples include oxygen scavengers to remove traces of oxygen from boiler feed water and H_2S scavengers to remove H_2S for safety before shipping or handling an oil.

SCBA Self-contained breathing apparatus

SCC Stress corrosion cracking

SCFD or scfd or scf/d Standard cubic feet per day (1 Atm, 60 °F)

SCFM or scfm or scf/m Standard cubic feet per minute (1 Atm, 60 °F)

SCO (1) Synthetic crude oil; (2) stabilized crude oil

SCR Selective catalytic reduction – NO_x control by reacting ammonia with NO_x over a catalyst in a fired heater flue gas stream

SCSR Self-contained self-rescuer

SD or S/D (1) Shutdown; (2) sustainable development; (3) stream day

SDWA US Safe Drinking Water Act

SE Stack effect – in a fired heater – drives the heater draft profile

Seal Device for preventing loss of fluids past a rotating or reciprocating shaft. Refer to the equipment chapter sections on pumps and compressors. Includes packing, mechanical seals, liquid seals, and gas seals.

SEC US Securities and Exchange Commission

Selective catalytic reduction (SCR) NO_x control by reacting ammonia with NO_x over a catalyst in a fired heater flue gas stream

Selective non-catalytic reduction (SNCR) NO_x control by thermally reacting ammonia with NO_x in a fired heater flue gas stream

Sensing element The field device that measures a physical or chemical property to be used in controlling the process. The most common elements measure things like flow, temperature, pressure, and level.

Sensor See "Sensing Element."

Separator A vessel used to effect separation of a mixture phases into separate phases, such as vapor separation from liquid or water separation from vapor and oil. Refer to the equipment chapter, section on drums and vessels.

Service factor (1) In refinery terms, the service factor is the percentage of the time that a given type of process unit is expected to be operable during a year. The service factor depends on the average amount of scheduled shutdown time the particular unit requires over 1 year for maintenance. Typical service factors are listed in Table S.1 below. (2) In electric motors, from a practical standpoint, the service factor indicates the amount of current a motor can draw over a sustained period and still maintain reasonable life. A service factor of 1.0 means the motor can only provide its design power reliably. A motor with a service factor of, say, 1.15 means the motor could draw about 15 % higher amps that rated without major damage.

Setpoint The desired value of a process variable that is provided to a controller. The controller tries to make the process measurement match the setpoint.

Set pressure Set pressure, in psig or barg, is the inlet pressure at which the pressure relief valve is adjusted to open under service conditions.

Setting (1) The setpoint of a controller (or act of changing the setpoint). (2) In a fired heater, all parts that form the coil supports and enclosure (housing). Refer to the equipment chapter, on fired heaters.

SEU Solvent extraction unit

Sewer systems Generally underground piping systems that collect wastewater or chemicals and transport them to treatment facilities. Sewers may include storm runoff, oily water, chemical wastes, and others. Refer to the discussions in the chapters on utilities, offsites, and environmental engineering.

SGP Saturate gas plant

SGR Saturate gas recovery

Shale crude or gas Not the same as shale oil. This is crude or gas produced by hydraulically fracturing tight shale formations to allow crude and gases in the formation to be recovered, e.g., Bakken or Eagle Ford crudes.

Shale oil Not the same as shale crude. This is oil produced by destructive distillation of a rock called kerogen, either in place (in situ) or after mining the rock

Table S.1 Some typical service factors for refinery processes

Unit	Service factor percent
Crude distillation unit	95–98
Light ends distillation	98
Vacuum crude distillation	95–98
Visbreaker and thermal cracker	90
Cat reformer	90–92
Naphtha hydrotreaters	90–92
Gas oil hydrotreaters	90
Fluid catalytic crackers	85–90
Hydrocrackers (distillate feed)	90–95
Hydrocrackers (residue feed)	80–82 (includes residue hydrotreating)
Cokers	85

(ex situ). This oil has low sulfur but high nitrogen, arsenic, and olefins. Not generally available on the market

SHE Safety, health, and environment

Shield tube See "Shock Tube."

Shell In a shell-and-tube exchanger, the chamber surrounding the tubes on the outside which contains the shell-side fluid

Shell-and-tube heat exchanger An exchanger consisting of an outer pressure shell containing multiple, small tubes inside the shell. A channel directs flow in and out of the tubes. Baffles direct flow around the outside of the tubes. Refer to the discussion in the chapter on equipment, section on heat exchangers.

Shell storage capacity The design capacity of a petroleum storage tank which is always greater than or equal to working storage capacity

Shock tube or coil The first rows of tubes in the convection section of a fired heater that is exposed to the highest temperature flue gases. Normally not finned. May be exposed directly to radiant heat, also

Short ton An avoirdupois measure of weight equal to 2,000 lb

Short-term exposure limit (STEL) The maximum average concentration of a compound which is considered a safe (negligible health hazard) exposure for a short period of time, normally 15 min

Shot coke Spherically shaped petroleum coke that is difficult to control in a coker. Causes equipment damage and potential safety issues

SIC US Standard Industrial Classification Code

Sidestream A stream that is pulled from a distillation tower between the overhead and bottoms. See also "Sidestream Stripping."

Sidestream stripping Sidestreams from multicomponent distillation towers are stripped free of entrained lighter products in "sidestream strippers" (or side strippers). Stripping may be accomplished by injection of steam through the hot sidestream in a trayed column, by injection of an inert gas instead of steam, or by reboiling the bottom product of the stripper tower. Steam stripping is probably the most common approach. The most common application is on the atmospheric crude distillation unit. Figure S.1 is an example of the bottom distillate sidestream stripper of a crude distillation unit.

Sieve tray A type of tray used in distillation, absorber, or stripper columns that promotes contact between vapor and liquid on the tray using holes distributed over the tray

SIF Safety instrumented function

SIL Safety integrity level

SimDist Distillation curve (vol% distilled vs. temperature) using ASTM D2887 method normally. This method is for all ranges of hydrocarbons and employs a gas chromatograph.

SIP In the United States, a State Implementation Plan for reduction of air emissions. See the environmental chapter.

SIS (1) Safety instrumented system; (2) safety interlock system; (3) safety integrity system

SIT Spontaneous ignition temperature. The temperature at which an oil ignites of its own accord in the presence of air or oxygen under standard conditions

Skirt Generally cylindrical supports for vertical vessels and towers. The skirt holds the vessel at the proper elevation and connects the vessel to the foundation. A skirt is normally about the same or slightly smaller diameter as the vessel or tower it supports.

Slack wax Mixture of oil and wax, a by-product from the refining of lubricating oil, from which paraffin wax can be recovered

Slagging Formation of hard deposits on boiler tubes and/or piston crowns, usually due to the presence of sodium, vanadium, and sulfur

SLC Single-loop controller

Slops Also called recovered oil and other site-specific terms. Oil that is collected from leaks, off-spec production, equipment flushing, etc., that does not meet any product specifications or may be contaminated. The oil may be reprocessed or otherwise managed. Refer to the chapter on utilities.

Sludge Deposits in fuel tanks and caused by the presence of wax, sand, scale, asphaltenes, tars, water, etc. The "sludge" formed in a #6 fuel oil storage tank is mostly composed of heavy hydrocarbons.

Slug flow Piping flow regime that is an extreme form of wave flow where the waves are periodically picked up and "slugged" through the pipe. This regime is undesirable as it causes excessive pipe movement and hammer that is potentially damaging. Refer to the equipment chapter, section on two-phase pressure drop.

Slurry oil The heaviest oil product from an FCC

Smoke point A measure of kerosene tendency to smoke. Measured in millimeters of flame height before smoking. Relates to turbine blade path temperatures in a jet engine. Specific ASTM test.

SMR Steam-methane reformer

SMTP Simple Mail Transfer Protocol (Internet e-mail)

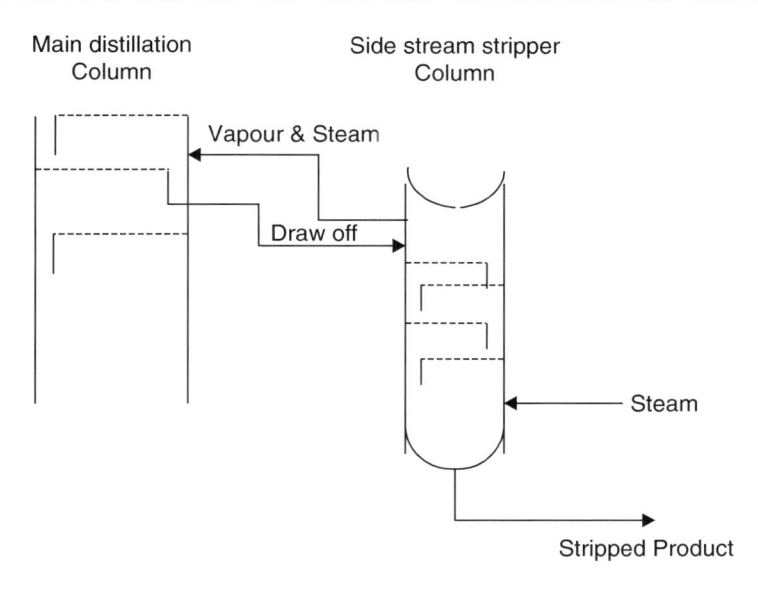

Fig. S.1 Sidestream steam stripper

SNCR Selective, non-catalytic reduction – NO_x control by thermally reacting ammonia with NO_x in a fired heater flue gas stream

Soaking volume factor (SVF) The design of a thermal cracker is keyed to the configuration and temperature profile across the heater and soaking drum or soaking coil. The degree of cracking is dependent on this temperature profile and the residence time of the oil under these conditions. The soaking volume factor (SVF) is related to product yields and the degree of conversion. Definition of these items is given in the chapter on "▶ Upgrading the Bottom of the Barrel" in this handbook. A design calculation using the SVF is given as Appendix 1 of that topic.

SOCMI Synthetic Organic Chemical Manufacturing Industry

Soda ash Sodium carbonate (Na_2CO_3)

Soft seat See "Resilient Seated."

SOL Safe operating limit – limit that, if exceeded for a short time, will result in failure including loss of containment

Solid waste A waste that is solid or semisolid (like a mud or gel)

Solomon Associates A company that does semiannual surveys of refining business and provides comparisons of refineries relative to each other

Soluble Matter or compounds capable of dissolving into a solution

Solvent A substance, normally a liquid, which is capable of absorbing another liquid, gas, or solid to form a homogeneous mixture

Soot blowing Removal of soot from the firebox side of a furnace by blowing high-velocity air through the furnace to loosen the soot. The released solids go out with the flue gas.

SOP Standard operating practice or procedure

SOR (1) Start of run; (2) statement of requirements

Sorbent Generic term for material used to remove specific components from a stream. May refer to absorbents (chemical or solution removal) or adsorbents (surface adsorption)

Sour crude Crude with high sulfur, typically >0.5–0.8 w% sulfur

Sour water stripper (SWS) A process which removes H_2S and ammonia from sour waters generated in a refinery by distillation. Refer to the environmental chapter section on air emissions.

SOW Scope of work

SO$_x$ Sulfur oxides – generically

SOX US Federal Sarbanes–Oxley Legislation

SP (1) Set pressure. (2) Setpoint

SPA Solid phosphoric acid

Spalling Removal of scale deposits and coke from heater tubes by forcing the materials to flake off the tubes using temperature changes

Sparing Providing more than one piece of equipment for a service to enable continued operation in the event of single equipment failure

SPC Statistical process control

SPCC Spill prevention, control, and countermeasures

Spec. sheet Specification sheet. Detailed information about a product or equipment

Special naphthas All finished products within the naphtha boiling range that are used as paint thinners, cleaners, or solvents. These products are refined to a specified flash point. Special naphthas include all commercial hexane and cleaning solvents conforming to ASTM Specification D1836 and D484, respectively. Naphthas to be blended or marketed as motor gasoline or aviation gasoline or that are to be used as petrochemical and synthetic natural gas (SNG) feedstocks are excluded.

Specific gravity The specific gravity of a liquid is the weight of a known volume of the liquid at a known temperature compared with water under the same conditions. The standard weight is taken as 1 g, and the standard temperature is usually 60 °F or 15 °C. The specific gravity of a petroleum compound is the basis for development of the material balance in design work and most measurements within the refinery. Abbreviated as Sp.Gr, SpGr, or SG

The basic specific gravities are given as an essential part of the crude assay. They are usually presented as a curve of specific gravities (usually quoted as °API) against midpoint distillation temperatures. API gravities are related to specific gravities by the equation: specific gravity $= 141.5/(131.5 + °API)$. The specific gravity of any petroleum compound may be calculated using the method provided in the topic "Introduction to Crude Oil and Petroleum Processing" in this handbook.

The specific gravity of crude oil and its products are obtained in the refinery using the test method described in the chapter on "▶ Quality Control of Products in Petroleum Refining" in this handbook. The method uses a properly calibrated hydrometer under laboratory conditions.

Specific heat The quantity of heat required to raise the temperature of a unit weight of a substance by 1°; usually expressed as calories/gram/°C or Btu/lb/°F

Specification (1) Detailed engineering definition of requirements for a product, equipment, or facility; (2) Term referring to the properties of a given crude oil or petroleum product, which are "specified" since they often vary widely even within the same grade of product. In the normal process of negotiation, seller will guarantee buyer that product or crude to be sold will meet certain specified limits and will agree to have such limits certified in writing. Generally, the major qualities of oil for which a buyer would demand a guarantee are API gravity (or specific gravity, in some cases), sulfur content, pour point, viscosity min/max, BS&W, etc.

SPL Sound pressure level. Refer to the environmental chapter section on noise.

Splitter, naphtha In all hydroskimming refineries, the key process next to the crude distillation process is the catalytic reformer. The correct design and subsequent operation of this process produces the hydrogen stream that is required by many refinery operations. Important to the efficient operation of this process is the correct boiling point range of the naphtha feed. This is ensured by the fractionation of the full range naphtha stream from the crude unit overhead distillate. This is accomplished as part of the light ends unit complex. Typically, the total overhead distillates plus, in some cases, other naphtha distillates (from thermal crackers) are first debutanized in the light ends debutanizer column. The bottom product from this column is the debutanized full range naphtha. This stream is delivered hot to a naphtha splitter fractionator which produces a light naphtha overhead and a heavy naphtha bottom product. The fractionation between these two products maximizes the naphthene content of the heavy naphtha. As this heavy naphtha is fed to the catalytic reformer, the amount of naphthenes in its composition will, to a large extent, determine the amount of hydrogen the unit will produce. Splitter towers contain between 25 and 35 actual distillation trays and operate at overhead reflux ratios of between 1.5 and 2.0. Further details and description are given in the topic "▶ Distillation of the "Light Ends" from Crude Oil in Petroleum Processing" in this handbook.

SPOC Single point of contact

SPR See "Strategic Petroleum Reserve."

Spot price The current value of any product on a volume basis on the open market

Spread In futures markets, applies to the difference between prices of futures contracts for different delivery months or to the difference in prices for different commodities. Spread traders try to capitalize on likely fluctuations in these relationships, and initial spread margins are often considerably lower than for outright positions.

SQC Statistical quality control
Squat A short, unplanned shutdown for repairs
SR1 Seconds Redwood # 1 at 100 °F. Measure of viscosity
SRD Safety and Reliability Directorate (UK Atomic Energy Authority)
SRHGO Straight-run heavy gas oil
SRU Sulfur recovery unit or sulfur plant
SRV Safety relief valve
SRxxx See "Straight Run."
SS Stainless steel
SSDC Single-station digital controller
SSF Seconds Saybolt Furol – a measure of viscosity. See "Saybolt Furol Viscosity."
SSSC Solid-state sequence controller
SSU Saybolt Seconds Universal. A viscosity measure. See "Saybolt Universal Viscosity."
ST Short ton = 2,000 lb.
Stabilize To convert a stock to a form that is stable to change or safe for storage. For example, stripping of light or sour gases from a hydrocarbon stock before sending it to tankage
Stack(s) Stacks are used to create an updraft of air from the firebox of a fired heater. The purpose of this is to cause a small negative pressure in the firebox and enable the introduction of combustion air from the atmosphere to the burners. This negative pressure also allows for the removal of the products of combustion from the firebox. The stack must have sufficient height to achieve these objectives and overcome the frictional pressure drop in the firebox and the stack itself. Refer to the handbook topic "▶ Process Equipment for Petroleum Processing" for the discussion of fired heater design.
Staged combustion Introduction of fuel to a burner in stages to reduce NO_x emissions
Static mixer A motionless mixer which has a series of fixed, geometric elements enclosed within a tubular housing. The internal elements impart flow division and radial mixing to materials flowing through the housing to produce a uniform mixture of the materials.
Steam and condensate systems In most plants, steam condensate, accumulated in the various processes, is collected into a single header and returned to the boiler or steam-generating plant. It is stored separately from the treated raw water because condensates may contain some oil contamination. A stream of treated water and condensate are taken from the respective storage tanks and pumped to a deaerator. The condensate stream passes through a simple filter on route to the deaerator to remove any oil contamination. Low-pressure steam is introduced immediately below a packed section of the deaerator and flows upwards countercurrent to the liquid stream to remove any entrapped air in the liquid. The deaerated boiler feed water (BFW) is pumped by the boiler feed water pumps into the steam drum of the steam generator. The steam drum is

located above the generator firebox. The liquid in the drum flows through the steam generator coils located in the firebox by gravity thermosyphon or is circulated by pump. A mixture of steam and water is generated in the coils and flows back to the steam drum. Here, the steam and water are separated with the steam leaving the drum to enter the superheater coil. This coil, located in the lower section of the convection side of the heater, superheats the saturated steam to the high-pressure refinery steam mains. Letdown stations may be located at various points in the refinery to create lower pressure main systems. Desuperheaters are used to establish the correct temperature levels in these lower-pressure mains. There are also normally waste heat recovery steam generation systems and cogeneration systems that use boiler feed water and contribute to the total plant steam production. The handbook topic "▶ Utilities in Petroleum Processing" gives details of about steam generation systems and condensate recovery.

Steam chest In a steam turbine or exchanger, the chamber where steam is initially admitted to the equipment

Steam drum A drum for separating steam from boiler water. The steam is produced from the drum and the boiler water is recirculated through one or more heat sources and back to the drum. There are internal steam separators inside the drum.

Steam-methane reforming (SMR) The most common process for making hydrogen or synthesis gas ($CO + H_2$) by reacting steam with hydrocarbon at very high temperatures and moderate pressure over a catalyst inside tubes suspended in a furnace. The hydrocarbon is converted to methane before or as it is process, hence the term steam-*methane* reforming.

Steam out Using steam to purge volatile materials and other contaminants from a piece of equipment or piping

Steam turbine A driver that uses steam pressure letdown and possible condensation to translate energy in the steam to rotational mechanical energy

STEL Short-term exposure limit

Stiff–Davis index (SDI) A measure of water fouling potential. See the utilities chapter.

Still gas (AKA refinery gas) Any form or mixture of gases produced in refineries by distillation, cracking, reforming, and other processes. The principal constituents are methane, ethane, ethylene, normal butane, butylenes, propane, propylene, etc. Still gas is used as a refinery fuel and a petrochemical feedstock. The usual conversion factor is 6.05 million Btus per fuel oil equivalent barrel.

Stoke The unit of kinematic viscosity

Stoke's law A mathematical relationship that describes the rate of movement of a particle or droplet through another continuous phase in laminar conditions. Refer to the environmental chapter section on water and the equipment chapter section on drums and separators.

Stop–check valve A valve normally used in steam systems on the steam drum which serves as a check valve and block valve for steam production off the drum

Storage facilities The term "storage facilities" generally refers to the tankage or tank farm areas of a petroleum processing facility. Please refer to the topic "▶ Off-Site Facilities for Petroleum Processing" in this handbook.

Stove oil Canadian term used to describe kerosene. Stove oil in Canada is the equivalent of US low-sulfur No. 1 oil or kerosene.

STP Standard temperature and pressure – in US refining, normally 60 °F and 1 Atm. In the metric system, normally 15 °C and 1 atm

STPD Short tons per day

Straight run or straight-run Refers to a petroleum product produced by the primary distillation of crude oil, free of cracked components. A crude cut "run straight from the still."

Strategic Petroleum Reserve (SPR) Petroleum stocks maintained by the US Federal Government for use during periods of major supply interruption

Stratification Phenomenon that occurs in mixed fuels or tanks that results in layers of materials in a tank with differing compositions

Stratified flow Flow regime that normally occurs in a horizontal flow when the gas and liquid separate as they flow through long horizontal pipes. The gas moves at a higher velocity than the liquid, with the liquid flowing along the bottom of the pipe. This is a regime that is a concern for vertical flow because it can cause slugging or pipe hammer. Refer to the equipment chapter, section on two-phase flow pressure drop.

Stream day Actual operating day where feedstock is being charged to a process unit. Stream days determine a unit's sizing capacity or rate. Stream days differ from calendar days by the amount of unit downtime. Stream days are always less than or equal to calendar days.

Stress relieve (or relief) Controlled heating and cooling of a metal to relieve residual stresses after forming or welding

Stripping steam ratio Normally, lb steam per barrel of product in a steam stripper

Stuffing box On rotating equipment, this is a chamber surrounding the shaft in which the packing or other sealing system is installed (or "stuffed" in the old vernacular). Refer to the chapter on equipment for details.

SU or S/U Startup

Sub-octane Usually applies to a gasoline that does not meet the 87-octane standard which most suppliers mandate for regular unleaded distinction. Sub-octanes are typically utilized by those using oxygenated components.

Sulfidation Reaction of sulfur with the surface of a metal

Sulfiding Activation of a hydroprocessing catalyst by reacting the catalytically active metals with sulfur to create the active sulfide form of the metals

Sulfiding chemical Refers to several types of sulfur-containing chemicals used to activate petroleum processing catalysts or temporarily suppress cracking activity in reformers. The most common sulfiding chemicals are DMS, DMDS, TBPS, TNPS, and H_2S.

Sulfinol A proprietary Shell Oil system that uses a regenerable solution containing both a chemical (DIPA) and physical solvent (sulfolane) to remove acid gases

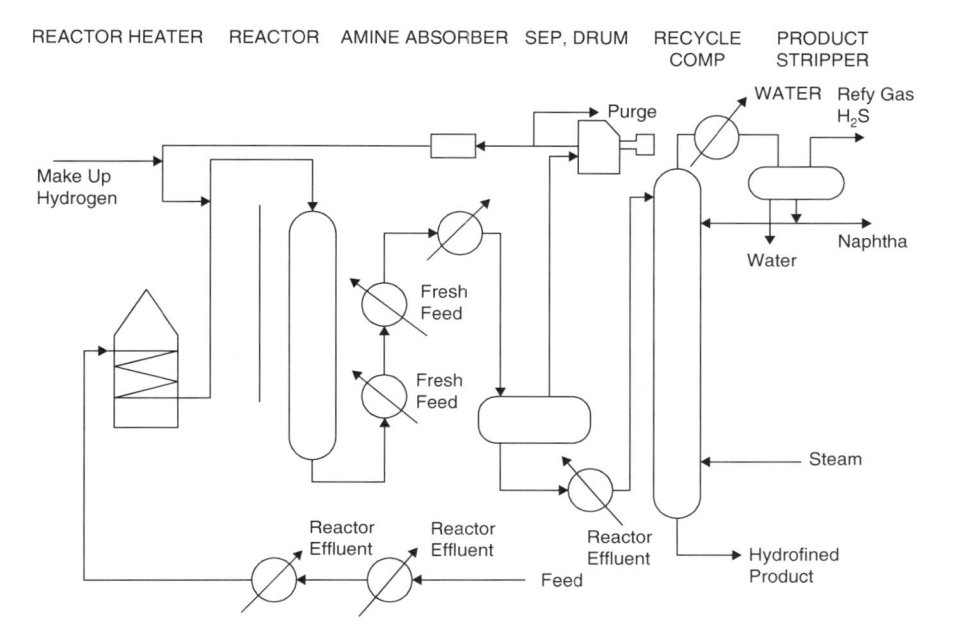

Fig. S.2 A typical middle distillate desulfurizer

(H_2S, CO_2) and some nonacidic gases (COS, RSH) from vapor streams. See the chapter on refinery gas processing.

Sulfolane (($CH_2)_4SO_2$) Tetrahydrothiophene-1-1-dioxide. A physical solvent component used for sulfur removal from foul gases. A component of sulfinol solution

Sulfur An element that is present in crude oil and natural gas as an impurity in the form of its various compounds.

Sulfur content Sulfur content of a petroleum product or cut is always quoted as a percent or parts per million by weight of the sample. One common laboratory test for sulfur content is "the lamp method ASTM D1266"; this is described in detail in the chapter on "▶ Quality Control of Products in Petroleum Refining" in this handbook. This method has been largely replaced by X-ray fluorescence analysis and other instrumental methods of analysis.

Sulfur removal Removal of sulfur from a petroleum stock. This is normally accomplished by hydrotreating, which converts the organic sulfur compounds to H_2S gas for separation from the hydrocarbons. Some sulfur compounds (notably mercaptans in light hydrocarbons) can also be removed by contact with caustic soda solutions (sweetening). Refer to the chapter "▶ Hydrotreating in Petroleum Processing" in this handbook for the hydrotreating discussion and the chapter "▶ Refinery Gas Treating Processes" for gas sweetening. Figure S.2 illustrates the flow sheet for a simple distillate hydrotreater for sulfur removal.

Sulfuric acid Common aqueous acid used in refineries for pH control and as a catalyst. H_2SO_4. Refer to the chapter on hazardous materials for additional information.

Sump Low spot for collection of a liquid and from which a pump suction may be taken. Sumps may collect liquid from vapors or from other, less dense, liquids.

Superheated A vapor that is heated to a temperature above its saturation temperature. For example, steam at atmospheric pressure and 450 °F is said to be superheated by 450–212 = 238 °F.

Superheater A coil or heat exchanger that provides additional sensible heat to steam above the boiling point

Superimposed back pressure Static pressure existing at the outlet of a pressure relief device at the time the device is required to operate. It is the result of pressure in the discharge system from other sources.

Supervisory control Basic control system for operating units

Surfactant Surface-active agent. The active agent in detergents that possesses a high cleaning ability. A surfactant enables hydrocarbons to be taken into solution or suspended in water by linking the hydrocarbon with the aqueous phases. A surfactant can also make emulsion breaking difficult.

Surge vessel A vessel intended to provide surge volume in a process. Refer to the chapters on equipment and controls.

Surge volume This is the volume of liquid between the normal liquid level (NLL) and the bottom (tan line) of a vessel. Refer to the equipment and controls chapters.

SUS Saybolt Universal Seconds viscosity measure. See "Saybolt Universal Viscosity."

SU/SD Start up/shut down

SV Safety valve – pressure relief valve

SVP Senior vice president

Sweet crude Crude with low sulfur, usually less than 0.5–0.8 w% sulfur

Swing capacity In crude planning, the total crude storage capacity plus seasonal storage. See the chapter on economics.

Swoopdown A short, unplanned shutdown for repairs

SWOT Strengths, weaknesses, opportunities, and threats analysis

SWP Safe work practice

SWPPP Stormwater Pollution Prevention Plan

SWS Sour water stripper or stripping – recovers H_2S and ammonia out of sour water by driving them out with heat. Refer to the environmental chapter section on air emissions control.

SWSOG Sour water stripper overhead gas

SynBit A blend of synthetic crude and bitumen. Some Canadian crudes fall in this category being blends of bitumen from tar/oil sands and partially refined bitumens.

Synthetic crude (or Syncrude) A crude-like hydrocarbon produced by methods that are not conventional. Basically an unconventional crude. Usually involves taking a raw, nonconventional crude and pre-refining it to remove the worst characteristics that would prevent its sale as a high-quality crude. Effort is made to give the Syncrude properties like a conventional crude. Can then be processed like a conventional crude

S Zorb Sulfur removal process for naphtha and distillates

T

T&C (or Ts and Cs) Terms and conditions in a contract
T&M Time and materials
T/A or TA Turnaround
TA (H, HH, L, LL) Temperature alarm high, high–high, low, or low-low. E.g., TAHH is a high–high temperature alarm.
Tag-Robinson colorimeter An instrument used to determine the color of oils. Also a scale of color values
Tail gas treating Generally refers to processing Claus sulfur recovery plant effluent gas to remove the final traces of sulfur compounds before emission to the air. Refer to the environmental chapter section on air emissions control.
Tallow Animal fat. See the discussions on renewables and unconventional stocks.
TAME Tertiary-amyl methyl ether – oxygenate blendstock for gasoline
TAN Total acid number. A measure of oil corrosivity due to carboxylic acids. Expressed as mg KOH per 100 g oil
Tank farm An installation used to store crude oil and products.
Tanker Vessel used to transport crude oil or petroleum products
Tar Tar is an ill-defined general term that describes heavy petroleum fractions that are solid or semisolid at room temperature. An alternative term is *bitumen* although the latter is better used to denote naturally occurring tar deposits, as in tar pits or tar sands. In addition to being very viscous or non-flowing materials (viscosity >10,000 cP), tars are also characterized by having relatively high densities lower than about 10 API degrees, which corresponds to a specific gravity (60/60) greater than 1.0. Depending on their physical properties, tars may be easily confused with asphalts. The main difference being that tars are usually either naturally occurring or unprocessed heavy fractions recovered as by-products from other sources (e.g., petroleum residues or coal processing), while asphalts are typically processed or manufactured materials, whether air-blown or solvent extracted. Sometimes, tars are further processed to recover the more volatile components. If so, the remaining very heavy residue is usually called *pitch* (Gerd Collin, "*Tar and pitch.*" *Ullmann's Handbook of Industrial Chemistry*, Wiley-VCH Verlag GmbH, 2002). See also the chapter on "▶ Upgrading the Bottom of the Barrel" in this handbook.
Tar sands Tar sands or bituminous sands or oil sands are several porous rock formations that contain highly viscous heavy hydrocarbon materials that cannot be recovered by conventional oil recovery methods, including enhanced oil recovery techniques. These hydrocarbon resources and the oils that come from them are discussed in the topic "▶ Unconventional Crudes and Feedstocks in Petroleum Processing."
Tariff A schedule of rates that a common carrier pipeline is permitted to charge to transport petroleum products or crude
TBA (1) Tertiary butyl alcohol; (2) tertiary butyl amine; (3) in marketing: tires, batteries, and accessories

TBN Total base number. ASTM D2896. This is measured in mg of KOH needed to neutralize an acidic solution through a reverse titration. TBN is the ability of the product to neutralize acid. In a motor oil, this is a property which allows the oil to neutralize acids from combustion that would otherwise degrade the oil.
TBP True boiling point
TC Temperature controller
TCC Thermal catalytic cracking unit
TCF Trillion cubic feet
TCU Thermal cracking unit
TDAF Tertiary dissolved air flotation
TEA See "Triethanolamine"
TEG Triethylene glycol
TEL Tetraethyl lead – octane booster – no longer used in United States. See "Tetraethyl Lead."
Tetraethyl lead (TEL) A former additive used as an antiknock additive to boost the octane number in gasoline. It was produced commercially from ethyl chloride. It was used at a dosage in 0–3 ml/gal range depending on the gasoline composition, sulfur content, and lead sensitivity. The use of tetraethyl lead has largely been discontinued worldwide, in particular, as a result of the introduction of catalytic converters to clean up the exhaust from internal combustion engines. The use of lead irreversibly poisons the oxidation catalysts used in such converters. Also, lead itself is a highly toxic substance that was present in automobile exhaust before it was banned with the advent of catalytic converters.
TEMA Tubular Exchanger Manufacturers Association – sets standards for heat exchangers
Tertiary-amyl methyl ether (TAME, $(CH_3)_2(C_2H_5)COCH_3$) An oxygenate blendstock formed by the catalytic etherification of isoamylene with methanol
Tertiary butyl alcohol (TBA, t-butanol, $(CH_3)_3COH$) An alcohol primarily used as a chemical feedstock, a solvent or feedstock for isobutylene production for MTBE; produced as a coproduct of propylene oxide production or by direct hydration of isobutylene
Tertiary storage Refers to petroleum storage tanks of end users, such as vehicle gasoline tanks or home heating oil storage
Test run A defined period of process unit operation where detailed performance data, sample analyses, and other information are recorded. The data are analyzed to define a process unit and equipment performance at a point in time. Test runs can be used to evaluate performance for guarantees, decisions on catalyst, or to determine alternate operations options, among many other uses. They also establish baseline equipment performance in new facilities.
TF Tank farm
TFE or PTFE Polytetrafluoroethylene
TGU or TGT Tail gas unit or tail gas treating unit. Refer to the chapter on environmental controls, air emissions section.

Thermal cracking Thermal cracking was the first commercial process used for the conversion of heavy petroleum fractions into more useful products. Though largely superseded by other processes (in particular catalytic cracking), thermal cracking was used for many years for the decomposition (cracking) of heavy, high-molecular-weight hydrocarbons into smaller molecules and is still used commercially in the processing of very heavy fractions, as in visbreaking or coking. Refer to the chapter on "▶ Upgrading the Bottom of the Barrel" in this handbook for additional discussion.

Thermal reforming Thermal reforming is similar to thermal cracking applied to gasoline boiling range hydrocarbons. The smaller molecules in naphtha are more difficult to crack and require higher severities than thermal cracking, with furnace outlet temperatures of up to about 600 °C. Good per-pass conversions and good octane improvements can be obtained while coke formation is limited because of the lighter nature of the feedstock. An excellent review of thermal processes, both cracking and reforming, can be found in the Petroleum Processing Handbook, edited by John J. McKetta, Marcel Dekker (1992). To a large extent, thermal reforming has been superseded by catalytic reforming and thermal cracking by fluidized catalytic cracking (FCC).

Thermal relief A pressure relief valve intended to avoid overpressure of a liquid-filled line due to external heating.

Thermal value Calories per gram or Btu per pound produced by burning fuels

Thermocouple The most common temperature measuring element. Takes advantage of the tiny voltage generated at a dissimilar metal junction that changes with temperature in a defined manner. See the chapter on process controls.

Thermofor The commercial name of a continuous moving bed process used either for catalytic cracking or catalytic reforming. A distinctive feature of this process is that the catalyst, usually chromia/alumina, flows down through the reactor concurrently with the hydrocarbons. A mechanical conveying system is used to circulate the catalyst back to the top of the reactors. The process is not common today. For more information, refer to James G. Speight and Baki Ö zu̇"m, *Petroleum Refining Processes*, Marcel Dekker, 2002.

Thermosiphon or thermosyphon An equipment arrangement, often used for reboilers, which uses the density difference created by heating a fluid to induce natural convection flow in the fluid. Refer to the section on heat exchangers in the chapter on equipment.

Thermowell A device inserted into a process in which the temperature measuring element is inserted. The thermowell protects the temperature element from damage by the process fluids.

Threshold limit value (TLV) The average concentration of a compound which is considered a safe (negligible health hazard) working environment over an 8 h period

Throttle valve A valve normally controlling steam rate entering a steam turbine or gas rate entering a compressor. Refer to the equipment chapter.

Throttling Controlling flow rate by a restriction, such as a valve

TIC Total indicated cost – in a project
Tight oil or tight shale oil See "Shale Crude and Gas." Oil produced from shale formations with negligible natural permeability. The rock must be fractured to create permeability before oil or gas can be produced.
TLV Threshold limit value
TMEL Targeted mitigating event likelihood – in LOPA
TOC Total organic carbon – in wastewater
Toluene ($C_6H_5CH_3$) Colorless liquid of the aromatic group of petroleum hydrocarbons, made by the catalytic reforming of petroleum naphthas containing methylcyclohexane. A high-octane gasoline-blending agent, solvent, and chemical intermediate base for TNT
Ton 2,000 lb in English engineering units
Tonne 1,000 kg or 1 metric ton. 2,205 lb. On average, there are 7 barrels of crude per metric ton.
Topped crude oil Oil from which the light ends have been removed by a simple refining process. Also referred to as "reduced crude oil"
Topping "Topping" or "skimming" is the name used for the distillation of crude oil to remove the lighter fractions. The crude oil with such fractions removed is sometimes called "topped crude." Some refineries are only designed to be topping units, but these facilities are severely disadvantaged by limited crude options and poor product quality. Very few simple topping plants are in operation any more.
Total existent sediment Combination of inorganic and hydrocarbon sediments existing in a stock as delivered
Tower Generally refers to a tall vessel with internals for fractionation, absorption, or stripping. Refer to the equipment chapter, section on vessel and tower design.
Tower fractionation A fractionation tower is a distillation column, typically with multiple trays or fractionation stages and with at least one feed and two product streams – top and bottom – but often having also provision for multiple feeds and multiple withdrawal points or side cuts. Design of these towers is discussed in the handbook topics on atmospheric and vacuum crude distillation units, light ends distillation, and process equipment design. The mechanical design of a fractionation tower is far more complex than just the specification of the operating conditions or the number of theoretical stages. Also needed are the specifications and designs for its internal components, such as dimensions, plates or packing, risers, downcomers, internal supports, distributor nozzles, reboilers, condensers, etc. Refer to the handbook topic "▶ Process Equipment for Petroleum Processing" for some details of tower internals.
Toxic Property of being poisonous to a living organism
Toxicity (1) The relative degree of being poisonous or toxic. A condition which may exist in wastes and will inhibit or destroy the growth or function of certain organisms. (2) Hazardous waste characteristic of being poisonous to living organisms. Refer to the environmental chapter section on hazardous wastes.
TPD Tons per day

TQ Threshold quantity

TR Temperature recorder

Trade secret Information that provides a company with an economic advantage. Proprietary information. Such information does not need to be disclosed except in an emergency.

Tramp compound A contaminant or undesired material in a process stream

Tramp air Air that enters a fired heater by leakage other than through the burners or stack cooling vents, wasting efficiency

Transesterification Reaction of a heavy polyorganic ester with a light alcohol to form lighter esters. See the discussion of renewables and unconventional stocks.

Transmix The interface material between shipments of different products through a pipeline. The interface does often not meet the specifications for either product, but may be blended into one product (generally downgrading) or reprocessed.

Transportation System used to move oil and gas from the field to the market

Trap A device for elimination of either condensed liquid from a vapor system (e.g., a steam trap to eliminate condensate from the steam header) or trapped gases from a liquid-filled system

TRC (1) Temperature recorder and controller; (2) total recordable cases (injuries)

TRI Toxic release inventory – environmental

Trickle filter A biological digestion unit for removal of BOD from wastewater by passing the water through a rock bed covered with organisms and exposed to air.

Triethanolamine (TEA, $(HOC_2H_4)_3N$) An early regenerable solvent used for removal of acid gases (mainly H_2S) from streams. Refer to the chapter on refinery gas treating. This material is not normally used for this service anymore because of the availability of better solvents.

Trip Unplanned shutdown caused by a process or equipment failure

True boiling point (TBP) The true boiling point distillation curve can be obtained in a laboratory apparatus. TBP distillation differs from an ASTM or *Engler* distillation. In process design, especially in fractionation, the TBP curve is critical. Various correlations exist to relate TBP, ASTM, and other laboratory distillation procedures. See the handbook topics "▶ Introduction to Crude oil and Petroleum Processing" and "▶ Atmospheric and Vacuum Crude Distillation Units in Petroleum Refineries." Another good reference is: Nelson, W. L. *Petroleum Refining Engineering*, McGraw-Hill, 4th edition, 1958.

TRV Thermal relief valve

TSAP EU Thematic Strategy for Air Pollution. Refer to the environmental chapter.

TSCA US Toxic Substances Control Act

TSV (1) Thermal or temperature safety valve – a relief valve for thermal expansion of fluids in a packed line; (2) total system value

TT Temperature transmitter

Tube hanger A device to support a tube in a fired heater

Tube sheet In a shell-and-tube exchanger, the plate where the tubes start and/or end and which isolates the tube-side fluid from the shell-side fluid

Tube skin The surface of a tube. Generally refers to furnace tube surface temperatures, which are normally the hottest part of the tube exposed to the firebox

Tube trailer A large trailer consisting of multiple high-pressure tubes (2,000–5,000 psig) used to supply relatively small specialty gas needs, such as hydrogen or nitrogen without requiring additional compression. Refer to the discussions in the chapters on utilities and hydrogen production.

Tube wall Literally, the wall of a tube. Usually applied in furnaces or exchangers

Turnaround A major, periodic shutdown maintenance period to inspect, repair, and renew equipment in a process unit

Turndown Reducing a unit's feed rate or production rate. Most units have a limited turndown capability before they become hard to control. Turndown may be described as % turndown; however, often what this number means is unclear: 33 % turndown sometimes means the plant can be run as low as 33 % of design rate or it can only be turned down to 67 % of design rate. It is always best to clarify what is meant.

Turnout gear In firefighting, these are the heavy, fire-resistant clothing and boots used to protect personnel

TUV Technischer Uberwachungs-Verein e.V. (Technical Inspection Association of Germany)

TW Thermowell

Two-phase Having two distinct phases, such as a mixture of liquid and vapor

U

Ubbehohde viscosimeter A suspended-level apparatus for accurately determining the viscosity of a liquid

UCARSOL A proprietary solution from Dow Chemical for absorbing acid gases. Refer to the environmental chapter on air emissions

UCC US Uniform Commercial Code

UFD Utility flow diagram

UL Underwriters Laboratory, Inc.

ULCC See "Ultra-Large Crude or Cargo Carrier."

Ullage The amount which a tank or vessel lacks of being full

ULNB Ultralow NO_x burner

ULS Ultralow sulfur – may apply to any product. ULSD = ultralow-sulfur diesel.

ULSD Ultralow-sulfur diesel – usually refers to diesel with <10 wppm sulfur

ULSK Ultralow-sulfur kerosene – usually refers to kerosene with <10 wppm sulfur

Ultra-large crude or cargo carrier (ULCC) The largest category of tanker, generally holding at least 3.5 million barrels of product

Unbranded A supply arrangement with a supplier that is usually not contractual and does not usually guarantee a specific amount of supply

Unconventional crude Crude that is produced by methods other than conventional – like shale oil that must be thermally decomposed from rocks or oil sands that must be thermally removed from sand via mining or special production methods. There are many types of unconventional crudes. Renewable fuels are in this category.

Underwood equation This correlation is used to estimate the minimum vapor flow above the feed in a distillation column. See the chapters on crude distillation, light ends processing, and equipment.

Unfinished oils All oils requiring further processing, except those requiring only mechanical blending. Unfinished oils are produced by partial refining of crude oil and include naphthas and lighter oils, kerosene and light gas oils, heavy gas oils, and residuum. AKA intermediates

Unfired vent stack See "Vent Stack."

UOP Originally Universal Oil Products – now UOP LLC, part of the Honeywell group. Independent refining process developer and licensor

UOP K (AKA Watson K) The *UOP K*, or the *Watson K*, or the *Watson characterization factor*, is a parameter identified by Kenneth Watson who defined it as

$$K = [T_B]^{1/3}/(Sp.Gr.)$$

where $[T_B]^{1/3}$ represents the cube root of the average molal boiling point of the hydrocarbon mixture $[T_B]$ in degrees Rankine ($°R = °F + 460$), divided by the specific gravity at 60 °F, relative to water at 60 °F. This is a correlation parameter based on the observation that $K \sim 12.5$ corresponds to paraffinic materials, while $K \sim 10.0$ indicates a highly aromatic material. It provides a means for roughly identifying the nature of a feedstock solely on the basis of two observable physical parameters. The characterization factor has also been related to viscosity, aniline point, molecular weight, critical temperature, percentage of hydrocarbons, etc., so it can be estimated using a number of laboratory methods. Refer also to (1) Characterization of petroleum fractions, *Ind. Eng. Chem.*, **27**, 1460, 1935 and (2) Nelson, W. L., Petroleum Refinery Engineering, McGraw-Hill, 4th edition, 1958.

UPS Uninterruptible power supply – a battery-backup power supply

Upstream Organization that finds and produces crude oil and natural gas

Urea dewaxing A process for making low pour point oils in which straight-chain paraffins are removed from the feedstock by complexing them with urea to form a crystalline adduct that can be separated by filtration. Refer to the chapter on "▶ Non-energy Refineries in Petroleum Processing" in this handbook and John J. McKetta (ed), *Petroleum Processing Handbook*, Marcel Dekker, 1992.

USD United States dollar

UST Underground storage tank

Utility Refers to fluids used by several pieces of equipment in refinery units – steam, air, water, cooling water, instrument air, power, nitrogen, etc.

Utility air See "Plant Air."

Utility station or utility drop A set of outlets provided at selected locations in a plant where utility connections are available. A utility station normally has plant air, plant water, and steam available. Some stations may have nitrogen or other utilities available. Refer to the chapter on utilities.
Utility water See "Plant Water."
UV Ultraviolet

V

V Vanadium

Vacuum distillation unit (VDU) Vacuum distillation unit refers to the further distillation of the residue portion of atmospheric distillation of the crude. The boiling curve range of this portion of the crude is too high to permit further vaporization at atmospheric pressure. Cracking of the residue would occur long before any temperature level for effective distillation would be reached. By reducing the pressure, the danger of cracking on further heating the residue oil for further distillation is reduced. Figure V.1 is a process diagram of a typical crude oil vacuum distillation unit. These units operate at overhead pressures as

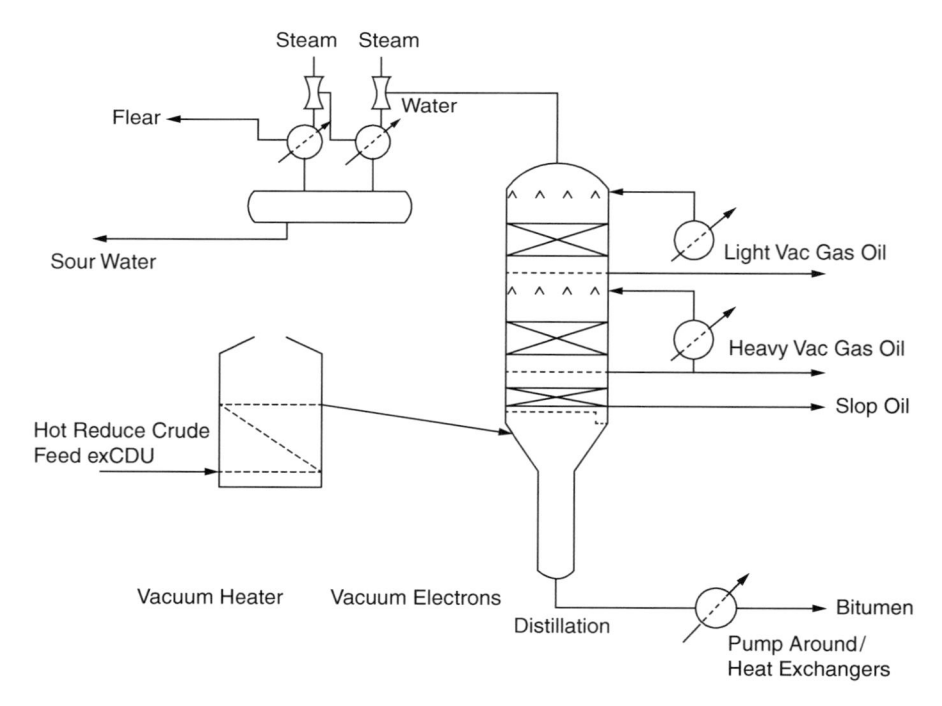

Fig. V.1 Vacuum distillation

low as 10 mmHg. This type of unit is further described and discussed in the handbook topics on atmospheric and vacuum distillation and process equipment design.

Vacuum pump A type of compressor used to create vacuum in a process system. These are normally liquid-ring or reciprocating compressors. They are often used as the last stage in a 3-stage vacuum system, with steam-jet ejectors used for the first two stages.

Vacuum resid See "Resid."

Value chain The string of organizations that bring something from raw material to finished product

Valve loader or lifter See "Loader or Lifter."

Valve tray A type of tray used in distillation, absorber, or stripper columns that promotes contact between vapor and liquid on the tray using simple, lifting valves placed over holes distributed over the tray. Refer to the chapter on process equipment.

Vanadium inhibitor An organic and/or inorganic metal bearing chemical intended to chemically and/or physically combine with the compounds formed during combustion of heavy fuel oil to improve the surface properties of the treated ash compounds

Vapor depressuring (or depressing) system Protective arrangement of valves and piping intended to provide for rapid reduction of pressure in equipment by release of vapors to a safe location, usually the flare, in emergencies.

Vapor disengaging drum Vessel for separation of vapor or gas from a mixture with liquid. Refer to the chapter on equipment, section on drums and separators.

Variable cost A cost that depends on processing rate. For example, energy costs that depend on number of barrels processed

Variable head-capacity pump A pump where the delivered flow rate depends on the back pressure, like a centrifugal pump

Variable speed A technique for controlling flow by changing the speed of a compressor or pump

VC Vertical cylindrical (or Can) heater style – also called petrochem heater – used extensively in refineries

VDU (1) See "Vacuum Distillation Unit"; (2) video display unit

Vena contracta The point of the narrowest flow streamline around an obstruction in a pipe or other flow channel. The pressure at this point has a specific relationship to the flow rate, as defined by the Bernoulli principle, and is used in flow metering.

Vent stack (or unfired vent stack) Elevated vertical termination of a disposal system which discharges vapors into the atmosphere without combustion or conversion of the relieved fluid

Venturi flow meter Similar to an orifice meter, a venturi meter creates a *vena contracta* with a smooth, nonturbulent channel that follows the fluid streamlines. This reduces pressure drop and improves accuracy.

Very large crude carrier (VLCC) A category of ocean-going tanker hauling from 1.5 million to 2.5 million barrels of product.

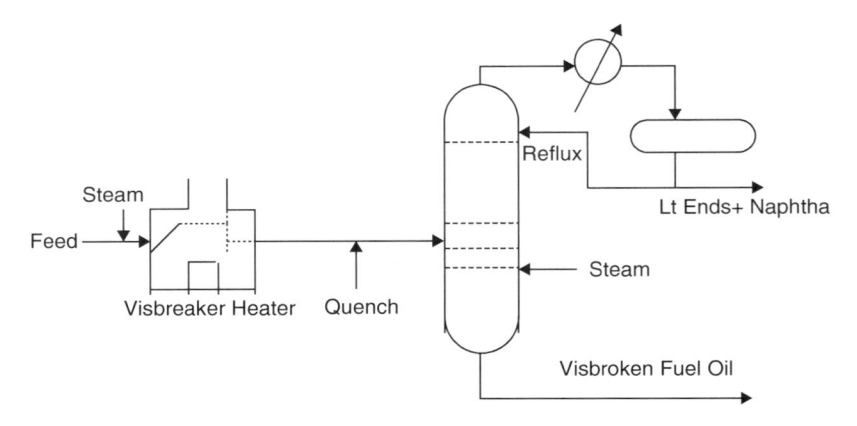

Fig. V.2 Typical visbreaker unit

Vetrocoke (AKA Giammarco-Vetrocoke) A proprietary process normally used for purification of product vapors from a gasifier to remove acid gases. This process uses either arsenite/arsenate or hot potassium carbonate solutions. It is not commonly found in refineries.

VGO Vacuum gas oil

VHAP Volatile hazardous air pollutant

VI Viscosity index

Virgin xxx See "Straight Run."

Visbreaking The visbreaking process is a mild thermal cracking of crude oil residues to reduce oil viscosity without blending. It is used especially to reduce the viscosity of vacuum residue to meet the fuel oil specification. The process configuration is very similar to the conventional, once through thermal cracker except for the routing of the recovery products from the fractionator. Figure V.2 shows the configuration of a typical visbreaker. Refer to the chapter on "▶ Upgrading the Bottom of the Barrel" in this handbook for a more detailed discussion.

Viscosimeter A device for determining the viscosity of oil. There are several methods or devices in general use. Basically, a fixed quantity of oil is allowed to pass through a fixed orifice at a specified temperature over a measured time span and the time is then compared to a standard liquid such as a calibration oil or water.

Viscosity This is a measure of the internal friction or resistance of an oil to flow. As the temperature of an oil is increased, its viscosity decreases, and it is therefore able to flow more readily. Viscosity is measured on several different scales, including Redwood No. 1 at 100 °F, Engler degrees, Saybolt Seconds, etc. The most common method for designation of viscosity is kinematic viscosity, measured in centistokes, cSt at 50 °C (see Saybolt Furol, Saybolt

Universal, Engler, Redwood, Kinematic). This measurement is important in many facets of process design and indeed is an essential quality of many finished products. The topic entitled "▶ Quality Control of Products in Petroleum Refining" describes some of the test methods for viscosity. There are two viscosity parameters normally defined: (1) dynamic or absolute viscosity and (2) kinematic viscosity. Both are related, since the kinematic viscosity may be obtained by dividing the dynamic viscosity by the mass density. The metric unit for viscosity is *poise* (P). The unit most often used in the petroleum industry for this measure is the *centipoise* (cP) which is the poise divided by 100. Thus, dimensions of the poise are grams/(cm × s). The kinematic viscosity dimension in English units is *square foot per second*. And in metric units is *square centimeter per second* called *the stoke*. In the petroleum refining industry, stokes divided by 100 called *the centistoke* (cSt) is the unit most often used. In blending for viscosity, a blending index concept must be used. You cannot blend viscosities directly using just proportions. The viscosity indices are given and discussed in the chapter titled "▶ Introduction to Crude Oil and Petroleum Processing" and again in the topic "▶ Petroleum Products and a Refinery Configuration."

Viscosity index (VI) Measure of the amount the viscosity of a lube oil stock will change with temperature. A high VI indicates less change with temperature. High VI is desired for premium lube oils. See the chapter titled "▶ Non-energy Refineries in Petroleum Processing."

VLCC Very large crude carrier

VOC Volatile organic compounds

Volatile A volatile substance is one that is capable of being evaporated or changed to a vapor at a relatively low temperature.

Volatile organic compound A combination of fugitive chemical pollutants that form ozone or smog. The Clean Air Act was designed in part to reduce VOCs in gasoline in order to reduce ozone pollution from gasoline exhaust and emissions.

Volatility (1) The degree to which an oil product will vaporize or turn from liquid to gas, when heated. (2) Range in price or frequency and magnitude of change in price for a product

VPP US OSHA Voluntary Protection Program

VRU Vapor recovery unit

VTB Vacuum tower bottoms – AKA vacuum resid. See "Resid."

VTU Vacuum tower unit

W

WABIT Weighted average bed inlet temperature – normally catalytic naphtha reformers

WABT Weighted average bed temperature – normally used for hydroprocessing reactors and other fixed bed units

Waste disposal facilities All process plants, including oil refineries, produce quantities of toxic and/or flammable materials during periods of plant upset or emergencies. Properly designed flare and slop handling systems are therefore essential to the plant operation. Facilities for recovery of hydrocarbon slops, safe disposal of excess gases, and environmentally responsible disposal of water effluents are discussed in the topics "▶ Off-Site Facilities for Petroleum Processing" and "▶ Environmental Control and Engineering in Petroleum Processing."

Waste minimization or waste reduction Practice of reducing the amount of net waste produced by a facility or process. Refer to the environmental chapter section on solid wastes.

Water–gas shift (WGS) The reaction of CO with steam to produce CO_2 and hydrogen. This reaction is used extensively in SMR hydrogen plants.

Water rate In a steam turbine, the steam required by a turbine for a given horsepower application. Units are lb/HP-hr. Refer to the chapter on equipment, section on steam turbines.

Water systems The major water systems in most petroleum processing plants include cooling water, treated boiler feed water, plant water, fire water, and potable water. The key elements of these systems are discussed in the topics "▶ Utilities in Petroleum Processing" and "▶ Fire Prevention and Firefighting in Petroleum Processing."

Wave or wavy flow Flow regime similar to stratified flow, but the gas velocity is higher, producing a wavy interface between gas and liquid layers. Again, this regime can give rise to problems when flow is vertical. Refer to the chapter on equipment, section on two-phase pressure drop.

Wax A solid or semisolid material consisting of a mixture of hydrocarbons obtained or derived from petroleum fractions, or through a Fischer–Tropsch-type process, in which the straight-chained paraffin series predominates. This includes all marketable wax, whether crude or refined, with a congealing point (ASTM D938) between 100 °F and 200 °F and a maximum oil content (ASTM D3235) of 50 wt%.

Waxy lube stock (WLS) An intermediate in lube oil production that has a high VI and can be finished into a lube oil blendstock

WBS Work breakdown structure in projects

WBT Web-based training

WC or "WC Inches of water column – a measure of draft in a heater firebox or vacuum

W.C.A. Water column absolute – normally inches of water pressure above vacuum

W.C.G. Water column gauge – normally inches of water pressure above atmospheric pressure

WCS (1) Western Canadian select crude oil blend; (2) worst case scenario

WDT Watchdog timer – in control systems

Weir A small "dam" used to create a controlled liquid level upstream by allowing overflow of excess liquid over the weir. In some typical applications, weirs are used to create liquid seals in distillation column downcomers and to control flow rates and distribution through wastewater treating units.

WFD EU Water Framework Directive

WGS (1) Wet gas scrubber – used for SO_x reduction in stacks, among others; (2) water–gas shift reaction (see "Water–Gas Shift")

What-if analysis A process hazards analysis method based on questions about possible impacts of process changes. Refer to the safety chapter.

WHB Waste heat boiler – recovers heat by generating steam. This will look like a heat exchanger.

WLS Waxy lube stock

Working capital This economic term includes capital items that require a front-end investment but are not part of the physical plant, accrued interest charges, accounts receivable, cash on hand, sorbents, catalysts, metals, warehouse, and spare parts inventories. See the chapter on economics.

Working storage capacity The difference in volume of a tank or other vessel between the maximum safe fill capacity and the quantity below which pump suction is ineffective (bottoms or tank heel/minimum level)

WPS Welding procedure specifications

WTI West Texas intermediate crude – a marker or price-setting crude oil

WTS West Texas sour crude oil

WWT (1) Wastewater treatment; (2) a Chevron process for recovering ammonia from sour wastewater

WWTP Wastewater treatment plant

WWTS Wastewater treatment standards

X

xC Controller – controls condition "x" where x = flow, temperature, pressure, level, etc.

xE Element – a device that detects condition "x" where x = flow, temperature, pressure, level, etc.

XHVGO Extra-heavy vacuum gas oil

xI Indicator – indicates condition "x" where x = flow, temperature, pressure, level, etc.

xIC Indicator–controller – indicates and controls condition "x" where x = flow, temperature, pressure, level, etc.

X-factor The sum of C_7+ hydrocarbons, C_6 naphthenes, and benzene in the feed to a light naphtha isomerization process unit. The higher the X factor, the more difficult the feed is to process.

X-grade Pipeline term for no. 2 fuel oil

XML Extensible Markup Language

xR Recorder – records condition "x" where x = flow, temperature, pressure, level, etc. – not used much anymore

XRD X-ray diffraction analysis method

XRF X-ray fluorescence analysis method

X-ray fluorescence analyzer In these analyzers, a sample is excited by X-rays which causes it to fluoresce or absorb in specific wavelengths that depend on composition. The fluorescence or absorption are detected and translated to composition. Refer to the chapter on controls.

xTL Shorthand for any carbon material converted to a liquid hydrocarbon. Refer to the chapters on unconventional feedstocks and chemicals from natural gas and coal.

xX Transmitter – transmits sensed condition "x" to control system (where x = flow, temperature, pressure, etc.)

Xylenes ($C_6H_4(CH_3)_2$) These are aromatic compounds consisting of two methyl groups attached to a benzene ring. There are three xylenes designated as meta, ortho, and para based on the positions of the two methyl groups around the aromatic ring. In ortho-xylene, the methyl groups are attached to adjacent carbon atoms. In meta-xylene, the methyl groups are separated by one carbon atom in the ring. Para-xylene has the methyl groups on opposite sides of the ring. Xylenes, coupled with benzene, toluene, and ethyl benzene, are the major products from a petrochemical petroleum refinery. These compounds as a whole are usually designated as BTX. Refer to the chapter "▶ Non-energy Refineries in Petroleum Processing" of this handbook for additional discussion.

Y

Yard air See "Plant Air."
Yard water See "Plant Water."
Y-grade Pipeline term for No. 1 Fuel Oil. The feedstocks that are sent to a fractionator in order to extract gas liquids
YTD Year to date

Z

"Z" factor The term "Z factor" is usually found in two contexts:
  • Compressibility factor for a gas
  • Pseudo heat transfer coefficient for a heat exchanger
  (1) *Gas compressibility factor*
  "Z" is often the symbol used for the compressibility factor of a gas. This may be derived from the equation:
$$PV = ZnRT$$
$$PV = Z(m/M)RT$$
$$PM = Z\rho RT$$
$$Z = PM/\rho RT$$

where:
V = volume of gas
m = mass of gas
R = gas constant
M = molecular weight of the gas
P = gas pressure (absolute)
T = absolute temperature
ρ = density of the gas at gas temperature and pressure
Typical values for R are:
8.3143 J/(mol × °K)
0.08205341 atm × m^3/(kmol × °K)
1.98716759 cal/(mol × °K)
10.7313 psia × ft^3/(lb mol × °K)

(2) *Pseudo heat transfer coefficient*
When limited data are available in a unit to track heat transfer in an exchanger, a Z factor can be tracked to infer exchanger fouling. The Z factor is developed in the following manner:

$$\Delta Q = U\ A\ \Delta T_{LM}\ \text{ and also }\ \Delta Q = C_{effective}\, DT$$

$$U\ A\, \Delta T_{LM} = C_{effective} \Delta T$$

$$U = C_{effective} \Delta T / A\, \Delta T_{LM}$$

Observe that the heat transferred per degree in an exchanger is roughly constant and the exchanger surface area is constant, assuming no major changes in flow rates on either side of the exchanger. Hence, assume that $C_{effective}/A$ is constant. So, $U\ \alpha\ \Delta T/\Delta T_{LM}$ which we define as the Z factor for the exchanger.

ΔQ = Heat transferred in the exchanger per unit time.
U = Overall heat transfer coefficient.
A = Exchanger surface area.
ΔT = Temperature difference for one of the fluids in the exchanger.
$C_{effective}$ = Heat transferred per degree for the fluid chosen as the ΔT basis.
ΔT_{LM} = Exchanger log-mean temperature difference between the hot and cold fluids.

A plot of Z over time will indicate the extent of fouling in an exchanger. Note that this approach assumes relatively constant flow conditions on both sides of the exchanger. It is confounded by large differences in flow rates. It is possible to adjust the Z for flow variations by introducing a correction factor developed from empirical tests.

Zeolite catalysts Zeolite catalysts contain acidic silica–alumina structures that promote cracking of hydrocarbons. They are used in catalytic cracking processes. This together with the technique of "riser cracking" revolutionized these processes (distillate feed crackers and residuum feed). This is described and discussed in detail in the chapter "▶ Fluid Catalytic Cracking (FCC) in Petroleum Refining." This type of catalyst is also incorporated into hydrocracking as part of the catalyst base. Discussion of hydrocracking is in the chapter on "▶ Hydrocracking in Petroleum Processing."

Part VI

Appendices

Examples of Working Flow Sheets in Petroleum Refining

Steven A. Treese

Contents

Appendix A: Example Flow Sheets	1830
A1a A Typical Process Flow Sheet	1830
A1b The Associated Material Balance	1830
A2 A Section of a Typical Mechanical Flow Sheet	1830
A3 A Section of a Typical Utility Flow Sheet	1830

Abstract

This appendix is a collection of working flow sheet examples that are developed for design and operation of refinery and other petroleum processing units. These are only used as illustrations of the format and content for these key documents. Included here are a Process Flow Diagram and accompanying Material Balance, a section of a Mechanical Flow Diagram (or P&ID), and a section of a Utility Flow Sheet.

Keywords

Process flow diagram • Mechanical flow diagram • Material balance • Utility flow diagram • P&ID

Steven A. Treese has retired from Phillips 66.

S.A. Treese (✉)
Puget Sound Investments LLC, Katy, TX, USA
e-mail: streese256@aol.com

© Springer International Publishing Switzerland 2015
S.A. Treese et al. (eds.), *Handbook of Petroleum Processing*,
DOI 10.1007/978-3-319-14529-7_32

Appendix A: Example Flow Sheets

A1a A Typical Process Flow Sheet

See Fig. 1.

A1b The Associated Material Balance

See Fig. 2.

A2 A Section of a Typical Mechanical Flow Sheet

See Fig. 3.

A3 A Section of a Typical Utility Flow Sheet

See Fig. 4.

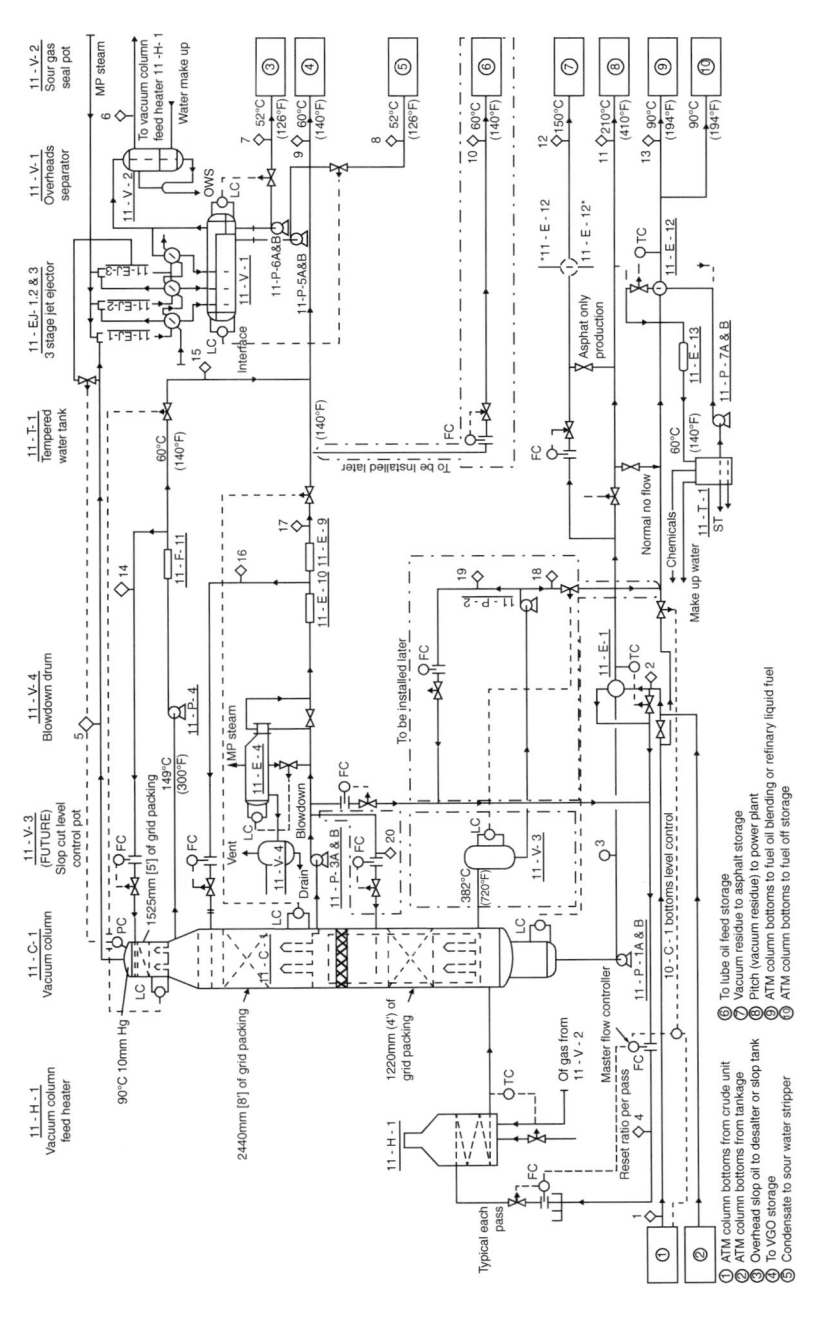

Fig. 1 A Typical Process Flow Sheet

Stream up	1	2	3	4	5
Stream identification	ATM column bottoms from crude column	Feed to vacuum column	Vacuum column residue	Feed to vacuum column	Vacuum column overhead vapour
Normal kg/h	470,269	391,836	245,982	391,836	2971
BPSD based on standard conditions	71,800	60,000	36,000	60,000	
Cut	371°C+	371°C+	80 pen	371°C+	94.1 (Av. mol/wt)
SG @ 15%	0.959	0.989	1.033	0.989	
API	11.5	11.5	5.5	11.5	
Max operating kg/h	470,300	457,070 (69,000 BPSD of 385°C+)	329,460 (48,500 BPSD) @ 800 pen)	457,070 (385°C+ resid)	3,000

Stream up	6	7	8	9	10
Stream identification	Vacuum column inerts exit stream	Vacuum column overhead slop	Condensate water from ejectors	Combined top side cut and heavy V.G.O.	Heavy vacuum gas oil
Normal kg/h	600	2,371		137,900	
BPSD based on standard conditions		407		22,693	
Cut	29 (Av. mol/wt)	218 (Av. mol/wt)	18		
SG @ 15%		0.883			
API		28.7		22.1	20.7
Max operating kg/h	600	2,371	7,710	137,901	

Stream up	11	12	13	14	15
Stream identification	Combined slop cut and vac. resid streams	Asphalt	ATM column bottoms to fuel blending	Top circulating reflux	Top cut TBP 370 to 427°C (700 to 800°F)
Normal kg/h	251,565		78,433	174,603	51,279
BPSD based on standard conditions	36,900		11,800		8,593
Cut			371°C+		
SG @ 15%			1.033	0.906	0.906
API	5.5	5.5	5.5	24.6	24.6
Max operating kg/h		11,900	457,000 (385°C+ resid)		51,280

Stream up	16	17	18	19	20
Stream identification	H.G.O. circulating reflux	Heavy vacuum gas oil	Slop cut vacuum column product	Vac. column bottoms recycle	Metals wash out
Normal kg/h	149,206	86,621	5,583	0 to 16,748	8,481
BPSD based on standard conditions		14,100	900	2,700	
Cut					
SG @ 15%	0.930	0.930	0.942	0.948	0.930
API	20.7	20.7	18.7	18.7	20.7
Max operating kg/h		86,621	16,748	16,748	15,964

Fig. 2 Material Balance Associated with the Typical Process Flow Sheet (Fig. 1)

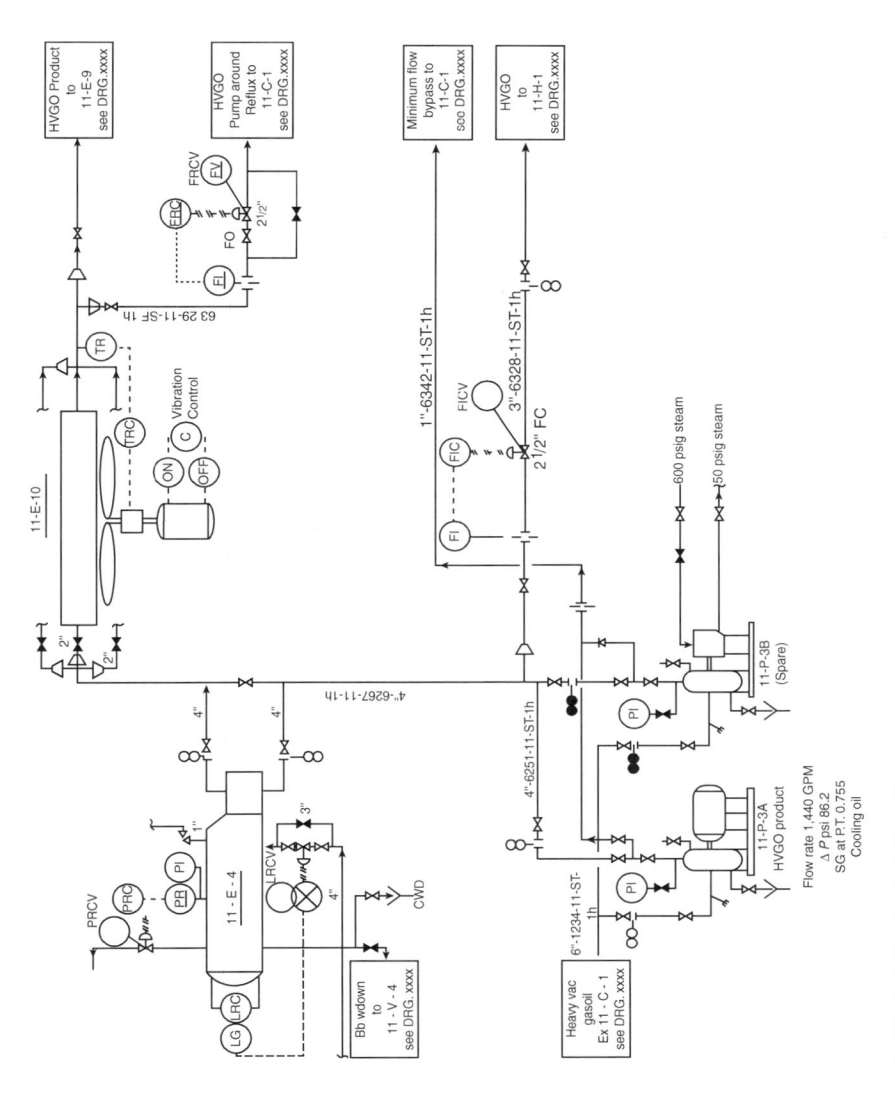

Examples of Working Flow Sheets in Petroleum Refining 1833

Fig. 3 A Section of a Typical Mechanical Flow Sheet

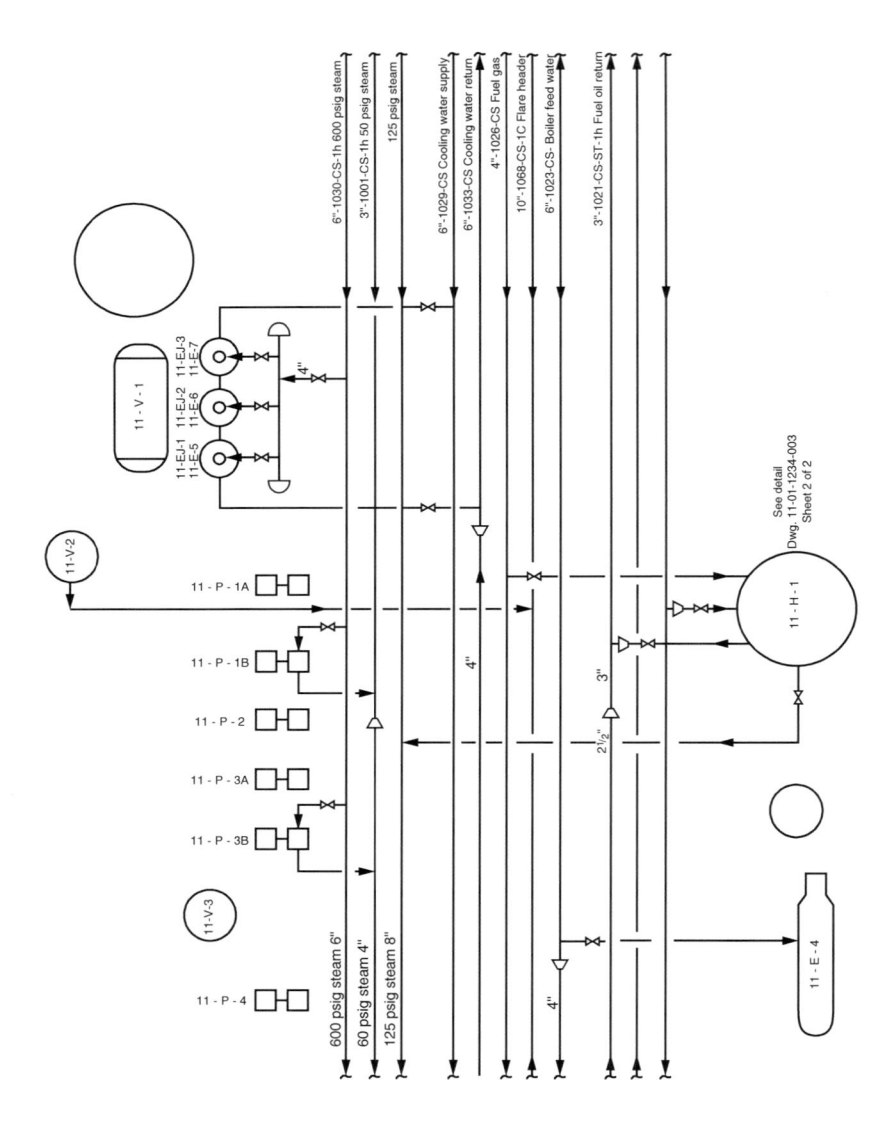

Fig. 4 A Section of a Typical Utility Flow Sheet

General Data for Petroleum Processing

Steven A. Treese

Contents

Appendix B: General Reference Data 1836
   B1 Hydrocarbon Viscosity Versus Temperature 1836
   B2 Hydrocarbon Specific Gravity Versus Temperature 1837
   B3a Relationship of SG, °API, and Lbs/Gal 1838
   B3b Relationship of SG, °API, and Lbs/Gal 1840
   B4 Relationship of Chords, Diameters, and Areas 1842
   B5 Boiling and Freezing Points of Normal Alkanes (Paraffins) 1842

Abstract
This appendix provides general data for petroleum processing which is used throughout this handbook. Included here are data on (1) viscosity and density/SG of stocks as a function of temperature; (2) relationships for chords, diameters, and areas of circular cross-sections that are used in vessel and tower design; and (3) boiling and freezing points for normal alkanes or paraffins up to a carbon number of 120.

Keywords
Refining • Viscosity • Specific gravity • API gravity • Boiling point • Freezing point • Hydrocarbon • Temperature response

---

Steven A. Treese has retired from Phillips 66.

S.A. Treese (✉)
Puget Sound Investments LLC, Katy, TX, USA
e-mail: streese256@aol.com

© Springer International Publishing Switzerland 2015
S.A. Treese et al. (eds.), *Handbook of Petroleum Processing*,
DOI 10.1007/978-3-319-14529-7_33

Appendix B: General Reference Data

B1 Hydrocarbon Viscosity Versus Temperature

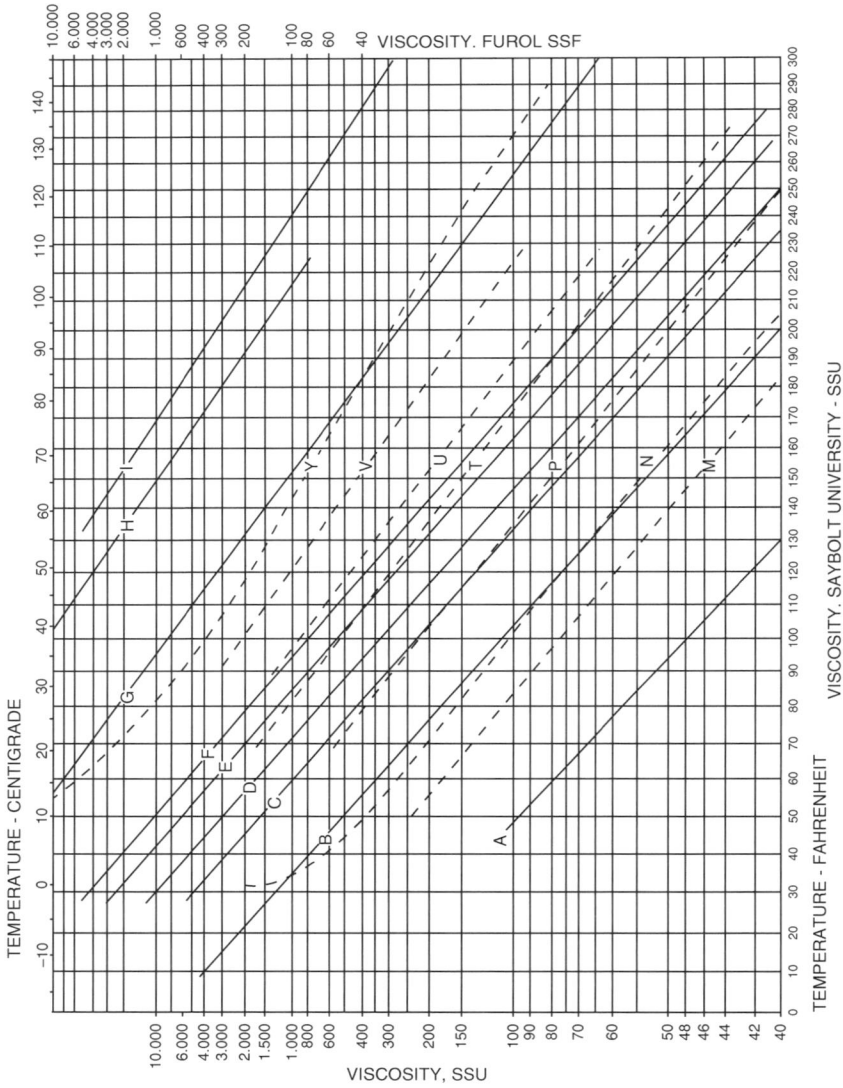

This chart may be used to determine the viscosity of an oil at any temperature provided its viscosity at two temperatures is known.

The lines on this chart show viscosities of representative oils.

Note: This chart is similar to ASTM tentative standard D341-32T which has a somewhat wider viscosity and temperature range.

Courtesy of Texaco, Inc.

General Data for Petroleum Processing 1837

B2 Hydrocarbon Specific Gravity Versus Temperature

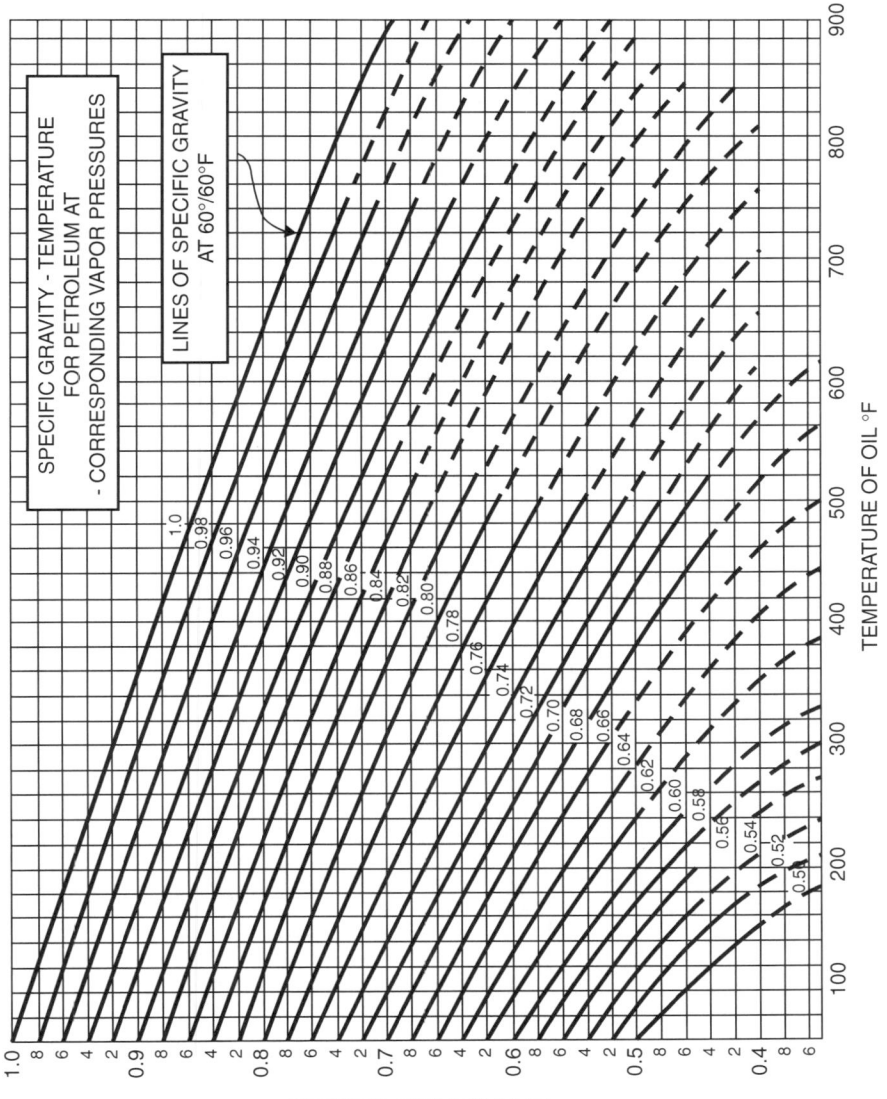

SPECIFIC GRAVITY AT °F

Specific Gravity - Referred to water at 60°F
Example: Oil with sp. gr. of 0.82 at 60°F will have sp. gr. of 0.64 at 500°F

Courtesy of Hydraulic Institute.

B3a Relationship of SG, °API, and Lbs/Gal

Pounds per gallon and specific gravities corresponding to degrees API at 60°F

Formula — $\text{sp gr} = \dfrac{141.5}{131.5 + °\text{API}}$

Deg API	\multicolumn{10}{c}{Tenths of Degrees}
	0	1	2	3	4	5	6	7	8	9
10	8.328	8.322	8.317	8.311	8.305	8.299	8.293	8.287	8.282	8.276
	1.0000	.9993	.9986	.9979	.9972	.9965	.9958	.9951	.9944	.9937
11	8.270	8.264	8.258	8.252	8.246	8.241	8.235	8.229	8.223	8.218
	.9930	.9923	.9916	.9909	.9902	.9895	.9888	.9881	.9874	.9868
12	8.212	8.206	8.201	8.195	8.189	8.183	8.178	8.172	8.166	8.161
	.9861	.9854	.9847	.9840	.9833	.9826	.9820	.9813	.9806	.9799
13	8.155	8.150	8.144	8.138	8.132	8.127	8.122	8.116	8.110	8.105
	.9792	.9786	.9779	.9772	.9765	.9759	.9752	.9745	.9738	.9732
14	8.099	8.093	8.088	8.082	8.076	8.071	8.066	8.061	8.055	8.049
	.9725	.9718	.9712	.9705	.9698	.9692	.9685	.9679	.9672	.9665
15	8.044	8.038	8.033	8.027	8.021	8.016	8.011	8.006	8.000	7.995
	.9659	.9652	.9646	.9639	.9632	.9626	.9619	.9613	.9606	.9600
16	7.989	7.984	7.978	7.973	7.967	7.962	7.956	7.951	7.946	7.940
	.9593	.9587	.9580	.9574	.9567	.9561	.9554	.9548	.9541	.9535
17	7.935	7.930	7.925	7.919	7.914	7.909	7.903	7.898	7.893	7.887
	.9529	.9522	.9516	.9509	.9503	.9497	.9490	.9484	.9478	.9471
18	7.882	7.877	7.871	7.866	7.861	7.856	7.851	7.846	7.841	7.835
	.9465	.9459	.9452	.9446	.9440	.9433	.9427	.9421	.9415	.9408
19	7.830	7.825	7.820	7.814	7.809	7.804	7.799	7.793	7.788	7.783
	.9402	.9396	.9390	.9383	.9377	.9371	.9365	.9358	.9352	.9346
20	7.778	7.773	7.768	7.762	7.757	7.752	7.747	7.742	7.737	7.732
	.9340	.9334	.9328	.9321	.9315	.9309	.9303	.9297	.9291	.9285
21	7.727	7.722	7.717	7.711	7.706	7.701	7.696	7.691	7.686	7.681
	.9279	.9273	.9267	.9260	.9254	.9248	.9242	.9236	.9230	.9224
22	7.676	7.671	7.666	7.661	7.656	7.651	7.646	7.641	7.636	7.632
	.9218	.9212	.9206	.9200	.9194	.9188	.9182	.9176	.9170	.9165
23	7.627	7.622	7.617	7.612	7.607	7.602	7.597	7.592	7.587	7.583
	.9159	.9153	.9147	.9141	.9135	.9129	.9123	.9117	.9111	.9106
24	7.578	7.573	7.568	7.563	7.558	7.554	7.549	7.544	7.539	7.534
	.9100	.9094	.9088	.9082	.9076	.9071	.9065	.9059	.9053	.9047
25	7.529	7.524	7.519	7.514	7.509	7.505	7.500	7.495	7.491	7.486
	.9042	.9036	.9030	.9024	.9018	.9013	.9007	.9001	.8996	.8990
26	7.481	7.476	7.472	7.467	7.462	7.458	7.453	7.448	7.443	7.438
	.8984	.8978	.8973	.8967	.8961	.8956	.8950	.8944	.8939	.8933
27	7.434	7.429	7.424	7.420	7.415	7.410	7.406	7.401	7.397	7.392
	.8927	.8922	.8916	.8911	.8905	.8899	.8894	.8888	.8883	.8877
28	7.387	7.383	7.378	7.373	7.368	7.364	7.360	7.355	7.350	7.346
	.8871	.8866	.8860	.8855	.8849	.8844	.8838	.8833	.8827	.8822
29	7.341	7.337	7.332	7.328	7.323	7.318	7.314	7.309	7.305	7.300
	.8816	.8811	.8805	.8800	.8794	.8789	.8783	.8778	.8772	.8767
30	7.296	7.291	7.287	7.282	7.278	7.273	7.268	7.264	7.259	7.255
	.8762	.8756	.8751	.8745	.8740	.8735	.8729	.8724	.8718	.8713
31	7.251	7.246	7.242	7.238	7.233	7.228	7.224	7.219	7.215	7.211
	.8708	.8702	.8697	.8692	.8686	.8681	.8676	.8670	.8665	.8660
32	7.206	7.202	7.198	7.193	7.188	7.184	7.180	7.176	7.171	7.167
	.8654	.8649	.8644	.8639	.8633	.8628	.8623	.8618	.8612	.8607
33	7.163	7.158	7.153	7.149	7.145	7.141	7.137	7.132	7.128	7.123
	.8602	.8597	.8591	.8586	.8581	.8576	.8571	.8565	.8560	.8555
34	7.119	7.115	7.111	7.106	7.102	7.098	7.093	7.089	7.085	7.081
	.8550	.8545	.8540	.8534	.8529	.8524	.8519	.8514	.8509	.8504
35	7.076	7.072	7.067	7.063	7.059	7.055	7.051	7.047	7.042	7.038
	.8498	.8493	.8488	.8483	.8478	.8473	.8468	.8463	.8458	.8453
36	7.034	7.030	7.026	7.022	7.018	7.013	7.009	7.005	7.001	6.997
	.8448	.8443	.8438	.8433	.8428	.8423	.8418	.8412	.8408	.8403
37	6.993	6.989	6.985	6.980	6.976	6.972	6.968	6.964	6.960	6.955
	.8398	.8393	.8388	.8383	.8378	.8373	.8368	.8363	.8358	.8353
38	6.951	6.947	6.943	6.939	6.935	6.930	6.926	6.922	6.918	6.914
	.8348	.8343	.8338	.8333	.8328	.8324	.8319	.8314	.8309	.8304
39	6.910	6.906	6.902	6.898	6.894	6.890	6.886	6.882	6.878	6.874
	.8299	.8294	.8289	.8285	.8280	.8275	.8270	.8265	.8260	.8256
40	6.870	6.866	6.862	6.859	6.854	6.850	6.846	6.842	6.838	6.834
	.8251	.8246	.8241	.8236	.8232	.8227	.8222	.8217	.8212	.8208
41	6.830	6.826	6.822	6.818	6.814	6.810	6.806	6.802	6.798	6.794
	.8203	.8198	.8193	.8189	.8184	.8178	.8174	.8170	.8165	.8160

(continued)

Pounds per gallon and specific gravities corresponding to degrees API at 60°F (Continued)

Deg API	\multicolumn{10}{c}{Tenths of Degrees}									
	0	1	2	3	4	5	6	7	8	9
42	6.790 .8155	6.786 .8151	6.782 .8146	6.779 .8142	6.775 .8137	6.771 .8132	6.767 .8128	6.763 .8123	6.759 .8118	6.756 .8114
43	6.752 .8109	6.748 .8104	6.744 .8100	6.740 .8095	6.736 .8090	6.732 .8086	6.728 .8081	6.724 .8076	6.720 .8072	6.716 .8067
44	6.713 .8063	6.709 .8058	6.705 .8054	6.701 .8049	6.697 .8044	6.694 .8040	6.690 .8035	6.686 .8031	6.682 .8026	6.679 .8022
45	6.675 .8017	6.671 .8012	6.667 .8008	6.663 .8003	6.660 .7999	6.656 .7994	6.652 .7990	6.648 .7985	6.645 .7981	6.641 .7976
46	6.637 .7972	6.633 .7967	6.630 .7963	6.626 .7958	6.622 .7954	6.618 .7949	6.615 .7945	6.611 .7941	6.607 .7936	6.604 .7932
47	6.600 .7927	6.596 .7923	6.592 .7918	6.589 .7914	6.585 .7909	6.582 .7905	6.578 .7901	6.574 .7896	6.571 .7892	6.567 .7887
48	6.563 .7883	6.560 .7879	6.556 .7874	6.552 .7870	6.548 .7865	6.545 .7861	6.541 .7857	6.537 .7852	6.534 .7848	6.530 .7844
49	6.526 .7839	6.523 .7835	6.520 .7831	6.516 .7826	6.512 .7822	6.509 .7818	6.505 .7813	6.501 .7809	6.498 .7805	6.494 .7800
50	6.490 .7796	6.487 .7792	6.484 .7788	6.480 .7783	6.476 .7779	6.473 .7775	6.469 .7770	6.466 .7766	6.462 .7762	6.459 .7758
51	6.455 .7753	6.451 .7749	6.448 .7745	6.445 .7741	6.441 .7736	6.437 .7732	6.434 .7728	6.430 .7724	6.427 .7720	6.423 .7715
52	6.420 .7711	6.416 .7707	6.413 .7703	6.410 .7699	6.406 .7694	6.402 .7690	6.399 .7686	6.396 .7682	6.392 .7678	6.389 .7674
53	6.385 .7669	6.381 .7665	6.378 .7661	6.375 .7657	6.371 .7653	6.368 .7649	6.365 .7645	6.360 .7640	6.357 .7636	6.354 .7632
54	6.350 .7628	6.347 .7624	6.344 .7620	6.340 .7616	6.337 .7612	6.334 .7608	6.330 .7603	6.326 .7599	6.323 .7595	6.320 .7591
55	6.316 .7587	6.313 .7583	6.310 .7579	6.306 .7575	6.303 .7571	6.300 .7567	6.296 .7563	6.293 .7559	6.290 .7555	6.287 .7551
56	6.283 .7547	6.280 .7543	6.276 .7539	6.273 .7535	6.270 .7531	6.266 .7527	6.263 .7523	6.259 .7519	6.256 .7515	6.253 .7511
57	6.249 .7507	6.246 .7503	6.243 .7499	6.240 .7495	6.236 .7491	6.233 .7487	6.229 .7483	6.226 .7479	6.223 .7475	6.219 .7471
58	6.216 .7467	6.213 .7463	6.209 .7459	6.206 .7455	6.203 .7451	6.199 .7447	6.196 .7443	6.193 .7440	6.190 .7436	6.187 .7432
59	6.184 .7428	6.180 .7424	6.177 .7420	6.174 .7416	6.170 .7412	6.167 .7408	6.164 .7405	6.161 .7401	6.158 .7397	6.154 .7393
60	6.151 .7389	6.148 .7385	6.144 .7381	6.141 .7377	6.138 .7374	6.135 .7370	6.132 .7366	6.129 .7362	6.125 .7358	6.122 .7354
61	6.119 .7351	6.116 .7347	6.113 .7343	6.109 .7339	6.106 .7335	6.103 .7332	6.100 .7328	6.097 .7324	6.094 .7320	6.090 .7316
62	6.087 .7313	6.084 .7309	6.081 .7305	6.078 .7301	6.075 .7298	6.072 .7294	6.068 .7290	6.065 .7286	6.062 .7283	6.059 .7279
63	6.056 .7275	6.053 .7271	6.050 .7268	6.047 .7264	6.044 .7260	6.040 .7256	6.037 .7253	6.034 .7249	6.031 .7245	6.028 .7242
64	6.025 .7238	6.022 .7234	6.019 .7230	6.016 .7227	6.013 .7223	6.010 .7219	6.007 .7216	6.004 .7212	6.000 .7208	5.997 .7205
65	5.994 .7201	5.991 .7197	5.988 .7194	5.985 .7190	5.982 .7186	5.979 .7183	5.976 .7179	5.973 .7175	5.970 .7172	5.967 .7168
66	5.964 .7165	5.961 .7161	5.958 .7157	5.955 .7154	5.952 .7150	5.949 .7146	5.946 .7143	5.943 .7139	5.940 .7136	5.937 .7132
67	5.934 .7129	5.931 .7125	5.928 .7121	5.925 .7118	5.922 .7114	5.919 .7111	5.916 .7107	5.913 .7103	5.910 .7100	5.907 .7096
68	5.904 .7093	5.901 .7089	5.898 .7086	5.895 .7082	5.892 .7079	5.889 .7075	5.886 .7071	5.883 .7068	5.880 .7064	5.877 .7061
69	5.874 .7057	5.871 .7054	5.868 .7050	5.866 .7047	5.863 .7043	5.860 .7040	5.857 .7036	5.854 .7033	5.851 .7029	5.848 .7026
70	5.845 .7022	5.842 .7019	5.839 .7015	5.836 .7012	5.833 .7008	5.831 .7005	5.828 .7001	5.825 .6998	5.823 .6995	5.820 .6991
71	5.817 .6988	5.814 .6984	5.811 .6981	5.808 .6977	5.805 .6974	5.802 .6970	5.799 .6967	5.796 .6964	5.793 .6960	5.791 .6957
72	5.788 .6952	5.785 .6950	5.782 .6946	5.779 .6943	5.776 .6940	5.773 .6936	5.771 .6933	5.768 .6929	5.765 .6926	5.762 .6923
73	5.759 .6919	5.757 .6916	5.754 .6913	5.751 .6909	5.748 .6906	5.745 .6902	5.743 .6899	5.740 .6896	5.737 .6892	5.734 .6889
74	5.731 .6886	5.728 .6882	5.726 .6879	5.723 .6876	5.720 .6872	5.718 .6869	5.715 .6866	5.712 .6862	5.709 .6859	5.706 .6856

B3b Relationship of SG, °API, and Lbs/Gal

Pounds per gallon and specific gravities corresponding to degrees API at 60°F (Continued)

Deg API	0	.1	2	3	4	5	6	7	8	9
75	5.703 .6852	5.701 .6849	5.698 .6846	5.695 .6842	5.693 .6839	5.690 .6836	5.687 .6832	5.685 .6829	5.682 .6826	5.679 .6823
76	5.676 .6819	5.673 .6816	5.671 .6813	5.668 .8809	5.665 .6806	5.662 .6803	5.660 .6800	5.657 .6796	5.654 .6793	5.652 .6790
77	5.649 .6787	5.646 .6783	5.643 .6780	5.641 .6777	5.638 .6774	5.635 .6770	5.632 .6767	5.630 .6764	5.627 .6761	5.624 .6757
78	5.622 .6754	5.619 .6751	5.617 .6748	5.614 .6745	5.611 .6741	5.608 .6738	5.606 .6735	5.603 .6732	5.600 .6728	5.598 .6725
79	5.595 .6722	5.592 .6719	5.590 .6716	5.587 .6713	5.584 .6709	5.582 .6706	5.579 .6703	5.577 .6700	5.574 .6697	5.571 .6693
80	5.568 .6690	5.566 .6687	5.563 .6684	5.561 .6681	5.558 .6678	5.556 .6675	5.553 .6671	5.550 .6668	5.548 .6665	5.545 .6662
81	5.542 .6659	5.540 .6656	5.537 .6653	5.534 .6649	5.532 .6646	5.529 .6643	5.526 .6640	5.524 .6637	5.522 .6634	5.519 .6631
82	5.516 .6628	5.514 .6625	5.511 .6621	5.508 .6618	5.506 .6615	5.503 .6612	5.501 .6609	5.498 .6606	5.496 .6603	5.493 .6600
83	5.491 .6597	5.489 .6594	5.486 .6591	5.483 .6588	5.480 .6584	5.477 .6581	5.475 .6578	5.472 .6575	5.470 .6572	5.467 .6569
84	5.465 .6566	5.462 .6563	5.460 .6560	5.458 .6557	5.455 .6554	5.453 .6551	5.450 .6548	5.448 .6545	5.445 .6542	5.443 .6539
85	5.440 .6536	5.437 .6533	5.435 .6530	5.432 .6527	5.430 .6524	5.427 .6521	5.425 .6518	5.422 .6515	5.420 .6512	5.417 .6509
86	5.415 .6506	5.412 .6503	5.410 .6500	5.407 .6497	5.405 .6494	5.402 .6491	5.400 .6488	5.397 .6485	5.395 .6482	5.392 .6479
87	5.390 .6476	5.387 .6473	5.385 .6470	5.382 .6467	5.380 .6464	5.377 .6461	5.375 .6458	5.372 .6455	5.370 .6452	5.367 .6449
88	5.365 .6446	5.363 .6444	5.361 .6441	5.358 .6438	5.356 .6435	5.353 .6432	5.351 .6429	5.348 .6426	5.346 .6423	5.343 .6420
89	5.341 .6417	5.338 .6414	5.336 .6411	5.334 .6409	5.331 .6406	5.329 .6403	5.326 .6400	5.324 .6397	5.321 .6394	5.319 .6391
90	5.316 .6388	5.314 .6385	5.312 .6382	5.310 .6380	5.307 .6377	5.305 .6374	5.302 .6371	5.300 .6368	5.297 .6365	5.295 .6362
91	5.293 .6360	5.291 .6357	5.288 .6354	5.286 .6351	5.283 .6348	5.281 .6345	5.278 .6342	5.276 .6340	5.274 .6337	5.271 .6334
92	5.269 .6331	5.266 .6328	5.264 .6325	5.262 .6323	5.260 .6320	5.257 .6317	5.254 .6314	5.252 .6311	5.250 .6309	5.248 .6306
93	5.245 .6303	5.243 .6300	5.241 .6297	5.238 .6294	5.236 .6292	5.234 .9289	5.232 .6286	5.229 .6283	5.227 .6281	5.225 .6278
94	5.222 .6275	5.220 .6272	5.217 .6269	5.215 .6267	5.213 .6264	5.211 .6261	5.208 .6258	5.206 .6256	5.204 .6253	5.201 .6250
95	5.199 .6247	5.196 .6244	5.194 .6242	5.192 .6239	5.190 .6236	5.187 .6233	5.185 .6231	5.183 .6228	5.180 .6225	5.179 .6223
96	5.176 .6220	5.174 .6217	5.172 .6214	5.170 .6212	5.167 .6209	5.164 .6206	5.162 .6203	5.160 .6201	5.158 .6198	5.156 .6195
97	5.154 .6193	5.151 .6190	5.149 .6187	5.146 .6184	5.144 .6182	5.142 .6179	5.140 .6176	5.138 .6174	5.136 .6171	5.133 .6168
98	5.131 .6166	5.129 .6163	5.126 .6160	5.124 .6158	5.122 .6155	5.120 .6152	5.118 .6150	5.116 .6147	5.113 .6144	5.111 .6141
99	5.109 .6139	5:107 .6136	5.104 .6134	5.102 .6131	5.100 .6128	5.098 .6126	5.096 .6123	5.093 .6120	5.091 .6118	5.089 .6115
100	5.086 .6112	5.09 .6110	5.09 .6107	5.08 .6104	5.08 .6102	5.08 .6099	5.08 .6097	5.07 .6094	5.07 .6091	5.07 .6089
101	5.07 .6086	5.07 .6083	5.06 .6081	5.06 .6078	5.06 .6076	5.06 .6073	5.05 .6070	5.05 .6068	5.05 .6065	5.05 .6063
102	5.05 .6060	5.04 .6058	5.04 .6055	5.04 .6052	5.04 .6050	5.04 .6047	5.03 .6044	5.03 .6042	5.03 .6039	5.03 .6037
103	5.02 .6034	5.02 .6032	5.02 .6029	5.02 .6026	5.02 .6024	5.01 .6021	5.01 .6019	5.01 .6016	5.01 .6014	5.01 .6011
104	5.00 .6008	5.00 .6006	5.00 .6003	5.00 .6001	4.99 .5998	4.99 .5996	4.99 .5993	4.99 .5991	4.99 .5988	4.98 .5986
105	4.98 .5983	4.98 .5981	4.98 .5978	4.98 .5976	4.97 .5973	4.97 .5970	4.97 .5968	4.97 .5965	4.97 .5963	4.96 .5960
106	4.96 .5958	4.96 .5955	4.96 .5953	4.95 .5950	4.95 .5948	4.95 .5945	4.95 .5943	4.95 .5940	4.94 .5938	4.94 .5935
107	4.94 .5933	4.94 .5930	4.94 .5928	4.93 .5925	4.93 .5923	4.93 .5921	4.93 .5918	4.93 .5916	4.92 .5913	4.92 .5911

(continued)

Pounds per gallon and specific gravities corresponding to degrees API at 60°F (Continued)

Deg API	\multicolumn{10}{c}{Tenths of Degrees}									
	0	1	2	3	4	5	6	7	8	9
108	4.92 .5908	4.92 .5906	4.92 .5903	4.91 .5901	4.91 .5898	4.91 .5896	4.91 .5893	4.91 .5891	4.90 .5888	4.90 .5886
109	4.90 .5884	4.90 .5881	4.90 .5879	4.89 .5876	4.89 .5874	4.89 .5871	4.89 .5869	4.89 .5867	4.88 .5864	4.88 .5862
110	4.88 .5859	4.88 .5857	4.87 .5854	4.87 .5852	4.87 .5850	4.87 .5847	4.87 .5845	4.87 .5842	4.86 .5840	4.86 .5837
111	4.86 .5835	4.86 .5833	4.85 .5830	4.85 .5828	4.85 .5825	4.85 .5823	4.85 .5821	4.84 .5818	4.84 .5816	4.84 .5813
112	4.84 .5811	4.84 .5909	4.83 .5806	4.83 .5804	4.83 .5802	4.83 .5799	4.83 .5797	4.82 .5794	4.82 .5792	4.82 .5790
113	4.82 .5787	4.82 .5785	4.82 .5783	4.81 .5780	4.81 .5778	4.81 .5776	4.81 .5773	4.81 .5771	4.80 .5768	4.80 .5766
114	4.80 .5764	4.80 .5761	4.80 .5759	4.79 .5757	4.79 .5754	4.79 .5752	4.79 .5750	4.79 .5747	4.78 .5745	4.78 .5743
115	4.78 .5740	4.78 .5738	4.78 .5736	4.77 .5733	4.77 .5731	4.77 .5729	4.77 .5726	4.77 .5724	4.76 .5722	4.76 .5719
116	4.76 .5717	4.76 .5715	4.76 .5713	4.76 .5710	4.75 .5708	4.75 .5706	4.75 .5703	4.75 .5701	4.75 .5699	4.74 .5696
117	4.74 .5694	4.74 .5692	4.74 .5690	4.74 .5687	4.73 .5685	4.73 .5683	4.73 .5680	4.73 .5678	4.73 .5676	4.73 .5674
118	4.72 .5671	4.72 .5669	4.72 .5667	4.72 .5665	4.72 .5662	4.71 .5660	4.71 .5658	4.71 .5655	4.71 .5653	4.71 .5651
119	4.70 .5649	4.70 .5646	4.70 .5644	4.70 .5642	4.70 .5640	4.69 .5637	4.69 .5635	4.69 .5633	4.69 .5631	4.69 .5628
120	4.69 .5626	4.68 .5624	4.68 .5622	4.68 .5620	4.68 .5617	4.68 .5615	4.67 .5613	4.67 .5611	4.67 .5608	4.67 .5606
121	4.67 .5604	4.67 .5602	4.66 .5600	4.66 .5597	4.66 .5595	4.66 .5593	4.66 .5591	4.65 .5588	4.65 .5586	4.65 .5584
122	4.65 .5582	4.65 .5580	4.64 .5577	4.64 .5575	4.64 .5573	4.64 .5571	4.64 .5569	4.64 .5566	4.63 .5564	4.63 .5562
123	4.63 .5560	4.63 .5558	4.63 .5556	4.62 .5553	4.62 .5551	4.62 .5549	4.62 .5547	4.62 .5545	4.62 .5542	4.61 .5540
124	4.61 .5538	4.61 .5536	4.61 .5534	4.61 .5532	4.61 .5530	4.60 .5527	4.60 .5525	4.60 .5523	4.60 .5521	4.60 .5519
125	4.59 .5517	4.59 .5514	4.59 .5512	4.59 .5510	4.59 .5508	4.59 .5506	4.58 .5504	4.58 .5502	4.58 .5499	4.58 .5497
126	4.58 .5495	4.57 .5493	4.57 .5491	4.57 .5489	4.57 .5487	4.57 .5484	4.57 .5482	4.56 .5480	4.56 .5478	4.56 .5476
127	4.56 .5474	4.56 .5472	4.56 .5470	4.55 .5468	4.55 .5465	4.55 .5463	4.55 .5461	4.55 .5459	4.54 .5457	4.54 .5455
128	4.54 .5453	4.54 .5451	4.54 .5449	4.54 .5446	4.53 .5444	4.53 .5442	4.53 .5440	4.53 .5438	4.53 .5436	4.53 .5434
129	4.52 .5432	4.52 .5430	4.52 .5428	4.52 .5426	4.52 .5424	4.51 .5421	4.51 .5419	4.51 .5417	4.51 .5415	4.51 .5413
130	4.51 .5411	4.50 .5409	4.50 .5407	4.50 .5405	4.50 .5403	4.50 .5401	4.50 .5399	4.49 .5397	4.49 .5395	4.49 .5393
131	4.49 .5390	4.49 .5388	4.49 .5386	4.48 .5384	4.48 .5382	4.48 .5380	4.48 .5378	4.48 .5376	4.48 .5374	4.47 .5372
132	4.47 .5370	4.47 .5368	4.47 .5366	4.47 .5364	4.47 .5362	4.46 .5360	4.46 .5358	4.46 .5356	4.46 .5354	4.46 .5352
133	4.46 .5350	4.45 .5348	4.45 .5346	4.45 .5344	4.45 .5342	4.45 .5340	4.45 .5338	4.45 .5336	4.44 .5334	4.44 .5332
134	4.44 .5330	4.44 .5328	4.44 .5326	4.43 .5324	4.43 .5322	4.43 .5320	4.43 .5318	4.43 .5316	4.43 .5314	4.42 .5312
135	4.42 .5310	4.42 .5308	4.42 .5306	4.42 .5304	4.42 .5302	4.41 .5300	4.41 .5298	4.41 .5296	4.41 .5294	4.41 .5292
136	4.41 .5290	4.40 .5288	4.40 .5286	4.40 .5284	4.40 .5282	4.40 .5280	4.40 .5278	4.39 .5276	4.39 .5274	4.39 .5272
137	4.39 .5270	4.39 .5268	4.39 .5266	4.38 .5264	4.38 .5262	4.38 .5260	4.38 .5258	4.38 .5256	4.38 .5254	4.37 .5252
138	4.37 .5250	4.37 .5249	4.37 .5247	4.37 .5245	4.37 .5243	4.36 .5241	4.36 .5239	4.36 .5237	4.36 .5235	4.36 .5233
139	4.36 .5231	4.35 .5229	4.35 .5227	4.35 .5225	4.35 .5223	4.35 .5221	4.35 .5219	4.35 .5218	4.34 .5216	4.34 .5214

B4 Relationship of Chords, Diameters, and Areas

R*	L*	A*	R*	L*	A*	R*	L*	A*	R*	L*	A*	R*	L*	A*	R*	L*	A*
0.070	0.511	0.0308	0.120	0.650	0.0680	0.170	0.751	0.113	0.220	0.828	0.163	0.280	0.898	0.230	0.390	0.977	0.361
1	0.514	0.0315	1	0.652	0.0688	1	0.753	0.114	1	0.829	0.164	5	0.903	0.236	5	0.979	0.367
2	0.517	0.0321	2	0.654	0.0697	2	0.755	0.115	2	0.831	0.165						
3	0.521	0.0328	3	0.657	0.0705	3	0.756	0.116	3	0.832	0.166	0.290	0.908	0.241	0.400	0.980	0.374
4	0.524	0.0335	4	0.659	0.0714	4	0.758	0.117	4	0.834	0.167	5	0.913	0.247	5	0.982	0.380
5	0.527	0.0342	5	0.661	0.0722	5	0.760	0.117	5	0.835	0.169	0.300	0.917	0.252	0.410	0.984	0.386
6	0.530	0.0348	6	0.663	0.0731	6	0.762	0.118	6	0.836	0.170	5	0.921	0.258	5	0.986	0.392
7	0.533	0.0355	7	0.665	0.0739	7	0.763	0.119	7	0.838	0.171						
8	0.536	0.0362	8	0.668	0.0748	8	0.765	0.120	8	0.839	0.172	0.310	0.925	0.264	0.420	0.987	0.398
9	0.539	0.0368	9	0.670	0.0756	9	0.786	0.121	9	0.841	0.173	5	0.930	0.270	5	0.989	0.405
0.080	0.542	0.0375	0.130	0.672	0.0765	0.180	0.768	0.122	0.230	0.842	0.174	0.320	0.933	0.276	0.430	0.991	0.412
1	0.545	0.0382	1	0.674	0.0774	1	0.770	0.123	1	0.843	0.175	5	0.937	0.282	5	0.993	0.418
2	0.548	0.0389	2	0.677	0.0782	2	0.772	0.124	2	0.845	0.176						
3	0.552	0.0396	3	0.679	0.0791	3	0.773	0.125	3	0.846	0.177	0.330	0.941	0.288	0.440	0.994	0.424
4	0.555	0.0403	4	0.682	0.0799	4	0.775	0.126	4	0.848	0.178	5	0.945	0.294	5	0.995	0.430
5	0.558	0.0410	5	0.684	0.0808	5	0.777	0.127	5	0.849	0.179	0.340	0.948	0.300	0.450	0.996	0.437
6	0.561	0.0418	6	0.686	0.0817	6	0.778	0.128	6	0.850	0.180	5	0.951	0.306			
7	0.564	0.0425	7	0.688	0.825	7	0.780	0.129	7	0.851	0.181				0.460	0.997	0.450
8	0.567	0.0432	8	0.691	0.0834	8	0.781	0.130	8	0.853	0.182	0.350	0.955	0.312			
9	0.570	0.0439	9	0.693	0.0842	9	0.783	0.131	9	0.854	0.183	5	0.958	0.318	0.470	0.998	0.462
0.090	0.573	0.0446	0.140	0.695	0.0851	0.190	0.784	0.132	0.240	0.855	0.184	0.360	0.961	0.324	0.480	0.998	0.475
1	0.576	0.0454	1	0.697	0.0860	1	0.786	0.133	5	0.860	0.190	5	0.964	0.330			
2	0.578	0.0461	2	0.699	0.0869	2	0.787	0.134							0.490	0.999	0.488
3	0.581	0.0469	3	0.700	0.0878	3	0.789	0.135	0.250	0.866	0.196	0.370	0.967	0.337			
4	0.583	0.0476	4	0.702	0.0887	4	0.790	0.136	5	0.872	0.202	5	0.969	0.343	0.500	1.0	0.50
5	0.586	0.0484	5	0.704	0.0896	5	0.792	0.137	0.260	0.878	0.207	0.380	0.971	0.348			
6	0.589	0.0491	6	0.706	0.0905	6	0.794	0.138	5	0.883	0.213	5	0.977	0.354			
7	0.592	0.0499	7	0.708	0.0914	7	0.795	0.139									
8	0.594	0.0506	8	0.710	0.0923	8	0.797	0.140	0.270	0.888	0.218						
9	0.597	0.0514	9	0.712	0.0932	9	0.798	0.141	5	0.893	0.224						
0.100	0.600	0.0521	0.150	0.714	0.0941	0.200	0.800	0.142									
1	0.603	0.0529	1	0.716	0.0950	1	0.802	0.143									
2	0.605	0.0537	2	0.718	0.0959	2	0.803	0.144									
3	0.608	0.0545	3	0.720	0.0969	3	0.805	0.145									
4	0.610	0.0555	4	0.722	0.0978	4	0.806	0.146									
5	0.613	0.0561	5	0.724	0.0987	5	0.808	0.148									
6	0.615	0.0568	6	0.726	0.0996	6	0.809	0.149									
7	0.618	0.0576	7	0.728	0.1005	7	0.810	0.150									
8	0.620	0.0584	8	0.729	0.1015	8	0.812	0.151									
9	0.623	0.0592	9	0.731	0.102	9	0.813	0.152									
0.110	0.625	0.0600	0.160	0.733	0.103	0.210	0.814	0.153									
1	0.628	0.0608	1	0.735	0.104	1	0.816	0.154									
2	0.630	0.0616	2	0.737	0.105	2	0.817	0.155									
3	0.633	0.0624	3	0.738	0.106	3	0.819	0.156									
4	0.635	0.0632	4	0.740	0.107	4	0.820	0.157									
5	0.638	0.0640	5	0.742	0.108	5	0.822	0.158									
6	0.640	0.0648	6	0.744	0.109	6	0.823	0.159									
7	0.643	0.0656	7	0.746	0.110	7	0.824	0.160									
8	0.645	0.0664	8	0.747	0.111	8	0.826	0.161									
9	0.648	0.0672	9	0.749	0.112	9	0.827	0.162									

* This table relates the downcomer area, the weir length, and the height of the circular segment formed by the weir

$$R = \frac{\text{*Downcomer rise}}{\text{Diameter}} = \frac{r}{\text{Dia}}$$

$$L = \frac{\text{*Weir length}}{\text{Diameter}} = \frac{l_0}{\text{Dia}}$$

$$A = \frac{\text{*Downcomer area}}{\text{Tower area}} = \frac{A_D}{A_S}$$

B5 Boiling and Freezing Points of Normal Alkanes (Paraffins)

Carbon number	Boiling point		Freeze point	
	°C	°F	°C	°F
3	−42	−44		
4	0	32	−138	−216
5	36	97	−129	−201
6	69	156	−96	−140

(*continued*)

General Data for Petroleum Processing

Carbon number	Boiling point		Freeze point	
	°C	°F	°C	°F
7	98	209	−91	−131
8	126	259	−57	−70
9	151	303	−53	−64
10	174	345	−29	−21
11	196	385	−25	−13
12	216	421	−9	15
13	235	455	−6	22
14	254	489	6	43
15	271	520	10	50
16	287	549	18	64
17	302	576	22	72
18	316	601	28	82
19	330	626	32	90
20	344	651	37	98
21	363	685	40	104
22	369	696	44	111
23	380	716	47	117
24	391	736	51	124
26	412	774		
28	431	808		
30	449	840		
32	466	871		
34	481	898		
36	496	925		
38	509	948		
40	522	972		
42	534	993		
44	545	1,013		
46	556	1,033		
48	566	1,051		
50	575	1,067		
52	584	1,083		
54	592	1,098		
56	600	1,112		
58	608	1,126		
60	615	1,139		
62	622	1,152		
64	629	1,164		
66	635	1,175		
68	641	1,186		
70	647	1,197		
72	653	1,207		
74	658	1,216		

(*continued*)

Carbon number	Boiling point		Freeze point	
	°C	°F	°C	°F
76	664	1,227		
78	670	1,238		
80	675	1,247		
82	681	1,258		
84	686	1,267		
86	691	1,276		
88	695	1,283		
90	700	1,292		
92	704	1,299		
94	708	1,306		
96	712	1,314		
98	716	1,321		
100	720	1,328		
110	735	1,355		
120	750	1,382		

Selection of Crude Oil Assays for Petroleum Refining

Steven A. Treese

Contents

Appendix C: A Selection of Crude Assays 1846

Abstract

This appendix provides a selection of several simple crude assays for stocks worldwide. It is intended to illustrate the range of variability and typical assay values. Caution: For final studies and definitive engineering, up-to-date assays from the crude oil suppliers should be used. The assays here are believed to be from reliable public sources, but they should be used only for informational purposes.

Keywords

Crude oil • Assay • Petroleum

Steven A. Treese has retired from Phillips 66.

S.A. Treese (✉)
Puget Sound Investments LLC, Katy, TX, USA
e-mail: streese256@aol.com

© Springer International Publishing Switzerland 2015
S.A. Treese et al. (eds.), *Handbook of Petroleum Processing*,
DOI 10.1007/978-3-319-14529-7_34

Appendix C: A Selection of Crude Assays

Alaskan North Slope, USA — Dec-10

Cut	Whole Crude	Light Ends	Light Naphtha	Heavy Naphtha	Kerosene	Diesel/Lt. Gasoil	Vacuum Gasoil	Vac Residuum
TBP Range, °F	-----	C_4–Minus	C_5–165	165-330	330-480	480-650	650-1000	1000+
Yield, v%	100.0	2.5	8.0	14.1	14.1	16.0	27.1	18.3
Yield, w%	100.0	1.7	6.1	12.4	13.4	16.1	29.0	21.4
Gravity, °API	32.1	113.7	83.77	55.02	41.1	31.4	21.2	6.9
Sulfur, w%	0.93		0.001	0.0107	0.0929	0.5304	1.21	2.34
Nitrogen, wppm	1490		0.01	0.15	1.1	75	1307	6322
CCR, w%	4.38					0	0.5	21.8
Viscosity	12.2 cSt @ 68°F	0.35 cSt @ 68°F	0.39 cSt @ 68°F	0.71 cSt @ 68°F	1.81 cSt @ 68°F	7.57 cSt @ 68°F	286 cSt @ 68°F	
Pour Point, °F	−3				−69	−2	88	121
TAN, mg/g	0.2					0.13	0.34	0.18
Ni, wppm	11					0	0	52
V, wppm	24.8					0	0	119
Paraffins, v%		100	85	43	36	32	19	
Naphthenes, v%		0	14	43	43	37	35	
Aromatics, v%		0	1	14	21	31	46	
RON Clear			71	38				
Freeze Point, °F					−52	14		
Aniline Point, °F					130	148	174	
Cetane Index	30	113	51	29	40	48		
Cloud Point, °F					−60	7		
Smoke Point, mm					22			

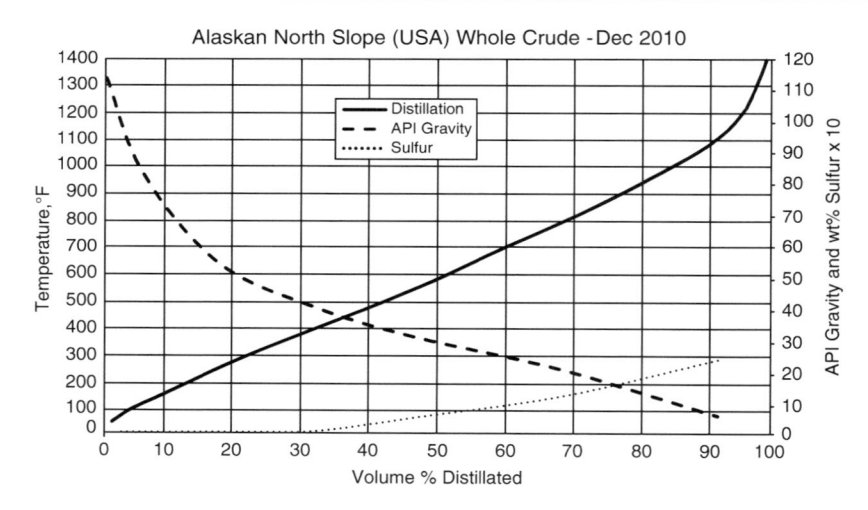

Selection of Crude Oil Assays for Petroleum Refining

Amna (high pour), Libya   Source: O&GJ

Cut	Whole Crude	Light Naphtha	Light Naphtha	Heavy Naphtha	Kerosene	Diesel/Lt. Gasoil	Vacuum Gasoil	Atm Residuum	Vac Residuum
TBP Range, °F	-----	C_5-120	120-250	250-330	330-443	443-600	600-850	655+	1000+
Yield, v%	100.0	2.4	7.7	6.9	9.4	15.0	23.3	55.4	21.8
Gravity, °API	36.1	93.6	69.5	57.4	49.9	42.3		25.4	16.7
Sulfur, w%	0.15	0.01	0.01	0.02	0.05	0.07	0.13	0.22	0.31
CCR, w%	3.7							6.22	14.5
Viscosity	13.7 cSt @ 100°F								
Pour Point, °F	75				−75	0	80	100	130
Salt, ptb	8.2								
Ni, wppm	5							8.5	20
V, wppm	0.6							1.1	2.5
Paraffins, v%		99	69.9	60.5	60.9				
Naphthenes, v%		1	26.5	33.6	29.9				
Aromatics, v%		0	3.7	5.9	9.2				
RON Clear		78							
Freeze Point, °F					−58				
Aniline Point, °F					150.6	169.1	202		
Diesel Index						75	71	69	
Smoke Point, mm					31				

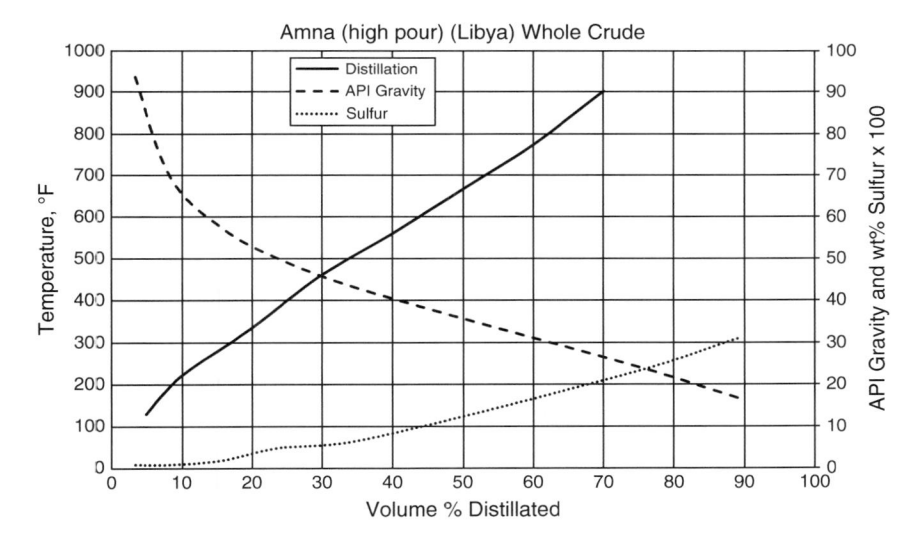

Amna (high pour) (Libya) Whole Crude — Distillation, API Gravity, Sulfur vs Volume % Distillated

Arabian Heavy, Saudi Arabia

Source: O&GJ

Cut	Whole Crude	Light Ends	Light Naphtha	Heavy Naphtha	Kerosene	Diesel/Lt. Gasoil	Vacuum Gasoil	Atm Residuum	Vac Residuum
TBP Range, °F	-----	X-68	68-212	212-302	302-455	455-650	650-1049	650+	1049+
Yield, v%	100.0	7.3	7.9	6.8	12.5	16.4	26.3	53.1	26.8
Gravity, °API	28.2		80.1	60.6	48.3	35.8	21.8	12.3	4.0
Sulfur, w%	2.84		0.0028	0.018	0.19	1.38	2.88	4.35	5.6
CCR, w%								13.2	24.4
Viscosity	18.9 cSt @ 100°F				1.12 cSt @ 100°F	3.65 cSt @ 100°F	62.5 cSt @ 100°F	106 cSt @ 210°F	13400 cSt @ 210°F
Pour Point, °F	-30					5	90	55	120
RVP, psi	8.5		10.2						
Ni, wppm								8.5	53
V, wppm								1.1	171
Paraffins, v%			89.6	70.3	58				
Naphthenes, v%			9.5	21.4	23.7				
Aromatics, v%			0.9	8.3	18.3				
RON Clear			59.7						
Freeze Point, °F					-64				
Aniline Point, °F					138	156	172		
Smoke Point, mm					26				

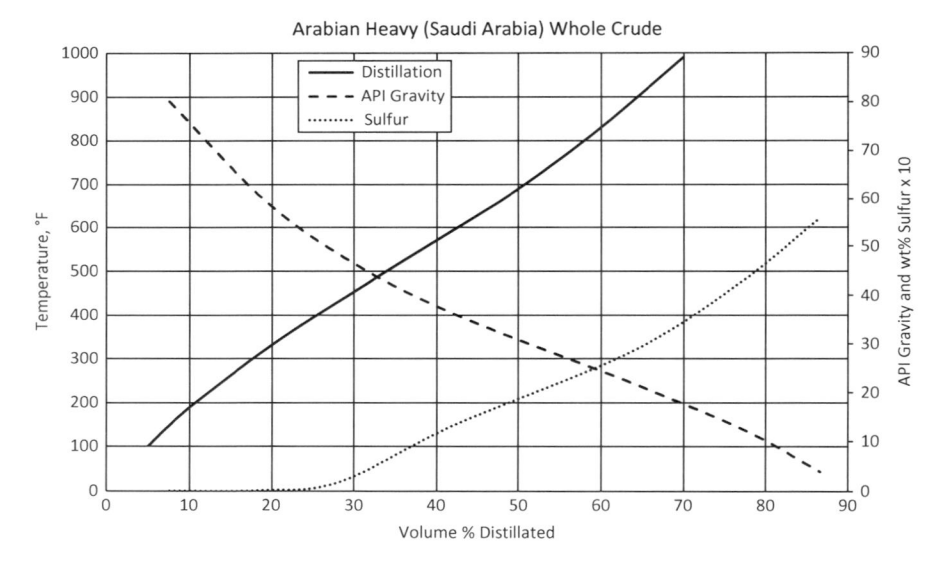

Arabian Heavy (Saudi Arabia) Whole Crude

Distillation, API Gravity, Sulfur vs Volume % Distillated

Selection of Crude Oil Assays for Petroleum Refining

Arabian Light, Saudi Arabia

Source: O&GJ

Cut	Whole Crude	Light Ends	Light Naphtha	Heavy Naphtha	Kerosene	Diesel/Lt. Gasoil	Vacuum Gasoil	Atm Residuum	Vac Residuum
TBP Range, °F	-----	X-68	68-212	212-302	302-455	455-650	650-1049	650+	1049+
Yield, v%	100.0	1.7	9.0	8.4	15.0	19.8	32.5	46.1	13.6
Gravity, °API	33.4		78.5	59.6	38.5	37.1	22.8	17.6	4.0
Sulfur, w%	1.8		0.024	0.027	0.094	1.05	2.46	3.08	6.5
CCR, w%								7.6	22.4
Viscosity	6.14 cSt @ 100°F					3.28 cSt @ 100°F	52.5 cSt @ 100°F	21.0 cSt @ 210°F	2017 cSt @ 210°F
Pour Point, °F	-30					0	90	40	115
RVP, psi	4.2		8.3						
Ni, wppm									22
V, wppm									94
Paraffins, v%			87.2	69.5					
Naphthenes, v%			10.4	18.2					
Aromatics, v%			2.4	12.3	20.4				
RON Clear			54.7						
Freeze Point, °F					-69				
Aniline Point, °F					135	156	179		
Smoke Point, mm					23				

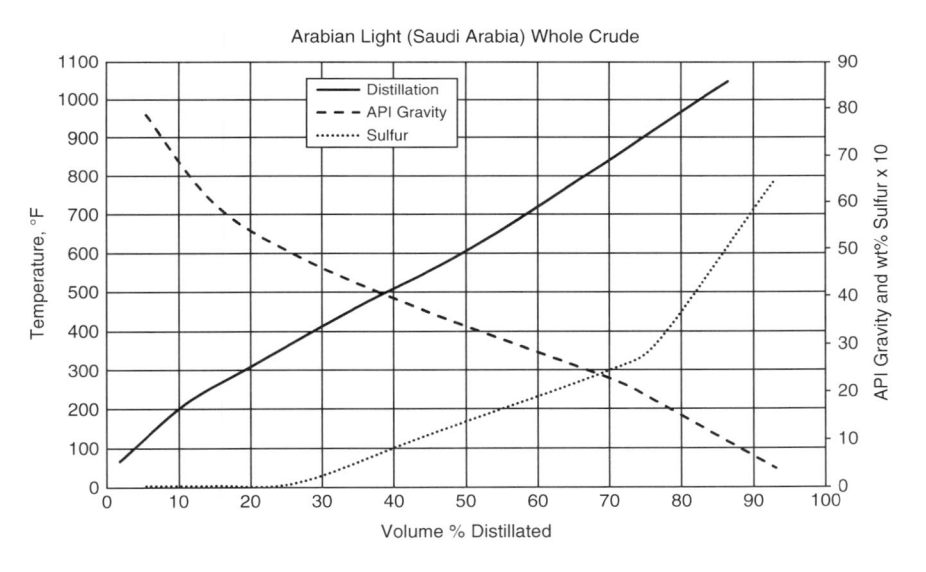

Arabian Light (Saudi Arabia) Whole Crude

Arabian Medium, Saudi Arabia

Source: O&GJ

Cut	Whole Crude	Light Ends	Light Naphtha	Heavy Naphtha	Kerosene	Diesel/Lt. Gasoil	Vacuum Gasoil	Atm Residuum	Vac Residuum
TBP Range, °F	-----	X-68	68-212	212-302	302-455	455-650	650-1049	650+	1049+
Yield, v%	100.0	1.2	8.9	7.7	14.5	18.1	30.9	49.6	18.7
Gravity, °API	30.8		77.7	59.1	48	36	21.9	15	4.9
Sulfur, w%	2.4		0.043	0.05	0.14	1.24	2.91	3.9	5.35
CCR, w%								9.9	23.3
Viscosity	9.41 cSt @ 100°F				1.13 cSt @ 100°F	3.53 cSt @ 100°F	49.2 cSt @ 100°F	36.0 cSt @ 210°F	3847 cSt @ 210°F
Pour Point, °F	5					15	100	55	115
RVP, psi	3.2		7.9						
Ni, wppm									32
V, wppm									96
Paraffins, v%			85.3	68.5					
Naphthenes, v%			12.3	18.7					
Aromatics, v%			2.4	12.7	20.6				
RON Clear			54.5						
Freeze Point, °F					-62				
Aniline Point, °F					136	157	176		
Smoke Point, mm					23				

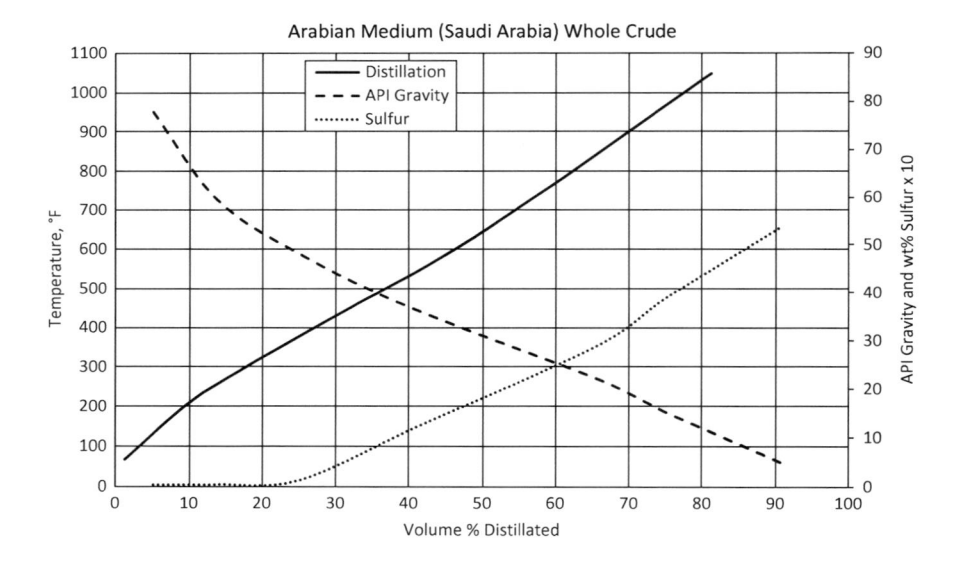

Arabian Medium (Saudi Arabia) Whole Crude

Selection of Crude Oil Assays for Petroleum Refining

Arjuna, Java, Indonesia — Source: O&GJ

Cut	Whole Crude	Light Ends	Light Naphtha	Heavy Naphtha	Distillate	Atm Gasoil	Vac Gasoil	Vac Residuum
TBP Range, °F	-----	X-50	50-230	230-380	380-640	640-780	780-1049	1049+
Yield, v%	100.0	3.3	13.8	17.9	28.7	10.7	20.2	5.4
Gravity, °API	37.7		68.2	48.5	33.0	31.3	25.8	5.6
Sulfur, w%	0.12		0.01	0.01	0.09	0.12	0.15	0.3
CCR, w%						0.08	0.42	
Viscosity	37.7 SUS @ 100°F							4100 cSt @ 275°F
Pour Point, °F	80				15	90	120	
RVP, psi	7		6.9					
Salt, ptb	< 10							
Ni, wppm							0.3	0.9
V, wppm							0	0.3
Paraffins, v%			41	43.8				
Naphthenes, v%			34.9	29.1				
Aromatics, v%			5.7	27.1				
RON Clear			72.2	61.5				
Aniline Point, °F						90.2	98.4	
Cetane Number					41			

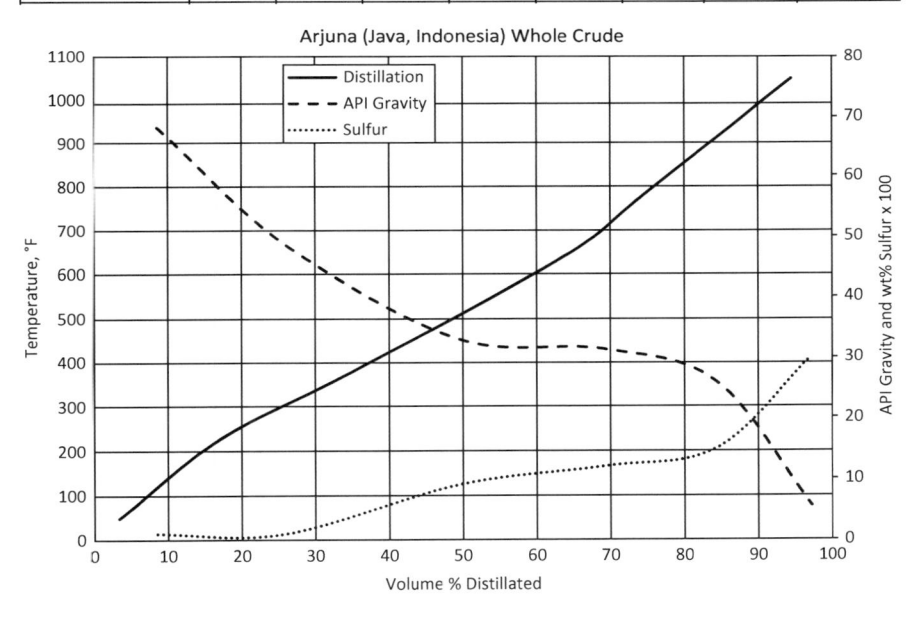

Arjuna (Java, Indonesia) Whole Crude

Bachequero (16.8°API) Heavy, Venezuela   Source: O&GJ

Cut	Whole Crude	Light Naphtha	Heavy Naphtha	Kerosene	Distillate	Atm Residuum
TBP Range, °F	-----	90-200	200-350	350-475	475-650	650+
Yield, v%	100.0	2.5	6.0	5.0	15.5	71.0
Gravity, °API	16.8	65.0	50.0	36.4	28.0	5.6
Sulfur, w%	2.4			0.48	0.99	3.0
Viscosity	1362 SUS @ 100°F					4100 cSt @ 275°F
Pour Point, °F	-10			-80		
RVP, psi		2.8				
Paraffins, v%		43.4		19.2		
Naphthenes, v%		51.9		54.8		
Aromatics, v%		4.7		26		
Aniline Point, °F				125	134	

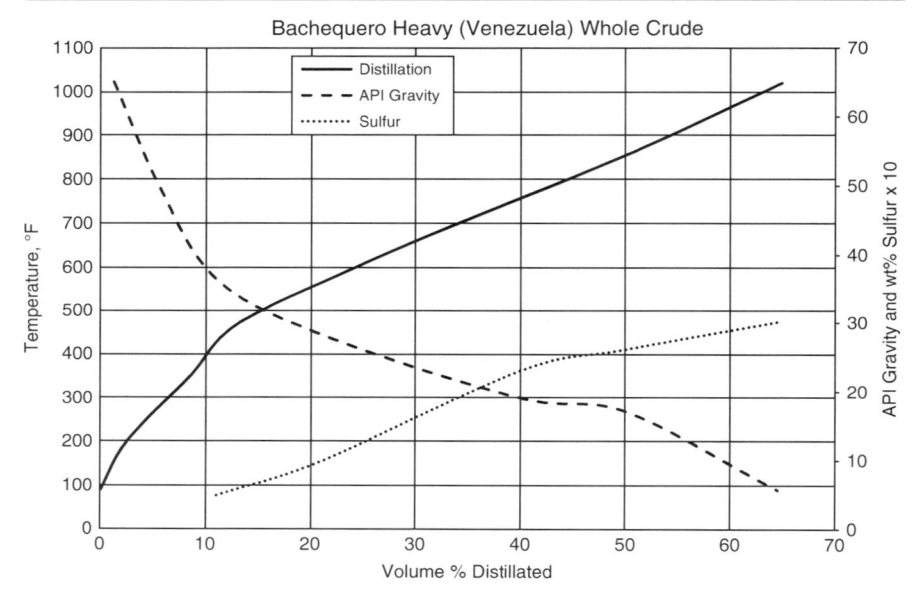

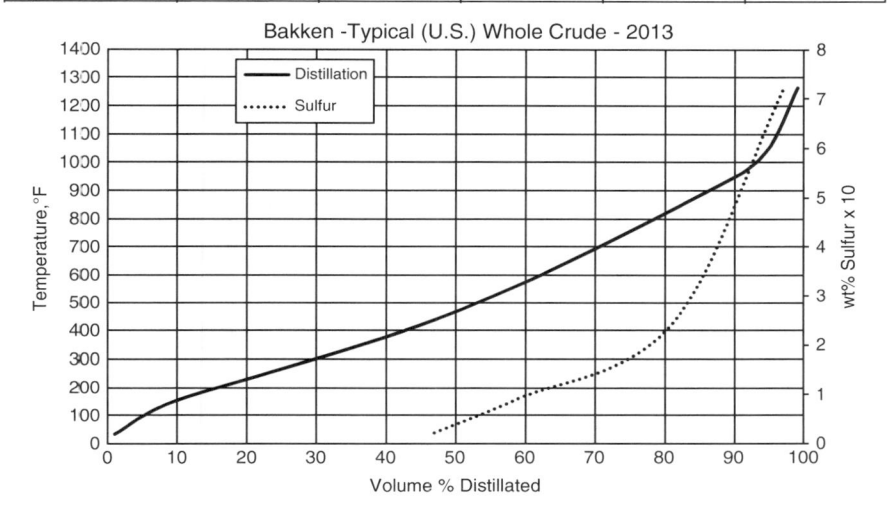

Selection of Crude Oil Assays for Petroleum Refining

Bakken - Typical (U.S.)   2013

Cut	Whole Crude	Kerosene	Diesel	Vacuum Gasoil	Resid
TBP Range, °F	-----	375-500	500-620	620-1050	1050+
Gravity, °API	41-42.1				14.0
Sulfur, w%	0.13-0.20	0.02	0.09	0.24	0.68-0.75
Pour Point, °F	−25				
Ni, wppm					7
V, wppm					2

Bakken -Typical (U.S.) Whole Crude - 2013

Bonney Light, Nigeria

Source: O&GJ

Cut	Whole Crude	Light Ends	Light Naphtha	Heavy Naphtha	Kerosene	Distillate	Vac Gasoil	Vac Residuum
TBP Range, °F	-----	X-60	60-167	167-347	347-482	482-662	662-977	977+
Yield, v%	100.0	2.2	6.4	22.0	15.4	23.2	23.1	7.7
Gravity, °API	37.6		79.9	53.6	40.2	33.2	25.4	11.8
Sulfur, w%	0.13		0.0002	0.003	0.03	0.13	0.21	0.39
Nitrogen, wppm							1150	
CCR, w%	1.1							12
Viscosity	36 SUS @ 100°F					40.3 SUS @ 100°F	48.1 SUS @ 210°F	2030 SUS @ 210°F
Pour Point, °F	5				−70	20	105	
Ni, wppm	4						<0.1	40
V, wppm	< 0.5						<0.1	3
Paraffins, v%			77	34				
Naphthenes, v%			21.5	55				
Aromatics, v%			1.5	11				
RON Clear			78					
Freeze Point, °F					−53			
Aniline Point, °F					120	136	190	
Diesel Index					55	53		
Cetane Index						51		
Smoke Point, mm					19			

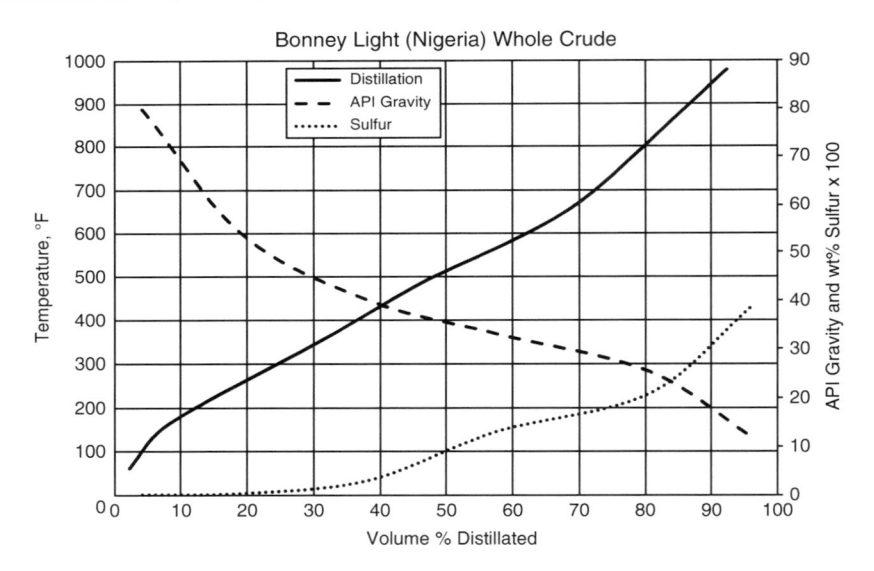

Bonney Light (Nigeria) Whole Crude

[Chart: Distillation, API Gravity, and Sulfur vs. Volume % Distilled. X-axis: Volume % Distilled (0–100). Left Y-axis: Temperature, °F (0–1000). Right Y-axis: API Gravity and wt% Sulfur × 100 (0–90).]

Selection of Crude Oil Assays for Petroleum Refining

Bonney Medium, Nigeria Source: O&GJ

Cut	Whole Crude	Light Ends	Light Naphtha	Heavy Naphtha	Kerosene	Distillate	Vac Gasoil	Vac Residuum
TBP Range, °F	-----	X-60	60-167	167-347	347-482	482-662	662-977	977+
Yield, v%	100.0	0.7	2.1	8.7	14.7	29.7	31.3	12.8
Gravity, °API	26		79.2	50.1	34.4	27.5	19.7	10.1
Sulfur, w%	0.23		0.001	0.01	0.063	0.18	0.31	0.48
Nitrogen, wppm								
CCR, w%	1.8							12.7
Viscosity	60.7 SUS @ 100°F					44.6 SUS @ 100°F	53.1 SUS @ 210°F	3690 SUS @ 210°F
Pour Point, °F	<-5					<-70	-15	80
Ni, wppm	7						<0.1	52
V, wppm	1						<0.1	7
Paraffins, v%			73	27.5				
Naphthenes, v%			24	58.5				
Aromatics, v%			3	14				
RON Clear			80					
Freeze Point, °F					-85			
Aniline Point, °F					110	123	163	
Diesel Index						42	37	
Cetane Index						40		
Smoke Point, mm					17			

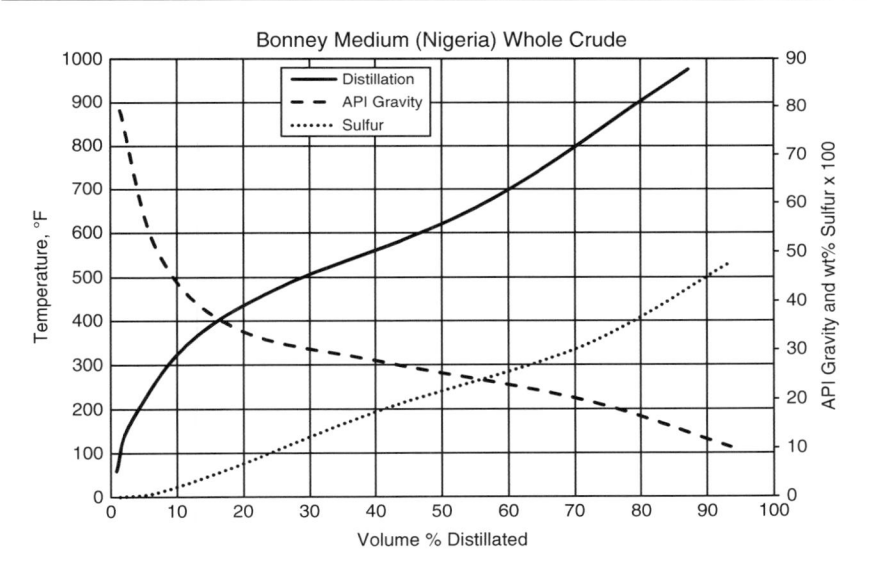

Bonney Medium (Nigeria) Whole Crude

[Graph showing Distillation (solid line), API Gravity (dashed line), and Sulfur (dotted line) versus Volume % Distillated on x-axis (0 to 100); Temperature, °F on left y-axis (0 to 1000); API Gravity and wt% Sulfur × 100 on right y-axis (0 to 90).]

Brega, Libya

Source: O&GJ

Cut	Whole Crude	Light Ends	Light Naphtha	Heavy Naphtha	Kerosene	Distillate	Vac Gasoil	Atm Residuum	Vac Residuum
TBP Range, °F	-----	X-68	68-212	212-302	302-455	455-650	650–1049	650+	1049+
Yield, v%	100.0	2.4	12.4	10.7	17.4	20.3	27.8	36.8	9.0
Gravity, °API	40.4		73.8	56.8	46.7	38.1	27.7	23.6	12.3
Sulfur, w%	0.21		0.014	0.02	0.035	0.10	0.30	0.41	0.69
Nitrogen, wppm									
CCR, w%								4.1	14.6
Viscosity	3.56 cSt @ 100°F				1.17 cSt @ 100°F	3.61 cSt @ 100°F	6.1 cSt @ 210°F	14.1 cSt @ 210°F	620 cSt @ 210°F
Pour Point, °F	30					15	105	95	>120
RVP, psi	6.4		7.4						
Ni, wppm									32
V, wppm									24
Paraffins, v%			72.5	53	51.2				
Naphthenes, v%			25.3	39.3	34.7				
Aromatics, v%			2.2	7.7	14.1				
RON Clear			59.9						
Freeze Point, °F						−68			
Aniline Point, °F						144	172	206	
Smoke Point, mm						27			

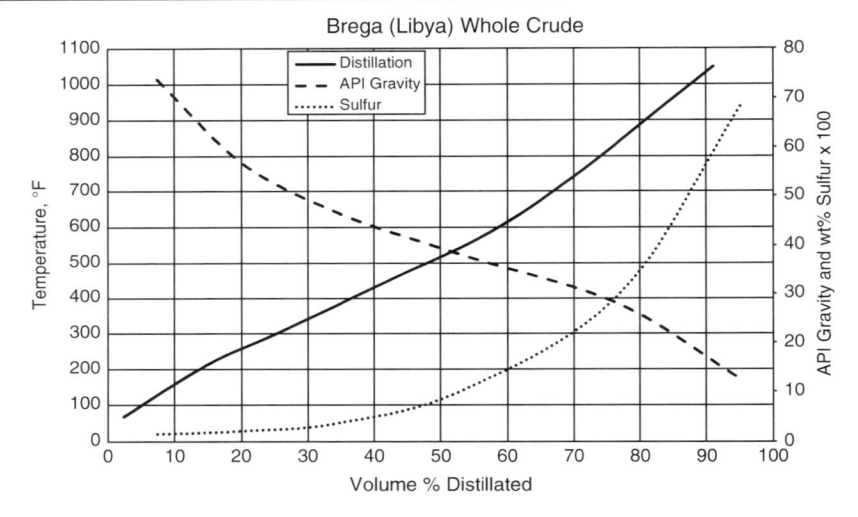

Brega (Libya) Whole Crude

Selection of Crude Oil Assays for Petroleum Refining

Brent (North Sea) 2009

Cut	Whole Crude	Light Naphtha	Medium Naphtha	Heavy Naphtha	Light Kerosene	Kerosene	Diesel	Vacuum Gasoil	Resid
TBP Range, °F	-----	X-158	158-212	212-374	374-455	455-536	536-649	649-1049	1049+
Yield, v%	100.0	7.6	6.3	18.6	7.0	9.1	11.1	27.1	12.3
Yield, w%	100.0	6.0	5.4	17.3	6.8	9.1	11.4	29.3	13.7
Gravity, °API	38.3	82.9	65.3	51.3	42.8	38.6	33.4	25.4	21.0
Sulfur, w%	0.4	0	0.001	0.002	0.012	0.07	0.327	0.603	1.303
CCR, w%	2.1							0.1	15.8
Viscosity	3.9 mm^2/s @ 100°F				1.4 mm^2/s @ 100°F	2.4 mm^2/s @ 100°F	5.0 mm^2/s @ 100°F		400 mm^2/s @ 200°F
Pour Point, °F	−44				−65	3	73		
TAN, mg/g	0.1								
Ni, wppm	1								7.8
V, wppm	6								46.6
Naphthenes, v%			28.6	35					
Aromatics, v%			9.6	17.5	22	22.9			
RON Clear		71.8	62.5	43.2					
Freeze Point, °F					−83	−27			
Aniline Point, °F					282	304	329	379	
Cetane Index					44	44.3	51.5		
Smoke Point, mm					20	17			

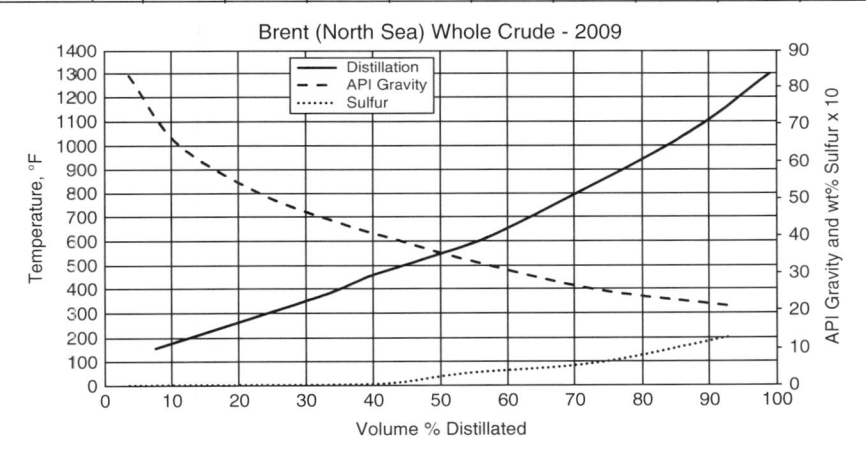

Brent (North Sea) Whole Crude - 2009

— Distillation
- - API Gravity
⋯⋯ Sulfur

X-axis: Volume % Distillated
Y-axis (left): Temperature, °F
Y-axis (right): API Gravity and wt% Sulfur × 10

Darius, Iran Source: O&GJ

Cut	Whole Crude	Light Ends	Light Naphtha	Heavy Naphtha	Kerosene	Distillate	Atm Gasoil	Vac Gasoil	Vac Residuum
TBP Range, °F	-----	X-60	60-200	200-350	350-540	540-620	620-690	690-1010	1010+
Yield, v%	100.0	2.8	10.0	15.3	17.6	7.0	6.4	22.2	18.9
Gravity, °API	33.9		83.3	56.9	42.7	33.4	26.4	23.4	5.3
Sulfur, w%	2.45		0.07	0.13	0.70	1.84	2.67	2.97	5.57
Nitrogen, wppm							0.055	0.093	
RCR, w%									18.9
Viscosity	40 SUS @ 130°F								620 cSt @ 210°F
Pour Point, °F	0				-23	57			
RVP, psi	6.4								
Ni, wppm								1.4 Equivalent	Ni+V 195
V, wppm									
Naphthenes, v%				N+A 30.2					
Aromatics, v%									
Arom. Carbon,w%							15.1	14.7	
RON Clear			57.2						
Cetane Index						51.5	54		

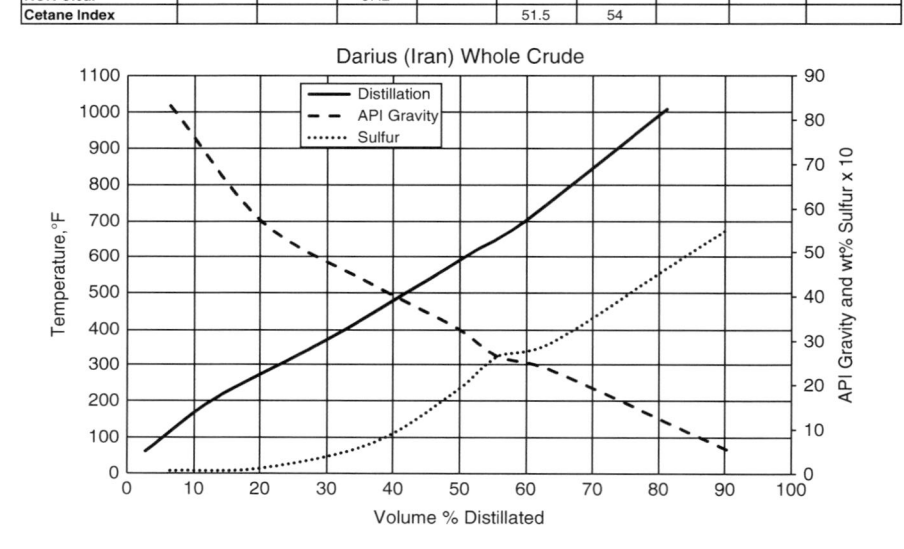

Darius (Iran) Whole Crude

[Chart: Distillation, API Gravity, and Sulfur curves vs. Volume % Distillated. X-axis: Volume % Distillated (0–100). Left Y-axis: Temperature, °F (0–1100). Right Y-axis: API Gravity and wt% Sulfur x 10 (0–90).]

Eagle Ford - Typical (U.S.) 2013

Cut	Whole Crude	Kerosene	Diesel	Vacuum Gasoil	Resid
TBP Range, °F	-----	375-500	500-620	620-1050	1050+
Gravity, °API	47-59				
Sulfur, w%	0.05-0.11	0.02	0.07	0.19	0.600
CCR, w%	0.07-0.24				
Viscosity	29-35 SSU @ 100°F				
Pour Point, °F	6-15				
RVP, psi	6.5-9.7				
Salt, ptb	0-1				
TAN, mg/g	0.02-0.03				
Ni, wppm	0.1-0.21				2-4
V, wppm	0.03-0.5				0.5-0.6

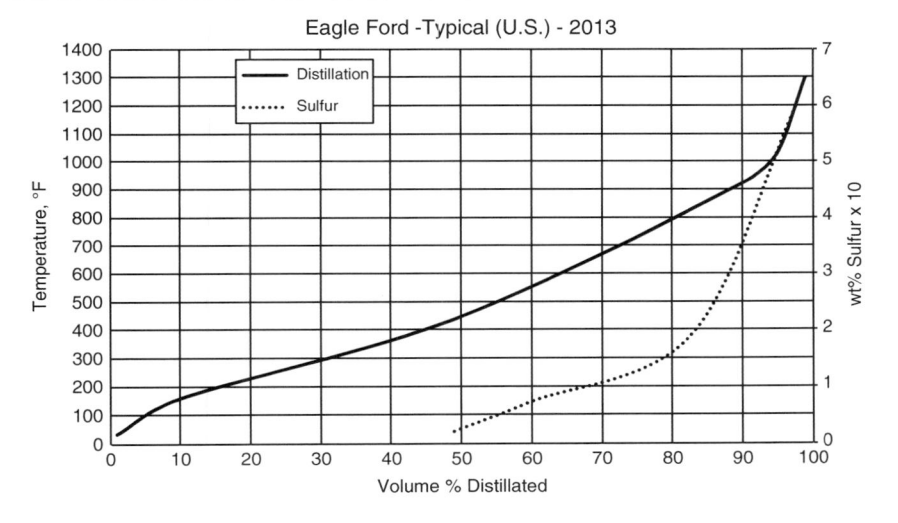

Eagle Ford -Typical (U.S.) - 2013

Ecuador Oriente, Ecuador   Source: O&GJ

Cut	Whole Crude	Light Ends	Light Naphtha	Heavy Naphtha	Kerosene	Distillate	Vac Gasoil	Atm Residuum
TBP Range, °F	-----	X-60	60-140	170-310	310-520	520-680	680-1000	680+
Yield, v%	100.0	0.9	3.2	11.9	18.4	18.9	27.7	44.94
Gravity, °API	30.4		90.6	58.2	41.9	32.0	23.3	15.2
Sulfur, w%	0.87		0.019	0.015	0.144	0.65	1.11	1.43
Nitrogen, wppm							1300	
CCR, w%								11.3
Viscosity	61.8 SUS @ 100°F					44.3 SUS @ 100°F	55.5 SUS @ 210°F	325 SUS @ 210°F
Pour Point, °F	20					25	105	95
RVP, psi	4.8							
Salt, ptb	5							
Ni, wppm	28						<0.1	56
V, wppm	61						<0.1	123
Paraffins					49.4			
Naphthenes					43.9			
Aromatics				6.7	18			
RON Clear			80.1					
Freeze Point, °F					-52			
Aniline Point, °F				131	137		198	
Diesel Index						51.2		
Cetane Index					46	52.4		
Smoke Point, mm					20			

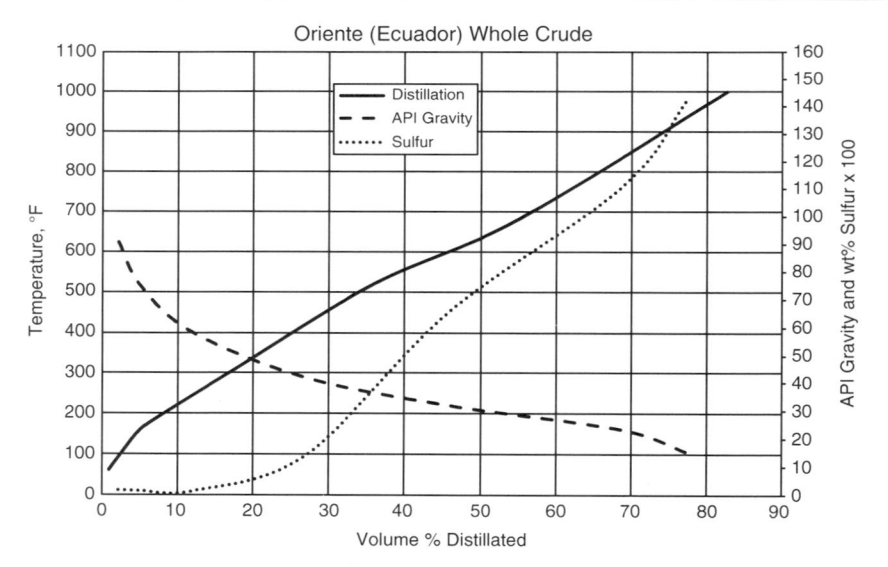

Oriente (Ecuador) Whole Crude

Ekofisk, Norway

Source: O&GJ

Cut	Whole Crude	Light Ends	Light Naphtha	Gasoline	Kerosene	Distillate	Atm Residuum
TBP Range, °F	-----	X-60	60-200	60-400	400-500	500-650	650+
Yield, v%	100.0	1.0	10.7	31.0	13.5	15.7	38.8
Gravity, °API	36.3		77.2	60.1	40.2	33.7	21.5
Sulfur, w%	0.21		0.003	0.0024	0.05	0.11	0.39
RCR, w%						0.08	4.0
Viscosity	42.5 SUS @ 100°F				32.3 SUS @ 100°F	43.8 SUS @ 100°F	80.3 SFS @ 122°F
Pour Point, °F	68					25	85
RVP, psi	5.1						
Salt, ptb	14.5						
BS&W, v%	1.0						
Ni, wppm							5.4
V, wppm							1.95
Paraffins, v%				56.5			
Naphthenes, v%				29.5			
Aromatics, v%				14.0	13.1		
RON Clear			74.4	52			
Freeze Point, °F					−38		
Aniline Point, °F					146	164	
Cetane Index						56.5	
Smoke Point, mm					21		

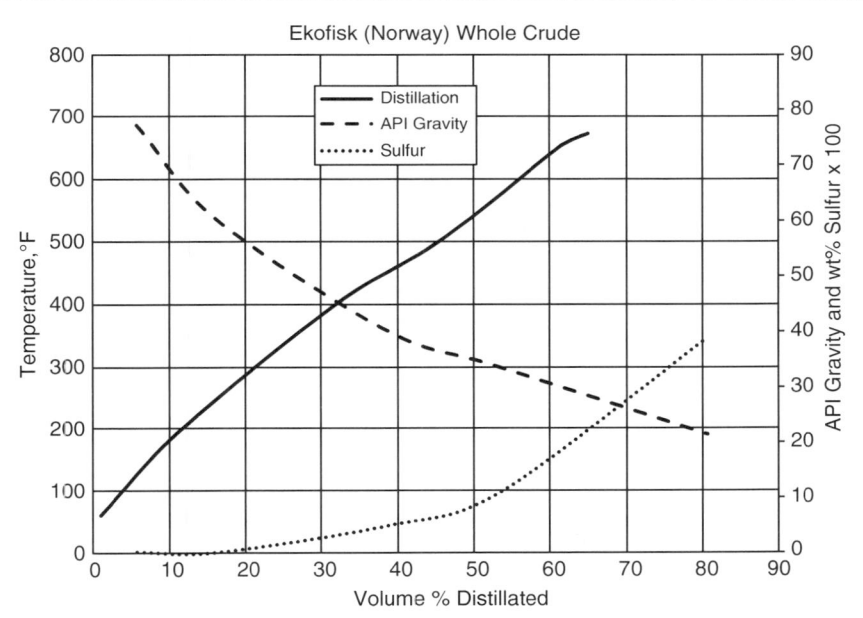

Ekofisk (Norway) Whole Crude

Escravos, Nigeria

Source: O&GJ

Cut	Whole Crude	Light Ends	Light Naphtha	Heavy Naphtha	Kerosene	Distillate	Vac Gasoil	Atm Residuum
TBP Range, °F	-----	X-60	60-140	170-310	310-520	520-680	680-1000	680+
Yield, v%	100.0	2.3	3.4	17.0	26.2	18.6	24.9	30.51
Gravity, °API	36.2		88.1	54	39.6	31.4	23.2	20.4
Sulfur, w%	0.16		0.008	0.006	0.057	0.17	0.25	0.3
Nitrogen, wppm							1200	
CCR, w%	1.3							3.62
Viscosity	38.0 SUS @ 100°F					44.1 SUS @ 100°F	56.4 SUS @ 210°F	93.2 SUS @ 210°F
Pour Point, °F	50					30	110	105
RVP, psi	4.5							
Salt, ptb	10							
Ni, wppm	5.1						1	15
V, wppm	0.2						<0.1	0.7
Paraffins, v%			89	38				
Naphthenes, v%			10	47.1				
Aromatics, v%			1	14.9	21.5			
RON Clear			81.2					
Freeze Point, °F					-48			
Aniline Point, °F				117	133		194	
Diesel Index						49.9		
Cetane Index					42	50.5		
Smoke Point, mm					19			

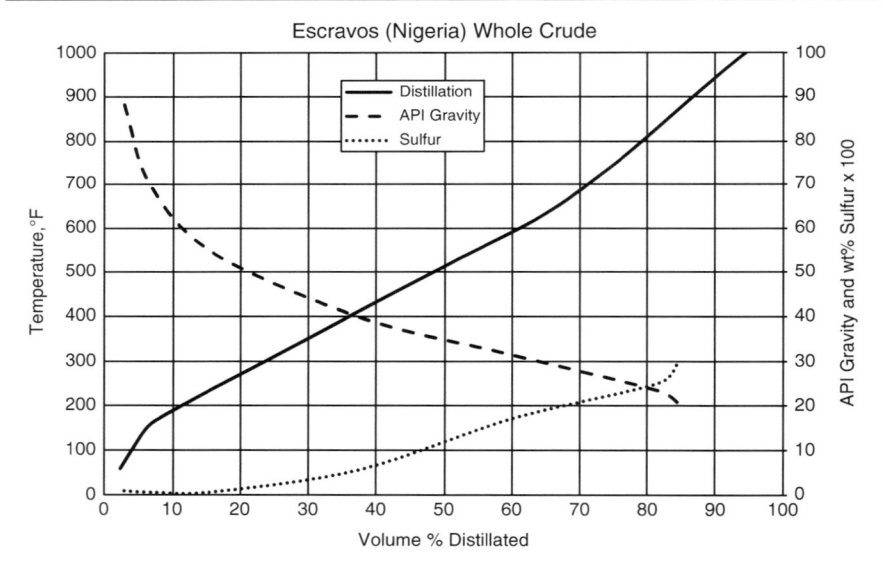

Escravos (Nigeria) Whole Crude

— Distillation
-- API Gravity
····· Sulfur

X-axis: Volume % Distillated
Left Y-axis: Temperature, °F
Right Y-axis: API Gravity and wt% Sulfur × 100

Selection of Crude Oil Assays for Petroleum Refining

Iranian Heavy, Iran **Source: O&GJ**

Cut	Whole Crude	Light Ends	Light Gasoline	Light Naphtha	Heavy Naphtha	Kerosene	Distillate	Vac Gasoil	Atm Residuum	Vac Residuum
TBP Range, °F	-----	X-60	60-200	200-300	300-400	400-500	500-650	650-1000	650+	1000+
Yield, v%	100.0	2.1	7.9	9.6	9.4	9.2	14.0	26.0	47.8	21.8
Gravity, °API	30.8		78.8	60	48.2	40.1	34.0	23.0	14.4	6.3
Sulfur, w%	1.6		0.1	0.13	0.22	0.44	1.1	1.8	2.5	3.2
RSH, wppm				340	340	100	14			
Nitrogen, wppm								1600	4700	8300
Ethane, v%		0.1								
Propane, v%		0.5								
I-Butane, v%		0.3								
N-Butane, v%		1.2								
CCR, w%	5									9.5
Viscosity	9.81 cSt @ 100°F							30.0 cSt @ 130°F	500 cSt @ 130°F	4250 cSt @ 210°F
Penetration, 77°F										47
Pour Point, °F	−5						−40	15	90	75	135
RVP, psi	6.6									
Ni, wppm	30									
V, wppm	88									
Paraffins, v%				53	50	27				
Naphthenes, v%				34	35	43				
Aromatics, v%				13	15	30				
RON Clear			66.1	49.6						
Freeze Point, °F							−35			
Aniline Point, °F					130	140	157	176		
Smoke Point, mm						23	19			

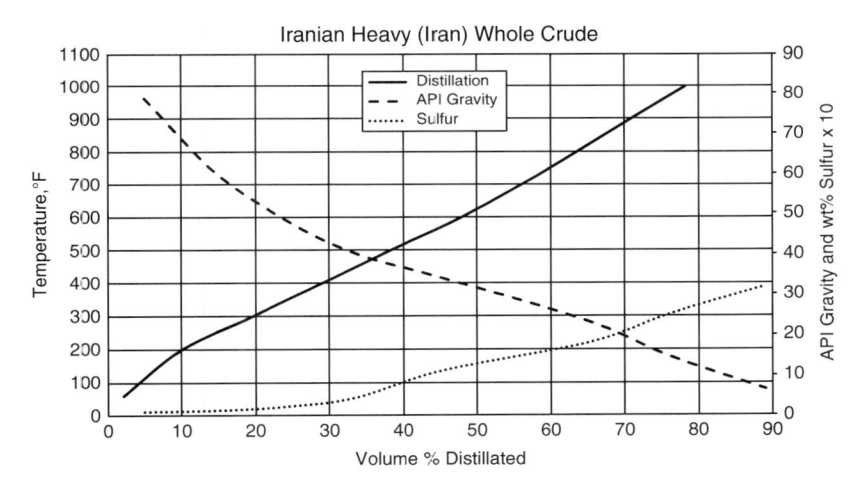

Iranian Heavy (Iran) Whole Crude

Chart showing Distillation, API Gravity, and Sulfur curves vs. Volume % Distillated (0-90). Left axis: Temperature, °F (0-1100). Right axis: API Gravity and wt% Sulfur x 10 (0-90).

Iranian Light, Iran

Source: O&GJ

Cut	Whole Crude	Light Ends	Light Gasoline	Light Naphtha	Heavy Naphtha	Kerosene	Distillate	Vac Gasoil	Atm Residuum	Vac Residuum
TBP Range, °F	-----	X-60	60-200	200-300	300-400	400-500	500-650	650-1000	650+	1000+
Yield, v%	100.0	1.9	8.1	10.0	10.1	10.5	14.0	26.8	45.4	18.6
Gravity, °API	33.5		75	57.8	49.6	40.9	34.4	23.7	17	11.0
Sulfur, w%	1.4		0.076	0.09	0.11	0.34	1	1.8	2.4	3.3
RSH, wppm			210	210	20	23				
Nitrogen, wppm								1300	2900	5100
Ethane, v%		0.1								
Propane, v%		0.4								
I-Butane, v%		0.3								
N-Butane, v%		1.1								
CCR, w%	3.4							6.8		
Viscosity	6.41 cSt @ 100°F							20.0 cSt @ 130°F	190 cSt @ 130°F	930 cSt @ 210°F
Penetration, 77°F										285
Pour Point, °F	-20						-35	20	90	75	100
RVP, psi	6.5									
Ni, wppm	13									
V, wppm	35									
Paraffins, v%					50	54	30			
Naphthenes, v%					33	30	41			
Aromatics, v%					17	16	29			
RON Clear			64.7	47.4						
Freeze Point, °F							-30			
Aniline Point, °F						135	146	162	183	
Smoke Point, mm						25	21			

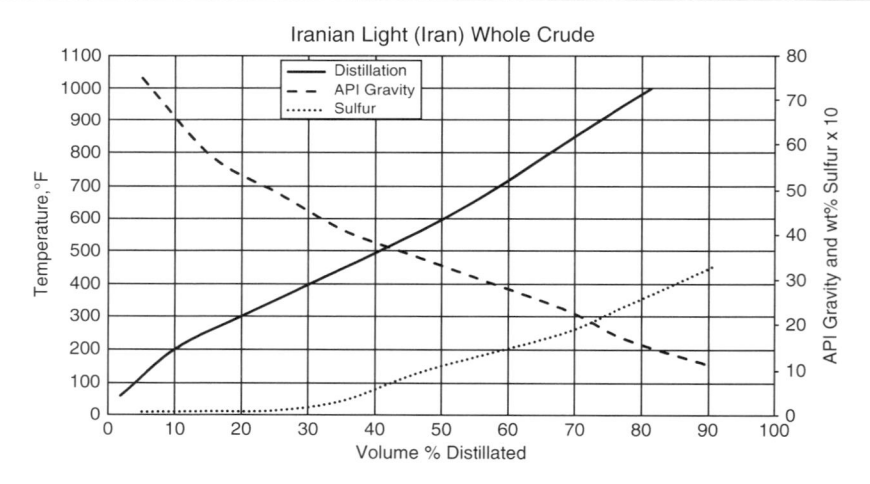

Iranian Light (Iran) Whole Crude

[Chart showing Distillation (Temperature, °F), API Gravity, and Sulfur (wt% Sulfur × 10) vs Volume % Distilled]

Kirkuk, Iraq

Source: O&GJ

Cut	Whole Crude	Light Ends	Light SR	Light Naphtha	Heavy Naphtha	Kerosene	Light Distillate	Heavy Distillate	Extra Hvy Distillate	Vac Residuum
TBP Range, °F	-----	X-60	60-149	149-212	212-302	302-392	392-482	572-662	662-698	698+
Yield, v%	100.0	1.4	6.1	6.4	10.0	10.5	9.9	8.8	3.1	34.4
Gravity, °API	35.9			70.4	58.2	49.4			28.6	15.4
Sulfur, w%	1.95			0.0333	0.0903	0.181	0.3	1.6	2.17	4
CCR, w%	3.8									10
Viscosity	4.61 cSt @ 100°F						1.68 cSt @ 100°F			43.9 cSt @ 210°F
Pour Point, °F	-33									86
Ni, wppm										<3
V, wppm										58
Paraffins, v%			97	80	69					
Naphthenes, v%				18	21					
Aromatics, v%				2	10					
RON Clear			73	52	38					
Aniline Point, °F						130.64				
Cetane Index							53			
Smoke Point, mm						24				
Wax Content, w%	3.6									
Asphaltenes, w%	1.5									

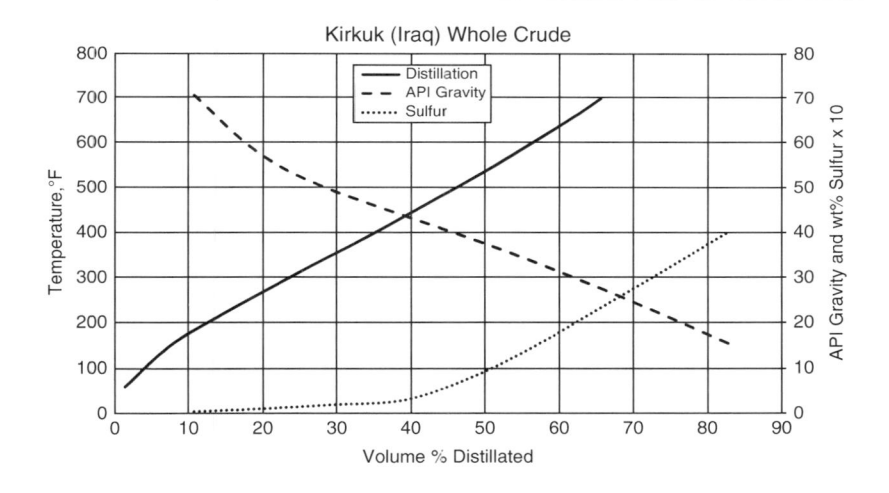

Kirkuk (Iraq) Whole Crude

Kuwait Crude (Kuwait)

Source: O&GJ

Cut	Whole Crude	Light Ends	Light SR	Light Naphtha	Heavy Naphtha	Kerosene	Light Distillate	Vac Gasoil	Vac Residuum
TBP Range, °F	-----	X-60	60-140	140-170	170-310	310-520	520-680	680-1000	680+
Yield, v%	100.0	0.8	5.5	1.9	12.0	18.2	14.1	26.6	47.5
Gravity, °API	31.2		94.6	78.2	62.2	45.9	33.7	21.7	14.0
Sulfur, w%	2.5		0.01	0.02	0.02	0.28	1.66	2.91	4.14
Nitrogen, wppm								950	
CCR, w%	5.3								9.37
Viscosity	58.7 SUS @ 100°F						41.3 SUS @ 100°F	53.9 SUS @ 210°F	267 SUS @ 210°F
Pour Point, °F	0						20	100	70
RVP, psi	5.4								
Salt, ptb	3								
Ni, wppm	9.6							0.1	18
V, wppm	31							0.4	59
Paraffins, v%			98.5	86.2	67.9				
Naphthenes, v%			1.4	12.5	22.1				
Aromatics, v%			0.1	1.3	10	19.7			
RON Clear			78.5	58.8					
Freeze Point, °F						−46			
Aniline Point, °F						133	143	178	
Diesel Index							53.6		
Cetane Index						52.3	55		
Smoke Point, mm						24			

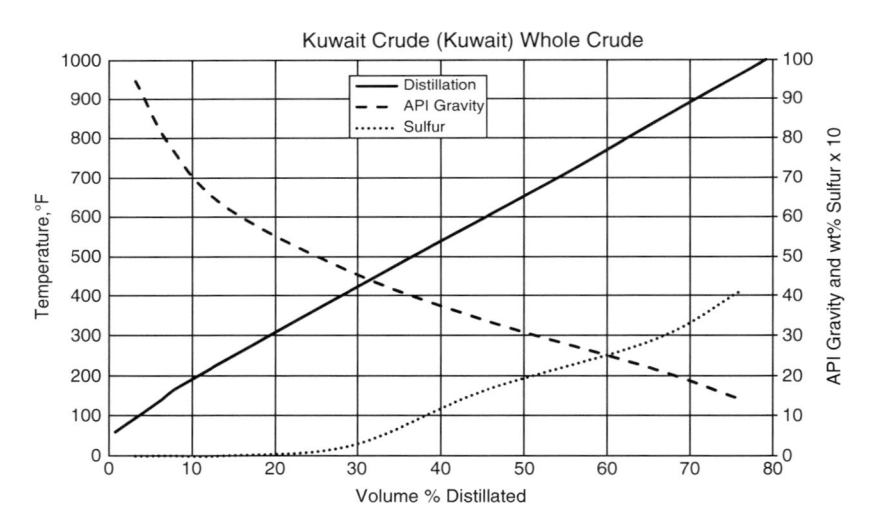

Kuwait Crude (Kuwait) Whole Crude

Selection of Crude Oil Assays for Petroleum Refining

Lloydminster Blend, Alberta, Canada — 2009

Cut	Whole Crude	Light Naphtha	Medium Naphtha	Heavy Naphtha	Light Kerosene	Kerosene	Diesel	Vacuum Gasoil	Resid
TBP Range, °F	-----	X-158	158-212	212-374	374-455	455-536	536-649	649-1049	1049+
Yield, v%	100.0	5.6	3.4	10.1	4.7	5.5	9.1	25.3	35.3
Gravity, °API	20.7	86.9	66.7	50.1	37.8	31.9	26.4	18.4	2.1
Sulfur, w%	3.15	0.022	0.046	0.191	0.426	0.61	0.89	2.9	5.620
CCR, w%	9.2							0.5	25.2
Viscosity	101 mm^2/s @ 100°F				1.6 mm2/s @ 100°F	2.9 mm2/s @ 100°F	6.4 mm2/s @ 100°F		
Pour Point, °F	−26				−150	−80	−9		
TAN, mg/g	0.78								
Ni, wppm	52.7							0	123
V, wppm	105							0.1	231
Naphthenes, v%			34.1	38.2					
Aromatics, v%			8.7	16.5	21.7	22.6			
RON Clear		70.5	57.6	55.2					
Freeze Point, °F					−92	−33			
Aniline Point, °F					257	262	273	324	
Cetane Index					36.1	36.1	41.2		
Smoke Point, mm					20.6	17.5			

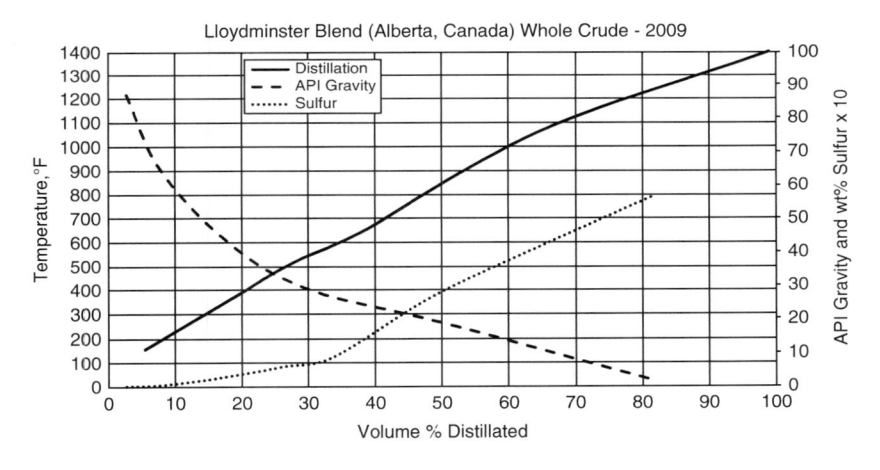

Lloydminster Blend (Alberta, Canada) Whole Crude - 2009

— Distillation
– – API Gravity
······ Sulfur

Louisiana Light Sweet (LLS), USA 2009

Cut	Whole Crude	Light Naphtha	Medium Naphtha	Heavy Naphtha	Light Kerosene	Kerosene	Diesel	Vacuum Gasoil	Resid
TBP Range, °F	-----	X-158	158-212	212-374	374-455	455-536	536-649	649-1049	1049+
Yield, v%	100.0	4.1	4.1	18.5	10.7	10.9	15.8	28.9	5.9
Yield, w%	100.0	3.2	3.5	16.9	10.3	10.8	16.0	31.1	8.2
Gravity, °API	36.1	83.9	66.4	51.8	43.4	38.0	33.6	24.3	-10.4
Sulfur, w%	0.45	0.018	0.022	0.031	0.074	0.2	0.361	0.792	1.343
CCR, w%	1.1							0.3	16.2
Viscosity	4.3 mm^2/s @ 100°F				1.5 mm^2/s @ 100°F	2.9 mm^2/s @ 100°F	5.7 mm^2/s @ 100°F		768 mm^2/s @ 200°F
Pour Point, °F	-35				-89	-17	50		
TAN, mg/g	0.58								
Ni, wppm	7.1							0.1	89
V, wppm	1.2							0	17
Naphthenes, v%			23	36					
Aromatics, v%			7.5	13.4	15.8	17.9			
RON Clear		73.3	66.6	49.8					
Freeze Point, °F					-74	7			
Aniline Point, °F					284	307	327	378	
Cetane Index					46	46.2	52		
Smoke Point, mm					22	17			

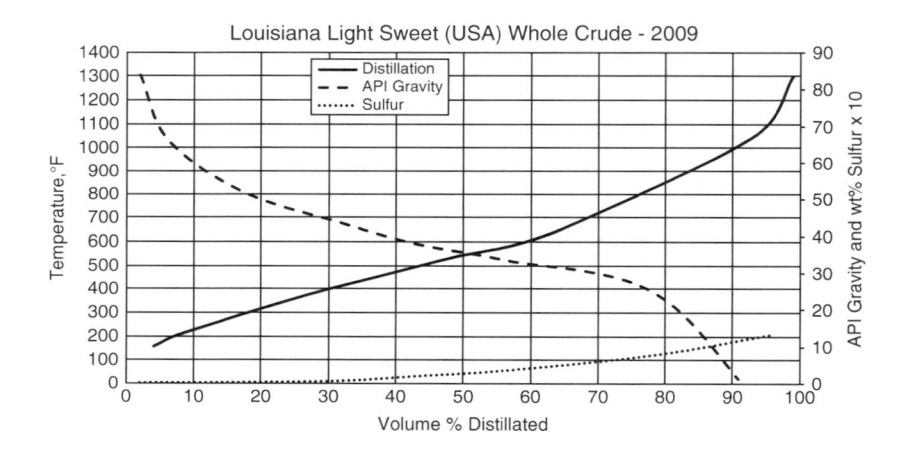

Louisiana Light Sweet (USA) Whole Crude - 2009

Selection of Crude Oil Assays for Petroleum Refining

Maya, Mexico — 2009

Cut	Whole Crude	Light Naphtha	Medium Naphtha	Heavy Naphtha	Light Kerosene	Kerosene	Diesel	Vacuum Gasoil	Resid
TBP Range, °F	-----	X-158	158-212	212-374	374-455	455-536	536-649	649-1049	1049+
Yield, v%	100.0	3.2	3.2	11.3	6.0	6.3	9.2	24.1	35.5
Yield, w%	100.0	2.4	2.5	9.3	5.3	5.8	8.8	24.4	40.5
Gravity, °API	22.2	76.0	66.4	54.9	43.2	36.8	30.0	20.3	3.4
Sulfur, w%	3.3	0.01	0.024	0.198	0.327	1.06	2.018	3.023	5.646
CCR, w%	12							0.1	25.2
Viscosity	102 mm^2/s @ 100°F				1.5 mm^2/s @ 100°F	2.6 mm^2/s @ 100°F	5.5 mm^2/s @ 100°F		306862 mm^2/s @ 200°F
Pour Point, °F	-33				-72	0	72		
TAN, mg/g	0.28								
Ni, wppm	52							0.6	122
V, wppm	314							0.1	735
Naphthenes, v%			21.7	27.9					
Aromatics, v%			28.1	29.2	19	21.6			
RON Clear		69.5	57.2	34.9					
Freeze Point, °F					-62	-15			
Aniline Point, °F					270	280	295	325	
Cetane Index					46	45.8	46.4		
Smoke Point, mm					15	10			

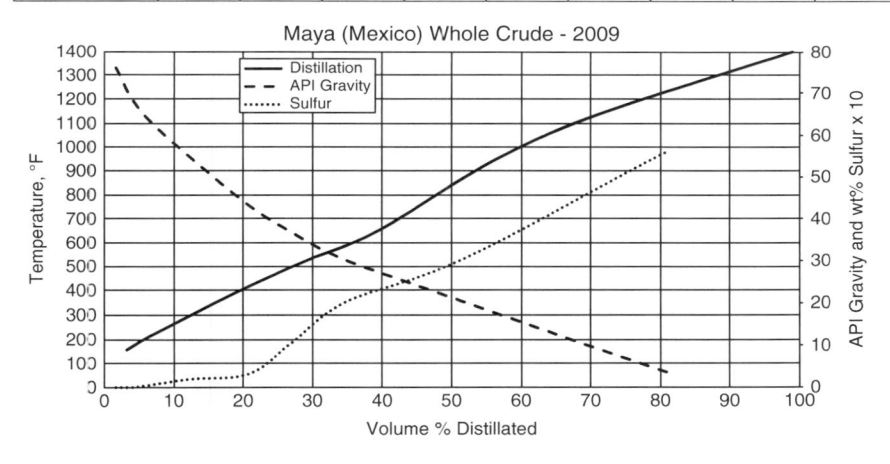

Maya (Mexico) Whole Crude - 2009

— Distillation
– – API Gravity
⋯⋯ Sulfur

X-axis: Volume % Distillated (0–100)
Left Y-axis: Temperature, °F (0–1400)
Right Y-axis: API Gravity and wt% Sulfur × 10 (0–80)

Murban, Abu Dhabi

Source: O&GJ

Cut	Whole Crude	Light Ends	Light SR	Heavy Naphtha	Kerosene	Light Distillate	Vac Gasoil	Atmos Residuum
TBP Range, °F	-----	X-60	60-167	167-347	347-482	482-572	572-662	662+
Yield, v%	100.0	1.7	6.8	21.2	16.1	10.4	9.2	34.5
Gravity, °API	39.4		82.2	56.9	45.4	37.8	33.6	22.6
Sulfur, w%	0.74		0.012	0.013	0.058	0.47	1.06	1.49
CCR, w%								3.6
Viscosity	5.0 cSt @ 70°F				1.8 cSt @ 68°F	4.2 cSt @ 68°F	9.5 cSt @ 68°F	104 cSt @ 100°F
Pour Point, °F	5						0	39	95
RVP, psi	5		10.1					
Salt, ptb	<5							
Ni, wppm	0.58							2
V, wppm	0.8							2
Wax Content, w%	8.0						17.5	19.5
Paraffins, v%				63				
Naphthenes, v%				20				
Aromatics, v%				17				
RON Clear			69					
Freeze Point, °F					−45			
Aniline Point, °F				124	143			
Diesel Index					65	59	58	
Cetane Index						54		
Smoke Point, mm					24			

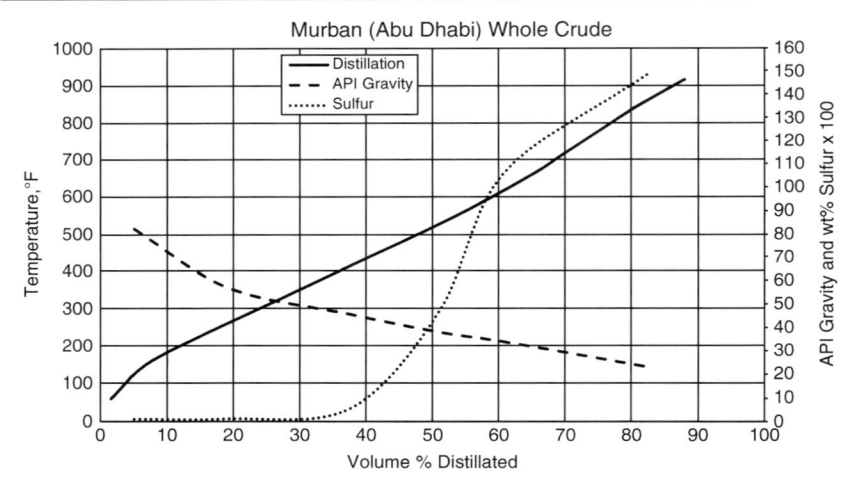

Murban (Abu Dhabi) Whole Crude

Selection of Crude Oil Assays for Petroleum Refining

Pennington, Nigeria **Source: O&GJ**

Cut	Whole Crude	Light Ends	Light SR	Heavy Naphtha	Kerosene	Distillate	Atmos Residuum
TBP Range, °F	-----	X-60	60-200	200-340	340-470	470-650	650+
Yield, v%	100.0	1.1	6.0	16.3	18.0	33.5	25.1
Gravity, °API	37.7		70.1	51.2	40.1	34.5	23.6
Sulfur, w%	0.076		0.001	0.004	0.018	0.058	0.19
CCR, w%							1.3
Viscosity	36 SUS @ 100°F						6.9 SUS @ 210°F
Pour Point, °F	37					10	88
Salt, ptb	2.9						
Paraffins, v%			74	32			
Naphthenes, v%			24	56.5			
Aromatics, v%			1.5	11			
Freeze Point, °F					-74		
Smoke Point, mm					21		

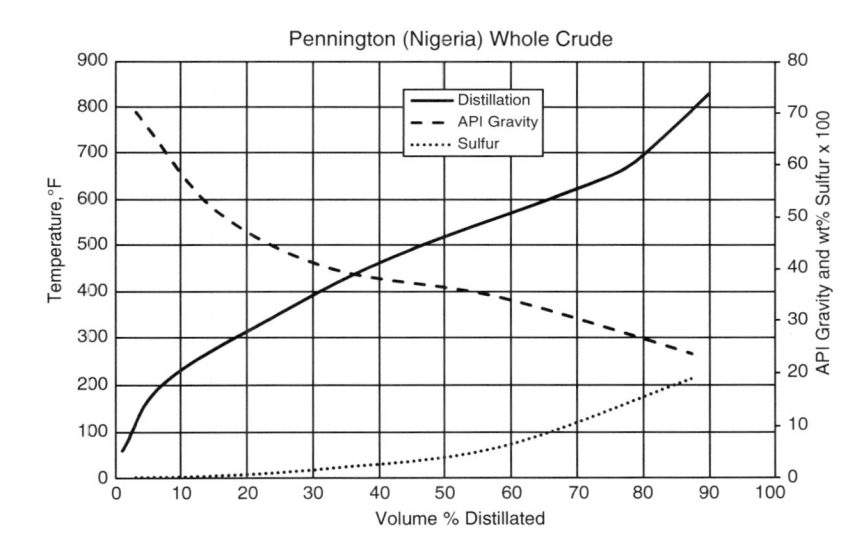

Pennington (Nigeria) Whole Crude

Qatar Marine, Qatar

Source: O&GJ

Cut	Whole Crude	Light Ends	Light SR	Heavy Naphtha	Distillate	Vac Gasoil	Atmos Residuum	Vac Residuum
TBP Range, °F	-----	X-113	113-220	220-390	390-680	680-1000	680+	1000+
Yield, v%	100.0	5.7	9.1	19.6	31.0	22.5	34.6	12.2
Gravity, °API	37.0		69.0	52.0		21.5	15.3	6.7
Sulfur, w%	1.5		0.04	0.07		2.24	2.69	3.32
Nitrogen, wppm						1400		
CCR, w%							9.27	21.6
Viscosity	42.8 SUS @ 80°F					58.8 SUS @ 210°F	236 SUS @ 210°F	
Penetration, 77°F								36
Pour Point, °F	25					100	85	127
RVP, psi	5.7			3.9				
Salt, ptb	9							
Ni, wppm						0.5	44	107
V, wppm						0.6	110	263
Paraffins, v%				71.3	59.2			
Naphthenes, v%				22.4				
Aromatics, v%				6.3				
Aniline Point, °F						176		

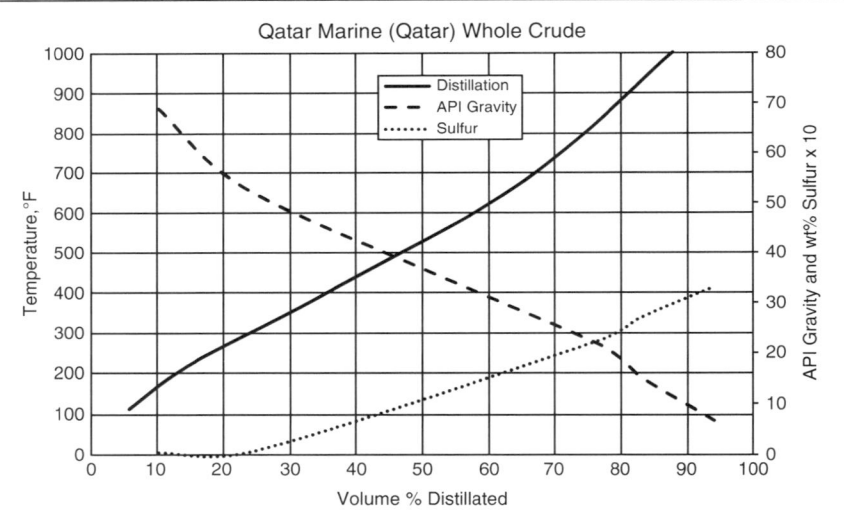

Qatar Marine (Qatar) Whole Crude

Selection of Crude Oil Assays for Petroleum Refining 1873

Russian Export Blend (REB), Russia — 2009

Cut	Whole Crude	Light Naphtha	Medium Naphtha	Heavy Naphtha	Light Kerosene	Kerosene	Diesel	Vacuum Gasoil	Resid
TBP Range, °F	-----	X-158	158-212	212-374	374-455	455-536	536-649	649-1049	1049+
Yield, v%	100.0	5.3	4.1	14.7	7.9	8.0	11.2	29.3	18.5
Yield, w%	100.0	4.0	3.3	13.0	7.4	7.7	11.2	31.2	21.2
Gravity, °API	31.8	84.2	69.8	53.7	42.8	37.6	32.3	21.8	11.1
Sulfur, w%	1.53	0.005	0.013	0.053	0.211	0.61	1.086	1.809	3.650
CCR, w%	3.9							0.7	18.5
Viscosity	8.3 mm^2/s @ 100°F				1.5 mm^2/s @ 100°F	2.6 mm^2/s @ 100°F	5.5 mm^2/s @ 100°F		1290 mm^2/s @ 200°F
Pour Point, °F	10				-89	-18	84		
TAN, mg/g	0.56								
Ni, wppm	14.7							0.2	73.4
V, wppm	46.7							0.2	230.6
Naphthenes, v%			25.2	37.1					
Aromatics, v%			3.7	11.7	21.8	27.8			
RON Clear		70.9	50.9	39.2					
Freeze Point, °F					-71	0			
Aniline Point, °F					277	288	300	334	
Cetane Index					45	44.9	49.7		
Smoke Point, mm					22	18			

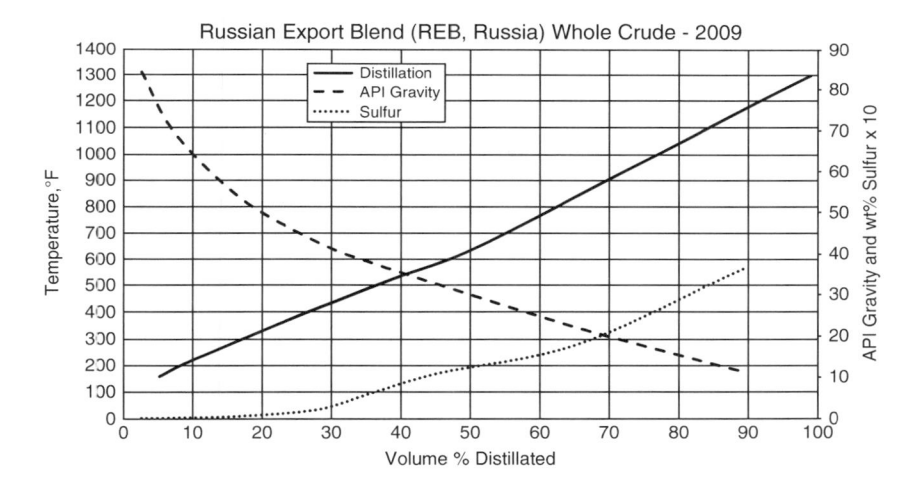

Russian Export Blend (REB, Russia) Whole Crude - 2009

Distillation, API Gravity, Sulfur vs Volume % Distillated

San Joaquin Valley Heavy, USA 2009

Cut	Whole Crude	Light Naphtha	Medium Naphtha	Heavy Naphtha	Kerosene	Diesel	Light Gasoil	Heavy Gasoil	Resid
TBP Range, °F	-----	160-250	250-325	325-375	375-500	500-620	620-800	800-1050	1050+
Yield, v%	100.0	0.3	0.7	1.1	7.5	11.9	21.9	26.2	30.5
Yield, w%	100.0	0.2	0.6	0.9	6.5	11.0	21.2	26.5	33.0
Gravity, °API	13.6					25	18.2	12.2	1
Sulfur, w%	1.38				0.33	0.72	1.17	1.52	1.88
CCR, w%									22.3
N+2A		72	82	79					
Cetane Index						33	32		

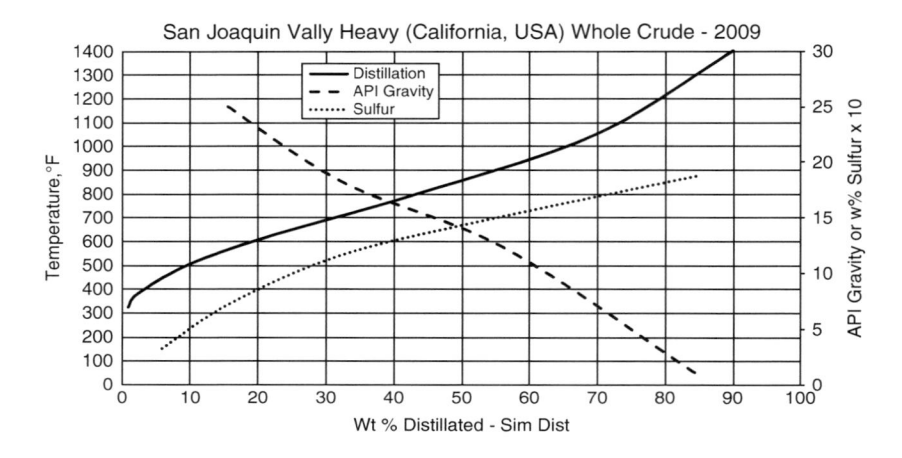

San Joaquin Vally Heavy (California, USA) Whole Crude - 2009
— Distillation
– – API Gravity
······ Sulfur

Wt % Distillated - Sim Dist

Selection of Crude Oil Assays for Petroleum Refining

Sassan, Iran
Source: O&GJ

Cut	Whole Crude	Light Ends	SR Gasoline	Naphtha	Kerosene	Distillate	Atmos Residuum	Vac Residuum
TBP Range, °F	-----	X-50	50-200	200-375	375-450	450-700	700+	1070+
Yield, v%	100.0	2.1	10.5	16.4	7.5	23.8	39.7	12.0
Gravity, °API	33.9		77.6	53.3	43.0	37.0	15.3	1.3
Sulfur, w%	1.91		0.081	0.082	0.19	1.2	3.5	5
RSH Sulfur, wppm			0.067	0.002				
Viscosity	44.2 SUS @ 100°F				2.05 cSt @ 60°F	7.52 cSt @ 60°F	154 SUS @ 210°F	8000 SUS @ 210°F
Pour Point, °F	-5					15	75	
Salt, ptb	1.76							
Ni, wppm							Ni+V	Ni+V
V, wppm							44	127
Naphthenes, v%				N+A				
Aromatics, v%				41				
RON Clear			61	35.4				
Freeze Point, °F					-50			
Aniline Point, °F					121.5	137.9	153	
Cetane Index							56.3	
Smoke Point, mm						20		
Wax Content, w%	1.9							

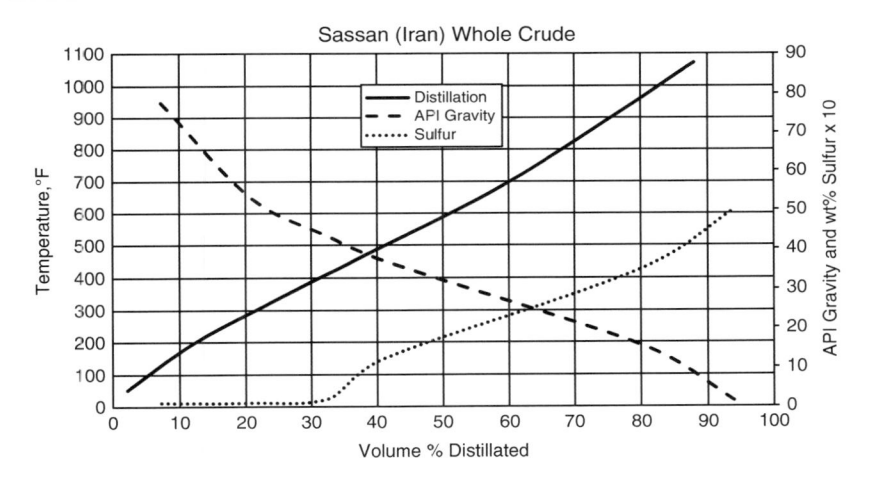

Sassan (Iran) Whole Crude

— Distillation
– – API Gravity
⋯⋯ Sulfur

Temperature, °F vs. Volume % Distillated; API Gravity and wt% Sulfur × 10

Tia Juana Pesado, Venezuela   2009

Cut	Whole Crude	Light Naphtha	Medium Naphtha	Heavy Naphtha	Light Kerosene	Kerosene	Diesel	Vacuum Gasoil	Resid
TBP Range, °F	-----	X-158	158-212	212-374	374-455	455-536	536-649	649-1049	1049+
Yield, v%	100.0	0.2	0.1	1.5	3.0	4.2	8.8	42.7	38.5
Yield, w%	100.0	0.2	0.1	1.2	2.6	3.8	8.1	41.9	41.1
Gravity, °API	12.1	81.6	61.8	44.5	34.2	28.9	23.7	14.8	2.8
Sulfur, w%	2.7	0.074	0.14	0.214	0.458	0.81	1.411	2.3	3.831
CCR, w%	11.2							1.2	25.6
Viscosity	3700 mm^2/s @ 100°F				1.9 mm^2/s @ 100°F	3.5 mm^2/s @ 100°F	10.1 mm^2/s @ 100°F		99000 mm^2/s @ 100°F
Pour Point, °F	30				−121	−69	−27		
TAN, mg/g	3.61								
Ni, wppm	38.5								91.4
V, wppm	284								674
Naphthenes, v%			52	72.3					
Aromatics, v%			2.8	11.1	18.1	24.4			
RON Clear		87.7	75.2	69.6					
Freeze Point, °F					−103	−51			
Aniline Point, °F					259	270	282	318	
Cetane Index					31.2	31.2	37.5		
Smoke Point, mm					16.5	10.4			

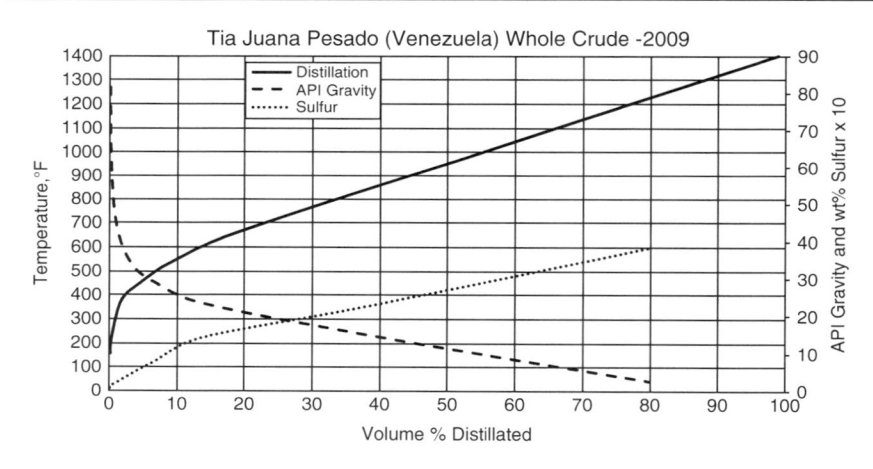

Tia Juana Pesado (Venezuela) Whole Crude -2009

Selection of Crude Oil Assays for Petroleum Refining

West Texas Intermediate (WTI), USA — 2009

Cut	Whole Crude	Light Naphtha	Medium Naphtha	Heavy Naphtha	Light Kerosene	Kerosene	Diesel	Vacuum Gasoil	Resid
TBP Range, °F	-----	X-158	158-212	212-374	374-455	455-536	536-649	649-1049	1049+
Yield, v%	100.0	6.4	6.0	23.2	9.1	8.7	10.9	24.4	10.3
Yield, w%	100.0	5.1	5.2	21.6	9.0	8.8	11.3	26.6	11.4
Gravity, °API	40.8	85.9	67.2	53.2	43.0	38.8	35.0	26.4	24.2
Sulfur, w%	0.34	0.011	0.027	0.044	0.088	0.17	0.274	0.444	1.374
CCR, w%	1.1							0.4	12.6
Viscosity	3.9 mm^2/s @ 100°F				1.5 mm^2/s @ 100°F	2.5 mm^2/s @ 100°F	5.5 mm^2/s @ 100°F		382 mm^2/s @ 200°F
Pour Point, °F	−20				−65	10	77		
TAN, mg/g	0.1								
Ni, wppm	1.6							0.1	15.3
V, wppm	1.6							0	15.2
Naphthenes, v%			37.6	38.5					
Aromatics, v%			4.3	13.2	13.7	13.7			
RON Clear		69.8	62.8	50.6					
Freeze Point, °F					−44	16			
Aniline Point, °F					289	316	345	396	
Cetane Index					45	45.1	54		
Smoke Point, mm					24	21			

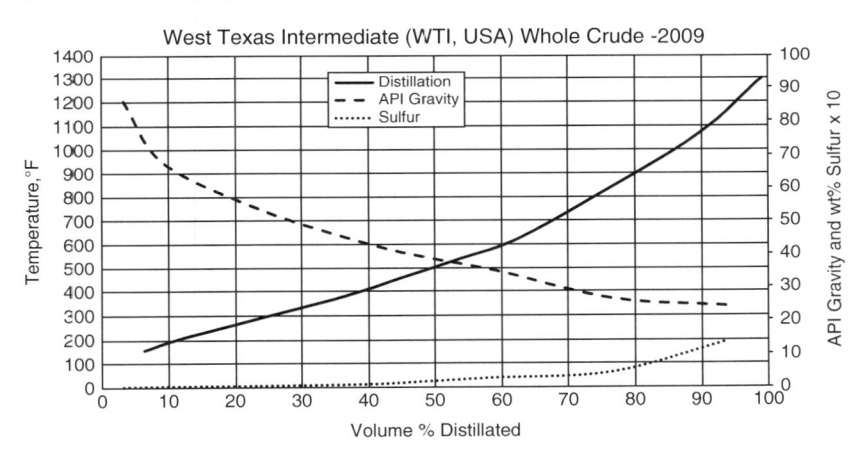

West Texas Intermediate (WTI, USA) Whole Crude -2009

West Texas Sour (WTS), USA — 2009

Cut	Whole Crude	Light Naphtha	Medium Naphtha	Heavy Naphtha	Light Kerosene	Kerosene	Diesel	Vacuum Gasoil	Resid
TBP Range, °F	-----	X–158	158–212	212–374	374–455	455–536	536–649	649–1049	1049+
Yield, v%	100.0	6.2	5.7	18.1	8.7	8.0	10.2	28.8	13.2
Yield, w%	100.0	4.7	4.8	16.4	8.3	7.9	10.4	31.1	15.4
Gravity, °API	34.1	87.9	65.9	51.3	42.1	37.0	31.5	22.0	9.9
Sulfur, w%	1.64	0.079	0.143	0.351	0.646	0.9	1.232	2.042	4.441
CCR, w%	3.3							0.7	22.6
Viscosity	4.6 mm^2/s @ 100°F				1.5 mm^2/s @ 100°F	2.5 mm^2/s @ 100°F	5.2 mm^2/s @ 100°F		7440 mm^2/s @ 200°F
Pour Point, °F	–51				–78	–18	54		
TAN, mg/g	0.11								
Ni, wppm	3.7							0.1	31.4
V, wppm	6.4							0.1	54.1
Naphthenes, v%			27.1	38.4					
Aromatics, v%			9	18	21.8	25.2			
RON Clear		69.1	59.3	43.1					
Freeze Point, °F					–63	1			
Aniline Point, °F					273	289	307	354	
Cetane Index					43	43.4	48.7		
Smoke Point, mm					20	16			

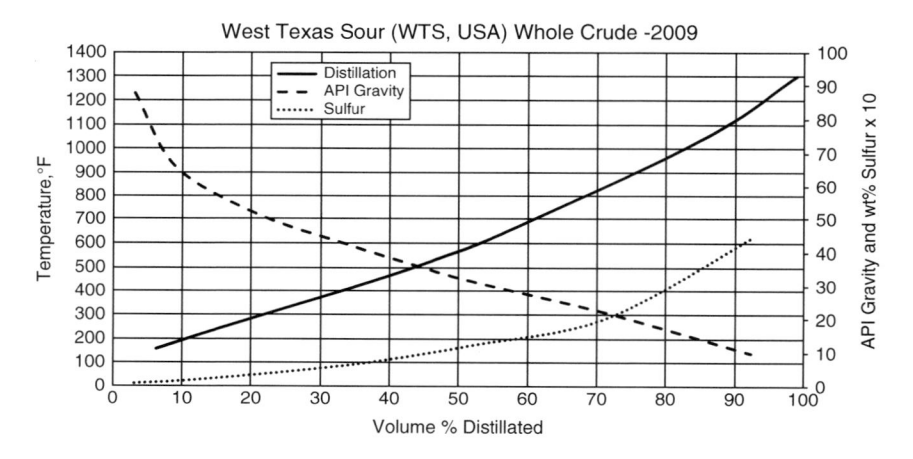

West Texas Sour (WTS, USA) Whole Crude -2009

Selection of Crude Oil Assays for Petroleum Refining

Western Candian Select, Canada Jan-13

Cut	Whole Crude	Full-Range Naphtha	Full-Range Distillate	Full-Range Gasoil	Residuum
TBP Range, °F	-----	IBP-374	374-649	649-981	981+
Yield, v%	100.0	20.1	17.3	26.6	36.0
Yield, w%	100.0	14.9	16.4	27.4	41.3
Gravity, °API	20.9	73.8	29.1	16.5	2.5
Sulfur, w%	3.48	0.044	1.295	2.924	5.684
Nitrogen, wppm	2481		28	1259	5150
MCR, w%	10.09		0.01	0.18	24.74
Viscosity					72,517 cSt @ 212°F
Salt, ptb	44				
BS&W, wppm	316				
TAN, mg KOH/g	0.76		0.25	1.26	0.61
Ni, wppm	46			0	
V, wppm	115			0	

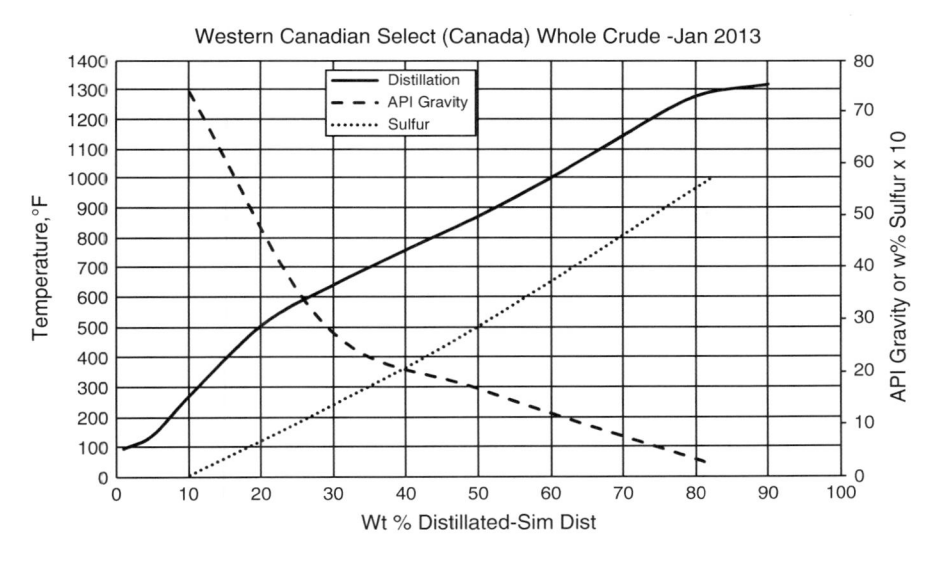

Western Canadian Select (Canada) Whole Crude -Jan 2013

Conversion Factors Used in Petroleum Processing

Steven A. Treese

Contents

Appendix D: Conversion Factors 1882
    D1 General Conversion Factors 1882
    D2 Pressure Conversion Table 1883
    Appendix D3 Viscosity Conversion Table 1883

Abstract

This appendix provides general conversion factors, pressure conversions, and viscosity conversions for petroleum processing. These factors are used throughout this handbook.

Keywords

Conversion Factors • Pressure • Viscosity

Steven A. Treese has retired from Phillips 66.

S.A. Treese (✉)
Puget Sound Investments LLC, Katy, TX, USA
e-mail: streese256@aol.com

Appendix D: Conversion Factors

D1 General Conversion Factors

To convert from	To	New unit abbreviation	Multiply by
Length			
Feet	Meter	m	0.304
Inch	Millimeter	mm	25.4
Statute mile	Kilometer	km	1.609
Area			
Square inches	Square millimeters	mm^2	645.2
Square inches	Square centimeters	cm^2	6.452
Square inches	Square meters	m^2	0.000645
Square feet	Square meters	m^2	0.0924
Acres	Hectare		0.4047
Volume			
Cubic inches	Cubic millimeter	mm^3	16,387
Cubic inches	Cubic centimeter	cm^3 or cc	16.387
Cubic inches	Cubic meter	m^3	0.00001639
Cubic feet	Cubic meter	m^3	0.0281
Fluid ounces	Milliliter	ml	29.57
Gallons (US)	Liter	l	3.785
Mass			
Pounds	Kilogram	kg	0.4536
Ton (short)	Metric ton	tonne or MT	0.9072
Ton (long)	Metric ton	tonne or MT	1.016
Pressure			
Pounds per square inch	Pascal	Pa	6,895
Pounds per square inch	Kilopascal	kPa	6.895
Kilograms per sq meter	Pascal	Pa	9.807
Bar	Kilopascal	kPa	100
Force			
Pounds force	Newton	N	4.448
Kilogram force	Newton	N	9.807
Work			
British thermal unit	Joule	J	1,055
Foot pound	Joule	J	1.356
Calorie	Joule	J	4.186
Power			
Btu/hour	Watt	w	0.293
Btu/sec	Watt	w	1,055
Horsepower	Kilowatt	kw	0.746
Flow rate			
Cubic feet per minute	Cubic meters/minute	m^3/m	0.0283

(continued)

To convert from	To	New unit abbreviation	Multiply by
Gallons (US)/minute	Liter/minute	l/m	3.785
Barrels (US) oil	Gallons oil (US)	gal	42
Specific energy. Latent heat			
Btu/pound	Joule/kilogram	J/kg	2,326
Specific heat, specific			
Entropy	Joule/kilogram – kelvin	J/kg-deg K	4,184
Btu/pound – deg F			
Miscellaneous			
(US) barrels per day	(US) gallons/hour	gph	1.75
(US) gallons	Imperial gallons		0.8326

D2 Pressure Conversion Table

To convert from ↓	Multiply by factor to convert to							
	Atm	psi	in. Hg	Ft H_2O (4 °C)	In. H_2O	mm Hg (torr)	Bar	kg/cm^2
Atm	1.00000	14.696	29.920	33.9000	406.800	760.00	1.0135	1.0333
psi	0.06804	1.000	2.036	2.3070	27.684	51.50	0.0690	0.0703
in. Hg	0.03340	0.491	1.000	1.1330	13.596	25.40	0.0339	0.0345
Ft H_2O (4 °C)	0.02950	0.434	0.883	1.0000	12.000	22.20	0.0299	0.0305
In. H_2O	0.00246	0.036	0.074	0.0833	1.000	1.85	0.0025	0.0025
mm Hg (torr)	0.00132	0.020	0.039	0.0446	0.535	1.00	0.0013	0.0014
Bar	0.98690	14.500	29.521	33.4484	401.381	749.88	1.0000	1.0195
kg/cm^2	0.96780	14.220	28.960	32.8100	393.690	735.51	0.9808	1.0000

Appendix D3 Viscosity Conversion Table

Given units	Multiply the given units by this factor to get						
	g/cm-s (Poise)	kg/m-s	lb$_m$/ft-s	lb$_f$-s/ft^2	Centipoise (cP)	lb$_m$/ft-h
g/cm-s (Poise)	1	10^{-1}	0.0672	2.09×10^{-3}	100	242
kg/m-s	10	1	0.672	0.0209	1,000	2,419
lb$_m$/ft-s	14.88	1.488	1	0.0311	1,488	3,600
lb$_f$-s/ft^2	4,788	47.88	32.17	1	47,880	1.158×10^5
Centipoise (cP)	0.01	0.001	6.72×10^{-4}	2.089×10^{-5}	1	2.4191
lb$_m$/ft-h	4.134×10^{-3}	4.134×10^{-4}	2.78×10^{-4}	8.634×10^{-6}	0.4134	1

Index

A
Abel flash points, 1686
Abel tester, 1686
Abnormal situation management (ASM), 1694
Absolute pressure, 1686
Absorbent, 1686
Absorber, 501–511, 1686
Absorption units, 1465–1470, 1686–1692
Accumulation, 1317, 1318, 1687, 1778
Accumulator, 1687
Acetic acid, 893
Acid gas, 500, 1687
Acid number, 944, 1687
Acid rain, 72
Acid-soluble oil(s) (ASO), 441, 1695
Acid suit, 1687
Activated sludge unit (ASU), 1688, 1696
Actuator, 1031, 1033, 1688
Adiabatic, 1688
Adiabatic plug flow reactor, 461–462
Adsorbent, 583–611, 1367–1368, 1688, 1785
Ad valorem tax, 732, 1688
Advanced process control (APC), 1011, 1688
Aerobic, 1270, 1688
A film-forming foams (AFFF), 1428
Air condensers, 1688
Air coolers, 1558, 1688
 description of, 1573
 thermal rating, 1573–1579
Air emissions from refinery, 1218–1221
Air flotation, 1270–1271
Air/fuel ratio, 1235, 1688
Air-line respirator, 1688
Air pollution, 71, 1218–1221
Air separation unit (ASU), 1696
Air systems, 1688–1689
Alcohol, 1689
Alcohol blend, 1689

Alcohol-resistant concentrate (ARC), 1428
Algal oil, 952, 1689
Aliphatic hydrocarbon, 1689
Alkylate, 41, 63–64, 1689, 1690
Alkylate product composition, 109, 452
Alkylation, 41, 63–64, 74, 435–456, 1338, 1689–1690
 unit feed composition, 109
 unit HF, 442–446, 1690
 unit H_2SO_4, 447–452, 1690
Alphabutol™, 475
American National Standards Institute, 1690
American Petroleum Institute (API), 1690
 base oil classifications, 849
 codes, 1693
 gravity, 10, 1694, 1837–1840
 oil-water separator, 1265–1267
 RP-530, 1598
 separator, 694, 1265–1267, 1301
 separator design, 1301–1302
 Standard 630, 1598
American Society for Testing and Materials (ASTM), 1690–1691
 distillation, 663, 1696
 distillation probability curves, 10, 18, 19
American Society of Mechanical Engineers (ASME), 1695
Amine, 500, 1691
 absorber, 508–513, 1692
 distribution, 1157
 solvents, 501–507, 1354–1356, 1691
 units, 500, 1691
Amine-based Claus tail gas process, 1232
Amine gas treating unit design, 500–529
 amine hold up, 518
 amine solution circulation rate, 514–515
 bottom tray conditions, 524
 contactor diameter, 516–518

Amine gas treating unit design (*cont.*)
  design specification, 514
  heat balance, 518, 522
  heat exchanger design, 518–519
  internal reflux, 521–523
  Kremser equation, 524–526
  material balance of stripper, 519–521
  number of trays, 515–516
  overhead product and reflux
    compositions, 521
  overhead vapor composition, 521
  preliminary process flow sheet, 528
  reboiler duty, 523
  reclaimer material balance and heat
    duty, 526
  stripper design, 519
  stripper tower dimensions, 526–529
  tower top pressure, 521
Ammonia (NH_3), 1154–1155, 1357–1358
Ammonia-burning claus plants, 1230
Ammonium salts, 368–369, 960, 1256
Amortization, 755
Anaerobic, 1270, 1692
Analyzer, 1076, 1692
  alarm, 1686
Ancillary equipment, 1692
Anhydrous, 1692
  ammonia, 1154–1155, 1357–1358, 1693
  hydrofluoric acid, 1377–1382, 1693
Aniline point, 1693
Annular flow, 1630–1631, 1693
Anti-backflow, 1693
Anti-cross-contamination, 1162–1163, 1693
Anti-surge, 1524–1525, 1693
Aqueous, 1694
  ammonia, 1154–1155, 1694
  waste treatment, 1256–1257
  waste water streams, 1251
Arabian American Oil Company
  (ARAMCO), 1694
Arbitrage, 1694
Arbor coil, 1694
Arch, 1694
Arithmetic progression, 770–771
Aromatics, 7, 47, 241, 872–875, 1694
  feedstocks, 47, 872–875
  reforming reactions in gasoline, 41–42, 244
Aromatics production
  dealkylation and benzene recovery, 874
  feed fractionation, 873
  recovery complex, 872–873
  xylene splitter and isomerization
    process, 873–874

Asphalt, 860–871, 1695
  products, 69
Asphalt blowing process, 863–871
  effect of oil level, 870
  flowsheet of bitumen manufacturing
    plant, 870–871
  rate of air injected, 864–866
  reaction temperature, 866–867
  retention and contact time, 867–870
  system pressure, 870–871
Asphaltene(s), 7, 944, 1695
Asphalt oxidizing mechanism, 861–862, 864
  asphalt factor *vs.* yield, 862–864
  asphalt yields, 862–863
Asphalt production, 860–871
  asphalt blowing process, 863–871
  asphalt oxidizing mechanism, 861–862, 864
  effect of oil level, 870
  flowsheet of bitumen manufacturing plant,
    870–871
  paving and liquid asphalt, 860–861
  rate of air injected, 864–866
  reaction temperature, 866–867
  retention and contact time, 867–870
  roofing asphalt, 861
  system pressure, 870–871
Aspiration, 1695
Assay, 8, 909–910, 1695–1696, 1845–1879
ASTM ON, 1696
Atmospheric
  discharge, 1697
  overhead distillate, 55
  residue, 58, 532
  straight run gas oils, 57
Atmospheric crude distillation unit, 33, 94,
    127–179, 1696–1697
  crude feed preheat exchanger system,
    155–158
  description, 127–129
  design characteristics, 134
  example, 158–179
  fractionator overhead equipment, 135–139
  material balance, 129–135
  side streams, 142–150
  tower dimensions, 150–155
Atmospheric gas oil (AGO), 1688
Atmospheric storage tanks
  cone roof tanks, 1169–1170
  floating roof tanks, 1170–1172
  nozzle arrangement and location
    considerations, 1172–1174
  sumps and drain-dry tanks, 1174–1175
  tank internals, 1174

Atmospheric tower unit (ATU), 1697. *See also* Atmospheric crude distillation unit
Automotive diesel specification, 76–82, 117, 654–659
Automotive grade gas oil, 656
Auto-thermal reforming (ATR), 618
Auxiliary, 1697. *See also* Atmospheric crude distillation unit
Aviation gasoline, 436–437, 653–654, 1697
　blending components, 1698
Aviation turbine gasoline (ATG), 654
Axens polynaphtha process, 476
Axial flow, 1514
　compressor, 1698

B
Bachequero pour point, 9
Back mix type catalyst cooler, 549–550
Back pressure, 1698
　sizing factor, 1328
BACT box, 1698
Baffle, 1698
Baker
　correlations, 1698
　parameters, 1633–1634
Balanced safety relief valves, 1317, 1698
Barge, 1187, 1698
Barrel (Bbl), 1698, 1699
Barrel of fuel oil equivalent (BFOE), 1699
Barrel of oil equivalent (BOE), 1700
　energy, 885
Base lube oils, 845–846, 1698
Base oils, 845
Base stock group statistics, 859
Base stock properties, 856–858
Basic process control system (BPCS), 1011, 1698, 1701
Batch blending, 1699
Battery limit (station), 1699
Bayonet heater, 1699
5-bed pressure swing adsorption unit, 626, 627
Benfield process, 506–507, 1699
BenSat, 1699
Benzene, 6, 872–875, 1359–1363, 1699
Berl saddle, 1462–1463
Bernoulli principle, 1020, 1702
Bimetal strip temperature measurement, 1044–1045
Biochemical oxygen demand (BOD), 1254, 1699, 1700
　reduction, 1205, 1261–1265

Biocide, 1699
Biocrude, 953, 1699
Biodiesel, 952–953, 973–1000, 1699
Bioethanol, 953, 970, 1699
Biofuel, 950–960, 966–1005, 1700
Biological oxygen demand (BOD), 1253–1254
Biomass fluid catalytic cracking (BFCC), 982
Biooil, 1700. *See also* Biocrude
Biorefining, 969
Bitumen, 930–935, 938, 942–945, 1700
　alkaline metals, 944
　API gravity, 942
　aromatics, 942
　asphaltenes, 944
　chlorides, 945
　coking/cracking propensity, 945
　cold production, 934
　contaminants, 944
　definitions, 930–931
　dumbbell distillation, 938, 942
　huff and puff, 933
　manufacturing plant, 871
　metals content, 943–944
　mining extraction, 932
　nitrogen content, 943
　oxidizer sizing, 875–882
　production locations, 931–932
　properties, 935–937
　steam-assisted gravity drainage, 933
　sulfur levels, 942–943
　total acid number, 944–945
　upgrading levels, 934
　viscosity, 945
Blank, 1700
Blanket, 1700
Bleach, 1700
Blend component, 1178–1180, 1700
Blind flange, 1700
Blocked discharge, 1700
Blocked-in, 1700
Blowdown, 1700
Boiler feed water (BFW), 1128–1138, 1701
　chemical treatment, 1137
　condensate recovery and recycle, 1138
　contamination prevention, 1138
　hot lime process, 1131
　impurities, 1128
　ion exchange processes, 1131–1133
　management, 1128–1131
　pretreatment process, 1135
　treatment requirements, 593, 1129, 1130, 1136

Boiling point
    analyzer, 1076, 1701
    boiling range, 126, 1701
Bottom(s), 1701
Bottom sediment and water (BS&W), 1702
Bourdon tube, 1052, 1701
Box cooler, 1558, 1701
BPCD and BPSD, 1701
Brake horsepower efficiency, 1494–1495
Breakeven, 1701
Bridgewall, 1597, 1701
Bright stock, 59, 1701
    lubes, 68, 846
British thermal unit (BTU), 1702
Bromine number, 678, 1702
Brown and Souder flood constant, 517
Bubble
    cap, 1702
    cap trays, 1442–1444
    flow, 1630–1631, 1702
    point, 13–14, 1702
Bulb thermometer, 1044
Bulk terminal, 1702
Bullet, 1702
Bunker fuel, 1702
Burner, 1604, 1702
    control system, 1113
    design, 1235–1239
    management system, 1702
    tip, 1702
Burst pressure, 1702
Butamer™ process, 481
Butane, 5
    equilibrium, 486
    LPG, 59

C
California Air Resources Board (CARB), 1703
Calorie, 1703
Calorific value, 1703
Capacity factored cost estimation, 723, 766, 1703
Capacity-factored estimate, 721–722, 1703
Capital cost estimation, 720–721
Capital plant cost, 709
CARB diesel, 1703
Carbon dioxide (CO_2), 1240
Carbon monoxide (CO), 72, 1239, 1363–1365, 1703
Carbon oxides, 1239
Carbon residue, 676, 1703
Carbonyl sulfide (COS), 1703
Carcinogen, 1703

Car seal, 1703
Cascade alkylation process, 447–448
Cascade control, 1011, 1704
Cash flow analysis, 745, 757–759
Catacarb process, 506–507, 1704
Catalyst, 1704
    activity, 329
    fines, 1704
    octane/activity curves, 491
    paraffin isomerization, 488–489
    regeneration, 424
    selectivity, 330
    stability, 330
    suppliers, 495
    support sintering, 423
Catalytic and non-catalytic NOx reduction, 1236–1238
Catalytic condensation process, 459–461
    diesel fuels, 471
    for gasoline production
        adiabatic plug flow reactor, 461–462
        oligomerization reactions, 463
        tubular reactor, 461–462
        typical polymer gasoline properties, 461, 463
    hydrogenated vs. non-hydrogenated polymer gasoline, 464–466
    selective vs. non-selective gasoline production, 466–468
Catalytic cooling vs. reactor yields, 549
Catalytic cracking, 460. See also Fluid catalytic cracking
    deep oil fluid, 542–545
        back mix type catalyst cooler, 549–550
        catalyst cooling vs. reactor yields, 549
        combustion air requirement vs. CO_2/CO ratio, 548
        enthalpy requirements, 548
        feed inlet enthalpy, 548
        flow through type catalyst cooler, 549–550
        heat of combustion vs. CO_2/CO ratio, 547
        lift gas technology, 550–552
        mix temperature control, 550–552
        two stage catalyst regeneration, 550–552
Catalytic naphtha reforming process, 35–36, 62, 229–260, 568–570
Catalytic olefin condensation, 458
    InAlk™ process, 468
        resin-catalyzed condensation, 468–469
        SPA-catalyzed condensation, 470–471
    Petrochemical operations (see Petrochemical)

Catalytic reformer operation, 106, 108, 229–260
Catalytic reformer unit, 35–36, 229–260
Catalytic reforming process, 62, 229–260
   alumina support, 247
   aromatic reaction, 244
   catalysts
      for CCR process units, 248–249
      for cyclic units, 248
      for semiregenerative units, 247–248
      suppliers, 249
   caustic scrubber, 257
   Chlorsorb system, 257
   continuous regeneration type reforming unit, 242
   deactivation mechanism, 249–250
   evolution, 232
   flow schemes
      continuous catalyst regeneration units, 255–256
      cyclic reforming units, 254
      fixed-bed SR reforming units, 253–254
      secondary recovery, 256
      unit improvement, 256
   history, 231–232
   hydrogen-to-hydrocarbon mole ratio, 253
   market segment
      aromatics, 241–242
      motor fuels, 238–240
   naphtha feedstocks (*see* Naphtha)
   naphthene reactions, 243–244
   octane number, 235–237
   paraffin reactions, 244–245
   platinum, 246–247
   pressure, 251–252
   rates and equilibrium, 245–246
   reaction sites, 243
   reformate composition, 235–237
   regeneration steps, 250
   SR regeneration, 250–251
   temperature and space velocity, 252–253
   unit capacity, 243
   vapor pressure, 237
Catalytic reforming unit, 1704
Caustic soda, 1156, 1255, 1370–1373, 1704
Caustic treating system, 1220
CBOB, 1704
Cellulose, 976
Celsius, 1705
Center for Chemical Process Safety (CCPS), 1704
Centigrade, 1705
Centipoise, 1705

Centistoke, 1705
Centrifugal compressors, 1705–1706
   description, 1515
   horsepower, 1515
   specification, 1497
   surge control, performance curves and seals, 1524
Centrifugal pump(s), 717, 1706
   capacity range for, 1492
   description, 1485–1486
   operating limits, 1493
   seal, 1505
   selection of, 1490
   specification, 1497
Centrifuge, 1706
Cetane, 1707
   index, 1707
   number, 78, 1707
C_6 fraction equilibrium, 487
CFR diesel testing unit, 1707
Channel, 1707
Characteristic curve, 1490–1494, 1525–1527, 1707
Charge capacity, 1707
Check valve, 1707
Chemical oxygen demand (COD), 1254, 1708
Chemical sewer, 1160–1161, 1707
Chemiluminescence analyzer, 1076, 1078, 1707
Chiksan joint, 1707
Chlorided alumina type catalysts, 489
Chlorsorb system, 257
Civil engineer (CE), 1705
Clarifier, 1708
Class A foams, 1429
Claus process, 1228–1230, 1708
Claus tailgas process, 1231–1233
Clean Air Act (CAA), 73, 1218, 1250
Clean Water Act (CWA), 1250
Clearance pocket, 1538, 1539, 1708
Cloud and pour points, 12, 666–667
Cloud point(s), 79, 666–667, 1708
Cloud point/pour point/ freeze point analyzer, 1076
Coagulation, 1708
Coal, 1708
   gasification, 886
   synthesis gas utilization, 889
Coal to liquids (CTL) conversion, 888
Cobalt-molybdenum catalysts, 415
Codimers
   octane number, 467
   properties, 466–467

Cogeneration, 1147–1148, 1708
Coke, 1709
  deposition, 422
Coker, 534–538, 1709
Coker naphtha hydrotreating unit, 396, 397
Coking process, 67, 534–535, 1709
Cold filter plugging point (CFPP), 79, 1707
Cold flash separator, 1709
Cold properties, 79, 1709, 1711
Color, 848
Column, 1711
  pressure, 1055
Combustion air requirement *vs.* CO_2/CO ratio, 548
Commingling, 1711
Commissioning, 824–825, 1711
Common carrier, 1711
Complex refinery, 87, 1711
Component balances, 1711
Composition measurement elements, 1076–1077
Compressible flow pressure drop, 1622–1624
Compressors, 802–803, 1513–1557
Comulling, 419
Condensate, 1711
Condensation, 458. *See also* Condenser
Condenser, 1579–1584, 1711
  duty, 214
Conductivity, 1712
Cone-roof tank, 1169–1170, 1712
Configuration, 82–93, 1712
Confined space, 1713
  hazards, 1342
  management, 1343–1344
Conradson carbon residue (ASTM D189), 676–677, 1713
Consolidated Omnibus Budget Reconciliation Act (COBRA), 1708
Construction plan, 818
Contactor, 501, 514–519, 1713
Continuous emissions monitoring (CEM), 1238
Contractors' bid evaluation, 808
Control element, 1713
Control loop, 1015–1016, 1713
Control system
  architecture, 1017, 1713
  hierarchy, 1014
Control valve, 1011, 1713
  characteristics, 1036, 1713
  rangeability, 1088–1089
  response, 1713
Convection
  section, 1596–1597, 1713
  tube, 1713

Conventional crude, 1713
Conventional safety relief valve, 1318, 1321, 1322, 1713
Cooling tower, 1119, 1124, 1714
Cooling water system, 1714
  chemistry and treatment, 1124–1126
  cooling tower safety, 1127–1128
  distribution system, 1119, 1121
  flowsheet and equipment, 1119–1124
  monitoring, 1126–1127
  pump sumps, 1123
  troubleshooting, 1105–1107, 1127, 1181–1184
Cool planet process, 983, 984
Coprocessing, 1714
Coproduct, 1714
Correlation coefficient, 1714
Corrosion, 1714
  management, 600, 1345
Corrosive fluids, 1561
Corrosivity, 1715
Corrugated plate interceptor, 1257, 1265, 1715
COS removal, 1224–1225
Cost estimating, 720–728, 1715
Cost-plus, 1715
  contract, 805
Counterflow cooling tower, 1122, 1124, 1715
Coupon, 1715
Crack, 1715
Cracked product properties, 99
Cracking. *See also* Hydrocracking and FCC
  propensity, 1709
  refinery, 84–87, 1715
Cradle, 1716
Crankcase oil, viscosity *vs.* temperature, 847
Crinkled wire mesh screens (CWMS), 1471
Criteria air pollutants, 71–73, 1218–1220
Crossflow cooling tower, 1124, 1716
Crossover, 1597, 1716
Crude distillation unit (CDU), 94, 125–179, 1716
Crude feed preheat exchanger system, 155–158
Crude oil, 5, 1716
  distillation process, 125–179
    atmospheric overhead distillate, 55
    atmospheric residue, 58
    atmospheric straight run gas oils, 57
    LPGs, 55
    naphthas, 57
    refinery gas, 55
    straight run kerosene, 57
    straight run products, 55
    vacuum residue, 58–59

program, 697, 760
viscosity, 12
Crude vacuum distillation unit, 34, 179–194
Cryogenic, 1716
Cryogenic hydrogen recovery process, 623, 634–635
CTL/GTL with CTO/GTO, integration of, 901–902
Cumulative cash flow, 729, 1716–1717
Cumulative present worth, 733–734, 1716–1717
Curing, 1717
Cut(s), 94, 97, 1717
Cut point, 14–15, 1717
Cutter stock, 1717
Cycloparaffin, 1717

D

DAF. *See* Dissolved air flotation (DAF)
Damper, 1596, 1718
Darcy equation, 1619, 1718
Dashboard, 1718
Data historian, 1012, 1014, 1017, 1718
DCF. *See* Discounted cash flow (DCF)
DCS. *See* Distributed control system (DCS)
DEA. *See* Diethanolamine (DEA)
Dead end system, 1503
Dead time, 1012, 1718
Deaeration and degasification, 1135–1137
Deaerator, 1718
Dearomatization process, 81, 1718
Deasphalted oil, 411
Deasphalting, 1718–1719
Debottlenecking, 1719
Debutanized reformate properties, 62–63
Debutanizer, 1719
Decarboxylation, 957, 994–996, 1719
Decarburization, 1719
Decoking, 1244–1245, 1719
Deep catalytic cracking (DCC) process, 264, 301
Deep oil fluid catalytic cracking
 back mix type catalyst cooler, 549–550
 catalyst cooling *vs.* reactor yields, 549
 combustion air requirement *vs.* CO_2/CO ratio, 548
 enthalpy requirements, 548
 feed inlet enthalpy, 548
 flow through type catalyst cooler, 549–550
 heat of combustion *vs.* CO_2/CO ratio, 547
 lift gas technology, 550–552

 mix temperature control, 550–552
 two stage catalyst regeneration, 550–552
Deethanizer, 201, 1719–1720
Definitive cost estimation, 727
Definitive estimate, 1720
Degasifier, 582, 594, 643, 1135–1137, 1720
Degree days, 1720
Degree of conversion, 540
Deisohexanizer, 1724
Deisopentanizer, 1724
Delayed coker, 536
Delayed coking process, 535–536
Deluge system, 1310, 1424–1426, 1720
Demethanizer, 1721
Demethylation, 1721
Demulsibility, 1721
Demurrage, 1721
Denitrogenation, 1721
 mechanism, 374–375
Dense bed catalyst coolers, 287
Density, 1721
Deoxygenation, 994–997
Department of Energy, 1721
Department of Transportation, 1721
Depreciation, 711, 736–737
Depropanizer, 201, 1721
Derivative control, 1085–1086, 1721
Desalter, 1721
Design
 basis, 1721
 pressure, 1721
 specifications, fire prevention, 1417
 temperature, 1721
Desulfurization, 36, 1722. *See also* Hydrotreating, units
Desuperheater, 1096, 1100, 1722
Detonation, 1722
Dewaxing, 849–850, 854–856, 983, 1722
Dew point, 1722
DGA. *See* Diglycolamine (DGA)
Diaphragm, 1722
 control actuator, 1033–1034
 pumps, 1487, 1722
Diene(s), 451
 reactor, 395
Diesel engines
 air introduction into cylinder, 77
 aromatic content of fuel, 78
 categories, 76
 cetane number, 78
 cloud point, 79
 distillation range, of diesel fuel, 79

Diesel engines (*cont.*)
  electronically controlled fuel injection
    timing, 77
  four stroke design, 76
  four valve cylinder heads, 77
  fuel density, 78
  fuel injection improvements, 77
  fuel viscosity, 79
  pour point, 79
  sulfur in diesel fuel, 78
Diesel feedstocks, 107
Diesel fuels, 79, 82
  cold flow properties, 81
  hydrogenated polymer product
    comparison, 471
  increasing cetane value, 80
  reducing aromatic content, 81
  sulfur content, 81
  zeolitic distillate, 471
Diesel hydrotreater, 707
  material, 105
Diesel index, 658, 1722
Diesel oil, 76–82, 656–658, 1722–1724
Diesel product, 76–82, 656–658
  distillation check, 118
  viscosity estimate, 119
Diethanolamine (DEA), 503, 1354, 1724
Differential pressure (dP), 1724
  dP cell, 1012, 1728
Diglycolamine (DGA), 505–506,
    1354–1356, 1724
Diisopropanol amine (DIPA), 507, 1724
Dilbit, 931, 1724
Dimerization
  of ethylene to 1-butene, 475
  of ethylene to n-butenes, 475
  of propylene and butenes, 475–476
Dimersol catalysts, 477
Dimersol E, 475
Dimersol G, 475
Dimersol™ process, 474
  dimerization
    of ethylene to 1-butene, 475
    of ethylene to n-butenes, 475
    of propylene and butenes, 475–476
  Dimersol E, 475
  Dimersol G, 475
  Dimersol X, 475
Dimethyl ether (DME), 898
DIPA. *See* Diisopropanol amine (DIPA)
Direct contact condenser, 1559, 1724
Discounted cash flow (DCF), 1724
  rate of return, 758

Disk and donut stripper, 758
  design, 278
  tray/stage efficiency, 278
Dispersed flow, 1630–1632, 1724
Dissolved air flotation (DAF),
    1248, 1718
Distillate, 1724
  cuts, 94, 97
  drum, 140–141
  hydrocracking, 1724–1725
Distillate hydrotreating
  analytical characterization, 381–382
  catalyst and reactions, 382–386
  diesel desulfurization, 380–381
  feedstocks, 388–389
  process conditions, 389–391
  process configuration, 389
Distillation, 1442–1470, 1725–1727
  analyzer, 1076
  trays, 1442–1446, 1643–1645
Distributed control system (DCS), 1012,
    1718, 1727
  controller, 1015
Distribution baffles, 1268
Distributor, 1727
DME. *See* Dimethyl ether (DME)
Doctor test, 1727
Dodecene, 461
Doppler meter, 1027
Double-pipe heat exchanger,
    1557–1558, 1728
Downcomer, 1728
Downstream, 1728
dP. *See* Differential pressure (dP)
Draft, 1055, 1108, 1114, 1608–1609, 1728
Drawoff, 1728
Drilled well, 1048
Drums, 1470–1476, 1728
Dry milling ethanol process, 972
Dumbbell distillation, 938, 942, 1728

E

Eagle Ford shale crude production,
    918, 921
Earned value (EV), 1734
Economic evaluation, 699–741, 1729
Economic life, 730, 1729
Edmister correlations, 19–20, 1729
Edmister method, 20
EDS. *See* Emergency depressuring system
    (EDS)
Eductor, 1729

Effluent, 1729
  refrigerated alkylation process, 448–449
  water, 1729
  water treating facilities, 1202–1204
EFV. *See* Equilibrium flash vaporization (EFV)
EHO. *See* Extra-heavy oil (EHO)
EIA. *See* Energy Information Administration (EIA)
EII. *See* Energy Intensity Index (EII)
Ejector, 1730
Electrical and instrument specifications, 1418
Electrical power distribution system, 1146–1147
Electrical resistance corrosion probe, 1733
Electric motor drivers, 1505
Electrolytic process, 1730
Electronic speed controllers, 1038
Electrostatic precipitator, 1730
Element, 1730
Embrittlement, 1731
Emergency depressuring system (EDS), 1335–1337, 1729
Emergency Planning and Community Right-to-Know Act (EPCRA), 1250
Emergency response plan (ERP), 1433–1435, 1733
Emergency response team (ERT), 1433–1435, 1733
Emergency shutdown system (ESD), 1433–1435, 1733
Emission, 1731
Emulsion, 1731
Endothermic reaction, 1732
End points, 15, 1731
Energy Information Administration (EIA), 1732
Energy Intensity Index (EII), 1732
Energy Policy Act (EPAct), 950
Engineering flow diagrams, 798, 1732
Engler viscosity, 1732
Enquiry document, 807–808
Enterprise control, 1014, 1017, 1732
Enthalpy requirements, fluid catalytic cracking, 548
Entrainment, 1732
Environmental Protection Agency (EPA), 71, 484, 951, 1218, 1732
Environmental work practice (EWP), 1734
EPA. *See* Environmental Protection Agency (EPA)
EPAct. *See* Energy Policy Act (EPAct)
EPCRA. *See* Emergency Planning and Community Right-to-Know Act (EPCRA)

Equilibrium flash calculation, 48–51
Equilibrium flash vaporization (EFV), 1733
Equipment-factored estimate, 722–724, 1733
Erected offsite cost, 748
Erected plant cost, 747
Ergun equation, 1733
ERP. *See* Emergency response plan (ERP)
Error, 1733
ERT. *See* Emergency response team (ERT)
ESD. *See* Emergency shutdown system (ESD)
Ethane, 1733
Ethanol, 1733
Ether, 1734
Etherification, 74, 1734
Ether production, 42–44
Ethylene, 1734
Ethyl mercaptan, 6
Ethyl tertiary-butyl ether, 1734
EuReDatA. *See* European Reliability Data Association (EuReDatA)
Euro fuel specifications, 484
Euro I-V gasoline specifications, 484
European Reliability Data Association (EuReDatA), 1734
European Union (EU), 1734
  regulations and directives, 1219, 1252, 1274
EV. *See* Earned value (EV)
EVP. *See* Executive vice president (EVP)
EWP. *See* Environmental work practice (EWP)
Excess air, 1601–1606, 1734
Exchanger reformer, 617–618
Executive vice president (EVP), 1734
Exothermic reaction, 1734
Exploration, 1734
Exports, 1734
Ex situ, 1734
Extra-heavy oil (EHO), 930, 1734. *See also* Bitumen
Extrudate, 1734
Exxon Flexicracking IIIR unit, 288
Eyes, sodium hydroxide as hazardous, 1371
Eyewash, 1734

F
FA. *See* Flow alarm (FA)
Facility siting hazards, 1344–1345
Fahrenheit, 1735
Fail open, 1037, 1090, 1740
Fail-safe, 1735
Failure position, 1735

Falldown, 1735
FAME. *See* Fatty acid methyl ester (FAME)
Fatigue, 1735
Fatty acid methyl ester (FAME), 952–953, 973–975, 1735
Fatty acids
 deoxygenation, 994–997
 hydroprocessing conversion, 986–987
 isomerization and cracking reactions, 997–999
 in plants and animals, 989
Faujasite zeolite structure, 294
FBP. *See* Final boiling point (FBP)
FCC. *See* Fluid catalytic cracking (FCC) process
Feed contaminants and catalyst poisons, 432–433
Feed/effluent exchangers, 367
Feed filters, 366
Feed-forward control, 1012, 1735
Feedstocks, 388–389, 408, 1736
FEL. *See* Front-end loading (FEL)
Fenceline, 1736
Fenske calculation, 211
Fermentation, 970–973
Filming amine, 594, 1103, 1736
Filtration, 1736
Final boiling point (FBP), 10, 1736
Final control element, 1031, 1736
Fines, 1736
Finished lube oil, 846
Finned air coolers, 1558
Finned tube, 1558, 1736
Firebox, 1737
Fire brick, 1606, 1736
Fired heaters, 1418, 1594, 1737
 burners, 1604–1606
 codes and standards, 1597–1599
 efficiency, 1601–1604
 refractories, stacks and stack emissions, 1606–1611
 specification, 1611–1617
 thermal rating, 1599–1601
 types, 1594–1597
Fire extinguisher, 1429–1433, 1736
Firefighting equipment, 1424–1426
Fire foam systems, 1426–1429, 1736
Fire main, 1423–1424, 1736
Fire point, 1736
Fireproofing, 1737
Fire relief, 1323–1325, 1737
Fire water system, 1140, 1737
Fischer assay, 909, 1737

Fischer-Tropsch (FT) synthesis, 858, 887, 946–949, 968–983, 1737
Fixed cost, 1737
Fixed equipment, 1737
Fixed fee contract, 805
Flame impingement, 1737
Flameout, 1737
Flame retardant, 1737
Flammability, 1737
 limits, 1737
Flare, 1195, 1738
Flare tip, 1195, 1196, 1199, 1738
Flashing, 1738
 liquids, 1088, 1331
Flash point, 11, 848, 1076, 1738
 test method, 664
Flash zone, 1738
Flexible thermowells, 1046
Flexicoking process™, 1738
 flow diagram, 536–537
 operating conditions and yields, 537–538
Flight scraper, 1738
Floating-roof tank, 1170–1172, 1738
Flocculation, 1139, 1204–1205, 1738–1739
Flooding, 1465–1466, 1739
Flow alarm (FA), 1735
Flow element, 1019–1028, 1735
Flow measurement and control
 controller processing, 1030
 flow element installation, 1028–1030
 flow signal, 1030
 magnetic flow meters, 1025
 measurement principle, 1019
 orifice meter, 1020–1025
 oscillatory flow meters, 1026
 output signal, 1030–1031
 PD meters, 1026–1027
 thermal meters, 1028
 turbine meter, 1027
 ultrasonic meters, 1027
 valves, 1031–1034
 variable area meter, 1027
 variable speed control, 1037–1038
 variable volume control, 1038–1039
Flow meter, 1028–1030, 1739
Flow regime, 1629–1632, 1739
Flow through type catalyst cooler, 549–550
Flue gas SOx scrubbing system, 1226
Flue gas tunnels, 1739
Fluid catalytic cracker, 711
Fluid catalytic cracker unit (FCCU), 1739
 material and sulfur balance, 105
 naphtha from, 62–63

Index

Fluid catalytic cracking (FCC) process,
    261–312, 898, 1735
  applications, 272–273
  catalysts, 269
  characteristics of mechanisms, 272
  development, 262, 264
  evolution, 263
  gas oil, 273–277. (see also Gas oil cracking
      technology)
  gasoline
    analytical characterization, 406–407
    catalysts and reactions, 407–408
    composition, 397
    desulfurization, 399–400
    diolefin saturation, 399
    feedstock characteristics, 403
    feedstocks, 408
    process conditions, 403, 409
    process configuration, 400–403,
        408–409
    process considerations, 399
    recombination, 398–399
    sulfur compounds, 406
  growth, 265
  innovations in process, 267
  for light olefins and aromatics, 297–305
  primary cracking reactions, 272
  reactor-regenerator, 264–266
  secondary reactions, 272
  yields
    DCC vs. FCCU, 266
    typical resid cracking, 264–265
Fluid coking method, 536–538, 661, 1740. See
    also Flexicoking process™
Fluidized-bed catalytic cracking process, 460
Fluoroprotein foam concentrates, 1428
Flushing, 1740
  oil, 1153–1154, 1740
  system, 1505
Flux, 1740
Foam generator, 1425, 1426, 1740
Foaming, 1740
FOB. See Free on board (FOB)
FOE. See Fuel oil equivalent (FOE)
FOEB. See Fuel oil equivalent barrels (FOEB)
Force
  draft, 1735, 1740
  majeure, 1740
Formaldehyde, 893–894
Fossil fuel resources, 884–886
Fouling, 1740
  liquids, 1561
FQD. See Fuel quality directive (FQD)

Fracking, 922, 1740–1741
Fraction, 1741
Fractionation, 369, 443–444, 450, 1741
Fractionator overhead equipment, 135–142
Framework order, 1740
Free on board (FOB), 1740
Fresh feed input, 1741
Front-end loading (FEL), 789–796, 1736, 1741
Froth flow, 1629–1632, 1741
FT process. See Fischer-Tropsch (FT)
    synthesis
Fuel coke, 661, 1742
Fuel ethanol, 1742
Fuel gas, 1742
  cleanup and burner fouling, 1115–1116
  system, 1108–1111
  tramp nitrogen compounds, 1234–1235
Fuel oil, 1742
  pour point blending, 117
  products, 67
  sulfur control, 1225
  system, 1111–1113
Fuel oil equivalent (FOE), 1742
Fuel oil equivalent barrels (FOEB), 1740
Fuel quality directive (FQD), 1219, 1740
Fuel rich, 1114, 1742
Fuel solvent deasphalting, 1742
Fuel system troubleshooting, 1116
Fugitive, 1221–1223, 1227, 1240–1241,
    1243–1247, 1742
Full-conversion/complex refinery, 87
Full-conversion refinery, 1742
Full-conversion refinery hydrogen
    management approach, 635–636
Fungible, 1742
Furfural, 1374–1377
  extraction process, 852–853
Furfuraldehyde, 1742
Furnace oil, 1742

G
Gallon, 1743
Galvanic corrosion, 1743
Gap control, 1085, 1743
Gaps and overlaps, 1743
Gas blanketed, 1169, 1170, 1744
Gas burners, 1607
Gas chromatograph, 1076, 1743
Gasification, 619–620, 883–909, 948,
    983–986, 1744
Gasohol, 1744–1745
Gas oil, 15, 1744

Gas oil cracking technology
    data for FCCU monitoring, 291–294
    FCC reliability and maintenance, 290–291
    FCC zeolite, 294–297
    hydrocracking, 317–359
    reaction technology
        dry gas production, 276
        feed injection system, 273–274
        impact of vapor quench on FCC yields, 277
        quenching, 276–277
        regenerated catalyst/feed contacting configurations, 274
        riser separator strippers, (RSS) 275
    regeneration technology, 280–283
    resid catalytic cracking (see Resid, catalytic cracking)
    stripping technology, 280
        disk and donut design, 277
        disk and donut stripper, 278
        packing designs, 279
        stripper packing designs, 279
Gasoline, 651–653
    blend component properties, 112
    blending components, 1745
    blending specifications, 111
    demand by region, 483
    engine
        carbon monoxide, 72
        deposits, 71
        hazardous air pollutants, 71
        lead emission, 72
        nitrogen dioxide, 72
        octane number, 70
        ozone, 73
        particulate matter, 72–73
        sulfur dioxide, 72
        thermal efficiency, 71
        volatility, 71
    manufacture, 75–76
        aromatic reduction, 73
        olefin reduction, 73
        process, 74
        sulfur content, 74
        tetra ethyl lead restriction, 73
    precursors, 61
    production
        adiabatic plug flow reactor, 461
        oligomerization reactions, 463
        selective vs. non-selective, 466–468
        tubular reactor, 461–462
Gas Processors Suppliers Association (GPSA), 1745

Gas-refinery fuel, 1108–1111, 1743
Gas-to-liquids (GTL), 44, 888, 901, 947–950, 1744
Gas treating processes, 499–529, 1744
Gas turbine drivers, 1554–1555
Gauge pressure, 1745
Gear pump, 1486
GHG. See Greenhouse gas (GHG)
GHV. See Gross heating value (GHV)
Gilliland correlation, 214, 1745
Glitsch Ballast trays, 1444
Gold cup measurement method, 610
GPSA. See Gas Processors Suppliers Association (GPSA)
Green coke, 67
Green diesel, 953, 1745
Green gasoline, 953, 1746
Greenhouse gas (GHG), 1745
Green olefins, 953, 1746
Grid, 1746
Gross, 1746
Gross estimation approach 748
Gross heating value (GHV), 1674, 1675, 1745
GTL. See Gas-to-liquids (GTL)
Guarantee test run, 828–830
    consumption guarantee, 832–833
    feedstock, 831
    hydraulic guarantee, 832
    product quality, 831–832
Gulf Coast spot market, 1746
Gum, 1746

H

Halide removal, 379–380
HAPs. See Hazardous air pollutants (HAPs)
Hazard and operability review (HAZOP), 1312, 1747
Hazardous air pollutants (HAPs), 71
Hazardous materials, 1354
Hazardous waste, 1275, 1746
HAZOP. See Hazard and operability review (HAZOP)
HCGO. See Heavy coker gas oil (HCGO)
Header, 1747
Heat of combustion
    vs. CO_2/CO ratio, 547
    gross, 1674–1675, 1747
    net, 1674–1675, 1747
Heat exchangers, 1557, 1748
    and coolers, 1604–1606, 1747–1748
    design considerations, 1559–1561
    surface area and pressure drop, 1562–1572

tube data, 1655
types, 1557–1559
Heating oil, 656, 1748
Heat loss and heater surface area, 1205–1209
Heat and material balance, 1747
Heat pump, 1747
Heated storage tanks, 1175–1176
Heater, 1594, 1747
    burners, 1604–1606, 1747–1748
    coil, 1748
    draft, 1055, 1108, 1114, 1608–1609
    efficiency, 1601, 1748
    noise, 1606
Heat recovery steam generator (HRSG), 1750
Heavy coker gas oil (HCGO), 410
Heavy crude, 1748
Heavy fuel oil blending, 116–117
Heavy fuel oil specification, 116
Heavy gas oil (HGO), 1748
Heavy naphtha, 57
Heavy oil cracking (HOC), 1748
Heavy vacuum gas oil (HVGO), 94, 97, 103
Height equivalent to a theoretical tray (HETP), 1467, 1748
Heptene *vs.* olefin conversion, 468
HETP. *See* Height equivalent to a theoretical tray (HETP)
HF alkylation, 445, isobutane
    acid regeneration, 444
    acid strength, 445
    acid-to-hydrocarbon volume ratio, 445
    feed contaminants, 445–446
    feed pretreatment, 442
    fractionation, 443–444
    KOH regeneration, 444
    maintenance, 446
    process flow description, 442
    reaction, 443
    reaction temperature, 445
    safety and volatility suppression, 446–447
HGO. *See* Heavy gas oil (HGO)
Higher heating value (HHV), 1749
High expansion foam concentrates, 1428–1429
High speed diesel engines, 76
High sulfur bunker fuel, 116, 117
HOC. *See* Heavy oil cracking (HOC)
Homogenizer, 1749
Horton sphere, 1749
Hose reel/station, 1424, 1749
Hot and cold flash separators, 1749
Hot lime process, 1428, 1749
Hot potassium carbonate process, 506–507
Hot soaking, 419–420

Hot work, 1749
HRSG. *See* Heat recovery steam generator (HRSG)
Huff and puff, 933, 1750
Human-machine interface (HMI), 1017, 1749
HVGO. *See* Heavy vacuum gas oil (HVGO)
Hydrant, 1424, 1425, 1750
Hydraulic analysis, 1679–1684
    of process systems, 1750
Hydraulic fracturing, 922, 1750
Hydraulic guarantee, 832
Hydraulic horsepower, 1501, 1750
Hydrocarbon, 1255
    octane values, 485
    pressure-temperature curves for, 1648–1650
Hydroclone, 1750–1751
Hydrocodimers, 461, 464
    properties, 466–467
Hydrocracker, 1751
Hydrocracking, 431–432, 856
    catalyst deactivation
        catalyst support sintering, 348
        coking, 347
        metal agglomeration, 347
        non catalyst metals deposition, 348
        poisoning, 347
        regeneration, 348
    catalyst manufacturing
        drying and calcination, 342
        impregnation, 342–343
    feedstock, 319
    hydroprocessing reactions, 331–333
    once-through hydrocracking unit, 323–324
    2013 OPEC world oil outlook, 320
    process, 38
    process variables
        boiling range, 352
        cracked feedstocks, 352
        fresh feed quality, 351
        hydrogen content, 351–352
        hydrogen purity, 356
        makeup hydrogen, 355–356
        nitrogen and methane content, 356
        nitrogen and sulfur compounds, 351
        permanent catalyst poisons, 352–353
        recycle gas rate, 355
    reaction
        evolution of, 333
        heavy polynuclear aromatics (HPNA), 336
        multi-ring aromatics, 335–336

Hydrocracking (cont.)
    steps involved for paraffins, 334
    thermodynamics, 335
    separate hydrotreating with two-stage, 327
    single stage with recycle, 324–326
    suppliers, 357–358
    two stage recycle, 326
    types of operating units, 323
Hydrodenitrogenation, 374–375, 1751
Hydrodesulfurization, 371–374, 1751
Hydrodewaxing/isomerization, 856
Hydrofinishing, 856
Hydrofluoric acid (HF, AHF), 1377–1383
Hydrogen, 1383–1385, 1751–1752
    distribution, 1156
    management, 568
    purification, 369
Hydrogenated polygasolines
    effect of propylene on RON, 464–465
    octane numbers, 464
Hydrogenated *vs.* non-hydrogenated polymer gasoline, 664–666
Hydrogenation, 1752
Hydrogenolysis, 488
Hydrogen once-through (HOT), 491–492
Hydrogen production
    catalytic naphtha reforming process, 567–570
    electrolytic process, 621
    partial oxidation process, 618–621
    steam-methane reforming process, 570–575
Hydrogen sulfide (H_2S), 1226–1228, 1385–1388, 1752
Hydrolysis, 1752
Hydrometer, 1752
Hydroprocessed esters of fatty acids (HEFA), 1000–1001
Hydroprocessing, 317–434, 855, 858, 1752
    base stock properties, 856–858
    hydrodewaxing/isomerization, 856
    hydrofinishing, 856
    hydrotreating/hydrocracking, 317–434, 856
    technology sources, 858
Hydroskimming refinery, 84, 1752
Hydrothermal liquefaction, 978–979
Hydrotreater, 361–434, 1752
Hydrotreating, 361–434, 856, 1224–1225, 1752–1753
    aromatic saturation, 377–378
    catalyst activation, 421–422
    catalyst deactivation, 422–424
    catalyst loading, 420–421
    catalyst manufacturers, 433
    catalyst performance, 417
    cobalt-molybdenum catalysts, 415
    distillate (*see* Distillate hydrotreating)
    FCC gasoline (*see* Fluid catalytic cracking (FCC) process, gasoline)
    feed/effluent exchangers, 367
    feed filters, 366–367
    feed quality and rate, 428–429
    fractionation section, 369
    gas-to-oil ratio, 430
    halide removal, 379–380
    history, 365–366
    hydrocracking, 410–441
    hydrogen partial pressure, 429–430
    hydrogen purification, 369
    liquid hourly space velocity, 430
    make-up hydrogen system, 367
    metals and non-metals removal, 378–379
    naphtha (*see* Naptha hydrotreating)
    nickel-molybdenum catalysts, 415
    olefin saturation, 376–377
    oxygen removal, 376
    reactor charge heater, 367
    reactor design and construction, 424–426
    reactor effluent water wash, 368–369
    reactor operation, 427
    reactor temperature, 427–428
    recycle gas purity, 430–431
    recycle gas scrubbing, 368
    recycle hydrogen system, 368
    residual oil hydrodesulfurization, 411–413
    sulfur removal, 371–374
    technology suppliers, 433
    type I *vs.* type II catalysts, 416
    units, 36
    vapor/liquid separation, 369
Hysomer™ catalyst, 489

I
i-component, 1753–1754
Ignitability, 1754
Immediately Dangerous to Life and Health (IDLH), 1754
Immediate oxygen demand (IOD), 1254, 1758
Impeller speeds (pumps), 1488, 1490, 1492, 1493, 1754
Impingement, 1754–1755
Imports, 1755
InAlk™ process, catalytic olefin condensation
    resin-catalyzed condensation, 468–469
    SPA-catalyzed condensation, 470–471
Incident command system (ICS), 1434, 1755

Incompatible, 1755
Incomplete combustion, 1241–1242
Incompressible flow pressure drop, 1617–1622
Incremental cost, 781
Indices, 1755
Induced, draft, 1755
Inert entry, 1340–1344, 1756
Inferred properties, 1756
Inflation, 745
Infrared analyzer, 1076–1077, 1756
Initial boiling point (IBP), 10, 1753, 1756
Inlet diffuser, 1756
Inlet guide vane, 1756
In-line blending, 1178–1180, 1756
In-line vs. batch blending, 1178–1179
Innage, 1756
INPNA, 1756
In situ, 1756
Instantaneous analysis, 741, 745
Instrument air system, 1141–1144, 1756
Instrumentation, 1013, 1756
Insulation, 1757
Intalox saddle, 1757
Integral control, 1086, 1087, 1757
Integrated gasification combined-cycle (IGCC) system, 887
Integrated oil company, 1757
Intellectual property (IP), 1758
Intercooling, 1538, 1757
Interface, 1757
Intermediate, 1757
  law, 1266, 1471, 1757
  stock report, 697, 763
Internal combustion engine, 1038
Internal rate of return (IRR), 758–759
Internals (vessels), 1119, 1132, 1174, 1442, 1471, 1484–1485, 1641–1642, 1757
International Energy Agency (IEA), 1757
International Maritime Organization (IMO) legislation, 1219
Interstate Commerce Commission (ICC), 1757
Ion exchange, 1131–1133, 1758
Ionic liquid alkylation, 454–456
Ionization, 1758
Ion-specific and pH electrodes, 1077
Iron sulfide (FeS), 1736
Isobutane, 1758
Isobutylene, 1758
Iso-container, 1758
Isohexane, 1758
Isomerate, 483

Isomerization, 74, 441, 997–999
  Butamer™ process, 481
  catalysts
    chlorided alumina, 489
    octane/activity curves, 490–491
    sulfated metal oxide catalysts, 490
    zeolitic type catalysts, 489–490
  flow scheme improvements, 482
  HF alkylation, 481
  history of, 481–482
  hydrogen once-through (HOT), 492
  isobutylene, 481
  light hydrocarbon (C4-C7) streams, 480
  light-naphtha, 480
  market trends, 482–484
  n-butane to isobutane, 481
  Oleflex™ process, 481
  paraffin, 480
    primary reaction pathways, 486–488
    process chemistry, 485–486
  Par-Isom process flow schemes, 491
  Penex™ process, 481
  Penex unit process flow scheme, 492
  process configurations and octane values, 482
  process economics, 494–495
  process requirements for catalysts, 491–492
  regenerable catalysts, 482
  regulation, 484–485
  skeletal isomerization of paraffins, 481
Itemized estimation approach, 748

J

Jet fuel, 653–654, 1759
Jet mixer, 1173, 1176, 1759
Jetties, 1759–1760
Jetty and dock facilities
  equipment, 1186–1187
  loading rates, 1187–1188
  ship ballast water, 1188
  size, access, and location, 1185–1186
  slop and spill facilities, 1189
Jetty on shore loading stations, 1422
Jobber, 1760
Job safety analysis (JSA), 1760
Joint venture, 1760
Joule-Thomson effect, 1760

K

Kellogg resid cracker, 287
Kerogen-oil, 907

Kerogen rock, 908–909
Kero operation, 105–106
Kerosenes, 653, 1760
Kerosene-type jet fuel, 653–654, 1760–1761
Kettle, 1761
    reboiler, 1584–1586
Kettle-type waste heat boiler, 1099–1100
Kick off meeting, 810
Kieselguhr, 459–460
Kinematic viscosity, 667–669, 1761
Knock/cetane engine, 1077
Knockout, 11, 681, 1761
    drum, 1470–1471, 1761
Kremser equation, 525, 527

L

Laboratory information system (LIMS), 1764
Langelier Saturation Index (LSI), 1125, 1762
Larkins, White, and Jeffery method,
    1639, 1762
Latent heat, 1762
Layering, 1762
Layers of protection analysis (LOPA),
    1312, 1762
Lead, 72
Leaded gasolines, 1762–1763
Leaks, 1246
Legionella bacteria, 1128, 1763
Level alarm (LA), 1762
Level bridles, 1067
Level control range, 1763
Level gauges/sight gauges, 1069
Level measurement and control
    elements and characteristics, 1063–1067
    level instrument installations, 1067–1069
    specific considerations, 1069–1070
Life cycle of petroleum fuels, 967–968
Lift gas technology, 550–553
Lifting, 1763
Light crude, 1763
Light cycle oil, 410
Light ends, 19, 34, 1763
    distillation, 1727
    condenser duty, 214–215
    description, 200–201
    material balance, 202–206
    operation conditions, 206–210
    reboiler duty, 215
    tower loadind and sizing, 216–217
    tower operation and performance, 225–227
    trays in, 211
    units, 34–35, 1763

Light gas oil, 1763
Light hydrocarbon (C_4-C_7) streams, 480
Light naphtha, 57
Light-naphtha isomerization, 480
Light vacuum gas oil (LVGO), 94, 101, 104
Lignin, 976, 977
Ligno-cellulosic conversion, 976–986
Linear programming, 55, 739–741,
    774–783, 1764
Liquefied petroleum gas(es) (LPG), 55, 651,
    1764–1765
Liquefied refinery gas(es) (LRG), 1765
Liquid alive bacteria (LAB), 1762
Liquid entrainment, 226
Liquid ring, 1170, 1515, 1764
Liquid separation drums, 1471–1473
Liter, 1765
Loader, 1765
Loan repayment, 769–770, 772–773
Lobular pump, 1486
Lockhart-Martinelli modulus, 1631, 1765
Logarithmic mean temperature difference,
    1564, 1765
Long ton, 1765
Lower heating value (LHV), 1763
Low-low heater draft, 1114
Low speed diesel engines, 76
Low-sulfur diesel (LSD), 81, 656–659, 1765
Low-sulfur gasoline (LSG), 81, 656–659, 1765
Lube base oil group classifications, 849
Lube base oils, 845
Lube base stock, 845
Lube oil properties, 847–848
    color, 848
    flash point, 848
    oxidation stability, 848
    pour point, 848
    thermal stability, 848
    viscosity, 848
    viscosity index, 848
    volatility, 848
Lube oil refineries, 44–46, 844–845
    asphalt production, 860
        asphalt blowing process, 863–864
        asphalt oxidizing mechanism, 861–862
        paving and liquid asphalt, 860–861
        roofing asphalt, 861
    base stock group statistics, 859
    bright stock lubes, 846
    finished lube oil, 846
    hydroprocessing, 855–856, 858
        base stock properties, 856–858
        hydrodewaxing/isomerization, 856

Index 1901

hydrofinishing, 856
hydrotreating/hydrocracking, 856
technology sources, 858
lube base stock, 845
lube slate, 845
naphthenic lube oils, 846
neutral lubes, 846
paraffinic lube oil, 846
refining process options, 850
re-refined base stocks, 846
traditional solvent-based lube production, 850–855
   furfural extraction process, 852–853
   MEK dewaxing process, 854–855
   propane deasphalting process, 850–852
virgin base lube oils, 846
Lube oil re-refining, 960–961
Lube oils, 660, 1766
   base stock, 68
   neutral lubes, 68
   performance specifications, 846–847
   properties, 69
   quality specifications, 846
   slate, 68
Lube slate, 845
Lubricants, 1766
Lubricity, 1766
Lump sum contract, 806
Lurgi MTP process, 898–899

M

Magnetic flow meters, 1025
Make-up hydrogen system, 367
Management by objectives (MBO), 1768
Manifold, 1767
Margin, 1767
Marginal crude oils, 689
Marine diesel
   fuel, 659
   oil, 1767
   specification, 116, 659
Market(ing), 1767
   department, 687
   trends, isomerization, 482–484
MARPOL, 1767
Mass flow meters, 1026
Mass spectrometry, 1077, 1767
Master construction plan, 818
Material balance, 1767
Material safety data sheet (MSDS), 1767–1768
Maxofin FCC, 304–305
Maxwell method, 21

McCabe Thiele-Graphical Method, 508, 510
Mechanical flow diagram, 793–794, 1768
Mechanical flow sheet, 794
   conference, 813–816
Mechanical integrity (MI), 1313, 1768, 1769
Mechanical specifications, fire
   prevention, 1417
Mechanical speed control, 1038
Medium curing cutbacks, 861
Medium speed diesel engines, 76
MEK dewaxing process, 854–855
Membranes, for hydrogen recovery, 634
Membrane unit, 1768
Merichem, 1768
Merox, 1768
Metal dusting, 599
Metals and non-metals removal, 378–379
Metals deposition, 423
Metals in crude oil, 1768
Metering pumps, 1487, 1769
Methanation, 584–587, 619, 1769
Methanol, 889–892, 1249
   derivatives, 893–897
   production economies, 891–892
   properties and specifications, 892
   technology licensors, 891
Methanol to gasoline technology (MTG), 888
Methanol-to-hydrocarbon, 897
Methanol-to-olefins, 897–898
Methyldiethanolamine (MDEA), 506, 1354–1356, 1769
Methyl ethyl ketone (MEK), 854–855, 1388–1391, 1769
Methyl tertiary-butyl ether (MTBE), 42–43, 890, 1769
Metric ton, 1769
Micron, 1769
Mid-barrel, 1769
Mid boiling point components, 15–16
Middle distillate, 1769
   products, 64–66
   schedule, 698
Midstream, 1770
Mid volume percentage point components, 16
Military jet fuel, 654
Million volt-amps (MVA), 1771
Mix temperature control, 550–552
Mobile Source Air Toxics (MSAT) rule, 484
Molecular sieve, 1770
Molecule, 1770
Mollier diagram, 1770
Monitor, 1770

Monoethanolamine (MEA), 500, 502–503, 1354, 1770
   preliminary process flow sheet, 528
   removing degradation impurity, 513–514
Mortgage formula, 772–773
Motor fuel alkylation
   alkylate properties, 452
   complex alkylation reactions, 439–441
   HF alkylation (see HF alkylation)
   history, 437–438
   ionic liquid alkylation, 454–456
   isomerization, 441
   primary alkylation reactions, 438–439
   solid catalyst alkylation, 453–454
   sulfuric acid (see Sulfuric acid alkylation)
Motor gasoline, 1770
Motor gasoline blending, 1178–1181, 1770
   components, 1180, 1771
Motor octane number, 651, 681, 1771
Multipass tray, 1455, 1771
Multiple single loop controls, 1016

N

Naphtha
   characterization, 233
   composition, 233–234
   consolidation and treating process, 106
   feed pretreatment, 234–235
   from fluid catalytic cracker unit, 62–63
   hydrotreater, 235
   straight run, 233
   stream properties, 63
Naphthalene, 1772
Naphtha-type jet fuel, 653–654, 1772
Naphthene(s), 5, 1772
   reactions, 243–244
Naphthenic lube(s), 69
   oils, 846
Naphtha hydrotreating
   diene reactor, 395
   gas-to-oil ratio, 395
   liquid hourly space velocity, 395
   naphtha disposition, 392
   post treat reactor, 395
   pressure, 395
   process configuration, 389, 390
   processing objectives and considerations, 392–394
   recombination with cracked stock, 394
   straight-run versus cracked naphtha, 391–392

National Council of Examiners for Engineering and Surveying (NCEES), 1773
National Pollutant Discharge Elimination System (NPDES), 1250, 1772
Natural draft, 1772
Natural gas, 1772
Natural gas liquids (NGL), 901, 1772, 1774
Natural gasoline and isopentane, 1773
Natural gas plant liquids, 1773
Near-infrared analyzer, 1076, 1773
Needle coke, 67, 1773
Netback, 1773
Net investment, 731, 1773
Net profit margin, 1773
Neutralization, 1773
   number, 1773
Neutralizing amine, 594, 1103, 1774
Neutral lubes, 68, 846
Newton's law, 1774
New York Mercantile Exchange (NYMEX), 1774
Nickel carbonyl, 1391–1394, 1774
Nickel-molybdenum catalysts, 415
Nitrogen compounds, 386–387
Nitrogen dioxide, 72, 1222, 1234, 1236–1239, 1611
Nitrogen oxides, 1222, 1234, 1236–1239, 1611
Nitrogen peroxide. See Nitrogen dioxide
Nitrogen removal, 374–375
Nitrogen system, 1149–1153
   utilities, 1149–1150
Noise control, 1280–1286
Noise pollution, 1278–1280
Noise program, 1286–1288
Non-energy product refineries, 1774
Non-energy refineries, 843–882
   lube oil properties, 847–848
      color, 848
      flash point, 848
      oxidation stability, 848
      pour point, 848
      thermal stability, 848
      viscosity index, 848
      volatility, 848
   lube oil refineries, 844–845, 849–850
      bright stock lubes, 846
      finished lube oil, 846
      lube base oil group classifications, 849
      lube base stock, 845
      lube oil properties, 847–848
      lube slate, 845
      naphthenic lube oils, 846
      neutral lubes, 846

paraffinic lube oil, 846
    quality and performance specifications, 846–847
    refining process options, 850
    re-refined base stocks, 846
    traditional solvent-based lube production, 850–855
    virgin base lube oils, 846
  lube oil specifications, 846
    performance specifications, 846–847
    quality specifications, 846
Nonenes, 474
Non-hazardous waste, 1275
NOx emissions monitoring and operations, 1238–1239
Nozzle meters, 1025

O

Octane enhancement processes, 41
Octane numbers, 11, 70, 464, 681, 1775
  hydrogenated polygasolines, 464–466
Octenes production, 459
Offsite systems, 1167–1213, 1776
Oil burners, 1114, 1604
Oil, Chemical, and Atomic Workers Union (OCAW), 1775
Oil dropping method, 418
Oil recovery sump, 1267–1268
Oil retention baffles, 1268
Oil sand crude, 930–946, 1776
Oil skim pipes, 1267
Oil-water separation, 1265
Oily water sewer, 1157–1159, 1776
Olefin(s), 1776
  condensation, 458. (*See also* Catalytic Olefin condensation)
  feedstock, 48
Olefin cracking byproduct recovery, 622–623
Olefin cracking process (OCP) integration, 902
Olefinic hydrocarbons, 7
Olefin saturation, 376–377
Oleflex™ process, 481
Oligomerization reactions, 463
Omnibus fee contract, 805
Once-through thermosiphon reboiler, 1586–1587
One-stage thermal cracker, 533
Opacity, 1776
Organic sulfur compounds, 1233–1234
Organometallic catalysts, 459
Orifice meter, 1020–1025, 1028–1030, 1776–1777

Original equipment manufacturer (OEM), 1776
Orsat, 1777
Oscillatory meters, 1026
Outage, 1777
Outlet collector, 1777
Overall tower heat balance, 216
Overflash, 1777
Overhead, 1777
Overlap, 1777
Overplus, 1777
Overpressure, 1778
Overspeed, 1038, 1778
Over-the-fence supply, 625, 1155–1156
Oxidation, 1262–1265, 1778
  mercaptans, 1262–1264
  pond, 1270, 1778
  stability, 848
  sulfides to thiosulfates, 1261–1262
  of sulfide to sulfate, 1264–1265
Oxidizing
  agent, 1778
Oxygenate
  gasoline, 41–42, 74–76
Oxygen deficient, 1778
Oxygen deficient environments
  hazards and potential exposures, 1338–1340
  management, 1340–1341
Oxygen enrichment, 1778
  supply, 1155–1156
Oxygen removal, 376
Ozonation, 1778
Ozone, 73, 1248, 1778

P

Packed tower, 1778–1779
Packing, 1779
Padding, 1779
Pall ring, 1463, 1779
Pantograph, 1174, 1779
Paraffin(s), 5, 1779
  isomerization, 480–481
    primary reaction pathways, 486–488
    process chemistry, 485–486
    skeletal isomerization, 481
  reactions, 244–245
Paraffinic lube(s), 69
  oil, 846
Par-Isom process flow schemes, 491–494
Partial oxidation (POX) process, 619–621, 1779
  and autothermal reforming, 618

Partial oxidation (POX) process (*cont.*)
  block flow diagram, 619
  costs, 621
  design and metallurgy, 620–621
  gasification, 619–620
  product purification and recovery, 620
  shift conversion, 620
Partial pressure, 1779
Particulate matter (PM), 72, 1782
Partition wall, 1597, 1779
Paving and liquid asphalt, 860–861
  medium curing cutbacks, 861
  rapid curing cutbacks, 860
  slow curing cutbacks, 861
Peened thermocouple, 1048
Penex™ process, 481
Penex unit process flow scheme, 492
Pensky-Martens (PMCC), 664, 1780
Pentane equilibrium, 486
Performance specifications, lube oil, 846–847
Permissible exposure limit (PEL), 1780
Personal protective equipment (PPE), 1784
Petrochemical(s), 297–307
  applications, 472
    heptenes, 472–474
    nonenes, 474
  DCC, 300–302
    diagram of process, 300
    light olefin yields, 302
    with naphtha recycle, 301
    vs. FCC, 300
    yields for atmospheric resid, 302
  DCC process, 299
  feedstocks, 1780
  heavy feed and naphtha propylene processes, 299
  HS-FCC and R2P process, 306
  Maxofin process, 304
  operations (*see* Petrochemical, applications)
  PetroFCC, 304
    vs. FCC, 304
  PetroFCC process, 303
  refinery, 872–875
    aromatics production, 872–874
    dealkylation unit, 875
    feed fractionation and xylene splitter, 874–875
    isomerization unit, 875
  selectivity of spent catalyst, 303
  Shell Milos and INDMAX processes, 307
PetroFCC, 303
  process, 303

Petroleum, 1780
  ash content, 1694
  coke, 67, 661, 1780
  process, 1215–1302
  process study approach, 700–703
  products, 1781
Petroleum Administration Defense District (PADD), 1780
pH, 1077, 1781
Phenol, 1255, 1781
Pickling, 826
Pilot, 1781
Pilot-operated pressure relief valve, 1321, 1781
Pinch technique, 1781
Pipe
  element, 1782
  fitting, 1781
  hanger, 1781
  layout specifications, 1418
  line, 1782
  support, 1781
Pitch, 1782
Planning refinery operations, 687–698, 1782
Plant air, 1141, 1782
Plant commissioning, 824–825, 1782
Plant life, 732
Plantrose™ process, 983, 985
Plant water system, 1138–1139, 1782
Play, 1782
Plenum, 1597, 1782
Plot plan development, 817
Plug flow, 1782
Plug valve, 1034–1035
Poison, 1783
Pollution, 1216, 1218, 1250, 1278, 1606, 1783
Pollution prevention design (PPD), 1784
Poly-alpha-olefins (PAOs), 845
  production, 858–859
Polyaromatic hydrocarbon, 1783
Polychlorinated biphenyls (PCBs), 1783
Poly gasolin. *See* Polymer, gasoline
Polymer, 1783
  gasoline, 461, 465–466
    blending octanes, 465–466
    hydrogenated *vs.* non-hydrogenated, 464–466
Polymerization, 1783
Polyvinylchloride (PVC), 460
Porosity, 909, 921–922, 1607, 1635, 1783
Positive displacement (PD)
  meters, 1026–1027
  pumps, 1111
Post treat reactor, 395

Index 1905

Potable water, 1139–1140, 1783
Pour point, 79, 666, 848, 1783–1784
Power recovery, 1147–1148, 1784
Practical (or Puckorius) Scale Index (PSI), 1126, 1784
Preheat and preheat exchanger train, 1784
Pre-ignition, 70
Premium gasoline grade blend, 111
Pre-refining, 1784
Pre-reformer, 617
Pressure alarm (PA), 1778
Pressure drop, 1631
  beds of solids, 1635–1643
  coefficients, 1456
  compressible flow, 1622–1624
  fittings and piping elements, 1625
  gas streams, 1677
  incompressible flow, 1617–1622
  overall, 1631
  two-phase, 1629–1634
Pressure measurement and control, 1054
  distillation column pressure, 1055
  elements and characteristics, 1051
  heater draft, 1055
  installations, 1051–1054
  pressure regulator, 1054
  tank blanket gas control, 1054–1055
Pressure regulator, 1014–1015, 1054, 1784
Pressure relief valve (PRV), 1319, 1785
Pressure safety
  balanced safety relief valve, 1321, 1323
  conventional safety relief valves, 1321
  definitions, 1317–1320
  orifice area sizing, 1326–1332
  pilot-operated safety relief valve, 1321
  relief capacity, 1322–1326
  relief cases, 1315–1317
  resilient-seated safety relief valve, 1321
  rupture disk, 1321
Pressure storage tanks, 1175
Pressure swing adsorption (PSA), 1785
  adsorption, 628
  5-bed PSA unit, 626–627
  blowdown, 628
  design and metallurgy, 629–630
  first equalization valve, 628
  hydrogen recovery, 624
  monitoring parameters, 631
  operating phases, 630
  purges, 628

repressurization, 628
troubleshooting, 631–633
Pressure vessels, 1419–1420
Pre-startup safety review (PSSR), 1314–1315, 1785
Pretreatment, 1785
Primary storage, 1785
Prime mover, 1505, 1550, 1785
Process chemistry
  paraffin isomerization, 485–486
Process configurations, 1785–1786
Process control, 1009–1092, 1785
Process controllers, 1084–1085
Process economics
  analysis, 699–728
  isomerization, 494–495
Process flow coefficient, 1087–1088
Process guarantees, 828–833, 1785
Process hazards analysis (PHA), 1312
Processing gain, 1786
Process safety management (PSM), 1311–1315
  compliance audits, 1314
  emergency planning and response, 1315
  employee participation, 1314
  foundational needs, 1315
  hot work, 1313
  incident investigation, 1314
  management of change, 1313
  mechanical integrity, 1313
  operating procedures, 1312–1313
  pre-startup safety review, 1314–1315
  process safety information, 1311–1312
  trade secrets, 1314
  training, 1313
Product blending, 1786
  facilities, 1178–1181
Product gravity, 23
Production, 1787
  capacity, 1787
  cost, 746
Product molecular weight, 28–33
Product properties, 1787
Product property prediction, 23–33, 1787
Product quality, 330, 650
Proesa process, 978
Programmable logic controller (PLC), 1013, 1018, 1787
Progressing cavity, 1787
Project duty specification, 796–803, 1787
Project management, 804–809
Project procedure manual, 811–813

Propane, 1787
  deasphalting process, 852
    product yield and properties, 852
    salient operating conditions, 852
  LPG, 59
Propant, 1787
Properties, lube oil, 847–848
  color, 848
  flash point, 848
  oxidation stability, 848
  pour point, 848
  thermal stability, 848
  viscosity index, 848
  volatility, 848
Proportional band, 1086, 1788
Proportional control, 1086, 1788
Proportioning pump, 1489, 1788
Propylene, 1788
  tetramer, 461
Protein foam concentrates, 1428
Pseudo components, 13, 130, 1788
Publically Owned Treatment Works (POTWs), 1265
Puckorius Scale Index (PSI), 1126
Pulsation, 1494
  drums, 1473–1476
Pulsation dampener (PD), 1555–1557, 1788
Pump, 1789
  centrifugal, 1486
  characteristic curve for, 1487–1489
  cooling water requirement, 1513
  diaphragm, 1487
  drivers and utilities, 1505–1508
  metering, 1487
  performance evaluation, 1494–1496
  reacceleration requirement, 1509
  reciprocating, 1493–1494
  rotary, 1493
  selection of, 1490–1494
  turbine, 1486
Pumparound, 1788–1789
Pumping temperature, 1498–1501
Purchase orders, 817
Purifier, 1789
Pyrolysis, 1789

Q

Quality specifications, 650
  asphalts, 661
  aviation turbine gasoline, 653
  fuel oil products, 659
  gas oils, 654
  gasolines, 651
  jet fuels, 653
  kerosene, 653
  LPG, 651
  lube oils, 660, 846
  petroleum coke, 661
  sulfur, 661
Quench zone, 368, 1332, 1641–1642, 1789
Quickie method, fractionation design, 1446–1449

R

Radiant tube, 1790
Raffinate, 1790
Ramsbottom coke, 1790
Rangeability, 1790
Rapid curing cutbacks, 860
Rapid thermal processing, 980–981
Raschig ring, 1462–1463, 1637, 1790
Raw hydrogen purification
  pressure swing adsorption, 583–584
  wet chemical/solvent approach, 584–587
Reacceleration, 1509, 1790
Reactivity, 1791
Reactor charge heater, 367
Reactor effluent water wash, 368–369
Reagent, 1791
Reboiler, 1791
  duty, 215
Reciprocating compressor, 1514, 1791
  control of, 1538–1540
  dampening facilities, 1556
  description, 1518
  horsepower, 1534–1537
  inter-cooling, 1538–1540
Reciprocating pumps, 1487, 1791
Recirculating thermosiphon reboiler, 1586–1587
Reclaiming, 513–514, 526, 1791–1792
Recommended exposure limit (REL), 1792
Recovery process, for hydrogen, 568
Recycle, 1792
  gas scrubbing, 368
  hydrogen system, 368
Reduced crude oil, 1793
Reducing agent, 1793
Redwood viscosity, 1793
Refinery, 1793. *See also* Non-energy refineries
  configuration, 82–93, 704
  emissions, 1219–1220
  gas, 55

Index   1907

gas recovery
    membranes, 623–624
    PSA, 624
gas treating process
    absorber tray size and design, 511
    amine circulation rate calculation, 507–508
    description, 500–501
    diethanol amine, 503–505
    diglycol amine, 505–506
    heat transfer area, 511–512
    methyl diethanolamine, 506
    monoethanol amine, 502–503, 513–514
    stripper design and performance, 512–513
    theoretical trays, in amine contractor, 508–510
hydrogen
    fresh hydrogen header, 636–637
    header pressures, 637–638
    management, 635–636
    overall planning, 638–639
    sour hydrogen, 637
    users management, 638
input-crude or total, 1793
operating program
    content of, 697–698
    monitoring, 699
    preparation of, 698–699
    and schedule, 761
process configuration, 53–124
    block flow diagram, 82, 121, 123
    comparison, 90–92
    cracking refinery, 84–87
    example, 93, 121
    full-conversion/complex refinery, 87
    geographic trends, 87–90
    hydroskimming refinery, 84
    topping refinery, 83–84
production, 1793
types, 1793
yield, 1793–1794
Refinery-grade butane, 651, 1793
Refining, 1794
    department, 688
Reflux, 1794
    drum, 1473, 1794
    ratio, 1794
Reformate, 1794
    catalysts, 246–249
    reactions, catalyst, 243–246
Reformulated gasoline, 41–42, 70–76, 651–653, 1794–1795

Refractory, 1606–1608, 1795
Refractory-shielded pads, 1047–1048
Regenerable catalysts, 482
Regeneration technology, 1795
    parameters, 281
    plug flow *vs.* backmixed regenerators, 280–281
    regenerator designs, 282
Register, 1604, 1795
Regular grade gasoline blend, 115
Regulation, isomerization, 484–485
Regulatory control, 1017, 1795
Reid vapor pressure (RVP), 669, 1795
Reliability, 1795
Reliability-centered maintenance (RCM), 1790
Reliability operating limit (ROL), 1797
Relief valves, 1245, 1321–1335, 1795
Relieving conditions, 1326–1331, 1795
Renewable, 950, 966, 1796
Renewable feedstocks
    ammonium bicarbonate deposition, 960
    catalyst poisoning, 959
    cold flow properties, 959
    corrosion, 959
    definitions, 952–953
    economics, 959
    ethanol, 951
    fatty acid esters, 951
    greenhouse gas emission, 950
    hydrogen consumption, 959
    hydrotreating, 956
    plugging, 959
    properties, 953, 954
    raw material supply, 958
    transesterification, 956
Renewable Fuels Standards, 950
Renewable Identification Number, 950
Re-refined base stocks, 846, 960–961
Re-refining, 960–961, 1796
Research octane number (RON), 681, 1796
Reset, 1087, 1796
Resid, 1796
    catalytic cracking
        additive coke, 283
        carbon *vs.* hydrogen in coke burns, 287, 289
        catalyst coolers, 289
        Exxon Flexicracking IIIR unit, 288
        Kellogg resid cracker, 287
        methods for coke making, 285
        percentage of Conradson carbon, 284
        shell resid crackers, 286
        sources of coke production, 283

Resid (*cont.*)
  Technip-S&W/Axens R2R, 285
  UOP resid designs, 286
  cracking, 264–265
Residual fuel oil, 1796
Residual oil hydrodesulfurization
  process conditions, 413
  process configuration, 413
Residue, 1796
  conversion, 1796
Residuum hydrocracking, 553–556
  catalysts and reactions, 412–413
  heavy metals effect on catalyst, 556
  yields, and product properties, 555–556
Resilient seat, 1321, 1796
Resin-catalyzed condensation, 468–469
Resistance temperature detector (RTD), 1041, 1043–1044, 1798
Retort, 908–911, 1796
Return bend, 1597, 1797
Return on investment (ROI), 711, 756
Reverse osmosis, 1133–1135, 1797
Road and rail loading facilities, 1184, 1797
Roofing asphalt, 861
Rotary compressor, 1515, 1797
Rotary pump, 1486, 1797
Rotating equipment, 1797
Runaway reaction, 1335–1338, 1798
Running plans, 688–692
Rupture disk, 1319, 1321, 1798
Ryznar Stability Index (RSI), 1125, 1798

S

SAE viscosity, for single-grade motor oil grades, 69
Safe operating limit (SOL), 1803
Safe park, 1798
Safety integrated system (SIS), 1013, 1798
Safety shower, 1798
Safety systems, 1307–1349, 1798–1799
Sanitary sewage treatment, 1205
Sanitary sewer, 1161, 1799
Saybolt Furol viscosity, 1799
Saybolt Universal viscosity, 1799
Scavenger, 1799
Screw pumps, 1486
Seal, 1800
Selective catalytic reduction (SCR), 1799–1800
  system, 1236–1237
  units, 597
Selective non-catalytic reduction (SNCR), 1800

Semi-definitive cost estimate, 726
Sensing element, 1800
Separator, 1470, 1800
Service factor, 121, 688, 1551, 1800
Setpoint, 1013, 1800
Set pressure, 1320, 1800
Setting, 1597, 1800
Sewer systems, 1157–1161, 1800
Shale crude and gas, 918, 921–929, 1800
  aluminum, potassium, and zirconium compounds, 927
  calcium and magnesium, 927
  corrosion, 928
  crude and product compatibility, 928
  definitions, 918
  glutaraldehyde, 929
  hydroprocessing units, 929
  impacts, 923, 927
  methods of production, 921–923
  naphtha reformers, 929
  phosphorous, 924–925
  production
    chemicals, 926
    locations, 918
    rates, 918
  properties, 923
  silicon, 925–926
Shale oil, 907–917, 1800–1801
Shell, 1801
  resid crackers, 286
  storage capacity, 1801
  tube heat exchanger, 1559, 1801
Short-term exposure limit (STEL), 1801
Shot coke, 1801
Side stream, 1801
  draw-off temperatures, 188
  filtration, 600
  stripping, 1801
Sieve tray, 1444, 1802
SimDist, 1802
Sinking funds, 768–769
Sizing drivers, 1550
Skeletal isomerization of paraffins, 481
Skirt, 1802
Slack wax, 1802
Slagging, 1802
Slops, 1802
  recovered oil, 1161–1162
Slow curing cutbacks, 861
Sludge, 1802
  sumps, 1268
Slug flow, 1629–1632, 1802
Slurry oil, 1802

Index

Smoke point, 673
Soaking volume concept, 538–540
Soaking volume factor (SVF), 540–543, 1803
    critical decomposition zone, 544
    degree of conversion, 541
    product yields, 542–543
Sodium hydroxide (NaOH), 1370–1374
Solid catalyst alkylation, 453–454
Solid waste, 1273–1274, 1803
Solomon Associates, 1803
Soluble, 1803
Solvent, 1803
    dewaxing, 81
Soot blowing, 1804
Sorbent, 1366–1370, 1804
Sour crude, 1804
Sour water stripper (SWS), 1230, 1804, 1810
    gas, 1230
SOx sources management, 1225
SPA catalyst, 460, 478
SPA-catalyzed condensation, 470–471
Spalling, 1804
Sparing, 121, 802–803, 943, 1804
Special naphthas, 1804
Specialty oligomerization processes, 476–477
Specification, 1805
Specific gravity, 663, 1804–1805
Specific heat, 1805
Spent caustic oxidation, 1261
Splitter, naphtha, 1805
Sponge coke, 67–68, 661
Spontaneous ignition temperature (SIT), 1802
Spot price, 1805
Spread, 1805
Stabilize, 1806
Stacks, 1608–1609, 1806
Staged combustion, 1235, 1806
State regulatory decisions, 1219
Static mixer, 1806
Steam and condensate systems, 1098, 1100–1101, 1104–105, 1806–1087
    boiler controls, 1102
    boiler feed water, 1098
    condensate chemical treatment, 1102–1103
    contamination prevention, 1102
    monitoring, 1104–1105
    pressure letdown and desuperheating, 1100
    refinery steam system, 1095–1097
    steam generation, 1098
    steam supply reliability, 1103–1104
    steam traps and condensate, 1100–1101
    utility steam stations, 1101
    waste heat boilers, 1099–1100
Steam-assisted gravity drain (SAGD), 933, 1799
Steam chest, 1807
Steam drum, 1016, 1098–1099, 1472, 1807
Steam-methane reforming (SMR), 570–618, 1807
    advantages, 571
    block flow diagram, 571–572
    burners, 595–596
    convection section, 596
    description, 570
    disadvantages, 571
    furnace draft control, 597
    LTS converter activity, 610
    methanator activity, 611
    monitoring parameters, 605
    pressure swing adsorption, 600
    PSA recovery, 609
    radiant tubes, 595
    wet chemical/solvent systems, 600
Steam out, 1807
Steam turbine, 1510–1511, 1553, 1555, 1807
    efficiencies, 1511, 1553
    operation of, 1509
    rated horsepower, 1553
Stiff-Davis Index (SDI), 1125, 1807
Still gas, 1807
Stoke's law, 1471, 1807
Stop-check valve, 1807
Storage bullet, 1702
Storage facilities, 1808
    atmospheric storage tanks, 1169–1175
    heated storage tanks, 1175–1176
    pressure storage tanks, 1175
    tank management and design considerations, 1176–1178
Storm sewer, 1159–1160
Storm surge pond, 1269
Stove oil, 1808
Straight run, 1808
    heavy gas oil, 65–66
    kerosene, 57
    kerosene products, 64–65
    light gas oil, 65
    naphtha, 233
    product streams, 59
Straight thermowells, 1046
Strategic Petroleum Reserve (SPR), 1808
Stratification, 1808

Stratified flow, 1630–1632, 1808
Stream day, 1808
Stress relieve, 1808
Stripping steam ratio, 1808
Stripping technology, 277–280
　disk and donut
　　design, 277
　　stripper, 278
　packing designs, 279
　stripper packing designs, 279
Stuffing box, 1502–1504, 1808
Sub-octane, 1808
Suction throttling, 1526
Suction valve lifters, 1539
Sulfated metal oxide catalysts, 490
Sulfidation, 1808
Sulfiding, 1808
　chemical, 1397–1403, 1808
Sulfinol, 1808–1809
　systems, 507
Sulfolane, 507, 1809
Sulfur, 661–662, 1809
　content, 12, 23–25, 1809
　dioxide, 72, 1220–1226
　removal, 1809
Sulfuric acid, 1156–1157, 1403–1405, 1810
Sulfuric acid alkylation, 447–456
　cascade alkylation process, 447–448
　effluent refrigerated alkylation process, 448–449
　effluent treating, 449–450
　feed pretreatment, 449
　fractionation, 450
　H_2SO_4 consumption and contaminants, 450–451
　H_2SO_4 regeneration, 451
　process variables, 450
Sump, 1810
Sump pump, 1268
Superheater, 1098, 1099, 1810
Superimposed back pressure, 1320, 1810
Supervisory control, 1013, 1017, 1810
Surfactant, 1810
Surge drums, 1473
Surge volume, 1013, 1810
Suspended solids, 1562
Sweet crude, 1810
Swing capacity, 693, 1810
Swoopdown, 1810
SynBit, 931, 935–937, 1810
Syncrudes, 946
Synthetic crude, 946, 1810
S Zorb, 1810

T

Tag-Robinson colorimeter, 1811
Tail gas system, 1233
Tail gas treating, 1231–1233, 1811
Tallow, 952, 954, 1811
Tank(s), 1421–1422
Tank farm, 1421–1422, 1811
　stocks, 961–963
Tapered thermowells, 1046
Tar, 1811
　sands, 932, 1811
Tariff, 1811
Taxable income, 732
Technip-S&W/Axens R2R, 285
Temperature alarm (TA), 1811
Temperature measurement and control, 1039, 1048
　reactor and vessel wall temperatures, 1048
　thermocouples, 1039–1043
Temperature mixing distances, 1048–1049
Temperature safety, 1332–1338
　acid/base reactions, 1337–1338
　alkylation unit acid runaway, 1338
　hydroprocessing unit, 1336–1337
　methanation, 1338
　polymerization reactions, 1337
　thermal relief, 1333–1335
Temperature *vs.* KT/ K800F ratio, 539
Terminal investment item, 733
Tertiary-amyl methyl ether (TAME), 42–43, 74, 953, 968, 1811–1812
Tertiary butyl alcohol (TBA), 1812
Tertiary storage, 1812
Test run, 830–832, 1812
Tetraethyl lead (TEL), 1812
Thermal conductivity, 1077
Thermal cracker, 706–708, 714, 833–834
　heater/reactor sizing, 538–540, 557–564
　product streams, 99
Thermal cracking, 533–534, 536, 538–553, 1813
　applications, 533–534, 536
　　coking process, 534–335
　　one-stage thermal cracker, 533
　　two-stage thermal cracker, 534
　　yield comparisons, 537
　soaking volume factor, 540–545
　　critical decomposition zone, 544
　　degree of conversion, 540
　　product yields, 540
　　test run data comparison, 545
　temperature *vs.* KT/ K800F ratio, 539
　units, 38–39

Thermal meters, 1028
Thermal rating
  air coolers, 1573–1579
  fired heaters, 1599–1601
Thermal reforming, 1813
Thermal relief, 1813
Thermal relief valves (TRVs), 1333–1335
Thermal safety valves (TSVs), 1333
Thermal stability, 848
Thermal value, 1813
Thermocouple, 1043, 1813
Thermodynamic equilibria, 486
  butane equilibrium, 486
  C_6 fraction equilibrium, 487
  pentane equilibrium, 486
Thermofor, 1813
Thermosiphon, 1813
Thermowell, 1046, 1813
Thiophenes, 7
Three element boiler control, 1016
Threshold limit value (TLV), 1813
Throttle valve, 1526, 1539, 1813
Throttling, 1813
Tight oil, 917, 1814. *See also* Shale crude and gas
Toluene, 1814
Topped crude oil, 1814
Topping, 1814
  refinery, 83–84
Total acid number (TAN), 1811
Total base number (TBN), 1812
Total existent sediment, 1814
Total vapor condensation processes, 1726
Tower, 1814
  flash zone, 134
  fractionation, 125–199, 1442–1470, 1814
  over head ejector system, 183–186
  top pumparound configuration, 138
Toxicity, 1814
Toxic pollutants, 1216–1218, 1220–1223, 1251, 1255–1256, 1275
Trade secret, 1815
Traditional shale oil, 907–917
  aromatic content, 911
  arsenic, 915
  cold properties, 916
  contaminants, 913
  direct heating, 909
  ex-situ retorting, 910–911
  fines, 913–915
  Fischer Assay, 909
  history and production rates, 908
  in-situ production, 909–910

  iron, 915
  nickel, 916
  nitrogen, 916
  oxygen, 916
  pre-refining, 913
  production locations, 907
  properties, 911
  vs. shale crudes, 907
  toxicity, 917
  unsaturated compounds, 913
  vanadium, 916
Traditional solvent-based lube production
  furfural extraction process, 852–853
  MEK dewaxing process, 854–855
  propane deasphalting process, 850–852
Tramp air, 1239–1240, 1815
Tramp compound, 1815
Tramp nitrogen compounds, 1234–1235
Transesterification, 1815
Transition metal sulfide (TMS) catalysts, 416–417
Transit time or time of flight meters, 1027
Transmix, 961, 1815
Transportation, 1815
  department, 687
Trap, 1097, 1100–1101, 1815
Tray spacing, 221, 1442–1444
Trickle filter, 1256, 1815
Triethanol amine (TEA), 500, 1354–1356, 1815
Triglycerides, 968, 986–1001
  deoxygenation, 994–997
  hydroprocessing conversion, 986–987
  isomerization and cracking reactions, 997–999
  oil production, 989–994
  structure of, 988
Trip, 1815
True boiling point (TBP), 1815
  curve, 10
Tube hanger, 1815
Tube sheet, 1420, 1562–1563, 1585, 1816
Tube skin, 1047, 1419, 1816
Tube trailer, 1816
  hydrogen supply, 626
Tube wall, 1816
Tubular reactor, 461–462
Turbine(s), 1038
  meters, 1027
  pumps, 1486
Turnaround, 1816
Turndown, 1816
Turnout gear, 1816

Two-bed hydrotreating reactor, 425–426
Two-phase, 1629, 1816
Two-phase flow, 1088, 1629
Two stage catalyst regeneration, 546–547
Two-stage thermal cracker, 534

U

Ubbehohde viscosimeter, 1816
Ullage, 1816
Ultra-large crude or cargo carrier (ULCC), 1816
Ultralow-sulfur diesel (ULSD), 76, 654, 1816
Ultralow-sulfur kerosene (ULSK), 1816
Ultrasonic meters, 1027
Ultraviolet spectrometer, 1077
Unbranded, 1816
Uncalcined coke, 68
Unconventional crude, 906, 1817
Underwood equation, 213, 1817
Unfinished oils, 1817
Unit design capacities, 121
Universal Oil Products (UOP), 459
UOP/HYDRO MTO™ process, 897, 899–901
UOP resid designs, 286
Upstream, 1817
Urea dewaxing, 1817
U.S. Energy Independence and Security Act (EISA), 950
Utility, 1094–1166, 1817
  flow sheet, 800
  station, 1818

V

Vacuum crude distillation unit, 179–193
  description, 181–182
  draw-off temperature, 188
  flash zone, 63–64, 187
  internal flow and pumparound duties, 189–191
  tower loading, 191–193
  tower over head ejector system, 183–186
Vacuum crude unit (VDU), 34, 94, 179–193
Vacuum distillation
  of atmospheric residue, 58, 179–193
  processes, 1727
Vacuum distillation unit (VDU), 1818
Vacuum gas oil (VGO), 410
Vacuum pump, 1819
Vacuum residue, 58–59
Value chain, 1819
Valve flow coefficient, 1089
Valve tray, 1444, 1819
Vanadium inhibitor, 1819
Vane pumps, 1487
Vapor depressuring, 1819
Vapor disengaging drum, 1470, 1819
Vaporizer systems, 1237
Vapor/liquid separation, 369
Vapor piping, 141
Variable
  area meter, 1027
  cost, 751, 1819
  head-capacity pump, 1489, 1819
  speed, 1819
Vena contracta, 1020, 1819
Vent stack, 1819
Venturi and butterfly control valves, 1035
Venturi flow meter, 1025, 1819
Vent valves, 1245
Vertical cooling water circulating pumps, 1122
Very large crude carrier (VLCC), 1819
Vessels and towers, 1448–1449
  bubble cap trays, 1442–1444
  liquid separation drums, 1471–1473
  packed towers and packed tower sizing, 1461–1465
  pressure vessel, 1476–1485
  pulsation drums, 1473–1476
  quickie method, 1446–1449
  rigorous method, 1449–1461
  sieve trays, 1444
  surge drums, 1473
  trayed tower sizing, 1442, 1444–1445
  valve trays, 1446–1448
  vapor disengaging drums, 1470–1471
Vetrocoke, 1820
Virgin base lube oils, 846
Visbreaker process, 534
Viscosimeter, 1820
Viscosity, 1820–1821
  prediction, 25–26
  vs. temperature, crankcase oil, 847
Viscosity index (VI), 848, 1821
Volatile, 1821
Volatile organic carbon (VOC), 1245
Volatile organic compound, 1245, 1821
Volatility, 848, 1821
Vortex separators (VSS), 275

W

Waste disposal facilities, 1822
Waste heat boiler (WHB), 1823

Waste hydrocarbon disposal facilities,
 1189–1202
 blowdown drums, 1191, 1193
 design, 1289–1301
 disposal regulations, 1272–1274
 elevated flare, 1195–1200
 integrated waste disposal system, 1189
 multi-jet flare, 1200–1202
 non-condensable blow-down drum, 1191
 water separation drum, 1194
Waste minimization, 1822
 techniques, 1276–1277
Water and oxygenates, 451
Water-gas shift (WGS), 580–582, 592,
 597–598, 995, 1822
Water quality guidelines, 1136
Water rate, 1822
Water systems, 1116–1118, 1822
 calcium carbonate equivalents, 1117–1118
Watson K, 1817
Wave flow, 1630–1632, 1822
Wax, 1822
Waxy lube stock (WLS), 845, 856, 1822
Weathering test, 671–672
Wedge/V meters, 1025
Weighted average bed inlet temperature
 (WABIT), 1821
Weighted average bed temperature
 (WABT), 1821
Weir, 1822–1823
Welded and threaded stud, 1048
Welded pads, 1047
West Texas Intermediate (WTI), 92, 1877

Wet chemical/solvent approach, 584–587
Wet gas scrubber (WGS), 1823
Wet milling ethanol process, 972
What-if analysis, 1823
Working capital, 750, 1823
Working storage capacity, 1168, 1823

X

X-factor, 494–495
X-ray fluorescence/absorption, 1077
X-ray fluorescence analyzer, 1824
"x"-to-liquids (xTL) feedstocks, 946–950
 definitions, 947
 methanol synthesis, 946
 production rates, 947
 properties, 949
 refining considerations, 949–950
 synthesis process, 947
Xylenes, 1824

Y

Y-grade, 1824

Z

Zeolite(s), 37
 catalysts, 1825–1826
Zeolitic type catalysts, 489–490
Z factor
 gas compressibility factor, 1824
 pseudo heat transfer coefficient, 1825

Printed by Printforce, the Netherlands